# Ring Theory

Volume II

This is volume 128 in
PURE AND APPLIED MATHEMATICS

H. Bass and S. Eilenberg, editors

A list of titles in this series appears at the end of this volume.

# Ring Theory

## Volume II

**Louis H. Rowen**
*Department of Mathematics and Computer Science*
*Bar Ilan University*
*Ramat Gan, Israel*

ACADEMIC PRESS, INC.
Harcourt Brace Jovanovich, Publishers
Boston  San Diego  New York
Berkeley  London  Sydney
Tokyo  Toronto

ACADEMIC PRESS, INC.
1250 Sixth Avenue, San Diego, CA 92101


*United Kingdom Edition published by*
ACADEMIC PRESS, INC. (LONDON) LTD.
24–28 Oval Road, London NW1 7DX


Library of Congress Cataloging-in-Publication Data

Rowen, Louis Halle.
  Ring theory.

  (Pure and applied mathematics; v. 127–128)
  Includes bibliographies and indexes.
  1. Rings (Algebra)  I. Title.  II. Series: Pure and
applied mathematics (Academic Press);  127–128.
QA3.P8  vol. 127–128  510 s  [512′.4]      87-14536
[QA247]
ISBN 0-12-599841-4 (v. 1)
ISBN 0-12-599842-2 (v. 2)


88 89 90 91    9 8 7 6 5 4 3 2 1
Printed in the United States of America

# Contents

# Introduction to Volume II

A comprehensive introduction was furnished in Volume I, but a few comments are in order about the specific nature of Volume II.

After a brief treatment of those techniques of homology and cohomology needed for rings, the book turns specifically to certain types of rings that have been objects of intense investigation in recent years—Azumaya algebras (5.3), rings with polynomial identity (Chapter 6), division rings, and, more generally, (finite dimensional) central simple algebras (Chapter 7), group rings (8.1 and 8.2), and enveloping algebras (8.3). Since the group ring contains all possible information about group representations and the enveloping algebra contains all possible information about Lie representations, these two types of rings have been unified in Chapter 8 under the heading, "Rings of Representation Theory." These rings often are Noetherian, and the last part of Chapter 8 deals with general aspects of Noetherian theory that have been used in studying these rings (and also with Hopf algebras, another generalization).

Although the general setup is the same as in Volume I, the pace of the text is somewhat faster than in the first volume. Indeed, the subjects of each chapter have had books devoted exclusively to them in the past, and to deal adequately with current research in any of these subjects would require several volumes. The goal in this volume has been to present a unified treatment to the reader and to give the reader some idea, at least, of research in ring theory of the 1980s. In certain fast-moving areas, such as polycyclic group algebras and primitive ideals of enveloping algebras, only a brief sketch has been given of recent advances; fortunately, excellent surveys in both subjects have been published within the last five years by Farkas and Passman (for the former); and by Borho, Joseph, and Rentschler (for the latter).

# Table of Principal Notation for Volume II

*Note:* ! after a page reference means the symbol is used differently in another part of the text.

# 5 Homology and Cohomology

One of the major tools of algebra is homology theory and its dual, cohomology theory. Although rooted in algebraic topology, it has applications in virtually every aspect of algebra, as explained in the classic book of Cartan and Eilenberg. Our goal here is to develop enough of the general theory to obtain its main applications to rings, i.e., projective resolutions, the homological dimensions, the functors $\mathscr{T}or$ and $\mathscr{E}xt$, and the cohomology groups (which are needed to study division rings).

Duality plays an important role in category theory, so we often wish to replace $R\text{-}\mathscr{M}od$ by a self-dual category, i.e., we use only those properties of $R\text{-}\mathscr{M}od$ whose duals also hold in $R\text{-}\mathscr{M}od$. For the most part, it is enough to consider abelian categories with "enough" projectives and injectives, i.e., for each object $M$ there is an epic $P \to M$ for $P$ projective and a monic $M \to E$ for $E$ injective.

## §5.0 Preliminaries About Diagrams

Certain easy technical lemmas concerning diagrams are used repeatedly in homology and cohomology. For the sake of further reference we collect them here. We shall assume all of the diagrams are commutative, with all

*1*

rows and columns exact; as usual a dotted line indicates a map to be found. Also please see the remark at the end.

(1) *The five lemma (repeated from proposition 2.11.15). In the diagram*

$$
\begin{array}{ccccccccc}
A & \xrightarrow{f} & B & \xrightarrow{g} & C & \xrightarrow{h} & D & \xrightarrow{j} & E \\
\downarrow{\scriptstyle\alpha} & & \downarrow{\scriptstyle\beta} & & \downarrow{\scriptstyle\gamma} & & \downarrow{\scriptstyle\delta} & & \downarrow{\scriptstyle\varepsilon} \\
A' & \xrightarrow{f'} & B' & \xrightarrow{g'} & C' & \xrightarrow{h'} & D' & \xrightarrow{j'} & E'
\end{array}
$$

*if every vertical arrow but $\gamma$ is an isomorphism then $\gamma$ is also an isomorphism.*
(2) *Given*

$$
\begin{array}{ccc}
M & \xrightarrow{\beta} & N \\
\downarrow{\scriptstyle f} & & \downarrow{\scriptstyle g} \\
M' & \xrightarrow{\gamma} & N''
\end{array}
$$

*then $f$ restricts to a map $\tilde{f}: \ker\beta \to \ker\gamma$, and $g$ induces a map $\bar{g}: \operatorname{coker}\beta \to \operatorname{coker}\gamma$ yielding the commutative diagram*

$$
\begin{array}{ccccccccc}
0 & \longrightarrow & \ker\beta & \longrightarrow & M & \xrightarrow{\beta} & N & \longrightarrow & \operatorname{coker}\beta & \longrightarrow & 0 \\
& & \downarrow{\scriptstyle\tilde{f}} & & \downarrow{\scriptstyle f} & & \downarrow{\scriptstyle g} & & \downarrow{\scriptstyle\bar{g}} & & \\
0 & \longrightarrow & \ker\gamma & \longrightarrow & M'' & \longrightarrow & N'' & \longrightarrow & \operatorname{coker}\gamma & \longrightarrow & 0
\end{array}
$$

*If $f$ and $g$ are isomorphisms then $\tilde{f}$ and $\bar{g}$ are also isomorphisms.*

**Proof:** If $\beta x = 0$ then $\gamma f x = g\beta x = 0$ implying $f(\ker\beta) \subseteq \ker\gamma$. On the other hand, $f(\beta M) = \gamma f M \subseteq \gamma M''$, so $g$ induces a map $\bar{g}: N/\beta M \to N''/\gamma M''$. The commutativity of the ensuing diagram is obvious. The last assertion is a special case of the five lemma. To wit, adding $0$ at the right so that $\bar{g}$ is in the middle of the appropriate diagram (with $f$ at the left side) shows $\bar{g}$ is an isomorphism; adding, instead, $0$ at the left so that $\tilde{f}$ is in the middle shows $\tilde{f}$ is an isomorphism.     Q.E.D.

(3) *The snake lemma. Given*

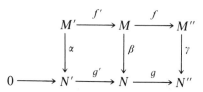

(i) $\ker \alpha \to \ker \beta \to \ker \gamma$ *is exact, where the maps are obtained as in* (2).

(ii) *If f is epic then there is an exact sequence*

$$\ker \alpha \xrightarrow{\tilde{f}'} \ker \beta \xrightarrow{\tilde{f}} \ker \gamma \xrightarrow{\partial} \operatorname{coker} \alpha \to \operatorname{coker} \beta \to \operatorname{coker} \gamma$$

*where $\partial$ is defined in the proof, and the other maps are in* (2).

**Proof:**

(i) The composition is certainly 0. On the other hand, if $z \in \ker \beta$ and $fz = 0$ then $z = f'x'$ for some $x'$ in $M'$; then $0 = \beta z = g'\alpha x'$, so $\alpha x' = 0$ and $x' \in \ker \alpha$ as desired.

(ii) Define $\partial$ as follows: Given $x''$ in $\ker \gamma$ we take $x$ in $M$ with $fx = x''$; then $g\beta x = 0$ so $\beta x \in \ker g = g'N'$, yielding $y'$ in $N'$ with $g'y' = \beta x$, and we take $\partial x''$ to be $y' + \alpha M'$. To check $\partial$ is well-defined, suppose we took instead $x_1$ with $fx_1 = x''$ and we took $y_1'$ with $g'y_1' = \beta x_1$. Then $f(x_1 - x) = 0$ so $x_1 - x \in f'M'$, and thus

$$g'(y_1' - y_1) = \beta(x_1 - x) \in \beta f'M' = g'\alpha M'.$$

Hence $y_1' - y_1 \in \alpha M'$ since $g'$ is monic.

Having defined $\partial$, which is surely a map, we note that the hypothesis is self-dual, so it is enough to show $\ker \alpha \xrightarrow{\tilde{f}'} \ker \beta \xrightarrow{\tilde{f}} \ker \gamma \xrightarrow{\partial} \operatorname{coker} \alpha$ is exact; in view of (i) we need only show $\ker \beta \xrightarrow{\tilde{f}} \ker \gamma \xrightarrow{\partial} \operatorname{coker} \alpha$ is exact. We shall use throughout the notation of the previous paragraph. First note that if $x'' \in \tilde{f}(\ker \beta)$ then in the definition of $\partial x''$ above we could take $x$ in $\ker \beta$, so $g'y' = \beta x = 0$ and thus $y' = 0$, proving $\partial \tilde{f} = 0$. On the other hand, if $\partial x'' = 0$ then $y' \in \alpha M'$ so writing $y' = \alpha x'$ we see $x - f'x' \in \ker \beta$ and $\tilde{f}(x - f'x') = f(x - f'x') = fx = x''$ proving $x'' \in \tilde{f}(\ker \beta)$.     Q.E.D.

(4) *The nine lemma (or 3 × 3 lemma). Given*

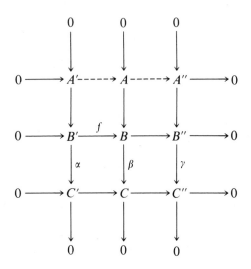

one can find dotted arrows making an exact row such that the above diagram is commutative.

**Proof:**   Immediate from the snake lemma, since

$$\ker \alpha \xrightarrow{\tilde{f}} \ker \beta \to \ker \gamma \xrightarrow{\partial} \operatorname{cok} \alpha = 0$$

is exact and $f$ is monic, yielding $0 \to A' \to A \to A'' \to 0$ as desired.     Q.E.D.

(5)   *The horseshoe lemma. If $P'$ and $P''$ are projective then the following diagram can be "filled in" where $\mu: P' \to P' \oplus P''$ is the canonical monic and $\pi: P' \oplus P'' \to P''$ is the projection:*

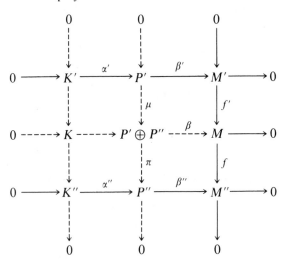

(*You get the horseshoe lying on its side by erasing the dotted lines and the maps to and from* 0.)

**Proof:** There is a map $h: P'' \to M$ lifting $f$, i.e., $\beta'' = fh$. Defining $\beta$ by $\beta(x', x'') = f'\beta'x' + hx''$ one sees at once that the top right square is commutative. Moreover, $f\beta(x', x'') = ff'\beta'x' + \beta''x'' = 0 + \beta''\pi(x', x'')$ so the bottom right square is commutative. Taking $K = \ker \beta$ we can draw the upper row by means of the nine lemma (turned on its side). Q.E.D.

(6)  *Turning the corner Given*

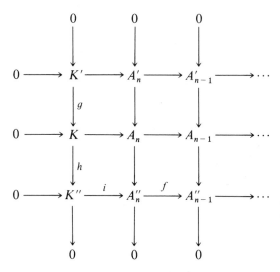

One has an exact sequence $0 \to K' \xrightarrow{g} K \xrightarrow{ih} A_n'' \xrightarrow{f} A_{n-1}'' \to \cdots$

**Proof:** $\ker f = iK'' = ihK$ since $h$ is epic; moreover, since $i$ is monic we have $\ker ih = \ker h = gK'$. Q.E.D.

(7)  *The pullback and pushout have been introduced, respectively, as inverse and direct limits in exercises* 1.8.1 *and* 1.8.1′. *We start over again here, obtaining some obvious properties. First, given*

*we define the pullback $P$ to be $\{(a_1, a_2) \in A_1 \oplus A_2 : fa_1 = ga_2\}$, and let $\pi_i\colon P_i \to A_i$ be the restrictions of the projection maps from $A_1 \oplus A_2$ to $A_i$. Note $\ker \pi_1 = \{(0, a_2): a_2 \in \ker g\} \approx \ker g$, yielding a commutative diagram*

$$
\begin{array}{ccccccc}
0 & \longrightarrow & K & \longrightarrow & P & \xrightarrow{\ \pi_1\ } & A_1 \\
 & & \downarrow{\scriptstyle 1_K} & & \downarrow{\scriptstyle \pi_2} & & \downarrow \\
0 & \longrightarrow & K & \longrightarrow & A_2 & \xrightarrow{\ g\ } & B
\end{array}
$$

*Furthermore, if $g$ is epic then $\pi_1$ is epic for if $a_1 \in A_1$ then $fa_1 = ga_2$ for some $a_2$ in $A_2$ implying $a_1 = \pi_1(a_1, a_2)$.*

*Dually given*

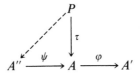

*we define the pushout $P' = (A_1' \oplus A_2')/N$ where $N = \{f'b, -g'b): b \in B'\}$, and let $\mu_1, \mu_2$ be the compositions of the canonical injections into $A_1' \oplus A_2'$ with the canonical map factoring out $N$. Then $\operatorname{coker} \mu_1 \approx \operatorname{coker} g'$ (proof dual to the dual assertion for the pullback), and if $g'$ is monic then $\mu_1$ is monic.*

(8) *The projective lifting lemma. If $P$ is projective and*

$$
\begin{array}{ccc}
 & P & \\
 \swarrow\ \ & \downarrow{\scriptstyle \tau} & \\
A'' \xrightarrow{\ \psi\ } & A & \xrightarrow{\ \varphi\ } A'
\end{array}
$$

*is exact with $\varphi\tau = 0$ then $\tau = \psi\sigma$ for some $\sigma\colon P \to A''$: (Indeed, replace $A$ by $\psi A''$. Since $\tau P \subseteq \ker \varphi \subseteq \psi A''$ we see $\psi$ lifts.)*

**Remark:** Although for concreteness we have framed the proofs in the category *R-Mod*, the proofs could be given in any abelian category. Indeed, the five lemma was already proven in general (exercise 1.4.21), and it is quite straightforward to modify the other diagram-chasing lemmas accordingly. The general pullback for abelian categories was described in exercise 1.4.22, and the general pushout is its dual. The facts concerning projectives were proved directly from the lifting property, which is categorical. Since "abelian" is self-dual we shall also claim the dual results of any theorems based on these facts.

## §5.1 Resolutions and Projective and Injective Dimension

In this section we introduce two new important dimensions, the *projective dimension* and its dual, the *injective dimension*. It turns out these dimensions have immediate application to ring theory, and we shall use these to develop the motivation of the subject.

### *Projective Resolutions*

**Definition 5.1.1:** A *projective resolution* $\mathscr{P}$ of a module $M$ is an exact sequence of the form

$$\cdots \to P_n \xrightarrow{f_n} P_{n-1} \to \cdots \to P_1 \xrightarrow{f_1} P_0 \xrightarrow{\varepsilon} M \to 0$$

where each $P_i$ is projective. We say $\mathscr{P}$ is f.g. (usually called *finite* in the literature) if each $P_i$ is f.g.; $\mathscr{P}$ is *free* if each $P_i$ is free. The smallest $n$ for which $P_n = 0$ is called the *length* of the resolution $\mathscr{P}$.

Several immediate observations are available.

**Remark 5.1.2:**

(i) Every module $M$ has a free resolution. Indeed, take an epic $f_0 : F_0 \to M$ with $F_0$ free and, inductively, given $f_{i-1} : F_{i-1} \to F_{i-2}$ take an epic $f_i : F_i \to$ ker $f_{i-1}$. We could view $f_i : F_i \to F_{i-1}$; then $f_i F_i = \ker f_{i-1}$ so the sequence $\cdots \to F_2 \xrightarrow{f_2} F_1 \xrightarrow{f_1} F_0 \to M \to 0$ is exact.

(ii) If $M$ is f.g. then we could take $F_0$ f.g. as well as free. If $R$ is left Noetherian then ker $f_0 \leq F_0$ is f.g., and, continuing by induction, we may assume each $F_i$ is f.g. Thus every f.g. module over a left Noetherian ring has an f.g. free resolution.

In order to study a resolution we shall repeatedly use the observation that we can "cut" an exact sequence $\cdots \to M'' \xrightarrow{f} M \xrightarrow{g} M' \to \cdots$. at $g$ to produce exact sequences

$$\to M'' \xrightarrow{f} M \to gM \to 0 \qquad \text{and} \qquad 0 \to \ker g \to M \xrightarrow{g} M' \to \cdots.$$

In particular, any projective resolution $\mathscr{P}$ can be cut at $f_n$ to yield

$$0 \to K \to P_n \xrightarrow{f_n} P_{n-1} \to \cdots \to P_0 \to M \to 0;$$

$K = \ker f_n$ is called the *$n$-th syzygy* of $M$. If $\mathscr{P}$ has length $n+1$ then $K \approx P_{n+1}$ is projective. Conversely, if $K$ is projective then we have produced a new projective resolution of length $n + 1$, which we shall compare to $\mathscr{P}$ by means of the next result.

**Proposition 5.1.3:**  *(Generalized Schanuel's lemma) Given exact sequences*

$$0 \to K \to P_n \xrightarrow{f_n} P_{n-1} \to \cdots \to P_1 \xrightarrow{f_1} P_0 \xrightarrow{\varepsilon} M \to 0 \tag{1}$$

$$0 \to K' \to P'_n \xrightarrow{g_n} P'_{n-1} \to \cdots \to P'_1 \xrightarrow{g_i} P'_0 \xrightarrow{\varepsilon} M \to 0 \tag{2}$$

*with each $P_i$ and $P'_i$ projective then*

$$K \oplus P'_n \oplus P_{n-1} \oplus P'_{n-2} \oplus \cdots \approx K' \oplus P_n \oplus P'_{n-1} \oplus P_{n-2} \oplus \cdots.$$

***Proof:***   We cut (1) at $f_1$ to get

$$0 \to K \to \cdots \to P_2 \to P_1 \to \ker \varepsilon \to 0 \tag{3}$$

$$0 \to \ker \varepsilon \to P_0 \to M \to 0 \tag{4}$$

and cut (2) at $g_1$ to get

$$0 \to K' \to \cdots \to P'_2 \to P'_1 \to \ker \varepsilon' \to 0 \tag{5}$$

$$0 \to \ker \varepsilon' \to P'_0 \to M \to 0 \tag{6}$$

Let $M' = P'_0 \oplus \ker \varepsilon \approx P_0 \oplus \ker \varepsilon'$ by Schanuel's lemma (2.8.26) applied to sequences (4) and (6). Then we can modify (3), (5) to get

$$0 \to K \to \cdots \to P_2 \to P_1 \oplus P'_0 \xrightarrow{f_1 \oplus 1} M' \to 0$$

$$0 \to K' \to \cdots \to P'_2 \to P'_1 \oplus P_0 \xrightarrow{g_1 \oplus 1} M' \to 0$$

and by induction on the length we get the desired conclusion.          Q.E.D.

**Definition 5.1.4:**   Two modules $N, N'$ are *projectively equivalent* if there are projectives $P, P'$ such that $N \oplus P \approx N' \oplus P'$; if, furthermore, $P, P'$ are f.g. free then $N, N'$ are *stably equivalent*. $N$ is *stably free* if $N$ is stably equivalent to a free module, i.e., $N \oplus R^{(n)}$ is free for suitable $n$.

   Note that any module $M$ projectively equivalent to a projective module is itself projective since $M$ is a summand of a projective module. In particular, all stably free modules are projective. (Compare to remark 2.8.4' and exercise 10.)

**Corollary 5.1.5:** *For any n the n-th syzygies of any two resolutions of M are projectively equivalent; the n-th syzygies of any two f.g. free resolutions of M are stably equivalent.*

We are ready to open an important new dimension in module theory.

**Definition 5.1.6:** *M* has *projective dimension n* (written $\text{pd}(M) = n$) if *M* has an f.g. projective resolution of length *n*, with *n* minimal such. In the literature the projective dimension is also called the *homological dimension*.

**Remark 5.1.7:** $\text{pd}(M) = 0$ iff *M* is projective.

**Proposition 5.1.8:** *The following are equivalent:*

(i) $\text{pd}(M) \leq n + 1$.

(ii) *The n-th syzygy of any projective resolution of M is projective.*

(iii) *Any projective resolution $\mathscr{P}$ of M can be cut at $f_n$ to form a projective resolution of length n + 1.*

**Proof:** (i) $\Rightarrow$ (ii) By corollary 5.1.5 the *n*-th projective syzygy is projectively equivalent to a projective module and thus is itself projective; (ii) $\Rightarrow$ (iii) clear; (iii) $\Rightarrow$ (i) by definition.     Q.E.D.

**Proposition 5.1.9:** *The following are equivalent:*

(i) *M has a f.g. free resolution of length $\leq n + 1$.*

(ii) *The n-th syzygy of any f.g. free resolution is stably free.*

(iii) *Any f.g. free resolution can be cut and modified to form a f.g. free resolution of length n + 1.*

**Proof:** (iii) $\Rightarrow$ (i) $\Rightarrow$ (ii) is clear. To see (ii) $\Rightarrow$ (iii) note that if $P \oplus R^{(n)}$ is free and $0 \to P \to F \to F'$ is exact with $F, F'$ free then the sequence

$$0 \to P \oplus R^{(n)} \to F \oplus R^{(n)} \to F'$$

is exact.     Q.E.D.

## Elementary Properties of Projective Dimension

Let us now compare projective dimensions of modules in an exact sequence. This can be done elegantly using the functor Ext of §5.2 or in a simple but ad hoc fashion by means of the following result:

**Proposition 5.1.10:** *Suppose* $0 \to M' \to M \to M'' \to 0$ *is exact. Given projective resolutions* $\mathscr{P}', \mathscr{P}''$ *of* $M', M''$ *respectively, we can build a projective resolution* $\mathscr{P}$ *of* $M$ *where each* $P_n = P'_n \oplus P''_n$ *and, furthermore, letting* $K'_n, K''_n$ *denote the n-th syzygies we have the following commutative diagram for each* $n$, *obtained by cutting* $\mathscr{P}', \mathscr{P},$ *and* $\mathscr{P}''$ *at* $f'_n, f_n,$ *and* $f''_n$, *respectively:*

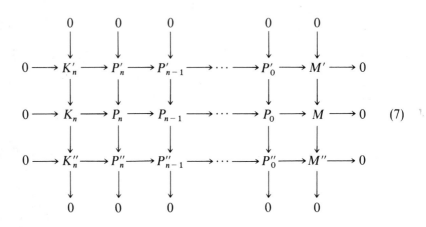

$$(7)$$

**Proof:** By induction on $n$. Suppose for $i \leq n-1$ we have $P_i = P'_i \oplus P''_i$ together with the diagram

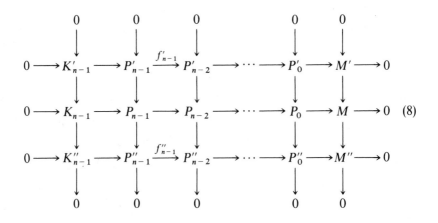

$$(8)$$

Note that $K'_{n-1} = \ker f'_{n-1} = f'_n P'_n$ and $K''_{n-1} = \ker f''_{n-1} = f''_n P''_n$. Cutting the resolutions $\mathscr{P}', \mathscr{P}''$ at $f'_n, f''_n$ yields the diagram

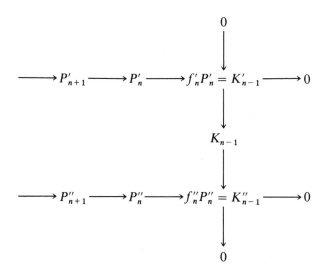

where the column is the $n - 1 -$ syzygy column from above. Now the horse-shoe lemma enables us to fill in the diagram

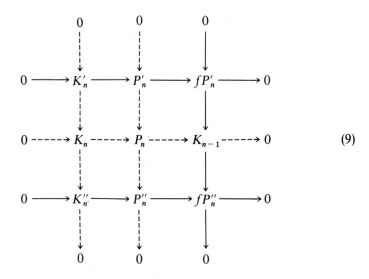

(9)

where $P_n = P'_n \oplus P''_n$, and we get (7) by pasting (9) to (8).     Q.E.D.

**Corollary 5.11:** *Suppose* $0 \to M' \to M \to M'' \to 0$ *is exact. If* $M', M''$ *have projective (resp. f.g. projective, resp. free, resp. f.g. free) resolutions of length* $\leq n$ *then so does* $M$. *In particular,* $\mathrm{pd}(M) \leq \max\{\mathrm{pd}(M'), \mathrm{pd}(M'')\}$. *If* $\mathrm{pd}(M') > \mathrm{pd}(M'')$ *then* $\mathrm{pd}(M) = \mathrm{pd}(M')$.

**Proof:** The first two assertions are immediate since $0 \oplus 0 \approx 0$. For the last assertion let us examine the exact sequence of $n$-th syzygies $0 \to K_n' \to K_n \to K_n'' \to 0$ for $n = \mathrm{pd}(M'')$. This splits since $K_n''$ is projective, so $K_n'$ is projectively equivalent to $K_n$, and we apply proposition 5.1.8.     Q.E.D.

**Corollary 5.1.12:** *Suppose* $0 \to M' \to M \to M'' \to 0$ *is exact. Suppose* $\mathscr{P}', \mathscr{P}''$ *are projective resolutions of* $M', M''$ *and let* $\mathscr{P}$ *be built as in proposition 5.1.10. Let* $K_m', K_m, K_m''$ *denote the respective* $m^{\text{th}}$ *syzygies. Then for any* $m$ *there is an exact sequence*

$$0 \to K_{m-1}' \to K_{m-1} \to P_{m-1}'' \to P_{m-2}'' \to \cdots \to P_0'' \to M'' \to 0.$$

*This displays* $K_{m-1}'$ *as an* $m$-*th syzygy of a projective resolution of* $M''$, *if* $K_{m-1}$ *is projective.*

   *Writing* $n', n, n''$ *for the respective pd of* $M', M,$ *and* $M''$ *we then have* $n'' \leq \max\{n', n\} + 1$. *Furthermore,* $n'' = n$ *if* $n' < n$, *and* $n'' = n' + 1$ *if* $n' > n$.

**Proof:** The first assertion is obtained by "turning the corner" in (7). Everything else follows by choosing $m$ properly. Taking $m = \max\{n', n\}$ we see $K_{m-1}'$ and $K_{m-1}$ are projective so $n'' \leq m + 1 = \max\{n', n\} + 1$.

   If $n' < n$ then we could choose $\mathscr{P}'$ so that $K_n' = 0$; taking $m = n + 1$ we have $n'' \leq n$ and thus $n'' = n$ by corollary 5.1.11.

   If $n' > n$ then taking $m = n'$ we see $K_{m-1}$ is projective but $K_{m-1}'$ is not, so $n'' > m$ by proposition 5.1.8, i.e., $n'' \geq n' + 1$; hence $n'' = n' + 1$ since $n'' \leq \max\{n', n\} + 1$.     Q.E.D.

**Summary 5.1.13:** *Suppose* $0 \to M' \to M \to M'' \to 0$ *is exact. Recapitulating corollary 5.1.12 where* $n' = \mathrm{pd}(M')$, $n = \mathrm{pd}(M)$, *and* $n'' = \mathrm{pd}(M'')$ *we have the following possibilities:*

*Case I.* $n < n'$. *Then* $n'' = n' + 1$.

*Case II.* $n = n'$. *Then* $n'' \leq n + 1$.

*Case III.* $n > n'$. *Then* $n'' = n$.

These formulas could be proved directly using induction, cf., Kaplansky [72B, p. 169]; we shall give a less ad hoc proof in §5.2.

**Example 5.1.14:** Suppose $f: F \to M''$ is epic with $F$ free and $M''$ not projective. Then $\mathrm{pd}(F) = 0$ so $\mathrm{pd}(M'') = \mathrm{pd}(\ker f) + 1$. In particular, if $Rr$ is *not* projective then $\mathrm{pd}(Rr) = \mathrm{pd}(\mathrm{Ann}\, r) + 1$.

## The (Left) Global Dimension

**Definition 5.1.15:** The *global dimension*

$$\mathrm{gl.\, dim}(R) = \sup\{\mathrm{pd}(M): M \in R\text{-}\mathcal{M}od\}.$$

**Example 5.1.16:**

(i) $R$ is semisimple Artinian iff every module is projective, iff $\mathrm{gl.\, dim}(R) = 0$.

(ii) If $R$ is hereditary then $\mathrm{gl.\, dim}(R) \leq 1$. (Indeed, every submodule of a projective is projective, so the first syzygy is always projective, implying $\mathrm{pd}(M) \leq 1$ for any module $M$.)

There is another way of viewing the global dimension, by dualizing everything.

**Definition 5.1.17:** An *injective resolution* $\mathbb{E}$ of a module $M$ is an exact sequence $0 \to M \xrightarrow{f_0} E_0 \xrightarrow{f_1} E_1 \to \cdots$ with each $E_i$ injective; the $n$-th *cosyzygy* is coker $f_n$. $\mathbb{E}$ has *length n* if $E_i = 0$ for all $i \geq n$, with $n$ minimal such. The *injective dimension* $\mathrm{id}(M)$ (if it exists) is the smallest $n$ for which $M$ has an injective resolution of length $n$.

Dually to proposition 5.1.8 we see $\mathrm{id}(M) = n$ iff every injective resolution can be cut to a resolution of length $n$.

In the next section we shall see $\mathrm{gl.\, dim}(R) = \sup\{\mathrm{id}(M): M \in R\text{-}\mathcal{M}od\}$, which is easy to believe since it holds for $R$ semisimple Artinian ($\mathrm{gl.\, dim}\, 0$) or hereditary ($\mathrm{gl.\, dim}\, 1$). Kirkland-Kuzmanovich [87] have constructed examples of Noetherian $R$ with $\mathrm{gl.\, dim}\, R < \mathrm{gl.\, dim}\, R/\mathrm{Nil}(R)$.

Our present interest is to see the connection between $R$ and $R[\lambda]$ for a commuting indeterminate $\lambda$. If $R$ is a field then $R[\lambda]$ is a PID and thus hereditary, so $\mathrm{gl.\, dim}\, R[\lambda] = \mathrm{gl.\, dim}(R) + 1$. Actually, this relation holds much more generally, and we shall verify it for the skew polynomial ring $R[\lambda; \sigma]$ where $R$ is arbitrary and $\sigma$ is an automorphism of $R$. (In particular, one could take $\sigma = 1$ and obtain the ordinary polynomial ring.) To see this we need to introduce some new module structure.

If $M$ is an $R$-module then we define $M[\lambda; \sigma]$ to be $R[\lambda] \otimes_R M$ as an additive group, i.e., the elements of $M[\lambda; \sigma]$ have the unique form $\sum \lambda^i x_i$ where $x_i \in M$. $M[\lambda; \sigma]$ is made into $R[\lambda; \sigma]$-module via scalar multiplication

$$r\lambda^j \sum \lambda^i x_i = \sum \lambda^{i+j} (\sigma^{-(i+j)} r) x_i.$$

**Remark 5.1.18:** Since $R[\lambda; \sigma]$ is free over itself we see that $F[\lambda; \sigma]$ is a free $R[\lambda; \sigma]$-module for any free $R$-module $F$; consequently, any projective (resp. free) resolution $0 \to P_n \to \cdots \to P_0 \to M \to 0$ of $M$ in $R\text{-}\mathcal{M}od$ yields a projective (resp. free) resolution $0 \to P_n[\lambda; \sigma] \to \cdots \to P_0[\lambda; \sigma] \to M[\lambda; \sigma] \to 0$ in $R[\lambda; \sigma]\text{-}\mathcal{M}od$. (This is really a fact about graded modules.) In particular, $\text{pd}_{R[\lambda; \sigma]} M[\lambda; \sigma] \leq \text{pd}_R M$; in fact, equality holds by exercise 2.

On the other hand, any $R[\lambda; \sigma]$-module $M$ is viewed naturally as $R$-module by forgetting $\lambda$, and then we can form $M[\lambda; \sigma]$ as above. To distinguish between the two $R[\lambda; \sigma]$-module structures we write $\lambda x$ to denote the original product in $M$, and $\lambda \cdot x$ to denote the new product in $M[\lambda; \sigma]$. We need one more module. Define $\sigma M$ to be the set of formal elements $\{\sigma x : x \in M\}$ made into a module under the operations $\sigma x_1 + \sigma x_2 = \sigma(x_1 + x_2)$ (so that $\sigma M \approx M$ as abelian groups) and $r\sigma x = \sigma x'$ where $x'$ is the product $(\sigma^{-1} r)x$ in $M$.

**Lemma 5.1.18':** (*Hochschild's trick*) *For any $R[\lambda; \sigma]$-module $M$ there is an exact sequence*

$$0 \to (\sigma M)[\lambda; \sigma] \to M[\lambda; \sigma] \xrightarrow{f} M \to 0$$

*where $f(\lambda^i \cdot x) = \lambda^i x$.*

**Proof:** Define $g: (\sigma M)[\lambda; \sigma] \to M[\lambda; \sigma]$ by $g(\sum \lambda^i \cdot \sigma x_i) = \sum (\lambda^i \cdot \lambda x_i - \lambda^{i+1} \cdot x_i)$. Clearly $g(\sigma M[\lambda; \sigma]) \subseteq \ker f$. Also

$$(r\lambda^j) \cdot \lambda^i \sigma x_i = \lambda^{i+j} \sigma^{-(i+j)} r\sigma x_i = \lambda^{i+j} \cdot \sigma((\sigma^{-(i+j+1)} r)x_i) \qquad \text{so}$$

$$g\left( r\lambda^j \sum_i \lambda^i \cdot \sigma x_i \right) = \sum_i (\lambda^{i+j} \cdot \lambda(\sigma^{-(i+j+1)} r)x_i - \lambda^{i+j+1} \cdot (\sigma^{-(i+j+1)} r)x_i)$$

$$= \sum_i r(\lambda^{i+j} \cdot \lambda x_i - \lambda^{i+j+1} x_i)$$

$$= r\lambda^j g\left( \sum \lambda^i \cdot \sigma x_i \right)$$

proving $g$ is indeed a map. Moreover, we can rewrite $g(\sum \lambda^i \cdot \sigma x_i)$ as $1 \cdot \lambda x_0 + \sum_{i \geq 1} \lambda^i \cdot (\lambda x_i - x_{i-1})$. Hence $\sum \lambda^i \cdot \sigma x_i \in \ker g$ iff $x_0 = 0$ and $\lambda x_i = x_{i-1}$ for all $i \geq 1$; but one has $x_i = 0$ for large $i$, so $0 = x_{i-1} = x_{i-2} = \cdots$, implying $g$ is monic.

Finally suppose $x' = \sum_{i=0}^m \lambda^i \cdot x_i' \in \ker f$. Put $x_m = 0$, $x_{m-1} = -x_m'$, and given $x_i$ put $x_{i-1} = \lambda x_i - x_i'$. Then $g(\sum_{i \geq 0} \lambda^i \cdot \sigma x_i) = x'$, proving exactness. Q.E.D.

We are ready for the first inequality.

**Proposition 5.1.19:** gl. dim $R[\lambda; \sigma] \leq$ gl. dim $R + 1$. *In fact*, $\mathrm{pd}_{R[\lambda;\sigma]} M \leq \mathrm{pd}_R \sigma M + 1$ *for any* $R[\lambda; \sigma]$-*module M*.

**Proof:** Let $n = $ gl. dim $R$. For any $R[\lambda; \sigma]$-module $M$ we have

$$\mathrm{pd}_{R[\lambda;\sigma]} M[\lambda; \sigma] \leq \mathrm{pd}_R M \leq n,$$

and $\mathrm{pd}_{R[\lambda;\sigma]}(\sigma M)[\lambda; \sigma] \leq \mathrm{pd}_R \sigma M \leq n$. Applying corollary 5.1.12 to lemma 5.1.18' thus shows $\mathrm{pd}_{R[\lambda;\sigma]} M \leq n + 1$. Q.E.D.

Our next goal is to see, in fact, that equality holds.

**Proposition 5.1.20:** $\mathrm{pd}(\bigoplus_{i \in I} M_i) = \sup\{\mathrm{pd}(M_i) : i \in I\}$.

**Proof:** Write $n_i = \mathrm{pd}(M_i)$ and $n = \mathrm{pd}(M)$; take projective resolutions $\mathscr{P}_i$ of $M_i$ of respective lengths $n_i$, and let $\mathscr{P} = \bigoplus \mathscr{P}_i$, i.e., $\mathscr{P}$ is formed by taking the direct sum of the respective terms. Then, clearly, length $(\mathscr{P}) = \sup(n_i)$, proving $n \leq \sup(n_i)$. On the other hand, the $(n - 1)$-st syzygy of $\mathscr{P}$ is projective, so the $(n - 1)$-st syzygy of each $\mathscr{P}_i$ is a summand of a projective and thus projective, proving each $n_i \leq n$. Q.E.D.

**Corollary 5.1.21:** $\mathrm{pd}(M^{(I)}) = \mathrm{pd}(M)$ *for any index set I*.

**Proposition 5.1.22:** *Suppose* $\varphi: R \to T$ *is a ring homomorphism. Viewing any T-module M as R-module by means of* $\varphi$ *we have*

$$\mathrm{pd}_R M \leq \mathrm{pd}_T M + \mathrm{pd}_R T.$$

*Moreover, if equality holds for all M such that* $\mathrm{pd}_T M \leq 1$ *then equality holds for all M having finite pd.*

**Proof:** Induction on $m = \text{pd}_T M$. Write $n = \text{pd}_R M$ and $t = \text{pd}_R T$; we want to show $n \leq m + t$. For $m = 0$ we have $M$ projective as $T$-module so $M \oplus M' = F$ for some free $T$-module $F$. This is also a direct sum as $R$-modules, so applying proposition 5.1.20 twice yields $n \leq \text{pd}_R F = t$, as desired.

For $m > 0$ take $0 \to M' \to F \to M \to 0$ exact where $F$ is free. Then $\text{pd}_T M' = m - 1$ by summary 5.1.13, so $\text{pd}_R M' \leq (m - 1) + t$ by induction. Thus

$$n \leq 1 + \max\{\text{pd}_R F, \text{pd}_R M'\} \leq 1 + \max\{t, (m - 1) + t\} = 1 + m - 1 + t = m + t$$

proving the first assertion. To prove the second assertion we note the same induction argument works if we have the case $m = 1$.     Q.E.D.

**Remark 5.1.23:** Suppose $Ra = aR$. Any $M$ in $R\text{-}\mathcal{M}\!od$ has the submodule $aM$, and $\bigoplus M_i / a(\bigoplus M_i) \approx \bigoplus (M_i / aM_i)$. In particular, if $F$ is a free (resp. projective) $R$-module then $F/aF$ is a free (resp. projective) $R/Ra$-module.

**Theorem 5.1.24:** ("*Change of rings*") *Suppose $a \in R$ is a regular noninvertible element, and $Ra = aR$. If $M$ is an $R/Ra$-module with $\text{pd}_{R/Ra} M = n < \infty$ then $\text{pd}_R M = n + 1$. In particular,* $\text{gl.}\dim R \geq \text{gl.}\dim(R/Ra) + 1$ *provided* $\text{gl.}\dim R/Ra$ *is finite.*

**Proof:** Write $\bar{R} = R/Ra$. First we show $\text{pd}_R \bar{R} = 1$. Right multiplication by $a$ gives an isomorphism $R \to Ra$, so applying summary 5.1.13 to the exact sequence $0 \to Ra \to R \to \bar{R} \to 0$ we see $\text{pd}_R \bar{R} \leq 1$. But $\bar{R}$ is not projective (for otherwise $Ra$ would be a summand of $R$, implying $Ra = Re$ for an idempotent $e \neq 0$, and then $a(1 - e) = 0$, contrary to $a$ regular.) Thus $\text{pd}_R \bar{R} = 1$.

By proposition 5.1.22 we have $\text{pd}_R M \leq n + 1$, and we need to show equality for the case $n \leq 1$. Assume $M$ is a counterexample; we may assume $\text{pd}_R M \leq 1$. By definition $aM = 0$. Hence $M$ cannot be a submodule of a free $R$-module. In particular, $M$ is not projective so $\text{pd}_R M = 1$. Hence $n = 1$ (since $M$ is a counterexample). Any exact sequence of $R$-modules

$$0 \to M' \to F \to M \to 0 \qquad (F \text{ free})$$

sends $aF$ to $aM = 0$, so we get the exact sequence

$$0 \to M'/aF \to F/aF \to M \to 0. \tag{10}$$

But this can be read in $\bar{R}\text{-}\mathcal{M}\!od$, in which $F/aF$ is free. Also we have

$$0 \to aF/aM' \to M'/aM' \to M'/aF \to 0. \tag{11}$$

Applying example 5.1.14 to (10), (11) we have $\mathrm{pd}_R M' = 0$ and $\mathrm{pd}_R M'/aF = 0$. But then $M'/aM'$ and $M'/aF$ are projective $\bar{R}$-modules. In particular, (11) splits, implying $aF/aM'$ is a projective $\bar{R}$-module. Thus $M \approx F/M' \approx aF/aM'$ is projective.     Q.E.D.

**Corollary 5.1.25:**   $\mathrm{gl.dim}\, R[\lambda;\sigma] = \mathrm{gl.dim}\, R + 1$.

**Proof:**   Apply the theorem, with $R[\lambda;\sigma]$ and $\lambda$ replacing, respectively, $R$ and $a$.     Q.E.D.

(For $\mathrm{gl.dim}\, R = \infty$ a separate argument is needed, which is an instant application of exercise 4. In fact, infinite gl. dim has led to serious errors in the literature, cf., exercise 16 and McConnell[77].) It is also useful to have a localization result.

**Proposition 5.1.26:**   *If $S$ is a left denominator set for $R$ and $T = S^{-1}R$ then* $\mathrm{pd}_T S^{-1}M \leq \mathrm{pd}_R M$ *for $M$ in $R$-$\mathcal{M}od$. In particular,* $\mathrm{gl.dim}\, T \leq \mathrm{gl.dim}\, R$.

**Proof:**   Since the localization functor is exact (theorem 3.1.20), any projective resolution of $M$ can be localized to a projective resolution of $S^{-1}M$ as $T$-module.     Q.E.D.

There is a right-handed version of projective dimension and thus of gl. dim, which also is 0 if $\mathrm{gl.dim}\, R = 0$ since "semisimple Artinian" is left-right symmetric. However, for $\mathrm{gl.dim}\, R = 1$ the symmetry fails because left hereditary rings need not be right hereditary, and Jategaonkar's example in §2.1 can be used to find rings of arbitrary left and right gl. dim $\geq 1$. A thorough discussion of global dimension and its peculiar connection to the continuum hypothesis can be found in Osofsky [73B].

## Stably Free and FFR

**Definition 5.1.27:**   A module $M$ has FFR (of length $n$) if $M$ has an f.g. free resolution of length $n$. (In particular $M$ is f.g.)

Our interest in FFR is derived from the next result.

**Lemma 5.1.28:**   *If $P$ is projective with FFR then $P$ is stably free.*

***Proof:*** Take an f.g. free resolution $0 \to F_n \xrightarrow{f_n} F_{n-1} \to \cdots \to F_1 \xrightarrow{f_1} F_0 \xrightarrow{\varepsilon} P \to 0$. Since $\varepsilon$ splits we can write $F_0 = P \oplus P'$ with $f_1 F_1 = \ker \varepsilon = P'$, so $P'$ has a f.g. free resolution $0 \to F_n \to F_{n-1} \to F_1 \to P' \to 0$ of length $n-1$. By induction $P'$ is stably free; write $P' \oplus R^{(n)} \approx F$ for $F$ free. Then $P \oplus F \approx P \oplus P' \oplus R^{(n)} \approx F_0 \oplus R^{(n)}$ is f.g. free, proving $P$ is stably free. Q.E.D.

Conversely, we have

**Lemma 5.1.29:** *If $M$ has an f.g. projective resolution of length $n$, with each $P_i$ stably free then $M$ has FFR of length $\leq n + 1$.*

***Proof:*** Take f.g. free $F, F'$ such that $P_0 \oplus F' \approx F$. If $n = 0$ then $M \approx P_0$ so $0 \to F' \to F \to M \to 0$ is the desired sequence. In general, we break the resolution $0 \to P_n \to \cdots \to P_0 \to M \to 0$ at $\varepsilon$ to get $0 \to P_n \to \cdots \to P_2 \to P_1 \to \ker f \to 0$ and $0 \to \ker f \to P_0 \to M \to 0$. By induction $\ker f$ has an f.g. free resolution $0 \to F_n \to \cdots \to F_0 \to \ker f \to 0$; piecing back together yields

$$0 \to F_n \to \cdots \to F_0 \to P_0 \to M \to 0, \qquad \text{or better yet}$$

$$0 \to F_n \to \cdots \to F_1 \to F_0 \oplus F' \to F \to M \to 0. \qquad \text{Q.E.D.}$$

***Definition 5.1.30:*** $K_0(R) = \mathscr{F}/\mathscr{R}$, where $\mathscr{F}$ is the free group whose generators correspond to the isomorphism classes of f.g. projective $R$-modules, and $\mathscr{R}$ is the subgroup generated by all $[P \oplus Q] - [P] - [Q]$ for $P, Q$ f.g. projective.

Since $\sum [P_i] = [\bigoplus P_i]$ we see every element of $K_0(R)$ can be written in the form $[P] - [Q]$ for $P, Q$ f.g. projective.

***Remark 5.1.30':*** $P$ is stably free iff $[P] = [R]^k$ for some $k$. (Proof: ($\Rightarrow$)) If $P \oplus R^{(m)} \approx R^{(n)}$ then $[P] = [R]^{n-m}$. ($\Leftarrow$) If $[P] = [R]^k$ then $P \oplus Q \approx R^{(k)} \oplus Q$ for a suitable f.g. projective module $Q$; writing $Q \oplus Q' \approx R^{(m)}$ we see $P \oplus R^{(m)} \approx R^{(m+k)}$.)

In view of remark 5.1.30' we see every projective is stably free iff $K_0(R) = \langle [R] \rangle$, and we shall use this characterization from now on.

**Proposition 5.1.31:** *Suppose $R$ is left Noetherian and* gl. dim $R < \infty$. *$K_0(R) = \langle [R] \rangle$ iff every f.g. projective $R$-module has FFR.*

***Proof:*** ($\Rightarrow$) by lemma 5.1.29. ($\Leftarrow$) by lemma 5.1.28. Q.E.D.

Our principal interest here is when $K_0(R) = \langle[R]\rangle$. One way of proving this is to show that every f.g. module has FFR. Thus we might hope for an analog of corollary 5.1.25, to tell us that $K_0(R[\lambda]) = \langle[R[\lambda]]\rangle$ if $K_0(R) = \langle[R]\rangle$. Unfortunately, we lack an f.g. analog of proposition 5.1.22, so we start over again.

**Proposition 5.1.32:** *Suppose $R$ is left Noetherian and* gl. dim $R < \infty$. *If $K_0(R) = \langle[R]\rangle$ then $K_0(S^{-1}R) = \langle[S^{-1}R]\rangle$ for any left denominator set $S$ of $R$.*

**Proof:** By lemma 5.1.28 it suffices to show every f.g. projective $S^{-1}R$-module $P$ has FFR. $P$ is naturally an $R$-module; take an f.g. $R$-submodule $M$ with $S^{-1}M$ maximal. If $S^{-1}M < P$ then taking $x$ in $P - S^{-1}M$ we would have $S^{-1}(M + Rx) > S^{-1}M$, contrary to choice of $M$; thus $S^{-1}M = P$. But now take an FFR for $M$. Tensoring each term by $S^{-1}R$ yields an FFR for $S^{-1}M = P$. Q.E.D.

This argument actually yields an epic $K_0(R) \to K_0(S^{-1}R)$, cf., exercise 12.

**Proposition 5.1.33:** *Suppose $R$ is left Noetherian and $0 \to M' \to M \to M'' \to 0$ is exact. If two of the modules have FFR then so does the third.*

**Proof:** In view of remark 5.1.2(ii) there are f.g. free resolutions $\mathscr{P}', \mathscr{P}''$ (of possibly infinite length) of $M', M''$, and we form the f.g. free resolution $\mathscr{P}$ of $M$ as in proposition 5.1.10. By lemma 5.1.29 it suffices to find some $n$ for which the $n$-th syzygies $K_n', K_n$, and $K_n''$ are stably free. Recall $0 \to K_n' \to K_n \to K_n'' \to 0$ is exact. If $M', M''$ have FFR then we may assume $\mathscr{P}', \mathscr{P}''$ are f.g., so for large enough $n$ we get $K_n' = K_n'' = 0$ and thus $K_n = 0$. Therefore, we may assume one of the two modules with FFR is $M$. Now the generalized Schanuel lemma implies $K_n$ is stably free for large enough $n$. But $M'$ or $M''$ has FFR so for $n$ large enough we may assume $K_n' = 0$ or $K_n'' = 0$; thus $K_n$ is isomorphic to the other syzygy which is thus stably free. Q.E.D.

**Theorem 5.1.34:** *(Serre's theorem) Suppose $R$ is a left Noetherian $\mathbb{N}$-graded ring, with* gl. dim $R < \infty$. *If $K_0(R_0) = \langle[R_0]\rangle$ then $K_0(R) = \langle[R]\rangle$.*

Before presenting the proof of this important result, we should note the same proof shows $K_0(R) \approx K_0(R_0)$. This result will be generalized further in appendix A. We follow Bass [68B] and start by noting various properties of graded modules, of independent interest. Let $R_+ = \bigoplus_{n>0} R_n$, an ideal of $R$. Then $R_0 \approx R/R_+$. We consider the category $R\text{-}\mathcal{G}\imath\text{-}\mathcal{M}od$ of definition 1.9.1.

**Remark 5.1.34':**   If $M \in R\text{-}\mathcal{G}\imath\text{-}\mathcal{M}od$ and $R_+M = M$ then $M = 0$. (Indeed, otherwise, take $n$ minimal such that $M_n \neq 0$. Then $M_n \cap R_+M = 0$, contradiction.)

**Remark 5.1.34'':**   If a projective module $P$ happens to be in $R\text{-}\mathcal{G}\imath\text{-}\mathcal{M}od$ then there is an epic $F \to P$ in $R\text{-}\mathcal{G}\imath\text{-}\mathcal{M}od$ for $F$ a suitable graded free $G(R)$-module (cf., remark 1.9.7), and the epic splits in $R\text{-}\mathcal{G}\imath\text{-}\mathcal{M}od$ by remark 1.9.6. In view of remark 1.9.5 we see if p.d. $M = t < \infty$ and $M$ is graded then $M$ has a projective resolution of length $t$ in $R\text{-}\mathcal{G}\imath\text{-}\mathcal{M}od$.

In view of remark 5.1.34'' there is no ambiguity in talking about the graded f.g. projective $R$-modules, which we call $\mathcal{G}\imath\text{-}\mathit{proj}(R)$, viewed as a full subcategory of $R\text{-}\mathcal{G}\imath\text{-}\mathcal{M}od$.

**Claim 1:**   *The functor $F = R \otimes_{R_0}$— (cf., remark 1.9.11) is a category equivalence $\mathcal{G}\imath\text{-}\mathit{proj}(R_0) \to \mathcal{G}\imath\text{-}\mathit{proj}(R)$.*

**Proof of Claim 1:**   Take the functor $G = R_0 \otimes_R$—$: R\text{-}\mathcal{G}\imath\text{-}\mathcal{M}od \to R_0\text{-}\mathcal{G}\imath\text{-}\mathcal{M}od$. Since $GF$ is naturally equivalent to 1 it suffices to show $P \approx FGP$ (graded) for every $P$ in $\mathcal{G}\imath\text{-}\mathit{proj}(R)$. Let $Q = GP \approx P/R_+P$ by example 1.7.21'; since $Q$ is projective the epic $P \to Q$ splits in $R_0\text{-}\mathcal{M}od$, and thus in $R_0\text{-}\mathcal{G}\imath\text{-}\mathcal{M}od$ by remark 1.9.6. Hence there is a graded monic $f: Q \to P$. Now $FGP = R \otimes_{R_0} Q$ so we can define a graded map $\varphi: FGP \to P$ such that $\varphi(r \otimes x) = rfx$ for $r$ in $R$ and $x$ in $Q$. Since $G$ right exact and $G\varphi$ is an isomorphism we see $G(\operatorname{coker}\varphi) = 0$, i.e., $\operatorname{coker}\varphi = R_+ \operatorname{coker}\varphi$, implying $\operatorname{coker}\varphi = 0$ by remark 5.1.34'. But then $\varphi$ is epic and thus split, so $G(\ker\varphi) = 0$, likewise implying $\ker\varphi = 0$. Thus $\varphi$ is an isomorphism, proving claim 1.

The next part of the proof of the theorem involves two clever tricks, the first of which is very cheap. Grade $R[\lambda]$ by putting $R[\lambda]_n = \bigoplus_{i \leq n} R_i \lambda^{n-i}$. Then $R[\lambda]_0 = R_0$, so $R$ is replaced by the larger graded ring $R[\lambda]$. The second trick is a "homogenization map" $\psi_d$, given by $\psi_d r = r\lambda^{d-n}$ for any $r$ in $R_n$. This defines the map $\psi_d: \bigoplus_{n \leq d} R_n \to R[\lambda]_d$, so $\psi_d$ homogenizes all elements of degree $\leq d$. Now define the functor $H: R[\lambda]\text{-}\mathcal{G}\imath\text{-}\mathcal{F}imod \to R\text{-}\mathcal{F}imod$ given by $HM = M/(1 - \lambda)M$, i.e., we specialize $\lambda$ to 1. Then $H$ is given by $R \otimes_{R[\lambda]}$—, where we view $R$ as the $R[\lambda]$-module $R[\lambda]/R[\lambda](1 - \lambda)$ as in example 1.7.21'. In particular, $H$ is right exact.

**Claim 2:** *For every f.g. R-module N there is an f.g. graded $R[\lambda]$-module M with $HM = N$.*

**Proof of Claim 2:** Write $N \approx R^{(n)}/K$ where $K = \sum_{i=1}^{t} Rx_i$. Writing $x_i = (x_{i1}, \ldots, x_{in})$ in $R^{(n)}$, we take $d$ greater than the maximum of the degrees of all the $x_{ij}$, and define $x_i' = (\psi_d x_{i1}, \ldots, \psi_d x_{in}) \in (R[\lambda]_d)^{(n)}$. Letting $K' = \sum_{i=1}^{t} R[\lambda]x_i'$, a graded submodule of $R[\lambda]^{(n)}$, we have $H(R[\lambda]^{(n)}/K') \approx H(R[\lambda]^{(n)})/HK' \approx R^{(n)}/K \approx N$ since $H$ is right exact.

**Claim 3:** *H is exact.*

**Proof of Claim 3:** We need to show $H$ is left exact, i.e., given a monic $f: N \to M$ in $R[\lambda]\text{-}\mathscr{G}\imath\text{-}\mathscr{F}\imath mod$ then $Hf: HN \to HM$ is monic. But the canonical map $N \to HM = M/(1 - \lambda)M$ has kernel $N \cap (1 - \lambda)M$, so it suffices to show $N \cap (1 - \lambda)M \le (1 - \lambda)N$. Suppose $x = \sum x_n \in M$ with $(1 - \lambda)x \in N$, where the $x_n$ are homogeneous. $x_0 = ((1 - \lambda)x)_0 \in N$, and $x_i - \lambda x_{i-1} = ((1 - \lambda)x)_i \in N$ for each $i \ge 1$, so by induction we see each $x_i \in N$. Thus $x \in N$, so $(1 - \lambda)x \in (1 - \lambda)N$ as desired.

**Proof of the Theorem:** Suppose $N$ is an f.g. R-module. Write $N = HM$ for $M \in R[\lambda]\text{-}\mathscr{G}\imath\text{-}\mathscr{M}od$, by claim 2. $R[\lambda]$ is Noetherian of finite gl. dim, by proposition 5.1.19. Thus $M$ has an f.g. projective resolution

$$0 \to P_n \to \cdots \to P_0 \to M \to 0,$$

which by remark 5.1.34″ can be taken to be graded. Then $0 \to HP_n \to \cdots \to HP_0 \to N \to 0$ is a projective resolution of $N$ by claim 3. On the other hand, $R[\lambda]_0 = R_0$ so claim 1 shows each $P \approx R[\lambda] \otimes_{R_0} Q_i$ for suitable f.g. projective $R_0$-modules $Q_i$, which by hypothesis are all stably free. Then $HP_i \approx R \otimes_{R_0} Q_i$ are stably free. Thus $K_0(R) = \langle [R] \rangle$ by proposition 5.1.31.

Q.E.D.

We now have quite a wide assortment of rings $R$ with $K_0(R) = \langle [R] \rangle$.

**Corollary 5.1.35:** *Suppose R is left Noetherian, gl. dim $R < \infty$. The hypothesis "every f.g. projective is stably free" passes from R to the following rings:*

(i) $R[\lambda; \sigma]$ *for any monic $\sigma: R \to R$.*
(ii) $R[\lambda_1, \ldots, \lambda_t]$ *for commuting indeterminates $\lambda_1, \ldots, \lambda_t$.*

In particular, if $R$ is a PLID then every f.g. projective $R[\lambda_1, \ldots, \lambda_t]$-module is stably free.

*Proof:*

(i) $R[\lambda;\sigma]$ is $\mathbb{N}$-graded according to degree and is left Noetherian (proposition 3.5.2) of finite gl. dim (corollary 5.1.25).

(ii) Apply (i) $t$ times.

To see the last assertion note $K_0(R) = \langle [R] \rangle$ and gl. dim $R = 1$ since $R$ is hereditary.     Q.E.D.

The hypothesis $R$ left Noetherian of finite gl. dim could be weakened throughout to "Every f.g. $R$-module has an f.g. projective resolution of finite length," without change in proof. Such rings are called (*homologically*) *regular.*

# O Supplement: Quillen's Theorem

One good theorem deserves another, and Quillen [73] proved a startling extension of Serre's theorem. Before stating Quillen's theorem let us make another definition.

**Definition 5.1.36:** A ring $R$ is *filtered* if $R$ has a chain of additive sub-groups $R_0 \subseteq R_1 \subseteq \cdots$ such that $R_i R_j \subseteq R_{i+j}$ for all $i, j$ and $\bigcup_{i \in \mathbb{N}} R_i = R$.

Note that this concept is not new to us; $R$ is filtered iff $R$ has a filtration over $(\mathbb{Z}, +)$ for which $R(1) = 0$. (Just take the new $R_i$ to be the old $R_{-i}$). Thus example 1.8.14(i) and the results applicable to that example can be used here, and, in particular, we can form the associated graded ring $G(R)$; also note that $R_0$ is a subring of $R$. The terminology introduced here has become standard when applied to enveloping algebras, and the results to be given here are fundamental in the theory of enveloping algebras, cf., §8.4.

**Quillen's Theorem:** *Suppose $R$ is filtered, such that the graded ring $G(R)$ is Noetherian and of finite* gl. dim. *If $G(R)$ is flat as $R_0$-module then the natural injection $R_0 \to R$ induces an isomorphism $\varphi: K_0(R_0) \to K_0(R)$.*

Quillen actually proved a much stronger result, that the corresponding isomorphism also holds in the "higher" $K$-theories. Our exposition (also, cf., the appendix and exercise 13) follows the very readable account of McConnell [85], which in turn relies on Roy [65].

The main object is to develop the homological module theory of filtered rings to the same stage that we developed the module theory of graded rings. To this end we assume throughout this discussion that $R$ is a filtered ring; we say an $R$-module $M$ is *filtered* if $M$ has a chain of subgroups $M_0 \subseteq M_1 \subseteq \cdots$ such that $R_i M_j \leq M_{i+j}$ and $\bigcup_{i \in \mathbb{N}} M_i = M$. $R$ is itself filtered as $R$-module in the natural way; on the other hand, any $R$-module $M$ can be *trivially filtered* by putting $M_0 = M$.

A map $f: M \to N$ of filtered $R$-modules will be called *filtered* if $fM_i \subseteq N_i$ for all $i$; $f$ is *strictly filtered* if $fM_i = N_i \cap fM$ for all $i$. If $N$ is a filtered module and $f: M \to N$ is an arbitrary map then $M$ can be filtered in such a way that $f$ is strictly filtered; namely, put $M_i = f^{-1}N_i$.

If $M$ is a filtered $R$-module we define the associated *graded module* $G(M) = \bigoplus(M_{i+1}/M_i)$, viewed naturally as $G(R)$-module by the rule

$$(r_i + R_{i+1})(x_j + M_{j+1}) = r_i x_j + M_{i+j+1}$$

for $r_i$ in $R_i$ and $x_i$ in $M_i$. Given $x$ in $M_i$ we shall write $\bar{x}$ for the corresponding element of $G(M)$.

Any filtered map $f: M \to N$ gives rise naturally to a map $G(f): G(M) \to G(N)$, so $G$ is a functor. The point of strictly filtered maps lies in the following "exactness" feature of this functor:

If $M'' \overset{f}{\to} M \overset{g}{\to} M'$ is exact with $f, g$ strictly filtered then

$$G(M'') \xrightarrow{G(f)} G(M) \xrightarrow{G(g)} G(M') \qquad \text{is exact.}$$

We say $M$ is *filtered-free* if $G(M)$ is graded free over $G(R)$, cf., remark 1.9.7.

**Remark 5.1.37:** Any filtered-free $R$-module is free, for if $\{\bar{e}_i : i \in I\}$ is a base of homogeneous elements of $G(M)$ then $\{e_i : i \in I\}$ is a base of $M$; in fact, writing $e_i \in M_{n(i)}$ we have

$$M_k = \sum_{n(i) \leq k} R_{k-n(i)} e_i.$$

Conversely, any graded free $G(R)$-module $\tilde{F}$ gives rise to a filtered-free $R$-module $F$ such that $G(F) = \tilde{F}$. (Proof. ($\Rightarrow$) Clearly the above holds for $k = 0$, and thus is easily seen to hold by induction for all $k$, by passing to $G(M)$ at each step; likewise, any dependence would be reflected in some $M_k$ and thus pass to $G(M)_k$. ($\Leftarrow$) Let $\{\bar{e}_i : i \in I\}$ denote a base of $\tilde{F}$ over $G(R)$ and define $F = R^{(I)}$. Labeling a base of $F$ as $\{e_i : i \in I\}$ use the above equation to filter $F$; then $G(F) = \tilde{F}$ by inspection.)

Now that we want to see that our object is really "free".

**Lemma 5.1.38:** *Suppose $M$ is a filtered $R$-module.*

(i) *If F is a filtered-free module then for every graded map $\tilde{f}: G(F) \to G(M)$ there is a filtered map $f: F \to M$ such that $G(f) = \tilde{f}$.*

(ii) *There exists a filtered-free module F and a strictly filtered epic $F \to M$.*

(iii) *If P is filtered and $G(P)$ is $G(R)$-projective then P is projective.*

### Proof:

(i) Let $\{e_i : i \in I\}$ be a base of $F$ taken as in remark 5.1.37, and define $f: F \to M$ by $\overline{f e_i} = \tilde{f} \bar{e}_i$. Then $G(f) = \tilde{f}$ by construction, and, obviously, $f$ is filtered.

(ii) Let $\tilde{F}$ be a graded-free $G(R)$-module, together with a graded epic $\tilde{f}: \tilde{F} \to G(M)$. By remark 5.1.37 we can write $\tilde{F}$ in the form $G(F)$, so by (i) we have a filtered epic $f: F \to M$. It remains to show $f$ is strictly filtered, i.e., $f: F_i \to M_i$ is epic for each $i$; we do this by induction on $i$, noting it is clear for $i = 0$ and thus follows for general $i$ by applying proposition 2.11.15 to the diagram

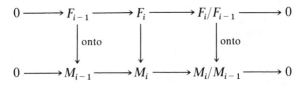

(iii) By (ii) we have an exact sequence $0 \to N \xrightarrow{f} F \xrightarrow{g} P \to 0$ where $F$ is filtered-free and $f, g$ are strictly filtered. Applying $G(\ )$ yields an exact sequence $0 \to G(N) \to G(F) \to G(P) \to 0$ which splits by remark 1.9.6.

We need to show $g$ is split, i.e. $g\mu = 1_P$ for some $\mu: P \to F$. Note $P_i/P_{i-1}$ is a summand of $G(P)$ and thus is projective; hence the epic $P_i \to P_i/P_{i-1}$ splits. But then $P_i \approx P_{i-1} \oplus (P_i/P_{i-1})$, so we can build $\mu$ inductively on $i$ (i.e., piecing together $P_i/P_{i-1} \hookrightarrow F_i/F_{i-1}$ and $P_{i-1} \hookrightarrow F_{i-1}$ to get $P_i \hookrightarrow F_i$).     Q.E.D.

**Proposition 5.1.39:**   *For any filtered module M we have $\operatorname{pd}_R M \le \operatorname{pd}_{G(R)} G(M)$.*

**Proof:**   Let $n = \operatorname{pd}_{G(R)} G(M)$. There is nothing to prove unless $n$ is finite, so we proceed by induction on $n$. Lemma 5.1.38(iii) gives the result for $n = 0$. For $n > 0$ take an exact sequence $0 \to N \to F \to M \to 0$ where $F$ is filtered-free and the maps are strictly filtered, cf., lemma 5.1.38(ii). Then we get an exact sequence $0 \to G(N) \to G(F) \to G(M) \to 0$. But $\operatorname{pd}_{G(R)} G(N) \le n - 1$ by summary 5.1.13, so $\operatorname{pd}_R N \le n - 1$ by induction, implying $\operatorname{pd}_R M \le n$ by summary 5.1.13.     Q.E.D.

**Corollary 5.1.40:** *If $R$ is a filtered ring then* gl. dim $R \leq$ gl. dim $G(R)$.

**Proof:** Any $R$-module $M$ can be filtered trivially, and then by the proposition $\mathrm{pd}_R M \leq \mathrm{pd}_{G(R)} G(M) \leq$ gl. dim $G(R)$. Thus gl. dim $R \leq$ gl. dim $G(R)$.
Q.E.D.

**Digression:** These results actually yield more. We could define the *graded* pd of a module and graded gl. dim of a ring, in terms of graded projective resolutions. Then we see at once the graded gl. dim of a graded ring equals its usual gl. dim.

We are ready for our watered-down version of Quillen's theorem.

**Theorem 5.1.41:** *Suppose $R$ is filtered and $G(R)$ is Noetherian of finite gl. dim. If $K_0(R_0) = \langle [R_0] \rangle$ then $K_0(R) = \langle [R] \rangle$.*

**Proof:** $R$ is Noetherian by corollary 3.5.32(i), and has finitely gl. dim by corollary 5.1.40. We want to show that every f.g. projective $R$-module $P$ is stably free. Viewing $P$ as filtered trivially, we see $G(P)$ is f.g. over $G(R)$ and thus we can build a graded resolution of f.g. graded free modules, as in remark 5.1.34''. By hypothesis the $n$-th syzygy $K_n$ is projective and is the kernel of a graded map, so is graded projective. By claim 1 (following remark 5.1.34'') there is a projective $R_0$-module $Q_0$ with $K_n \approx G(R) \otimes_{R_0} Q_0$. But $Q_0$ is stably free by hypothesis, so $K_n$ is stably free as $G(R)$-module. Thus $K_n$ is "graded" stably free by remark 5.1.34'', so just as in lemma 5.1.29 we have a graded FFR of $G(P)$ having finite length. Using lemma 5.1.38 we have the resolution in the form $0 \to G(F_n) \to G(F_{n-1}) \to \cdots \to G(P) \to 0$, and this "pulls back" to an FFR $0 \to F_n \to F_{n-1} \to \cdots \to P \to 0$ in $R$-$\mathcal{Mod}$, as needed.

## Euler Characteristic

The FFR property has an important tie to topology.

**Definition 5.1.42:** Suppose $R$ has IBN. The *Euler characteristic* of a module $M$ with an f.g. free resolution $0 \to F_n \to \cdots \to F_0 \to M \to 0$ is defined as $\chi(M) = \sum_{i=0}^{n} (-1)^i \mathrm{rank}(F_i)$.

**Remark 5.1.43:** $\chi(M)$ is independent of the FFR because of the generalized Schanuel lemma. In particular, if $0 \to F_n \to \cdots \to F_0 \to 0$ is exact then $\sum (-1)^i \mathrm{rank}(F_i) = 0$.

**Remark 5.1.44:** Suppose $S$ is a left denominator set of $R$ and $S^{-1}R$ also has IBN. If an $R$-module has FFR then $S^{-1}M$ has FFR in $S^{-1}R\text{-}\mathcal{M}od$, and $\chi(S^{-1}M) = \chi(M)$. (Just localize the resolution, since the localization functor is exact and preserves "free" and the rank.)

The name "Euler characteristic" comes from Euler's observation that $\#$ faces $-\ \#$ edges $+\ \#$ vertices of a simplicial complex is a topological invariant. The standard reference on properties of the Euler characteristic is Bass [76].

**Theorem 5.1.45:** *(Walker [72]) If $R$ is a prime left Noetherian ring of finite gl. dim with $K_0(R) = \langle [R] \rangle$ then $R$ is a domain.*

**Proof:** By Goldie's theorem $R$ is a left order in a simple Artinian ring $S^{-1}R \approx M_n(D)$. We claim $n = 1$, which would prove $R$ is a domain. Indeed, $L = R \cap M_n(D)e_{11}$ is a left ideal of $R$ which is nonzero since $M_n(D)$ is an essential extension of $R$; thus $S^{-1}L \leq M_n(D)e_{11}$ so $0 < [S^{-1}L\!:\!D] \leq n$. On the other hand, take an FFR $0 \to F_m \to \cdots \to F_0 \to L \to 0$. Then each $S^{-1}F_i$ is a free $S^{-1}R$-module so $n^2$ divides $[S^{-1}F_i\!:\!D]$. Hence remark 5.1.43 shows $n^2$ divides $[S^{-1}L\!:\!D]$, a contradiction unless $n = 1$.     Q.E.D.

Walker's theorem is more general, with "semiprime" replacing "prime", cf., exercise 14. However, the proof given here will be relevant when we consider group rings later.

## O  Supplement: Serre's Conjecture

One of the outstanding recent problems in algebra has been to verify *Serre's conjecture*: Every f.g. projective module over the polynomial ring $R = F[\lambda_1, \ldots, \lambda_t]$ is free, for any field $F$. This is trivial for $n = 0$ and is easy for $n = 1$ since $R$ is a PID (and so every submodule of a free module is free, by corollary 2.8.16). Attempts to verify Serre's conjecture for larger $n$ spurred the development of algebraic $K$ theory, a deep and useful subject which also ties in to results about division algebras; Serre's conjecture was finally established by Quillen [76] and Suslin [76], two masters of $K$-theory, although the solutions were (surprisingly) not so difficult. Following Rotman [79B] we present Vaserstein's simplification of Suslin's proof.

The solution is comprised of two parts:

(i) f.g. projective modules are stably free (by corollary 5.1.35);
(ii) Stably free modules are free, which we consider now.

The techniques are not homological, but instead involve linear algebra over PIDs. Thus this part of the proof lies in the realm of noncommutative algebra, although the theorem itself is over a commutative ring. We say a vector $v = (r_1, \ldots, r_n) \in R^{(n)}$ is *unimodular* if $\sum_{i=1}^{n} r_i r_i' = 1$ for suitable $r_i'$ in $R$; $v$ is *strongly unimodular* if $v$ is the first row of an invertible matrix (in $GL(n, R)$). Any strongly unimodular vector certainly is unimodular, and we are interested in the converse. In the following discussion let $e_1, \ldots, e_n$ be a standard base of $R^{(n)}$.

### Remark 5.1.46:

(i) $v = (r_1, \ldots, r_n)$ is unimodular iff there is an epic $f: R^{(n)} \to R$ with $fv = 1$. (Proof: ($\Rightarrow$) Given $r_1, \ldots, r_n'$ with $\sum r_i r_i' = 1$ define $f: R^{(n)} \to R$ by $f(x_1, \ldots, x_n) = \sum x_i r_i'$. ($\Leftarrow$) $1 = fv = f(\sum r_i e_i) = \sum r_i f e_i$, so take $r_i' = f e_i$.)

(ii) $v$ is strongly unimodular iff there is an invertible transformation $T: R^{(n)} \to R^{(n)}$ for which $T e_1 = v$ (seen at once by viewing $T$ as a matrix).

*Digression 5.1.47:* Remark 5.1.47(i) enables us more generally to define an element $x$ of an arbitrary $R$-module $M$ to be *unimodular* iff $Rx$ is a summand of $M$ isomorphic to $R$.

We shall describe "stably free implies free" in terms of the *unimodular vector property*, which is defined as the property that all unimodular vectors are strongly unimodular.

*Remark 5.1.48:* Suppose $P = R^{(n)}/Rv$ where $v$ is a strongly unimodular vector of $R^{(n)}$. Then $P \approx R^{(n-1)}$. (Indeed, take an invertible transformation $T: R^{(n)} \to R^{(n)}$ with $T e_1 = v$. Then $P = R^{(n)}/Rv = TR^{(n)}/T(Re_1) \approx T(R^{(n)}/Re_1) \approx R^{(n-1)}$.)

**Proposition 5.1.49:** *Suppose $R$ has IBN (invariant base number). Every f.g. stably free $R$-module is free iff $R$ has the unimodular vector property. (Note: "f.g." actually is superfluous by exercise 10.)*

*Proof:* ($\Rightarrow$) Suppose $v_1$ is a unimodular vector. Taking $f: R^{(n)} \to R$ epic with $fv_1 = 1$ we note $R^{(n)} \approx R \oplus \ker f$ since $R$ is projective; by hypothesis $\ker f$ is free with some base $\{v_2, \ldots, v_n\}$. Thus we have an invertible transformation $T: R^{(n)} \to R^{(n)}$ given by $T e_i = v_i$, corresponding to an invertible matrix whose first row is $v_1$.

($\Leftarrow$) First assume $P \oplus R \approx R^{(n)}$, i.e., the canonical map $f: R^{(n)} \to P$ has kernel $Rv \approx R$ for suitable $v$ in $R^{(n)}$. Since $f$ splits we see by remark 5.1.46 that

$v$ is unimodular and thus strongly modular, so $P$ is free by remark 5.1.48. By induction on $m$ we now see that if $P \oplus R^{(m)} \approx R^{(n)}$ then $P$ is free.     Q.E.D.

We see from proposition 5.1.49 ($\Rightarrow$) that local commutative rings have the unimodular vector property, since all projectives are free.

**Proposition 5.1.50:** (*Horrocks*) *Suppose $C$ is a local commutative ring, and $v = (f_1,\ldots,f_n)$ is a unimodular vector in $C[\lambda]^{(n)}$. If some $f_i$ is monic then $v$ is strongly unimodular.*

**Proof:** Reordering the $f_i$ we may assume $f_1$ is monic. Suppose $m = \deg(f_1)$. We can use $f_1$ to cancel higher degree terms in the other entries; since this involves elementary transformations and thus affects neither unimodularity nor strong unimodularity, we may assume $\deg(f_i) < m$ for all $i > 1$. For $m = 0$ we have $v \in C^{(n)}$ is strongly unimodular. Thus we may induct on $m > 0$.

For $n = 1$ the assertion is vacuous. For $n = 2$ we have $(g_1,g_2)$ in $C[\lambda]^{(2)}$ such that $f_1 g_1 + f_2 g_2 = 1$, so $\begin{pmatrix} f_1 & f_2 \\ -g_2 & g_1 \end{pmatrix}$ is invertible. For $n \geq 3$, letting $^-$ denote the canonical image in the field $\bar{C} = (C/\mathrm{Jac}(C))$, we see $\bar{v}$ is unimodular in $\bar{C}[\lambda]^{(n)}$. But $\bar{f}_1 = \lambda^m + \cdots$ is not invertible in $\bar{C}[\lambda]$ since $m > 0$, so $\bar{f}_i \neq 0$ for some $i \geq 2$. Assume $\bar{f}_2 \neq 0$. Then some coefficient of $f_2$ is not in $\mathrm{Jac}(C)$ and thus is invertible. If $C[\lambda]f_1 + C[\lambda]f_2$ contains a monic polynomial of degree $m - 1$ then using elementary transformations we can change $f_3$ to a monic of degree $m - 1$, and are done by induction. Thus it remains to prove the next result.     Q.E.D.

**Lemma 5.1.51:** (*Suslin*) *Suppose $f = \lambda^m + \sum_{i=1}^m a_i \lambda^{m-i}$ and $g = \sum_{i=1}^m b_i \lambda^{m-i}$ in $C[\lambda]$. Then for any $j < m$ the ideal $C[\lambda]f + C[\lambda]g$ has a polynomial $h$ of degree $\leq m - 1$ and leading coefficient $b_j$.*

**Proof:** Induction on $j$. For $j = 1$ we take $h = g$. In general, let $A = C[\lambda]f + C[\lambda]g$ and define $g' = \lambda g - b_1 f = \sum_{i=1}^m b_i' \lambda^{m-i} \in A$, where $b_i' = b_{i+1} - b_1 a_i$. By induction $b_{j-1}'$ is the leading coefficient of a polynomial $h'$ in $A$ having degree $m - 1$. But now $h' + a_{j-1}g$ has leading coefficient $b_j - b_1 a_{j-1} + a_{j-1}b_1 = b_j$, as desired.     Q.E.D.

Next we need the following passage from $C[\lambda]^{(n)}$ to $C[\lambda]$.

**Lemma 5.1.52:**   (*Suslin*) *Suppose* $v = (f_1(\lambda), \ldots, f_n(\lambda))$ *is a unimodular vector with some* $f_i$ *monic. Then, viewing* $v$ *as* $1 \times n$ *matrix, we have* $v = wA$ *for some* $A \in GL(n, C[\lambda])$ *and* $w \in C^{(n)}$.

**Proof:**   Write $G = GL(n, C[\lambda])$. Given any $h$ in $C[\lambda]$ let $v(h)$ denote the vector $(f_i(h)) \in C[\lambda]^{(n)}$. (Thus $v = v(\lambda)$.) Clearly $v(0) \in C^{(n)}$, and we shall prove the result by showing $v \in v(0)G$. Let $I = \{c \in C : v \in v(g + c\lambda h)G$ for all $g, h$ in $C[\lambda]\}$, an ideal of $C$. If $I = C$ we are done by taking $g = \lambda$, $c = 1$, and $h = -1$. Thus we assume $I \neq C$ and aim for a contradiction.

Taking a maximal ideal $P \supset I$ of $C$ we work in $C_P[\lambda, \mu]$ where $\mu$ is another indeterminate commuting with $\lambda$ and $C$. By proposition 5.1.50 $v(\lambda + \mu)$ is the first row of some matrix $A = A(\lambda + \mu)$ in $GL(n, C_P(\lambda + \mu))$, i.e., $v(\lambda + \mu) = e_{11}A$. Write $A_0$ for the image of $A$ under $\mu \to 0$, also an invertible matrix. Then $B = A^{-1}A_0 \in GL(n, C_P[\lambda, \mu])$ satisfies

$$v(\lambda + \mu)B = e_{11}A_0 = v. \tag{12}$$

View $B = B(\lambda, \mu)$ as a matrix whose entries are functions in $\lambda, \mu$; and likewise for $B^{-1} = B^{-1}(\lambda, \mu)$. Then $B(\lambda, 0) = A_0^{-1}A_0$ is the identity matrix, so the coefficients for $\lambda^i$ for $i > 0$ must all be 0. There is some $c \notin P$ such that $cB$ and $cB^{-1}$ are in $C[\lambda, \mu]$. It follows that $B(\lambda, c\mu) \in M_n(C[\lambda, \mu])$ with inverse $B^{-1}(\lambda, c\mu)$, and (12) implies $v = v(\lambda + c\mu)B(\lambda, c\mu)$. Specializing $\lambda \to g$ and $\mu \to h\lambda$ shows $c \in I$, contrary to $c \notin P$, the desired contradiction.

Q.E.D.

**Theorem 5.1.53:**   *Every f.g. projective* $F[\lambda_1, \ldots, \lambda_t]$*-module is free for any field F.*

**Proof:**   Induction on $t$; $t = 0$ is obvious. We need to verify the unimodular vector property. Let $v = (f_i(\lambda_1, \ldots, \lambda_t))$ be a unimodular vector with $f_1 \neq 0$. By the Noether Normalization Theorem (cf., Lang [64B, p. 260]) we have some $\alpha$ in $F$ such that $\alpha f_1$ is monic over $\mu_1, \ldots, \mu_{t-1}$ where the $\mu_i$ are algebraically independent polynomials. By Suslin's lemma $\alpha v \in GL(n, K[\lambda_1, \ldots, \lambda_t])w$ for some $w$ in $F[\mu_1, \ldots, \mu_{t-1}]$. Clearly, $w$ is a unimodular vector and thus by induction is strongly unimodular. Hence $\alpha v$ is strongly unimodular, and so $v$ also is strongly unimodular, as desired.   Q.E.D.

Of course, theorem 5.1.53 is the solution to Serre's conjecture. This proof illustrates one of the features of the Russian school—translate a problem to matrices and tackle it there. Quillen's solution is more structural and also has several very interesting ideas, cf., Quillen [76]. An excellent treatment of Serre's conjecture is to be found in Lam [78B].

## M Supplement: Cancellation and Its Consequences

Unfortunately, the Quillen-Suslin theorem does not generalize directly to noncommutative rings, in view of the fact that $D[\lambda_1, \lambda_2]$ has nonfree projectives for any noncommutative division ring $D$ (cf., exercise 9). Nevertheless, there are noncommutative results to be had, when the projective module is required to have sufficiently large rank. Suslin [79] lists a host of interesting theorems, including the following:

Suppose $R$ is f.g. as a module over a central Noetherian ring $C$ of classical Krull dimension $n$. Let $T$ be any ring obtained by localizing $R[\lambda_1, \ldots, \lambda_n]$ by a submonoid generated by various $\lambda_i$. Then any projective $T$-module of rank $> \max(1, n)$ has the form $T \otimes_R M$ for a suitable $R$-module $M$.

In particular, if $D$ is a division algebra of dimension $2^2 = 4$ over its center then every projective $D[\lambda_1, \ldots, \lambda_n]$-module of rank $\geq 5$ is free. Actually, by using theorem 3.5.72 we can obtain some rather sweeping results which imply, for example, that "big enough" f.g. projective modules over any Noetherian ring $R$ with $K_0(R) = \langle [R] \rangle$ are free. Recall $s(R)$ from definition 3.5.71.

**Lemma 5.1.54:** *Suppose $P \oplus R \approx R^{(n)}$ and $s(R) < n$. Then $P$ is free.*

**Proof:** More precisely, $P \oplus Rv = R^{(n)}$ where $v$ is unimodular, cf., the proof of proposition 5.1.49 ($\Leftarrow$). Writing $v = (a_1, \ldots, a_n)$ in $R^{(n)}$ we thus have $\sum Ra_i = R$, so the stable range condition yields $r_i$ in $R$ for which

$$\sum_{i < n} R(a_i + r_i a_n) = R,$$

i.e., $v$ is a unimodular vector in $R^{(n-1)}$ (viewed in $R^{(n)}$) under the corresponding change of base. Hence $Rv$ is a summand of $R^{(n-1)}$, so writing $R^{(n-1)} = Rv \oplus P'$ we get

$$R^{(n)} \approx (Rv \oplus P') \oplus R = Rv \oplus (P' \oplus R) = Rv \oplus R^{(n-1)}.$$

Factoring out $Rv$ yields $P = R^{(n)}/Rv \approx R^{(n-1)}$.        Q.E.D.

(This is sharp even for $R$ commutative.)

**Proposition 5.1.55:** *Suppose $s(R) = k$. If $P$ is an f.g. stably free module which cannot be generated by $k$ elements then $P$ is free.*

***Proof:*** Write $P \oplus R^{(m)} \approx R^{(n)}$ and induct on $m$, the case $m = 1$ given by the lemma (since, clearly, $n > k$).     Q.E.D.

**Corollary 5.1.56:** *If $R$ is left Noetherian with $K_0(R) = \langle [R] \rangle$, and if $K$-dim $R = k$, then every f.g. projective which cannot be generated by $k$ elements is free.*

***Proof:*** Apply theorem 3.5.72′ to the proposition.     Q.E.D.

Projective modules of low rank need not be free, as we mentioned earlier, also, cf., §8.4. The following two properties obviously are related to the freeness of projectives:

*Property 1.* (Cancellation) If $M \oplus R^{(n)} \approx N \oplus R^{(n)}$ then does $M \approx N$?

*Property 2.* Does a given module have $R$ as a summand?

These properties hold under much more general conditions than the Quillen-Suslin theorem. The exposition here follows Stafford [82]. We recall the notation of the normalized reduced rank

$$\hat{g}(M, P) = \rho_{R/P}(M/PM)/\rho_{R/P}(R/P) \qquad \text{where } \rho \text{ is the reduced rank.}$$

**Lemma 5.1.57:** *Suppose $M, N$ are modules over a left Noetherian ring $R$, and $K$ is an f.g. projective $R$-module such that $\hat{g}(K, P_j) \geq \hat{g}(M, P_j)$ for given prime ideals $P_1, \ldots, P_m$ of $R$. Given maps $f: K \to M$ and $f': N \to M$ we can find $h: K \to N$ for which*

$$\hat{g}(M/(f + f'h)K, P_j) = \hat{g}(M/(fK + f'N), P_j) \qquad \text{for each } 1 \leq j \leq m.$$

***Proof:*** First assume $m = 1$ and let $P = P_1$. Since $K$ is projective any map $\bar{h}: K/PK \to N/PN$ lifts to a map $h: K \to N$ via the diagram

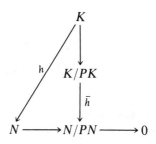

Thus we can assume that $P = 0$, i.e., $R$ is prime left Noetherian and so an order in a simple Artinian ring $Q$. Thus we are given $\rho(M) \leq \rho(K)$ and, letting $M' = fK + f'N$ we want to find $h: K \to N$ for which $\rho((f + f'h)K) = \rho(M')$. Since $(f + f'h)K \leq M'$ we can find $h: K \to N$ such that $\rho((f + f'h)K)$ is maximal. Replacing $f$ by $f + f'h$ we may assume $\rho(fK) \geq \rho((f + f'h)K)$ for all $h: K \to N$; we want to show $\rho(fK) = \rho(M')$, or by proposition 3.5.7 that $M'/fK$ is singular.

First assume $\rho(\ker f) > 0$. If, on the contrary, $M'/fK$ is nonsingular then there is some $x$ in $f'N$ such that $(fK)x^{-1}$ is *not* large and thus misses a left ideal $L$ of $R$; in other words, $L \approx (Lx + fK)/fK$. Since $K$ is a summand of $R^{(n)}$ for some $n$ we see by means of tensoring by $Q$ that there is a nonzero map $h_0: K \to L$ such that $h_0(\ker f) \neq 0$. Writing $x = f'a$ for suitable $a$ in $N$ we have a map $h: K \to N$ given by $hy = (h_0 y)a$. Note $f'hK = f'((h_0 K)a) = (h_0 K)f'a = h_0 Kx$; thus $\rho(f'hK) = \rho(h_0 Kx) = \rho(h_0 K) > 0$.

We claim the contradiction $\rho((f + f'h)K) > \rho(fK)$ or, equivalently, $\rho(\ker(f + f'h)) < \rho(\ker f)$. Indeed, if $y \in \ker(f + f'h)$ then $fy = f'hy \in fK \cap Lx = 0$ so $y \in \ker f$. Thus $\ker(f + f'h) \leq \ker f$. On the other hand, $(f + f'h)\ker f = f'h \ker f = h_0(\ker f)x$ and so by construction

$$0 \neq \rho(h_0(\ker f)) = \rho(h_0(\ker f)x) = \rho((f + f'h)\ker f),$$

proving $\rho(\ker(f + f'h)) = \rho(\ker f) - \rho((f + f'h)\ker f) < \rho(\ker f)$, as desired. This contradiction arose from the assumption $\rho(\ker f) > 0$, so we have $\rho(\ker f) = 0$.

But now by hypothesis $\rho(M) \leq \rho(K) = \rho(fK) \leq \rho(M') \leq \rho(M)$, so equality holds and we are done for $m = 1$.

In general we apply the same argument inductively. Namely, we arrange the primes $P_1, \ldots, P_m$ such that if $i < j$ then $P_i \nsubseteq P_j$ (by starting with the largest ones and working down) and assume by induction the result has been proved for $m - 1$. Let $A = P_1 \cap \cdots \cap P_{m-1}$. Now $A \nsubseteq P_m$ by arrangement of the $P_i$, so the canonical image $\bar{A}$ of $A$ in $R/P_m$ is nonzero. But now we reduce to the case $m = 1$ by passing to $R/P_m$, $K/P_m K$, $M/P_m M$, and $AN/P_m N$ since this has the same reduced rank as $N/P_m N$.     Q.E.D.

Recall $b(M, P)$ and $b(M, S)$ from definition 3.5.67, and define $b(M)$ to be $b(M, \text{J-Spec}(R))$ where $\text{J-Spec}(R)$ is the set of prime, semiprimitive ideals of $R$. Remark 3.5.68(iii) shows $b(M) \leq \max \{K\text{-dim } R/P + g(M, P): P \in \text{J-Spec}(R)\}$.

**Proposition 5.1.58:** *Suppose $R$ is left Noetherian with $n = K\text{-dim } R < \infty$, and $K$ is an f.g. projective $R$-module for which $\hat{g}(K, P) \geq K\text{-dim } R/P + 1$*

*for all $P$ in* J-Spec$(R)$. *If* $b(R/f_1 K) \leq n$ *for some* $f_1 : K \to R$ *then there are* $f_2, \ldots, f_{n+1}$ *in* $\text{Hom}(K, R)$ *such that* $b(R^{(n+1)}/(f_i)K) \leq n$ *(viewing* $(f_i)$ *as* $(f_1, \ldots, f_{n+1}) : K \to R^{(n+1)})$.

**Proof:** We shall prove by induction on $t$ that for all $t \leq n + 1$ there are $f_2, \ldots, f_t$ for which $b(R^{(t)}/(f_i)K) \leq n$, where $(f_i) = (f_1, \ldots, f_t)$; the assertion is given for $t = 1$. Let $N = R^{(t)}/(f_i)K$. By induction we may assume that we have $f_2, \ldots, f_t$ with $b(N) \leq n$; our first task is to find $f_{t+1}$. Let $\mathscr{S} = \{P \in \text{J-Spec}(R) : b(N, P) = n \text{ or K-dim } R/P = n\}$. $\mathscr{S}$ is finite by proposition 3.5.70 and corollary 3.5.45'.

Let $R_{t+1}$ denote the submodule $(0, \ldots, 0, R)$ of $R^{(t+1)}$ and view $R^{(t)} < R^{(t+1)}$ via the first $t$ coordinates. Letting $f$ be the composition $(f_i) : K \to R^{(t)} < R^{(t+1)}$ and letting $f' : R_{t+1} \to R^{(t+1)}$ be the canonical inclusion, we have by lemma 5.1.57 some map $h : K \to R_{t+1}$ for which

$$\hat{g}(R^{(t+1)}/(f + f'h)K, P) = \hat{g}(R^{(t+1)}/(fK + f'R_{t+1}, P)$$

$$= \hat{g}(R^{(t)}/fK, P) = \hat{g}(N, P)$$

for all $P$ in $\mathscr{S}$. We take $f_{t+1}$ to be the composition $K \to R_{t+1} \to R$ and $\bar{f} = (f_1, \ldots, f_{t+1}) = f + f'h$.

It remains to show $b(R^{(t+1)}/\bar{f}K, P) \leq n$ for all $P$ in J-Spec$(R)$. If $P \in \mathscr{S}$ we have

$$b(R^{(t+1)}/\bar{f}K, P) \leq g(R^{(t+1)}/\bar{f}K, P) + \text{K-dim}(R/P) = g(N, P) + \text{K-dim } R/P \leq n.$$

(The last inequality is clear if $g(N, P) = 0$, and if $g(N, P) > 1$ then $g(N, P) + \text{K-dim } R/P = b(N, P) \leq n$.) On the other hand, if $P \notin \mathscr{S}$ we use the exact sequence

$$0 \to R_{t+1}/(R_{t+1} \cap (\bar{f}K + PR^{(t+1)})) \to R^{(t+1)}/(\bar{f}K + PR^{(t+1)})$$

$$\to R^{(t)}/((f_i)K + PR^{(t)}) \to 0;$$

since $\hat{g}(M, P) \leq 1$ for any cyclic module, the additivity of reduced rank yields

$$\hat{g}(R^{(t+1)}/\bar{f}K, P) \leq 1 + \hat{g}(N, P).$$

Thus $b(R^{(t+1)}/\bar{f}K, P) \leq \text{K-dim}(R/P) + g(N, P) + 1 \leq b(N, P) + 1 \leq n$ since $P \notin \mathscr{S}$, so we have proved $b(R^{(t+1)}/\bar{f}K) \leq n$ as desired. Q.E.D.

We are finally ready for our main result, followed by the promised applications.

**Theorem 5.1.59:**  *Suppose $R$ is a left Noetherian ring of K-dim $n < \infty$, and $K$ is an f.g. projective $R$-module with $\hat{g}(K, P) \geq$ K-dim $R/P + 1$ for all $P$ in $\mathrm{Spec}(R)$. Given $r \in R$ and $f: K \to R$ for which $R = fK + Rr$, we can find some $h': K \to R$ such that $R = (f + \tilde{r}h')K$, where $\tilde{r}$ denotes the right multiplication map by $r$.*

**Proof:**  Let $P_1, \ldots, P_m$ be the primes of $R$ for which K-dim $R/P_j = n$, cf., corollary 3.5.45'. By lemma 5.1.57 there is $h$ in $\mathrm{Hom}(K, R)$ for which $g(R/(f + \tilde{r}h)K, P_j) = g(R/(fK + Rr), P_j) = 0$ by hypothesis. Let $f_1 = f + \tilde{r}h$. Note $g(R/f_1K, P) \leq 1$ for each $P$ in $\mathrm{Spec}(R)$, since $R/f_1K$ is cyclic. Hence proposition 5.1.58 gives us $f_2, \ldots, f_{n+1}$ in $\mathrm{Hom}(K, R)$ such that $b(R^{(n+1)}/\hat{K}) \leq n$, where $\hat{K} = (f_i)K = \{(f_1x, \ldots, f_{n+1}x): x \in K\}$. Let $e_1, \ldots, e_{n+1}$ be the standard base of $R^{(n+1)}$, and write $\bar{f} = (f_i)$. Clearly $Re_1 = f_1Ke_1 + Rre_1$ so

$$R^{(n+1)} = \hat{K} + Rre_1 + \sum_{i=2}^{n+1} Re_i.$$

But $s(R^{(n+1)}/\hat{K}) \leq n$ by theorem 3.5.72. Hence we have $r_i$ in $R$ such that

$$R^{(n+1)} = \hat{K} + \sum_{i=2}^{n+1} R(e_i + r_ire_1).$$

In particular, $e_1 = (\bar{f_i})x + \sum_{i=2}^{n+1} a_i(e_i + r_ire_1)$ for suitable $x$ in $K$ and $a_i$ in $R$. Matching coefficients of $e_i$ shows

$$0 = f_ix + a_i \qquad \text{for } 2 \leq i \leq n+1$$

$$1 = f_1x + \sum_{i=2}^{n+1} a_ir_ir$$

$$= f_1x - \sum_{i=2}^{n+1} (f_ix)r_ir$$

$$= (fx + (hx)r) - \sum_{i=2}^{n+1} (f_ix)r_ir.$$

Putting $h' = h - \sum_{i=2}^{n+1} \tilde{r_i}f_i$ we have $1 = (f + \tilde{r}h')x$, implying $R = (f + \tilde{r}h')K$ as desired.    Q.E.D.

**Corollary 5.1.60:**  *(Serre's theorem generalized) Suppose $R$ is a left Noetherian ring, and $K$ is an f.g. projective $R$-module satisfying $\hat{g}(K, P) \geq$ K-dim $R/P + 1$ for each $P$ in $\mathrm{J}\text{-}\mathrm{Spec}(R)$. Then $R$ is a summand of $K$.*

**Proof:**  Taking $f = 0$ and $r = 1$ in the theorem, we have an epic $K \to R$ which splits since $R$ is projective.    Q.E.D.

**Corollary 5.1.61:** (*"Cancellation"*) *Hypotheses as in corollary 5.1.60, any isomorphism* $\sigma: K \oplus R \to M \oplus R$ *yields an isomorphism* $K \to M$.

**Proof:** Let $\pi_2: M \oplus R \to R$ be the projection; then there are $f_1$ in $\mathrm{Hom}(K, R)$ and $f_2$ in $\mathrm{Hom}(R, R)$ for which $\pi_2\sigma(k, r) = f_1 k + f_2 r$ for all $k$ in $K$ and $r$ in $R$. But $f_2 = \tilde{r}_2$ for some $r_2$ in $R$ and $R = f_1 K + R r_2$ since $\pi\sigma$ is onto, so theorem 5.1.59 provides $h': K \to R$ for which $R = (f_1 + \tilde{r}_2 h')K$. In particular, there is $x_1$ in $K$ such that $1 - r_2 = (f_1 + \tilde{r}h')x_1$. Define $h: K \oplus R \to K \oplus R$ by

$$h(x, r) = (x + rx_1, r + h'(x + rx_1)),$$

an isomorphism since $h^{-1}$ is given by $h^{-1}(x, r) = (x - rx_1 + (h'x)x_1, r - h'x)$. Then

$$\pi_2\sigma h(0, r) = \pi_2\sigma(rx_1, r + h'(rx_1)) = f_1(rx_1) + f_2 r + f_2 h'(rx_1)$$

$$= r(f_1 x_1 + f_2 h' x_1) + f_2 r = r(1 - r_2) + rr_2 = r,$$

implying $\sigma h(0, r) = (vr, r)$ for some $v: R \to M$. Defining an isomorphism $j: M \oplus R \to M \oplus R$ by $j(x, r) = (x - vr, r)$ we see $j\sigma h$ restricts to the identity on $0 \oplus R$, so we have a natural chain of isomorphisms

$$K \to \frac{(K \oplus R)}{R} \xrightarrow{j\sigma h} \frac{(M \oplus R)}{R} \approx M. \qquad \text{Q.E.D.}$$

This result should be compared to exercise 2.9.16. In certain instances the bounds can be lowered, e.g., polynomial rings over simple Noetherian rings and Weyl algebras. The interested reader should consult Stafford [77, Sections 5,6,7] for such results.

As mentioned earlier, more sweeping results are to be found in Stafford [81] and in Coutinho [86]. One of Coutinho's results involves a sweeping generalization of the Eisenbud-Evans "basic element" theorem. Write $M^*$ for $\mathrm{Hom}(M, R)$. An element $x$ of $M$ is called pre-*basic* if the left ideal $\{fx: f \in M^*\}$ of $R$ contains a nonzero ideal; $x$ is *basic* if its image in the $R/P$-module $M/PM$ is pre-basic for every $P$ in J-Spec$(R)$. Also we say $r$-$rk(M) \geq n$ if for any given $x_1, \ldots, x_{n-1}$ in $M$ and $P$ in $\mathrm{Spec}(R)$ there are $x_n \in M$ and $f \in M^*$ such that $0 = fx_1 = \cdots = fx_{n-1}$ but $fx_n$ is regular modulo $P$. Coutinho [86, theorem 4.6.1] proves a "descent" for basic elements which yields the conclusions of corollaries 5.1.60 and 5.1.61 under the weaker hypothesis that $r$-$rk(K) \geq$ K-dim $R + 1$. His result is very strong in view of the following:

(i) $K$ is not required to be f.g.

(ii) $K$ is not required to be projective!

(iii) The group $K_0(R)$ is easily seen to be generated by those projectives whose $r$-$rk$ $\leq$ K-dim $R$.

As the reader can readily surmise, Coutinho's dissertation is requisite for anyone interested in projective modules over noncommutative rings.

## Large Projective Modules are Free

Since the theory of f.g. projectives is so complicated, one might consider it even harder to study non-f.g. projectives. However, the opposite is true, as Bass [63] so skillfully pointed out.

**Remark 5.1.62:** Suppose there is some infinite cardinal X for which the projective module $M$ is generated by X elements and contains a summand isomorphic to $R^{(X)}$. Then $M$ is free. (Indeed, writing $M \approx M_0 \oplus R^{(X)}$ we have $M_0 \oplus M' \approx R^{(X)}$ for some projective $M'$, so $M_0 \oplus R^{(X)}$ is free by Eilenberg's trick (remark 2.8.4').)

Thus to prove $M$ is free we need merely find a "big" free summand $R^{(X)}$. The first step, of course, is to find when $R$ is isomorphic to a summand of $M$; since $R$ is projective it suffices to find an epic $f: M \to R$. This implies the trace ideal $T(M) = R$, and we have the following converse.

**Lemma 5.1.63:** *Suppose $R/\mathrm{Jac}(R)$ is left Noetherian. If $M = \bigoplus_{i \in I} M_i$ is projective with $I$ infinite and $T(M) = R$ then $R$ is a summand of $M$.*

**Proof:** Let $J = \mathrm{Jac}(R)$. Given $L_t < R$ there is $f_t: M \to R$ with $f_t M \nsubseteq L_t$; hence $f_t M_i \nsubseteq L_t$ for some $i = i(t)$. Now we define the $L_t$ inductively by $L_0 = J$ and $L_{t+1} = L_t + f_t M_{i(t)}$. Then $L_0 < L_1 < \cdots$ so some $L_n = R$ since, otherwise, $R/J$ has an infinite ascending chain of left ideals, which is impossible. Thus $1 = a + \sum_{t=1}^{n} r_t$ for suitable $a$ in $J$ and $r_t$ in $f_t M_{i(t)}$, implying $\sum_{t=1}^{n} r_t$ is invertible so $\bigoplus f_t: \bigoplus_{t=1}^{n} M_{i(t)} \to R$ is epic, as desired.     Q.E.D.

Recall a module is said to be *countable* if it is generated by a countable set of elements.

**Corollary 5.1.64:** *Suppose $R/\mathrm{Jac}(R)$ is left Noetherian and $T(M) = R$ for every projective R-module M. Then every infinite direct sum of countable projectives is free.*

***Proof:*** Take an epic $\pi\colon M \to R$. Writing $1 = \pi x$ we see $x$ lies in $M' = \bigoplus_{i \in I} M_i$ for some finite $I' \subset I$; letting $I'' = I - I'$ and proceeding inductively in $M'' = \bigoplus_{i \in I''} M_i$ we have $T(M'') = R$ by hypothesis, and continuing inductively we have $R^{(X)}$ is a summand of $M$, where $X = |I|$, so $M$ is free by remark 5.1.62.    Q.E.D.

Note the hypothesis of corollary 5.1.64 was rather stronger than needed. We say a module $M$ is *uniformly big* if $M$ is generated by X elements for some cardinal X such that for all $A \lhd R$, $M/AM$ is *not* generated by $<$X elements. Clearly, if $M \approx F \oplus N$ where $F$ is generated by $<$X elements then $N$ is uniformly big also. Furthermore, any uniformly big projective has $T(M) = R$ since $T(M)M = M$ by the dual basis lemma. Thus the argument of corollary 5.1.64 yields

**Proposition 5.1.65:** *Suppose $R/\mathrm{Jac}(R)$ is left Noetherian. Then every uniformly big projective which is an infinite direct sum of countable projectives is free.*

This result gains power when confronted with a theorem of Kaplansky that every projective $M$ is a direct sum of countables. (Kaplansky's theorem is an immediate consequence of exercise 2.9.2.) If $M$ is not countable then by matching cardinality we see $M$ is an infinite direct sum of countables, yielding

**Proposition 5.1.66:** *Suppose $R/\mathrm{Jac}(R)$ is left Noetherian. Any uniformly big projective $M$ which is not countable is free.*

To show any uniformly big projective $M$ is free, we may therefore assume $M$ is countable, and need find a summand $R^{(\mathbb{N})}$ of $M$. Bass [63] accomplishes this for $R/\mathrm{Jac}(R)$ left Noetherian, by a delicate manipulation of infinite matrices. His argument becomes much clearer if we do not have to worry about $\mathrm{Jac}(R)$, so we prove the result for left Noetherian rings, leaving the more general case for exercises 25ff using an elegant reduction of Beck. When K-dim $R$ is finite the next theorem also follows from Coutinho's work cited above.

**Theorem 5.1.67:** *If $R$ is left Noetherian then every uniformly big projective $M$ is free.*

***Proof:*** By proposition 5.1.66 we may assume $M$ is countable. Write $M \oplus M' = R^{(\mathbb{N})}$ which has a standard base $\{e_1, e_2, \ldots\}$, and write the generators of $M$ as $x_i = \sum r_{ij} e_j$ for $i = 1, 2, \ldots$. Given any $x = \sum r_j e_j$ in $M$ we define $\operatorname{supp} x = \{j : r_j \neq 0\}$, a finite subset of $\mathbb{N}$. $\sum R r_{i1}$ is a left ideal of $R$ and thus f.g. by elements of the form $b_k = \sum_i a_{ik} r_{i1}$ where $a_{ik} \in R$. Note that only a finite number of $i$ are involved in the $b_k$; i.e., there is $n = n(1)$ for which $b_k$ is the coordinate of $e_1$ in $\sum a_{ik} x_i$. Thus subtracting $R$-linear combinations of the $x_i$ for $i < n$ from the other $x_i$ we may assume $1 \notin \operatorname{supp}(x_i)$ for all $i \geq n$. Continuing in this way for each $j$ we have $n(j)$ for which $\operatorname{supp}(x_i) \cap \{1, \ldots, j\} = \varnothing$ for all $i \geq n(j)$.

We now modify the idea of lemma 5.1.63 to build elements $y_t$ in $M$, left ideals $L_t$, and maps $f_t : M \to R$ for $t \in \mathbb{N}$, by induction on $t$ as follows:

Suppose we have $y_u$ and $f_u$ for all $u < t$. Let $L_t = \sum_{u=0}^{t-1} R f_u y_u$. Let $t'$ be the largest number in $\bigcup_{u < t} \operatorname{supp} y_u$ and let $n = n(t')$. Then $\operatorname{supp} y_u \cap \operatorname{supp} x_m = \varnothing$ for all $u < t$ and all $m \geq n$. For each $m \geq n$ let $A_m = \sum_j R r_{mj} R$, and $A = \sum_{m \geq n} A_m$. Then $M/AM$ is (finitely) generated by the images of $x_1, \ldots, x_n$, so $A = R$ since $M$ is uniformly big. Hence $A_m \nsubseteq L_t$ for some $m \geq n$, implying $\sum_j r_{mj} R \nsubseteq L_t$ for suitable $j$, none of which is in $\operatorname{supp} y_u$ for any $y \leq t$. Pick elements $a_j$ in $R$ for which $\sum_{\text{finite}} r_{mj} a_j \notin L_t$, and let $\rho_j$ denote the right multiplication map by $a_j$. Taking $f_t = \sum_{\text{finite}} \rho_j \pi_j$ where $\pi_j : R^{(\mathbb{N})} \to R$ is the projection onto the $j$-th coordinate, we have

$$f_t x_m = f_t\left(\sum r_{mj} e_j\right) = \sum \rho_j r_{mj} = \sum r_{mj} a_j \notin L_t$$

and we take $y_t = x_m$.

By construction $f_i y_u = 0$ whenever $i \neq u$. Also the $L_i$ are a strictly ascending sequence so some $L_{i+1} = R$ and thus $1 = \sum_{u=0}^{i} r_u f_u y_u$ for suitable $r_u$ in $R$. But taking $f = f_1 + \cdots + f_i$ we then have $1 = f\left(\sum_{u=0}^{i} r_u y_u\right)$ so $\sum_{u=0}^{i} R y_u$ contains a copy of $R$ which is a summand of $M$.

Now we could repeat the whole procedure starting with $j$ greater than any element in $\bigcup_{u \leq i} \operatorname{supp} y_u$ and get a new copy of $R$ which is a summand of $M$ and independent of the first copy; continuing in this way we see inductively that $R^{(\mathbb{N})}$ is isomorphic to a summand of $M$, so $M$ is free by remark 5.1.62.

Q.E.D.

Bass' results are applied to commutative rings in exercise 32.

## §5.2 Homology, Cohomology, and Derived Functors

In this section we develop (as quickly as possible) the theory of derived functions with special emphasis on $\mathscr{T}or$ and $\mathscr{E}xt$, the derived functors of $\otimes$ and $\mathscr{H}om$. Our reason is twofold: (1) We get much deeper insight into homological dimensions; (2) tools are forged which apply to diverse subjects in the sequel. To do this we view projective resolutions more categorically, building homology and cohomology. We shall draw on the basic results of §5.0. To keep the discussion explicit, we work in the category $R\text{-}\mathcal{M}od$; however, see note 5.2.5.

**Definition 5.2.1:** A (chain) *complex* $(\mathbb{A};(d_n))$ is a sequence (not necessarily exact) of maps

$$\cdots \to A_{n+1} \xrightarrow{d_{n+1}} A_n \xrightarrow{d_n} A_{n-1} \to \cdots$$

for all $n$ in $\mathbb{Z}$, such that $d_n d_{n+1} = 0$ for all $n$. The maps $d_n$ are called *differentiations*; when unambiguous $\mathbb{A}$ is used to denote the complex $(\mathbb{A};(d_n))$. We call the complex $\mathbb{A}$ *positive* if $A_n = 0$ for all $n < 0$; $\mathbb{A}$ is *negative* if $A_n = 0$ for all $n > 0$. We shall also find it convenient to call a complex *almost positive* if $A_n = 0$ for all $n < -1$; in this case the map $d_0: A_0 \to A_{-1}$ has a special role and is called the *augmentation map*, designated as $\varepsilon$. To unify notation we shall retain $d_0$. There is a category $\mathscr{Comp}$ whose objects are the complexes and whose morphisms $f: \mathbb{A} \to \mathbb{A}'$ are $\mathbb{Z}$-tuples $(f_i)$ of morphisms $f_i: A_i \to A_i'$ for each $i$, such that the following diagram commutes:

$$
\begin{array}{ccccccccc}
\cdots & \longrightarrow & A_{n+1} & \xrightarrow{d_{n+1}} & A_n & \xrightarrow{d_n} & A_{n-1} & \longrightarrow & \cdots \\
& & \downarrow{\scriptstyle f_{n+1}} & & \downarrow{\scriptstyle f_n} & & \downarrow{\scriptstyle f_{n-1}} & & \\
\cdots & \longrightarrow & A'_{n+1} & \xrightarrow{d'_{n+1}} & A'_n & \xrightarrow{d'_n} & A'_{n-1} & \longrightarrow & \cdots
\end{array}
$$

These morphisms $f$ are called *chain maps*.

**Example 5.2.2:** Suppose $(\mathbb{A};(d_n))$ is a complex. We can form a new complex $(\mathbb{A}';(d'_n))$ where $A'_n = A_{n-1}$ and $d'_n = d_{n-1}$; we can view $d: \mathbb{A} \to \mathbb{A}'$ as a chain map. Note that $\mathbb{A}'$ here is almost positive iff $\mathbb{A}$ is positive.

Although originating in topology, complexes are very relevant to the study of sequences of modules. Every exact sequence is obviously a complex. In particular, any projective resolution is an almost positive complex, and any injective resolution is an almost negative complex. Although functors do not necessarily preserve exactness they *do* preserve complexes. Thus if $\mathbb{A}$ is a complex and $F: \mathscr{C} \to \mathscr{D}$ is a covariant functor then $F\mathbb{A}$ is a complex of $\mathscr{D}$ given by

$$\cdots \to FA_{n+1} \xrightarrow{Fd_{n+1}} FA_n \xrightarrow{Fd_n} FA_{n-1} \to \cdots.$$

There is a more concise way of describing complexes, in terms of graded objects. Suppose $G$ is a given abelian group and $M, M'$ are $G$-graded modules. A map $M \to M'$ has *degree* $h$ for suitable $h$ in $G$ if $fM_g \subseteq M'_{g+h}$ for all $g$ in $G$. Taking $G = \mathbb{Z}$ and $A = \bigoplus_{i \in \mathbb{Z}} A_i$ in definition 5.2.1 we see $d: A \to A$ is a map of degree $-1$, leading us to the following alternate definition:

**Definition 5.2.3:**  A *complex* is a $\mathbb{Z}$-graded module $A$ together with a map $d: A \to A$ of degree $-1$ satisfying $d^2 = 0$. *Comp'* is then the category of $\mathbb{Z}$-graded modules, whose morphisms are those maps $f: A \to A'$ of degree 0 satisfying $fd = d'f$.

As we just saw, any complex $(\mathbb{A}; (d_n))$ of definition 5.2.1 gives rise to the complex $(A; d)$ of definition 5.2.3 where $A = \bigoplus A_n$ and $d = \bigoplus d_n$; and conversely. Thus *Comp* and *Comp'* are easily seen to be isomorphic categories, and we shall use the definitions interchangeably. One should note that definition 5.2.3 can be easily generalized, using different groups $G$ in place of $(\mathbb{Z}, +)$; $G = (\mathbb{Z}^{(2)}, +)$ is a useful candidate.

**Definition 5.2.4:**  Dually to definition 5.2.1, define a *cochain complex* $\mathbb{A}'$ to be a sequence $\cdots \leftarrow A^{n+1} \xleftarrow{d^n} A^n \xleftarrow{d^{n-1}} A^{n-1} \leftarrow \cdots$ with each $d^n d^{n-1} = 0$ where, by convention, one writes superscripts instead of subscripts. Then one can define *Cocomp* analogously.

In fact, there is an isomorphism *Comp* $\to$ *Cocomp* given by sending the complex $\cdots \to A_{n+1} \xrightarrow{d_{n+1}} A_n \to \cdots$ to the cochain complex $\mathbb{A}'$ obtained by replacing $n$ by $-n$, i.e., $(A')^n = A_{-n}$ and $(d')^n = d_{-n}$.

**Note 5.2.5:**  If one worked with an arbitrary abelian category $\mathscr{C}$ instead of $R\text{-}\mathcal{M}od$ one could formally define the corresponding category *Comp*$(\mathscr{C})$ of chain complexes. There is a formal isomorphism between *Comp*$(\mathscr{C}^{op})$ and *Cocomp*$(\mathscr{C})^{op}$ given by reversing arrows, so we see that *Comp*$(\mathscr{C})^{op}$ and *Comp*$(\mathscr{C}^{op})$ are isomorphic categories. Aiming for the best of both worlds, we work in $R\text{-}\mathcal{M}od$ but keep the duality in mind, often skipping dual proofs (which can be filled in easily).

## Homology

We saw before that functors preserve complexes, and this is a good reason to consider complexes in preference to exact sequences. But then we should know when a complex is exact (as a sequence). Obviously the complex $\cdots \to A_{n+1} \xrightarrow{d_{n+1}} A_n \xrightarrow{d_n} A_{n-1} \to \cdots$ is exact at $A_n$ iff $\ker d_n / \operatorname{im} d_{n+1} = 0$, leading us to the following important definition:

**Definition 5.2.6:** Suppose $(A, d)$ is a complex, notation as in definition 5.2.3. Define $Z = Z(A) = \ker d$, $B = B(A) = dA$ (the image of $d$), and $H = H(A) = Z/B$. $B$ is called the *boundary* and $Z$ the *cycle*, and $H$ is the *homology* of $A$.

**Remark 5.2.7:** $Z$ and $B$ are graded submodules of $A$, where $Z_n = \ker d_n$ and $B_n = d_{n+1}A_{n+1}$. Moreover, since $d^2 = 0$ we have a canonical injection $i: B \to Z$, so $H$ is a graded module and is coker $i$. Dually one can define the *coboundary* $Z' = \operatorname{coker} d = A/B$, and the *cocycle* $B' = A/Z$ (the coimage of $d$). Then there is a canonical epic $p: Z' \to B'$, and $\ker p \approx Z/B = H$. Thus $H$ is "self-dual." Moreover, $d$ induces a map $\bar{d}: Z' \to Z$, and we have the exact sequence

$$0 \to H \to Z' \xrightarrow{\bar{d}} Z \to H \to 0. \tag{1}$$

In view of the above discussion we have $H \approx \bigoplus_{n \in \mathbb{N}} H_n$ where $H_n = Z_n/B_n$. Thus $H_n = 0$ *iff the complex is exact at* $A_n$. We write $H_n(A)$ for $H_n$ when $A$ is ambiguous. Actually $Z( \ )$ and $B( \ )$ can be viewed as functors. Indeed if $f: (A; d) \to (A'; d')$ is a map of complexes then $f$ restricts to graded maps $Zf: Z \to Z'$ and $Bf: B \to B'$, by fact 5.0.2 applied to each $f_n$. Consequently we get a graded map $f_*: H(A) \to H(A')$.

## The Long Exact Sequence

We will often want to compare the homology of different modules, especially those in an exact sequence.

**Proposition 5.2.8:** *The functor* $Z( \ )$ *is a left exact functor.*

**Proof:** If $0 \to A' \xrightarrow{f} A \xrightarrow{g} A''$ is exact then $ZA' \xrightarrow{Zf} ZA \xrightarrow{Zg} ZA''$ is exact by the snake lemma (5.0.3(i)); applying the same argument to $0 \to 0 \to A' \to A$ shows $Zf$ is monic. Q.E.D.

**Remark 5.2.9:** By duality $Z'$ is a right exact functor.

Now we have all the necessary requirements for a focal result.

**Theorem 5.2.10:** ("*Exact homology sequence*") If $0 \to \mathbb{A}' \xrightarrow{f} \mathbb{A} \xrightarrow{g} \mathbb{A}'' \to 0$ is *an exact sequence of complexes, then there is a long exact sequence*

$$\cdots \to H_{n+1}(A') \xrightarrow{f_*} H_{n+1}(A) \xrightarrow{g_*} H_{n+1}(A'') \xrightarrow{\partial} H_n(A') \xrightarrow{f_*} H_n(A) \xrightarrow{g_*} H_n(A'') \xrightarrow{\partial}$$

$$H_{n-1}(A') \to \cdots$$

*where $\partial: H_{n+1}(A'') \to H_n(A')$ is obtained by applying the snake lemma to the exact sequence*

$$Z'_{n+1}(A') \xrightarrow{f_{n+1}} Z'_{n+1}(A) \xrightarrow{g_{n+1}} Z'_{n+1}(A'') \longrightarrow 0$$

$$0 \longrightarrow Z_n(A') \xrightarrow{f_n} Z_n(A) \xrightarrow{g_n} Z_n(A'')$$

**Proof:** The rows of the sequences are exact by proposition 5.2.8 and remark 5.2.9, so we are done by the snake lemma and (1).    Q.E.D.

One could rephrase the conclusion of theorem 5.2.10 by saying the following graded triangle is exact:

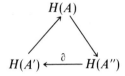

## Cohomology

The same procedure applied to cochain complexes yields a *cohomology* functor from $\mathcal{C}\!o\!c\!o\!m\!p(R\text{-}\mathcal{M}\!o\!d)$ to $\{\mathbb{Z}\text{-graded } R\text{-modules}\}$, and theorem 5.2.10 has the following dual theorem:

**Theorem 5.2.11:**   (*Exact cohomology sequence*) *If $0 \to A' \xrightarrow{f} A \xrightarrow{g} A'' \to 0$ is an exact sequence of cochain complexes then there is a long exact sequence*

$$\to H^n(A') \xrightarrow{f^*} H^n(A) \xrightarrow{g^*} H^n(A'') \xrightarrow{\delta} H^{n+1}(A') \xrightarrow{f^*} H^{n+1}(A) \to \cdots$$

*where the maps are defined as in theorem 5.2.10.*

Cohomology turns out to be particularly useful in algebra, and we digress a bit to present an example of paramount importance for §7.2. First we transfer some terminology from algebras to groups.

**Definition 5.2.12:**   Suppose $G$ is a group. A *G-module* is an abelian group $M$ together with a scalar multiplication $G \times M \to M$ satisfying the axioms $g(x_1 + x_2) = gx_1 + gx_2, (g_1 g_2)x = g_1(g_2 x)$, and $1x = x$, for all $g_i$ in $G$ and $x_i$ in $M$.

Formally we have merely written down whichever module axioms make sense. Looking at the $G$-module axioms from a different perspective, we could view the elements of $G$ as maps from $M$ to itself. In particular, any group $M$ has the *trivial* $G$-module action obtained by taking $gx = x$ for all $g$ in $G$. A deeper example: If $G$ is a group of automorphism of a field $F$ then the multiplicative group $M = F - \{0\}$ (rewritten additively) is a $G$-module under the given action of $G$.

**Example 5.2.13:** Suppose $M$ is a $G$-module. There is a very useful positive cocomplex $\mathbb{A}$ of abelian groups (i.e., $\mathbb{Z}$-modules): $A^n = \{$functions $f: G^{(n)} \to M\}$ where $G^{(0)} = \{1\}$, and $d^n: A^n \to A^{n+1}$ is defined by

$$(d^n f)(g_1, \ldots, g_{n+1}) = g_1 f(g_2, \ldots, g_{n+1}) + (-1)^{n+1} f(g_1, \ldots, g_n)$$

$$+ \sum_{i=1}^{n} (-1)^i f(g_1, \ldots, g_i g_{i+1}, \ldots, g_{n+1}). \qquad (2)$$

The cohomology groups are denoted $H^n(G, M)$, and are torsion if $G$ is a finite group, cf., exercise 2.

A key example: Any exact sequence of groups $1 \to M \to E \to G \to 1$ with $M$ abelian determines a $G$-module structure on $M$, by the following rule: Suppose $x \in M$ and $g \in G$. View $M \subseteq E$ (and thereby use multiplicative notation). Letting $h$ be a preimage of $g$ in $E$, define $gx$ to be $h^{-1}xh$ (which is clearly in $M$). Note that this action is independent of the choice of $h$, for if $h_1 \in Mh$ then writing $h_1 = hx_1$ we have $h_1^{-1}xh_1 = x_1^{-1}(h^{-1}xh)x_1 = h^{-1}xh$ since $M$ is abelian.

This set-up is called a *group extension* of $G$ by $M$, and there is a correspondence given in Jacobson [80B, pp. 363–366] between these group extensions and $H^2(G, M)$. In exercises 8.2.7ff. we shall see how group extensions tie in with division algebras.

## Derived Functors

**Proposition 5.2.14:** *Suppose* $\mathbb{P} = \cdots \to P_n \overset{f_n}{\to} P_{n-1} \to \cdots \to P_0 \overset{f_0}{\to} M \to 0$ *is a projective resolution, and* $\mathbb{A} = \cdots \to A_n \overset{d_n}{\to} A_{n-1} \to \cdots \to A_0 \overset{d_0}{\to} A_{-1} \to 0$ *is an arbitrary exact sequence. Then any map* $g: M \to A_{-1}$ *can be lifted to maps* $g_i: P_i \to A_i$ *such that* $(g_n): \mathbb{P} \to \mathbb{A}$ *is a chain map, i.e., there is a commutative diagram*

**Proof:** $d_0: A_0 \to A_{-1}$ is epic so we can lift $gf_0: P_0 \to A_{-1}$ to a map $g_0: P_0 \to A_0$ such that

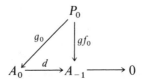

commutes, i.e., $gf_0 = d_0g_0$. Now, inductively, suppose we have $g_{n-1}f_n = d_ng_n$. Applying 5.0.8 with $P = P_{n+1}$, $\tau = g_nf_{n+1}$, $\varphi = d_n$, and $\psi = d_{n+1}$, we see $\varphi\tau = d_ng_nf_{n+1} = g_{n-1}f_nf_{n+1} = 0$, so $g_nf_{n+1} = \tau = d_{n+1}\sigma$ for some $\sigma: P_{n+1} \to A_{n+1}$, and we are done taking $g_{n+1} = \sigma$.          Q.E.D.

**Definition 5.2.15:** The $n$-th *derived* functor $L_nT$ of a given functor $T$: $R\text{-}\mathcal{M}od \to R'\text{-}\mathcal{M}od$ is defined as follows: Take a projective resolution

$$\mathbb{P} = \cdots \to P_n \xrightarrow{f_n} P_{n-1} \xrightarrow{f_{n-1}} \cdots \to M \to 0$$

viewed as an almost positive complex, and put $(L_nT)M = H_n(T\mathbb{P}) = \ker Tf_n/\operatorname{im} Tf_{n+1}$; given $g: M \to M'$ define $(L_nT)g$ as follows: First take a projective resolution $\mathbb{A}$ of $M'$, i.e., $A_{-1} = M'$; extend $g$ to a chain map as in proposition 5.2.14, then apply $T$, and finally take the $n$-th homology map.

We need to prove this is well-defined, i.e., independent of the choice of projective resolution. To this end we make the following definition:

**Definition 5.2.16:** Suppose $(\mathbb{A};(d_n))$ and $(\mathbb{A}';(d_n'))$ are complexes. Two chain maps $g: \mathbb{A}' \to \mathbb{A}$ and $h: \mathbb{A}' \to \mathbb{A}$ are *homotopic* if there are $s_n: A_n' \to A_{n+1}$ for each $n$ such that $h_n - g_n = d_{n+1}s_n + s_{n-1}d_n'$.

**Proposition 5.2.17:** *Homotopic chain maps induce the same maps on the homology modules.*

**Proof:** Suppose $g: \mathbb{A}' \to \mathbb{A}$ and $h: \mathbb{A}' \to \mathbb{A}$ are homotopic. For any cycle $z$ in $Z_n$ we have $h_nz - g_nz = d_{n+1}s_nz + s_{n-1}d_n'z = d_{n+1}s_nz \in B_n$, so the induced actions on $H_n$ are equal.          Q.E.D.

One easy application of this result is a method of checking that a given sequence is indeed exact.

**Proposition 5.2.18:** *Suppose we are given a sequence of R-modules*

$$\mathbb{A} = \cdots \to A_n \xrightarrow{d_n} A_{n-1} \to \cdots \to A_0 \xrightarrow{d_0} A_{-1} \to 0$$

*with $d_0$ epic and $d_0d_1 = 0$. If there are group homomorphisms $s_i: A_i \to A_{i+1}$ such that each $s_iA_i$ spans $A_{i+1}$ and $d_0s_{-1} = 1_{A_{-1}}$ and $d_{i+1}s_i + s_{i-1}d_i = 1_{A_i}$ for all $i \geq 0$ then $\mathbb{A}$ is exact.*

**Proof:** $d_id_{i+1}s_i = d_i(1 - s_{i-1}d_i) = d_i - (1 - s_{i-2}d_{i-1})d_i = s_{i-2}d_{i-1}d_i = 0$ by induction, so $d_id_{i+1} = 0$ by hypothesis on $s_i$. Thus $\mathbb{A}$ is a complex. Now we see the chain map $1_\mathbb{A}: \mathbb{A} \to \mathbb{A}$ is homotopic to the zero chain map, implying the homology of $\mathbb{A}$ is 0, i.e., $\mathbb{A}$ is exact.  Q.E.D.

We are also interested in the opposite direction. To illustrate what we are aiming for, here is a partial converse to proposition 5.2.18.

**Proposition 5.2.18′:** *Suppose $\cdots \to P_n \overset{d_n}{\to} P_{n-1} \to \cdots \to P_0 \overset{d_0}{\to} P_{-1} \to 0$ is an exact sequence of projective R-modules. Then there are maps $s_i: P_i \to P_{i+1}$ such that $d_0s_{-1} = 1_{P_{-1}}$ and $d_{i+1}s_i + s_{i-1}d_i = 1_{P_i}$ for all $i \geq 0$.*

**Proof:** Since $d_0$ is split epic we can find $s_{-1}$ with $d_0s_{-1} = 1_{P_{-1}}$. Now we merely proceed by induction on $i$, applying 5.0(8) to the diagram

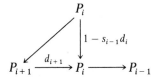

to obtain $s_i: P_i \to P_{i+1}$ satisfying $d_{i+1}s_i = 1 - s_{i-1}d_i$, or $d_{i+1}s_i + s_{i-1}d_i = 1$.  Q.E.D.

The same idea applies to homotopy.

**Lemma 5.2.19:** *(Comparison lemma) Notation as in proposition 5.2.14. Suppose $\mathbb{P}$ is a projective resolution of an R-module $M$, and $(\mathbb{A}; (d_n))$ is exact and almost positive. If $g: M \to A_{-1}$ is lifted to chain maps $(g_n), (h_n): \mathbb{P} \to \mathbb{A}$ as in proposition 5.2.14 then $(g_n)$ and $(h_n)$ are homotopic.*

**Proof:** Let $q_n = g_n - h_n$, which lifts $g - g = 0$. We use 5.0.8 repeatedly. First note $d_0q_0 = (g - g)f_0 = 0$, so $q_0P_0 \subset d_1A_1$, and thus there is $s_0: P_0 \to A_1$ such that $q_0 = d_1s_0$. Inductively, suppose we have defined $s_{n-1}: P_{n-1} \to A_n$ such that $q_{n-1} = d_ns_{n-1} + s_{n-2}f_{n-1}$. We need to define $s_n$ such that $d_{n+1}s_n = q_n - s_{n-1}f_n$. To do this it suffices to show $d_n(q_n - s_{n-1}f_n) = 0$. But $d_nq_n - d_ns_{n-1}f = q_{n-1}f_n - (q_{n-1} - s_{n-2}f_{n-1})f_n = s_{n-2}f_{n-1}f_n = 0$, as desired.  Q.E.D.

**Proposition 5.2.20:**   *Definition 5.2.15 is well-defined.*

**Proof:**   First $(L_n T)M$ is well-defined since for any two projective resolutions $\mathbb{P}, \mathbb{P}'$ of $M$ we have chain maps $(g_n): \mathbb{P} \to \mathbb{P}'$ and $(h_n): \mathbb{P}' \to \mathbb{P}$ lifting $1_M$, implying $(h_n g_n): \mathbb{P} \to \mathbb{P}$ lifts $1_M$. But the identity map $1_{\mathbb{P}}$ lifts $1_M$, so $h_n g_n$ is homotopic to 1 by lemma 5.2.19, implying $T h_n T g_n$ is homotopic to $T1 = 1$, and thus $(Tg_n)$ induces an isomorphism of the homology modules, by proposition 5.2.17. The well-definedness of $(L_n T)g$ is proved analogously.
                                                                                              Q.E.D.

There is one case in which it is easy to determine $L_n T$.

**Remark 5.2.21:**   If $P$ is projective then $(L_n T)P = 0$ for all $n$. (Indeed, $0 \to P \to P \to 0$ is a projective resolution where $P_0 = P$, and $0 \to TP \to TP \to 0$ is exact.)

Let us now restate theorem 5.2.10 for left derived functors.

**Theorem 5.2.22:**   *If $T$ is a right exact functor from $R\text{-}\mathcal{M}od$ to $R'\text{-}\mathcal{M}od$ then for any exact sequence $0 \to M' \xrightarrow{f'} M \xrightarrow{f} M'' \to 0$ there is a corresponding long exact sequence*

$$\cdots \to (L_{n+1} T)M'' \to (L_n T)M' \xrightarrow{(L_n T)f'} (L_n T)M \xrightarrow{(L_n T)f} (L_n T)M''$$
$$\to (L_{n-1} T)M' \to \cdots \to (L_1 T)M'' \to TM' \xrightarrow{Tf'} TM \xrightarrow{Tf} TM'' \to 0.$$

**Note:**   If $T$ is not right exact then the conclusion becomes true when we replace $TM'$, $TM$, and $TM''$ by $(L_0 T)M'$, $(L_0 T)M$, and $(L_0 T)M''$, but we shall not have occasion to use this generality. Before applying the theorem let us discuss it briefly. As usual we note that had the results above been stated more generally for abelian categories, the dualization would be automatic, so let us indicate briefly how this could have been done. As remarked at the end of §5.0 the results there hold for any abelian category $\mathscr{C}$. Assuming $\mathscr{C}$ has "enough projectives", which we recall means for any object $A$ there is an epic $P \to A$ with $P$ projective, we can form projective resolutions for any object. On the other hand, the definition of homology is applicable in any abelian category (and actually could be formulated in terms of "graded objects" by means of coproducts). Thus the left derived functors could be defined for any abelian category with enough projectives, thereby yielding a general version of theorem 5.2.22. We leave the details to the reader, although we shall need the following consequence.

**Corollary 5.2.23:** *Suppose* $T: \mathscr{C} \to \mathscr{D}$ *is a right exact (covariant) functor of abelian categories, and* $\mathscr{C}$ *has enough projectives.* $T$ *is exact iff the left derived functor* $L_1 T = 0$.

**Proof:** ($\Leftarrow$) For an exact sequence $0 \xrightarrow{f'} M' \xrightarrow{f} M \to M'' \to 0$ we have $\cdots \to (L_1 T)M \to (L_1 T)M'' \to TM' \xrightarrow{Tf'} TM \to TM'' \to 0$ exact. By hypothesis $(L_1 T)M'' = 0$ so clearly $T$ is exact.

($\Rightarrow$) Given any $M''$ we could take an exact sequence with $M$ projective; then $(L_1 T)M = 0$ and $Tf'$ is monic by hypothesis, implying $(L_1 T)M'' = 0$ for all $M''$. Hence $L_1 T = 0$.      Q.E.D.

There are two ways to dualize the discussion; we could either switch the direction of the resolution or replace $T$ by a contravariant functor. Let us describe these procedures respectively, assuming the underlying category has "enough injectives."

**Definition 5.2.24:** The $n$-th *right derived* functor $R^n T$ of the functor $T$ is given by $(R^n T)M = H^n(T\mathbb{E})$, the $n$-th cohomology group of $T$ applied to an injective resolution $\mathbb{E}$ of $M$; likewise, $(R^n T)f$ is the $n$-th cohomology map of $T(f^n)$, where $(f^n)$ is the extension of $f: M \to M'$ to injective resolutions of $M$ and $M'$.

**Theorem 5.2.25:** *If* $T$ *is a left exact functor then for any exact sequence* $0 \to M' \to M \to M'' \to 0$ *there is a long exact sequence*

$$0 \to TM' \to TM \to TM'' \to (R^1 T)M' \to (R^1 T)M \to \cdots$$

$(R^1 T)E = 0$ *for every injective* $E$; $T$ *is exact iff* $R^1 T = 0$.

**Proof:** Dual to corollary 5.2.23.      Q.E.D.

**Definition 5.2.26:** The $n$-th *right derived* functor $R^n T$ of a *contravariant* functor $T$ is defined by $(R^n T)M = H^n(T\mathbb{P})$ where $\mathbb{P}$ is a projective resolution of $M$. (Note we use cohomology since $T\mathbb{P}$ is a negative complex.) Given $f: M \to M'$ we lift $f$ to a map $(f_n): \mathbb{P} \to \mathbb{P}'$ of projective resolutions and define $(R^n T)f$ to be the $n$-th cohomology map of $T(f_n)$.

**Theorem 5.2.26':** *If* $T$ *is a left exact contravariant functor then for any exact sequence* $0 \to M' \to M \to M'' \to 0$ *there is a long exact sequence*

$$0 \to TM'' \to TM \to TM' \to (R^1 T)M'' \to (R^1 T)M \to \cdots.$$

$T$ *is exact iff* $R^1 T = 0$.

***Proof:***   As in theorem 5.2.25.     Q.E.D.

## Tor and Ext

At last we are ready to define the functors which will help us to measure homological dimensions of rings.

***Definition 5.2.27:***   $\operatorname{Tor}_n^R(\underline{\phantom{x}}, M)$ is defined as $L_n T: \mathscr{M}\!od\text{-}R \to \mathscr{A}\!b$ where $T = \underline{\phantom{x}} \otimes_R M$.

$\operatorname{Ext}_R^n(M, \underline{\phantom{x}})$ is defined as $R^n T: R\text{-}\mathscr{M}\!od \to \mathscr{A}\!b$ where $T = \operatorname{Hom}_R(M, \underline{\phantom{x}})$.
$\operatorname{Ext}_R^n(\underline{\phantom{x}}, M)$ is defined as $R^n T: R\text{-}\mathscr{M}\!od \to \mathscr{A}\!b$ where $T = \operatorname{Hom}_R(\underline{\phantom{x}}, M)$.

We delete the "$R$" when the ring $R$ is understood. These definitions make sense because $\underline{\phantom{x}} \otimes M$ is right exact, $\operatorname{Hom}(M, \underline{\phantom{x}})$ is left exact, and $\operatorname{Hom}(\underline{\phantom{x}}, M)$ is contravariant left exact. Let us review theorem 5.2.22, 5.2.25, and 5.2.26' for these particular functors.

***Summary 5.2.28:***   If $0 \to N' \to N \to N'' \to 0$ is exact in $\mathscr{M}\!od\text{-}R$ or $R\text{-}\mathscr{M}\!od$ according to the context then there are long exact sequences

(i)   $\cdots \to \operatorname{Tor}^2(N'', M) \to \operatorname{Tor}^1(N', M) \to \operatorname{Tor}^1(N, M) \to \operatorname{Tor}^1(N'', M)$

$\to N' \otimes M \to N \otimes M \to N'' \otimes M \to 0$

(ii)   $0 \to \operatorname{Hom}(M, N') \to \operatorname{Hom}(M, N) \to \operatorname{Hom}(M, N'') \to \operatorname{Ext}^1(M, N')$

$\to \operatorname{Ext}^1(M, N) \to \operatorname{Ext}^1(M, N'') \to \operatorname{Ext}^2(M, N') \to \cdots$

(iii)   $0 \to \operatorname{Hom}(N'', M) \to \operatorname{Hom}(N, M) \to \operatorname{Hom}(N', M) \to \operatorname{Ext}^1(N'', M)$

$\to \operatorname{Ext}^1(N, M) \to \operatorname{Ext}^1(N', M) \to \operatorname{Ext}^2(N'', M) \to \cdots$

There is a fundamental tie from $\operatorname{Ext}^1$ to projectives and injectives.

***Proposition 5.2.29:***   $\operatorname{Ext}^1(P, \underline{\phantom{x}}) = 0$ iff $P$ is projective; dually $\operatorname{Ext}^1(\underline{\phantom{x}}, E) = 0$ iff $E$ is injective.

***Proof:***   We prove the first assertion. Let $T = \operatorname{Hom}(P, \underline{\phantom{x}})$. In view of corollary 2.11.6 we want to show $R^1 T = 0$ iff $T$ is exact. ($\Rightarrow$) is immediate by theorem 5.2.25. ($\Leftarrow$) For any module $N$ we can take an exact sequence $0 \to N \to E \to E/N \to 0$ with $E$ injective, so $0 \to TN \to TE \to T(E/N) \to (R^1 T)N \to (R^1 T)E = 0$ is exact. Hence $(R^1 T)N \approx \operatorname{coker}(TE \to T(E/N)) = 0$ since $T$ is exact, proving $R^1 T = 0$.     Q.E.D.

There is some ambiguity in notation since $\text{Ext}^n(M, N)$ could denote either $\text{Ext}^n(M, \_)$ applied to $N$ or $\text{Ext}^n(\_, N)$ applied to $M$. We have seen these are both 0 for $n = 1$ when $M$ is projective (by proposition 5.2.29 and remark 5.2.21), thereby motivating the next theorem, that the two interpretations of $\text{Ext}^n(M, N)$ are the same. First we show this for $n = 1$.

**Proposition 5.2.30:**   *For this result write* $\overline{\text{Ext}}(M, \_)$ *instead of* $\text{Ext}(M, \_)$. *Then* $\overline{\text{Ext}}^1(M, N) \approx \text{Ext}^1(M, N)$ *for all modules* $M, N$.

**Proof:**   Take exact sequence $0 \to K \to P \to M \to 0$ and $0 \to N \to E \to K' \to 0$ with $P$ projective and $E$ injective. Then we have the commutative diagram

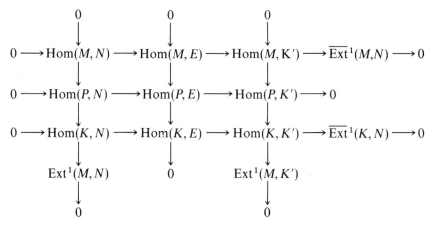

The zeros appear on the right and bottom because of remark 5.2.21 and proposition 5.2.29. Applying the snake lemma to the second and third rows gives the exact sequence

$$\text{Hom}(M, N) \to \text{Hom}(M, E) \to \text{Hom}(M, K') \to \text{Ext}^1(M, N) \to 0.$$

Thus $\text{Ext}^1(M, N) \approx \text{coker}(\text{Hom}(M, E) \to \text{Hom}(M, K')) \approx \overline{\text{Ext}^1(M, N)}$ as desired.

Q.E.D.

**Remark 5.2.31:**   In the lower right-hand corner of the above diagram there is an isomorphism $f: \overline{\text{Ext}}^1(K, N) \to \text{Ext}^1(M, K')$, sending any $x$ in $\overline{\text{Ext}}^1(K, N)$ to its preimage in $\text{Hom}(K, K')$ and then to the image in $\text{Ext}^1(M, K')$. (The proof is by diagram chasing. First note that if $y, y'$ are both preimages of $x$ in $\text{Hom}(K, K')$ then $y - y'$ has a preimage in $\text{Hom}(K, E)$ and thus in $\text{Hom}(P, E)$; travelling down the other side of the square shows $y - y'$ goes to 0 in $\text{Ext}^1(M, K')$. This proves $f$ is well-defined, and the same analysis going in the other direction enables us to construct $f^{-1}$.)

In order to generalize proposition 5.2.30 to arbitrary $n$ we need a way of shifting down.

**Lemma 5.2.32:** *If $\mathbb{P}$ is a projective resolution of $M$ and $K_n$ is the $n$-th syzygy for each $n$ then $(L_{n+1}T)M \approx (L_iT)K_{n-i}$ for any right exact functor $T$, for each $1 \leq i \leq n$.*

**Proof:** By induction it suffices to prove $(L_{n+1}T)M \approx (L_nT)K_0$. But we can cut $\mathbb{P}$ at $P_0$ to get a projective resolution $\cdots \to P_2 \to P_1 \to K_0 \to 0$ of $K_0$, which we call $\mathbb{P}'$. By definition $(L_nT)K_0 = H_n(T\mathbb{P}') = H_{n+1}(T\mathbb{P}) = (L_{n+1}T)M$ as desired.    Q.E.D.

**Proposition 5.2.33:** *Suppose $K_i$ is the $i$-th syzygy of some projective resolution of $M$. Then $\mathrm{Tor}_{n+1}(\underline{\quad}, M)$ and $\mathrm{Tor}_{n-i}(\underline{\quad}, K_i)$ are naturally isomorphic functors; also $\mathrm{Ext}^{n+1}(\underline{\quad}, M)$ and $\mathrm{Ext}^{n-i}(\underline{\quad}, K_i)$ are naturally isomorphic functors. Dually suppose $M$ has an injective resolution $0 \to M \xrightarrow{f_0} E_0 \xrightarrow{f_1} E_1 \cdots$ and let $K_i' = \mathrm{coker}\, f_i = E_i/f_iE_{i-1}$, called the $i$-th cosyzygy. Then $\mathrm{Ext}^{n+1}(M, \underline{\quad})$ and $\mathrm{Ext}^{n-i}(K_i', \underline{\quad})$ are naturally isomorphic functors.*

**Proof:** Apply the lemma and its analog for maps; this yields the first assertion, and the others follow in the same way.    Q.E.D.

**Theorem 5.2.34:** *For this result write $\overline{\mathrm{Ext}}(M, \underline{\quad})$ instead of $\mathrm{Ext}(M, \underline{\quad})$. Then $\mathrm{Ext}(M, N) \approx \overline{\mathrm{Ext}}(M, N)$ for all modules $M, N$, and all $n$.*

**Proof:** (Zaks) For $n = 0$ this is obvious and for $n = 1$ it is proposition 5.2.30. In general, take a projective resolution $\mathbb{P}$ of $M$, denoting the $n$-th syzygy by $K_n$, and take an injective resolution $\mathbb{E}$, denoting the $n$-th cosyzygy as $K_n'$. Applying remark 5.2.31 to the exact sequences $0 \to K_0 \to P_0 \to M \to 0$ and $0 \to K_{n-2}' \to E_{n-1}' \to K_{n-1}' \to 0$ yields $\mathrm{Ext}^1(M, K_{n-1}') \approx \mathrm{Ext}^1(K_0, K_{n-2}')$. Iterating remark 5.2.31 we have $\mathrm{Ext}^1(M, K_{n-1}') \approx \mathrm{Ext}^1(K_0, K_{n-2}') \approx \mathrm{Ext}^1(K_1, K_{n-3}') \approx \cdots \approx \mathrm{Ext}^1(K_{n-1}, N)$. Applying proposition 5.2.33 at both ends yields

$$\overline{\mathrm{Ext}}^{n+1}(M, N) \approx \overline{\mathrm{Ext}}^1(M, K_{n-1}') \approx \mathrm{Ext}^1(M, K_{n-1}')$$

$$\approx \mathrm{Ext}^1(K_{n-1}, N) \approx \mathrm{Ext}^{n+1}(M, N)$$

as desired.    Q.E.D.

We are now in a position for a basic result.

**Corollary 5.2.35:**  *The following are equivalent for a module P:*

(i) *P is projective.*
(ii) $\text{Ext}^1(P, \underline{\phantom{x}}) = 0$.
(iii) $\text{Ext}^n(P, \underline{\phantom{x}}) = 0$ *for all n.*

**Proof:**  (ii) $\Rightarrow$ (i) by proposition 5.2.29, and (i) $\Rightarrow$ (iii) by remark 5.2.21. (iii) $\Rightarrow$ (ii) is obvious.    Q.E.D.

Let us dualize all of this for Tor. Given a right module $M$ we can define $\text{Tor}_n^R(M, \underline{\phantom{x}})$ as $L_n T : R\text{-}\mathcal{M}od \to \mathcal{Ab}$ where $T$ is the functor $M \otimes_R \underline{\phantom{x}}$. By definition $M$ is flat iff $T$ is exact. Thus proposition 5.2.29 dualizes to

**Proposition 5.2.36:**  $\text{Tor}_1(F, \underline{\phantom{x}}) = 0$ *iff F is a flat right module; analogously,* $\text{Tor}_1(\underline{\phantom{x}}, F) = 0$ *iff F is a flat left R-module.*

But any projective module $P$ is flat, so an analogous proof to proposition 5.2.30 (but now using projective resolutions on both sides) shows $\text{Tor}_1(M, N)$ is well-defined for any right module $M$ and left module $N$. The analogous use of proposition 5.2.33 enables one to prove

**Theorem 5.2.37:**  $\text{Tor}_n(M, N)$ *is well-defined for any right module M and left module N.*

Tor has fundamental connections to tensor products, and we shall need the following one in Chapter 6:

**Example 5.2.38:**  (i) Suppose $0 \to K \to F \to M \to 0$ is an exact sequence of $R$-modules with $F$ free, and $A \lhd R$. Then $\text{Tor}_1^R(R/A, M)$ is canonically identified with $(K \cap AF)/AK$. Indeed, applying $\text{Tor}_1(R/A, \underline{\phantom{x}})$ yields the exact sequence

$$0 = \text{Tor}_1(R/A, F) \to \text{Tor}_1(R/A, M) \to K/AK \to F/AF \to M/AM$$

since $K/AK \approx (R/A) \otimes K$ and $F/AF \approx (R/A) \otimes F$ by example 1.7.21'. Hence we can identify $\text{Tor}_1(R/A, M)$ with $\ker(K/AK \to F/AF) = (K \cap AF)/AK$, as needed.

(ii) If $A, B \lhd R$ then $\text{Tor}_1(R/A, R/B)$ is canonically identified with $A \cap B/AB$, as seen by taking $K = B$ and $F = R$ in (i).

## Digression: Ext and Module Extensions

Our objective in introducing Ext and Tor was to understand the homological dimensions better. However, there are more concrete interpretations of these functors, and we shall digress a bit to discuss Ext, leaving Tor for exercise 8ff.

Given modules $K, N$ we say an *extension* of $K$ by $N$ is an exact sequence $0 \to K \xrightarrow{f} M \xrightarrow{g} N \to 0$. Sometimes we denote this extension merely as $M$, with f,g understood. Two extensions $M, M'$ of $K$ by $N$ are *equivalent* if there is a map $\varphi: M \to M'$ such that

$$
\begin{array}{ccccccccc}
0 & \longrightarrow & K & \xrightarrow{f} & M & \xrightarrow{g} & N & \longrightarrow & 0 \\
  &                 & \downarrow 1_M &  & \downarrow \varphi &  & \downarrow 1_N &  & \\
0 & \longrightarrow & K & \xrightarrow{f'} & M' & \xrightarrow{g'} & N & \longrightarrow & 0
\end{array}
$$

commutes. Note by the 5 lemma that $\varphi$ must be an isomorphism. The set of equivalence classes of extensions of $K$ by $N$ is denoted $e(N, K)$.

**Proposition 5.2.39:**  *There is a $1 - 1$ correspondence $e(N, K) \to \operatorname{Ext}^1(N, K)$.*

**Proof:**   Given an extension $0 \to K \to M \to N \to 0$ we shall define an element of $\operatorname{Ext}^1(N, K)$ as follows: Take a projective resolution $\mathbb{P}$ of $N$, and using proposition 5.2.14 build the commutative diagram

$$
\begin{array}{ccccccccccc}
\cdots & \longrightarrow & P_2 & \xrightarrow{d_2} & P_1 & \xrightarrow{d_1} & P_0 & \xrightarrow{d_0} & N & \longrightarrow & 0 \\
       &                 & \downarrow &  & \downarrow f_1 &  & \downarrow f_0 &  & \downarrow 1_N &  & \\
\cdots & \longrightarrow & 0 & \longrightarrow & K & \longrightarrow & M & \longrightarrow & N & \longrightarrow & 0
\end{array} \qquad (1)
$$

Now $0 = f_1 d_2 = d_2^* f_1$ so $f_1$ yields an element in $\ker d_2^* / \operatorname{im} d_1^* \in \operatorname{Ext}^1(N, K)$, and this is the element of $\operatorname{Ext}^1(N, K)$ corresponding to our original sequence. By lemma 5.2.19 and proposition 5.2.20 this is well-defined and sends equivalent extensions to the same element of $\operatorname{Ext}^1(N, K)$. Thus we have a functor $\Phi: e(N, K) \to \operatorname{Ext}^1(N, K)$.

Now we want to determine $\Phi^{-1}$. Given a cocycle $f_1: P_1 \to K$ we want to define a suitable extension. This will be done by means of the pushout (cf., 5.0.7). We are given $0 = d_2^* f_1 = f_1 d_2$, so we have a map $\bar{f}: P_1 / d_2 P_2 \to K$, giving rise to the sequence

$$
\begin{array}{ccccccc}
0 & \longrightarrow & P_1 / d_2 P_2 & \xrightarrow{\bar{d}_1} & P_0 & \longrightarrow & N & \longrightarrow & 0 \\
  &                 & \downarrow \bar{f} &  &  &  &  & \\
  &                 & K &  &  &  &  &
\end{array}
$$

with $\bar{d}_1$ monic. Taking $M$ to be the pushout of the upper left corner yields

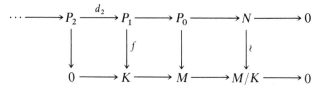

$$(2)$$

But (2) could also be rewritten as

which is the same as (1). Any extension satisfying (1) could be viewed in this manner as a pushout, so the universality of the pushout shows the extension is determined uniquely up to equivalence. Thus we have, indeed, created $\Phi^{-1}$.    Q.E.D.

**Corollary 5.2.39′:**    $\text{Ext}^1(K, N) = 0$ *iff every extension of $K$ by $N$ splits.*

**Proof:**    Every extension must be equivalent to the extension $0 \to K \xrightarrow{\mu} K \oplus N \xrightarrow{\pi} N \to 0$.    Q.E.D.

## Homological Dimension and the Derived Functors

We are finally ready to apply all this machinery to homological dimension; the proofs become much clearer intuitively.

**Theorem 5.2.40:**    *The following are equivalent for a module $M$:*

(i) $\text{pd } M \le n$;
(ii) $\text{Ext}^k(M, \_\_) = 0$ *for all $k > n$;*
(iii) $\text{Ext}^{n+1}(M, \_\_) = 0$;
(iv) *The $(n-1)$ syzygy of any projective resolution of $M$ is projective.*

**Proof:**    (i) $\Rightarrow$ (ii) By definition of $\text{Ext}^k(\_\_, N)$ we see $\text{Ext}^k(M, N) = 0$ for any $k > n$ since M has a projective resolution of length n. Hence $\text{Ext}^k(M, \_\_) = 0$ by theorem 5.2.34.

(ii) $\Rightarrow$ (iii) is a fortiori.

(iii) $\Rightarrow$ (iv) Let $K_{n-1}$ be the $(n-1)^{\text{st}}$ syzygy. Proposition 5.2.33 shows $\text{Ext}^1(K_{n-1}, \_\_) \approx \text{Ext}^{n+1}(M, \_\_) = 0$, so $K_{n-1}$ is projective by corollary 5.2.35.

(iv) $\Rightarrow$ (i) By proposition 5.1.8.    Q.E.D.

One has the dual theorem for injective resolutions:

**Theorem 5.2.40':** *The following are equivalent for a module $N$:*

(i) *$N$ has injective dimension $\leq n$;*
(ii) *$\operatorname{Ext}^k(\underline{\phantom{x}}, N) = 0$ for all $k > n$;*
(iii) *$\operatorname{Ext}^{n+1}(\underline{\phantom{x}}, N) = 0$;*
(iv) *The $(n-1)$ cosyzygy of every injective resolution is injective.*

**Corollary 5.2.41:** gl. dim $R = \sup\{$*injective dimensions of the $R$-modules*$\}$.

*Proof:* Each side $= \sup\{k: \operatorname{Ext}^k(M, N) \neq 0$ for all $R$-modules $M$ and $N\}$, by theorem 5.2.40(ii) and 5.2.40' (iii).    Q.E.D.

Using Ext one can show gl. dim depends only on the cyclic modules.

**Theorem 5.2.42:** (*Auslander*) gl. dim $R = \sup\{\operatorname{pd} R/L: L < R\}$.

*Proof:* Let $n = \sup\{\operatorname{pd} R/L: L < R\}$; then $\operatorname{Ext}^{n+1}(R/L, \underline{\phantom{x}}) = 0$. We need to show for any injective resolution $\mathbb{E}$ that the $(n-1)$ cosyzygy $K$ is injective. Well, $\operatorname{Ext}^1(R/L, K) \approx \operatorname{Ext}^{n+1}(R/L, N) = 0$, so $\operatorname{Hom}(R, K) \to \operatorname{Hom}(L, K) \to \operatorname{Ext}^1(R/L, K) = 0$ is exact by summary 5.2.28(iii). By Baer's criterion $K$ is injective, as desired.    Q.E.D.

**Corollary 5.2.43:** *$R$ is left hereditary iff* gl. dim$(R) = 1$.

Treating Tor analogously we discover a new dimension. In view of proposition 5.2.36 it makes sense to work with flat modules.

*Definition 5.2.44:* A *flat resolution* of a module $M$ is an exact sequence

$$\cdots \to F_n \xrightarrow{d_n} F_{n-1} \xrightarrow{d_{n-1}} \cdots \to F_0 \xrightarrow{d_0} M = 0$$

where each $F_n$ is flat. The *flat dimension of $M$* is the smallest $n$ for which $M$ has a flat resolution of length $n$; the *weak dimension*, written w. dim $R$, is the supremum of the flat dimensions of all $R$-modules.

**Proposition 5.2.45:** *For any ring* w. dim $R \leq$ gl. dim $R$.

*Proof:* Any projective resolution is a flat resolution, since projective modules are flat.    Q.E.D.

Of course w. dim $R = 0$ iff every $R$-module is flat, iff $R$ is (von Neumann) regular. Since left Noetherian regular rings are semisimple Artinian, this raises the hope that w. dim $R = $ gl. dim $R$ if $R$ is left Noetherian. In fact this is true, and implies gl. dim is left-right symmetric for Noetherian rings. These are other basic facts about w. dim are given in exercises 13ff; of course $\mathcal{T}or$ enters heavily in the proofs.

## Dimension Shifting

Dimension shifting is a rather straightforward technique which lifts properties from $\text{Ext}^n$ to $\text{Ext}^{n+1}$ and from $\text{Tor}_n$ to $\text{Tor}_{n+1}$ and thus transfers properties of Hom to Ext, and of $\otimes$ to Tor. The idea is rather simple.

**Remark 5.2.46:** "Dimension shifting". Suppose $K \leq M$. If $\text{Ext}^n(M, \_\_) = 0$ (for example, if $M$ is projective) then $\text{Ext}^{n+1}(M/K, \_\_) \approx \text{Ext}^n(K, \_\_)$ in the sense they are naturally isomorphic functors. (Indeed, $\text{Ext}^{n+1}(M, \_\_) = 0$ by theorem 5.2.40, so summary 5.2.28(iii) yields the exact sequence

$$0 = \text{Ext}^n(M, N) \to \text{Ext}^n(K, N) \to \text{Ext}^{n+1}(M/K, N) \to \text{Ext}^{n+1}(M, N) = 0,$$

implying $\text{Ext}^{n+1}(M/K, N) \approx \text{Ext}^n(K, N)$ for all modules $N$.)
Similarly, if $\text{Ext}^n(\_\_, M) = 0$ then $\text{Ext}^{n+1}(K, \_\_) \approx \text{Ext}^n(M/K, \_\_)$.
If $\text{Tor}_n(\_\_, M) = 0$ then $\text{Tor}_{n+1}(M/K, \_\_) \approx \text{Tor}_n(K, \_\_)$.

The principle of dimension shifting has already been used implicitly; this simple technique produces instant results.

**Remark 5.2.47:** Reproof of summary 5.1.13. Suppose $0 \to M' \to M \to M'' \to 0$ is exact and $n = $ pd $M$. $\text{Ext}^{n+1}(M, \_\_) = 0$, so by dimension shifting $\text{Ext}^{m+1}(M'', \_\_) \approx \text{Ext}^m(M', \_\_)$ for all $m > n$. This yields cases I and II $(n \leq n')$ at once. For case III $(n > n')$ we need a similar but modified argument. Taking $m > n'$ we have

$$0 = \text{Ext}^m(M', N) \to \text{Ext}^{m+1}(M'', N) \to \text{Ext}^{m+1}(M, N) \to \text{Ext}^{m+1}(M', N) = 0$$

for all $N$, so $\text{Ext}^{m+1}(M, \_\_) \approx \text{Ext}^{m+1}(M'', \_\_)$. Thus $n = n''$.

Our immediate interest in dimension-shifting is to extend certain basic properties from Hom to Ext, and from $\otimes$ to Tor.

**Theorem 5.2.48:** *Suppose $F: R\text{-}\mathcal{M}od^I \to R\text{-}\mathcal{M}od$ and $G: \mathcal{A}b \to \mathcal{A}b$ is a pair* *of right exact functors which commutes with* Hom *in the sense that there is a natural isomorphism* $\text{Hom}(\_, F(N_i)) \approx G(\text{Hom}(\_, (N_i))$ *(as functors from* $R\text{-}\mathcal{M}\!\mathit{od}$ *to* $\mathcal{A}\!\mathit{b}$*). Then* $F$ *and* $G$ *commutes with* $\text{Ext}^n$ *for all* $n \in \mathbb{N}$, *in the same sense.*

**Proof:** Take an exact sequence $0 \to K \to P \to M \to 0$ with $P$ free. Since $\text{Ext}_R^n(P, \_) = 0$ for all $n \geq 1$ we have the commutative diagram

$$\text{Ext}^{n-1}(P, F(N_i)) \longrightarrow \text{Ext}^{n-1}(K, F(N_i)) \longrightarrow \text{Ext}^n(M, F(N_i)) \longrightarrow 0$$

$$\downarrow \qquad\qquad\qquad\qquad \downarrow \qquad\qquad\qquad\qquad \downarrow$$

$$G(\text{Ext}^{n-1}(P, N_i)) \longrightarrow G(\text{Ext}^{n-1}(K, N_i)) \longrightarrow G(\text{Ext}^n(M, N_i)) \longrightarrow 0$$

Identifying $\text{Ext}^0$ with Hom we may assume by induction on $n$ that the first two vertical maps are isomorphisms. (In fact, for $n > 1$ the left-hand terms are 0.) By property 5.0.2 the right-hand map exists and is an isomorphism, as desired. This is all natural, by inspection.     Q.E.D.

**Theorem 5.2.49:** *Suppose* $F: R\text{-}\mathcal{M}\!\mathit{od}^{(I)} \to R\text{-}\mathcal{M}\!\mathit{od}$ *and* $G: \mathcal{A}\!\mathit{b} \to \mathcal{A}\!\mathit{b}$ *is a pair of left exact functors for which there is a natural isomorphism of functors* $\_ \otimes F(N_i)$ *and* $G(\_ \otimes (N_i))$. *Then there is a natural isomorphism of functors* $\text{Tor}_n(\_, F(N_i))$ *and* $G(\text{Tor}_n(\_, (N_i)))$.

**Proof:** Dual to theorem 5.2.48.     Q.E.D.

**Corollary 5.2.50:** *Suppose* $N_i$ *are* $R$-*modules.* $\text{Ext}^n(M, \prod N_i) \approx \prod \text{Ext}^n(M, N_i)$ *for every* $R$-*module* $M$, *and* $\text{Tor}_n(M, \coprod N_i) \approx \coprod \text{Tor}_n(M, N_i)$ *for every right* $R$-*module* $M$. *(We write* $\coprod$ *instead of* $\bigoplus$ *to stress duality.)*

**Proof:** It suffices to prove $\prod$ is right exact, and dually $\coprod$ is left exact. But given $K_i \to L_i \to N_i \to 0$ exact we see at once $\coprod L_i \to \coprod N_i$ is epic with kernel $\coprod K_i$.     Q.E.D.

Let us improve this result.

**Lemma 5.2.51:** *If* $I$ *is a directed set then* $\varinjlim$ *is left exact, in the sense that if* $(A_i'; \varphi_{ij}')$, $(A_i; \varphi_{ij})$ *and* $(A_i''; \varphi_{ij}'')$ *are systems indexed by* $I$ *and* $0 \to A_i' \to A_i \to A_i''$ *is exact then* $0 \to \varinjlim A_i' \to \varinjlim A_i \to \varinjlim A_i''$ *is exact.* Indeed, left $f: \varinjlim A_i \to \varinjlim A_i''$ be the given map. Obviously, $\varinjlim A_i' \subseteq \ker f$. Conversely, suppose $(a^{(i)}) \in \ker f$. By the construction in theorem 1.8.7 this means for suitable $k \geq i$ we have $\varphi_i^k a^{(i)} \in A_k'$. But defining $a'^{(i)}$ to be $\varphi_i^k a^{(i)}$ we see easily that $(a'^{(i)})$ equals $(a^{(i)})$ in $\varinjlim A_i$, proving $\ker f = \varinjlim A_i'$.

**Proposition 5.2.52:** $\operatorname{Tor}_n(M, \varinjlim N_i) \approx \varinjlim \operatorname{Tor}_n(M, N_i)$ *whenever the index set is directed.*

**Proof:** Apply lemma 5.2.51 to theorem 5.2.49.    Q.E.D.

**Remark 5.2.53:** We should record some variants of these results; the proofs are analogous.

   (i) $\operatorname{Ext}^n(\coprod M_i, N) \approx \prod \operatorname{Ext}^n(M_i, N)$.

   (ii) $\operatorname{Tor}_n(\coprod M_i, N) \approx \coprod \operatorname{Tor}_n(M_i, N)$.

   (iii) $\operatorname{Tor}_n(\varinjlim M_i, N) \approx \varinjlim \operatorname{Tor}(M_i, N)$ for any directed index set $I$.

Other results of this genre are given in exercises 29–35; a detailed treatment is given in the classic Cartan-Eilenberg [56B]. There are many papers in the literature which study the homology of particular rings, but we should like to point the reader to Rinehart-Rosenberg [76] and Rosenberg-Stafford [76], who by a careful use of projective resolutions calculate explicitly the homological dimensions of modules over rings of differential polynomials and over Weyl algebras.

## Acyclic Complexes

Homology gives us an important connection between complexes and projective resolutions by means of the following definition:

**Definition 5.2.54:** A positive complex $(A, d)$ is *acyclic* if $H_n(A) = 0$ for all $n > 0$.

**Remark 5.2.55:** The acyclicity condition means

$$\to A_{n+1} \xrightarrow{d_{n+1}} A_n \xrightarrow{d_n} \cdots \xrightarrow{d_1} A_0 \to 0$$

is exact at $A_n$ for each $n > 0$. On the other hand, $H_0(A) = A_0/d_1 A_1$, so we have the exact sequence

$$\cdots \to A_{n+1} \xrightarrow{d_{n+1}} A_n \xrightarrow{d_n} \cdots \xrightarrow{d_1} A_0 \to H_0(A) \to 0.$$

(This is why we have been considering almost positive complexes). Thus we have a projective resolution of $H_0(A)$ if $A_n$ is projective for each $n \geq 0$, or, equivalently, if $A = \bigoplus_{n \geq 0} A_n$ is projective. Reversing the argument shows a projective resolution can be viewed as a positive acyclic complex which is projective.

**Proposition 5.2.56:** *Suppose $(A, d)$ and $(A', d')$ are complexes of vector spaces over a field F. Let $\tilde{A} = A \otimes_F A'$, which is graded via $\tilde{A}_n = \bigoplus_{u+v=n} A_u \otimes_F A'_v$, and define a differentiation $\tilde{d}$ by $\tilde{d}_n(a \otimes a') = d_u a \otimes a' + (-1)^u a \otimes d'_v a'$ for $a \in A_u$ and $a' \in A'_v$. Then*

(i) *$(\tilde{A}, \tilde{d})$ is a complex.*
(ii) *If $A$ and $A'$ are acyclic then so is $\tilde{A}$, and $H_0(\tilde{A}) = H_0(A) \otimes H_0(A')$ canonically.*

**Proof:** For purposes of calculation we shall take $a \in A_u$ and $a' \in A'_v$, with $u + v = n$.

(i) 
$$\tilde{d}_{n-1}\tilde{d}_n(a \otimes a') = \tilde{d}_{n-1}(d_u a \otimes a' + (-1)^u a \otimes d'_v a')$$
$$= d_{u-1}d_u a \otimes a' + (-1)^{u-1}d_u a \otimes d'_v a' + (-1)^u d_u a \otimes d'_v a'$$
$$+ (-1)^u a \otimes d'_{v-1}d'_v a' = 0$$

since the middle two terms cancel and the end terms are each 0.

(ii) We verify the condition of proposition 5.2.18 for $\tilde{A}$. Let $\varepsilon: A_0 \to H_0(A)$ and $\varepsilon': A'_0 \to H_0(A')$ be the canonical maps, and $\tilde{d}_0 = \varepsilon \otimes \varepsilon'$. By proposition 5.2.18' we have $s_{-1}: H_0(A) \to A_0$ and $s_i: A_i \to A_{i+1}$ for $i \geq 0$ satisfying $\varepsilon s_{-1} = -1$, $d_1 s_0 + s_{-1}\varepsilon = 1$, and $d_{i+1}s_i + s_{i-1}d_i = 1$ for all $i > 0$. Likewise, we have $s'_{-1}: H_0(A') \to A'_0$ and $s'_i: A'_i \to A'_{i+1}$ for $i \geq 0$ satisfying $\varepsilon' s'_{-1} = -1$, $d'_1 s'_0 + s'_{-1}\varepsilon' = 1$, and $d'_{i+1}s'_i + s'_{i-1}d'_i = 1$ for all $i > 0$. Define

$$\tilde{s}_{-1} = s_{-1} \otimes s'_{-1} \quad \text{and} \quad \tilde{s}_n = s_{-1}\varepsilon \otimes s'_n + \sum_{i=0}^{n} s_i \otimes 1 \quad \text{for } n \geq 0.$$

Then $\tilde{d}_0 \tilde{s}_{-1} = 1 \otimes 1 = 1$. It remains to show $\tilde{d}_{n+1}\tilde{s}_n + \tilde{s}_{n-1}\tilde{d}_n = 1$ for all $n \geq 0$, which will be seen by evaluating it on $a_u \otimes a_v$.

*Case I.* $u > 1$. Then

$$(\tilde{d}_{n+1}\tilde{s}_n + \tilde{s}_{n-1}\tilde{d}_n)(a \otimes a') = \tilde{d}_{n+1}s_u a \otimes a' + \tilde{s}_{n-1}(d_u a \otimes a' + (-1)^u a \otimes d'_v a')$$
$$= d_{u+1}s_u a \otimes a' + (-1)^{u+1}s_u a \otimes d'_v a'$$
$$+ s_{u-1}d_u a \otimes a' + (-1)^u s_u a \otimes d'_v a'$$
$$= (d_{u+1}s_u + s_{u-1}d_u)a \otimes a' = a \otimes a'.$$

*Case II.* $u = 1$. Same calculation as in case I, except with the extra term $s_{-1}\varepsilon d_1 a \otimes s'_n a' = 0$ (since $\varepsilon d_1 = 0$).

*Case III.* $u = 0$.

$$(\tilde{d}_{n+1}\tilde{s}_n + \tilde{s}_{n-1}\tilde{d}_n)(a \otimes a') = \tilde{d}_{n+1}(s_{-1}\varepsilon a \otimes s'_n a' + s_0 a \otimes a') + \tilde{s}_{n-1}(a \otimes d'_n a')$$

$$= s_{-1}\varepsilon a \otimes d'_{n+1}s'_n a' + d_1 s_0 a \otimes a' - s_0 a \otimes d'_n a'$$

$$+ s_{-1}\varepsilon a \otimes s'_{n-1}d'_n a' + s_0 a \otimes d'_n a'$$

$$= s_{-1}\varepsilon a \otimes (d'_{n+1}s'_n + s'_{n-1}d'_n)a' + d_1 s_0 a \otimes a'$$

$$= (s_{-1}\varepsilon + d_1 s_0)a \otimes a' = a \otimes a'. \qquad \text{Q.E.D.}$$

**Corollary 5.2.57:** *If $R$ is an $F$-algebra and $\mathbb{P}_k$ is a projective resolution of $M_k$ for $k = 1, \ldots, m$ then $M_1 \otimes_F \cdots \otimes_F M_n$ has the projective resolution $\tilde{P} = \bigotimes_{k=1}^m \mathbb{P}_k$ where $\tilde{P}_n = \bigoplus_{u_1 + \cdots + u_m = n}(P_1)_{u_1} \otimes \cdots \otimes (P_m)_{u_m}$ and $\tilde{d}_n : \tilde{P}_n \to \tilde{P}_{n-1}$ is given by*

$$\tilde{d}(a_1 \otimes \cdots \otimes a_m) = \sum_{i=1}^m (-1)^{u(i)} a_1 \otimes \cdots \otimes da_i \otimes \cdots \otimes a_m$$

*where each $a_i \in (A_i)_{u_i}$ and $u(i) = \sum_{k=1}^{i-1} u_k$.*

*Proof:* View each $\mathbb{P}_k$ as a direct sum of projectives, which is thus projective; hence $\mathbb{P}$ is projective by the easy remark 5.3.0. Furthermore, $\tilde{P}$ is a projective resolution of $M_1 \otimes_F \cdots \otimes_F M_m$, by the proposition (since everything can be viewed naturally as vector spaces over $F$.)     Q.E.D.

Some remarks about these results. Proposition 5.2.56 is a special case of the Kunneth formulas of algebraic topology. The reader's eyebrows may have been raised by the sign in the formula for the differentiation. This sign arises naturally in the definition of tensor products of morphisms of $\mathbb{Z}$-graded modules and is discussed in detail in MacLane [63B, Chapter VII, esp. pp. 190–191].

## §5.3 Separable Algebras and Azumaya Algebras

As noted several times earlier, one can bypass the asymmetry between left and right modules by dealing with bimodules. On the other hand, we want the powerful techniques of module theory at our disposal, so we view $R - R$ bimodules as $R \otimes R^{\text{op}}$-modules, cf., proposition 1.7.31. Rather than taking tensors over $\mathbb{Z}$, we shall consider the more general case where $R$ is an algebra over an arbitrary commutative ring $C$, and shall write $R^e$ for $R \otimes_C R^{op}$. Furthermore, we assume throughout that $M$ is an $R - R$ bimodule respecting the algebra structure, i.e., $cx = xc$ for all $c$ in $C$, $x$ in $M$. Thus we shall view $M$ canonically as $R^e$-module, by the action $(r_1 \otimes r_2)x = r_1 x r_2$; also $(c \otimes 1)x = (1 \otimes c)x$ for all $c$ in $C$. Now we can define the *Hochschild* homology and cohomology groups

$$H_n(R, M) = \text{Tor}_n(R, M) \quad \text{and} \quad H^n(R, M) = \text{Ext}^n(R, M)$$

where the base ring is $R^e$. Although there is a lovely theory developed in Hochschild's articles, Cartan-Eilenberg [56B], and the series "On the dimension of modules and algebras" in the Nagoya Journal (1950s), we focus on the special case where $R$ is a projective $R^e$-module. This provides a modern and natural setting for the classical theory of separable algebras, cf., definition 2.5.28, and, in particular, for Wedderburn's principal theorem. We shall also deal with "Azumaya algebras," in preparation for Chapters 6 and 7. Our treatment follows Knus-Ojanguran [74B], Jacobson [80B], and Demeyer-Ingraham [71B].

**Remark 5.3.0:**  If $P_i$ are projective $C$-modules then $P_1 \otimes_C P_2$ is projective. (Indeed, if $F_i \approx P_i \oplus P_i'$ is free then

$$P_1 \otimes P_2 \oplus (P_1' \otimes P_2 \oplus P_1 \otimes P_2' \oplus P_1' \otimes P_2') \approx (P_1 \oplus P_1') \otimes (P_2 \oplus P_2')$$

$$\approx F_1 \otimes F_2$$

is free.)

**Remark 5.3.0':**  If $P$ is a projective $C$-module and $R$ is a $C$-algebra then $R \otimes_C P$ is a projective $R$-module. (For if $P \oplus P' = F$ then $R \otimes P \oplus R \otimes P' \approx R \otimes F$ is free by corollary 1.7.16.)

**Remark 5.3.1:**  $R$ is a cyclic $R^e$-module (spanned by 1), and there is an epic $p: R^e \to R$ in $R^e$-$\mathcal{M}od$ given by $p(\sum r_{1i} \otimes r_{2i}) = \sum r_{1i} r_{2i}$. Letting $J = \ker p$ we have the exact sequence of $R^e$-modules $0 \to J \to R^e \to R \to 0$. The notation $J$ and $p$ has become standard, and we shall use it without further ado.

**Proposition 5.3.2:**  $J = \sum_{r \in R} R^e (r \otimes 1 - 1 \otimes r)$.

**Proof:**  ($\supseteq$) is clear. ($\subseteq$) If $a = \sum r_{i1} \otimes r_{i2} \in J$ then $\sum r_{i1} r_{i2} = 0$ so $a = \sum_i (r_{i1} \otimes 1)(1 \otimes r_{i2} - r_{i2} \otimes 1)$ as desired.     Q.E.D.

**Remark 5.3.3:** The exact sequence of remark 5.3.1 yields an exact sequence

$$0 \to \operatorname{Hom}_{R^e}(R, M) \to \operatorname{Hom}_{R^e}(R^e, M) \to \operatorname{Hom}_{R^e}(J, M) \to \operatorname{Ext}^1_{R^e}(R, M) \to 0$$

since $\operatorname{Ext}^1_{R^e}(R^e, R) = 0$.

To interpret the other groups we need some definitions. First define $M^R = \{x \in M : rx = xr \text{ for all } r \text{ in } R\}$, viewing $M$ as $R - R$ bimodule; translated to $R^e\text{-}\mathcal{M}od$ the condition reads $(r \otimes 1)x = (1 \otimes r)x$, i.e., $Jx = 0$. Next define $\operatorname{Der}(M) = \{\partial : R \to M : \partial(r_1 r_2) = (\partial r_1)r_2 + r_1 \partial r_2\}$, the set of *derivations of $R$ in $M$*. Translated to $R^e\text{-}\mathcal{M}od$ this condition reads as $\partial(r_1 r_2) = (1 \otimes r_2)\partial r_1 + (r_1 \otimes 1)\partial r_2$. We can view $\operatorname{Der}(M)$ as $C$-module by defining $(c\partial)x$ to be $c(\partial x)$. For any $x$ in $M$ we define the *inner derivation* $\partial_x$ by $\partial_x r = (r \otimes 1 - 1 \otimes r)x$. When $M = R^e$ we write $\delta$ for $\partial_{1 \otimes 1}$, i.e., $\delta r = r \otimes 1 - 1 \otimes r$. In fact, $\delta \in \operatorname{Der}(J)$ by proposition 5.3.2. Let us view this fact in context.

**Proposition 5.3.4:** $\operatorname{Hom}_{R^e}(R, M) \approx M^R$ *via* $f \mapsto f1$; $\operatorname{Hom}_{R^e}(J, M) \approx \operatorname{Der}(M)$ *via* $f \mapsto f\delta$. *Consequently, we have an exact sequence*

$$0 \to M^R \to M \to \operatorname{Der}(M) \to \operatorname{Ext}^1_{R^e}(R, M) \to 0$$

**Proof:** The last assertion is obtained by matching the given isomorphisms to remark 5.3.3, noting also $\operatorname{Hom}(R^e, M) \approx M$ by $f \mapsto f1$. Thus it remains to verify the two given isomorphisms.

Define $\varphi : \operatorname{Hom}_{R^e}(R, M) \to M^R$ by $\varphi f = f1$; note $f1 \in M^R$ since

$$(r \otimes 1 - 1 \otimes r)f1 = f(r - r) = f0 = 0.$$

Thus $\varphi$ is a map whose inverse $M^R \to \operatorname{Hom}_{R^e}(R, M)$ is given by sending $x$ to right multiplication by $x$.

Next define $\psi : \operatorname{Hom}_{R^e}(J, M) \to \operatorname{Der}(M)$ by $\psi f = f\delta$; $f\delta$ is in $\operatorname{Der}(M)$ since

$$f\delta(r_1 r_2) = f((1 \otimes r_2)\delta r_1 + (r_1 \otimes 1)\delta r_2) = (1 \otimes r_2)f\delta r_1 + (r_1 \otimes 1)f\delta r_2.$$

$\ker \psi = 0$ since $f\delta = 0$ implies $fJ = 0$ by proposition 5.3.2. It remains to show $\psi$ is onto. Given $\partial \in \operatorname{Der}(M)$ we define $\bar{g} : R^e \to M$ by means of the balanced map $g : R \times R^{op} \to M$ given by $\bar{g}(a, b) = -(a \otimes 1)\partial b$, and let $f$ be the restriction of $\bar{g}$ to $J$. Then $f \in \operatorname{Hom}(J, M)$, for if $\sum a_i \otimes b_i \in J$ then

$$f((r_1 \otimes r_2)\sum a_i \otimes b_i) = \sum f(r_1 a_i \otimes b_i r_2) = -\sum(r_1 a_i \otimes 1)\partial(b_i r_2)$$

$$= -\sum((r_1 a_i \otimes 1)(1 \otimes r_2)\partial b_i + (b_i \otimes 1)\partial r_2)$$

$$= -\sum((r_1 a_i \otimes r_2)\partial b_i + (r_1 \sum a_i b_i \otimes 1)\partial r_2)$$

$$= -(r_1 \otimes r_2)\sum(a_i \otimes 1)\partial b_i + 0 = (r_1 \otimes r_2)f(\sum a_i \otimes b_i).$$

But $f\delta r = f(r \otimes 1 - 1 \otimes r) = -(r \otimes 1)\partial 1 + (1 \otimes 1)\partial r = 0 + \partial r$, proving $f\delta = \partial$ as desired. Thus $\psi$ is onto.      Q.E.D.

**Corollary 5.3.4′:**

(i) $\mathrm{Hom}_{R^e}(R, R) \approx Z(R)$.

(ii) $\mathrm{Hom}_{R^e}(R, R^e) \approx \mathrm{Ann}' J$, *the right annihilator of $J$ in $R^e$, under the correspondence $f \mapsto f1$.*

**Proof:**   (i) Take $M = R$. (ii) Take $M = R^e$. Then $\mathrm{Hom}_{R^e}(R, M) \approx M^R \approx \{x \in M : (r \otimes 1)x = (1 \otimes r)x \text{ for all } r \text{ in } R\} = \mathrm{Ann}'\{r \otimes 1 - 1 \otimes r : r \in R\} = \mathrm{Ann}' J$.
                                                                                 Q.E.D.

**Remark 5.3.5:**   In the exact sequence of proposition 5.3.4 the map $M \to \mathrm{Der}(M)$ is given by sending $x$ in $M$ to the inner derivation $\partial_x$. Thus $\mathrm{Ext}^1(R, M) = 0$ iff every derivation of $R$ in $M$ is inner.

## Separable Algebras

**Definition 5.3.6:**   $R$ is a *separable* $C$-algebra if $R$ is projective as $R^e$-module.

There are several nice criteria for separability.

**Theorem 5.3.7:**   *The following conditions are equivalent*:

(i) *$R$ is separable.*

(ii) *$\mathrm{Ext}^1_{R^e}(R, M) = 0$ for every $R$-module $M$.*

(iii) *There is an element $e$ in $(R^e)^R$ such that $pe = 1$ where $p: R^e \to R$ is the canonical epic.*

(iv) *The epic $p: R^e \to R$ restricts to an epic $(R^e)^R \to Z(R)$,*

(v) *Every derivation of $R$ in $M$ is inner, for every module $M$.*

**Proof:**   (i) $\Leftrightarrow$ (ii) by corollary 5.2.35. (i) $\Rightarrow$ (iii) $p$ splits so there is a monic $f: R \to R^e$ for which $pf = 1_R$; take $e = f1$.

(iii) $\Rightarrow$ (iv) $p((z \otimes 1)e) = z$ for any $z$ in $Z(R)$.

(iv) $\Rightarrow$ (iii) trivial. (iii) $\Rightarrow$ (i) Define $f: R \to R^e$ by $fr = (r \otimes 1)e$; we see $f((r_1 \otimes r_2)r) = f(r_1 r r_2) = (r_1 r r_2 \otimes 1)e = (r_1 r \otimes 1)((r_2 \otimes 1)e) = (r_1 r \otimes r_2)e = (r_1 \otimes r_2)fr$, proving $f$ is a map. Hence $p$ is split so $R$ is a summand of $R^e$ and thus projective.

(ii) $\Leftrightarrow$ (v) by the exact sequence of proposition 5.3.4.      Q.E.D.

I hope no confusion will arise from the use of the symbol $e$ both as part of the notation $R^e$ and as an element thereof.

**Remark 5.3.8:** The element $e$ of theorem 5.3.7(iii) is called a *separability idempotent* of $R$ and is indeed idempotent since writing $e = \sum r_{j1} \otimes r_{j2}$ we have

$$e^2 = \sum (r_{j1} \otimes 1)(1 \otimes r_{j2})e = \sum (r_{j1} \otimes 1)(r_{j2} \otimes 1)e.$$
$$= \left(\sum r_{j1} r_{j2} \otimes 1\right)e = (1 \otimes 1)e = e.$$

Obviously $C$ is separable over itself since $C \approx C^e$. $M_n(C)$ is separable since $M_n(C) \otimes M_n(C)^{op} \approx M_n(C) \otimes M_n(C) \approx M_t(C)$ where $t = n^2$. Other examples are had by means of the separability idempotent, so let us record some of its properties.

**Remark 5.3.9:** Suppose $e$ is a separability idempotent of $R$ and let $Z = Z(R)$.

(i) $eR^e \subseteq (R^e)^R$ since $(r \otimes 1)ex = ((r \otimes 1)e)x = ((1 \otimes r)e)x = (1 \otimes r)ex$ for $x$ in $R^e$;

(ii) $p(eR^e) = Z$ by (i) and theorem 5.3.7(iv); in fact

$$p(e(z \otimes 1)) = p((z \otimes 1)e) = z$$

for $z$ in $Z$;

(iii) $Z$ is a summand of $R$ as $Z$-module since $r \mapsto p(e(r \otimes 1))$ defines a projection $R \to Z$ fixing $Z$, by (ii);

(iv) If $A \lhd Z$ then $A = Z \cap AR$ by (iii) and sublemma 2.5.22'

**Proposition 5.3.10:**

(i) If $R_i$ *are separable $C$-algebras then* $R_1 \times R_2$ *is a separable $C$-algebra.*

(ii) If $R_i$ *are separable $C_i$-algebras where $C_i$ are commutative $C$-algebras then* $R_1 \otimes_C R_2$ *is a separable $C_1 \otimes_C C_2$-algebra and*

$$Z(R_1 \otimes R_2) = Z(R_1) \otimes Z(R_2).$$

**Proof:** Let $e_i$ be a separability idempotent of $R_i$.

(i) Viewing $R_1^e \times R_2^e \subset (R_1 \times R_2)^e$ canonically we see by inspection $(e_1, e_2)$ is a separability idempotent.

(ii) $R_1 \otimes R_2$ has the separability idempotent $e = e_1 \otimes e_2$, and thus is separable. Let $Z = Z(R_1 \otimes R_2)$. Clearly $Z(R_1) \otimes Z(R_2) \subseteq Z$. Conversely, if

$z = \sum r_{i1} \otimes r_{i2} \in Z$ then by remark 5.3.9(ii) we have

$$z = p(e(z \otimes 1)) = \sum p(e_1(r_{i1} \otimes 1) \otimes e_2(r_{i2} \otimes 1)) \in Z(R_1) \otimes Z(R_2)$$

as desired.        Q.E.D.

**Example 5.3.11:**

(i) The direct product of copies of $C$ is separable.

(ii) If $S$ is a submonoid of $C$ and $R$ is separable then $S^{-1}R \approx S^{-1}C \otimes_C R$ is separable (with the same separability idempotent).

(iii) If $R$ is separable then $M_n(R) \approx M_n(C) \otimes R$ is separable.

(iv) (extension of scalars) If $R$ is separable and $C_1$ is a commutative $C$-algebra then $C_1 \otimes_C R$ is a separable $C_1$-algebra.

**Proposition 5.3.12:**   *If $f: R \to R'$ is a surjection of $C$-algebras and $R$ is separable then $R'$ is separable and $Z(R') = fZ(R)$.*

**Proof:**   Let $p: R^e \to R$ and $p': (R')^e \to R$ be the usual epics. Let $\tilde{f} = f \otimes f$: $R^e \to (R')^e$. Then $fp = p'\tilde{f}$. Letting $e$ be a separability idempotent of $R$, we see $\tilde{f}e$ is a separability idempotent of $R'$, proving $R'$ is separable. Furthermore $f(Z(R)) = fp(eR^e) = p'\tilde{f}(eR^e) = p'((\tilde{f}e)(R')^e) = Z(R')$ by remark 5.3.9(ii).
                                                                                                          Q.E.D.

**Proposition 5.3.13:**   *Suppose $C'$ is a commutative $C$-algebra and $R$ is a $C'$-algebra.*

(i) *If $R$ is separable over $C$ then $R$ is separable over $C'$.*

(ii) *If $R$ is separable over $C'$ and $C'$ is separable over $C$ then $R$ is separable over $C$.*

**Proof:**

(i) The canonical epic $R \otimes_C R^{op} \to R$ factors through $R \otimes_{C'} R^{op}$, so we get a separability idempotent.

(ii) By definition 5.3.6 $C'$ is a summand of $(C')^e = C' \otimes_C C'$. Tensoring by $R^e$ over $(C')^e$ yields $R \otimes_{C'} R^{op}$ is a summand of $R^e$. But $R$ is a summand of $R \otimes_{C'} R^{op}$ by hypothesis, so $R$ is a summand of $R^e$.        Q.E.D.

To make separability "descend," we consider the following set-up: $C'$ is a commutative $C$-algebra and $R, R'$ are algebras over $C, C'$, respectively, such that $C'$ is a summand of $R'$. Write $(R')^e$ for $R' \otimes_{C'} (R')^{op}$. As $C'$-algebras

$(R \otimes_C R')^e \approx R^e \otimes_C (R')^e$ so any $(R \otimes R')^e$-module can be viewed naturally as $R^e$-module via the map $R^e \to R^e \otimes 1$. In particular, if $R \otimes R'$ is separable as $C'$-algebra then $R \otimes R'$ is a summand of $(R \otimes R')^e$ as module over $(R \otimes R')^e$ and thus as $R^e$-module. To conclude $R$ is separable over $C$ we merely need to show

(i) $R$ is a summand of $R \otimes R'$, and
(ii) $(R \otimes R')^e$ is $R^e$-projective.

(i) is clear if $C \cdot 1$ is a summand of $R'$ as $C$-module, since then we tensor on the left by $R$. On the other hand, (ii) follows if $(R')^e$ is projective as $C$-module since then we tensor on the left by $R^e$. In case $R'$ is projective as $C$-module we thus have (ii) if $R' = C'$ (trivially), or if $C' = C$ by remark 5.3.0. This provides the following two important instances of descent:

**Proposition 5.3.14:**

(1) *If $R \otimes_C R'$ is separable over $C$ and $R'$ is faithfully projective over $C$ then $R$ is separable over $C$.*

(2) *If $R \otimes_C C'$ is separable over $C'$ for a commutative algebra $C'$ which is faithfully projective over $C$ then $R$ is separable over $C$.*

**Proof:** By proposition 2.11.29 if $R'$ is faithfully projective over $C$ then $C \cdot 1$ is a summand of $R'$. Hence (i) of the above discussion is satisfied, and (ii) also holds since $C = C'$ in (1) and $R' = C'$ in (2). Q.E.D.

**Proposition 5.3.15:** *If $R$ is separable over $C$ and $R$ is faithfully projective over a commutative $C$-algebra $C'$ then $C'$ is separable over $C$.*

**Proof:** $R$ is a summand of $R^e$ which is projective over $(C')^e$, so $R$ is projective over $(C')^e$. But by proposition 2.11.29 $C'$ is a summand of $R$ as $C'$-module and thus as $C'$-bimodule since $C'$ is commutative (so we copy out the same scalar multiplication on the right also), and thus as $(C')^e$-module. Hence $C'$ is projective over $(C')^e$. Q.E.D.

We are ready to justify the name "separable", also, cf., theorem 5.3.18.

**Proposition 5.3.16:** *Suppose $R$ is a finite field extension of a subfield $F$. This field extension is separable iff $R$ is separable as $F$-algebra.*

**Proof:** In view of propositions 5.3.13 and 5.3.15 we may assume $R$ is a simple field extension, i.e., $R = F[r] \approx F[\lambda]/\langle f \rangle$ for some irreducible polynomial $f$ in $F[\lambda]$.

($\Rightarrow$) Let $K$ be a splitting field of $f$ over $F$. Then $K \otimes R \approx R^{(n)}$ where $n = \deg f$, by the Chinese Remainder Theorem, cf., after remark 2.2.7, so $K \otimes R$ is separable as $R$-algebra, implying $K$ is separable as $F$-algebra by proposition 5.3.14.

($\Leftarrow$) Suppose the field extension were not separable. Then $\mathrm{char}(F) = p > 0$ and $R \approx L[\lambda]/\langle \lambda^n - a \rangle$ for some field $L \subset R$ and $a \in L$. Then differentiation with respect to $\lambda$ yields a nontrivial derivation of $R$, which cannot be inner since $R$ is commutative, so $R$ is not separable by theorem 5.3.7(v).     Q.E.D.

**Proposition 5.3.17:** *Suppose $R$ is a separable $C$-algebra and is projective as $C$-module. Then $R$ is f.g. as $C$-module.*

**Proof:** Take a separability idempotent $e = \sum_{\text{finite}} r_{j1} \otimes r_{j2}$. Picking a dual base $\{(x_i, f_i): i \in I\}$ of $R$ we have $f_i r_{j1} = 0$ for almost all $i$, and we claim the $x_i r_{j2}$ span $R$. Indeed, for any $r$ in $R$ we have

$$r = p((r \otimes 1)e) = p\left(\sum_j rr_{j1} \otimes r_{j2}\right) = p\left(\sum_{j,i} f_i(rr_{j1})x_i \otimes r_{j2}\right) = \sum_{j,i} f_i(rr_{j1})x_i r_{j2}$$

as desired.     Q.E.D.

**Theorem 5.3.18:** *An algebra $R$ over a field $F$ is separable in the sense of this section iff it is separable in the sense of definition 2.5.38.*

**Proof:** ($\Rightarrow$) First we show $R$ is semisimple Artinian. It suffices to show $\mathrm{Hom}_R(M, \_)$ is exact for any $R$-module $M$, by corollary 2.11.6. But $\mathrm{Hom}_F(M, \_)$ is exact. Furthermore, $\mathrm{Hom}_F(M, N)$ is an $R^e$-module (i.e., $R - R$ bimodule) by remark 1.5.18; the explicit action is $((r_1 \otimes r_2)f)x = r_1 f(r_2 x)$ for $f$ in $\mathrm{Hom}(M, N)$. Then

$$\mathrm{Hom}_R(M, N) = (\mathrm{Hom}_F(M, N))^R \approx \mathrm{Hom}_{R^e}(R, \mathrm{Hom}_F(M, N))$$

by proposition 5.3.4. Thus $\mathrm{Hom}_R(M, \_)$ is the composite of the exact functors $\mathrm{Hom}_F(M, \_)$ and $\mathrm{Hom}_{R^e}(R, \_)$ (noting $R$ is $R^e$-projective), so $R$ indeed is semisimple Artinian.

Let $K$ be the algebraic closure of $F$. Then $R \otimes_F K$ is separable, so by the first paragraph is semisimple Artinian and f.d. over $K$ by proposition 5.3.17 implying $R \otimes_F K$ is split. Now for any field extension $L$ of $F$ let $L'$ be the composition of $K$ and $L$; then $R \otimes L'$ is split, so $R \otimes_F L$ is semisimple Artinian. Hence $R$ is separable in the sense of definition 2.5.38.

($\Leftarrow$) Let $K$ be a splitting field of $R$. Then $R \otimes_F K$ is separable over $K$ by example 5.3.11, implying $R$ is separable over $F$.    Q.E.D.

## Hochschild's Cohomology

Having seen that this notion of separable generalizes the classical notion and is more elegant, our next objective is to redo the classical theory of separable algebras in this context, in particular to recast Wedderburn's principal theorem in this setting. First we want Hochschild's explicit description of his cohomology groups.

**Construction 5.3.19:** Assuming $R$ is projective as $C$-module, we shall construct an explicit cochain complex for any $R^e$-module $M$. Define $C^n(R, M) = \text{Hom}_C(R^{(n)}, M)$ (so $C^0(R, M) = M$) and define the cochain map $\delta^n : C^n(R, M) \to C^{n+1}(R, M)$ by

$$\delta^n f(r_1, \ldots, r_{n+1}) = r_1 f(r_2, \ldots, r_{n+1})$$

$$+ \sum_{i=1}^{n} (-1)^i f(r_1, \ldots, r_{i-1}, r_i r_{i+1}, r_{i+2}, \ldots, r_{n+1}) + (-1)^{n+1} f(r_1, \ldots, r_n) r_{n+1}$$

for $r_i$ in $R$. Let $B^n = \delta^{n-1} C^{n-1}(R, M)$ and $Z^n = \ker \delta^n$.

We want to identify $H^n(R, M)$ with $Z^n / B^n$. To do this we must identify this cochain complex with a cochain complex arising from $\text{Hom}(\_\_, M)$ applied to a projective resolution of $R$ in $R^e\text{-}\mathcal{M}od$.

Define $P_n = R^{\otimes(n+2)}$ for each $n \geq -1$, viewed naturally as $R - R$ bimodules (i.e., $R^e$-modules). Then $P_0 \approx R^e$ and the $P_n$ are all projective $C$-modules by remark 5.3.0, implying $P_n \approx R^e \otimes_C P_{n-2}$ is a projective $R_e$-module, by remark 5.3.0'. Now by the adjoint isomorphism

$$\text{Hom}_{R^e}(P_n, M) \approx \text{Hom}_{R^e}(P_{n-2} \otimes R^e, M) \approx \text{Hom}_C(P_{n-2}, \text{Hom}_{R^e}(R^e, M))$$

$$\approx \text{Hom}_C(P_{n-2}, M) \approx C^n(R, M)$$

the last isomorphism arising from viewing an $n$-linear map as a balanced map. It remains to define the differentiations $\partial_n : P_n \to P_{n-1}$, by

$$\partial_n(r_1 \otimes \cdots \otimes r_{n+2}) = \sum_{i=1}^{n+1} (-1)^{i-1} r_1 \otimes \cdots \otimes r_i r_{i+1} \otimes \cdots \otimes r_{n+2}.$$

We have a diagram

$$\cdots \to P_2 \xrightarrow{\partial_2} P_1 \xrightarrow{\partial_1} P_0 \xrightarrow{p} R \to 0 \qquad (\mathbb{P})$$

where $p$ is as in remark 5.3.1. Define $s_n: P_n \to P_{n+1}$ by $s_n(x_1 \otimes \cdots \otimes x_{n+2}) = 1 \otimes x_1 \otimes \cdots \otimes x_{n+2}$. Clearly $s_n P_n$ spans $P_{n+1}$, $ps_{-1} = 1_R$, and $\partial_{n+1} s_n + s_{n-1} \partial_n = 1$ on each $P_n$, by direct computation. Hence $\mathbb{P}$ is a projective resolution by proposition 5.2.18. Moreover, we have a commutative diagram

$$\begin{array}{ccc}
\mathrm{Hom}_{R^e}(P_n, M) & \xrightarrow{\partial^{\#}} & \mathrm{Hom}_{R^e}(P_{n+1}, M) \\
\downarrow & & \downarrow \\
C^n(R, M) & \xrightarrow{\quad\delta\quad} & C^{n+1}(R, M)
\end{array}$$

which yields the desired isomorphism of $H^n(R, M)$ with the cohomology group of $(C^n; \delta)$, i.e., with $Z^n/B^n$.

Let us look at certain relevant cocycles and coboundaries.

A map $f: R \to M$ is in $B^1$ iff $f = \delta^0 x$ for some $x \in C^0(R, M) \approx M$, i.e., $f$ is the inner derivation with respect to $x$.

$$f: R \to M \text{ is in } Z^1 \text{ iff } r_1 f r_2 - f(r_1 r_2) + (f r_1) r_2 = 0 \qquad \text{for all } r_i \text{ in } R,$$

i.e., $f$ is a derivation of $R$ in $M$. This shows $H^1(R, M)$ is the derivations modulo the inner derivations, a result implicit in the proof of proposition 5.3.4.

A map $f: R^{(2)} \to M$ is in $B^2$ iff there is $g: R \to M$ for which $f(r_1, r_2) = r_1 g r_2 - g(r_1 r_2) + (g r_1) r_2$ for all $r_i$ in $R$. Now we have a much more intuitive proof of a strengthening of Wedderburn's principal theorem (2.5.37).

**Theorem 5.3.20:** (*Wedderburn's principal theorem revisited*) *Suppose $R$ is a finite dimensional algebra over a field $F$, with $H^2(R, \_\_) = 0$. Then $R$ has a semisimple Artinian subalgebra $S$ for which $R \approx S \oplus \mathrm{Jac}(R)$ as vector spaces over $F$.*

**Proof:** As in the "general case" of the proof of theorem 2.5.27 we can reduce readily to the case $N^2 = 0$ where $N = \mathrm{Jac}(R)$. Let $V$ be a complementary vector space to $N$ in $R$. Then $V \approx R/N$ as vector spaces, and this would be an algebra isomorphism if $V$ were closed under multiplication in $R$, implying $V$ were semisimple Artinian, as desired. Hence we want to modify $V$ by suitable elements of $N$ to make $V$ multiplicatively closed. Letting $\pi: R \to V$ be the vector space projection define $f: R^{(2)} \to N$ by

$$f(r_1, r_2) = \pi(r_1 r_2) - \pi r_1 \pi r_2.$$

Then

$$\delta f(r_1, r_2, r_3) = r_1 f(r_2, r_3) - f(r_1 r_2, r_3) + f(r_1, r_2 r_3) - f(r_1, r_2) r_3$$

$$= r_1 \pi(r_2 r_3) - r_1 \pi r_2 \pi r_3 - \pi(r_1 r_2 r_3) + \pi(r_1 r_2) \pi r_3$$

$$+ \pi(r_1 r_2 r_3) - \pi r_1 \pi(r_2 r_3) - \pi(r_1 r_2) r_3 + \pi r_1(\pi r_2) r_3$$

$$= (r_1 - \pi r_1)(\pi(r_2 r_3) - \pi r_2 \pi r_3) + (\pi r_1 \pi r_2 - \pi(r_1 r_2))(r_3 - \pi r_3) = 0$$

since $N^2 = 0$. Thus $f \in Z^2(R, N) = B^2(R, N)$ by hypothesis, so there is some $g: R \to N$ for which $f = \delta g$. Hence $f(r_1, r_2) = r_1 g r_2 - g(r_1 r_2) + (g r_1) r_2$. Let $S = \{\pi r + g r : r \in R\}$. $R = S + N$, so $[V : F] = [S : F]$ and $R = S \oplus N$ as vector spaces. Furthermore, $\pi r_1 g r_2 = r_1 g r_2$ since $(r_1 - \pi r_1) g r_2 \in N^2 = 0$, and, likewise, $g r_1 \pi r_2 = (g r_1) r_2$, yielding

$$(\pi r_1 + g r_1)(\pi r_2 + g r_2) = \pi r_1 \pi r_2 + g r_1 \pi r_2 + \pi r_1 g r_2 + 0$$

$$= \pi(r_1 r_2) - f(r_1, r_2) + (g r_1) r_2 + r_1 g r_2$$

$$= \pi(r_1 r_2) + g(r_1 r_2)$$

proving $S$ is multiplicatively closed, as desired.     Q.E.D.

This also yields a direct approach to a theorem of Malcev concerning the uniqueness of $S$. Note first by proposition 2.5.6(i) that for any unit $u$ in $R$ we have $u^{-1} N u = N$ where $N = \text{Jac}(R)$, so whenever $R = S \oplus N$ we also have $R \approx u^{-1} S u \oplus N$.

**Theorem 5.3.21:** *Suppose $R$ is a finite dimensional $F$-algebra with $S \oplus N = S' \oplus N = R$ as vector spaces, where $N = \text{Jac}(R)$ with $R/N$ separable and $S \approx R/N \approx S'$. Then there is $a \in N$ for which $S' = (1 + a)^{-1} S(1 + a)$.*

**Proof:**  Let $\varphi: S \to S'$ be the composition of the isomorphisms $S \approx R/N \approx S'$. Let $\bar{e} \in (R/N)^e$ be a separability idempotent of $R/N$, written as the image of a suitable element $e$ of $R^e$ (not necessarily idempotent), and let $e' \in S \otimes (S')^{\text{op}}$ be the image of $\bar{e}$ under the isomorphism $(R/N)^e \to S \otimes (S')^{\text{op}}$. Then for any $s$ in $S$ we have $(s \otimes 1 - 1 \otimes \varphi s) e' = 0$. Passing up to $R^e$ and applying the canonical maps $p: R^e \to R$ and $R \to R/N$ gives $spe = (pe)\varphi s$ for all $s$ in $S$. Since the image of $e'$ in $R/N$ is 1 we see $pe \in 1 + N$ is invertible and thus $\varphi s = (pe)^{-1} s p e$ for all $s$ in $S$, as desired.     Q.E.D.

This result can be formulated in the much broader scope of "inertial sub-algebras," cf., Ingraham [74].

## Azumaya Algebras

Separable algebras have been seen to have many properties in common with simple algebras. The next one is particularly noteworthy.

**Lemma 5.3.22:**  *Suppose $R$ is a separable $Z(R)$-algebra. Then any maximal ideal $A$ has the form $A_0 R$ for suitable maximal $A_0 \lhd Z(R)$. In fact $A_0 = A \cap Z(R)$.*

**Proof:**  Let $Z = Z(R)$ and $A_0 = A \cap Z$. By proposition 5.3.12 $R/A$ is separable with center $Z/A_0$. But $R/A$ is simple so $Z/A_0$ is a field and $A_0$ is a maximal ideal. But now $R/A_0 R$ is separable over the field $Z/A_0$ so is central simple by theorem 5.3.18. Hence $A_0 R$ is already a maximal ideal, implying $A_0 R = A$.    Q.E.D.

**Definition 5.3.23:**  A $C$-algebra $R$ is *Azumaya* if the following two conditions are satisfied:

   (i) $R$ is faithful and f.g. projective as $C$-module, and
   (ii) $R^e \approx \operatorname{End}_C R$ under the correspondence sending $r_1 \otimes r_2$ to the map $r \to r_1 r r_2$.

By theorem 2.3.27 every central simple algebra is Azumaya, and this is the motivating example of our discussion. In fact, we shall see that Azumaya algebras possess properties in common with central simple algebras, because of the following connection to separability.

**Theorem 5.3.24:**  *The following are equivalent for a $C$-algebra $R$:*
   (i) *$R$ is Azumaya.*
   (ii) *$C = Z(R)$ and $R$ is separable over $C$.*
   (iii) *There is a category equivalence $C\text{-}\mathcal{M}od \to R^e\text{-}\mathcal{M}od$ given by $R \otimes \underline{\ \ }$.*
   (iv) *$R^e \approx \operatorname{End}_C R$ and $R$ is a progenerator of $C\text{-}\mathcal{M}od$.*
   (v) *$R^e \approx \operatorname{End}_C R$ and $R$ is a generator of $R^e\text{-}\mathcal{M}od$.*
   (vi) *$R^e \approx \operatorname{End}_C R$ and $R^e \operatorname{Ann}' J = R^e$ (where $\operatorname{Ann}' J$ denotes the right annihilator of $J$ in $R^e$).*

Before proving this theorem let us note that conditions (i), (ii), (iii) interconnect Azumaya algebras with separability theory and Morita theory: furthermore, we shall tie Azumaya algebras to polynomial identity theory in theorem 6.1.35. Conditions (iv), (v), (vi) are more technical, and all involve $R^e \approx \operatorname{End}_C R$ plus another condition. It may well be that $R^e \approx \operatorname{End}_C R$ already implies $R$ is Azumaya! Indeed Braun [86] has shown this is the case if $R$ has an anti-automorphism fixing $Z(R)$, cf., exercises 6–9. Another criterion for $R$ to be Azumaya is given in exercise 11.

***Proof of Theorem 5.3.24:***   We show (i) $\Rightarrow$ (iv) $\Rightarrow$ (iii) $\Rightarrow$ (ii) $\Rightarrow$ (i) and (ii) $\Rightarrow$ (vi) $\Rightarrow$ (v) $\Rightarrow$ (i).

(i) $\Rightarrow$ (iv) by remark 4.1.9 (since $R^e = R \otimes R^{\mathrm{op}}$).

(iv) $\Rightarrow$ (iii) by Morita's theorem.

(iii) $\Rightarrow$ (ii) $R = R \otimes C$ is a progenerator in $R^e\text{-}\mathcal{M}od$. By proposition 4.1.19 we can build a Morita context, thereby yielding a category equivalence in the other direction given by the functor $\mathrm{Hom}_{R^e}(R, R^e) \otimes_{R^e}$ —, which by the adjoint isomorphism is naturally equivalent to $\mathrm{Hom}_{R^e}(R, \_)$. Thus $C \approx \mathrm{Hom}_{R^e}(R, R) \approx Z(R)$ by corollary 5.3.4'.

(ii) $\Rightarrow$ (i)   $C = Z(R) \approx \mathrm{Hom}_{R^e}(R, R)$ so applying example 4.1.16 to the projective $R^e$-module $R$ gives the Morita context

$$(R^e, C, R, \mathrm{Hom}_{R^e}(R, R^e), \tau, \tau') \tag{1}$$

where $\tau'$ is epic by proposition 4.1.19 since $R$ is cyclic projective.

To conclude by Morita's theorem we need to show $\tau$ is epic; by remark 4.1.9 and proposition 4.1.19 it is enough to show $R$ is faithfully projective over $R^e$. Suppose $A \lhd R^e$ and $AR = R$. We may assume $A$ is maximal. By proposition 5.3.10 $R^e \approx R \otimes R^{\mathrm{op}}$ is separable with center $C$, so $A = A_0 R^e$ for some maximal $A_0 \lhd C$, by lemma 5.3.22. Now $A_0 R = R$, which is impossible by remark 5.3.9(iv).       Q.E.D.

(ii) $\Rightarrow$ (vi) As in the proof of (ii) $\Rightarrow$ (i) we can set up the Morita context (1) with $\tau'$ epic. Thus $R^e \approx \mathrm{End}_C R$ and we need to show $R^e \mathrm{Ann}' J = R^e$. $\tau'$ epic implies every element of $R^e$ has the form $\sum_i f_i r_i = \sum r_i f_i 1 = \sum (r_i \otimes 1) f_i 1 \in R^e \mathrm{Ann}' J$ since each $f_i 1 \in \mathrm{Ann}' J$ by corollary 5.3.4(ii); here each $f_i \in \mathrm{Hom}_{R^e}(R, R^e)$.

(vi) $\Rightarrow$ (v) Reverse the previous argument to show $\tau'$ is epic in (1), so $R$ is a generator in $R^e\text{-}\mathcal{M}od$ by Lemma 4.1.7.

(v) $\Rightarrow$ (i) Setting up the Morita context (1) we see $R$ is f.g. projective over $\mathrm{End}_{R^e} R \approx Z(R)$ by remark 4.1.17'.       Q.E.D.

***Remark 5.3.24':***   The reverse equivalence $R^e\text{-}\mathcal{M}od \to C\text{-}\mathcal{M}od$ is given by $M \mapsto M^R$. (Indeed, as shown in the proof of (iii) $\Rightarrow$ (ii) it is given by $\mathrm{Hom}_{R^e}(R, \_)$, so we appeal to proposition 5.3.4.)

**Corollary 5.3.25:**   *If $R$ is an Azumaya $C$-algebra then there is a 1:1* *correspondence between* $\{ideals\ of\ C\}$ *and* $\{ideals\ of\ R\}$ *given by* $A_0 \mapsto A_0 R$ (*and the reverse correspondence* $A \mapsto A \cap C$).

**Proof:** Apply proposition 4.1.18, noting the $R^e$-submodules of $R$ are the ideals of $R$.     Q.E.D.

On the other hand, let us consider the consequences of viewing $R$ as a projective $C$-module. In view of theorem 2.12.22 we can write $C = \prod_{i=1}^{t} C_i$ where each $C_i = Ce_i$ for a suitable idempotents $e_i$ of $C$, and $R_i = Re_i$ has constant rank as projective module over $Ce_i$. But each $R_i$ is Azumaya as $C_i$-algebra by theorem 5.3.24(ii), and $R \approx \prod R_i$, so we have proved

**Proposition 5.3.26:** *Every Azumaya algebra is a finite direct product of Azumaya algebras of constant rank* (*as projective module over the center*).

One further reduction is worth noting, for this is what gives Azumaya algebras the connection to the original definition of Azumaya [50]. We say $R$ is *proper maximally central of rank n* over a subring $C$ of $Z(R)$ if $R$ is a free $C$-module of rank $n$ and $R^e \approx \text{End}_C R$.

**Proposition 5.3.27:** *Suppose $R$ is Azumaya over $C$ of constant* rank $n$.

(i) $R_P$ *is proper maximally central over $C_P$ of* rank $n$, *for every* $P \in \text{Spec}(C)$.

(ii) $R/P'$ *is central simple of dimension $n$ over $C/(P' \cap C)$, for every maximal ideal $P'$ of $R$. In particular, $n$ is square.*

**Proof:**

(i) By proposition 5.3.10(ii) $R_P \approx C_P \otimes_C R$ is separable over $C_P$ with $Z(R_P) = C_P$. $C_P$ is local so $R_P$ is free of rank $n$ by definition 2.12.21.

(ii) Let $P = P' \cap C$. Then $P' = PR$ by lemma 5.3.22, so $P'_P = P_P R_P$. Thus by Nakayama's lemma the image of a base of $R_P$ (over $C_P$) must be a base of $R_P/P'_P$ over $C_P/P_P$, the center of $R_P/P'_P$ by proposition 5.3.12. We conclude by remark 2.12.12, which shows $R_P/P'_P \approx R/P'$ and $C_P/P_P \approx C/P$.     Q.E.D.

The converse results are in Knus-Ojanguren [74B, pp. 79 and 95], and reduce the study of Azumaya algebras to the local case. Recently Azumaya algebras have become a key tool in PI-theory, and we close this discussion with results of Braun to be used in §6.3.

**Remark 5.3.28:** If $R = C\{r_1, \ldots, r_t\}$ and $S$ is a submonoid of $Z(R)$ then

$Z(S^{-1}R) = S^{-1}Z$ where $Z = Z(R)$. (Indeed if $s^{-1}r \in Z(S^{-1}R)$ then $s^{-1}[r, r_j] = 0$ for $1 \le j \le t$ so there is $s'$ in $S$ for which $0 = s'[r, r_j] = [s'r, r_j]$ for each $j$, implying $s'r \in Z$; thus $s^{-1}r = (s's)^{-1}s'r \in S^{-1}Z$.) Hence $(S^{-1}R)^e \approx S^{-1}R^e$.

**Proposition 5.3.29:** *(R arbitrary) Suppose* $x = \sum a_i \otimes b_i \in R^e$.

(i) *If* $x \in \text{Ann}' J$ *then* $\sum a_i r b_i \in Z(R)$ *for all $r$ in $R$.*

(ii) *If* $R = C\{r_1, \ldots, r_t\}$ *and* $\{r_i \otimes 1 - 1 \otimes r_i)x = 0$ *for all* $1 \le i \le t$ *then* $x \in \text{Ann}' J$ *(so the conclusion of* (i) *holds).*

**Proof:** For all $r'$ in $R$ we have $r' \otimes 1 - 1 \otimes r' \in J$ so

$$0 = (r' \otimes 1 - 1 \otimes r')x = \sum(r'a_i \otimes b_i - a_i \otimes b_i r').$$

Taking the image under the canonical map $R^e \to \text{End}_C R$ and applying it to $r$ yields

$$0 = \sum_i (r'a_i r b_i - a_i r b_i r') = [r', \sum a_i r b_i] \qquad \text{for all } r' \text{ in } R,$$

implying $\sum a_i r b_i \in Z(R)$.

(ii) By proposition 5.3.2 it suffices to prove $(r \otimes 1 - 1 \otimes r)x = 0$ for all $r$ in $R$. By hypothesis we may assume $r = r_{i_1} \cdots r_{i_t}$ for some $t$. By induction on $t$ we have $(r_{i_1} \cdots r_{i_{t-1}} \otimes 1 - 1 \otimes r_{i_1} \cdots r_{i_{t-1}})x = 0$, for all $i_1, \ldots, i_{t-1}$. Hence

$$(r_{i_1} \cdots r_{i_t} \otimes 1)x = (r_{i_1} \otimes 1)(r_{i_2} \cdots r_{i_t} \otimes 1)x = (r_{i_1} \otimes 1)(1 \otimes r_{i_2} \cdots r_{i_t})x$$

$$= (1 \otimes r_{i_2} \cdots r_{i_t})(r_{i_1} \otimes 1)x = (1 \otimes r_{i_2} \cdots r_{i_t})(1 \otimes r_{i_1})x$$

$$= (1 \otimes r_{i_1} r_{i_2} \cdots r_{i_t})x$$

as desired .    Q.E.D.

**Corollary 5.3.30:** *Suppose* $R = C\{r_1, \ldots, r_t\}$ *and* $S^{-1}R$ *is Azumaya for a given submonoid $S$ of $Z(R)$. If* $1^{-1}\sum_i a_i r b_i \in Z(S^{-1}R)$ *for all $r$ in $R$ then there is $s$ in $S$ satisfying* $\sum s a_i r b_i \in Z(R)$ *for all $r$ in $R$.*

**Proof:** Let $\psi: (S^{-1}R)^e \to \text{End } S^{-1}R$ be the canonical isomorphism. For any $s^{-1}r$ in $S^{-1}R$ and any $1 \le j \le t$ we have

$$0 = s^{-1}\left[r_j, \sum_i a_i r b_i\right] = s^{-1}\sum_i(r_j a_i r b_i - a_i r b_i r_j)$$

$$= \psi\left((1^{-1}r_j \otimes 1 - 1 \otimes 1^{-1}r_j)\sum_i(1^{-1}a_i \otimes 1^{-1}b_i)\right)s^{-1}r.$$

Thus $1^{-1}(r_j \otimes 1 - 1 \otimes r_j)\sum(a_i \otimes b_i) = 0$ in $(S^{-1}R)^e \approx S^{-1}R^e$ (cf., remark 5.3.28) so for some $s$ in $S$ we have

$$0 = (s \otimes 1)(r_j \otimes 1 - 1 \otimes r_j) \sum a_i \otimes b_i$$

$$= (r_j \otimes 1 - 1 \otimes r_j) \sum sa_i \otimes b_i \qquad \text{for } 1 \le j \le t.$$

Now proposition 5.3.29 implies $\sum sa_i r b_i \in Z(R)$ for all $r$ in $R$.           Q.E.D.

In analogy with the results presented earlier, one can use Tor to define the *weak dimension* of an algebra. This was carried out by Villamayor [59] who managed to find an intuitive proof of a strengthened version of corollary 2.5.33. Let us briefly sketch his approach. Noting that $M$ is flat iff $\mathrm{Tor}(N, M) = 0$ for every *cyclic* right module $N$, he determines an elementary condition for a cyclic left module to be flat, and deduces that $R$ is flat as $R^e$-module iff for every *finite* subset $\{r_1, \ldots, r_n\}$ of $R$ there is a suitable element $e$ of $R^e$ satisfying $(r_i \otimes 1 - 1 \otimes r_i)e = 0$ for each $i$, such that $pe = 1$. Thus for finite dimensional algebras over a field this notion corresponds with separability. He proves finally that $\mathrm{Jac}(R_1 \otimes_F R_2) = \mathrm{Jac}(R_1) \otimes_F R_2$ whenever $R_i$ are algebras over a field with $R_2$ flat as $R_2^e$-module. He also characterizes which left Artinian rings satisfy this property and, as another application, completes a characterization of regular group rings.

## Appendix to Chapter 5: The Grothendieck Group, and Quillen's Theorem Revisited

**Definition 5A.1:**   Let $\mathscr{C}$ be an abelian category. The *Grothendieck group* $\mathscr{G}_0(\mathscr{C})$ (also denoted $K_0(\mathscr{C})$ in Bass [68B, Chapter 8] and in exercise 5.1.11) is the abelian group $F/H$, where $F$ is the free abelian group whose base is the isomorphism classes of objects of $\mathscr{C}$, and $H$ is the subgroup of $F$ generated by all

$$[M] - [M''] - [M']$$

for all exact sequences $0 \to M'' \to M \to M' \to 0$. Thus in $\mathscr{G}_0(\mathscr{C})$ we have $[M] = [M''] + [M']$. Write $\mathscr{G}_0(R)$ for $\mathscr{G}_0(\mathscr{C})$ when $\mathscr{C} = R\text{-}\mathscr{F}imod$. Since any exact sequence of projective modules splits, we have a natural map $K_0(R) \to \mathscr{G}_0(R)$, which turns out to be an isomorphism when $R$ is left Noetherian of finite gl. dim., by the *resolution theorem*, cf., exercise 5.1.11. In fact, the results of 5.1.34ff can be formulated more generally when we replace $K_0$ by $\mathscr{G}_0$, and we shall now give this more general version, for the graded case, following an exposition of Cliff (as recorded and kindly provided by Passman); we shall want this result in appendix 8B.

**Theorem 5A.2:** *Suppose $R$ is left Noetherian and $\mathbb{N}$-graded, $R$ is flat as right $R_0$-module, and $R_0$ has finite pd as right $R$-module. Then $R \otimes_{R_0}$ ___ induces an isomorphism $\psi: \mathscr{G}_0(R_0) \to \mathscr{G}_0(R)$.*

**Proof:** We give Cliff's short proof. All undecorated tensors are to be taken over $R_0$. Note the functor $R \otimes$ ___ is exact since $R$ is flat over $R_0$; hence $\psi$ is a well-defined group map. Let $m = \text{pd}_R R_0$ (as right module).

*Step I.* $\text{Tor}_n^R(R_0, M) = 0$ for all $n > m$ and all $M$ in $R$-$\mathscr{M}\!od$, in analogy to theorem 5.2.40 (or by exercise 5.2.13, noting all projectives are flat).

*Step II.* If $N \in R_0$-$\mathscr{M}\!od$ then $\text{Tor}_n^R(R_0, R \otimes N) = 0$ for all $n \geq 1$. (Indeed take a projective resolution $\mathbb{P}$ of $N$. Tensoring each term in $\mathbb{P}$ by $R$, we obtain a projective resolution $\mathbb{P}'$ of $R \otimes N$ since $R \otimes$ ___ is exact. Now applying $R_0 \otimes_R$ ___ to $\mathbb{P}'$ gives $\mathbb{P}$ again since any $R_0$-module $M$ satisfies $R_0 \otimes_R (R \otimes_{R_0} M) \approx (R_0 \otimes_R R) \otimes_{R_0} M \approx M$ canonically; hence the homology is 0 on $\mathbb{P}'$ as a complex, implying $\text{Tor}_n^R(R_0, R \otimes N) = 0$ for all $n \geq 1$.)

*Step III.* $\psi$ is 1:1. (Indeed define $\varphi: \mathscr{G}_0(R) \to \mathscr{G}_0(R_0)$ by $[M] \to \sum_{i=0}^{m}(-1)^i \cdot [\text{Tor}_i^R(R_0, M)]$. $\varphi$ is a homomorphism, as seen by the long exact Tor sequence. Furthermore, $\varphi\psi[N] = \varphi[R \otimes N] = [N]$ by step II, for all $N$ in $R_0$-$\mathscr{M}\!od$, proving $\psi$ is 1:1.)

At this stage it only remains to prove $\psi$ is onto.

*Step IV.* (Serre's reduction) It is enough to show $R \otimes$ ___ induces an onto map $\psi: \mathscr{G}_0(R_0) \to \mathscr{G}_0(R\text{-}\mathscr{G}\!\imath\text{-}\mathscr{F}\!imod)$. (Indeed, as in the proof of theorem 5.1.34 we grade $R[\lambda]$ by putting $R[\lambda]_n = \bigoplus_{i \leq n} R_i \lambda^{n-i}$. $R[\lambda]$ is left Noetherian and free as module over $R[\lambda]_0 = R$, and $\text{pd}_{R[\lambda]} R \leq 1$ in view of proposition 5.1.19 (taking $\sigma = 1$ and $M = R$, viewed as $R[\lambda]$-module by annihilating $\lambda$). Hence by hypothesis $\mathscr{G}_0(R_0) \approx \mathscr{G}_0(R[\lambda]\text{-}\mathscr{G}\!\imath\text{-}\mathscr{F}\!imod)$. Now specializing $\lambda \mapsto 1$ sends $\mathscr{G}_0(R[\lambda]\text{-}\mathscr{G}\!\imath\text{-}\mathscr{F}\!imod)$ onto $\mathscr{G}_0(R\text{-}\mathscr{F}\!imod)$ by claim 2 of the proof of theorem 5.1.34).

Having proved step IV, we put $\mathscr{C} = \{M \in R\text{-}\mathscr{G}\!\imath\text{-}\mathscr{F}\!imod: \text{Tor}_n^R(R_0, M) = 0$ for all $n \geq 1\}$. By step II we see $\psi(\mathscr{G}_0(R_0)) \subseteq \mathscr{C}$.

*Step V.* Suppose $M \in R\text{-}\mathscr{G}\!\imath\text{-}\mathscr{F}\!imod$. If $M = RM_t$ for some $t$ and $\text{Tor}_1^R(R_0, M) = 0$ then $M \approx R \otimes M_t$. (Indeed, let $f: R \otimes M_t \to M$ be the natural graded map. Take any $x \neq 0$ in $\ker f$. Writing $x = \sum_{n \in \mathbb{N}} \sum_i r_{in} \otimes x_{in}$

where $r_{in} \in R_n$ and $x_{in} \in M_t$, we have

$$0 = fx = \sum r_{in}x_{in} \in M_{n+t}.$$

The component in $M_t$ is $0 = \sum r_{i0}x_{i0}$, so $0 = 1 \otimes \sum r_{i0}x_{i0} = \sum r_{i0} \otimes x_{i0}$. Let $R^+ = \bigoplus_{n>0} R_n$. We have proved $x \in R^+ \otimes M_t$; thus $\ker f \subseteq R^+ \otimes M_t$.

Applying $R_0 \otimes_R \underline{\ \ }$ to the exact sequence $0 \to \ker f \to R \otimes M_t \to M \to 0$ and using example 1.7.21' yields

$$\ldots \to \operatorname{Tor}_1^R(R_0, M) \to \frac{\ker f}{R^+ \ker f} \to \frac{R \otimes M_t}{R^+ \otimes M_t} \to M/R^+M \to 0.$$

Since $\operatorname{Tor}_1^R(R_0, M) = 0$ by hypothesis and $\ker f \subseteq R^+ \otimes M_t$ we conclude $\ker f/R^+ \ker f = 0$, i.e., $\ker f = R^+ \ker f$, which implies $\ker f = 0$ as desired.)

*Step VI.* If $0 \to M' \to M \to M'' \to 0$ is exact and $M, M'' \in \mathscr{C}$ then $M' \in \mathscr{C}$, seen via the long exact sequence for Tor (5.2.28).

*Step VII.* $\psi(\mathscr{G}_0(R_0)) = \mathscr{C}$. (Indeed, suppose $M \in \mathscr{C}$. Write $M = \sum_{n=0}^t RM_n$ for suitable $t$, and put $M' = \sum_{n=0}^{t-1} RM_n$. If $M' = 0$ then $M = RM_t \approx R \otimes M_t$ by step V, so we are done. Thus we may assume $M' \neq 0$. Let $M'' = M/M'$. Applying $R_0 \otimes_R \underline{\ \ }$ to the exact sequence $0 \to M' \to M \to M'' \to 0$ yields the long exact sequence

$$\cdots \to \operatorname{Tor}_1^R(R_0, M) \to \operatorname{Tor}_1^R(R_0, M'') \to M'/R^+M'$$

$$\to M/R^+M \to M''/R^+M'' \to 0.$$

But it is easy to see $R^+M' = M' \cap R^+M$, by writing $M' = \sum_{n=0}^{t-1}(R_0 + R^+)M_n$ and matching components. Thus the map $M'/R^+M' \to M/R^+M$ is monic; furthermore, $\operatorname{Tor}_1^R(R_0, M) = 0$ by hypothesis, so we must have $\operatorname{Tor}_1^R(R_0, M'') = 0$.

Clearly $M'' = RM_t''$, so step V implies $M'' \in \mathscr{C}$. Thus $M' \in \mathscr{C}$ by step VI, so by induction on $t$ we see $[M'] \in \psi(\mathscr{G}_0(R_0))$. It follows that $[M] = [M'] + [M''] \in \psi(\mathscr{G}_0(R_0))$, as desired.)

*Step VIII.* $\psi$ is onto. By step I it suffices to show for all $t$ that if $\operatorname{Tor}_n^R(R_0, M) = 0$ for all $n > t$ then $M \in \psi(\mathscr{G}_0(R_0))$. The proof will be by induction on $t$; the case $t = 0$ is step VII, so assume $t \geq 1$. Take $0 \to K \to F \to M \to 0$ with $F$ f.g. free. Then $\operatorname{Tor}_n^R(R_0, K) = \operatorname{Tor}_{n+1}^R(R_0, M) = 0$ for all $n > t - 1$, so by induction $[K] \in \psi(\mathscr{G}_0(R_0))$. Thus $[M] = [K] - [F] \in \psi(\mathscr{G}_0(R_0))$, as desired.

$$\text{Q.E.D.}$$

**Remark 5A.3:** If $S$ is a left denominator set of $R$ then there is an epic

$\mathscr{G}_0(R) \to \mathscr{G}_0(S^{-1}R)$ given by $[M] \mapsto [S^{-1}M]$. (This is clear since the localization functor is exact.)

# Exercises

## §5.1

1. Schanuel's lemma fails for flat modules. (Hint: Let $F_0$ be any localization of $\mathbb{Z}$; take $F$ free with $f: F \to F_0/\mathbb{Z}$ epic. Then $0 \to \mathbb{Z} \to F_0 \to F_0/\mathbb{Z} \to 0$ but $F_0 \oplus \ker f \not\approx \mathbb{Z} \oplus F$ unless $F_0$ is projective.)
2. $\mathrm{pd}_{R[\lambda]} M[\lambda] = \mathrm{pd}_R M$. (Hint: corollary 5.1.21, viewing $M[\lambda]$ as $R$-module.)
3. If each $R$-module has finite pd then gl. dim $R < \infty$. (Hint: proposition 5.1.20.)
4. (I. Cohen) If $R$ is a subring of $T$ such that $R$ is a summand of $T$ in $R\text{-}\mathscr{M}od\text{-}R$ then gl. dim $R \leq$ gl. dim $T$ + $\mathrm{pd}_R(T)$. In particular, if $T$ is projective as $R$-module then gl. dim $R \leq$ gl. dim $T$. (Hint: Writing $T = R \oplus N$ as bimodule yields $\mathrm{Hom}_R(T, M) = M \oplus \mathrm{Hom}_R(N, M)$ for any $R$-module $M$. Let $M' = \mathrm{Hom}_R(T, M)$. Then $\mathrm{pd}_R M \leq \mathrm{pd}_R M'$. Apply proposition 5.1.22 to $M'$.)
5. gl. dim $M_n(R) =$ gl. dim $R$. (Hint: Morita.)
6. If $R$ is left Noetherian and $0 \to M' \to M \to M'' \to 0$ is exact with two of the modules having Euler characteristic then so does the third, and $\chi(M) = \chi(M') + \chi(M'')$.
7. If $M > M_1 > \cdots > M_t = 0$ and each factor has FFR then $M$ has FFR.
8. Fields [70] Prove proposition 5.1.19 under the weaker assumption $\sigma: R \to R$ is an injection. (Hint: Consider $\sigma R$-modules.) Thus gl. dim $R \leq$ gl. dim $R[\lambda; \sigma] \leq 1 +$ gl. dim $R$. There are examples for strict inequality at either stage. Furthermore, these results hold for $R[\lambda; \sigma, \delta]$.

### F.g. Projectives vs. Free

9. (Ojanguren-Sridhan) For any noncommutative division ring $D$ there is an f.g. projective $D[\lambda_1, \lambda_2]$-module which is not free. (Hint: Pick $a, b$ in $D$ with $ab \neq ba$. The vector $v = (\lambda_1 + a, -\lambda_2 + b)$ in $D[\lambda_1, \lambda_2]^{(2)}$ is unimodular since $(\lambda_2 - b) \cdot (\lambda_1 + a) + (\lambda_1 + a)(-\lambda_2 + b) = [a, b]$, but $v$ is not strongly unimodular.)
10. (Gabel) If $P$ is stably free and not f.g. then $P$ is free. (Hint: Suppose $P \oplus R^{(m)} = F$ is free, and $\pi: F \to R^{(m)}$ be the projection with $\pi x_j = e_j$ where $e_1, \ldots, e_m$ is the standard base. Take a base $\{y_i\}$ and let $F'$ be the f.g. free module spanned by all $y_i$ involved in $x_1, \ldots, x_m$. Let $P' = P \cap F'$. Then $P/P'$ is free, so $0 \to P' \to P \to P/P' \to 0$ and $0 \to P' \to F' \to R^{(m)} \to 0$ are split; also there is a (split) epic $P'/P \to R^{(m)}$ whose kernel $F''$ is free; conclude $P \approx P' \oplus (P/P') \approx F' \oplus F''$.)

### $K_0$-Theory

11. Define more generally $K_0(R\text{-}\mathscr{F}imod)$ to be the free abelian group whose generators are isomorphism classes of f.g. $R$-modules, modulo the relation $[M] = [M'] + [M'']$ if there is an exact sequence $0 \to M'' \to M \to M' \to 0$. If $R$ is left

Noetherian of finite gl. dim show $K_0(R\text{-}\mathscr{F}imod) \approx K_0(R)$ canonically. (Hint: Split an f.g. projective resolution into short exact sequences.)

12. Using exercise 11 and the proof of proposition 5.1.32 show there is a canonical epic $K_0(R\text{-}\mathscr{F}imod) \to K_0(S^{-1}R\text{-}\mathscr{F}imod)$ whose kernel is generated by those modules with $S$-torsion.

13. Prove that $\varphi$ of Quillen's theorem is epic. (Hint: Let $R'$ be the "Rees subring" $\sum_{n \in \mathbb{N}} R_n \lambda^n$ of $R[\lambda]$, which takes the place of $R[\lambda]$ used in proving Serre's theorem. Grading $R'$ according to degree in $\lambda$ one has $R_0 = (R')_0$. But the summands $R_{n+1}/R_n$ of $G(R)$ are flat as $R_0$-modules so each $R_0, R_1, \ldots$ is flat. Hence $R$ is flat; also $R'$ is flat since $R'_n \approx R_n$ is flat. On the other hand, $R'$ is also filtered by $(R')_n = R_n[\lambda] \cap R'$; the associated graded ring is isomorphic to $G(R)[\lambda]$ under the grade $G(R)[\lambda]_n = \bigoplus_{i \leq n} R_i \lambda^{n-i}$ so is left Noetherian with finite gl. dim. Thus these properties pass to $R'$. Now factor $\varphi$ as $K_0(R) \to K_0(R') \to K_0(R)$, the first first arrow of which is an isomorphism by Serre's theorem. It remains to show the second arrow is onto, seen just as in the proof of Serre's theorem by viewing $R$ as $R'$-module via specializing $\lambda$ to 1 and applying the functor $H$ given by $HM = M/(1 - \lambda)M$.)

14. (Walker [72]) If $R$ is semiprime Noetherian and gl. dim $R < \infty$ and $K_0(R) = \langle [R] \rangle$ then $R$ is a domain. (Hint: Use the uniform dimension instead of the Euler characteristic in the proof of theorem 5.1.45.)

15. State and prove explicitly the dual to Schanuel's lemma.

## Change of Rings

16. Theorem 5.1.24 can fail if gl. dim $R/Ra$ is infinite. (Example: $R = \mathbb{Z}$ and $a = 4$.)

17. Suppose $a$ is a normalizing element of $R$ and $M$ is a module with $ax \neq 0$ for all $x \neq 0$ in $M$. Then any exact sequence $0 \to K \to L \xrightarrow{f} M \to 0$ yields an exact sequence (of $R/Ra$-modules) $0 \to K/aK \to L/aL \xrightarrow{\bar{f}} M/aM \to 0$. (Hint: ker $\tilde{f} = (K + aL)/aL \approx K/(K \cap aL)$ so it suffices to show $K \cap aL = aK$. But if $ax \in K$ then $0 = f(ax) = afx$.)

18. Second "change of rings" theorem. Hypotheses as in exercise 17 we have $\text{pd}_{R/aR} M/aM \leq \text{pd}_R M$. (Hint: Induction on $n = \text{pd}_R M$. For $n > 0$ take $0 \to K \to F \to M \to 0$ exact, with $F$ free. Then $\text{pd}_R K/aK \leq n - 1$ by induction; conclude using exercise 17.)

19. Third "change of rings" theorem. Hypotheses as in exercise 17, if, moreover, $R$ is left Noetherian, $M$ is f.g. and $a \in \text{Jac}(R)$, then $\text{pd}_{R/aR} M/aM = \text{pd}_R M$. (Hint: Using an induction procedure analogous to that of exercise 18, one is left with the case $\text{pd}_{R/aR} M/aM = 0$. Take $0 \to K \to F \to M \to 0$ with $F$ free. By exercise 17 one has $0 \to K/aK \to F/aF \to M/aM \to 0$ which thus splits. It suffices to show $K \oplus M$ is free; replacing $M$ by $K \oplus M$ assume $M/aM$ is free. But any base $x_1 + aM, \ldots, x_t + aM$ of $M/aM$ lifts to a base of $M$. Indeed, the $x_i$ span by Nakayama; if $\sum r_i x_i = 0$ take such $r_i$ with $Rr_1$ maximal and note each $r_i \in Ra = aR$, and writing $r_i = ar'_i$ conclude $r_1 \in Rar'_1$, contrary to $a \in \text{Jac}(R)$.) A more homological approach taken by Strooker [66] is given in Exercise 5.2.24ff.

20. If gl. dim $R \leq 2$ then $R$ is a PP ring. (Hint: By exercise 3.1.20 it suffices to show Ann $a$ is projective for any $a$ in $R$. But $0 \to \text{Ann } a \to R \to R \to R/Ra \to 0$ is exact where the second arrow is given by right multiplication by $a$.)

21. Generalizing the argument of exercise 20, show that if gl. dim $R \leq 2$ then every left annihilator of $R$ is projective. This result is used by Faith [73B] to characterize simple Noetherian rings of gl. dim $\leq 2$; in fact, Bass [60] characterized Noetherian rings of gl. dim $\leq 2$ along these lines.

22. (Zaks [71]) Suppose $L$ is a left ideal containing $P_1 \cdots P_t$ such that each $P_i \lhd R$ is f.g. projective as left module and each $R/P_i$ is simple Artinian. Then $L$ is projective. (Hint: Suppose $L$ is not projective. Let $A = P_1 \cdots P_t$. $R/A$ is Artinian, so one may take $L$ maximal with respect to not projective, and also there is $L < L' \leq R$ with $L'/L$ simple. Hence Ann$(L'/L)$ is some $P_i$. But $\mathrm{pd}_R R/P_i \leq 1$ so pd $L'/L \leq 1$. $L'$ is projective; hence $L$ is projective.)

23. (Zaks [71]) Suppose $R$ is left bounded, all proper images are Artinian, and all maximal ideals are projective as left modules. Then $R$ is left hereditary. (Hint: If $L < R$ then exercise 22 shows $L \oplus L'$ is projective where $L'$ is an essential complement of $L$) This is one of a host of his results about when a ring must be left hereditary.

24. If $R$ is left Noetherian then the direct limit of modules of injective dimension $\leq n$ has injective dimension $\leq n$.

## Free Summands of Large Projectives

The following results of Beck and Trosberg give interesting criteria for a module to have a free direct summand and lead to a reduction result strengthening parts of Bass [63].

25. Suppose $I$ is a set. Given $S \subseteq I \times I$ there is $I' \subseteq I$ with a well-ordering such that (i) if $i_1 < i_2$ in $I'$ then $(i_1, i_2) \in S$; (ii) if $I' \times \{i\} \subseteq S$ then $i \in I'$. (Hint: Consider all sets $I'$ with well-orderings satisfying (i), and apply the maximal principal.)

26. Consider $F = R^{(I)}$ with the canonical base $\{e_i : i \in I\}$ and projections $\pi_i : F \to R$, with $I$ infinite. If there are $x_i$ in $F$ which satisfy $\pi_i x_i = 1$ for each $i$ in $I$ then there is some $I' \subseteq I$ for which $|I'| = |I|$ and $\{x_i : i \in I'\} \cup \{e_i : i \in I - I'\}$ is a base of $F$. (Hint: Take $T = \{(i, j) \in I \times I : j \notin \mathrm{supp}(x_i)\}$. Take $I'$ as in exercise 25, and let $\psi: F \to \sum_{i \in I'} Re_i$ be the canonical projection. If $\sum_{i \in I'} r_i x_i \in \ker \psi$ then applying $\psi$ shows each $r_i = 0$, so the $x_i$ are independent for $i \in I'$ and $\ker \psi \cap \sum_{i \in I'} Rx_i = 0$. But an easy transfinite induction now shows $e_i \in \sum_{i \in I'} R\psi x_i$ for each $i$ in $I'$, so $\ker \psi + \sum_{i \in I'} Rx_i = F$.)

27. If $F$ is a free module and $M \leq F$ with $M + \mathrm{Jac}(R)F = F$ then $M$ has a summand isomorphic to $F$. (Hint: Let $F$ have a base $\{e_i : i \in I\}$; one may assume $I$ is infinite. Write $e_i = y_i + y_i'$ for $y_i$ in $M_i$ and $y_i'$ in $\mathrm{Jac}(R)F$. Then $y_i = \sum r_{ij} e_j$ where $r_{ii}$ is invertible, so apply exercise 26 to the $x_i = r_{ii}^{-1} y_i$ to get a free submodule $F'$ of $M$. The projection $F \to F'$ restricts to an epic $M \to F'$.)

28. If $R/\mathrm{Jac}(R)$ is left Noetherian then every uniformly big projective $R$-module is free, by applying exercise 27 to theorem 5.1.67.

29. Over any ring $R$, a non-f.g. projective $P$ is free iff there is a dual base $\{(x_i, f_i) : i \in I\}$ such that there is $I' \subseteq I$ with $|I'| = |I|$ and $f_i$ epic for all $i$ in $I'$. (Hint: ($\Leftarrow$) Define $f: P \to R^{(I)}$ by $fx = \sum (f_i x) e_i$ where $\{e_i : i \in I\}$ are the standard base for $R^{(I)}$; as in exercise 26 show $R^{(I)}$ is a summand of $P$.)

## Big Projectives are Free (Bass)

**30.** If $R = R_1 \times R_2$ then $R_1^{(\mathbb{N})} \times 0$ is a non-f.g., non-free projective.

**31.** Say $R$ is *S-connected* if for any prime ideals $P, Q$ there is a sequence $P_0 = P, P_1, \ldots, P_k = Q$ such that $P_{i-1} \subseteq P_i$ or $P_i \subseteq P_{i-1}$ for each $i$. Show $R/\mathrm{Nil}(R)$ is a finite direct product of $S$-connected rings if there are only a finite number of minimal primes. (Hint: Let $A_i$ be the intersection of all minimal primes in the $i$-th $S$-connected component. The $A_i$ are comaximal, so apply the Chinese Remainder Theorem.)

**32.** If $R$ is commutative Noetherian and $\mathrm{Spec}(R)$ is connected then every non-f.g. projective $M$ is free. (Hint: One may assume $M$ is countable by Kaplansky's theorem. $R$ is $S$-connected, so the rank function on $M$ is constant, cf., §2.12. If $M/PM$ were f.g. for each minimal prime $P$ then building a chain $M > P_1 M > P_2 P_1 M > \cdots > 0$ would show $M$ is f.g. Hence some $M/PM$ is not f.g. One wants to show $M$ is uniformly big. Otherwise, passing to $R/P$ and $M/PM$ one may assume $R$ is an integral domain, and $M/AM$ is f.g. for some maximal ideal $A$. Let $n = [M/AM : R/A] = \mathrm{rank}(M)$. $M$ has a free submodule $F$ of rank $n$ since $R$ is a domain. Hence $M/F$ is singular. Write $M \oplus M' = F'$ free, and write $F' = F_0 \oplus F_1$ where $F_0$ is f.g. free containing $F$. Then $M/F \leq \mathrm{Sing}((F_0 \oplus F_1)/F) = \mathrm{Sing}(F_0/F)$ so $M \leq F_0$. Hence $M$ is f.g., contradiction.)

## §5.2

**1.** The *ring of dual numbers* $\tilde{R}$ is defined as $R[\lambda]/\langle \lambda^2 \rangle$ (also, cf., exercise 1.9.7). Any $R$-module $M$ can be viewed as $\tilde{R}$-module since $R$ is a homomorphic image of $\tilde{R}$. Thus $Z(\ ) \approx \mathrm{Hom}_{\tilde{R}}(R, \underline{\ \ })$ which also proves left exactness of $Z$.

## Group Cohomology

**2.** If $t = |G|$ is finite then each $H^n(G, M)$ is $t$-torsion (notation as in example 5.2.13). (Hint: Let $h(g_1, \ldots, g_{n-1}) = \sum_{g \in G} f(g_1, \ldots, g_{n-1}, g)$. Summing over (2) of example 5.2.13, as $g_n$ runs over $g$ yields $\sum_{g \in G} (d^n f)(g_1, \ldots, g_n) = (-1)^{n+1} t f(g_1, \ldots, g_n) - (d^n h)(g_1, \ldots, g_n)$. If $d^n f = 0$ then $tf \in B^n$ as desired.)

**3.** Let $G$ be a group. Define $P_0 = \mathbb{Z}[G]$ and inductively $P_{n+1} = P_n \otimes \mathbb{Z}[G]$, a free $\mathbb{Z}[G]$-module with base $\{1 \otimes g_1 \otimes \cdots \otimes g_n : g_i \in G\}$. Also define maps $d_n : P_n \to P_{n-1}$ by $d_n(1 \otimes g_1 \otimes \cdots \otimes g_n) = g_1 \otimes g_2 \otimes \cdots \otimes g_n + (-1)^n \otimes g_1 \otimes \cdots \otimes g_{n-1} + \sum_{1 \leq i \leq n-1} (-1)^i \otimes g_1 \otimes \cdots \otimes g_i g_{i+1} \otimes \cdots \otimes g_n$. Then $(P_n, d_n)$ is a projective resolution of $\mathbb{Z}$ as $\mathbb{Z}[G]$-module with the trivial $G$-action. (Hint: Define $s_{-1} : \mathbb{Z} \to P_0$ by $s_{-1} 1 = 1$ and $s_n : P_n \to P_{n-1}$ by $s_n(g_0 \otimes \cdots \otimes g_n) = 1 \otimes g_0 \otimes \cdots \otimes g_n$, and use proposition 5.2.18.) Conclude that $H^n(G, M) \approx \mathrm{Ext}^n_{\mathbb{Z}[G]}(\mathbb{Z}, M)$ for any $\mathbb{Z}[G]$-module $M$ by looking at the isomorphisms $\mathrm{Hom}(P_n, M) \approx H^n(G, M)$.

**4.** Prove in detail the assertions made after theorem 5.2.22 concerning derived functors of Abelian categories, to obtain corollary 5.2.23.

## Basic Properties of $\mathcal{T}or$ (also, cf., Exercises 23ff.)

**5.** Use Tor to prove quickly the following facts about an exact sequence $0 \to M' \to$

$M \to M'' \to 0$: (1) If $M', M''$ are flat then $M$ is flat; (2) If $M''$ is flat the sequence is pure.

**6.** Suppose $0 \to K \to F \to N \to 0$ with $F$ flat then $\text{Tor}_n(M, N) \approx \text{Tor}_{n+1}(M, K)$.

**7.** If $M$ is a $T\text{-}\mathcal{M}od\text{-}R$ bimodule then $\text{Tor}_n^R(M, \_)$ can be viewed as a functor from $R\text{-}\mathcal{M}od$ to $T\text{-}\mathcal{M}od$. Verify similar assertions for $\text{Tor}(\_, M)$ and for Ext.

## $\mathcal{T}or$ as Torsion Module

**8.** Define the *torsion submodule* $tM = \{x \in M : \text{Ann}_R x \text{ contains a regular element}\}$. If $R$ is left Ore then $tM \leq M$ and $t(M/tM) = 0$; we say $M$ is *torsion* if $tM = M$. If $R$ is semiprime Goldie then $tM = \text{Sing}(M)$.

In Exercises 9 through 12 we assume $R$ is left Ore with ring of fractions $Q$, and $M$ is an arbitrary $R$-module.

**9.** $Q \otimes_R tM = 0$. Conclude $\text{Tor}_1(Q/R, tM) \approx tM$. (Hint: Apply $\text{Tor}(\_, tM)$ to $0 \to R \to Q \to Q/R \to 0$, noting $Q$ is $R$-flat.)

**10.** $\text{Tor}_n(Q/R, M) = 0$ for all $n \geq 2$. (Hint: As in exercise 9, noting $Q$ and $R$ are flat.)

**11.** The functors $t(\ )$ and $\text{Tor}_1(Q/R, \_)$ are naturally isomorphic. (Hint: Apply $\text{Tor}(Q/R, \_)$ to the exact sequence $0 \to tM \xrightarrow{i} M \to M/tM \to 0$ to get $0 \to \text{Tor}_1(Q/R, tM) \xrightarrow{i_*} \text{Tor}_1(Q/R, M) \to 0$ by exercises 9, 10. Conclude by composing $i_*$ with the isomorphism $tM \to \text{Tor}_1(Q/R, tM)$ of exercise 9.)

**12.** If $R$ is an integral domain then $\text{Tor}_n(M, N)$ is torsion for all $R$-modules $M, N$ and for all $n \geq 1$, and $\text{Tor}_0(M, N)$ also is torsion if $N$ is torsion. (Hint: First assume $N$ is torsion. Then $\text{Tor}_0(M, N) \approx M \otimes N$ is torsion. For $n \geq 1$ apply $\text{Tor}(\_, N)$ to a sequence $0 \to K \to F \to M \to 0$ with $F$ free and thus flat. Finally for arbitrary $N$ apply $\text{Tor}(M, \_)$ to $0 \to tN \to N \to N/tN \to 0$ to get $\text{Tor}_n(M, tN) \xrightarrow{f} \text{Tor}_n(M, N) \xrightarrow{g} \text{Tor}_n(M, N/tN)$ exact, and note $\ker f$ and $\text{im } g$ are torsion.)

## Weak Dimension and G1. Dim

**13.** In analogy to the syzygy of a projective resolution define the $n$-th *yoke* of a flat resolution of $M$ to be $\ker d_n$ and prove the following are equivalent: (i) $M$ has a flat resolution of length $n$. (ii) $\text{Tor}_k(\_, M) = 0$ for all $k \geq n + 1$. (iii) $\text{Tor}_{n+1}(\_, M) = 0$. (iv) The $(n-1)$ yoke of any flat resolution of $M$ is flat.

**14.** Show w. dim $R$ is the same as the right-handed version, by applying theorem 5.2.37 to the right-handed version of exercise 13. In this way w. dim is "better" than gl. dim, which is not left-right symmetric.

**15.** If $R$ is left Noetherian then pd $M =$ the flat dimension of $M$, for any f.g. module $M$. (Hint: Suppose $M$ has flat dimension $n$. We need to find a projective resolution of length $n$. Take any f.g. projective resolution. This is also a flat resolution so the $(n-1)$ syzygy $K$ is flat. But $K$ is f.g. and thus finitely presented, and so $K$ is projective by exercise 2.11.8.)

**16.** If $R$ is left Noetherian then gl. dim $R =$ w. dim $R$. (Hint: By theorem 5.2.42 one need only check the f.g. modules, but this is exercise 15.) Consequently, gl. dim $R$ is left-right symmetric for Noetherian rings. For non-Noetherian rings one can sometimes find a bound for the difference of the left and right gl. dim, as shown in Osofsky [73B, p. 57].

17. (i) A module $F$ is flat iff $\mathrm{Tor}_1(R/I, F) = 0$ for every right ideal $I$. (Hint: Use $0 \to I \to R \to R/I \to 0$.) (ii) w.dim($R$) is the supremum of the flat dimensions of the cyclic modules. (Hint: Dual of Auslandler's theorem.)

18. The following are equivalent for a ring $R$: (i) Every left ideal is flat. (ii) w. dim $R \leq 1$. (iii) Every right ideal is flat. Consequently, if $R$ is left or right semihereditary then w.dim($R$) $\leq 1$ and every submodule of a flat module is flat.

## Global Dimension for Semiprimary and/or Hereditary Rings

19. (Eilenberg-Nagao-Nakayama [56]) Suppose $R$ is hereditary and $A \lhd R$ such that $A^n = A^{n+1}$. Then gl. dim $R/A \leq 2n - 1$. (Hint: Let $B/A \lhd R/A$. The series $B \geq A \geq AB \geq A^2 \geq A^2B \geq \cdots$ yields an exact sequence

$$\cdots \to AB/A^2B \to A/A^2 \to B/AB \to B/A \to 0.$$

Each term has the form $P/AP$ where $P \lhd R$ and is thus projective, so $P/AP$ is projective as $R/A$-module. Since $A^n/A^{n+1} = 0$ conclude pd($B/A$) $\leq 2n - 1$ as desired.) Using exercise 2.7.17 conclude every homomorphic image of a hereditary semiprimary ring has finite global dimension, bounded by twice the index of nilpotence of the radical.

20. If $T = \begin{pmatrix} R & M \\ 0 & R' \end{pmatrix}$ then gl. dim $T \leq \max\{\text{gl. dim } R, \text{ gl. dim } R'\} + 1$. (Hint: By example 1.1.9 pd $L \leq \max\{\text{gl. dim } R, \text{gl. dim } R'\}$ for $L \leq T$.)

21. (Mostly Chase [61]; also, cf., Zaks [67]) Suppose $R$ is semiprimary with $J = \mathrm{Jac}(R)$. We say $R$ is *generalized upper triangular* with respect to a complete set of orthogonal idempotents $\{e_1, \ldots, e_n\}$ if $e_i J e_j = 0$ for all $i \geq j$. (In particular, $e_i R e_i$ is a division ring.) Prove the following are equivalent: (i) $R$ is generalized upper triangular with respect to *every* complete set of orthogonal idempotents. (ii) $R$ is generalized upper triangular with respect to *some* complete set of orthogonal idempotents. (ii) $R$ is either semisimple Artinian or has the form $\begin{pmatrix} R' & M \\ 0 & R'' \end{pmatrix}$ where $R'$ is generalized upper triangular (with respect to a smaller complete set of orthogonal idempotents) and $R''$ is semisimple Artinian. (iv) $R$ is the homomorphic image of a hereditary semiprimary ring. (v) gl. dim $R/A < \infty$ for every $A \lhd R$. (vi) gl. dim $R/J^2 < \infty$. In each case if $t$ is the number of isomorphism classes of simple $R$-modules we have gl. dim $R < t$ and $\mathrm{Jac}(R)^t = 0$. (Hint: Use exercise 2.7.17. (ii) $\Rightarrow$ (iii) If $t = 1$ then $J = 0$ and $R$ is semisimple Artinian. For $t > 1$ take $u$ minimal possible for which $e_u J = 0$ for $u < i \leq n$. Let $e = e_{u+1} + \cdots + e_n$ and $e' = 1 - e$. Then $eRe' = 0$ by exercise 2.7.11, and $eRe$ is semisimple Artinian. Note that (iii) implies the last assertion since $R'$ has $\leq t - 1$ classes of simple modules so we can apply induction to $R'$. (iii) $\Rightarrow$ (iv) Let $T$ be the tensor ring of $J/J^2$ as $R/J$-module. $T$ is hereditary semiprimary, and by induction applied to $R'$ one sees $R$ is an image of $T$. (iv) $\Rightarrow$ (v) by exercise 19. (vi) $\Rightarrow$ (i) First assume $J^2 = 0$. Let $N_i = Re_i/Je_i$. Each $Je_i$ is simple since $J^2 = 0$. If $e_i J e_i \neq 0$ then $N_i$ is a summand of $N_j$ so pd $N_i < $ pd $N_j$. Thus one merely need arrange the $e_i$ such that pd $N_i$ are increasing. For $J^2 \neq 0$ show by induction that $e_i J e_j \subseteq e_i J^u e_j$ for all $i \geq j$; the induction step is $e_i J e_j \subseteq e_i J^u e_j = \sum e_i J^{u-1} e_k J e_j \subseteq e_i J^{u+1} e_j$ since $i \geq k$ or $k \geq j$.)

22. (Auslander [55]) If $R$ is semiprimary and not semisimple Artinian then gl. dim $R = \text{pd}_R R/J = \text{pd}_R J + 1$ where $J = \text{Jac}(R)$. (Hint: Let $n = \text{pd}_R R/J$ and suppose $M \in R\text{-}\mathcal{M}od$. Every simple $R$-module is a summand of $R/J$ so $\text{Ext}_R^n(N, M) = 0$ for all simple $N$, implying $\text{Ext}_R^n(N, M) = 0$ for all $N$ satisfying $JN = 0$. Thus $\text{Ext}_R^n = 0$ by exercise 2.11.1 proving the first equality.) Auslander also showed this is the weak global dimension of $R$, thereby achieving left-right symmetry.

## Tor Module Theory

23. If $0 \to K \to F \to M \to 0$ is exact and $\text{Tor}_1^R(R/A, M) = 0$ then $0 \to K/AK \to F/AF \to M/AM \to 0$ is exact.
24. (Strooker [66]) Covers of projectives are projective. Explicitly suppose $R$ is left Noetherian and $J \subseteq \text{Jac}(R)$. If $P$ is an f.g. projective $R/J$-module and $M$ is a cover of $P$ with $\text{Tor}_1^R(R/J, M) = 0$ then $M$ is uniquely determined up to isomorphism, and $M$ is f.g. projective with $M/JM \approx P$. (Hint: Part of this is exercise 2.8.19. Finish by chasing diagrams using exercise 23).
25. If $R$ is left Noetherian and semilocal and $\text{Tor}_1^R(R/\text{Jac}(R), P) = 0$ then $P$ is projective.
26. View exercise 5.1.19 as a corollary of exercise 24.
27. Suppose $R$ is left Noetherian and $a$ is a regular normalizing element of $R$ with $a \in \text{Jac}(R)$. If gl. dim $R/Ra = n < \infty$ then gl. dim $R = n + 1$. (Hint: It suffices to show pd $M \leq n + 1$ for every f.g. module $M$. Take $0 \to K \to F \to M \to 0$ with $F$ f.g. free. Then $\text{pd}_R K = \text{pd}_{R/aR} K/aK \leq n$ by exercise 5.1.19, so $\text{pd}_R M \leq n + 1$.)

## Functorial Properties of Tor

28. Suppose $R, T$ are rings and $F$ is a flat right $R$-module and $M \in R\text{-}\mathcal{M}od\text{-}T$. Then $\text{Tor}_n^T(F \otimes_R M, \_) \approx F \otimes_R \text{Tor}_n^T(M, \_)$, cf., exercise 7. (Hint: It holds for $\text{Tor}_0 = \otimes \_.$)
29. If $S$ is a submonoid of $Z(R)$ and $T = S^{-1}R$ then $S^{-1}\text{Tor}_n^R(M, N) \approx \text{Tor}_n^T(S^{-1}M, S^{-1}N)$ as $S^{-1}Z(R)$-modules. If $M, N$ are f.g. and $R$ is left Noetherian then $S^{-1}\text{Ext}_R^n(M, N) \approx \text{Ext}_T^n(S^{-1}M, S^{-1}N)$. (Hint: exercise 2.11.15.) Improve this for left denominator sets.
30. If $M$ is f.g. and $R$ is left Noetherian then $\text{Tor}_n(M, \prod N_i) \approx \prod \text{Tor}_n(M, N_i)$. (Hint: Take $0 \to K \to P \to M \to 0$ with $P$ f.g. projective.)
31. If gl. dim $R = d$ then the functors $\text{Tor}_d(M, \_)$ and $\text{Tor}_d(\_, N)$ are left exact for any $M$ in $\mathcal{M}od\text{-}R$ and $N$ in $R\text{-}\mathcal{M}od$.
32. (Rosenberg-Stafford [76]) Suppose $R$ is left Noetherian and gl. dim $R = d$. Then the canonical map $\varphi \colon \text{Tor}_d(M, \prod N_i) \to \prod \text{Tor}_d(M, N_i)$ is monic. (Hint: By proposition 5.2.52 one may assume $M$ is f.g., so use exercise 30.)
33. (Rosenberg-Stafford [76]) Hypotheses as in exercise 32, if each $N_i \leq N$ then $\text{Tor}_d(M, \bigcap N_i) = \bigcap \text{Tor}_d(M, N_i)$. (Hint: Let $F$ denote the left exact functor $\text{Tor}_d(M, \_)$ and consider the commutative diagram

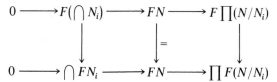

## §5.3

1. If $\partial \in \text{Der}(M)$ and $f: M \to N$ is a map in $R^e\text{-}\mathcal{M}od$ then $f\partial \in \text{Der}(N)$, so we have a commutative diagram each of whose vertical arrows arises from $f$:

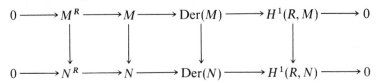

2. A map $f: J \to M$ extends to a map $R^e \to M$ iff $f\partial$ is an inner derivation of $R$ in $M$).
3. If $G$ is a group of order $n$ and $1/n \in C$ then the group algebra $C[G]$ is separable over $C$. (Hint: $\sum_{\sigma \in G}(1/n)\sigma \otimes \sigma^{-1}$ is a separability idempotent.)
4. $R^e$ has an involution given by $r_1 \otimes r_2 \mapsto r_2 \otimes r_1$.

**Braun's Characterization of Azumaya Algebras** Assume in exercises 5–8 that $R^e \approx \text{End}_C R$ canonically.

5. $Z(R^e) = C \otimes_C C \approx C$. (Hint: If $\sum x_i \otimes y_i \in Z(R^e)$ then $(r \otimes 1)\sum x_i \otimes y_i = (\sum x_i \otimes y_i)(r \otimes 1)$ implying $r\sum x_i y_i = \sum x_i r y_i = \sum x_i y_i r$; so $\sum x_i y_i \in C$, thereby yielding $\sum x_i \otimes y_i = \sum x_i y_i \otimes 1 \in C \otimes_C C$.)
6. If $R$ has an anti-automorphism (∗) fixing $C$ and $v(\sum x_i \otimes y_i) = (∗)$ then $(\sum x_i \otimes y_i) \cdot (\sum y_i \otimes x_i)$ is an invertible element $c$ of $C$, and $(∗)^{-1} = v(c^{-1}y_i \otimes x_i)$. (Hint: Compute $r_1^* r_2^* r_3^* = (r_3 r_2 r_1)^*$ to get $v((r_1^* \otimes r_3^*)(\sum x_i \otimes y_i))r_2 = v((\sum x_i \otimes y_i)(r_3 \otimes r_1))r_2$ and thus $(r_1^* \otimes r_3^*)\sum x_i \otimes y_i = (\sum x_i \otimes y_i)r_3 \otimes r_1$. Using the "switch" of exercise 4 conclude $(\sum x_i \otimes y_i)(\sum y_i \otimes x_i) \in Z(R^e) \approx C$. Identifying $c$ with $\sum(x_i \otimes y_i)\sum(y_i \otimes x_i)$ one sees $1 \in Rc$ so $c$ is invertible.)
7. Notation as exercise 6, and writing (∗′) for $(∗)^{-1}$ show $\sum y_i \otimes x_i^{*'}$ and $\sum x_i \otimes y_i^*$ are in $\text{Ann}'J$. (Brief hint: This follows from computing $(r_1 r^{*'})^* = r_2 r_1^*$ and manipulating with the maps $\alpha: R \otimes R \to R^e$ and $\beta: R \otimes R \to R^e$ defined by $\alpha(r_1 \otimes r_2) = r_1 \otimes r_2$ and $\beta(r_1 \otimes r_2) = r_1 \otimes r_2^*$.)
8. Applying theorem 5.3.24(vi) to exercise 7 and similar manipulations, show that if $R$ has an anti-automorphism (∗) fixing $C$ and $R^e \approx \text{End}_C R$ then $R$ is Azumaya.
9. $R$ is Azumaya iff the canonical map $R^e \otimes (R^e)^{\text{op}} \to \text{Hom}(R^e, R^e)$ is an isomorphism. (Hint: $R^e$ is Azumaya by exercises 4, 8; thus $R$ is Azumaya.)
10. (Schelter [84] Suppose $\varphi: C' \to C$ is a surjection of commutative rings with $(\ker \varphi)^2 = 0$, and $C$ is local. Given an Azumaya $C$-algebra $R$ there is an Azumaya $C$-algebra $R'$ and a map $R' \to R$ which extends $\varphi$. (Hint: $R \approx C^{(n)}$ as $C$-module, so let $R' = (C')^{(n)}$ and extend $\varphi$ to $\tilde{\varphi}: R' \to R$ componentwise. It suffices to find a multiplication on $R'$ for which $\tilde{\varphi}$ is a ring homomorphism. Letting $e_1, \ldots, e_n$ be the canonical base of $R$ write $e_i e_j = \sum c_{ijk} e_k$ and picking $c'_{ijk}$ in $C'$ such that $\varphi c'_{ijk} = c_{ijk}$ define multiplication on $R'$ by $e_i e_j = \sum c'_{ijk} e_k$. Let $N = (\ker \varphi)^{(n)} \subset R'$. Then $f: (R')^{(3)} \to N$ defined by $f(r'_1, r'_2, r'_3) = r'_1(r'_2 r'_3) - (r'_1 r'_2)r'_3$ induces a map $\bar{f}: R^{(3)} \to N$ obtained by defining $\bar{f}(\tilde{\varphi}r'_1, \tilde{\varphi}r'_2, \tilde{\varphi}r'_3) = f(r'_1, r'_2, r'_3)$ because $N^2 = 0$; $\bar{f} \in Z^3(R, N) = B^3(R, N)$ so $\bar{f} = \delta g$. Use $g$ to define a new multiplication.)

**11.** (Braun) $R$ is Azumaya iff there are elements $a_1, \ldots, a_t, b_1, \ldots, b_t$ in $R$ such that $\sum a_i b_i = 1$ and $\sum a_i r b_i \in Z(R)$ for all $r$ in $R$. (Hint (Dicks): ($\Rightarrow$) Write a separability idempotent $e$ as $\sum a_i \otimes b_i$ and apply proposition 5.3.29(i). ($\Leftarrow$) Let $e = \sum a_i \otimes b_i$ in $R^e$. First prove $R^e e R^e = R^e$. For, otherwise, let $P$ be a maximal ideal of $R$ containing $\{r \in R : r \otimes 1 \in R^e e R^e\}$ and let $\bar{R} = R/P$; then $0 \neq \bar{e}$ in $\bar{R} \otimes \bar{R}^{op}$, a simple ring, but $\bar{e}$ generates a proper ideal, contradiction. Now check for any $vew$ in $R^e e R^e$ that $\sum a_i \otimes (vew \cdot b_i) \in \mathrm{Ann}' J$ implying $e = \sum a_i \otimes b_i \in \mathrm{Ann}' J$ so $e$ is a separability idempotent.)

# 6 Rings with Polynomial Identities and Affine Algebras

In this chapter we shall treat two theories each of which in recent years has served as a way of attracting directly some of the techniques of commutative algebras to study suitably conditioned noncommutative algebras. Both theories have their roots in specific questions outside of "pure algebra" but have blossomed spectacularly in recent years. The two subjects complement each other beautifully and often intertwine both in hypotheses and in mathematical history, so we shall treat them together.

§6.1 is a resume of "elementary" theory of *rings with polynomial identity* (PI), focusing on central polynomials and their most direct applications. In §6.2 we introduce *affine algebras* and the Gelfand-Kirillov dimension, which perhaps has become the most prominent dimension of recent research; the springboard is the Golod-Shafarevich counterexample to Kurosch's problem. In §6.3 we study the structure of affine PI-algebras, including the positive answer to Kurosch's problem using Shirshov's techniques; one of the key features is Schelter's important "trace ring" construction which has been the arena of many significant recent advances. In §6.4 we consider relatively free PI-algebras, both as examples of rings and as a launching pad for the "quantitative" PI-theory. However, we barely scrape the surface of this area.

## §6.1 Rings with Polynomial Identities

Suppose one were to look for the most naive way possible to generalize the property of commutativity of rings. One might begin by observing that the commutative law for $R$ reads "$X_1 X_2 = X_2 X_1$ for all substitutions of $X_1, X_2$ in $R$," or, equivalently, $X_1 X_2 - X_2 X_1$ is "identically" 0 in $R$. Similarly, we could view the associative law as $(X_1 X_2) X_3 - X_1 (X_2 X_3)$ vanishing identically, and likewise for the distributive law. Thus it is natural to examine formal polynomials, say as element of $\mathbb{Z}\{X\}$, and see what happens when some polynomial vanishes for all substitutions for $R$. Some immediate examples:

*Example 6.1.1:*

(i) Any Boolean ring satisfies the identity $X_1^2 - X_1$.
(ii) Any finite field having $m$ elements satisfies the identity $X_1^m - X_1$ by Fermat's little theorem.
(iii) The algebra $R$ of upper triangular $n \times n$ matrices over a commutative ring satisfies the identity $[X_1, X_2][X_3, X_4] \cdots [X_{2n-1} X_{2n}]$.

These examples are not nearly as important as two deeper examples, finite dimensional algebras and algebraic algebras of bounded degree, which will be discussed a bit later. There is another obvious example—any algebra of characteristic $p$ satisfies the identity $pX_1$. In order to exclude such a huge class of algebras, we want to dismiss such an identity as "trivial." We are ready for a formal definition. Throughout $X = \{X_1, X_2, \ldots\}$ is a countable set of non-commuting indeterminates over a commutative ring $C$, and a *polynomial* is an element $f$ of $C\{X\}$, written $f(X_1, \ldots, X_m)$ to indicate that no $X_i$ occurs in $f$ for $i > m$. But now $f$ can be viewed as a function $f: R^{(m)} \to R$ acting by substitution, i.e., $f(R^{(m)}) = \{f(r_1, \ldots, r_m): r_i \in R\}$.

When $m$ is not of concern to us we shall write $f(R)$ in place of $f(R^{(m)})$. It is also convenient to deal with $f(R)^+$, defined as the additive subgroup of $R$ generated by $f(R)$. Obviously, $f(R) = 0$ iff $f(R)^+ = 0$.

*Remark 6.1.1':*   If $f$ is a polynomial and $\sigma$ is a $C$-algebra automorphism of $R$ then $f(\sigma r_1, \ldots, \sigma r_m) = \sigma f(r_1, \ldots, r_m)$; consequently, $f(R)$ and $f(R)^+$ are invariant under all automorphisms of $R$.

*Definition 6.1.2:*   $f$ is an *identity* of $R$ if $f(R) = 0$; an identity $f$ is a *polynomial identity* if one of the monomials of $f$ of highest (total) degree has coefficient 1. $R$ is a PI-*algebra* if $R$ satisfies a polynomial identity; $R$ is a PI-*ring* if $R$ satisfies a polynomial identity with $C = \mathbb{Z}$.

The reason for the somewhat technical definition of polynomial identity is to exclude identities such as $[X_1, X_2] + 2[X_1^2, X_2]$, since one usually wants to work with the monomials of highest degree.

The subject of most of this chapter is the "qualitative" PI-theory, i.e., the structure theory that can be developed for a ring on the mere assumption that it satisfies a polynomial identity. PI-rings were introduced by Dehn [22] who was searching for an algebraic framework for axiomatic Desargiuan geometry. He was missing a key ingredient, the use of inverses, and his project receded into oblivion until Amitsur [66] brought it to a successful completion. With a few exceptions mostly of historical interest, PI-rings remained in a state of suspended animation until Jacobson, Kaplansky, and Levitzki became interested in them, largely in connection with Kurosch's problem (to be discussed in §6.2). When Kaplansky proved every primitive PI-algebra is simple and finite dimensional over its center, the PI-theory began to breathe anew, especially in the hands of Amitsur. Another breakthrough came when Posner [60] proved every prime PI-ring is Goldie, yielding an instance where Goldie's theory could be applied to non-Noetherian rings. Deep applications were found to Azumaya algebras, geometry, and division algebras. But the best was yet to come. In response to an old problem of Kaplansky, Formanek [72] and Razmyslov [73] constructed polynomials on arbitrary $n \times n$ matrices which are not identities yet take only scalar values. Although the question had at first been considered mostly as a curiosity, it soon became clear that these "central polynomials" provided enough commutativity to permit direct application of standard techniques from the commutative theory, and the entire PI-theory was reworked from the bottom up to give central polynomials their proper dominant role.

During the 1970s researchers began to turn to basic matrix theory techniques to study PI-rings and expanded the rings under consideration by adjoining coefficients of the characteristic polynomials of various elements. On the one hand, this tied PI-theory to classical invariant theory, and brought in the "quantitative" PI-theory; on the other hand, Schelter and others used this technique to prove striking new results about prime PI-rings and later to develop the rudiments of a noncommutative algebraic geometry, cf, the papers of Artin-Schelter.

This section concerns that part of the PI structure theory which is obtained as a direct consequence of the existence of central polynomials for matrices. In the later sections we shall obtain the other basic PI-structure theorems, concentrating on the deeper theory of prime PI-rings. As hinted in the previous paragraphs, the structure theory hinges on what can be said about matrices, in particular on the two most basic properties of matrices—the determinant and the Hamilton-Cayley theorem. The determinant is linear and alternating in its rows; to study the analogous situation for polynomials we introduce the notation $f(X_i \mapsto h)$ to denote that $h$ is substituted throughout for $X_i$: an alternate longer notation is $f(X_1, \ldots, X_{i-1}, h, X_{i+1}, \ldots, X_m)$.

**Definition 6.1.3:**  A polynomial $f$ is *linear in $X_i$* if $X_i$ occurs exactly once (of degree 1) in every monomial of $f$; $f$ is *t-linear* if $f$ is linear in $X_1, \ldots, X_t$. $f$ is *t-alternating* if $f(X_i \mapsto X_j) = 0$ for all $1 \le i < j \le t$. A polynomial which is $t$-linear and $t$-alternating is called *t-normal*.

For example $[X_1, X_2] = X_1 X_2 - X_2 X_1$ is 2-normal. The value of $t$-linear polynomials lies in the following observation:

**Remark 6.1.4:**  If $f(X_1, \ldots, X_m)$ is $t$-linear then

$$f\left(\sum c_{i1} r_{i1}, \ldots, \sum c_{it} r_{it}, r_{t+1}, \ldots, r_m\right)$$

$$= \sum_{i_1, \ldots, i_t} c_{i_1 1} \cdots c_{i_t t} f(r_{i_1 1}, \ldots, r_{i_t t}, r_{t+1}, \ldots, r_m)$$

for all $c_{ij}$ in $C$ and $r_{ij}$ in $R$. In particular, if $B$ spans $R$ as $C$-module then $f(R^{(m)}) = f(B^{(t)} \times R^{(m-t)})$. Likewise, if $f$ is 1-linear then $f(R)^+$ is a $C$-submodule of $R$.

**Example 6.1.5:**  The 3-linear polynomial $[[X_1, X_2], X_3]$ is an identity of each exterior algebra, seen by checking monomials.

**Proposition 6.1.6:**  *If $f(X_1, \ldots, X_m)$ is t-normal and $R$ is spanned by fewer than $t$ elements over $C$ then $f$ is an identity of $R$.*

**Proof:**  Suppose $R = CB$ where $|B| \le t - 1$. Then $f(b_1, \ldots, b_t, r_{t+1}, \ldots, r_m) = 0$ by definition whenever the $b_i \in B$, since two $b_i$ must be the same. Consequently $f(B^{(t)} \times R^{(m-t)}) = 0$, so $f$ is an identity by remark 6.1.4.   ∎

It is a simple enough matter to construct normal polynomials.

**Definition 6.1.7:**  Define the *Capelli polynomial*

$$C_{2t}(X_1, \ldots, X_{2t}) = \sum_{\pi \in \mathrm{Sym}(t)} (\mathrm{sg}\,\pi) X_{\pi 1} X_{t+1} X_{\pi 2} X_{t+2} \cdots X_{\pi(t-1)} X_{2t-1} X_{\pi t} X_{2t}$$

and the *standard polynomial*

$$S_t(X_1, \ldots, X_t) = C_{2t}(X_1, \ldots, X_t, 1, \ldots, 1) = \sum_{\pi \in \mathrm{Sym}(t)} (\mathrm{sg}\,\pi) X_{\pi 1} \cdots X_{\pi t}$$

**Proposition 6.1.8:** $C_{2t}$ *and* $S_t$ *are* $t$-*normal and thus are identities of any* $C$-*algebra spanned by* $<t$ *elements.*

**Proof:** The terms in $C_{2t}(X_i \mapsto X_j)$ (or $S_t(X_i \mapsto X_j)$) subdivide into pairs of the same monomials appearing with opposite sign (one corresponding to a permutation $\pi$ and the other corresponding to $\pi$ composed with the transposition $(ij)$ and thus having opposite sign), so the polynomial is sent to 0.    Q.E.D.

For our final example we say $R$ is *integral of bounded degree* $n$ if every element of $R$ is integral of degree $\leq n$ over $C$, with $n$ minimal such.

**Example 6.1.9:** If $R$ is integral of bounded degree $\leq n$ and $f(X_1, \ldots, X_m)$ is $n$-normal then $S_n([X_1^n, X_2], \ldots, [X_1, X_2])$ is a 2-variable identify of $R$.

Indeed, for any $r_i$ in $R$ we can write $r_1^n = \sum_{i=0}^{n-1} c_i r_1^i$ so $[r_1^n, r_2] = \sum_{i=1}^{n-1} c_i[r_1^i, r_2]$ implying $S_n([r_1^n, r_2], \ldots, [r_1, r_2]) = 0$ by remark 6.1.4, so $f$ is an identity.

As a special case, consider $M_n(C)$ over an arbitrary commutative ring $C$. On the one hand, taking $t = n^2 + 1$ we see $S_t$ and $C_{2t}$ are identiities of $M_n(C)$ by proposition 6.1.8. On the other hand, by the Hamilton-Cayley theorem (which holds for arbitrary commutative rings, cf., Jacobson [85B, page 203]) we see $M_n(C)$ is integral of bounded degree $n$ so $S_n([X_1^n, X_2], \ldots, [X_1, X_2])$ is an identity of $M_n(C)$.

Given these basic examples of PI-rings, we want to see how broad the class of PI-rings really is.

**Remark 6.1.10:**

(i) If $f$ is an identity of $R$ then $f$ is also an identity of every homomorphic image of $R$ and of every subalgebra of $R$.

(ii) If $f$ is an identity of $R_i$ for each $i$ in $I$ then $f$ is an identity of $\prod\{R_i : i \in I\}$, and thus of any subdirect product of the $R_i$. (Indeed (i) is *a fortiori* and (ii) is by checking components and then applying (i).)

A polynomial $f(X_1, \ldots, X_m)$ is *multilinear* if $X_i$ has degree 1 in each monomial of $f$, for each $1 \leq i \leq m$. Multilinear identities play a special role because of the next observation (compare with theorem 6.1.44 on page 106).

**Proposition 6.1.11:**  *Any multilinear identity $f$ of $R$ is an identity of each central extention $R'$ of $R$.*

**Proof:**  If $R' = RZ$ then remark 6.1.4 shows $f(R') = f(R)Z = 0$.     Q.E.D.

To utilize this result we need a method of *multilinearizing* polynomial identities. Given $f(X_1, \ldots, X_m)$ define $\Delta_i f(X_1, \ldots, X_{m+1})$ for $1 \leq i \leq n$ by

$$\Delta_i f = f(X_i \mapsto (X_i + X_{m+1})) - f - f(X_i \mapsto X_{m+1}).$$

For example, if $f = X_1^2 X_2$ then $\Delta f = (X_1 + X_3)^2 X_2 - X_1^2 X_2 - X_3^2 X_2 = X_1 X_3 X_2 + X_3 X_1 X_2$ and $\Delta_2 f = X_1^2 (X_2 + X_3) - X_1^2 X_2 - X_1^2 X_3 = 0$.

The multilinearization process only works on the assumption that any $X_i$ appearing in $f$ appears in *all* the monomials of $f$; such a polynomial will be called "blended." All the polynomials we wrote down thus far are blended, and exercise 1 shows any identity is a sum of blended identities, so this assumption is very mild.

**Remark 6.1.12:**  Suppose $f(X_1, \ldots, X_m)$ is blended. Then $\Delta_i f$ is blended (in $X_1, \ldots, X_{m+1}$). Furthermore, any monomial $h$ in $f$ of degree $d \geq 1$ in $X_i$ produces $2^d - 2$ monomials in $\Delta_i f$, each having the same coefficient as $h$ and of degree $< d$ in $X_i$. If $f$ is an identity then $\Delta_i f$ is identity. (Indeed, the last assertion is obvious; furthermore, $h(X_i \mapsto X_i + X_{m+i})$ produces $2^d$ monomials, all of which are blended except $h$ and $h(X_i \mapsto X_{m+i})$, and these two drop out by definition of $\Delta_i$.)

**Remark 6.1.13:**  Using remark 6.1.12 one can turn any blended polynomial identity into a multilinear polynomial identity of the same degree. (Indeed, take a monomial $h$ of highest (total) degree having coefficient 1. If $h$ is linear in each $X_i$ then $f$ is multilinear since $f$ is blended. If $h$ is not linear in $X_i$ then $\Delta_i h$ is blended and has smaller degree in $X_i$; repeating the procedure we eventually reach a multilinear identity. Note that all monomials of smaller degree than $h$ have vanished in the multilinearization procedure.)

To see how this works let us consider the identities of example 6.1.1.

(i) $\Delta_1(X_1^2 - X_1) = \Delta_1(X_1^2) = X_1 X_2 + X_2 X_1$ which is thus an identity of every Boolean algebra.

(ii) Applying $\Delta_1$ $(m - 1)$ times to $X_1^m$ yields $\sum_{\pi \in \text{Sym}(m)} X_{\pi 1} \cdots X_{\pi m}$, which is a multilinear identity of any field of $m$ elements and is called the *symmetric identity.*

(iii) $[X_1 X_2] \cdots [X_{2n-1}, X_{2n}]$ is already multilinear.

## Central Polynomials and Identities of Matrices

We want now to examine identities of matrices more carefully, for this will pay handsome dividends in structure theoretical applications. If $x_1$ and $x_2$ are in $M_2(C)$ then $[x_1, x_2]$ has trace 0 so $[x_1, x_2]^2$ is a scalar matrix by the Hamilton-Cayley theorem. This led Kaplansky to ask whether arbitrary $n \times n$ matrix rings have nonidentities taking on scalar values. We shall see soon the answer is "yes." (Also see exercise 22.)

**Definition 6.1.14:**  $f(X_1, \ldots, X_m)$ is a *central polynomial* for $R$, otherwise called $R$-*central*, if $0 \neq f(R) \subseteq Z(R)$, i.e., if $f$ is *not* an identity of $R$ but $[X_{m+1}, f]$ is an identity of $R$.

**Example 6.1.15:**  $[X_1, X_2]^2$ is $M_2(C)$-central for any commutative ring $C$, as we just saw. $[X_1, X_2]$ is a central polynomial for the exterior algebra, in view of example 6.1.5.

Our aim is to construct multilinear $M_n(C)$-central polynomials. We follow the exposition of Rowen [80B] (based on Razmyslov-Bergman-Amitsur) since the ideas also apply to §6.3. The study of identities and central polynomials of matrices are closely related. Our first concern is to distinguish $M_n(C)$ from $M_{n-1}(C)$ by means of their identities.

**Proposition 6.1.16:**  For $t = n^2$, the Capelli polynomial $C_{2t}$ is not an identity of $R = M_n(H)$ for any commutative ring $H$ (although $C_{2(t+1)}$ is an identity, as noted earlier). In fact $C_{2t}(R)^+ = R$.

**Proof:**   Order the matric units $\{e_{ij}: 1 \leq i, j \leq n\}$ lexicographically on the subscripts, i.e., $e_{11} < e_{12} < \cdots < e_{1n} < e_{21} < \cdots < e_{nn}$, and write $r_k$ for the $k$-th matric unit on this list. Let us evaluate $C_{2t}(r_1, \ldots, r_t, r_1, \ldots, r_t)$. Taking $\pi$ in Sym($t$) let $a_\pi = r_{\pi 1} r_1 r_{\pi 2} r_2 \cdots r_{\pi t} r_t$. Then $r_1 r_{\pi 2} r_2 = e_{11} r_{\pi 2} e_{12}$ so $a = 0$ unless $r_{\pi 2} = e_{11}$; likewise, $r_2 r_{\pi 3} r_3 = e_{12} r_{\pi 3} e_{13}$ is 0 unless $r_{\pi 3} = e_{21}$. Continuing in this way we have precisely one choice of $r_{\pi 2}, \ldots, r_{\pi t}$ for $a$ to be nonzero. Since $e_{n1}$ has not yet been selected we take $r_{\pi 1} = e_{n1}$; then $a_\pi = e_{nn}$ for this particular $\pi$, and all other $a_\pi = 0$; proving $C_{2t}(r_1, \ldots, r_t, r_1, \ldots, r_t) = \pm e_{nn}$. By symmetry each $e_{ii} \in C_{2t}(R)$. For $i \neq j$ we have $(1 + e_{ij})^{-1} = 1 - e_{ij}$ so $(1 + e_{ij})^{-1} e_{ii} (1 + e_{ij}) = e_{ii} + e_{ij} \in C_{2t}(R)$ by remark 6.1.1'. Hence each $e_{ij} \in C_{2t}(R)^+$, proving $C_{2t}(R)^+ = R$.     Q.E.D.

**Remark 6.1.16':**  For any field $F$, elements $\{b_i: 1 \leq i \leq n^2\}$ of $M_n(F)$ form a base iff there are $\{r_i \colon 1 \le i \le n^2\} \subset M_n(F)$ such that $C_{2t}(b_1,\ldots,b_t,r_1,\ldots,$ $r_t) \ne 0$ where $t = n^2$. (Indeed by remark 6.1.4 we may assume the $b_i$ are matric units, so apply the proposition.)

**Digression 6.1.17:**   The standard polynomial $S_{2n}$ actually is an identity of $M_n(C)$ for any $n$. This celebrated theorem of Amitsur-Levitzki is proved in exercise 10 as a fairly direct consequence of the Hamilton-Cayley theorem, translated to traces by means of "Newton's formulas." Actually Razmyslov [74] has proved that *all* identities of $n \times n$ matrices can be obtained from the Hamilton-Cayley theorem by using traces formally in PI-theory. Nevertheless, $S_{2n}$ has the smallest degree of all the identities of $M_n(C)$, as we see now.

**Remark 6.1.18:**   No polynomial $f \ne 0$ of degree $\le 2n - 1$ is an identity of $M_n(C)$. (Indeed, by multilinearizing we could assume $f = \sum \alpha_\pi X_{\pi 1} \cdots X_{\pi m}$ with $\alpha_{(1)} \ne 0$; then $f(e_{11},e_{12},e_{22},\ldots) = \alpha_{(1)} e_{11} e_{12} \cdots \ne 0$.)

One might have hoped $S_n([X_1^n, X_2],\ldots,[X_1,X_2])$ to be the minimal 2-variable identity, but for $n \ge 8$ the polynomial identity $S_{2n}(X_1, X_2, X_1 X_2,$ $X_2 X_1, X_1^2 X_2,\ldots)$ has lower degree.

Our key to finding central polynomials for $M_n(C)$ is the following lemma which mimics the known properties of determinants.

**Proposition 6.1.19:**   *Suppose $R \in C\text{-}\mathcal{Alg}$, $x_1,\ldots,x_t$ are arbitrary elements of $R$, and $T \colon \sum_{i=1}^t C x_i \to \sum_{i=1}^t C x_i$ is a map (in $C\text{-}\mathcal{Mod}$). Viewing $T$ as an image of a $t \times t$ matrix $T'$ (cf., theorem 1.5.13) we have the following formulas, for any $t$-normal polynomial $f(X_1,\ldots,X_k)$:*

(i) $(\det T') f(x_1,\ldots,x_k) = f(Tx_1,\ldots,Tx_t,x_{t+1},\ldots,x_k)$,

(ii) $\det(\lambda - T') f(x_1,\ldots,x_k) = f(\lambda x_1 - Tx_1,\ldots,\lambda x_t - Tx_t, x_{t+1},\ldots,x_k)$

*for a commuting indeterminate $\lambda$ (viewed as a scalar $t \times t$ matrix)*

(iii) *If $\lambda^t + \sum_{i=1}^t (-1)^i c_i \lambda^{t-i}$ is the characteristic polynomial of $T'$ then each*

$$c_i f(x_1,\ldots,x_k) = \sum f(T^{j_1} x_1,\ldots,T^{j_t} x_t, x_{t+1},\ldots,x_k)$$

*summed over $(j_1,\ldots,j_t)$ in $\{0,1\}^{(t)}$ such that $j_1 + \cdots + j_t = i$;*

(iv) $(\operatorname{tr} T') f(x_1,\ldots,x_k) = \sum_{i=1}^t f(x_1,\ldots,x_{i-1}, Tx_i, x_{i+1},\ldots,x_k)$.

*Proof:*

(i) Write $Tx_j = \sum_{i=1}^{t} c_{ij}x_i$. Using the $t$ normality of $f$ we have

$$f(Tx_1,\ldots,Tx_t,x_{t+1},\ldots,x_k) = f\left(\sum_{i=1}^{t} c_{i1}x_i,\ldots;\sum_{i=1}^{t} c_{it}x_i,x_{t+1},\ldots,x_k\right)$$

$$= \sum_{\pi \in \text{Sym}(t)} c_{\pi 1,1}\cdots c_{\pi t,t} f(x_{\pi 1},\ldots,x_{\pi t},x_{t+1},\ldots,x_k)$$

$$= \sum_{\pi}(\text{sg }\pi)c_{\pi 1,1}\cdots c_{\pi t,t} f(x_1,\ldots,x_k) = (\det T')f(x_1,\ldots,x_k)$$

(ii) Follows from (i), working in $R[\lambda]$ and using $\lambda 1 - T$ in place of $T$ (where 1 is the identity map).

(iii) Match coefficients of $\lambda^{t-i}$ in (ii).

(iv) Take $i = 1$ in (iii).     Q.E.D.

**Theorem 6.1.20:**   *There is a multilinear polynomial which is $M_n(H)$-central for every commutative ring $H$.*

*Proof:*   Put $t = n^2$, and write

$$\sum_{i=1}^{t} C_{2t}(X_1,\ldots,X_{i-1},X_{2t+1}X_iX_{2t+2},X_{i+1},\ldots,X_{2t})$$

$$= \sum_{j=1}^{m} \sum_{i=1}^{t} h_{ij1}X_{2t+1}X_iX_{2t+2}h_{ij2}$$

for suitable $m$ and multilinear monomials $h_{ij1},h_{ij2}$ (in $X_1,\ldots,X_{i-1},X_{i+1},\ldots,X_{2t}$). Pick arbitrary $a,b,r_1,\ldots,r_{2t}$ in $M_n(H)$; viewing $M_n(H)$ as an $n^2$-dimensional module over $H$ having base $e_{ij}$, $1 \le i,j \le n$, we define the map $T: M_n(H) \to M_n(H)$ given by $Tx = axb$. Writing $a = (a_{ij})$ and $b = (b_{ij})$ we have $Te_{ij} = \sum_{u,v=1}^{n} a_{ui}b_{jv}e_{uv}$, whose coefficient of $e_{ij}$ is $a_{ii}b_{jj}$. Hence as $n^2 \times n^2$ matrix, $T$ has trace $\sum_{u,v=1}^{n} a_{ii}b_{jj} = \text{tr}(a)\text{tr}(b)$ so by proposition 6.1.19(iv), putting $w_{iju} = h_{iju}(r_1,\ldots,r_{i-1},r_{i+1},\ldots,r_{2t})$ we have

$$\text{tr}(a)\text{tr}(b)C_{2t}(r_1,\ldots,r_{2t}) = \sum_{i=1}^{t} C_{2t}(r_1,\ldots,r_{i-1},ar_ib,r_{i+1},\ldots,r_{2t})$$

$$= \sum_{j=1}^{m} \sum_{i=1}^{t} w_{ij1}ar_ibw_{ij2}.$$

Put $w = C_{2t}(r_1,\ldots,r_{2t})$. Taking traces of both sides yields

$$\text{tr}(\text{tr}(a)\text{tr}(w)b) = \text{tr}(a)\text{tr}(w)\text{tr}(b) = \text{tr}(\text{tr}(a)\text{tr}(b)w) = \sum_{i,j}\text{tr}(w_{ij1}ar_ibw_{ij2})$$

$$= \sum_{i,j}\text{tr}(w_{ij2}w_{ij1}ar_ib) = \text{tr}\left(\sum_{i,j}w_{ij2}w_{ij1}ar_i\right)b\right),$$

implying $\mathrm{tr}((\mathrm{tr}(a)\,\mathrm{tr}(w) - \sum_{i,j} w_{ij2} w_{ij1} a r_i) b) = 0$ for all $b$; nondegeneracy of the trace yields $\mathrm{tr}(a)\,\mathrm{tr}(w) - \sum_{i,j} w_{ij2} w_{ij1} a r_i = 0$, i.e., $\sum w_{ij2} w_{ij1} a r_i$ equals the scalar $\mathrm{tr}(a)\,\mathrm{tr}(w)$. By proposition 6.1.16 we can pick $r_1,\dots,r_{2t}$ such that $\mathrm{tr}(a)\,\mathrm{tr}(w) \neq 0$. Thus we define $f = \sum_{j=1}^m \sum_{i=1}^t h_{ij2} h_{ij1} X_{2t+1} X_i$, which takes only scalar values and is not an identity, i.e., $f$ is $M_n(H)$-central.          Q.E.D.

Actually we want an $n^2$-normal central polynomial, but this is rather easy to find.

**Definition 6.1.21:**   $g_n = f(C_{2n^2}(X_1,\dots,X_{2n^2})X_{2n^2+1},\ X_{2n^2+2},\dots,\ X_{2n^2+2t+1})$ where $f(X_1,\dots,X_{2t+1})$ is the central polynomial for $n \times n$ matrices we found in the theorem.

**Corollary 6.1.22:**   $g_n$ is an $n^2$-normal polynomial which is $M_n(C)$-central for all commutative rings $C$.

**Proof:**   Clearly, $g_n$ is $n^2$-normal and is either $M_n(C)$-central or an identity. But $f$ is not an identity, so $g_n$ is not an identity, by proposition 6.1.17.          Q.E.D.

We carry the notation $g_n$ throughout this chapter and shall denote its degree as $m$.

**Digression 6.1.23:**   We made little effort to limit the degree, and indeed there are multilinear polynomials of considerably lower degree. Formanek's central polynomials have degree $n^2$, but this is not the best possible. Also Formanek [86] verified a central polynomial of degree $2n^2$ (conjectured by Regev) which is of considerable importance in the quantitative theory, cf., §6.4.

Halpin [83] has simplified Razmyslov's original construction of a central polynomial. Let us outline the main ideas, which are of considerable interest. We say $f$ is a *weak identity* of $M_n(C)$ if $f$ vanishes on all matrices of trace 0. The connection between identities and weak identities is found in the *Razmyslov transformations* $T_k$ and $T'_k$ defined on multilinear polynomials as follows: It is enough to define them for multilinear monomials $h = h_1 X_k h_2$; define $T_k h = h_2 X_k h_1$ and $T'_k h = h_2 h_1 = T_k h(X_k \mapsto 1)$. One verifies at once

(i)   $T_k^2 f = f$.

(ii)   $T_k f = (T'_k(X_{m+1} f))(X_{m+1} \mapsto X_k)$.

(iii)   $T_k[X_{m+1}, f] = (T_k f)(X_k \mapsto [X_k, X_{m+1}])$.

By a careful accounting of the diagonal contributions, one can show if a multilinear polynomial $f$ is a weak identity of $M_n(C)$ then $T'_k f$ is either $R$-central or an identity; if $f$ is an identity then $T'_k f$ is an identity.

Applying (ii), one sees $f$ is an identity iff $T_k f$ is an identity. Using commutators, it is now easy to show that any weak identity of degree $d$ which is *not* an identity produces a central polynomial of degree $\leq 2d$. The obvious such weak identity is the Capelli polynomial $C_{2n^2}$. However, noting the trace is the second coefficient of the Hamilton-Cayley polynomial, one sees the multilinearization of

$$S_{n-1}([X_1^n, X_2], [X_1^{n-2}, X_2], [X_1^{n-3}, X_2], \ldots, [X_1, X_2])$$

is a weak identity which is not an identity, cf., exercise 11.

Here is another result used in verifying identities and central polynomials.

**Remark 6.1.23':** Suppose $R$ is central simple over an infinite field $F$. To prove a continuous map $f: R^{(m)} \to T$ satisfies $f(r, r_2, \ldots, r_m) = 0$ for all $r$ in $R$ (where $r_2, \ldots, r_m$ are fixed) it suffices to show this holds for a Zariski open subset. Using proposition 2.3.35 we may assume $\deg r = n$ and the roots of the minimal polynomial are distinct. Thus $r$ is diagonalizable in $M_n(\bar{F}) \approx R \otimes \bar{F}$, where $\bar{F}$ is the algebraic closure of $F$. Suppose $f(ara^{-1}, ar_2a^{-1}, \ldots, ar_ma^{-1}) = af(r, r_2, \ldots, r_m)a^{-1}$ (e.g., $f$ is a polynomial or a trace polynomial). Diagonalizing $r$, it *suffices to assume $r$ is diagonal with distinct eigenvalues* (in $M_n(\bar{F})$).

## Structure Theory for PI-Rings

We can now build a structure theory for semiprime PI-rings. The cornerstone is Kaplansky's theorem, that any primitive PI-ring $R$ is simple and finite dimensional over the center. The proof is twofold. First one notes $R$ is simple Artinian by Jacobson's density theorem, and then one shows the underlying division ring is finite dimensional by means of some splitting technique. The splitting technique used here is a special case of example 2.12.28, spelled out for the reader's convenience.

**Proposition 6.1.24:** *Suppose $D$ is a division algebra over an algebraically closed field $F$ and $D$ has a (possibly infinite) base over $F$ of cardinality $< |F| - 1$. Then $D = F$. (More generally, if $F$ is not algebraically closed then $D$ is algebraic over $F$.)*

**Proof:** If $d \in D$ were not algebraic over $F$ then $\{(d - \alpha)^{-1} : \alpha \in F\}$ is $F$-independent by proposition 2.5.21, contrary to hypothesis. Thus $F[d]$ is a finite field extension of $F$, so $d \in F$, i.e., $D = F$. Q.E.D.

We say $R$ has PI-*degree* $n$ if $R$ satisfies all multilinear identities of $M_n(\mathbb{Z})$ and $g_n$ (of corollary 6.1.22) is $R$-central. This was called PI-*class* $n$ in Rowen [80B] to avoid confusion wih the degree of a minimal polynomial identity, which is $2n$ by the Amitsur-Levitzki theorem, but the terminology PI-degree has become standard.

**Theorem 6.1.25:** (*Kaplansky's Theorem*) *Suppose $R$ is a primitive ring satisfying a polynomial identity $f$ of degree $d$. Then $R$ has some PI-degree $n \leq [d/2]$, and $R \approx M_t(D)$ for a division ring $D$ (unique up to isomorphism) with $n^2 = [R:Z(R)] = t^2[D:Z(D)]$.*

**Proof:** We may assume $f$ is multilinear and $X_d \cdots X_1$ has nonzero coefficient $\alpha$ in $f$. Let $M$ be a faithful simple $R$-module, and $D = \text{End}_R M$. We claim $R \approx M_t(D)$ for some $t \leq d$. Otherwise, taking any $x_1$ in $M$ we take $r_1$ such that $r_1 x_1 \notin x_1 D$, and put $x_2 = r_1 x_1$; inductively, given $r_1, \ldots, r_{i-1}$ and $x_1, \ldots, x_{i-1}$, take $r_i$ such that $r_i x_j = 0$ for all $j < i$ and $r_i x_i \notin \sum_{j<1} x_i D$, and put $x_i = r_i x_i$. Then $f(r_1, \ldots, r_d)x_1 = \alpha r_d \cdots r_1 x_1 = \alpha x_d \neq 0$.

Let $F = Z(D) = Z(R)$ and take an algebraically closed field $K$ of cardinality $> 1 + [R:F]$ (possibly infinite). Then $R_1 = R \otimes_F K$ is a simple $K$-algebra and satisfies the identity by proposition 6.1.11, so as above $R_1 \approx M_n(D_1)$ for some $n$ and some $K$-division algebra $D_1$ with $[D_1:K] \leq [R_1:K] = [R:F] < |K| - 1$. Hence $D_1 = K$ by proposition 6.1.25 so $R_1 \approx M_n(K)$ and $n \leq [d/2]$ by remark 6.1.18. But then $n = [R:F] = t^2[D:F]$, and clearly $R$ has PI-degree $n$.    Q.E.D.

To utilize Kaplansky's theorem most effectively we need an embedding procedure into nicer rings. We say $R$ is *embeddible in $n \times n$ matrices* if there is an injection from $R$ into $\prod_{k<n} M_k(H_k)$ where $H_k$ is commutative.

**Theorem 6.1.26:** *If $R$ satisfies a polynomial identity of degree $d$ then $R/N(R)$ is embeddible in $n \times n$ matrices for $n = [d/2]$. (Recall $N(R)$ from definition 2.6.25.) In fact, each $H_k$ can be taken as a direct product of fields.*

**Proof:** In the notation of theorem 2.6.27 we have an injection of $R/N(R)$ into $R_1$ with $\text{Nil}(R_1) = 0$. Multilinearizing we can pass the PI to $R_1$. But $R_1[\lambda]$ is semiprimitive by Amitsur's theorem 2.5.23, so can be injected into a direct product of primitive rings, each of which by Kaplansky's theorem is simple of dimension $\leq n^2$ over its center. Splitting each of these primitive components to inject it into $M_k(F_{ik})$ for a suitable field $F_{ik}$ and $k \leq n$, we conclude by taking $H_k = \prod_i F_{ik}$.    Q.E.D.

This embedding procedure, foreshadowed in §2.6, is extremely useful. We give the principal application here and leave the others for exercises 14, 15. (Note that an arbitrary PI-ring need not be embeddible in matrices, by example 3.2.48. We pursue this question further in §6.3.)

**Theorem 6.1.27:** *Suppose R satisfies a PI of degree d. If R is semiprime then* $\text{Nil}(R) = 0$. *In general, for every nil weakly closed subset A of R we have* $A^{[d/2]} \subseteq N(R)$.

**Proof:** Let $\bar{A}$ be the image of $A$ in $R/N(R)$. By theorem 6.1.26 $R/N(R)$ is embeddible in $n \times n$ matrices for $n = [d/2]$, so $\bar{A}^n = 0$, i.e., $A^n \subseteq N(R)$, proving the second assertion. In particular, $\text{Nil}(R)^n \subseteq N(R) = 0$ for $R$ semiprime, so $\text{Nil}(R) = 0$.     Q.E.D.

Although exercise 12 shows this theorem is sharp insofar as $n$ is concerned, one can replace $N(R)$ by $L_1(R)$, the sum of the nilpotent ideals, as expounded in Rowen [80B, theorem 1.6.36(i) and exercise 1.6.6]; consequently the nil-radical is reached in only two steps. Instead of pursuing this matter further, we push on to a central result.

**Lemma 6.1.27':** *If f is an identity or central polynomial of $M_n(C)$ then f is an identity of $M_{n-1}(C)$.*

**Proof:** Matrices $x_i = \sum_{i,j=1}^{n-1} c_{ij}e_{ij}$ can be reread as elements of $M_n(C)$ so $f(x_1, \ldots, x_m)$ is some scalar $\alpha$; but $\alpha e_{nn} = f(x_1, \ldots, x_m)e_{nn} = 0$ so $\alpha = 0$.
                                                                          Q.E.D.

**Theorem 6.1.28:** *Every semiprime PI-ring R has PI-degree n for suitable n, and every ideal of R intersects the center nontrivially.*

**Proof:** Let $d$ be the degree of a PI of $R$; let $Z = Z(R)$ and $0 \neq A \lhd R$. We shall show $R$ has PI-degree $n \leq [d/2]$, and $A \cap Z \neq 0$.

*Case I.* $R$ is semiprimitive. $R$ is a subdirect product of primitive $\{R_i : i \in I\}$, and by Kaplansky's theorem each $R_i$ is central simple over $Z(R_i)$ of degree $n_i \leq [d/2]$. Let $\pi : R \to R_i$ denote the canonical projection and let $A_i = \pi A_i$. Each $A_i$ is an ideal of $R_i$ so $A_i = 0$ or $A_i = R_i$. Let $I' = \{i \in I : A_i \neq 0\}$ and $n = \max\{n_i : i \in I'\}$. Then $g_n(A_i) \subseteq Z(R_i)$ for $i \in I'$ and $g_n(A_i) = 0$ for $i \notin I'$ by lemma 6.1.27'. But $g_n(A_i) \neq 0$ for $i$ such that $n_i = n$, so $0 \neq g_n(A) \subseteq A \cap Z$. Furthermore, taking $A = R$ we have each $A_i = R_i$ so $I' = I$ and $g_n$ is $R$-central, proving $R$ has PI-degree $n$.

*Case II.* $R$ semiprime. Then $\mathrm{Nil}(R) = 0$ by theorem 6.1.27 so $R[\lambda]$ is semi-primitive by Amitsur's theorem. Now case I is applicable. $0 \neq A[\lambda] \cap Z[\lambda] = (A \cap Z)[\lambda]$, proving $A \cap Z \neq 0$; also $R[\lambda]$ has some PI-degree $n$, implying $R$ also has PI-degree $n$.     Q.E.D.

**Corollary 6.1.29:** *If $R$ is semiprime PI and $Z(R)$ is a field $F$ then $R$ is a central simple $F$-algebra.*

*Proof:* Every nonzero ideal of $R$ contains a unit in $F$, so $R$ is simple, and Kaplansky's theorem is applicable.     Q.E.D.

**Theorem 6.1.30:** *If $R$ is a prime PI-ring and $S = Z(R) - \{0\}$ then $S^{-1}R$ is central simple over $S^{-1}Z(R)$ of degree $n$, where $n = \mathrm{PI\text{-}deg}(R)$.*

*Proof:* Let $Z = Z(R)$. $S^{-1}R$ is prime by proposition 2.12.9' and has PI-degree $n$ since $S^{-1}R$ is a central extension of $R$. But proposition 1.10.13 shows $Z(S^{-1}R) = S^{-1}Z$, a field, so $S^{-1}R$ is central simple by corollary 6.1.29.
         Q.E.D.

**Corollary 6.1.31:** *If $R$ is a semiprime PI-ring then* $\mathrm{Sing}(R) = 0$.

*Proof:* If $L$ is a large left ideal of $R$ and $Lz = 0$ for $z$ in $Z(R)$ then $z \in \mathrm{Ann}\, L = 0$. Thus $Z(R) \cap \mathrm{Sing}(R) = 0$, implying $\mathrm{Sing}(R) = 0$.     Q.E.D.

*Digression:* We now see the theory of nonsingular rings is applicable to semiprime PI-rings. An alternate proof of theorem 6.1.28 comes by means of rings without 1, cf., exercises 18–20, and shows, furthermore, that $L < R$ is large iff $L \cap Z(R)$ is large in $Z$ (exercise 21), leading to an easy description of the maximal quotient ring, cf., Rowen [74]. Continuing in the general structure theory, let us exploit the $n^2$-normality of $g_n$.

**Lemma 6.1.32:** *Suppose $R$ has PI-degree $n$. For each $r_1, \ldots, r_{m+1}$ in $R$ we have*

$$g_n(r_1, \ldots, r_m) r_{m+1} = \sum_{i=1}^{t} (-1)^{i+1} g_n(r_{m+1}, r_1, \ldots, r_{i-1}, r_{i+1}, \ldots, r_m) r_i$$

*where* $t = n^2$.

*Proof:* Define $\tilde{g}(X_1, \ldots, X_{m+1})$ as $\sum_{i=1}^{t+1} (-1)^i g_n(X_1, \ldots, X_{i-1}, X_{i+1}, \ldots, X_{m+1}) X_i$. Then $\tilde{g}$ is $(t+1)$-alternating by inspection since $\tilde{g}(X_i \mapsto X_j)$ has exactly two nonzero parts, which appear with opposite signs. Hence $\tilde{g}$ is an identity of $R$, so $\tilde{g}_n(r_{m+1}, r_1, \ldots, r_m) = 0$ yielding the desired equation.  Q.E.D.

**Theorem 6.1.33:** *Suppose $R$ has PI-degree $n$. If there are elements $r_1, \ldots, r_m$ in $R$ for which $g_n(r_1, \ldots, r_m) = 1$ then $R$ is a free $Z(R)$-module with base $r_1, \ldots, r_t$ where $t = n^2$.*

**Proof:** Let $Z = Z(R)$. By lemma 6.1.32 we have $r = \sum_{i=1}^{t} (-1)^{i+1} g(r, r_1, \ldots, r_{i-1}, r_{i+1}, \ldots, r) r_i \in \sum Z r_i$ for each $r$ in $R$, so $r_1, \ldots, r_t$ span $R$. If $\sum_{i=1}^{t} z_i r_i = 0$ for $z_i$ in $Z$ then for each $j \leq t$ we have

$$0 = g_n\left(r_1, \ldots, r_{j-1}, \sum_{i=1}^{t} z_i r_i, r_{j+1}, \ldots, r_m\right) r_j$$

$$= \sum_{i=1}^{t} g_n(r_1, \ldots, r_{j-1}, z_i r_i, r_{j+1}, \ldots, r_m) r_j$$

$$= g_n(r_1, \ldots, z_j r_j, \ldots, r_m) = z_j$$

proving $r_1, \ldots, r_t$ are independent.  Q.E.D.

## The Artin-Procesi Theorem

The last result hints $R$ may be Azumaya over $Z(R)$, leading us to another major result; the equivalence (i') $\Leftrightarrow$ (iv) below is called the *Artin-Procesi* theorem.

**Lemma 6.1.34:** *Suppose $R$ has a multilinear central polynomial $g$. Then $g(R)^+$ is an ideal of $Z(R)$; furthermore, if $g(R)R = R$ then $g(R)^+ = Z(R)$.*

**Proof:** Let $Z = Z(R)$. If $z \in Z$ then $zg(r_1, \ldots, r_m) = g(zr_1, \ldots, r_m) \in g(R)$, proving $g(R)^+$ is an ideal of $Z$. Suppose $g(R)R = R$. If $g(R)^+ \neq Z$ then take a maximal ideal $P$ of $Z$ containing $g(R)$ and let $\bar{R} = R_P/B$ where $B$ is a maximal ideal of $R_P$. Then $1 \in g(R_P)R_P$ and $g(R_P) \subseteq g(R)_P \subseteq Z_P$, implying $g$ is $\bar{R}$-central. Since $\bar{R}$ is simple we see $g(\bar{R})^+ = Z(\bar{R})$, so

$$(Z_P + B)/B \subseteq Z(\bar{R}) = g(\bar{R})^+ = (g(R_P)^+ + B)/B \subseteq (Z_P + B)/B$$

so equality holds and $1 = s^{-1}z + b$ for suitable $z$ in $g(R)^+$, $s$ in $Z - P$, and $b$ in $B$. Thus $b = 1 - s^{-1}z \in B \cap Z_P \subseteq P_P$ since $Z_P$ is local, implying the absurdity $1 \in P_P$. The only way out is $g(R)^+ = Z$.  Q.E.D.

For the purpose of the next result we say an identity $f$ of $M_n(\mathbb{Z})$ is *critical* if $f$ is not an identity of $M_{n+1}(F)$ for any field $F$.

**Theorem 6.1.35:**   *The following conditions are equivalent:*

(i) *$R$ satisfies all the multilinear identities of $M_n(\mathbb{Z})$, and each critical multilinear identity $f$ of $M_{n-1}(\mathbb{Z})$ is not satisfied by any homomorphic image of $R$.*

(i') *$R$ satisfies all the multilinear identities of $M_n(\mathbb{Z})$, and there is some multilinear identity $f$ of $M_{n-1}(\mathbb{Z})$ not satisfied by any homomorphic image of $R$.*

(ii) *$R$ has PI-degree $n$ and $1 \in g_n(R)R$.*

(iii) *$R$ has PI-degree $n$ and $1 \in g_n(R)^+$.*

(iv) *$R$ is Azumaya of constant rank $n^2$.*

**Proof:**   Let $Z = Z(R)$. (i) $\Rightarrow$ (i') is obvious. (i') $\Rightarrow$ (ii) Otherwise, take a maximal ideal $P$ containing $g_n(R)R$. Then $g_n$ is an identity of $R/P$, implying $R/P$ has PI-degree $< n$, and thus $R/P$ satisfies $f$, contradiction.

(ii) $\Rightarrow$ (iii) by lemma 6.1.34.

(iii) $\Rightarrow$ (iv) Write $1 = \sum_{i=1}^{k} g_n(r_{i1}, \ldots, r_{im})$. First we show $R$ is an f.g. projective $Z$-module by means of the dual basis lemma. Indeed, define $f_{ij} : R \to Z$ by

$$f_{ij}r = (-1)^{j+1} g_n(r, r_{i1}, \ldots, r_{i,j-1}, r_{i,j+1}, \ldots, r_{im}).$$

Then lemma 6.1.32 implies $r = \sum_{i=1}^{k} \sum_{j=1}^{n^2} (f_{ij}r)r_{ij}$, so $\{(r_{ij}, f_{ij}) : 1 \le j \le n^2\}$ is the desired dual basis. $\operatorname{rank}(R) = n^2$ by theorem 6.1.33, since $g_n(R_P)$ contains an invertible element of $Z_P$ for any $P$ in $\operatorname{Spec}(Z)$.

Clearly $R$ is faithful over $Z$ since $1 \in R$, so it suffices to prove the canonical homomorphism $\psi : R \otimes R^{\mathrm{op}} \to \operatorname{End}_Z R$ is an isomorphism. To this end write $g_n = \sum_u h_{u1} X_1 h_{u2}$ for suitable polynomials $h_{uv}(X_2, \ldots, X_m)$. Let $a_{iju} = (-1)^{j+1} h_{u1}(r_{i1}, \ldots, r_{i,j-1}, r_{i,j+1}, \ldots, r_{im})$ and $b_{iju} = h_{u2}(r_{i1}, \ldots, r_{i,j-1}, r_{i,j+1}, \ldots, r_{im})$.

To show $\psi$ is onto let $\beta \in \operatorname{End}_Z R$. We shall show $\beta = \psi\left(\sum_{i,j,u} a_{iju} \otimes b_{iju}\beta r_{ij}\right)$. Indeed,

$$\beta r = \beta(1r) = \beta\left(\sum_i g_n(r_{i1}, \ldots, r_{im})r\right)$$

$$= \beta\left(\sum_{i,j}(-1)^{j+1} g_n(r, r_{i1}, \ldots, r_{i,j-1}, r_{i,j+1}, \ldots, r_{im})r_{ij}\right)$$

$$= \sum_{i,j}(-1)^{j+1} g_n(r, r_{i1}, \ldots, r_{i,j-1}, r_{i,j+1}, \ldots, r_{im})\beta r_{ij}$$

$$= \sum_{i,j,u} a_{iju}rb_{iju}\beta r_{ij}.$$

as desired.

To show $\psi$ is 1:1 suppose $\sum_v x_v \otimes y_v \in \ker \psi$ for suitable $x_v$ in $R$, $y_v$ in $R^{op}$.

$$\sum x_v \otimes y_v = \sum_{i,v} g_n(r_{i1}, \ldots, r_{im}) x_v \otimes y_v$$

$$= \sum_{i,j,v} (-1)^{j+1} g_n(x_v, r_{i1}, \ldots, r_{i,j-1}, r_{i,j+1}, \ldots, r_{im}) r_{ij} \otimes y_v$$

$$= \sum_{i,j,v} (-1)^{j+1} r_{ij} \otimes g_n(x_v, r_{i1}, \ldots, r_{i,j-1}, r_{i,j+1}, \ldots, r_{im}) y_v$$

$$= \sum_{i,j,u,v} r_{ij} \otimes (a_{iju} x_v b_{iju}) y_v$$

$$= \sum_{i,j} r_{ij} \otimes \sum_u a_{iju} \left( \sum_v x_v b_{iju} y_u \right) = 0$$

since $\sum_v x_v b_{iju} y_v = (\sum x_v \otimes y_v) b_{iju} = 0$. Thus $\ker \psi = 0$.

(Once upon a time this was the "hard direction" of the Artin-Procesi theorem; see exercise 13 for an even shorter proof.)

(iv) $\Rightarrow$ (i) First we claim no homomorphic image $R/A$ of $R$ satisfies $f$. Indeed, otherwise, taking $P$ maximal containing $A$ we see $R/P$ satisfies $f$; but $R/P$ is central simple over its center and splitting by a maximal subfield we see $f$ is an identity of $M_{n'}(F)$ for some field $F$, where $n'$ is the PI-degree of $R/P$. Since $f$ is critical we have $n' < n$, so $R/P$ has dimension $\leq (n-1)^2$ over its center $Z/(Z \cap P)$, contrary to proposition 5.3.27.

It remains to show $R$ satisfies all identities of $M_n(\mathbb{Z})$. In view of proposition 2.12.15(i) we may assume $Z$ is local, with maximal ideal $P$; then $PR$ is the unique maximal ideal of $R$, and $\bar{R} = R/PR$ is simple of dimension $n^2$ over $Z/P$, by proposition 5.3.27.

We need to quote the following fact from the theory of central simple algebras, to be proved below as proposition 7.1.22. There are elements $r_j, a$ in $R$ for which $\{\bar{a}^{i-1} \bar{r}_j : 1 \leq i, j \leq n\}$ are a base for $\bar{R}$ over $\bar{Z}$, where $^-$ denotes the canonical image in $\bar{R}$. Now $\{a^{i-1} r_j : 1 \leq i, j \leq n\}$ span $R$ over $Z$ by Nakayama's lemma, implying $r_1, \ldots, r_n$ span $R$ as module over the commutative ring $Z[a]$. The regular representation enables us to embed $R$ into $\text{End}_{Z[a]} R$, which by theorem 1.5.13 is a homomorphic image of a subring of $M_n(Z[a])$; since $Z[a]$ is a homomorphic image of the free commutative ring $\mathbb{Z}[\lambda]$ we see $R$ satisfies all multilinear identities of $M_n(\mathbb{Z}[\lambda])$ and thus of $M_n(\mathbb{Z})$. Q.E.D.

The critical identity of $M_{n-1}(\mathbb{Z})$ usually used in (i') is $S_{2n-2}$, in view of the Amitsur-Levitzki theorem. The Artin-Procesi theorem is a very explicit computational tool for studying Azumaya algebras of rank $n$, and actually provides a new invariant, the minimal number $k$ for which $1 = \sum_{i=1}^{k} z_i$ for suitable $z_i$ in $g_n(R)$. If $k = 1$ then $R$ is free over $Z(R)$, by theorem 6.1.33.

**Remark 6.1.35′:**   Examining the proof of lemma 6.1.32 and theorem 6.1.35, we see that the condition in (iii) that $R$ has PI-degree $n$ could be weakened to "$\tilde{g}$ is an identity of $R$."

## The n-Spectrum

One way to achieve $1 \in g_n(R)$ is by localization.

**Corollary 6.1.36:**   *Suppose $R$ has PI-degree $n$ with center $Z$ and $s \in g_n(R)$. Then $R[s^{-1}]$ is Azumaya of rank $n^2$ over $Z[s^{-1}]$, and, in fact, is free over $Z[s^{-1}]$.*

**Proof:**   $1 \in g_n(R[s^{-1}])$ so lemma 6.1.34 shows $Z(R[s^{-1}]) = g_n(R[s^{-1}])^+ \subseteq Z[s^{-1}] \subseteq Z(R[s^{-1}])$; thus equality holds throughout. Now $R[s^{-1}]$ is Azumaya by theorem 6.1.35 and is free by theorem 6.1.33.     Q.E.D.

This provides a most powerful tool, since it is applicable whenever $g_n(R) \nsubseteq P$. Accordingly we make the following definition.

**Definition 6.1.37:**   Suppose $Z = Z(R)$. $\mathrm{Spec}_n(R) = \{P \in \mathrm{Spec}(R): g_n(R) \nsubseteq P\}$ and $\mathrm{Spec}_n(Z) = \{P \in \mathrm{Spec}(Z): g_n(R) \nsubseteq P\}$.

$g_n(R)$ is often called the *Formanek center*; the primes in $\mathrm{Spec}_n(R)$ are called *identity-faithful* or *regular*. The following theorem and subsequent results have more elementary proofs in Rowen [80B, p. 76–78].

**Theorem 6.1.38:**   *Suppose $P \in \mathrm{Spec}_n(Z(R))$. There is a unique $P'$ in $\mathrm{Spec}(R)$ lying over $P$, and $P'$ contains every ideal $A$ of $R$ for which $A \cap Z(R) \subseteq P$. In other words, $R_P$ is local with maximal ideal $P_P R_P$.*

**Proof:**   Pass to $R_P$, which is Azumaya and thus local with maximal ideal $P_P R_P$, and translate this back to $R$.     Q.E.D.

**Corollary 6.1.39:**   LO, GU, *and* INC *hold from* $Z(R)$ *to* $R$ *in* $\mathrm{Spec}_n$.

This permits us to build an inductive procedure on the PI-degree, and there is also a pleasant correspondence of maximal ideals.

**Lemma 6.1.40:** *If $P' \in \mathrm{Spec}_n(R)$ is a maximal ideal of $R$ then $Z(R) \cap P'$ is a maximal ideal of $Z(R)$.*

**Proof:** Let $Z = Z(R)$ and $\bar{R} = R/P'$, a simple ring. Then $1 \in g_n(\bar{R})^+$ by theorem 6.1.34 so

$$Z(\bar{R}) = g(\bar{R})^+ = \overline{g(R)^+} \subseteq \bar{Z} \subseteq Z(\bar{R})$$

implying $Z(\bar{R}) = \bar{Z} \approx Z/(P' \cap Z)$.　　Q.E.D.

These results show that the "difficult" part of $\mathrm{Spec}(R)$ is the complement of the $n$-spectrum. Nevertheless, there are some interesting results, most notably the Bergman-Small "additivity principle" to be discussed in §6.3.

## The Algebra of Generic Matrices

We shall now deal with a PI-ring which is a very important "test ring." Given a commutative ring $C$ we let $\Lambda = \{\lambda_{ij}^{(k)} : 1 \leq i, j \leq n, k \in \mathbb{N}\}$ be a set of commuting indeterminates over $C$, and let $Y_k$ denote the matrix $(\lambda_{ij}^{(k)})$ in $M_n(C[\Lambda])$. Each entry of $Y_k$ is a different indeterminate; accordingly, we call $Y_k$ a *generic matrix* and define the algebra of *generic matrices*, denoted $C_n\{Y\}$, to be the $C$-subalgebra of $M_n(C[\Lambda])$ generated by the $Y_k$.

**Proposition 6.1.41:** *If $C$ is an integral domain then the algebra of generic matrices $C_n\{Y\}$ is prime.*

**Proof:** Let $K$ be the field of fractions of $C[\Lambda]$ and let $R' = C_n\{Y\}K \subseteq M_n(K)$. Clearly the Capelli polynomial of degree $2n^2$ is nonvanishing on generic matrices (seen by specializing the indeterminates), so $[R':K] \geq n^2$. But this means $R' = M_n(K)$; hence $M_n(K)$ is a central extension of $C_n\{Y\}$, which thus is prime by proposition 2.12.39 (taking $P' = 0$).　　Q.E.D.

The significance of $C_n\{Y\}$ lies in the following result:

**Proposition 6.1.42:**

　(i) *There is a canonical surjection $\varphi: C\{X\} \to C_n\{Y\}$ given by $\varphi X_i = Y_i$, and $\ker \varphi = \{identities\ of\ M_n(C[\Lambda])\}$.*

　(ii) *$C_n\{Y\}$ is free in the class $\mathscr{C}$ of algebras satisfying the identities of $M_n(C[\Lambda])$.*

*Proof:*

(i) $\varphi$ exists because $C\{X\}$ is free. If $f(X_1, \ldots, X_m)$ is an identity of $M_n(C[\Lambda])$ then $f(Y_1, \ldots, Y_m) = 0$ so $f \in \ker \varphi$. To finish the proof of (i) it suffices to prove if $f(Y_1, \ldots, Y_m) = 0$ then $f(X_1, \ldots, X_m)$ is an identity of $M_n(C[\Lambda])$. Given any matrices $a_k = (c_{ij}^{(k)})$ in $M_n(C[\Lambda])$ we could extend the map $\lambda_{ij}^{(k)} \mapsto c_{ij}^{(k)}$ to a homomorphism $C[\Lambda] \to C[\Lambda]$ and thus to a homomorphism $M_n(C[\Lambda]) \to M_n(C[\Lambda])$ sending $Y_k = (\lambda_{ij}^{(k)})$ to $a_k$; then $f(a_1, \ldots, a_m)$ is the image of $f(Y_1, \ldots, Y_m) = 0$.

(ii) This argument will be repeated more generally in §6.4. Suppose $R \in \mathscr{C}$. Given any map $Y_i \mapsto r_i$ for $i = 1, 2, \ldots$ we have a homomorphism $\psi : C\{X\} \to R$ sending $X_i \mapsto r_i$ for each $i$. We shall show $\ker \psi \supseteq \ker \varphi$, for then we can factor $\psi$ through $C_n\{Y\}$ to get the desired homomorphism $\bar{\psi} : C_n\{Y\} \to R$ satisfying $\bar{\psi} Y_i = r_i$, and clearly $\bar{\psi}$ is unique such. If $f \in \ker \varphi$ then $f$ is an identity of $M_n(C[\Lambda])$ by (i), so $f(r_1, \ldots, r_m) = 0$ since $R \in \mathscr{C}$, implying $f \in \ker \psi$.       Q.E.D.

## The Identities of a PI-Algebra

Proposition 6.1.42 heightens our interest in which rings satisfy *all* the identities of $M_n(C[\Lambda])$. Accordingly we define $\mathscr{Z}(R) = \{$identities of $R\}$, and we say $C$-algebras $R$ and $T$ are PI-*equivalent* if $\mathscr{Z}(R) = \mathscr{Z}(T)$. Let us start by asking when $R$ and $R \otimes_C H$ are PI-equivalent, recalling proposition 6.1.11. Now we must require some restriction on $C$, as evidenced by the example that $\mathbb{Z}/2\mathbb{Z}$ satisfies the identity $f = X_1^2 - X_1$ but no other field satisfies $f$.

**Lemma 6.1.43:** *Suppose $C$ is an infinite integral domain, and the polynomial $p \in C[\lambda_1, \ldots, \lambda_k]$ satisfies $p(c_1, \ldots, c_k) = 0$ for all $c_i$ in $C$. Then $p = 0$.*

*Proof:* This is well-known, but we review it here. The case $k = 1$ follows at once from the fact that any nonzero $p(\lambda_1)$ has only a finite number of zeroes in the field of fractions of $C$, so some $p(c) \neq 0$. In general, $p = \sum_{i=0}^t p_i(\lambda_1, \ldots, \lambda_{k-1})\lambda_k^i$ and for each $\mathbf{c} = (c_1, \ldots, c_{k-1})$ in $C^{(k-1)}$ write $p_\mathbf{c} = \sum_{i=0}^t p_i(c_1, \ldots, c_{k-1})\lambda_k^i \in F[\lambda_k]$. Then as seen above $p_\mathbf{c} = 0$ so $p_\mathbf{c} = 0$ for each $i$ and all $c$ in $C^{(k-1)}$, implying each $p_i = 0$ by induction on $k$; hence $p = 0$.       Q.E.D.

**Theorem 6.1.44:** *Suppose $R$ is torsion-free over an infinite integral domain $C$. Then $R$ is PI-equivalent to $R \otimes_C H$ for any torsion-free commutative $C$-algebra $H$.*

**Proof:** Identify $R$ with $R \otimes 1 \subseteq R \otimes H$. Then every identity of $R \otimes H$ is an identity of $R$, so it remains to show every identity $f$ of $R$ is an identity of $R \otimes H$. Let $F$ be the field of fractions of $C$. Replacing $H$ by $H \otimes_C F$ we may assume $H$ is an $F$-algebra. $R \otimes F$ has a base $\{r_i : i \in I\}$ over $F$, where each $r_i \in R$. Given $x_1, \ldots, x_m$ in $R \otimes H$ write $x_j = \sum_i r_i \otimes a_{ij}$ where the $a_{ij} \in H$. Then $f(x_1, \ldots, x_m) = \sum_i r_i \otimes p_i(\mathbf{a})$ where each $p_i \in C[\lambda_1, \ldots, \lambda_m]$ and $p_i(\mathbf{a})$ is the evaluation at various $a_{ij}$. If all our $a_{ij} \in C$ then each $x_1, \ldots, x_m \in R$ so $f(x_1, \ldots, x_m) = 0$ and thus each $p_i(\mathbf{a}) = 0$; therefore, each $p_i = 0$ by lemma 6.1.43. But then $p_i(\mathbf{a}) = 0$ for arbitrary $a_{ij}$ in $H$, implying $f(x_1, \ldots, x_m) = 0$ for all $x_1, \ldots, x_m$ in $R \otimes H$. Q.E.D.

**Corollary 6.1.45:** *If $R$ is a torsion-free algebra over an infinite integral domain $C$ then $R$ is PI-equivalent to any central extension $T$ of $S^{-1}R$, where $S = C - \{0\}$.*

**Proof:** Let $H = Z(T)$. Then $R$ is PI-equivalent to $R \otimes_C H$. But $T$ is a homomorphic image of $R \otimes_C H$ so using the theorem we have

$$\mathscr{L}(R) = \mathscr{L}(R \otimes_C H) \subseteq \mathscr{L}(T) \subseteq \mathscr{L}(R)$$

since $R \subseteq T$; hence equality holds and $R$ is PI-equivalent to $T$. Q.E.D.

**Corollary 6.1.46:** *Suppose $R$ is a semiprime PI-algebra over an infinite field $F$. Then $R$ is PI-equivalent to $M_n(F)$ for some $n$.*

**Proof:** $R$ is PI-equivalent to $R[\lambda]$, which is semiprimitive by Amitsur's theorem since $\mathrm{Nil}(R) = 0$. Hence we may assume $R$ is semiprimitive, by theorem 2.5.23. But $\mathscr{L}(M_{n-1}(F)) \supseteq \mathscr{L}(M_n(F))$ by lemma 6.1.27′, so passing to primitive images we may assume $R$ is primitive and thus central simple by Kaplansky's theorem. Taking a splitting field $K$ we see $R$ is PI-equivalent to $R \otimes K \approx M_n(K) \approx M_n(F) \otimes_F K$, which is PI-equivalent to $M_n(F)$. Q.E.D.

**Corollary 6.1.46′:** *Suppose $F$ is a field, and $R$ is a simple $F$-algebra of PI degree $n$, with $Z = Z(R)$ infinite. Then $R$ is PI-equivalent to $F_n\{Y\}$.*

**Proof:** We have the following sequence of PI-equivalent algebras: $R$, $M_n(Z)$, $M_n(Z[\Lambda])$, $M_n(F[\Lambda])$, $F_n\{Y\}$. Q.E.D.

**Corollary 6.1.47:** *Suppose $C$ is an integral domain. Every semiprime $C$-algebra $R$ of PI-degree $n$ satisfies all the identities of $M_n(C[\Lambda])$ and is thus a homomorphic image of the generic matrix algebra $C_n\{Y\}$.*

*Proof:* By proposition 6.1.42(ii) it is enough to show $R$ satisfies all identities of $M_n(C[\Lambda])$; since $R$ is a subdirect product of prime images we may assume $R$ is prime. Replacing $C$ by its image $C \cdot 1$ in $R$, we may assume $C \subseteq R$. We may replace $R$ by $R[\Lambda]$, which in turn is PI-equivalent to its algebra of central quotients, which is an algebra over the (infinite) field of fractions of $C[\Lambda]$, so we are done by corollary 6.1.46.     Q.E.D.

*Remark 6.1.47':* The same proof shows: If every prime homomorphic image of $C$ is infinite then every semiprime $C$-algebra $R$ of PI-degree $n$ satisfies all identities of $M_n(C)$.

## Noetherian PI-Rings

The subject of Noetherian PI-rings relies heavily on the following observation, which permits us at once to lift many of the results directly from §3.5.

**Proposition 6.1.48:**  *Every PI-ring $R$ is left and right fully bounded.*

*Proof:* We may assume $R$ is prime, and by symmetry need to show any large left ideal $L$ contains a two-sided ideal. But letting $S = Z(R) - \{0\}$ we show $S^{-1}L$ is a large left ideal of the simple Artinian ring $Q = S^{-1}R$, so $S^{-1}L = Q$ and thus some $s \in S \cap L$. Now $0 \neq Rs \lhd R$, and clearly every nonzero ideal of a prime ring is large.     Q.E.D.

**Theorem 6.1.49:** *If $R$ is a Noetherian PI-ring then $\bigcap_{i \in \mathbb{N}} \mathrm{Jac}(R)^i M = 0$ for all f.g. $R$-modules $M$.*

*Proof:* By theorem 3.5.28, noting the hypotheses hold by proposition 6.1.48 and Kaplansky's theorem.     Q.E.D.

**Theorem 6.1.50:** *In a left Noetherian PI-ring every prime ideal has finite height.*

*Proof:* By theorem 3.5.13, noting that every nonzero ideal of a prime PI-ring has a nonzero central element which thus is normalizing.     Q.E.D.

To determine when $R$ is Noetherian we can use the following property of arbitrary prime rings of PI-degree $n$.

**Proposition 6.1.51:** *Suppose $R$ is prime of PI-degree n. Then any left ideal $L$*

*is isomorphic to a $Z(R)$-submodule of a free $Z(R)$-module $M$ of* rank $\leq n^2$ *contained in $L$. Explicitly there is $z \neq 0$ in $Z(R)$ for which $zL \subseteq M \subseteq L$.*

**Proof:** Let $Z = Z(R)$ and take $k$ maximal possible for which there are $r_1, \ldots, r_m$ such that $z = g_n(r_1, \ldots, r_m) \neq 0$ with $k$ of the $r_i$ from $L$. Passing to the ring of fractions of $R$ we see the $r_i$ are $Z$-independent; relabeling the $r_i$ from $L$ as $a_1, \ldots, a_k$ we take $M = \sum_{i=1}^{k} Za_i \approx Z^{(k)}$. It remains to show $zL \subseteq M$. But lemma 6.1.32 yields

$$za = \sum_{j=1}^{t} (-1)^{j+1} g_n(a, r_1, \ldots, r_{j-1}, r_{j+1}, \ldots, r_m) r_j$$

and the $j$-th summand is 0 by maximality of $k$ unless $r_j \in L$, implying each nonzero summand is in $M$. Q.E.D.

**Corollary 6.1.52:** *(Formanek) Suppose $R$ is prime PI. If $Z(R)$ is Noetherian then $R$ is f.g. as $Z(R)$-module and hence is Noetherian. If $Z(R)$ has Krull dimension $\alpha$ then $R$ has Krull dimension $\leq \alpha$.*

**Proof:** $R$ is a submodule of an f.g. $Z(R)$-module. Q.E.D.

A cute variant of this idea yields *Cauchon's theorem*, that any semiprime PI-ring with ACC(ideals) is Noetherian, cf., exercise 24.

Formanek's result raised the hope that the structure of a prime Noetherian PI-ring is closely tied to the structure of its center. Although this is true in certain special cases, e.g., when $R$ hereditary (cf., Robson-Small [74]), the general situation is the opposite; the center of a prime Noetherian PI-ring can be rather different. Perhaps the most efficient counterexample (i.e., most pathology for least work) is exercise 27. In desperation it was asked whether a prime PI-ring must be f.g. as module over a suitable commutative subring. Building on examples by Cauchon and Bergman, Sarraille [82] produced a counterexample, cf., exercise 6.3.2.

If $R$ is Noetherian PI then $K$-dim$(R) \leq$ gl. dim $R$ when the latter exists. This was proved more generally for fully bounded Noetherian rings under the restriction that the center contains an uncountable domain, by Brown-Warfield [84]; Goodearl-Small [84] finished the result for Noetherian PI-rings by passing to the Laurent series ring $R((\lambda))$, whose center $Z((\lambda))$ is uncountable. To accomplish this reduction they had to show gl. dim $R((\lambda)) =$ gl. dim $R$ and K-dim $R \leq$ K-dim $R((\lambda))$. The latter fact is easy since there is a lattice injection $\mathscr{L}(R((\lambda))) \to \mathscr{L}(R)$ given by $L \mapsto L'$, where $L' = \{a \in R : a$ is the lowest order coefficient of some element of $L$ (viewed as Laurent series)$\}$.

Vamos [77] showed the completion $\hat{R}$ of a semilocal Noetherian PI-ring $R$ is Noetherian, and $\hat{R} \otimes \underline{\quad}$ is an exact functor which on $R\text{-}\mathscr{F}imod$ is naturally equivalent to the $I$-adic completion.

Much recent work on Noetherian PI-rings has focused on the "trace ring" $T(R)$, so we postpone the continuation of this discussion until proposition 6.3.45ff.

## Digression: Verifying PI-Rings

How does one spot that a ring is PI? This is usually obvious, for we are usually given the polynomial identity. However, there are instances when the PI is not clear. Usually it is easy to determine whether or not a semiprime ring is PI, for this is equivalent to its being embeddible in matrices, cf., theorem 6.1.26. The reduction to the semiprime case is given by the following useful result of Amitsur.

**Theorem 6.1.53:**   *Suppose $\mathscr{C}$ is a class of algebra closed under direct powers, i.e., if $R \in \mathscr{C}$ then $\prod_{i \in I} R_i \in \mathscr{C}$ whenever each $R_i \approx R$. If $f$ is an identity of $R/\mathrm{Nil}(R)$ for each $R$ in $\mathscr{C}$ then every $R$ in $\mathscr{C}$ satisfies a suitable power of $f$. If, furthermore, $\mathscr{C}$ is closed under direct products then there is $k$ in $\mathbb{N}$ such that $f^k$ is an identity of all $R$ in $\mathscr{C}$.*

**Proof:**

(i) Write $f = f(X_1, \dots, X_m)$. For each $\mathbf{r} = (r_1, \dots, r_m)$ in $R^{(m)}$ let $R_\mathbf{r}$ denote a copy of $R$ and let $R_1 = \prod_\mathbf{r} R_\mathbf{r}$. Let $\hat{r}_i$ be the element of $R_1$ whose $\mathbf{r}$-component is $r_i$. Then $f(\hat{r}_1, \dots, \hat{r}_n) \in \mathrm{Nil}(R_1)$ by hypothesis so matching the $\mathbf{r}$-components shows $f(r_1, \dots, r_m)^k = 0$ for each $\mathbf{r}$, i.e., $f^k$ is an identity of $R$.

(ii) So far $k$ depends on $R$. Suppose $\mathscr{C}$ is closed under direct products. We are done unless for each $k$ in $\mathbb{N}$ there is some $R_k$ for which $f^k$ is not an identity of $R_k$. Then no power of $f$ is an identity of $\prod_{k \in \mathbb{N}} R_k$, contrary to (i).                                                                                      Q.E.D.

**Corollary 6.1.54:**   *If $R$ is a PI-ring then $M_t(R)$ is a PI-ring for any $t$.*

**Proof:**   First note if $R$ is semiprime of PI-degree $n$ then $R$ is embeddible in $n \times n$ matrices so $M_t(R)$ is embeddible in $tn \times tn$ matrices and thus has PI-degree $tn$. Letting $\mathscr{C} = \{M_t(R): R$ satisfies a given polynomial identity $f\}$, we see $\mathscr{C}$ satisfies the criteria of the theorem.      Q.E.D.

There are many similar uses, cf., exercises 28, 29, 34, 35.

## Digression: Generalized Identities and Rational Identities

The polynomial identity is an example of a universal atomic sentence and has rather powerful consequences as we have seen. This raises the question of whether weaker atomic sentences would also be of use. There is an assortment of such results in the literature, but many are of limited use because of the difficulty of verifying the given condition. However, one such theory has proved applicable in a variety of situations and, not surprisingly, is related to the PI-theory. Logically it involves universal atomic sentences which arise when we permit assignment of constant symbols to elements of the ring. For example $M_n(F)$ satisfies the sentence $\forall x_1 \forall x_2 ([e_{11} x_1 e_{11}, e_{11} x_2 e_{11}] = 0)$.

The correct algebraic framework is to fix a ring $W$ and deal with the category of $W$-*rings*, i.e., those rings $R$ for which there is a ring homomorphism $\varphi: W \to R$ such that $\varphi(Z(W)) \subseteq Z(R)$. Writing $Z$ for $Z(W)$ we see the free $W$-ring is merely the free product of $W$ and $Z\{X\}$ as $Z$-algebras; its elements are called *generalized polynomials*, or $W$-*polynomials*, and are sums of terms of the form $w_1 X_{i_1} w_2 X_{i_2} \cdots w_t X_{i_t} w_{t+1}$ where the $w_j \in W$. Normally one takes $W = R$.

**Definition 6.1.55:** A *generalized identity* (or GI) of a $W$-ring $R$ is a generalized polynomial which vanishes for all substitutions in $R$.

The multilinearization procedure enables us to reduce to the case of multilinear GI's. However, it is not immediately apparent what kind of GI is appropriate to require for a ring; any nonprime ring $R$ satisfies the GI $aX_1 b$ for suitable $a, b$ (viewed as $R$-ring). The key lies in defining a *generalized monomial* to be a generalized polynomial in which the indeterminates of each monomial occur in the same order, e.g., $X_2 w_1 X_1 X_3 - X_2 X_1 w_2 X_3 w_3$. Every multilinear generalized polynomial can be written uniquely as the sum of generalized monomials.

**Definition 6.1.56:** A multilinear GI of $R$ is *improper* if its generalized monomials each are GI's of $R$. A GI is *proper* if it is not improper.

We say a primitive ring $R$ is *strongly primitive* if $R$ is dense in End $M_D$ where the division ring $D$ is PI, and soc($R$) $\neq 0$.

**Remark 6.1.57:** Any strongly primitive ring $R$ satisfies a proper GI. (Indeed let $e$ be a primitive idempotent of $R$. Then $D = eRe$ satisfies some standard identity $S_k$, and $S_k(eX_1 e, \ldots, eX_k e)$ is a GI of $R$ which is proper since $R$ is prime.)

This remark motivates the two main structure theorems of the subject:

**Amitsur's Theorem for GI's:**   *A primitive ring is strongly primitive iff it satisfies a proper* GI.

**Martindale's Theorem:**   *If R is prime with proper* GI *then its central closure is strongly primitive.*

Full accounts of these theorems are given in Rowen [80B, Chapter 7]; we shall indicate a fast proof of Amitsur's theorem in exercises 30, 31. However, the full power of the GI lies in the more technical fact that the GI can be rewritten such that each of its monomials contains a coefficient in the socle, cf., Rowen [80B, theorem 7.2.2].

Multilinear GI's are very effective in the study of "elementary" properties, because they pass to tensor extensions (as in proposition 6.1.11), and thus often permit one to pass to matrices over an algebraically closed field. An example of their application is the Martindale-Miers [85] exposition of Herstein's theorem on Lie and Jordan structure.

Note also that exercise 5.3.11 is really a theorem concerning GI's.

A different universal atomic sentence is obtained if instead of expanding the assignments of constants we expand the unary operations of the language. One possibility is to allow the formal introduction of inverses into the identities; this defines *rational identities*. One example of a rational identity is Hua's famous identity $(X^{-1} + (Y^{-1} - X)^{-1})^{-1} - (X - XYX)$ which arises in the fundamental theorem of projective geometry. But Hua's identity vanishes for all substitutions in every ring (whenever one can take the inverses), so must be called "trivial"; we see that it is difficult to identity the "nontrivial" rational identities. Amitsur [66] showed that rational identities are precisely what we get from intersection theorems in Desarguian geometries and proved that any division algebra satisfying a nontrivial rational identity is actually a PI-ring. The theory of rational identities is treated in Rowen [80B, Chapter 8] and relies heavily on applications by Bergman [76] of the Bergman-Small theorem discussed in §6.3.

## P Supplement: Identities and Generalized Identities with Involution

One can handle identities in rings with involution, called (*)-identities merely by expanding the language to include the involution. Thus we say

$f(X_1, X_1^*, \ldots, X_m, X_m^*)$ is an *identity* of $(R, *)$ if $f(r_1, r_1^*, \ldots, r_m, r_m^*) = 0$ for all $r_i$ in $R$. More generally, one can define $(*)$-GI's. The question arises at once if this theory is any richer than the theory without $(*)$. To this end we say a $(*)$-identity $f(X_1, X_1^*, \ldots, X_m, X_m^*)$ is $(R, *)$-*special* if $f(X_1, X_2, \ldots, X_{2m-1}, X_{2m})$ is an identity of $R$; likewise for $(*)$-GI's.

### Example 6.1.58:

(i) Let $(*)$ be the transpose involution on $R = M_2(F)$. Then $[X_1 - X_1^*, X_2 - X_2^*]$ is a non-special identity of $(R, *)$.

(ii) Let $(*)$ be the symplectic involution on $M_2(F)$. Then every symmetric element is central, so $[X_1 + X_1^*, X_2]$ is a non-special identity of $(R, *)$.

(iii) If $R \oplus R^{\mathrm{op}}$ has the exchange involution $(*)$ then every $(*)$-identity is special, for if $f(X_1, X_1^*, \ldots, X_m, X_m^*)$ is a $(*)$-identity we have

$$(f(r_1, \ldots, r_{2m}), 0) = (1, 0) f((r_1, r_2), (r_2, r_1), (r_3, r_4), (r_4, r_3), \ldots)$$

$$= (1, 0) f((r_1, r_2), (r_1, r_2^*), \ldots) = (0, 0).$$

Note that the minimal degree of a $(*)$-identity might be smaller than the minimal degree of a PI, so one may wonder whether the existence of a $(*)$-identity implies the ring is PI. This is true by a theorem of Amitsur, cf., exercise 36.

**Theorem 6.1.59:** *Suppose $R$ is a semiprime PI ring which has an involution $(*)$. Then every nonzero ideal $A$ of $(R, *)$ intersects $Z(R, *)$ nontrivially.*

**Proof:** By theorem 6.1.28 there is $0 \neq z \in A \cap Z(R)$. Then $z + z^* \in A \cap Z(R, *)$ so we are done unless $z = -z^*$. But now $0 \neq -z^2 = zz^* \in A \cap Z(R, *)$. Q.E.D.

**Corollary 6.1.60:** *If $R$ is a semiprime PI-ring with involution $(*)$ and $Z(R, *)$ is a field then $(R, *)$ is simple.*

**Theorem 6.1.61:** *If $R$ is a PI-ring and $(R, *)$ is prime then taking $S = Z(R, *) - \{0\}$ we have $S$ is regular and $(S^{-1}R, *)$ is simple.*

**Proof:** Apply proposition 2.13.35 and corollary 6.1.60 to $(S^{-1}R, *)$, noting $Z(S^{-1}R, *) = S^{-1}Z(R, *)$. Q.E.D.

## §6.2 Affine Algebras

**Definition 6.2.1:** $R$ is an *affine* $C$-algebra if $R$ is generated *as algebra* by a finite number of elements $r_1, \ldots, r_t$; in this case we write $R = C\{r_1, \ldots, r_t\}$. (In the literature $C$ is often required to be a field.)

Commutative affine algebras over fields are obviously Noetherian, and we record the following useful result.

**Noether (–Bourbaki) Normalization Theorem:** (*cf.*, Bourbaki [72B, theorem v.3.1]) *Suppose* $A_1 \subseteq \cdots \subseteq A_k$ *are ideals in a commutative affine algebra* $C$ *over a field* $F$. *There is a transcendence base* $c_1, \ldots, c_m$ *of* $C$ *for which* $C$ *is integral over* $F[c_1, \ldots, c_m]$, *and such that there are* $n(1) \le n(2) \le \cdots \le n(k)$ *such that* $A_i \cap F[c_1, \ldots, c_m] = \sum_{u=1}^{n(i)} Cc_u$ *for* $1 \le i \le k$.

On the other hand, the free algebra in two noncommuting indeterminates is far from Noetherian and so affine algebras offer an alternative to Noetherian rings in building a general geometric-flavored ring theory. We shall start with some general results concerning affine algebras and then delve into the Gelfand-Kirillov dimension.

**Remark 6.2.2:** If $R$ is an affine $C$-algebra and $A \lhd R$ then $R/A$ is affine over $C/(C \cap A)$.

Recall from theorem 2.5.22 that the weak Nullstellensatz holds for affine algebras over uncountable fields, so the starting point of our inquiry might be whether this is true in general, i.e., if $\text{Jac}(R)$ is nil for $R$ affine over a field. This question was open for a long time but was settled negatively by Beidar using a very easy method which we describe below now (with some Small modifications). *Note* f.g. *always means as module* in what follows.

**Example 6.2.3:** (Affinization) If $T$ is an affine $C$-algebra and $L$ is a f.g. left ideal of $T$ then

$$R = \begin{pmatrix} C + L & T \\ L & T \end{pmatrix}$$

is affine over $C \cdot 1$, and $Z(R) \subseteq C + L$. Indeed if $T = C\{a_1, \ldots, a_t\}$ and $L = \sum_{i=1}^{n} Ta_i'$ for $a_i', a_j$ in $T$ then $R = C\{e_{11}, e_{12}, a_i'e_{21}, a_je_{22} : 1 \le i \le n, 1 \le j \le t\}$.

Note $Z(R)$ can be very poorly behaved, and even not affine, e.g., take $T = \mathbb{Q}[\lambda]$ and $L = T\lambda$. A related instance was given in exercise 6.1.28. This example also is used in exercises 3, 4. However, we want to see what happens when $T$ is the free associative ring.

**Example 6.2.4:** (Failure of the Nullstellensatz for affine algebras.) Build $R$ as in example 6.2.3, where $C$ is a countable field $F$, $T = F\{X_1, X_2\}$, and $L = TX_2$. Then $R$ has a prime ideal $P$ for which $R/P$ is right Noetherian affine over $F$ but $\text{Jac}(R/P) \neq 0 = \text{Nil}(R/P)$. To see this let

$$T' = \left\{ \begin{pmatrix} a & 0 \\ 0 & a \end{pmatrix} : a \in F + L \right\} \approx F + L = F\{X_1^i X_2 : i \in \mathbb{N}\},$$

which is isomorphic to the free algebra in countably many indeterminates. Let $H$ be a countable Noetherian local integral domain, such as the localization of $F[\lambda]$ at the prime ideal $\lambda$. Then $H \approx T'/A$ for some $A$ in $\text{Spec}(T')$. $R = \begin{pmatrix} T' & T \\ L & T \end{pmatrix}$; hence $RAR = \begin{pmatrix} A & AT \\ LA & TAT \end{pmatrix}$ is contained in an ideal $P$ of $R$ maximal with respect to $P \cap T' = A$, and $P \in \text{Spec}(R)$ by remark 2.12.42. Let $\bar{R} = R/P$ and let $e$ denote the image of $e_{11}$ in $R/P$. Then $e\bar{R}e \approx T'/A \approx H$ so $0 \neq \text{Jac}(e\bar{R}e) = e\,\text{Jac}(\bar{R})e$ by proposition 2.5.14, proving $\text{Jac}(\bar{R}) \neq 0$.

$\bar{R}$ is left Noetherian since right multiplication by the image of $X_2 \cdot 1$ gives a lattice injection from the left ideals $\mathscr{L}(\bar{R})$ to $\mathscr{L}(M_2(H))$. In particular, $\text{Nil}(\bar{R}) = 0$ since $\bar{R}$ is prime and left Noetherian.

In practice many affine algebras do satisfy the weak Nullstellensatz and more, as we shall see in §6.3 and §6.4, and there is no known answer to

**Question 6.2.4′:** If $R$ is Noetherian and affine over a field then is $\text{Jac}(R)$ necessarily nil? (An affirmative answer would imply $R$ is Jacobson by passing to homomorphic images.)

Incidentally, the relationship between affine and Noetherian is mysterious; it is not even known if every affine subalgebra of the Weyl algebra $\mathbb{A}_1(F)$ is Noetherian.)

Let us try to salvage something concrete from the wreckage of example 6.2.4. First we note that there are cases when the center is affine.

**Proposition 6.2.5:** (*Artin-Tate*) *If $C$ is commutative Noetherian and $R$ is an affine $C$-algebra and $Z$ is a $C$-subalgebra of $Z(R)$ over which $R$ is f.g. then $Z$ is also affine.*

**Proof:**   Write $R = C\{r_1, \ldots, r_t\}$ and $R = \sum_{k=1}^{t} Zx_k$ for suitable $x_k$ in $R$. Then we have elements $z_{ijk}, z'_{uk}$ in $Z$ for which

$$x_i x_j = \sum_k z_{ijk} x_k \quad \text{and} \quad r_u = \sum_k z'_{uk} x_k.$$

Let $Z_1 = C[z_{ijk}; z'_{uk} : 1 \leq i, j, k \leq m \text{ and } 1 \leq u \leq t]$, which is commutative affine and thus Noetherian. Then one can check at once $R = \sum_{k=1}^{m} Z_1 x_k$ is f.g. over $Z_1$, so $R$ is a Noetherian $Z_1$-module and thus its submodule $Z$ is f.g. over $Z_1$, implying $Z$ is affine.      Q.E.D.

Other results verifying when an algebra is affine are given in exercises 1, 2; a sweeping result can be found in Lorenz [84].

## Kurosch's Problem and the Golod-Shafarevich Counterexample

Having seen that affine algebras exist in abundance, we might wonder whether something can be said if extra restrictions are imposed.

**Problem 6.2.6:**   (Kurosch's problem) If $R$ is affine over a field $F$ and algebraic over $F$ then is $R$ finite dimensional as a vector space over $F$?

This famous problem has two key special cases:

**Kurosch's Problem for Division Rings:**   If a division algebra $D$ is affine and algebraic over $F$ then is $D$ central simple? (If so then $D$ is f.g. over $Z(D)$, implying $Z(D)$ is affine by Artin-Tate and thus $[Z(D):F] < \infty$ by corollary 6.3.2 below, and thence $[D:F] < \infty$.)

**Levitzki's Problem:**   If $N$ is a nil, affine $F$-algebra without 1 then is $N$ nilpotent? This is cleary a special case of Kurosch, since adjoining 1 formally would give an affine algebraic algebra $R$ and if $[R:F] < \infty$ then $N = \text{Nil}(R)$ is nilpotent.

There is a cute reduction of Kurosch's problem to the prime case, cf., exercise 5. Kurosch for division rings is still open in general, but has a positive solution for $F$ uncountable, by proposition 6.1.24; Levitzki's problem, however, is false even for $F$ uncountable, by a celebrated example of Golod [64] and Shafarevich. Recently Bergman [86] has studied local finiteness (and the Golod-Shafarevich example) in terms of the Jacobson radical of extension rings, and sharper examples have recently appeared. Let us study the original Golod-Shafarevich example.

**Main Computational Lemma 6.2.7:** *Grade the free associative algebra $F\{X_1, \ldots, X_m\}$ by the (total) degree. Let $A$ be an ideal generated as an ideal by a countably infinite set $S$ of homogeneous elements of degree $\geq 2$. Writing $S_k = \{elements\ of\ S\ of\ degree\ k\}$, suppose there are numbers $m_k \geq |S_k|$ for which the coefficients of the power series*

$$p(\lambda) = \left(1 - m\lambda + \sum_{k=2}^{\infty} m_k \lambda^k\right)^{-1}$$

*are all non-negative. Then $R = F\{X_1, \ldots, X_m\}/A$ is infinite dimensional over $F$.*

**Proof:** (Vinberg) We write power series $\sum a_j \lambda^j \leq \sum b_j \lambda^j$ to indicate that each $a_j \leq b_j$. Let $n_j = [R_j : F]$. The aim of the proof is to show the series $\sum_{j \in \mathbb{N}} n_j \lambda^j \geq p(\lambda)$. This would finish the proof since $p$ has infinite support, i.e., an infinite number of nonzero coefficients. (Indeed, if $p$ had finite support then $(1 + \sum_{k \geq 2} m_k \lambda^k) p$ would have infinite support since all the coefficients of $p$ are non-negative; but $(1 - m\lambda + \sum_{k \geq 2} m_k \lambda^k) p = 1$, so $-m\lambda p$ must have infinite support, implying $p$ has infinite support.)

Write $T$ for $F\{X_1, \ldots, X_m\}$, and $T_j$ for the homogeneous part of degree $j$. Then $T_1 = \sum F X_i$ and $TT_1 = \bigoplus_{j > 0} T_j$ so $T = TT_1 \oplus F$. Let $B_j$ be a vector subspace of $T_j$ complementary to $A_j$, and $B = \bigoplus B_j$. Then $T = A \oplus B$ so

$$A = TST = TSTT_1 + TSF = AT_1 + TS = AT_1 + (A + B)S = AT_1 + BS.$$

Taking homogeneous parts yields $A_k = A_{k-1} T_1 + \sum_j B_{k-j} S_j$; computing dimensions (noting $[A_j : F] + [B_j : F] = [T_j : F] = m^j$ and $[B_j : F] = [R_j : F] = n_j$) we have

$$m^k - n_k \leq (m^{k-1} - n_{k-1})m + \sum_j n_{k-j} d_j$$

where $d_j = |S_j|$, implying

$$0 \leq n_k - n_{k-1}m + \sum_j n_{k-j} d_j \leq n_k - n_{k-1}m + \sum_j n_{k-j} m_j$$

for all $k \geq 1$. Letting $h(\lambda) = (\sum_j n_j \lambda^j)(1 - m\lambda + \sum_u m_u \lambda^u)$ our inequality shows $h \geq 1$. But then $\sum_j n_j \lambda^j = hp = p + (h-1)p \geq p$ as desired, since $(h-1)p$ has only non-negative coefficients.    Q.E.D.

As final preparation we need a standard observation about counting.

**Remark 6.2.8:** The number of ways of selecting $n$ objects from $q$ allowing repetitions is $\binom{q+n-1}{n}$. (Indeed, since $\binom{q+n-1}{n}$ is the number of ways of selecting $n$ from $q + n - 1$ without repetition, we wish to find a 1:1 correspondence between these selections and selections of $n$ from $q$ with repetitions. Arranging selections in ascending order, we define the "bottom" of a selection of $n$ from $q + n - 1$ to be those numbers $\leq q$. We shall conclude by showing for any $t$ that the number of selections of $n$ from $q + n - 1$ having a particular bottom $(i_1, \ldots, i_t)$ is the same as the number of selections of $n$ from $q$ which are $i_1, \ldots, i_t$ with repetitions. Indeed, in the first case we are selecting $n - t$ objects from the $n - 1$ objects $q + 1, \ldots, q + n - 1$ without repetition, and in the latter case we are selecting $n - t$ objects from the $t$ objects $i_1, \ldots, i_t$ with repetition, which by induction on $n$ has $\begin{pmatrix} t + n - t - 1 \\ n - t \end{pmatrix} = \begin{pmatrix} n - 1 \\ n - t \end{pmatrix}$ possibilities, so indeed these numbers are the same).

**Theorem 6.2.9:** *(Golod [66]) Given any field F and any $n \geq 2$ there is an infinite dimensional graded F-algebra $R_0$ without 1, generated as algebra without 1 by m elements, such that every subset of $<m$ elements of $R_0$ is nilpotent. In particular, $R_0$ is nil and is a counterexample to Levitzki's and Kurosch's problem. Furthermore $\bigcap_{n \in \mathbb{N}} R_0^n = 0$.*

**Proof:** Since we want to use lemma 6.2.7 we first look for a general instance for which the coefficients of the power series $(1 - m\lambda + \sum_{k=2}^{\infty} m_k \lambda^k)^{-1}$ are non-negative. We claim this holds for

$$m_k = \varepsilon^2 (m - 2\varepsilon)^{k-2} \qquad \text{where } 0 < 2\varepsilon < m.$$

Indeed we compute

$$1 - m\lambda + \sum_{k \geq 2} \varepsilon^2 (m - 2\varepsilon)^{k-2} \lambda^k = 1 - m\lambda + \frac{\varepsilon^2 \lambda^2}{1 - (m - 2\varepsilon)\lambda}$$

$$= \frac{((1 - m\lambda)(1 - (m - 2\varepsilon)\lambda) + \varepsilon^2 \lambda^2)}{1 - (m - 2\varepsilon)\lambda}$$

$$= \frac{(1 - (m - \varepsilon)\lambda)^2}{1 - (m - 2\varepsilon)\lambda},$$

so its reciprocal is

$$\frac{1 - (m - 2\varepsilon)\lambda}{(1 - (m - \varepsilon)\lambda)^2} = (1 - (m - 2\varepsilon)\lambda)(1 + (m - \varepsilon)\lambda + (m - \varepsilon)^2 \lambda^2 + \cdots)^2$$

$$= (1 - (m - 2\varepsilon)\lambda)\left(1 + \sum_{k \geq 1} (k + 1)(m - \varepsilon)^k \lambda^k\right)$$

$$= 1 + \sum((k + 1)(m - \varepsilon)^k - (m - 2\varepsilon)k(m - \varepsilon)^{k-1})\lambda^k$$

$$= 1 + \sum_{k \geq 1} (m - \varepsilon)^{k-1}(m + (k - 1)\varepsilon)\lambda^k$$

and the coefficients are visibly positive.

Next we shall construct inductively a set $S$ of homogeneous elements $\{s_1, s_2, \ldots\} \subset T$ having the following properties:

(i) There are $\leq \varepsilon^2(m - 2\varepsilon)^{k-2}$ elements of $S$ having degree $k$, for each $k$;

(ii) If $f_1, \ldots, f_{m-1}$ are arbitrary in $T$ having constant term 0 then $\{f_1, \ldots, f_{m-1}\}^n \subseteq \langle S \rangle$ for suitable $n = n(f_1, \ldots, f_{m-1})$.

Clearly $R_0 = T_1 T/\langle S \rangle$ would be the desired algebra. Assume by induction that for given $k_0$ there is $d = d(k_0)$ for which homogeneous $s_1, \ldots, s_d$ of degree $k_0$ have been selected and such that whenever each $f_i$ has degree $\leq k_0$ we have $\{f_1, \ldots, f_{m-1}\}^n \subseteq \langle s_1, \ldots, s_d \rangle$ for suitable $n$ depending on $f_1, \ldots, f_{m-1}$. We want to select the next bunch of $s_j$, each of degree $> k_0$. To do this formally write

$$f_i = \sum \alpha_{iw} w \qquad \text{for } 1 \leq i \leq m - 1$$

where $\alpha_{iu} \in F$ and $w$ runs over the words in $X_1, \ldots, X_m$ having degree between 1 and $k_0$. Now any product of length $n$ (to be determined) of the $f_i$ has total degree $\leq nk_0$ and is a linear combination of homogeneous elements whose coefficients are products of the $\alpha_{iu}$. Taking $s_{d+1}, s_{d+2} \cdots$ to be these homogeneous elements we formally have satisfied condition (ii).

It remains to verify (i), which we claim is automatic if $n$ is large enough. First of all taking $n > d$ we see that each of the new $s_j$ have degree $> d$; the number of possible new $s_j$ is at most the number of commutative monomials in the $\alpha_{iw}$, and this can be estimated as $\leq d'd''$ where

$d'$ is the number of commutative monomials having length $n$ in the $\alpha_{1w}$;

$d'' = (m - 1)^n$ is the number of possible ways to replace a product $\alpha_{1w_1} \cdots \alpha_{1w_n}$ by $\alpha_{i_1 w_1} \cdots \alpha_{i_n w_n}$ where $1 \leq i_u \leq m - 1$ for each $u$.

Note $d'$ is the number of ways of choosing $n$ from

$$q = (m - 1)(m + m^2 + \cdots + m^{k_0})$$

(the number of words of degree $\leq k_0$) allowing repetitions, which by remark 6.2.8 is

$$\binom{q + n - 1}{n} \leq (q + n - 1)^{q-1}.$$

Let $c = (m - 2\varepsilon)/(m - 1)$. Taking $\varepsilon < \frac{1}{2}$ we may assume $c > 1$. For sufficiently large $n$ we have $(q + n - 1)^{q-1} \leq \varepsilon^2 c^n/(m - 2\varepsilon)^2$ since the exponential rises faster than the polynomial; thus for such $n$ we see

$$d'd'' \leq (q + n - 1)^{q-1}(m - 1)^n \leq (\varepsilon^2 c^n/(m - 2\varepsilon)^2)(m - 1)^n$$
$$= \varepsilon^2(m - 2\varepsilon)^n/(m - 2\varepsilon)^2 = \varepsilon^2(m - 2\varepsilon)^{n-2}.$$

Since the new monomials all have degree $\geq n > k_0$ we see (i) holds, as desired.    Q.E.D.

*Note 6.2.9':*    The algebra of theorem 6.2.9 is considerably more than a counterexample to Kurosch's problem. For example, taking $m = 4$ we get a counterexample $R_0$ for which $M_2(R_0)$ also is nil. If we are interested in finding the quickest counterexample to Kurosch's problem we could require merely that each $s_j$ has a power in $\langle S \rangle$; thus one can work with one $j$ at a time and streamline the above proof considerably, cf., Herstein [69B] which produces a quick application of lemma 6.2.7 for $F$ countable and $m = 3$.

## Growth of Algebras

The crux of the Golod-Shafarevich example is that $S$ was "sparse" enough that the algebra $R$ although nil grew exponentially in terms of the generators. Thus it makes sense to require that our algebras do not grow so fast. This leads us to the following situation. For the remainder of this section *we assume $R$ is an algebra over a field $F$.* If $S = \{r_1, \ldots, r_m\}$ we write $F\{S\}$ for $F\{r_1, \ldots, r_m\}$.

*Definition 6.2.10:*    Suppose $R = F\{S\}$. Let $V_n(S)$ be the subspace $\sum_{k=0}^{n} FS^k$, and define the *growth function* $G_S(n)$ of $R$ by

$$G_S(n) = [V_n(S) : F].$$

We say $R$ has *exponential growth* if $G_S(n) \geq t^n$ for some $t > 1$ and all $n \in \mathbb{Z}$; otherwise, $R$ has *subexponential growth*. $R$ has *polynomially bounded growth* if $G_S(n) \leq cn^t$ for suitable $c, t$ in $\mathbb{N}$, for all $n$ in $\mathbb{N}$.

*Examples 6.2.11:*

(i) If $R$ is the free algebra $F\{S\}$ where $S = \{X_1, \ldots, X_m\}$ then $G_S(n) = \sum_{k=0}^{n} m^k$ so $R$ has exponential growth.

(ii) If $R = F[\lambda_1, \ldots, \lambda_m]$ then as noted in remark 6.2.8 we have $\binom{m+n-1}{n}$ elements in $\{\lambda_1, \ldots, \lambda_m\}^n$, and thus $G_S(n) = \sum_{k=0}^{n} \binom{m+k-1}{k} = \binom{m+n}{n}$

$\left(\text{since } \binom{m+n}{n} = \binom{m+n-1}{n} + \binom{m+n-1}{n-1}\right)$ and apply induction to $\binom{m+n-1}{n-1}$. But $\binom{m+n}{n} = \frac{(n+m)\cdots(n+1)}{m\cdots 1} \le 2n^m$ whenever $n > 1$. Consequently $G_S(n)$ has polynomially bounded growth. Note for $m \ge 1$ and $n > 4$ that $G_S(m) \le n^m$.

(iii) The Weyl algebra $\mathbb{A}_n(F)$ "grows" at the same rate as $F[\lambda_1, \ldots, \lambda_n]$, so $G_S(n) \le 2n^m$.

It would be rather embarrassing if the type of growth changed according to the generating set, so we need the following fact.

**Remark 6.2.12:** If $R = F\{S\} = F\{S'\}$ for $S'$ finite then $S' \subseteq S^{n'}$ for some $n'$. (Each element of $S'$ lies in suitable $V_{n'}(S)$ so take the maximal such $n'$.) Consequently $G_{S'}(n'n) \ge G_S(n)$ for all $n$, thereby implying exponential, subexponential, and polynomially bounded growth are each well-defined.

This remark leads us to the following definition.

**Definition 6.2.13:** Suppose $R, R'$ are affine $F$-algebras where $R = F\{S\}$ and $R' = F\{S'\}$. $R'$ grows *at least as fast* as $R$, written $G(R) \le G(R')$, if there are positive integers $n_0, n_1$ for which $G_S(n) \le n_1 G_{S'}(n_0 n)$ for all $n$ in $\mathbb{N}$. We say $R$ and $R'$ have the *same* growth if $G(R') \le G(R)$ and $G(R') \le G(R)$.

In view of remark 6.2.12 this definition is independent of the choice of generating sets $S$ and $S'$. Thus we have the following observations.

**Remark 6.2.14:**

(i) If $R$ is a homomorphic image of $R'$ then $G(R) \le G(R')$. In particular, if $R = F\{r_1, \ldots, r_m\}$ then $G(R) \le G(F\{X_1, \ldots, X_m\})$.

(ii) If $R$ is an affine subalgebra of $R'$ then $G(R) \le G(R')$. (Indeed, if $S, S'$ are generating sets of $R, R'$, respectively, then replace $S'$ by $S \cup S'$ and the assertion is obvious.)

(iii) If $R = F\{S\}$ and $R' = F\{S'\}$ then $R \otimes_F R'$ has generating set $S'' = (S \otimes 1) \cup (1 \otimes S')$, so $V_n(S) \otimes V_n(S') \subseteq V_{2n}(S'') \subseteq V_{2n}(S) \otimes V_{2n}(S')$. In particular, if $[R':F] < \infty$ then $R \otimes_F R'$ and $R$ have the same growth. As a special case $R$ and $M_n(R) \approx R \otimes_F M_n(F)$ have the same growth.

(iv) If $R \subseteq R'$ and $R'$ is f.g. as $R$-module then $R'$ and $R$ have the same growth, by (iii) and the regular representation of $R'$ as an $F$-subalgebra of $M_t(R)$.

(v) The Noether normalization theorem shows if $C$ is $F$-affine commutative of transcendence degree $t$ then $C$ has the same growth as $F[\lambda_1, \ldots, \lambda_t]$.

(vi) Viewing the $\{V_n(S): n \in \mathbb{N}\}$ as a filtration of $R$, we see by comparing dimensions that $R$ has the same growth as the associated graded algebra.

There are domains having subexponential but not polynomially bounded growth, cf., exercise 8.3.14. Having various examples at our disposal, let us now show that subexponential growth has important structural implications.

**Proposition 6.2.15:** (*Jategaonkar*) *Suppose $R$ is a domain not necessarily with 1, and $R$ does not contain a copy of the free algebra $F\{X_1, X_2\}$ without 1. (In particular, this latter hypothesis holds when $R$ has subexponential growth). Then $R$ is left and right Ore.*

**Proof:** Given $0 \neq r_1, r_2$ in $R$ take $f \in F\{X_1, X_2\}$ of minimal total degree such that $f(r_1, r_2) = 0$. Writing $f = f_1(X_1, X_2)X_1 + f_2(X_1, X_2)X_2$ we have $f_i(r_1, r_2) \neq 0$ for $i = 1, 2$ by assumption on $i$, so $f_1(r_1, r_2)r_1 = -f_2(r_1, r_2)r_2 \neq 0$, proving $Rr_1 \cap Rr_2 \neq 0$. The right Ore condition is proved analogously.
                                                                                    Q.E.D.

This does not generalize to arbitrary prime algebras of subexponential growth, cf., exercise 7. However, Irving-Small [83] do have a generalization.

**Corollary 6.2.16:** *Suppose $R$ is an affine prime $F$-algebra having subexponential growth. If $R$ has a left ideal $L$ for which $\mathrm{Ann}' L$ is a maximal right annihilator then $L$ is uniform.*

**Proof:** Let $I = \mathrm{Ann}' L$. Then $I \cap L \lhd L$ as rng, and $\bar{L} = L/(I \cap L)$ is a domain; indeed, if $a_1, a_2 \in L - (I \cap L)$ then $La_1 \neq 0$ implying $(La_1)a_2 \neq 0$ (for, otherwise, $\mathrm{Ann}' La_1 \supset I$) and so $a_1a_2 \notin I \cap L$. By the proposition $\bar{L}$ is left and right Ore. To prove $L$ is uniform suppose $0 \neq L_1, L_2 < L$. Then their images $\bar{L}_1$ and $\bar{L}_2$ in $\bar{L}$ are nonzero since $L_iL \neq 0$ (for $R$ is prime). Hence $\bar{L}_1$ and $\bar{L}_2$ have a common element $\bar{a} \neq 0$, and $0 \neq La \subseteq L_1 \cap L_2$.     Q.E.D.

**Theorem 6.2.17:** (*Irving-Small*) *Suppose $R$ is an affine, semiprime $F$-algebra satisfying ACC on left and right annihilators. If $R$ has subexponential growth then $R$ is Goldie.*

**Proof:** By theorem 3.2.27 we may assume $R$ is prime. But ACC(right annihilators) implies DCC(left annihilators). Let $L$ be a minimal left annihilator. By corollary 6.2.16 $L$ is uniform. $R$ is prime so $\operatorname{Ann} L = 0$ and consequently there are $a_1, \ldots, a_n$ in $L$ for which $\operatorname{Ann}\{a_1, \ldots, a_n\} = 0$. The map $r \mapsto (ra_1, \ldots, ra_n)$ gives us a monic $R \to \bigoplus_{i=1}^{n} Ra_i \subseteq L^{(n)}$ implying $R$ has finite uniform ( = Goldie) dimension. Since $R$ has ACC(Ann) by hypothesis we conclude $R$ is Goldie.    Q.E.D.

(The hypothesis ACC on *left* annihilators is superfluous, cf., exercise 8; however, some chain condition is needed in light of exercise 7.) Having whetted our appetites for studying subexponential growth and, in particular, polynomially bounded growth, we want a more precise measure of growth.

## Gelfand-Kirillov Dimension

**Definition 6.2.18:** The *Gelfand-Kirillov* dimension of an affine algebra $R$, written $\operatorname{GK} \dim(R)$, is $\overline{\lim}_{n \to \infty} \log_n(G_S(n)) = \overline{\lim}_{n \to \infty}(\log G_S(n)/\log n)$.
    The point of this definition is that if $G_S(n) \le cn^t$ then $\operatorname{GK} \dim(R) \le \overline{\lim}_{n \to \infty}(\log_n c + t) = t$. Thus one sees $R$ has finite GK dim if $R$ has polynomially bounded growth. Similarly we have

**Remark 6.2.18':** $\overline{\lim} \log_n G_S(n) = \overline{\lim} \log_n G_S(n_0 n)$, for fixed $n_0 > 0$, since $\log_n G_S(n_0 n) = \log G_S(n_0 n)/\log n = (\log G_S(n_0 n)/\log n_0 n)((\log n_0 n)/(\log n))$    and $(\log n_0 n)/\log n = 1 + (\log n_0/\log n) \to 1$ as $n \to \infty$. Consequently if $G(R) \le G(R')$ then $\operatorname{GK} \dim R \le \operatorname{GK} \dim R'$.

When $R$ is not affine we could still define $\operatorname{GK} \dim(R)$ as $\sup\{\operatorname{GK} \dim(R')$: $R'$ is an affine subalgebra of $R\}$. A related notion (nowadays called GK tr deg) was introduced by Gelfand-Kirillov [66] to show that the Weyl algebras $A_n(F)$ and $A_m(F)$ have nonisomorphic rings of fractions for $n \ne m$ and to frame the Gelfand-Kirillov conjecture (cf., §8.3). Within a decade GK dim was seen to be a general tool in ring theory. A thorough, readable treatment of GK dim is given in Krause-Lenagan [85B], which is the basis of much of the discussion here.
    By example 6.2.11 we see that $\operatorname{GK}(F[\lambda_1, \ldots, \lambda_m]) = m$, and $\operatorname{GK}(A_m(F)) = 2m$. On the other hand, K-dim $F[\lambda_1, \ldots, \lambda_m] = m$ by exercise 3.5.23, but K-dim $A_m(F) = m$ by exercise 3.5.25. See exercise 9 for a worse example.

**Remark 6.2.19:** If $R \subseteq R'$ or $R$ is a homomorphic image of $R'$ then $\operatorname{GK} \dim(R) \le \operatorname{GK} \dim(R')$ by remark 6.2.14(i), (ii).

***Remark 6.2.20:***   If $R'$ is f.g. as module over a subring $R$ then GK dim$(R')$ = GK dim$(R)$ by remark 6.2.14(iv). This simple observation has the following consequences:

(i) If $C$ is commutative and $F$-affine then GK dim $C$ = tr deg$(C/F)$ by Noether normalization and example 6.2.11(ii).

(ii) GK dim$(M_n(R))$ = GK dim$(R)$.

(iii) If $R$ is locally finite over $F$ then GK$(R) = 0$. (Indeed one may assume $R$ is affine and thus finite dimensional over $F$, so GK dim$(R)$ = GK dim$(F) = 0$.)

As typified by these results, the GK dimension is usually difficult to compute from scratch but is very amenable because closely related rings often have the same GK dimension. Another instance is

**Proposition 6.2.21:**   *If $R$ is a subdirect product of $R_1,...,R_t$ then* GK dim $R$ = $\max\{$GK dim$(R_i)$: $1 \le i \le t\}$.

***Proof:***   "$\ge$" is clear by remark 6.2.19, and to prove "$\le$" we note $R \subseteq \prod_{i=1}^t R_i$, so it suffices to assume $R = \prod R_i$. Let $S_i$ be a generating set for $R_i$, with $1 \in S_i$ for each $i$. Then $S = S_1 \times \cdots \times S_n$ is certainly a generating set for $R$, and $V_n(S) = \prod V_n(S_i)$; writing $G_i(n)$ for the generating function of $S_i$ in $R_i$ we have

$$G_S(n) = \sum_{i=1}^t G_i(n) \le t \max\{G_i(n): 1 \le i \le t\}.$$

Since $\log_n t \to 0$ we see GK dim$(R)$ = max GK dim$(R_i)$, as desired.

$\hspace{10cm}$ Q.E.D.

**Proposition 6.2.22:**   GK dim$(R[\lambda])$ = $1$ + GK dim$(R)$.

***Proof:***   Given $S = \{1,r_1,...,r_t\}$ let $S' = S \cup \{\lambda\} \subset R[\lambda]$; then $V_{2n}(S') \supset \sum_{i=0}^n V_n(S)\lambda^i$ so in view of remark 6.2.18' we see

$$\text{GK dim}(R[\lambda]) \ge \overline{\lim} \log_n(n + 1)G_S(n) = 1 + \overline{\lim} \log_n G_S(n)$$

proving ($\ge$).

To prove ($\le$) note that any affine subalgebra $T$ of $R[\lambda]$ is contained in $R_1[\lambda]$ where $R_1$ is the affine subalgebra of $R$ generated by the coefficients of the generators of $T$. Thus we may assume $R$ is affine. Now taking $S, S'$ as above we have $V_n(S') \subset \sum_{i=0}^n V_n(S)\lambda^i$ so

$$\text{GK dim}(R[\lambda]) \le \overline{\lim} \log_n(n + 1)G_S(n) = \overline{\lim} \log_n(n + 1) + \overline{\lim} \log_n G_S(n)$$

$$= 1 + \text{GK dim}(R). \quad \text{Q.E.D.}$$

**Remark 6.2.22':** The same proof shows that $GK \dim(T) \geq 1 + GK \dim(R)$ whenever there is $x$ in $T$ such that $\{x^i : i \in \mathbb{N}\}$ are $R$-independent.

The opposite inequality is trickier and can fail for Ore extensions; cf., theorem 8.2.16 and exercise 8.3.13. However, it holds for differential polynomial extensions which are affine, cf., exercise 10.

**Proposition 6.2.23:** $GK \dim(S^{-1}R) = GK \dim(R)$ *for any regular submonoid $S$ of $Z(R)$.*

**Proof:** Any finite subset of $S^{-1}R$ can be written in the form $s^{-1}r_1, \ldots, s^{-1}r_t$, and thus $s^n \{s^{-1}r_1, \ldots, s^{-1}r_t\}^n \subseteq \{r_1, \ldots, r_t\}^n$, from which we see at once that $R$ grows at least as fast as $S^{-1}R$. Thus $GK \dim(S^{-1}R) \leq GK \dim(R)$, and equality follows by remark 6.2.14(ii) since $R \subseteq S^{-1}R$ (because $S$ is regular).

Q.E.D.

On the other hand, this result does not hold for arbitrary Ore sets $S$, in view of Makar-Limanov's result that the division algebra of fractions of $A_1(F)$ does not have GK dim, cf., example 7.1.46.

**Proposition 6.2.24:** *Suppose an ideal $A$ of $R$ contains a left regular element $a$. Then $GK \dim(R) \geq 1 + GK \dim(R/A)$.*

**Proof:** Take any finite set $S$ of $R$ containing 1 and $a$. Let $V'_n = A \cap V_n(S)$ and let $V''_n$ be a complementary subspace of $V'_n$ in $V_n(S)$. Since $V''_n \cap Ra = 0$ we see

$$V''_n, aV''_n, \ldots, a^n V''_n$$

are independent subspaces all contained in $V_{2n}(S)$, so

$$G_S(2n) \geq n[V''_n : F] = nG_{\bar{S}}(n)$$

where $\bar{S}$ is the image of $S$ in $R/A$. Thus by remark 6.2.18'

$$GK \dim(R) \geq \overline{\lim} \log_n G_S(2n) \geq \overline{\lim} \log_n(nG_{\bar{S}}(n))$$

$$= 1 + \overline{\lim} \log_n(G_{\bar{S}}(n)) = 1 + GK \dim(R/A). \qquad \text{Q.E.D.}$$

**Theorem 6.2.25:** *Suppose the prime images of $R$ are left Goldie. (In particular, this is true if $R$ is PI or if $R$ is left Noetherian.) Then*

$$GK \dim(R) \geq GK \dim(R/P) + \text{height}(P) \qquad \text{for any } P \text{ in } \text{Spec}(R).$$

***Proof:***   For any chain $P = P_0 \supset P_1 \supset \cdots \supset P_m$ we need to show GK dim($R$) $\geq$ GK($R/P$) + $m$, which would certainly be the case if GK($R/P_i$) $\geq$ GK($R/P_{i-1}$) + 1 for each $i$; but this latter assertion follows from proposition 6.2.24 since $P_{i-1}/P_i$ is a nonzero ideal of the prime Goldie ring $R/P_i$ and thus contains a regular element.   Q.E.D.

***Note:***   It follows that if GK dim($R/P_1$) = GK dim($R/P_2$) for $P_1, P_2$ in Spec($R$) then $P_1$ and $P_2$ are incomparable.

**Corollary 6.2.25′:**   *Under hypothesis of theorem 6.2.25, the little Krull dimension of $R$ is at most* GK dim($R$).

**Proposition 6.2.26:**   GK dim($R_1 \otimes R_2$) $\leq$ (GK dim $R_1$)(GK dim $R_2$).

***Proof:***   We may assume $R_1, R_2$ are affine. Then the inequality follows from remark 6.2.14(iii).   Q.E.D.

A useful related result involves changing the base field.

***Remark 6.2.26′:***   If $K$ is a field extension of $F$ then GK dim $R =$ GK dim $R \otimes_F K$ (as $K$-algebra) for any affine $F$-algebra $R$. (Indeed, if $S$ generates $R$ then $S \otimes 1$ generates $R \otimes_F K$, and $G_S(n) = \sum_{k \leq n} [FS^k : F] = \sum_{k \leq n} [FS^k \otimes K : K] = G_{S \otimes 1}(n)$, so appeal to the definition.)

One may wonder, "When does equality hold in proposition 6.2.26?", i.e., when does $\alpha_1 \alpha_2 \leq$ GK dim($R_1 \otimes R_2$) when $\alpha_i =$ GK dim $R_i$? The hitch in proving equality in general via remark 6.2.14(iii) is that when taking a sequence $n_1, n_2 \ldots$ such that $\log G_{S_2}(n_i)/\log n_i \to \alpha_2$, one need not have $\log G_{S_1}(n_i)/\log n_i \to \alpha_i$. Of course, this occurs if also $\alpha_i = \underline{\lim} \log G_{S_1}(n)/\log n$, which we shall temporarily call the *lower* GK-dim of $R_1$. In certain cases the GK-dim and lower GK-dim agree and are integervalued; for example, when $R_1$ is either commutative (remark 6.2.20(i)), affine PI (theorem 6.3.41), or the homomorphic image of an enveloping algebra of a finite dimensional Lie algebra (remark 8.3.37). Also see exercise 6.3.24.

Indeed, it is not easy to come up with an example for which GK dim $R$ is not an integer, but Borho-Kraft [76] showed for any real $\alpha$ between 2 and 3 that there is a homomorphic image of $F\{X_1, X_2\}/\langle X_2^3 \rangle$ which has GK dim $\alpha$. In view of proposition 6.2.22 GK dim $R$ thus can take on any real value $\geq 2$.

Warfield [84] showed equality need not hold in proposition 6.2.26, and

Krempa-Okninski [86] extend Warfield's example to an algebra $R$ whose GK-dim and lower GK-dim take on arbitrary values $\alpha > \alpha' \geq 2$. For GK-dim $\leq 2$, however, the story is completely different, as we shall now see.

## Low GK Dimension

**Remark 6.2.27:** The GK dimension of $R$ cannot be strictly between 0 and 1; in fact, $\mathrm{GK}(R) \geq 1$ if $R$ is not locally finite over $F$. (Indeed, by remark 6.2.20(iii) we may assume $R$ is affine but not finite dimensional over $F$. Then $V_n(S) \neq R$ for any $n$, where $S$ is a generating set; hence $V_n(S) \subset V_{n+1}(S)$ for each $n$, implying $G_S(n) \geq n$ and thus $\mathrm{GK}(R) \geq 1$.)

Bergman showed that the GK dimension also cannot be strictly between 1 and 2. His technique shows how words affect the study of affine algebras, and so we make a slight digression to study them.

**Definition 6.2.28:** Let $M$ denote the word monoid in $\{1, \ldots, t\}$, cf., example 1.2.4, and given the total order of example 1.2.18. We say $w'$ is a *subword* of $w$ if there are words $v', v''$ (possibly blank) for which $w = v'w'v''$; $w'$ is *initial* if $v'$ is blank; $w'$ is *terminal* if $v''$ is blank.

**Remark 6.2.29:** Given $R = F\{r_1, \ldots, r_t\}$ define the monoid homomorphism $M \to R$ given by $i \mapsto r_i$, and write $^-$ for the canonical image. For example, if $w = 231$ then $\bar{w} = r_2 r_3 r_1$. It is convenient to define the set $W = W(R) \subset M$ as follows: List the elements of $M$ in ascending order $w_1 = 1$, $w_2 = 2, \ldots$, $w_t = t$, $w_{t+1} = 11$, etc., and, inductively, say $w_k \in W$ iff $\bar{w}_k \notin \sum_{j<k} F\bar{w}_j$. Then $\bar{W}$ is an $F$-base of $R$. (Indeed, obviously $\bar{W}$ spans. But also $\bar{W}$ is $F$-independent for, otherwise, take $\sum_{j \leq k} \alpha_j \bar{w}_j = 0$ with $\alpha_k \neq 0$, where we took $w_j$ only from $W$; then $\bar{w}_k = -\sum_{j<k} \alpha_k^{-1} \alpha_j w_j \in \sum_{j<k} F\bar{w}_j$, contrary to definition of $W$.) In the foregoing write $\varphi(n)$ for the number of words of $W(R)$ having length $n$.

**Remark 6.2.30:** If $w_k \in W$ then every subword of $w_k$ is in $W$. (Indeed, otherwise, writing $w_k = w'w_m w$ where $\bar{w}_m \in \sum_{j<m} F\bar{w}_j$ note $w_k \in \sum_{j<m} F\bar{w}'\bar{w}_j\bar{w} \subseteq \sum_{j<k} F\bar{w}_j$ since $w'w_jw < w'w_mw = w_k$.)

We need some facts about repeated words.

**Lemma 6.2.31:** *Given a word* $w = i_1 \cdots i_n$ *suppose for some* $n' \leq n$ *that* $i_1 \cdots i_n i_1 \cdots i_{n'} = v^k v'$ *where* $v$ *is an initial subword of* $w$ *and* $\mathrm{length}(v) = m \leq n'$.

*Then* $v = \tilde{v}^t$ *for a suitable word* $\tilde{v}$ *whose length divides both m and n, where* $t = m/length(\tilde{v})$.

**Proof:** Note $i_j = i_{j+m}$ for all $j$. Taking $k$ minimal for which $j + km > n$ we see $i_j = i_{j+km} = i_{j+km-n}$; but the greatest common divisor $d$ of $m$ and $n$ can be written in the form $k_1 m + k_2 n$ and working back and forth we see $i_j = i_{j+d}$ for all $j$. Letting $\tilde{v} = i_1 \cdots i_d$ we have $v = \tilde{v}^t$ where $t = m/d$.     Q.E.D.

**Lemma 6.2.32:** *Assume there is* $n_0$ *in* $\mathbb{N}$ *for which* $\varphi(n_0) \leq n_0$. *Then* $\varphi(n) \leq n_0^3$ *for all* $n > n_0$.

**Proof:** Claim. Any $w$ in $W(R)$ of length $\geq 2n_0$ has the form $w'v^k v'w''$ where length$(v) \leq n_0$, $w'$ and $w''$ have length $\leq n_0 - $ length$(v)$, and $v'$ is an initial subword of $v$.

Before proving the claim let us show how it implies the lemma. Note any word in $W(R)$ of length $\leq 2n_0$ is determined by its initial and terminal subwords of length $n_0$, each of which is in $W(R)$ by remark 6.2.30 and thus has $\leq n_0$ possibilities, implying $\varphi(n) \leq n_0^2$ for $n \leq 2n_0$.

Thus we may assume $n \geq 2n_0$. We shall now show any word $w$ of length $n$ is determined by its initial subword $y$ of length $2n_0$ and its terminal subword $y'$ of length $n_0$; thereby implying $\varphi(n) \leq n_0^2 n_0 = n_0^3$. For $n \leq 3n_0$ this is obvious, so assume $n \geq 3n_0$. By the claim write $w = w'v^k v'w''$. We see $y$ has a terminal subword of the form $v^k v'$ where length$(v^k v') = 2n_0 - $ length$(w') \geq n_0 + $ length$(v)$, and $v'$ is an initial subword of $v$; we choose this terminal subword of $y$ such that first $v$ and then $v'$ have minimal possible length. Now $v$ is uniquely determined by $y$, and we "fill in" the missing entries of $w$ by repeating $v$. (This procedure determines $w$ uniquely in view of lemma 6.2.31.)

*Proof of Claim.* We shall actually prove the additional conclusion that $k \geq 2$ and length$(v^k v') \geq n_0 + $ length$(v)$.

*Case I.* $n = 2n_0$. Note $w$ has $n_0 + 1$ subwords of length $n_0$, so two must be equal, i.e., $w = w'vv'v_1 w''$ where $vv' = v'v_1$ has length $n_0$. Thus $w = w'v^2 v'w''$ and length$(w') + $ length$(w'') = 2n_0 - $ length$(v^2 v') = n_0 - $ length$(v)$, so we are done.

*Case II.* $n > 2n_0$. Write $w = u_1 u$ where $u_1$ has length 1. By induction $u = u'v^k v'u''$ where $k \geq 2$, and letting $m = $ length$(v)$, we know $u'$ and $u''$ have length $\leq n_0 - m$ and length$(v^k v') \geq n_0 + m$. Now $w = u_1 u'v^k v'u''$ so we are done unless length$(u_1 u') > n_0 - m$, i.e., length$(u') = n_0 - m$. Applying case I to the initial subword of $u$ of length $2n_0$ yields

$$u = q'p^{k'}p''q''$$

where $\text{length}(q') \leq n_0 - \text{length}(p)$, $\text{length}(p^{k'}p'') \geq n_0 + \text{length}(p)$, and $p''$ is an initial subword of $p$.

Note we have copies of $v$ appearing in $w$, starting in positions $n_0 - m + 2$ and $n_0 + 2$, and one of these two appears as a *noninitial* subword of $p^{k'}p''$. On the other hand, letting $m' = \text{length}(p)$ we see $p^{k'}p''$ starts no earlier than position 2 and extends through position $n_0 + m' + 1$. Thus there are $m'$ positions of $p^{k'}p''$ which occur in the subsequent copy of $v$, and lemma 6.2.31 implies that $p$ can be written in the form $(\tilde{p})^j$ where $\text{length}(\tilde{p}) \mid m$. Hence the last digit appearing in $v$ is clearly the last digit of $u'$. Writing $u_1u'$ as $\tilde{u}i$ and $v = \tilde{v}i$ for suitable $i$, we see

$$w = u_1u'v^kv'u'' = \tilde{u}i(\tilde{v}i)^kv'u'' = \tilde{u}(i\tilde{v})^k(iv')u''.$$

Now $\text{length}(\tilde{u}) = n_0 - m$ and $\text{length}(i\tilde{v}) = m$, and $iv'$ is an initial subword of $i\tilde{v}$, so we are done.     Q.E.D.

**Theorem 6.2.33:**     (*Bergman*) *If* $1 \leq \text{GK dim}(R) \leq 2$ *then* $\text{GK dim}(R)$ *is an integer.*

**Proof:**     We may assume $R = F\{r_1, \ldots, r_n\}$ is affine. Taking $S = \{r_1, \ldots, r_n\}$ we have $[FS^n : F] \leq n$ for some $n$, for, otherwise, $G_S(n) \geq \sum_{k=1}^{n} k = n(n+1)/2$ for all $n$, implying $\text{GK dim}(R) \geq \overline{\lim}_{n \to \infty} \log_n(n^2/2) = 2$, contrary to hypothesis. But $[FS^n : F] = \varphi(n)$, as defined above, so some $\varphi(n_0) \leq n_0$ implying $\varphi(n) \leq n_0^3$ for all $n > n_0$, by the lemma. Hence

$$G_S(n) \leq G_S(n_0) + (n - n_0)n_0^3$$

which is linear in $n$, so $\text{GK dim}(R) \leq 1$.     Q.E.D.

Small-Stafford-Warfield [84] continue this analysis and show that if $R$ is affine and $\text{GK dim}(R) = 1$ then $R$ is f.g. over $Z(R)$, and, in particular, is a PI-ring. Then Artin-Tate says $Z(R)$ is commutative affine of GK dimension 1, so we know by remark 6.2.20(i) that $Z(R)$ is integral over a subring isomorphic to $F[\lambda]$.

*Recapitulation 6.2.34:*     Let $\alpha = \text{GK dim}(R)$ for $R$ affine. $\alpha = 0$ iff $R$ is finite dimensional over $F$. $\alpha = 1$ iff $R$ is f.g. over a central subring isomorphic to $F[\lambda]$. Otherwise $\alpha \geq 2$, and any such $\alpha$ can appear.

Returning to Kurosch's problem, we see that if GK dimension $= 0$ or 1 then Kurosch's problem has a trivial solution. Thus we are led to ask,

**Question 6.2.35:**   If $R$ is algebraic over $F$ and $\mathrm{GK}\dim(R) < \infty$ then is $\mathrm{GK}(R) = 0$?

## The GK Dimension of a Module

Although the GK dimension is used predominantly for algebras, it is also applicable to modules.

**Definition 6.2.36:**   If $R$ is affine and $M = \sum_{i=1}^{t} Rx_i \in R\text{-}\mathscr{F}imod$ then $\mathrm{GK}\dim(M) = \overline{\lim}_{n\to\infty} \log_n[\sum V_n(S)x_i : F]$, notation as in definition 6.2.10. More generally, for any module $M$ over an $F$-algebra $R$ define $\mathrm{GK}\dim M = \sup\{\mathrm{GK}$ dimension of f.g. submodules over affine subalgebras of $R\}$.

It is not difficult to show this definition is well-defined. Clearly $\mathrm{GK}\dim(R)$ as $R$-module reduces to the definition of $\mathrm{GK}\dim(R)$ as algebra, and the GK module dimension has ridden on the shoulders of the algebraic-theoretic dimension. Its main disadvantage in general is that is not *exact*, i.e., one does not always have the formula $\mathrm{GK}\dim M = \max\{\mathrm{GK}\dim(N), \mathrm{GK}\dim(M/N\}$ for $N < M$, cf., exercise 13. We do have the following partial results.

**Remark 6.2.37:**   $\mathrm{GK}\dim M \geq \max\{\mathrm{GK}\dim(N), \ \mathrm{GK}\dim(M/N)\}$ for any $N < M$, as an immediate consequence of the definition. Furthermore, $\mathrm{GK}\dim M = \mathrm{GK}\dim(M^{(t)})$ for any $t$ since taking $x_i = (0,\dots,1,\dots,0)$

$$\mathrm{GK}\dim M^{(t)} = \overline{\lim} \log_n\left[\sum_{i=1}^{t} V_n(S)x_i : F\right] = \overline{\lim} \log_n(t[V_n(S)x_1 : F])$$

$$= \overline{\lim}(\log_n t + \log_n[V_n(S)x_1 : F] = \mathrm{GK}\dim M.$$

In particular, for any f.g. $R$-module $M$ we have $\mathrm{GK}\dim M \leq \mathrm{GK}\dim R^{(t)} = \mathrm{GK}\dim R$ (for suitable $t$).

There are two instances when the GK module dimension is exact. The first is when $R$ is Noetherian PI, cf., exercise 6.3.27, although in this case there are other dimensions and techniques readily available. The second, more significant, instance is when $R$ is an enveloping algebra. The exactness of the GK module dimension then follows from the next result.

**Theorem 6.2.38:** *Suppose R is a filtered affine F-algebra such that the associated graded algebra $\tilde{R}$ is affine and left Noetherian. Then GK module dimension is exact for f.g. R-modules.*

**Proof:** Suppose $M = \sum_{k=1}^{t} Rx_k$. We modify the associated graded module construction as follows:

Let $M_i = \sum_{k=1}^{t} R(i)x_k$ and $\tilde{M} = \bigoplus_{i \in \mathbb{N}} M_i/M_{i-1}$, viewed naturally as module over $\tilde{R} = \bigoplus R(i)/R(i-1)$ by

$$(r_i + R(i-1))(y_j + M_{j-1}) = r_i y_j + M_{i+j-1} \qquad \text{for } r_i \text{ in } R(i), y_j \text{ in } M_j.$$

(Note $M(0) = 0$ and $\sum_k Fx_k \subseteq M(1)$ so we can view the $x_k$ in $\tilde{M}$.)

Suppose now $K \leq M$ and $N = M/K$. Defining $K_i = K \cap M_i$ and $N_i = (K + M_i)/K$ for $i$ in $\mathbb{N}$ we also can form $\tilde{R}$-modules $\tilde{K} = \bigoplus K_i/K_{i-1}$ and $\tilde{N} = \bigoplus N_i/N_{i-1}$, and have an exact sequence

$$0 \to \tilde{K} \to \tilde{M} \to \tilde{N} \to 0.$$

Of course $\tilde{M}$ is generated by $x_1, \ldots, x_t$ so is Noetherian, implying $\tilde{K}$ and $\tilde{N}$ are Noetherian and thus f.g. Taking appropriate generating sets of homogeneous elements and noting

$$[M_i:F] = [K_i:F] + [N_i:F],$$

we see $\text{GK dim } \tilde{M} = \overline{\lim}_{n \to \infty} \log_n[M_n:F] = \overline{\lim}_{n \to \infty} \log_n([K_n:F] + [N_n:F]) \leq \max\{\text{GK dim } \tilde{K}, \text{GK dim } \tilde{N}\}$. But comparing dimensions in definition 6.2.36 we see at once $\text{GK dim } \tilde{M} = \text{GK dim } M$, and, likewise, $\text{GK dim } \tilde{K} \leq \text{GK dim } K$ and $\text{GK dim } \tilde{N} \leq \text{GK dim } N$ so we conclude $\text{GK dim } M \leq \max\{\text{GK dim}(K), \text{GK dim}(N)\}$, as desired.     Q.E.D.

There is a way of making this result even more explicit.

**Definition 6.2.39:** Suppose $R$ is an $\mathbb{N}$-graded algebra over a field $F$, and $M$ is a graded $R$-module. The *Hilbert series* is $\sum_{i \in \mathbb{N}} [M_i:F] \lambda^i$.

This is also called the *Poincare series*, and has arisen in the Golod-Shafarevich example (lemma 6.2.7); for this reason Cohn calls it the *Gocha*. One of the goals of current research is to determine the Hilbert series of a graded algebra as a closed rational expression; for example, the Hilbert series of $F\{X_1, \ldots, X_m\}$ is $\sum_{i \in \mathbb{N}} m^i \lambda^i = (1 - m\lambda)^{-1}$. See exercise 16 for another example. Hilbert proved the Hilbert series has a closed expression in case $R$ is commutative Noetherian (cf., Jacobson [80B, p. 442]). In fact, it is enough to have $\tilde{R}$ commutative, as explained in [Krause-Lenagan 85B, Chapter 7], and this is the case for enveloping algebras, cf., §8.3. Hilbert series have also come into prominence in the study of relatively free PI-algebras, as we shall indicate in §6.4.

## §6.3 Affine PI-Algebras

We start this section by verifying that for affine PI-algebras the general questions raised in §6.2 actually have positive solutions. In fact, using Shirshov's positive solution to Kurosch's problem we can describe the structure of prime affine PI-algebras extremely precisely, leading to many important results such as the catenarity of affine PI-algebras and the nilpotence of the Jacobson radical. Several techniques obtained thereby, most notably the use of the trace ring, are applicable for arbitrary prime PI-rings, and we shall note their impact on the Bergman-Small theorem and on localization (but without proof). Often the base ring $C$ is required to be Noetherian and/or Jacobson; this is certainly the case when $C$ is a field.

### The Nullstellensatz

There are several ways of proving the Nullstellensatz for affine PI-algebras, as we shall see; possibly the fastest method is Duflo's proof using generic flatness. In order to achieve the result when the base ring is arbitrary Jacobson we need another version of the Artin-Tate Lemma.

**Proposition 6.3.1:**   *Suppose $R$ is $C$-affine and is f.g. free as $Z$-module with a base containing 1, for suitable $Z \subseteq Z(R)$. Then $Z$ is affine.*

**Proof:**   Let $R = C\{r_1,\ldots,r_n\}$ and let $b_1 = 1,\ldots,b_t$ be a base of $R$ as $Z$-module. As in the proof of proposition 6.2.5 write $b_i b_j = \sum_{k=1}^{t} z_{ijk} b_k$ and $r_u = \sum_k z_{uk} b_k$, and let $C' = C[z_{ijk}, z_{uk}: 1 \le i,j,k \le t; 1 \le u \le n]$. Then $C'$ is affine. Moreover, $\sum_{k=1}^{t} C' b_k$ is a subalgebra of $R$ containing each $r_i$ and so equals $R$. It follows $C' = Z$, for if $\sum c'_k b_k = z \in Z$ then matching coefficients of 1 shows $z = c'_1 \in C'$, as desired.         Q.E.D.

**Corollary 6.3.2:**   *Suppose $R$ is $C$-affine primitive PI. Then $C$ is a G-domain, i.e., there is $s \neq 0$ in $C$ for which $C[s^{-1}]$ is a field $F$; furthermore, $[R:F] < \infty$.*

**Proof:**   Let $F$ be the field of fractions of $C$. By Kaplansky's theorem $R$ is finite dimensional over $Z(R)$, which is affine over $F$ by proposition 6.3.1. Write $Z(R) = F[z_1,\ldots,z_t]$.

We claim any field of the form $F[z_1, \ldots, z_t]$ is algebraic over $F$. This is seen by induction on the transcendence degree over $F$. Suppose, on the contrary, $z_t$ is transcendental over $F$, and let $F_1$ be the field of fraction of $F[z_1, \ldots, z_{t-1}]$ taken inside $Z(R)$. Then $Z(R) = F_1[z_t]$ is a field, implying $z_t$ is algebraic over $F_1$ (for, otherwise, $F_1[z_t] \approx F_1[\lambda]$ is not a field) and thus $[Z(R):F_1] < \infty$ implying $F_1$ is affine by proposition 6.3.1. But tr $\deg(F_1) <$ tr deg $Z(R)$ so by induction $F_1$ is algebraic over $F$, yielding the claim.

Thus $[Z(R):F] < \infty$ so $[R:F] < \infty$. Hence $F$ is $C$-affine by proposition 6.3.1. Write $F = C[s^{-1}c_1, \ldots, s^{-1}c_t]$ for $s$, $c_i$ in $C$; then $F = C[s^{-1}]$ as desired.     Q.E.D.

**Theorem 6.3.3:** (*Amitsur-Procesi*) *Suppose $R$ is an affine PI-algebra over a commutative Jacobson ring $C$. Then the generic flatness hypotheses of theorem 2.12.36 are satisfied, and consequently*

(i) *$R$ is Jacobson.*

(ii) *$R$ satisfies the maximal Nullstellensatz.*

(iii) *If $P$ is a primitive ideal of $R$ then $C \cap P$ is a maximal ideal of $C$, and $R/P$ is simple and finite dimensional over $C/C \cap P$.*

*Proof:* (Duflo[73]) Since any image of $R[\lambda]$ is also affine PI, we need only to show that $R/PR$ satisfies generic flatness over $C/P$ for every $P$ in Spec($C$); passing to $P/PR$ over $C/P$ we may assume $C$ is a domain. Let $M$ be a simple $R$-module; by remark 2.12.33(i) we may assume $M$ is faithful over $C$. But then $R/\text{Ann}_R M$ is $C$-affine, so $C[s^{-1}]$ is a field for some $s$, by corollary 6.3.2. $M[s^{-1}]$ is free over this field, so we have verified generic flatness.

Now conditions (i) and (ii) follows from theorem 2.12.36. The first part of (iii) comes from remark 2.12.33″, taking $P$ the annihilator of a suitable simple module. But then $R/P$ is affine over the field $F = C/C \cap P$ and is primitive PI so we are done by corollary 6.3.2.     Q.E.D.

Another proof of theorem 6.3.3 comes from the powerful "trace ring" techniques of the latter part of this section; a far-reaching generalization of McConnell [84a] is discussed in §8.4.

The weak Nullstellensatz shows Jac($R$) is nil for an affine PI-algebra $R$ over a field, so the next question is whether Nil($R$) is nilpotent. The story of this question involves most of the important ideas of affine algebras. Razmyslov [74] (and later Schelter who had been unaware of Razmylov's work) proved the nilpotence of the radical for homomorphic images of prime affine PI-algebras. To obtain the maximal value from this result we need an affine version of generic matrices. Let $F_n\{Y_1, \ldots, Y_m\}$ denote the subalgebra of the algebra of generic matrices $F_n\{Y\}$.

**Remark 6.3.4:** $F_n\{Y_1, \ldots, Y_m\}$ is $F$-affine and PI, and is prime by the argument of proposition 6.1.41. If $R$ satisfies all the identities of $M_n(F[\Lambda])$ then $R$ is a homomorphic image of $F_n\{Y_1, \ldots, Y_m\}$, by proposition 6.1.42(ii).

Thus Razmyslov's theorem (once proved) yields the nilpotence of the nilradical for every affine PI-algebra which satisfies the identities of $n \times n$ matrices. Since virtually all affine PI-algebras are visibly seen to satisfy the identities of $n \times n$ matrices, this formulation is quite satisfactory; furthermore, Lewin proved the converse, that nilpotence of the nilradial implies satisfaction of the identities of $n \times n$ matrices.

Razymslov also showed in characteristic 0 that it is enough to assume the affine algebra satisfies a Capelli identity. Kemer showed that every affine PI-algebra of characteristic 0 indeed satisfies a Capelli identity, thereby verifying the nilpotence of the radical of arbitrary affine PI-algebras in characteristic 0. Nevertheless, the PI community was relieved when Braun proved the nilpotence of the nilradical in complete generality, in arbitrary characteristic, using a different technique.

We shall discuss this subject rather leisurely, following a semihistorical trek along the main directions of inquiry concerning the affine PI-theory. Since most problems of structure become rather easy for rings embeddible in matrices (over a commutative ring), the first question one might ask is, "Which affine rings are embeddible in matrices?" This ultimately turns out to be a blind alley, but, nevertheless, is an interesting excursion.

## Digression: The Embedding Question for Affine PI-Algebras

There has been considerable interest in finding "reasonable" conditions which would characterize embeddibility into matrices, i.e., $R \subset M_n(C)$ for suitable $C$. (One could match this with the definition preceding 6.1.26 by noting that if $R \subseteq \prod_{k \leq n} M_k(H_k)$ then $R \subseteq M_t(C)$ where $t = n!$ and $C = \prod H_k$.) Theorem 6.1.26 showed any semiprime PI-ring is embeddible in matrices. On the other hand, arbitrary PI-rings need not be embeddible, cf., example 3.2.48, so we are led to narrow the scope to affine PI-algebras. Even in this restricted case we shall see that there is no obvious necessary and sufficient criterion for embeddibility in matrices, but as consolation we shall find a different kind of embeddibility, due to Lewin, which is quite satisfactory.

Let us start by listing criteria for embeddibility. We assume throughout that $F$ is a field and $R = F\{r_1, \ldots, r_m\}$ is affine PI.

**Reduction procedure 6.3.5:** Assume $R \subseteq M_n(C)$ for suitable commutative $C$. Then we may further assume either (but not both) of the following:

(i) $C$ is an affine (commutative) $F$-algebra. (Indeed, $F \subseteq R$ so $C$ contains a subring $H$ isomorphic to a subring of $F$; replacing $C$ by $C \otimes_H F$ we may assume $C$ is an $F$-algebra. Write each $r_k$ as a matrix $(c_{ij}^{(k)})$ for $1 \le k \le m$. Replacing $C$ by $F[c_{ij}^{(k)}: 1 \le i, j \le n, 1 \le k \le m]$ we may assume $C$ is affine.)

(ii) $C$ is an Artinian $F$-algebra. (Indeed, by (i) we may assume $C$ is $F$-affine and thus Noetherian; but then $C$ is embeddible on an Artinian ring, by theorem 3.2.53.)

**Remark 6.3.6:** Any $R$ embeddible in matrices satisfies the following properties not necessarily holding in arbitrary PI-algebras:

(i) $R$ satisfies all multilinear identities of $M_n(\mathbb{Z})$ (since these pass to $M_n(\mathbb{Z}[\lambda_i: i \in I])$) and thus to $M_n(C)$);

(ii) Every nil subring of $R$ is nilpotent of bounded index, by theorem 3.2.38 and reduction 6.3.5;

(iii) There is some $k \in \mathbb{N}$ which bounds the length of any chain of left (or right) annihilators (since this condition is inherited by subrings of Artinian rings, and we may assume by reduction 6.3.5(ii) that $M_n(C)$ is Artinian).

(iv) Any condition necessary for embeddibility must also be satisfied by $M_t(R)$ for all $t$, since $M_t(R)$ is embeddible in $M_{tn}(C)$.

Suppose a ring $R$ satisfies conditions (i)–(iv) of remark 6.3.6. Can $R$ be embedded into a simple Artinian ring? One method to answer this question negatively would be to dredge out another sufficient condition not dependent on (i)–(iv); however, we could then throw this condition into the list and ask the same question again.

A different approach initiated by Lewin is a counting argument. If $\alpha = \max(|F|, \aleph_0)$ then $|F[\lambda_1, \ldots, \lambda_m]| \le \alpha$ for any $m$; since each ideal of $F[\lambda_1, \ldots, \lambda_m]$ is f.g. there are at most $\alpha$ ideals and thus at most $\alpha$ homomorphic images $C$ of $F[\lambda_1, \ldots, \lambda_m]$. Summing over all $m$ in $\mathbb{N}$ there are at most $\alpha$ commutative affine $F$-algebras $C$, up to isomorphism. For any of these $C$ we have $|M_n(C)| \le \alpha$, so there are at most $\alpha$ finite subsets and thus there are $\le \alpha$ affine subalgebras of $M_n(C)$. This proves that the number of embeddible affine $F$-algebras (up to isomorphism) is $\le \alpha$.

**Proposition 6.3.7:** (Lewin) Suppose $\{A_i: i \in I\}$ are ideals of an $F$-affine algebra $R$ with $|I| > \alpha = \max(|F|, \aleph_0)$. Then there are $|I|$ nonisomorphic $R/A_i$ which are not embeddible in matrices.

***Proof:*** We claim that at most $\alpha$ of the $R/A_i$ are isomorphic to each other; this would yield $|I|$ isomorphism classes of $n$ nonisomorphic $R/A_i$, only $\alpha$ of which are embeddible, so throwing these away would yield the proposition. The claim follows at once from

***Remark 6.3.8:*** There are at most $\alpha$ algebra endomorphisms of $R = F\{r_1,\ldots,r_t\}$, since any algebra endomorphism is defined by sending $r_i \mapsto \sum_{\text{finite}} c_{i_p} p(r_1,\ldots,r_m)$ where $p$ ranges over the monomials of $F\{X_1,\ldots,X_m\}$; we are choosing a finite set of $c_{i_p}$ from $F$, and so there are at most $\alpha$ choices.

By constructing $R$ satisfying proposition 6.3.7 one easily concludes there are an uncountable number of nonembeddible affine algebras, cf., exercise 5. Furthermore, applying proposition 6.3.7 to the affine algebra $R = \mathbb{Q}\{X_1, X_2\}/ \langle X_1^2, X_2 X_1 X_2 \rangle$, Irving-Small [86] proved that there are an uncountable number of nonembeddible homomorphic images of $R$ each satisfying all the conditions of remark 6.3.6; cf., exercise 6 for details. Lest the reader think that the subject of embeddibility is finally closed, let us consider

***Question 6.3.9:*** Is every Noetherian affine PI-algebra embeddible in matrices? (The counting technique fails in this case, by exercise 8.)

There is a positive result for f.g. modules over the center, cf., exercise 3. Nevertheless, the embeddibility theory of affine algebras into matrices has proved a disappointment except as a source of interesting counterexamples.

## Lewin's Embedding Procedure

The "correct" kind of embedding procedure turns out to be a general method of Lewin [74] which deals with the generalized triangular matrix rings of example 1.1.9. In the following discussion we assume $A, B$ are given ideals of a $C$-algebra $R$, and $M$ is an $R/A - R/B$ bimodule (and, in particular, a $C$-module). We want to embed $R$ into $T = \begin{pmatrix} R/A & M \\ 0 & R/B \end{pmatrix}$. First we consider arbitrary homomorphisms $\varphi: R \to T$ and then ask when $\ker \varphi = 0$. Our treatment stems from Bergman-Dicks [75].

***Remark 6.3.10:*** Define $\varphi: R \to T$ by $\varphi r = \begin{pmatrix} r + A & \delta r \\ 0 & r + B \end{pmatrix}$ where $\delta: R \to M$ is some $C$-module map. $\varphi$ is a $C$-algebra homomorphism iff $\delta$ is a derivation, i.e., $\delta(r_1 r_2) = (\delta r_1)\bar{r}_2 + \bar{r}_1 \delta r_2$ where $\bar{r}_1 = r_1 + A$ and $\bar{r}_2 = r_2 + B$; this is seen by computing $\varphi(r_1 r_2)$.

Now assume $\bar{M}$ is a homomorphic image of $M$ as bimodule. Then we have a derivation $R \to \bar{M}$ composing $\delta$ with $M \to \bar{M}$, and thus we have a homomorphism $\bar{\varphi}: R \to \bar{T} = \begin{pmatrix} R/A & \bar{M} \\ 0 & R/B \end{pmatrix}$. Clearly $\ker \bar{\varphi} \supseteq \ker \varphi$, so $\ker \varphi$ is expected to be smallest possible when $\delta: R \to M$ is "initial," in the sense that any derivation from $R$ to an $R/A - R/B$ bimodule factors through $\delta$. Lewin calls $M$ the *universal bimodule* and $\delta$ the *universal derivation*. The universal derivation can actually be described in terms of tensor products. Let $K \subseteq R \otimes_C R$ be the kernel of the canonical map $R \otimes_C R \to R$ defined by multiplication in $R$, and

$$M = R/A \otimes_R K \otimes_R R/B. \tag{1}$$

Let $\delta: R \to M$ be given by

$$\delta r = 1 \otimes (r \otimes 1 - 1 \otimes r) \otimes 1. \tag{1'}$$

It is enough for us to have this $M$ in our hands, although one can readily check $\delta$ is indeed the universal derivation, cf., exercise 10. The neatest way to describe $\ker \varphi$ is by means of $\mathrm{Tor}_1^R$ of homology (denoted below as $\mathrm{Tor}_1$); in fact, $\mathrm{Tor}_1$ is a natural and useful tool in connection with tensor products, and we need hardly more than its definition.

**Theorem 6.3.11:** *Let $K$, $M$, $\delta$ be as in* (1), *and let* $\varphi: R \to \begin{pmatrix} R/A & M \\ 0 & R/B \end{pmatrix}$ *be as in remark 6.3.10. Then $AB \subseteq \ker \varphi \subseteq A \cap B$, and $\ker \varphi = AB$ if $C$ is a field.*

**Proof:** Obviously $AB \subseteq \ker \varphi \subseteq A \cap B$, so it remains to prove $\ker \varphi = AB$ when $C$ is a field. Note $\ker \delta \supseteq AB$ since $\delta(ab) = a\delta b + (\delta a)b = 0 + 0 = 0$, so $\delta$ induces a derivation $\delta: R/AB \to M$. We aim to show $((A \cap B)/AB) \cap \ker \bar{\delta} = 0$ for then $\ker \varphi = A \cap B \cap \ker \delta = AB$. First we use the canonical epic $R \otimes_C R \to R$ to build the exact sequence of bimodules

$$0 \to K \xrightarrow{i} R \otimes_C R \to R \to 0. \tag{2}$$

$R$ is projective as right $R$-module so (2) is split in $\mathscr{M}od$-$R$; tensoring by $R/B$ yields the exact sequence

$$0 \to K \otimes_R R/B \xrightarrow{i'} R \otimes_C R/B \to R/B \to 0 \tag{3}$$

and this can also be viewed naturally in $R$-$\mathscr{M}od$. Note $i'$ is given by

$(r_1 \otimes r_2) \otimes \bar{1} \mapsto r_1 \otimes \bar{r}_2$ where $^-$ denotes the canonical image in $R/B$. Applying $\operatorname{Tor}(R/A, \_)$ yields

$$\operatorname{Tor}_1(R/A, R \otimes_C R/B) \to \operatorname{Tor}_1(R/A, R/B) \xrightarrow{h} M \to R/A \otimes_C R/B$$

$$\to R/A \otimes_R R/B \to 0 \tag{4}$$

Note $\operatorname{Tor}_1(R/A, R/B) \approx (A \cap B)/AB$ by example 5.2.38(ii). On the other hand, applying $\operatorname{Tor}(\_, R \otimes_C R/B)$ to the exact sequence

$$0 \to A \to R \to R/A \to 0$$

yields (cf., proposition 5.2.36)

$$0 \to \operatorname{Tor}_1(R/A, R \otimes_C R/B) \to A \otimes_C R/B \to R \otimes_C R/B$$

$$\to R/A \otimes_C R/B \to 0. \tag{5}$$

Since $C$ is a field we see $A \otimes_C R/B \to R \otimes_C R/B$ is monic so $\operatorname{Tor}_1(R/A, R \otimes_C R/B) = 0$. Making these two substitutions in (4) yields

$$0 \to (A \cap B)/AB \xrightarrow{h} M \to R/A \otimes_C R/B \to R/A \otimes_R R/B \to 0. \tag{6}$$

In fact, $h: (A \cap B)/AB \to M$ is given by $h(x + AB) = 1 \otimes (x \otimes 1 - 1 \otimes x) \otimes 1$. This can be seen computationally or by the following easy homological argument:

Defining $f: B \to K \otimes R/B$ by $fb = (b \otimes 1 - 1 \otimes b) \otimes \bar{1}$ and $g: R \to R \otimes_C R/B$ by $gr = r \otimes \bar{1}$, we have the commutative diagram

$$
\begin{array}{ccccccccc}
0 & \longrightarrow & B & \longrightarrow & R & \longrightarrow & R/B & \longrightarrow & 0 \\
& & \downarrow f & & \downarrow g & & \| & & \\
0 & \longrightarrow & K \otimes_R R/B & \xrightarrow{i'} & R \otimes_C R/B & \longrightarrow & R/B & \longrightarrow & 0
\end{array}
\tag{7}
$$

where the bottom row is (3); the diagram is commutative since

$$(i'f)b = i'((b \otimes 1 - 1 \otimes b) \otimes \bar{1}) = b \otimes \bar{1} - 1 \otimes \bar{b} = b \otimes \bar{1} - 0 = gb$$

for all $b$ in $B$. Applying $\operatorname{Tor}(R/A, \_)$ as above we have

$$
\begin{array}{ccc}
\operatorname{Tor}_1(R/A, R/B) = (A \cap B)/AB & \longrightarrow & R/A \otimes B = B/AB \\
\downarrow = & & \downarrow 1_A \otimes f \\
\operatorname{Tor}_1(R/A, R/B) & \xrightarrow{h} & M = R/A \otimes K \otimes R/B
\end{array}
$$

where the bottom row is from (4). But $1_A \otimes f$ is the restriction of $\bar{\delta}$ to $B/AB$

by (1'), so we have factored the monic $h$ as

$$(A \cap B)/AB \hookrightarrow B/AB \overset{\bar{\delta}}{\to} M.$$

This proves the restriction of $\bar{\delta}$ to $(A \cap B)/AB$ is monic, as desired. Q.E.D.

**Digression 6.3.12:** The only place in theorem 6.3.11 we used that $C$ was a field was in showing $\text{Tor}_1(R/A, R \otimes_C R/B) = 0$. Thus one can generalize theorem 6.3.11, cf., exercise 11. As usual, Bergman-Dick's paper delves even deeper into the result and its ramifications.

**Corollary 6.3.13:** *Suppose $R$ is an algebra over a field and $A, B \lhd R$ with $AB = 0$. Then there is an $R - R$ bimodule $M$ and an injection*

$$R \hookrightarrow \begin{pmatrix} R/A & M \\ 0 & R/B \end{pmatrix}.$$

**Proof:** Immediate from theorem 6.3.11. Q.E.D.

In order to obtain full mileage from this result we shall now take $R$ to be the free algebra $F\{X\}$.

**Theorem 6.3.14:** *Suppose $A, B \lhd F\{X\}$, $F$ a field, such that $F\{X\}/A$ and $F\{X\}/B$ are embeddible, respectively, in $m \times m$ and $n \times n$ matrices. Then $F\{X\}/AB$ is embeddible in $(m + n) \times (m + n)$ matrices.*

**Proof:** Taking $R = F\{X\} = F\{X_k : k \in I\}$ we have an embedding

$$R/AB \to \begin{pmatrix} R/A & M \\ 0 & R/B \end{pmatrix}$$

as in corollary 6.3.13. We want to embed this further into matrices. Let $\hat{M}$ be the free $R/A - R/B$ bimodule (i.e., free $(R/A) \otimes (R/B)^{\text{op}}$-module) with base which we denote as $\{\hat{X}_k : k \in I\}$. There is a derivation $\hat{\delta} : R \to \hat{M}$ given by $\delta X = \hat{X}$; thus we could replace $M$ by $\hat{M}$. But now $\hat{M}$ is flat as $R/A - R/B$ bimodule. View $R/A \subseteq M_m(C_1)$ and $R/B \subseteq M_n(C_2)$. Letting $C = C_1 \otimes_F C_2$ we can embed $R/A$ in $M_m(C)$ and $R/B$ in $M_n(C)$, so letting $N = M_m(C) \otimes_{R/A} \hat{M} \otimes_{R/B} M_n(C)$ we see

$$\begin{pmatrix} R/A & \hat{M} \\ 0 & R/B \end{pmatrix} \hookrightarrow \begin{pmatrix} M_m(C) & N \\ 0 & M_n(C) \end{pmatrix}. \tag{8}$$

Let $\Lambda$ be a set of commuting indeterminates $\{\lambda_{ij}^{(k)}: m < i \le m+n, n < j \le m+n, k \in I\}$. Writing $M_{mn}(C[\Lambda])$ for the set of $m \times n$ matrices with entries in the polynomial ring $C[\Lambda]$, viewed naturally as $M_m(C) - M_n(C)$ bimodule we let $N'$ be the submodule generated by the "generic" elements $X'_k = \{(\lambda_{ij}^{(k)}): m < i \le m+n, n < j \le m+n\}$. $N'$ is readily seen to be free as $M_m(C) - M_n(C)$ bimodule since any dependence relation translates to linear equations in the $\lambda_{ij}^{(k)}$, which can only have a trivial solution. On the other hand, there is a homomorphism

$$R/AB \to \begin{pmatrix} M_m(C) & N' \\ 0 & M_n(C) \end{pmatrix} \tag{9}$$

given by $X_k + AB \to \begin{pmatrix} X_k + A & X'_k \\ 0 & X_k + B \end{pmatrix}$, and whose kernel is 0 in view of (8) since $N'$ is free. The right hand side of (9) is visibly in $M_{m+n}(C[\Lambda])$, so $R/AB$ is embeddible in $M_{m+n}(C[\Lambda])$.    Q.E.D.

**Corollary 6.3.15:**   *If $F\{X\}/A$ is embeddible in $n \times n$ matrices then $F\{X\}/A^t$ is embeddible in $nt \times nt$ matrices for all $t$ in $\mathbb{N}$.*

The reader should note this result cannot be generalized to arbitrary affine PI-rings; any nonembeddible affine PI-algebra would be a counterexample, since we shall prove the nilpotence of its nilradical. Although involving neither affine nor PI explicitly, the corollary has the following immediate application:

**Theorem 6.3.16:**   *Suppose $R$ is a PI algebra over a field $F$. If $\mathrm{Nil}(R)$ is nilpotent then $R$ satisfies the identities of $k \times k$ matrices for some $k$. Explicitly if $R/\mathrm{Nil}(R)$ has PI degree $n$ and $\mathrm{Nil}(R)^t = 0$ then $R$ is a homomorphic image of an algebra embeddible into $nt \times nt$ matrices, and $R$ satisfies all the identities of $M_{nt}(F[\Lambda])$.*

**Proof:**   Let $A = \{\text{identities of } R/\mathrm{Nil}(R)\}$. Take a surjection $\varphi: F\{X\} \to R$, and let $\psi: F\{X\} \to R/\mathrm{Nil}(R)$ be the composite of $\varphi$ with $R \to R/\mathrm{Nil}(R)$. Then $\varphi A \subseteq \mathrm{Nil}(R)$ so $\varphi A^t = 0$, yielding a surjection $F\{X\}/A^t \to R$. On the other hand, by proposition 6.1.42 $F\{X\}/A$ is the algebra of generic $n \times n$ matrices and thus embeddible in $M_n(F[\Lambda])$, so $F\{X\}/A^t$ is embeddible in $nt \times nt$ matrices and thus satisfies all its identities; therefore, $R$ also satifies these identities.    Q.E.D.

**Corollary 6.3.17:**   *If we can prove the radical of an affine PI-algebra $R$ over a field is nilpotent then $R$ satisfies the identities of $n \times n$ matrices for suitable $n$.*

## Shirshov's Theorem

Having established an interesting converse to our question of nilpotence of the radical, we return to our casual pursuit of the original question (or better, its solution). There is a surprising connection to Kurosch's problem, which has a positive solution for affine PI-algebras; in fact, the solutions by Kaplansky and Levitzki were the first in the long list of major theorems in PI theory. There is a fast structural solution, cf., exercise 12, but we shall use a solution of Shirshov [57] based on combinatoric techniques on words which only requires the algebraicity of a finite (albeit huge) number of elements; this becomes very important in the applications. We start with definition 6.2.28 and its total order on words; the idea is to use remark 6.2.29 to pass from words to monomials in the generators $r_1, \ldots, r_t$.

**Definition 6.3.18:** A word $w$ is $m$-decomposable with $m$-decomposition $w = w_1 \cdots w_m$ for suitable words $w_1, \ldots, w_m$ if $w_{\pi 1} \cdots w_{\pi m} < w$ for each $\pi \neq 1$ in $\mathrm{Sym}(m)$.

The $m$-decomposition will enable us to utilize a polynomial identity $f(X_1, \ldots, X_m)$ to provide an inductive procedure based on the order of words.

**Remark 6.3.19:** If $w = w_1 w'$ and $w'$ has an $(m - 1)$-decomposition $w_2 \cdots w_m$ where $w_j < w_1$ for each $j \geq 2$ then $w_1 w_2 \cdots w_m$ is an $m$-decomposition of $w$.

Now for our main tool.

**Lemma 6.3.20:** There is a function $\beta \colon \mathbb{N}^{(3)} \to \mathbb{N}$ such that for every $k \geq \beta(t, u, m)$, every word of length $k$ and height $t$ has either (i) an $m$-decomposable subword or (ii) a subword of the form $w_0^u$ for suitable $w_0$.

**Proof:** Induction on $m$ and $t$. Every word is 1-decomposable so take $\beta(t, u, 1) = 1$; on the other hand, we can take $\beta(1, u, m) = u$ since $1^k$ is the only word of height 1 and length $k$. Fixing $u, m$ and $t$, suppose we have found $\beta(t', u, m - 1)$ for all $t'$ in $\mathbb{N}$, and also $\beta(t - 1, u, m)$. Putting $t' = ut^{\beta(t-1, u, d)}$ define

$$\beta(t, u, m) = (\beta(t - 1, u, m) + u - 1)\beta(t', u, m - 1)$$

We shall show any word $w$ of height $t$ not satisfying (i) or (ii) has length $\leq \beta(t, u, m)$.

Write $w$ as $1^{\varphi(1)} v_1 1^{\varphi(2)} v_2 \cdots 1^{\varphi(j)} v_j$ where each $\varphi(i) \in \mathbb{N}$, $\varphi(i) > 0$ for $i \geq 2$, and 1 does not occur in any $v_i$. For example, if $w = 54112314$ then $\varphi(1) = 0$,

$v_1 = 54, \varphi(2) = 2, v_2 = 23, \varphi(3) = 1$, and $v_3 = 4$. We are done unless $w$ satisfies the following two properties:

length$(v_i) < \beta(t - 1, u, m)$ for each $i$

$\varphi(i) < u$ for each $i$

In general, there are $t^{\beta(t-1,u,m)}$ ways of building a word $v$ of length $\leq$ $\beta(t - 1, u, m)$ in the "letters" $2, \ldots, t$, since in each position we could put $2, \ldots, t$ or nothing at all. Thus there are $t'$ possibilities for a word $\tilde{v} = 1 \cdots 1v$ where 1 occurs at most $u$ times at the beginning. We order these $t'$ words lexicographically and name them $\tilde{v}_1, \tilde{v}_2, \ldots, \tilde{v}_{t'}$. Thus each $1^{\varphi(i)}v_i = \tilde{v}_{\psi(i)}$ where $1 \leq \psi(i) \leq t'$.

Consider the word $\hat{w} = \psi(2) \cdots \psi(j)$, and suppose $j \geq \beta(t', u, m - 1) + 1$. By the induction on $m$, $\hat{w}$ satisfies either (i) or (ii), as a word of length $j - 1$ and height $t'$. But (ii) says $\hat{w}$ has a subword $\hat{w}_0^u$ which, when translated back to $w$ by means of $\tilde{v}$, gives a subword $w_0^u$ of $w$, so we would be done. Thus we may assume $\hat{w}$ has an $(m - 1)$-decomposable subword $\psi(i) \cdots \psi(i')$ for suitable $2 \leq i < i' \leq j$, i.e., the word $1^{\varphi(i)}v_i \cdots 1^{\varphi(i')}v_{i'}$ has an $(m - 1)$-decomposition $w_2 \cdots w_m$ where $w_2, \ldots, w_m$ each starts with 1. But $v_{i-1}$ does not start with 1 so is greater than each of $w_2, \ldots, w_m$; hence by remark 6.3.19 $v_{i-1}w_2 \cdots w_m$ is an $(m - 1)$-decomposition of $v_{i-1}1^{\varphi(i)}v_i \cdots 1^{\varphi(i')}v_{i'}$.

Thus we are done unless $j \leq \beta(t', u, m - 1)$. But each $v_i$ was assumed to have length $< \beta(t - 1, u, m)$, implying

$$k < (\beta(t - 1, u, m) + u - 1)\beta(t', u, m - 1) = \beta(t, u, m). \qquad \text{Q.E.D.}$$

We can improve this further.

**Lemma 6.3.21:** *If $w$ has length $\geq m$ and $w$ is not of the form $\tilde{v}^k$ for any $k > 1$ then $w^{2m}$ has a terminal $m$-decomposable subword.*

**Proof:** Writing $w = i_1 \cdots i_m w'$ for suitable $w'$ define $v_j = i_{j+1} \cdots i_m w' i_1 \cdots i_j$ for $1 \leq j < m$, and $v_m = w$. We claim $v_1, \ldots, v_m$ are distinct. Indeed, if $v_j = v_{j'}$ for $j < j'$ then putting $u = j' - j$ we have $w = v_u = i_{u+1} \cdots i_m w' i_1 \cdots i_u$ and so $i_{u+1} \cdots i_m$ is an initial subword of $w$; cancelling $i_{u+1} \cdots i_m$ repeatedly and applying lemma 6.2.31 we get $i_{u+1} \cdots i_m = \tilde{w}^t$ for some $\tilde{w}$ of length dividing $m$, implying $w = \tilde{w}^k$ for some $k$, contrary to hypothesis.

Thus the claim is established, so for some $\pi$ in Sym$(m)$ we have $v_{\pi 1} > v_{\pi 2} > \cdots > v_{\pi m}$. Each $v_j$ is obviously a subword of $w^2$. Writing $w^2 = w'_j v_j w''_j$

we have the $m$-decomposition

$$(v_{\pi 1} w''_{\pi 1} w'_{\pi 2})(v_{\pi 2} w''_{\pi 2} w'_{\pi 3}) \cdots (v_{\pi m} w''_{\pi m})$$

of a terminal subword of $w$, as desired.     Q.E.D.

**Proposition 6.3.22:**  (*Shirshov*) *There is a function* $\beta \colon \mathbb{N}^{(3)} \to \mathbb{N}$ *such that for all* $u \geq 2m$ *and all* $k \geq \beta(t, u, m)$, *every word of length* $k$ *and height* $m$ *has either*

(i) *a subword of the form* $v^u$ *with length* $(v) < m$ *or*
(ii) *a* $d$-*decomposable subword.*

**Proof:**   Assume (ii) fails. By lemma 6.3.20 $w$ contains a subword $w_0^u$; taking $w_0$ of minimal possible length $n$ we have $n < m$ for, otherwise, $w_0 = v^k$ by lemma 6.3.21, so $w_0^u = v^{ku}$ with length$(v) < n$, contradiction. But this is (i).
                                                                            Q.E.D.

**Theorem 6.3.23:**  (*Levitzki-Kaplansky-Shirshov*) *Suppose* $R = C\{r_1, \ldots, r_t\}$ *satisfies a multilinear polynomial identity* $f(X_1, \ldots, X_m)$, *and each monomial in the* $r_i$ *of length* $< m$ *is integral over* $C$. *Then* $R$ *is f.g. as* $C$-*module, spanned by the monomials in the* $x_i$ *of length* $\leq \beta(t, u, m)$ *where* $u = \max\{2m,\ \text{degree of integrality of each monomial in the}\ r_i\ \text{of length} \leq m\}$.

**Proof:**   Let $B = \{\text{monomials in the } r_i \text{ of length} \leq \beta(t, u, m)\}$. We shall prove the theorem by showing in the notation of remark 6.2.29 that each $\bar{w}_k$ is spanned by $B$. We shall prove this by induction on $k$. We are done by hypothesis unless length $(w_k) > \beta(t, u, m)$, since otherwise $\bar{w}_k \in B$. Thus (i) or (ii) of proposition 6.3.22 holds.

If (i) holds then $w_k$ has a subword $v^u$. By induction hypothesis $\bar{v}^u = \sum_{i=0}^{u-1} c_i \bar{v}^i$. Writing $w_k = w' v^u w''$ we have $\bar{w}_k = \sum_{i=0}^{u-1} c_i \bar{w}' \bar{v}^i \bar{w}''$; but each $w' v^i w'' < w_k$ since the length is smaller, so by induction each $\bar{w}' \bar{v}^i \bar{w}''$ is spanned by $B$, so $\bar{w}_k$ is spanned by $B$.

If (ii) holds then $w_k = w' v w''$ where $v = v_1 \cdots v_m$ is an $m$-decomposition of $v$. We may write

$$f = X_1 \cdots X_m + \sum_{\pi \neq 1} c_\pi X_{\pi 1} \cdots X_{\pi m}.$$

Then $\bar{v} = -\sum_{\pi \neq 1} c_\pi \bar{v}_{\pi 1} \cdots \bar{v}_{\pi m}$ so $\bar{w} = \sum_{\pi \neq 1} c_\pi \bar{w}' \bar{v}_{\pi 1} \cdots \bar{v}_{\pi m} \bar{w}''$. But each $w' v_{\pi 1} \cdots v_{\pi m} w'' < w' v_1 \cdots v_m w'' = w_k$, so by induction each $\bar{w}' \bar{v}_{\pi 1} \cdots \bar{v}_{\pi m} \bar{w}''$ is spanned by $B$; hence $\bar{w}_k$ is spanned by $B$.     Q.E.D.

Schelter noted that the same combinatorial argument actually gives a stronger result, cf., exercise 14 and Rowen [80B, theorem 4.2.8]. Shirshov proved the following stronger result than theorem 6.3.23.

**Theorem 6.3.24:** (*Shirshov*) *Suppose* $R = C\{r_1, \ldots, r_t\}$ *satisfies a multilinear polynomial identity* $f(X_1, \ldots, X_m)$. *Let* $w_1, \ldots, w_{m'}$ *denote the words of length* $\leq m$, *and let* $\bar{w}_k$ *denote the image of a word $w$ in $R$, as in remark 6.2.29. There is a number* $k = k(m, t)$ *such that for any word $w$ of length* $\geq \beta(t, m, m)$ *we have*

$$\bar{w} \in \sum C \bar{v}_{i_1} \cdots \bar{v}_{i_k}$$

*where each* $i_j \in \{1, \ldots, m'\}$, $v_i$ *denotes a power of* $w_i$, *and each word* $v_{i_1} \cdots v_{i_k}$ *is a rearrangement of the letters of $w$.*

***Proof:*** Let $k = (m')^2 m \beta(t, m, m)$. Proof is by induction on the order of $w$ (cf., example 1.2.18). In particular, we may assume $w$ does not have an $m$-decomposable subword since then we could use the identity $f$ to express $\bar{w}$ in terms of lower order monomials, just as in the proof of theorem 6.3.23 case (ii). Write

$$\bar{w} = \bar{v}_{i_1} \cdots \bar{v}_{i_d}$$

where each $v_i$ denotes a power of $w_i$ and each $i_j \in \{1, \ldots, m'\}$, such that $d$ is minimal possible, temporarily we call $d$ the *breadth* of $w$. We are done unless $w$ has breadth $> k$, so, in particular, length$(w) > k$.

*Claim.* Any subword $v$ of $w$ of breadth $> \beta(t, m, m)$ contains a subword $\tilde{w}^m \tilde{w}'$ where $\tilde{w}, \tilde{w}'$ have length $\leq m$ and each of $\tilde{w}, \tilde{w}'$ is *not* an initial subword of the other.

*Proof of claim.* A fortiori $v$ cannot have an $m$-decomposable subword, so by proposition 6.3.22 $v$ contains a subword $\tilde{w}^m$ with length$(\tilde{w}) \leq m$. We choose $\tilde{w}$ such that the initial subword $v' \tilde{w}^m$ has length$(v')$ minimal possible. Write $v = v' \tilde{w}^u v''$ for $u \geq m$, such that $\tilde{w}$ is not an initial subword of $v''$. We are done unless $v''$ is an initial subword of $\tilde{w}$. Write $\tilde{w} = v'' a$ for a suitable subword $a$, and note length$(v'') <$ length$(\tilde{w}) \leq m$. Now

$$v = v'v''(av'')^u.$$

Since length$(av'') = $ length$(\tilde{w}) \leq m$ we see breadth$(v'v'') \geq$ breadth$(v) - 1 \geq \beta(t, m, m)$, so again proposition 6.3.22 shows $v'v''$ has a subword $\hat{w}^m$ for suitable $\hat{w}$ of length $\leq m$. But then $v'v''$ has an initial subword $a'\hat{w}^m$ where length$(a') \leq$ length$(v') +$ length$(v'') - m <$ length$(v')$; since $a'\hat{w}^m$ is also an initial subword of $v$ we have contradicted the minimality of length$(v')$.

Having established the claim we have a subword $\tilde{w}^m\tilde{w}'$ where neither $\tilde{w}$ nor $\tilde{w}'$ is initial in the other. But the number of choices for the pair $(\tilde{w}, \tilde{w}')$ is $\leq (m')^2$; since $k = (m')^2 m\beta(t, m, m)$ we can apply the claim successively to the remaining subwords to assure that some $\tilde{w}^m\tilde{w}'$ recurs in $m$ distinct places. Thus $w$ has a subword

$$\tilde{w}^m\tilde{w}'a_1\tilde{w}^m\tilde{w}'a_2\cdots\tilde{w}^m\tilde{w}'a_{m-1}\tilde{w}^m\tilde{w}'.$$

If $\tilde{w} > \tilde{w}'$ this has an $m$-decomposable subword

$$(\tilde{w}^m\tilde{w}'a_1\tilde{w})(\tilde{w}^{m-1}\tilde{w}'a_2\tilde{w}^2)\cdots(\tilde{w}\tilde{w}');$$

if $\tilde{w} < \tilde{w}'$ we use instead the $m$-decomposable subword

$$(\tilde{w}'a_1\tilde{w}^{m-1})(\tilde{w}\tilde{w}'a_2\tilde{w}^{m-2})\cdots(\tilde{w}^{m-2}\tilde{w}'a_{m-1}\tilde{w})(\tilde{w}^{m-1}\tilde{w}').$$

In either case $w$ has an $m$-decomposable subword, so as noted above, we are done.    Q.E.D.

This result is rather technical and was slow in being incorporated into the theory. However, it has some far-reaching consequences. The first one is due to Berele, using a different proof.

**Theorem 6.3.25:** *Any affine PI-algebra over a field has finite Gelfand-Kirillov dimension.*

***Proof:*** (Amitsur; Drensky) Take $S = \{\bar{w}_1, \ldots, \bar{w}_{m'}\}$ of theorem 6.3.24. In calculating $G_S(n)$ of definition 6.2.10 for $n$ large we see $V_n(S)$ is spanned by the $\bar{v}_{i_1}\cdots\bar{v}_{i_k}$. One could formally replace $i_j$ by $j$ and obtain $\bar{v}_1\cdots\bar{v}_k$, in this way creating a monomial which could be viewed in the commutative polynomial ring $C[\lambda_1, \ldots, \lambda_n]$. Since at most $k^k$ monomials $\bar{v}_{i_1}\cdots\bar{v}_{i_k}$ correspond to $\bar{v}_1\cdots\bar{v}_k$, we see by example 6.2.11(ii) that

$$G_S(n) \leq k^k\binom{k+n}{n} \leq k^k n^k = (kn)^k.$$

Hence the GK-dimension is bounded by

$$\lim_{n\to\infty}\log_n((kn)^k) = \lim k(1 + \log_n k) = k.    \text{Q.E.D.}$$

## The Theory of Prime PI-Rings

Many problems about an arbitrary PI-ring $R$ can be reduced to the prime case by means of the structure theory. Furthermore, prime PI-rings occupy a position of prominence because of

**Remark 6.3.26:** If $R$ satisfies the identities of $n \times n$ matrices then $R$ is a homomorphic image of the algebra of generic matrices, which is prime (cf., proposition 6.1.41 and 6.1.42).

Thus the theory of prime PI-rings should bear heavily on the PI-theory in general. So far we have several powerful tools at our disposal.

**Summary 6.3.27:** Suppose $R$ is a prime PI-ring with center $Z$, and $S = Z - \{0\}$. Then

(i) $R$ has some PI-degree $n$ (theorem 6.1.28)

(ii) $S^{-1}R$ is central simple over $S^{-1}Z$ of degree $n$ (theorem 6.1.30)

(iii) There exists $z$ in $Z$ for which $R_{<z>}$ is a free $Z_{<z>}$-module with a base of $n^2$ elements, and $R_{<z>}$ is Azumaya over $Z_{<z>}$. (In fact, any $z \neq 0$ in $g_n(R)$ will do; corollary 6.1.36.)

Despite these results, $Z(R)$ can still be rather poorly behaved, as we shall see when studying Noetherian PI-rings, and even localizing at a single element can be too crude a tool to work with certain delicate questions. There is one more construction which can be used in almost every situation.

**Construction 6.3.28:** *(Schelter's Trace Ring) Let a C-algebra R be prime of PI-degree n and $Q = S^{-1}R$ where $S = Z(R) - \{0\}$. We shall expand R to a certain subring $T(R)$ of Q satisfying the following properties:*

(i) *$T(R) = RC'$ for a suitable C-subalgebra $C'$ of $Z(Q)$; in particular, $T(R)$ is prime.*

(ii) *$Rg_n(R) \neq 0$ is a common ideal of R and $T(R)$. In fact $C'g_n(R) = g_n(R)^+$.*

(iii) *$T(R)$ is integral over $C'$.*

(iv) *If R is C-affine then $C'$ is C-affine and $T(R)$ is f.g. over $C'$; in particular, when C is Noetherian $C'$ and $T(R)$ are Noetherian.*

To construct $T(R)$ we first embed $R$ in $M_n(F)$ by splitting $Q$, and taking a base $b_1, \ldots, b_k$ of $M_n(F)$ where $k = n^2$ we view $R \subseteq M_k(F)$ by the regular representation. Writing $\hat{r}$ for the $k \times k$ matrix corresponding to $r$ we have the Hamilton-Cayley polynomial $\det(\lambda \cdot 1 - \hat{r}) = \sum (-1)^i \alpha_i \lambda^{k-i}$ for suitable $\alpha_i$, $0 \leq i \leq k$, where $\alpha_0 = 1$. We shall call these $\alpha_i$ the *characteristic coefficients* of $r$. Note $\alpha_i$ are independent of the choice of the base $b_1, \ldots, b_k$. Furthermore, choosing the base $b_1, \ldots, b_k$ in $Q$ it is easy to see that each $\alpha_i \in Q$.

Let $C'$ be the $C$-subalgebra of $Q$ generated by the characteristic coefficients of "enough" elements of $R$, and let $T(R) = RC' \subseteq Q$. "Enough" means "all" unless $R$ is affine, i.e., $R = C\{r_1, \ldots, r_t\}$ in which case the monomials in $r_1, \ldots, r_t$ of length $\leq \beta(t, 2n^2, d)$ are "enough" (where $d$ is the degree of a PI of $R$). We now have (i) and shall prove (iii), (iv), and (ii).

***Proof of (iii):*** Any element $a$ of $T(R)$ has the form $\sum_{i=1}^{u} a_i c_i$ where $a_i \in R$ and $c_i \in C'$. Hence $a \in C'\{a_1, \ldots, a_u\}$. But all the monomials in the $a_i$ are in $R$ and thus integral over $C'$ since $C'$ contains enough characteristic coefficients. Thus $a$ is integral over $C'$, by theorem 6.3.23.

***Proof of (iv):*** As in proof of (ii), since now each $a_i$ can be taken to be a monomial in $r_1, \ldots, r_t$, and $C'$ contains enough characteristic coefficients by construction; the last assertion follows from elementary properties of Noetherian rings.

***Proof of (ii):*** It suffices to show for any characteristic coefficient $\alpha$ of $r$ and any $b_1, \ldots, b_m$ in $R$ that $\alpha g_n(b_1, \ldots, b_m) \in g_n(R)^+$ where $m$ is the degree of $g_n$. This is obvious unless $z = g_n(b_1, \ldots, b_m) \neq 0$, in which case $b_1, \ldots, b_k$ is a base of $M_n(F)$, seen by applying theorem 6.1.33 to $z^{-1}b_1, \ldots, b_k$. Thus we could use $b_1, \ldots, b_t$ as our base in forming $\hat{r}$. By proposition 6.1.19(iii) we have

$$\alpha g_n(b_1, \ldots, b_n) \in g_n(R)^+, \text{ proving (ii).} \qquad \text{Q.E.D.}$$

This construction was called the *characteristic closure* in Rowen [80B], but the misnomer *trace ring* has stuck since for $\mathbb{Q}$-algebras all the characteristic coefficients can be obtained from the traces. Technically our ring $T(R)$ might be called the $n^2$-trace ring since we took the $n^2$-Hamilton-Cayley polynomial; Amitsur-Small [80] analogously built the $n$-trace ring using the $n$-Hamilton-Cayley polynomial. One advantage of the $n^2$-trace ring is that proposition 6.1.19 is itself a useful tool; for example, it could be used in the proof of exercise 15, due to Amitsur, which implies that if $T(R)$ is defined functorially by adjoining all characteristic coefficients of elements of $R$ then $T(T(R)) = T(R)$.

When $R$ is a $\mathbb{Q}$-algebra the $n$-trace ring and $n^2$-trace ring are the same. A detailed proof of this fact would rely on the reduced trace (cf., definition 2.3.32 ff), but essentially boils down to the observation that if $a$ is an $n \times n$ matrix then the trace of the $n^2 \times n^2$ matrix $\begin{pmatrix} a \cdot \\ & \ddots \\ & & a \end{pmatrix}$ is $n \operatorname{tr}(a)$, and $n^{-1} \in \mathbb{Q}$.

***Remark 6.3.29:*** $T(R)$ is isomorphic to an $R$-submodule of $R$. (Indeed, take $0 \neq z \in g_n(R)$ and note $T(R)z \subseteq Rg_n(R) \subseteq R$, so right multiplication by $z$ embeds $T(R)$ into $R$.)

In order to utilize $T(R)$ we must be able to pass prime ideals from $R$ to $T(R)$ and back, so the following observations are crucial:

**Theorem 6.3.30:** *Suppose a PI-ring $R$ is integral over $C \subseteq Z(R)$. Then* GU, LO *and* INC *hold from $C$ to $R$.*

***Proof:*** Any $C$-affine subalgebra of $R$ is f.g. by theorem 6.3.23, so satisfies GU and LO by theorem 2.12.48 and INC by theorem 3.4.13. Hence $C \subseteq R$ satifies GU and LO by proposition 2.12.49. To prove INC suppose $P_1 \subset P_2$ in Spec($R$) lie over the same prime ideal of $C$. Passing to $R/P_1$ and $C/(P_1 \cap C)$ we may assume $P_1 = 0$ and $P_2 \cap C = 0$.

In particular, $Z(R)$ is a domain. Take an affine subalgebra $Z'$ of $Z(R)$ containing some $0 \neq z \in P_2 \cap Z(R)$. Then $Z'z \cap C = 0$ so $Z'z$ is contained in a prime ideal of $Z'$ lying over 0, by remark 2.12.42. This contradicts INC for $Z'$ over $C$.     Q.E.D.

**Theorem 6.3.31:** *Suppose $R$ is prime of PI-degree $n$.*

(i) LO($P$) *and* GU$(\_\_, P)$ *hold from $R$ to $T(R)$, for all $P$ in* Spec$_n(R)$ *and for all $P$ minimal over a prime in* Spec$_n(R)$ *(i.e., for which there is $P_0$ in* Spec$_n(R)$ *satisfying height $(P/P_0) = 1$ in $R/P_0$.)*

(ii) *The strong version of* INC($P$) *holds from $R$ to $T(R)$ for all $P$ in* Spec$_n(R)$*, in the sense that if $P'_1, P'_2 \in$ Spec $T(R)$ with $P'_1 \cap R \subseteq P'_2 \cap R = P$ then $P'_1 \subseteq P'_2$.*

(iii) LO, GU, *and* INC *hold from $C'$ to $T(R)$.*

***Proof:***

(i) Let $A = Rg_n(R)$ and apply proposition 2.12.45 to 6.3.28(i),(ii).

(ii) By proposition 2.12.46, as in (i).

(iii) Theorem 6.3.30 is applicable by 6.3.28(iii).     Q.E.D.

Thus many questions about Spec($R$) can be translated to $T(R)$, which can be a suitable arena for computation even when $R$ is not affine. One of the key computations, due to Braun [85], is a Hamilton-Cayley type expression for $T(R)$ which passes to certain homomorphic images; Braun uses this in a short proof of

**Theorem (Bergman-Small):**   *Suppose $R$ is prime of PI-degree $n$, and $P \in$ Spec($R$). If $m = PI\text{-}degree(R/P)$ then $n - m$ is a sum of PI-degrees (with possible repetitions) of simple homomorphic images of $R$.*

In particular, if $R$ is local and $P$ is the unique maximal ideal then $m \mid n$. The reduction to $T(R)$ is easy, and is given in exercise 17. We shall use $T(R)$ here to polish off much of the theory of affine PI-algebras and then discuss its use for Noetherian rings. Let us start by reproving the weak Nullstellensatz.

**Remark 6.3.32:**   If $r \in R$ is regular and is algebraic over a central subring $Z$ then $Z \cap Rr \neq 0$. (Indeed, take $\sum_{i=0}^{t} z_i r^i = 0$ with $z_t \neq 0$, $t$ minimal such, and let $r' = \sum_{i=1}^{t} z_i r^{i-1} \neq 0$ by minimality of $t$. Then $0 \neq -r'r = z_0 \in Z \cap Rr$, as desired.)

**Application 6.3.33:**   Easy proof of the weak Nullstellensatz for affine PI-algebras over Jacobson $C$: It is enough to assume $R$ is prime and to prove Jac($R$) = 0. Using construction 6.3.28(iv) we see $C'$ is Jacobson, by the commutative Nullstellensatz, so Jac($C'$) = 0. By proposition 2.5.33 we have $0 = C' \cap \text{Jac}(T(R))$; but $Z(R)$ is integral over $C'$ and so $Z(R) \cap \text{Jac}(T(R)) = 0$ by remark 6.3.32 (since all of its elements are central and thus regular). Hence Jac($T(R)$) = 0 by theorem 6.1.28.

Now $g_n(R)\text{Jac}(R) \subseteq \text{Jac}(R)$ so is quasi-invertible and is an ideal of $T(R)$ by 6.3.28(ii), implying $g_n(R)\text{Jac}(R) = 0$. Since $R$ is prime we conclude Jac($R$) = 0.       Q.E.D.

**Theorem 6.3.34:**   *(Schelter) If $R$ is affine PI over a Noetherian ring $C$ then $R$ satisfies* ACC(*prime ideals*)

**Proof:**   Take a chain $P_1 \subset P_2 \subset \cdots$, and take $m$ for which PI-deg($R/P_m$) is minimal. Replacing $R$ by $R/P_m$ and replacing the original chain by the chain $P_{m+1}/P_m \subset P_{m+2}/P_m \subset \cdots$ we may assume $R$ is prime of some PI-degree $n$, and each $P_i \in \text{Spec}_n(R)$. But by theorem 6.3.31(i) $T(R)$ has an infinite ascending chain of prime ideals, contrary to $T(R)$ being Noetherian by 6.3.28(iv).       Q.E.D.

Of course when $C$ is a field the previous result follows from theorems 6.2.25 and 6.3.25. We shall obtain a stronger result in corollary 6.3.36', to follow.

## Nilpotence of the Jacobson Radical

**Theorem 6.3.35:** (*Razmyslov*) *Suppose R is affine over a Noetherian Jacobson domain C. (In particular, one could take C to be a field.) If R satisfies all the identities of $M_n(C[\Lambda])$ where $\Lambda$ is an infinite set of commuting indeterminates then* Jac(R) *is nilpotent.*

**Proof:** Since $R$ is a homomorphic image of an affine algebra of generic matrices, which is prime by remark 6.3.4, this follows at once from the weak Nullstellensatz and the following result:

**Theorem 6.3.36:** (*Schelter*) *Suppose R is prime PI and affine over a Noetherian ring C. If $A \triangleleft R$ then* Nil(R/A) *is a finite intersection of prime ideals of R/A and is nilpotent in R/A.*

**Proof:** In view of theorem 6.3.34 we may apply Noetherian induction to the prime images of $R$ and thereby assume the theorem is true in $R/P$ for every $0 \neq P \in \mathrm{Spec}(R)$. Pass to $T(R)$, which is Noetherian by 6.3.28(iv). Let $n = \mathrm{PI} \deg(R)$. Then $g_n(R)A \triangleleft T(R)$ by 6.3.28(ii), so there are a finite number of prime ideals $Q_1, \ldots, Q_t$ of $T(R)$ minimal over $g_n(R)A$. Since $(\bigcap Q_i)/g_n(R)A = \mathrm{Nil}(T(R)/g_n(R)A)$ is nilpotent, we have $(\bigcap Q_i)^k \subseteq g_n(R)A$ for some k.

Let $P_i = Q_i \cap R \neq 0$. Then $R/(P_i + A)$ is an image of $R/P_i$ so there are a finite set of prime ideals $P_{i1}, \ldots, P_{im}$ each containing $P_i + A$, for which $(\bigcap P_{ij})^{k'} \subseteq P_i + A$; technically $m$ and $k'$ depend on $P_i$ but there are only a finite number of the $P_i$, so we can take $m$ and $k'$ to be the same for all $i$. Now

$$\left(\left(\bigcap_{j=1}^m P_{1j}\right)^{k'} \cdots \left(\left(\bigcap_{j=1}^m P_{tj}\right)^{k'}\right)\right)^k \subseteq ((P_1 + A) \cdots (P_t + A))^k \subseteq g_n(R)A \subseteq A,$$

implying at once $\mathrm{Nil}(R/A) = \left(\bigcap_{i=1}^t \bigcap_{j=1}^m P_{ij}\right)/A$ and is nilpotent.      Q.E.D.

**Corollary 6.3.36':** *Any semiprime PI-algebra affine over a Noetherian ring C has a finite number of minimal prime ideals and satisfies* ACC(*semiprime ideals*).

**Proof:** This is shown directly in Rowen [80B, §4.5], but our proof here is based on a reduction to theorem 6.3.34. By exercise 3.5.34, it is enough to prove 0 is a finite intersection of prime ideals (for by passing to homomorphic images we may then assume any semiprime ideal is a finite intersection of prime ideals.) Since $C$ has only finitely many minimal primes $P_1, \ldots, P_t$ we may pass to $C/P_i$ and assume $C$ is prime. But by corollary 6.1.47 $R$ is now a homomorphic image of the generic matrix algebra, which is a prime ring, so Schelter's theorem applies.    Q.E.D.

We now aim for the nilpotence of the radical in general.

**Lemma 6.3.37:**  *(Latyshev) Suppose $R = C\{r_1,\ldots,r_t\}$ is PI. If $N \lhd R$ is nil and $f.g.$ as two-sided ideal, then $N$ is nilpotent.*

**Proof:**   Writing $N = \sum_{i=1}^{k} Ra_i R$, we incorporate $\{a_1,\ldots,a_k\}$ into the generating set $\{r_1,\ldots,r_t\}$ for $R$. We shall appeal to theorem 6.3.24, with its notation. Let $n$ be the largest index of nilpotency of $\{$monomials in $r_1,\ldots,r_t$ of length $\leq m$ which contains some $a_i\}$. (Certainly these are all elements of $N$ and are thus nilpotent.) Let $n' = \max\{\beta(t,m,m), kmn + 1\}$ where $k = k(m,t)$ was obtained in theorem 6.3.24.

We shall prove $N^{n'} = 0$ by showing any monomial $\bar{w}$ in $N^{n'}$ is in fact 0. By theorem 6.3.24 we may assume $\bar{w} = \bar{v}_{i_1} \cdots \bar{v}_{i_k}$, and $n'$ elements of $N$ appear in $\bar{v}_{i_1} \cdots \bar{v}_{i_k}$. But $n' > kmn$ so one of these $\bar{v}_i$ contains $> mn$ elements of $N$. On the other hand, theorem 6.3.24 also says $v_i = w_i^u$ for some $u$ and $u \geq n$ since length $w_i \leq m$. Hence $\bar{v} = \bar{w}_i^u = 0$.    Q.E.D.

**Remark 6.3.37':**   Latyshev's result already shows that $\mathrm{Nil}(R)$ is the sum of nilpotent ideals of $R$ since $\mathrm{Nil}(R) = \sum_{a \in \mathrm{Nil}(R)} RaR$.

**Remark 6.3.38:**   Suppose $R$ is affine over Noetherian $C$ and f.g. as $Z(R)$-module. Then $R$ is Noetherian so $\mathrm{Nil}(R)$ is nilpotent. (Indeed $Z(R)$ is affine over $C$ by Artin-Tate (proposition 6.2.5) so is Noetherian by the Hilbert basis theorem; thus $R$ is Noetherian.fl

**Theorem 6.3.39:**   *(Braun [84]) If $C$ is a commutative Noetherian ring then $\mathrm{Nil}(R)$ is nilpotent for any PI-algebra $R = C\{r_1,\ldots,r_t\}$.*

**Proof:**   Let $C'$ be the localization of the polynomial ring $C[\lambda]$ at the monoid generated by all $\lambda^i - \lambda^j$ for $i > j$. Since $\mathrm{Nil}(R)$ is locally nilpotent we may replace $R$ by $R \otimes C'$ and $C$ by $C'$, and thereby assume each homomorphic image of $C$ has an infinite number of distinct elements (the images of the powers of $\lambda$).

Let us collect some reductions which will be needed throughout. Write $N = \mathrm{Nil}(R)$ and $Z = Z(R)$. We reformulate the theorem as:

For any $A \lhd R$ and any semiprime ideal $B$ of $R$ such that $B/A = \mathrm{Nil}(R/A)$ we have $B^k \subseteq A$ for suitable $k$ (depending on $A, B$).

Suppose this assertion is false. Take a counterexample $R$ for which $R/N$ has minimal PI-degree; by corollary 6.3.36' there is a semiprime ideal $B$ maximal with respect to $B/A = \text{Nil}(R/A)$ and $B^k \nsubseteq A$ for all $k$. Passing to $R/A$ we may assume $A = 0$ and $B = N$.

**Reductions:** (1) (Amitsur) By the last paragraph, if $B \supset N$ is a semiprime ideal and $A \lhd R$ with $B/A$ nil then $B^k \subseteq A$ for some $k$.

(2) If $A$ is a non-nil ideal then taking $B = A + N$ in (1) we see $(A + N)^k \subseteq A$ for suitable $k$.

(3) If $z + N$ is central in $R/N$ then we may assume $z \in Z$. (Indeed, each $[z, r_i] \in N$ so $N_1 = \sum_{i,j=1}^{t} R[r_i, z]R$ is nilpotent by Latyshev's lemma; we obtain the desired reduction when we replace $R$ by $R/N_1$.)

(4) If $z \in Z - N$ then it is enough to prove the theorem for $R/\text{Ann}\, z$. (Indeed, suppose $N^{k(1)} \subseteq \text{Ann}\, z$. By reduction 2 we have $N^{k(2)} \subseteq Rz$ for some $k(2)$, so $N^{k(1)+k(2)} \subseteq (\text{Ann}\, z)Rz = 0$.)

*Case I.* (Folklore). $R/N$ is commutative. Applying reduction 3 to each of $r_1, \ldots, r_t$ in turn, we may assume each $r_i \in Z(R)$, i.e., $R$ is commutative, so we are done by remark 6.3.38.

*Case II.* $R/N$ is not prime, i.e., there are ideals $A_1, A_2 \supset N$ with $A_1 A_2 \subseteq N$. Since $R/N$ is semiprime we can take $z_j$ for $j = 1, 2$ such that $0 \neq z_j + N \in (A_j/N) \cap Z(R/N)$. By reduction 3 we may assume each $z_j \in Z$ so by reduction 2 there are numbers $k(1), k(2)$ such that $N^{k(j)} \subseteq Rz_j$. But $z_1 z_2 \in A_1 A_2 \subseteq N$ so $0 = (z_1 z_2)^m = z_1^m z_2^m$ for some $m$, and we are done by reductions 1 and 4.

*Case III.* Thus we may assume $R/N$ is prime of PI degree $> 1$. Write $^-$ for the canonical image in $\bar{R} = R/N$. Then $\bar{C} \subseteq Z(\bar{R})$ is an integral domain, so $N \cap C = \text{Nil}(C)$ is nilpotent. Replacing $R$ by $R/\text{Nil}(C)R$ we may assume $C$ is an integral domain, which as noted above is infinite. Our object is to prove "enough monomials" in $r_1, \ldots, r_t$ are integral over $Z$ to permit us to apply Shirshov's theorem (6.3.23), for then we would be done by remark 6.3.38.

In the above paragraph it is important to note "enough" is finite. Thus we want a procedure of modifying $R$ such that a given monomial in $r_1, \ldots, r_t$ is integral over $Z$. Such a procedure for prime affine algebras was in fact the key in constructing $T(R)$, but now $R$ need not be prime. However $\bar{R}$ is prime, and we want to find some way to adjoin characteristic coefficients in this slightly more general setting. The hurdle is that we do not know that the map from $R$ to its localization need be 1:1, and the kernel may contain a non-nilpotent part of $N$. The remainder of the proof will be to use several high-powered techniques to "create" enough characteristic coefficients to show {evaluations of the identities of $M_d(C)$ in $R$} is nilpotent, so that we can then appeal to theorem 6.3.36 by means of the algebra of generic matrices.

Our first task is to carry central localization as far as we can. Take $x_1, \ldots, x_m$ in $R$ such that $z = g_n(x_1, \ldots, x_m) \notin N$. By lemma 6.1.32 we have

$$\bar{z}\bar{r} = \sum_{u=1}^{n^2} (-1)^{u+1} g_n(\bar{r}, \bar{x}_1, \ldots, \bar{x}_{u-1}, \bar{x}_{u+1}, \ldots, \bar{x}_m)\bar{x}_u \qquad (10)$$

for any $r$ in $R$. Writing $z_{iju} = (-1)^{u+1} g_n(x_i x_j, x_1, \ldots, x_{u-1}, x_{u+1}, \ldots, x_m)$ we see $\bar{z}_{iju} \in Z(\bar{R})$ so by reduction 3 we may assume $z \in Z$ and each $z_{iju} \in Z$. Furthermore, $z x_i x_j - \sum z_{iju} x_u \in N$ so modding out by $\sum R(z x_i x_j - \sum z_{iju} x_u)R$ we may assume by Latyshev's lemma that

$$z x_i x_j = \sum_u z_{iju} x_u$$

for each $1 \le i, j \le n^2$. Likewise, we may assume $z r_i = \sum_u z_{iu} x_u$ for $z_{iu}$ in $Z$.

Let $B = \{x_1, \ldots, x_n\}$ and let $\tilde{g}$ be as in the proof of lemma 6.1.32. By Latyshev's lemma we also assume $\tilde{g}(B) = 0$ and $[g_n(B), x_u] = 0$ for each $u$. But $R[z^{-1}] = \sum_{u=1}^{n^2} Z[z^{-1}] x_u$, so $\tilde{g}$ is an identity of $R[z^{-1}]$ and $g_n$ is $R[z^{-1}]$-central, yielding

$$1^{-1} z r = 1^{-1} \sum_{u=1}^{n^2} (-1)^{u+1} g_n(r, x_1, \ldots, x_{u-1}, x_{u+1}, \ldots, x_m) x_u. \qquad (11)$$

We aim for a similar equation in $R$ itself. To this end write $g_n = \sum h_{1i} X_1 h_{2i}$ where the indeterminate $X_1$ appears in neither $h_{1i}$ nor $h_{2i}$. Letting $a_{iu} = h_{1i}(x_1, \ldots, x_{u-1}, x_{u+1}, \ldots, x_m)$ and $b_{iu} = h_{2i}(x_1, \ldots, x_{u-1}, x_{u+1}, \ldots, x_m)$ we see $g_n(r, x_1, \ldots, x_{u-1}, x_{u+1}, \ldots, x_m) = \sum_i a_{iu} r b_{iu} = \psi(\sum_i a_{iu} \otimes b_{iu})r$ where $\psi : R \otimes R^{\text{op}} \to \text{End } R$ is the canonical map. Thus (11) becomes

$$1^{-1} z \otimes 1 = \sum_{u,i} 1^{-1} (-1)^{u+1} a_{iu} \otimes 1^{-1} b_{iu}$$

in $R[z^{-1}]^e \approx R^e[z^{-1}]$ (cf., remark 5.3.28), so there is some $d$ such that

$$z^{d+1} \otimes 1 = \sum_{i,u} (-1)^{u+1} z^d a_{iu} \otimes b_{iu}$$

in $R^e$, and thus, for all $r$ in $R$,

$$z^{d+1} r = \sum_{i,u} (-1)^{u+1} z^d a_{iu} r b_{iu} = z^d \sum_{u=1}^{n^2} (-1)^{u+1} g_n(r, x_1, \ldots, x_{u-1}, x_{u+1}, \ldots, x_m) x_u.$$

$$(12)$$

Increasing $d$ if necessary, we may also assume by corollary 5.3.30 for all $r$ in $R$ that

$$z^d g_n(r, x_1, \ldots, x_{u-1}, x_{u+1}, \ldots, x_m) \in Z. \tag{13}$$

Just to clean up notation, we may assume $d = 0$ by passing to $R/\text{Ann}\, z^d$ (by reduction 4, noting by (13) that $[g_n(r, x_1, \ldots, x_{u-1}, x_{u+1}, \ldots, x_m), R] \subseteq \text{Ann}\, z^d$.) Thus (12) becomes $zr = \sum_{u=1}^{n^2} (-1)^{u+1} g_n(r, x_1, \ldots, x_{u-1}, x_{u+1}, \ldots, x_m) x_u$, and (13) becomes $g_n(r, x_1, \ldots, x_{u-1}, x_{u+1}, \ldots, x_m) \in Z$.

*Reduction 5.* Any *finite* set of homogeneous identities of $\bar{R}$ may be assumed to be identities for $R$. (An identity is *homogeneous* if the total degree of each monomial is the same.) Indeed suppose $f(X_1, \ldots, X_v)$ has degree $j$ in each monomial. Let $B = \{X_i : 1 \le i \le n^2\}$; passing to $R/Rf(B^{(v)})R$ via Latyshev's lemma we may assume $f(B^{(v)}) = 0$. But for any $r'_1, \ldots, r'_v$ in $R$ we have each $zr'_i \in ZB$, so

$$z^j f(r'_1, \ldots, r'_v) = f(zr'_1, \ldots, zr'_v) \in Zf(B^{(v)}) = 0,$$

i.e., $f(R) \in \text{Ann}\, z^j$, which by reduction 4 we may assume is 0.

One particular identity of $\bar{R}$ actually contains enough information to encode the Hamilton-Cayley theorem. Let $n' = n^2$ and, taking the Capelli polynomial $C_{2n'}$ define

$$f(X_1, \ldots, X_{2n'+1}) = X_{n'+1} C_{2n'}(X_1, \ldots, X_{n'}, X_{n'+2}, \ldots, X_{2n'+1})$$

$$= \sum_{\pi \in \text{Sym}(n')} (\text{sg}\, \pi) X_{n'+1} X_{\pi 1} X_{n'+2} \cdots X_{\pi n'} X_{2n'+1}.$$

Consider $n'$-tuples $\mathbf{j} = (j(1), \ldots, j(n'))$ where each $j(v) \in \{0, 1\}$ for $1 \le v \le n'$, and write $|\mathbf{j}|$ for the number of $j(v)$ which are 1, i.e., $|\mathbf{j}| = j(1) + \cdots + j(n')$. Picking $y_j$ in $R$ and letting $y = f(y_1, \ldots, y_{2n'+1})$ we have

$$c_u \bar{y} = \sum_{|\mathbf{j}| = u} \overline{f(r^{j(1)} y_1, \ldots, r^{j(n')} y_{n'}, y_{n'+1}, \ldots, y_{2n'+1})}$$

for any $r$ in $R$ by proposition 6.1.19, where $(-1)^u c_u$ denotes the appropriate "characteristic coefficient" of the matrix $T$ corresponding to left multiplication by $\bar{r}$ in $\bar{R}$. But $\sum (-1)^u c_u T^{n'-u} = 0$ so we have

$$0 = \sum (-1)^u c_u \overline{r^{n'-u} y} = \sum_u \overline{(-1)^u r^{n'-u} \sum_{|\mathbf{j}| = u} f(r^{j(1)} y_1, \ldots, r^{j(n')} y_{n'}, y_{n'+1} \cdots)}. \tag{14}$$

Thus $\bar{R}$ satisfies the identity

$$\tilde{f} = \sum_u \sum_{|\mathbf{j}| = u} (-1)^u X_0^{n'-u} f(X_0^{j(1)} X_1, \ldots, X_0^{j(n')} X_{n'}, X_{n'+1}, \ldots, X_{2n'+1})$$

which is homogeneous of degree $3n' + 1$.

*Reduction 6.* We may assume $\tilde{f}$ is an identity of $R$, by reduction 5.

The last ingredient in the proof is based on an ideal of Razmyslov. Since we have trouble working in $R$, we switch to the free algebra $C\{X, Y\} = C\{X_1, \ldots, X_t, Y_1, \ldots, Y_{n'}\}$ in noncommuting indeterminates $X_i, Y_i$ over $C$, and let $W = C\{X_1, \ldots, X_t\} \subset C\{X, Y\}$. We want to "create" characteristic coefficients of the elements of $W$. Namely, let $\mu = \{\mu_{u,h}: 1 \leq u \leq n', h \in W\}$ be a set of indeterminates which commute with $W$ but not with each other. (The reason the $\mu_{u,h}$ initially are not to commute with each other is to ensure that (16) below is well-defined.) Let $A$ be the ideal in $W\{\mu\}$ generated by $\{\sum_u (-1)^u \mu_{u,h} h^{n'-u}: h \in W\} \cup \{[\mu_{u,h}, \mu_{u',h'}]: 1 \leq u, u' \leq n'$ and $h, h' \in W\}$. Note the $\mu$ are commutative modulo $A$. Thus $(W + A)/A$ is integral of bounded degree $\leq n'$ over the integral domain which is the image of $C\{\mu\}$, and thus by example 6.1.9 satisfies some PI of some degree $m'$, which we multilinearize. Let $\tilde{\mu} = \{\mu_{u,h}: 1 \leq u \leq n', h$ is a word in the $X_i$ of degree $\leq m'\}$, and let $T = W\{\tilde{\mu}\}/(A \cap W\{\tilde{\mu}\})$. Then $T$ satisfies the same PI, so by Shirshov's theorem is f.g. over the Noetherian domain $C\{\tilde{\mu}\}/A \cap C\{\tilde{\mu}\}$. Hence $T$ is Noetherian by remark 6.3.38, implying $\text{Nil}(T)^k = 0$ for some $k$.

$T/\text{Nil}(T)$ has some PI-degree $d$. Let $\mathscr{L}$ be the set of identities of $M_d(C)$. $\mathscr{L}$ are all identities of $T/\text{Nil}(T)$ by remark 6.1.47'; letting $\mathscr{L}(W)$ be the set of evaluations of $\mathscr{L}$ in $W$ we see $\mathscr{L}(W)^k \subseteq A$. To utilize this result we let $L$ be the $C$-submodule of $C\{X, Y\}$ spanned by all $\{f(Y_1, \ldots, Y_{n'}, h_1, \ldots, h_{n'+1}): h_j \in W\}$. $L$ is naturally a $W$-module since

$$X_u f(Y_1, \ldots, Y_{n'}, h_1, \ldots, h_{n'+1}) = f(Y_1, \ldots, Y_{n'}, X_u h_1, \ldots, h_{n'+1}). \qquad (15)$$

On the other hand, each $\mu_{u,h}$ acts on $L$ via

$$\mu_{u,h} f(Y_1, \ldots, Y_{n'}, h_1, \ldots, h_{n'+1}) = \sum_{|\mathbf{j}| = u} f(h^{j(1)} Y_1, \ldots, h^{j(n')} Y_{n'}, h_1, \ldots, h_{n'+1}). \qquad (16)$$

To see the right-hand side is indeed in $L$ we must observe

$$\sum_{|\mathbf{j}| = u} f(h^{j(1)} Y_1, \ldots, h^{j(n')} Y_{n'}, h_1, \ldots, h_{n'+1})$$

$$= \sum_{\pi \in \text{Sym}(n')} \sum_{|\mathbf{j}| = u} (\text{sg } \pi) h_1 h^{j(\pi 1)} Y_{\pi 1} \cdots h_{n'} h^{j(\pi n')} Y_{\pi n'} h_{n'+1}$$

$$= \sum_{\pi \in \text{Sym}(n')} \sum_{|\mathbf{j}| = u} (\text{sg } \pi) h_1 h^{j(1)} Y_{\pi 1} \cdots h_{n'} h^{j(n')} Y_{\pi n'} h_{n'+1}$$

$$= \sum_{|\mathbf{j}| = u} f(Y_1, \ldots, Y_{n'}, h_1 h^{j(1)}, \ldots, h_{n'} h^{j(n')}, h_{n'+1}) \in L.$$

(The middle equality is obtained by noting that $(j(\pi 1), \ldots, j(\pi n'))$ and $(j(1), \ldots, j(n'))$ each have precisely $u$ entries which are 1.)

By iterating (16) we extend the action to monomials in $\mu$ and thus get a $W\{\mu\}$-module action on $L$. Also note $(-1)^u \mu_{u,h} f(Y_1, \ldots, Y_{n'}, h_1, \ldots, h_{n'+1})$ is the coefficient of $\lambda^{n'-u}$ in $f((\lambda - h)Y_1, \ldots, (\lambda - h)Y_{n'}, h_1, \ldots, h_{n'+1})$.

There are a finite number of polynomials $[\mu_{u,h}, \mu_{u',h'}]f$ for $\mu_{u,h}, \mu_{u'h'} \in \tilde{\mu}$; by reduction 5 we may assume they all are identities of $R$. Now let $I$ be the ideal of $C\{X, Y\}$ generated by all evaluations of $\tilde{f}$ and all evaluations of these $[\mu_{u,h}, \mu_{u',h'}]f$. Then (15) shows $I \supseteq AL$; we saw above $AL \supseteq \mathscr{L}(W)^k L$. Hence $I \supseteq \mathscr{L}(W)^k L$. Specializing down to $R$ then yields $0 \supseteq \mathscr{L}(R)^k f(R) \supseteq I(R)^k g_n(R)$, since $g_n$ is clearly a formal consequence of the Capelli polynomial, cf., exercise 6.1.4. But $g_n(R) \not\subseteq N$, so by reduction 4 we may assume $\mathscr{L}(R)^k = 0$. On the other hand, $R/\mathscr{L}(R)$ satisfies the identities of $d \times d$ matrices, so $N/\mathscr{L}(R)$ is nilpotent by theorem 6.3.36. Hence $N$ is nilpotent, as desired.     Q.E.D.

Crucial ingredients in this proof were Shirshov's theorem, the Artin-Procesi theorem, the properties of the Capelli polynomial which lead to the construction of the central polynomial, the Razmyslov-Schelter theorem (6.3.36), the construction of $\tilde{f}$ to incorporate a Hamilton-Cayley theorem into a polynomial identity, and the generic matrix algebras. It is safe to say this proof requires all the machinery of PI-theory developed here.

## Dimension Theory of Affine PI-Algebras

Although we have so far obtained a considerable amount of information about affine PI-rings, we have not yet built up a dimension theory. One obvious candidate is the GK dimension, which we have seen exists, by theorem 6.3.25. Thus we can draw on the theory of GK dimension, especially in view of the following result:

**Proposition 6.3.40:**    *If $R$ is a prime PI-algebra (not necessarily affine) over a field $F$ then* GK dim $R = \operatorname{tr} \deg Z(R)/F$ *(the transcendence degree).*

*Proof:*    By proposition 6.2.23 we can pass to the ring of central quotients of $R$, and assume $R$ is central simple over the field $Z = Z(R)$. But then GK dim $R =$ GK dim $Z = \operatorname{tr} \deg Z/F$ by remark 6.2.20.     Q.E.D.

When $R$ is affine as well this is an integer by theorem 6.3.25; however, GK dim $R$ need not be an integer for $R$ non-semiprime affine PI. Let us bring in now the classical Krull dimension (the maximal length of a chain of primes), which *is* always an integer.

**Theorem 6.3.41:** *If $R$ is prime affine PI over a field $F$ then* GK dim $R =$ cl. K-dim $R = \operatorname{tr deg} Z(R)/F$. *Furthermore, a chain of prime ideals of maximal length can be obtained from* $\operatorname{Spec}_n(R)$ *where $n = $ PI deg $R$.*

**Proof:** Induction on $d = $ GK dim $R$. We just saw $d = \operatorname{tr deg} Z(R)/F$, and $d \geq$ cl. K-dim $R$ by theorem 6.2.25. It remains to show that there is a chain $0 \subset P_1 \subset \cdots \subset P_d$ in $\operatorname{Spec}_n(R)$. If $d = 0$ then there is nothing to prove, so assume $d \geq 1$. We claim $g_n(R)$ contains an element $z_1$ transcendental over $F$. Indeed, this is clear unless each $r$ in $g_n(R)$ is algebraic over $F$; but then $F \cap \langle r \rangle \neq 0$ by remark 6.3.32 so $r$ invertible and thus $1 \in g_n(R)$, so $g_n(R)^+ = Z(R)$ and the claim follows at once.

Expand $z_1$ to a transcendence base $z_1, \ldots, z_d$ of $Z(R)$ over $F$, and let $S = \{ p(z_1, \ldots, z_{d-1}) : 0 \neq p \in F[\lambda_1, \ldots, \lambda_{d-1}] \}$. Then $S^{-1}R$ is affine over the ring $S^{-1}F$. If $S^{-1}R$ were simple then by corollary 6.3.2 $S^{-1}R$ would be algebraic over $S^{-1}F$, contrary to $d \geq 1$. Thus $S^{-1}R$ has a nonzero maximal ideal $S^{-1}P$, where $P \in \operatorname{Spec}(R)$ with $P \cap S = \varnothing$. In particular, $P \in \operatorname{Spec}_n(R)$. Let $\bar{R} = R/P$, which contains the algebraically independent elements $\bar{z}_1, \ldots, \bar{z}_{d-1}$. Thus

$$d - 1 \leq \frac{\operatorname{tr deg} Z(\bar{R})}{F} = \operatorname{GK dim} \bar{R} \leq d - 1$$

by induction and prosition 6.2.24. Hence equality holds, and by induction there is a chain $0 \subset \bar{P}_1 \subset \cdots \subset \bar{P}_{d-1}$ in $\operatorname{Spec}_n(\bar{R})$. This lifts to a chain $0 \subset P \subset P_1 \subset \cdots \subset P_{d-1}$ in $\operatorname{Spec}_n(R)$, of length $d$ as desired.          Q.E.D.

**Corollary 6.3.42:** *($R$ prime affine PI)* cl K-dim $R = $ cl K-dim $T(R)$ *where $T(R)$ is as in 6.3.28(iv).*

**Proof:** Chains from $\operatorname{Spec}_n$ lift, by theorem 6.3.31(i), and these determine cl K-dim, by the theorem.          Q.E.D.

Let us apply these results to obtain a lovely theorem of Schelter.

**Theorem 6.3.43:** *(Schelter) Catenarity of affine PI-algebras. If $R$ is affine PI over a field $F$, and $P \in \operatorname{Spec}(R)$ then*

$$\operatorname{cl. K-dim} R = \operatorname{cl. K-dim}\left(\frac{R}{P}\right) + \text{height } P,$$

*i.e., $P$ can be put into a chain of primes of maximal length.*

***Proof:*** Note $\geq$ is clear, so we only need to prove $\leq$. Let $d = $ cl. K-dim $R$, and $h = $ height($P$). If $P$ contains a nonzero prime ideal $P_1$ then by induction on $d$ applied to $\bar{R} = R/P_1$ we would have

$$\text{cl. K-dim } \bar{R} - \text{height } \bar{P} = \text{cl. K-dim}\left(\frac{\bar{R}}{\bar{P}}\right) = \text{cl. K-dim}\frac{R}{P},$$

and by induction on $h$ we would have

$$\text{cl. K-dim } R = \text{cl. K-dim } \bar{R} + \text{height } P_1$$

$$= \left(\text{cl. K-dim}\frac{\bar{R}}{\bar{P}} + \text{height } \bar{P}\right) + \text{height } P_1$$

$$\leq \text{cl. K-dim}\frac{R}{P} + \text{height } P$$

as desired.

Thus we may assume $P$ is minimal prime, i.e., $h = 1$, and we want to prove $d \leq$ cl. K-dim $R/P + 1$. Now either $P$ is in $\text{Spec}_n(R)$ where $n = $ PI degree($R$), or $P$ is minimal over $g_n(R)$; thus there is $P'$ in $T(R)$ lying over $P$, by theorem 6.3.31(i). Take $P'$ maximal such. Then every nonzero ideal of $T(R)/P'$ intersects $R/P$ nontrivially.

We claim cl. K-dim $T(R)/P' = $ cl. K-dim $R/P$. Indeed letting $R' = T(R)/P'$ and $S = Z(R/P) - \{0\}$, we see any nonzero ideal of $R'$ intersects $S$ nontrivially, so $S^{-1}R'$ is simple and by construction is affine over the field $F' = S^{-1}Z(R/P)$. Hence $S^{-1}R'$ is algebraic over $F'$ by corollary 6.3.2, so $\text{tr deg } Z(R')/F = \text{tr deg } Z(R/P)/F$ and hence cl. K-dim $R' = $ cl. K-dim $R/P$ by theorem 6.3.41, as desired.

In view of the claim we need to show $d \leq$ cl. K-dim $T(R)/P' + 1$. Note $T(R)$ is integral over $C'$ which is affine over $F$. Let $A_2 = P' \cap C'$ and $A_1 = A_2 \cap g_n(T(R))^+$, nonzero ideals of $C'$ since $Z(T(R))$ is integral over $C'$. By the Noether-Bourbaki normalization theorem quoted at the beginning of §6.2, we have a transcendence base $c_1, \ldots, c_d$ of $C'$ over $F$ and $n(1)$, $n(2)$ in $\mathbb{N}$, such that taking $C_0 = F[c_1, \ldots, c_d]$ we have $A_i \cap C_0 = \sum_{u=1}^{n(i)} C_0 c_u$ for $i = 1, 2$, and $C'$ is integral over $C_0$.

$T(R)$ is integral over $C_0$, which is "normal" in the terminology of commutative algebra, so one can modify the standard commutative proof to obtain a version of "going down," cf., exercise 18. Hence taking $A_0 = \sum_{u=2}^{n(2)} C_0 c_u$ we have some $P'_0 \neq 0$ in $\text{Spec}(T(R))$ lying over $A_0$ with $P'_0 \subset P'$. Moreover,

$P_0' \in \mathrm{Spec}_n(T(R))$ since $c_1 \notin P_0'$, and using theorem 6.3.30 we have

$$\mathrm{cl.\,K\text{-}dim} \frac{T(R)}{P_0'} = \mathrm{cl.\,K\text{-}dim}\left(\frac{C_0}{A_0}\right) = d + 1 - n(2)$$

$$= \mathrm{cl.\,K\text{-}dim} \frac{C_0}{C_0 \cap P} + 1 = \mathrm{cl.\,K\text{-}dim} \frac{T(R)}{P'} + 1.$$

Thus it remains to show $d \leq \mathrm{cl.\,K\text{-}dim}\, T(R)/P_0'$. But $R \cap P_0'$ is a prime ideal of $\mathrm{Spec}_n(R)$ contained in $P$; applying theorem 6.3.31(ii) (strong INC) to $R \cap P_0'$ we see $R \cap P_0' \subset P$, so $R \cap P_0' = 0$ implying $P_0' = 0$. Hence cl. K-dim $T(R)/P_0' = $ cl. K-dim $T(R) = d$ by theorem 6.3.40, as desired.    Q.E.D.

Having seen that the affine PI-theory is so close to the commutative theory, one might wonder whether a meaningful algebraic geometry can be developed for affine PI-algebras. This has been accomplished to some extent in a major series of papers of Artin-Schelter.

***Digression 6.3.44:***  Gabriel dimension. Suppose $R$ is prime PI of Krull dimension $\alpha$. Then K-dim $T(R) \leq \alpha$ by remark 6.3.29, and for $R$ affine we thus have  K-dim $T(R) = $ K-dim $C' = $ cl. K-dim $C' = $ tr deg $C'/F = $ tr deg $Z(R)/F = $ GK dim $R$. Thus one might expect the noncommutative Krull dimension to be a useful tool for studying affine PI-algebras. Unfortunately it may fail to exist! Indeed we saw in exercise 3.5.29 that an affine PI-algebra satisfying the identities of $2 \times 2$ matrices may lack K-dim, and thus the affine algebra of generic $2 \times 2$ matrices is prime PI but lacks K-dim.

When $R$ is affine PI and has K-dim then applying exercise 3.5.33 to the proof of proposition 3.5.51 one has K-dim $R = $ cl. K-dim $R = $ cl. K-dim $T(R) = $ cl. K-dim $C' = $ GK dim $R$ as above, so we have the peculiar situation that K-dim may fail to exist, but equals GK dim when it does exist. Of course this motivates us to consider the Gabriel dimension, which we recall is very close to K-dim except that it exists more often. Gordon-Small [84] proved every affine PI-algebra over a field has Gabriel dimension, by reducing to the prime case and using the theory of left-bounded Goldie rings, cf., exercise 19.

## Digression: Noetherian PI-Rings

For the remainder of this section we survey some of the recent research in (non affine) Noetherian PI-rings. We start by considering $T(R)$ for $R$ prime.

**Proposition 6.3.45:**   *If $R$ is left Noetherian, prime PI then $T(R)$ is a finite central extension of $R$ and thus is left Noetherian.*

**Proof:**   $T(R)$ is Noetherian as left $R$-module by remark 6.3.29, so $T(R) = \sum R\alpha$ summed over a finite number of characteristic coefficients $\alpha$ of elements of $R$, i.e., $T(R)$ is a finite central extension of $R$.     Q.E.D.

In particular, LO, GU, and INC hold from $R$ to $T(R)$, so $R$ and $C'$ (cf., construction 6.3.28) are intimately related, and the theory of Noetherian PI-rings has been seen recently to be an arena where a suitable general Noetherian structure theory can be developed. Before discussing this further, let us glance at noncommutative Krull dimension for arbitrary PI-rings.

**Digression 6.3.46:**   Dimension   theory.   We   *assume   throughout   $R$   is Noetherian* PI.

(1)  Recall from proposition 3.5.51 that K-dim $R =$ cl. K-dim $R$.

(2)  (Brown-Warfield [84]; Goodearl-Small [84]) K-dim $R \leq$ gl. dim $R$. Inequality need *not* hold, cf., Resco-Small-Stafford [82, example, 1.8]. We discussed this result in §6.1, but would like to add that some useful partial results are readily available using $T(R)$, cf., exercises 20–23.

(3)  (Lorenz-Small [82]) If $R$ is an algebra over a field then there is $P$ in Spec$(R)$ for which GK-dim $R =$ GK-dim $R/P$.

The Lorenz-Small theorem is proved in Krause-Lenagan [85B, pp. 150–154] and has the following consequences, proved in exercises 24–27:

(i)  GK-dim $R$ is an integer if it is finite (this is obvious).

(ii)  GK-dim $R \otimes T =$ GK-dim $R +$ GK-dim $T$ for any $F$-algebra $T$.

(iii)  If $A_1, A_2 \lhd R$ with $A_1 A_2 = 0$ then GK-dim $R = \max\{$GK-dim $R/A_1$, GK-dim $R/A_2\}$.

(The following results concern GK module dimension.)

(iv)  GK-dim $M =$ GK-dim$(R/$Ann $M)$ for any f.g. $R$-module $M$.

(v)  GK-dim $M = \max\{$GK-dim $N$, GK-dim $M/N\}$ for any $N \leq M$ in $R$-ℱ𝒾𝓂𝓸𝒹.

## *Digression: Localization in PI-Rings*

A classical question due to Goldie is, "Which prime ideals of a PI-ring are (left) localizable," i.e., letting $\mathscr{C}(P) = \{$elements regular modulo $P\}$, when is $\mathscr{C}(P)$ a left or right denominator set? When $R$ has PI-degree $n$ and $P \in$

$\mathrm{Spec}_n(R)$ then we only need central localization, cf., theorem 6.1.38. Thus we are first led to check whether there are other cases in which central localization is "good enough":

The first case one might check is when $R$ is f.g. over $Z = Z(R)$. Note $R$ is a PI-ring. Clearly $R_P/P_P R_P$ is finite dimensional over the field $Z_P/P_P$ so is Artinian.

**Proposition 6.3.47:** *Suppose $R$ is f.g. over $Z = Z(R)$ and $P \in \mathrm{Spec}(Z)$. Then $N = \bigcap \{$ primes of $R$ containing $P\}$ is localizable, in fact, by central localization in the sense that any element in $\mathscr{C}(N)$ is invertible modulo $R_P R_P$ in $R_P$.*

**Proof:** $R_P/N_P$ is Artinian. We shall conclude by showing every element of $\mathscr{C}(N)_P$ is invertible. Suppose $a \in \mathscr{C}(N)$. First we claim $1^{-1}a \in \mathscr{C}(N_P)$. Indeed, if $(s^{-1}r)(1^{-1}a) \in N_P$ then $ra \in N_P$ so $s'ra \in N$ for suitable $s'$ in $Z - P$; hence $s'r \in N$ implying $s^{-1}r \in N_P$, proving the claim. Thus $1^{-1}a + N_P$ is regular and thus invertible in $R_P/N_P$. But $N_P/P_P R_P = \mathrm{Jac}(R_P/P_P R_P)$ so $1^{-1}a$ is invertible in $R_P/P_P R_P$.    Q.E.D.

Mueller [76, theorem 7] goes on to prove that the following assertions are equivalent for $R$ f.g. over $Z$ and $P' \in \mathrm{Spec}(R)$

(i) $P'$ is the unique prime ideal of $R$ lying over $P' \cap Z$.

(ii) $P'$ can be localized by central localization.

(iii) $P'$ is left and right localizable.

Mueller's proof involves building a convergent sequence of idempotents in the completion $\hat{R}$, which thus must converge to 0 or 1. The reader is invited to simplify the proof. Of course the question remains how to verify these conditions. In this regard some examples of Small-Stafford [81] are relevant, in which $R$ is actually f.g. over its *Noetherian* center:

A prime ideal can be left but not right localizable, cf., exercise 30.

Even when $R$ is prime, a prime ideal need not be localizable, cf., exercise 29.

Given these negative results, it is surprising that Braun-Warfield [86] found a localizability criterion for arbitrary prime Noetherian PI-rings.

**Braun-Warfield Theorem:** *Suppose $R$ is prime Noetherian PI, and $B$ is an arbitrary semiprime ideal of $R$. Write $B = P_1 \cap \cdots \cap P_t$ where $P_i$ are the $B$-minimal prime ideals of $R$. The following are equivalent:*

(i) *$B$ is left localizable.*

(ii) *$B$ is right localizable.*

(iii) *$B$ is left and right localizable (so that Mueller's results are applicable).*

(iv) *Braun-Warfield localizability criterion. Let* $\mathscr{S} = \{prime\ ideals\ obtained$
*by lifting the* $P_i$ *to* $T(R)\}$ *If* $Q \in \mathrm{Spec}(T(R))$ *and* $Q \cap Z(T(R)) = Q_j \cap Z(T(R))$
*for some* $Q_j$ *in* $\mathscr{S}$ *then* $Q \in \mathscr{S}$.

Of course, everything would follow from proving (i) $\Leftrightarrow$ (iv) since (iv) is
left-right symmetric. The hard direction (iv) $\Rightarrow$ (i) is shown by means of
$T(R)$ with an application of Jategaonkar's localization techniques to finite
centralizing extensions of Noetherian PI-rings. Braun-Warfield also showed
their localizability results hold for f.g. height 1 primes in prime affine PI-rings.
When $P$ is a maximal ideal, Braun has found a trace ring criterion, formally
weaker than (iv), which characterizes when $\bigcap_{i \in \mathbb{N}} P^i = 0$.

## §6.4 Relatively Free PI-Rings and $T$-Ideals

Until now we have largely studied the properties of rings satisfying *one*
polynomial identity. In this section we shift emphasis to the set of *all* identities
of a given ring (or class of rings). This is best done by studying the *relatively
free*, or *generic*, ring with respect to these identities. Relatively free PI-rings
often serve as "test cases" for many proposed theorems, either yielding short
proofs or providing counterexamples. This theory applies, in particular, to the
generic matrix rings of §6.1. We start by examining the structure of the set of
identities of an algebra.

*Definition 6.4.1:* $A \lhd R$ is a $T$-*ideal* if $\varphi A \subseteq A$ for every surjection $\varphi: R \to R$.

*Remark 6.4.2:* If $A \lhd R$ and $\varphi$ is a surjection then $\varphi A \lhd R$. Hence $\mathrm{Jac}(R)$,
$\mathrm{Nil}(R)$, $N_\alpha(R)$ for all $\alpha$, and Prime rad$(R)$ are $T$-ideals of $R$. (Indeed, $\mathrm{Jac}(R)$
is a $T$-ideal by proposition 2.5.6(i); the other assertions follow from the
properties of nil sets.)

*Remark 6.4.3:* If $A$ is a $T$-ideal of $C\{X\}$ where $X$ is an infinite set of
indeterminates then $\varphi A \subseteq A$ for every homomorphism $\varphi: C\{X\} \to C\{X\}$.
(Indeed, suppose $f(X_1, \ldots, X_n) \in A$. Define $\psi: C\{X\} \to C\{X\}$ by $\psi X_i = \varphi X_i$
for $1 \leq i \leq n$ and $\psi X_i = X_{i-n}$ for $i > n$. Then $\psi$ is a surjection so $\varphi f = \psi f \in A$.)

*Definition 6.4.4:* Let $\mathscr{Z}(R) = \{$identities of $R\}$. A $C$-algebra $U$ is *relatively
free* if $U$ is free in the class of $C$-algebras $R$ for which $\mathscr{Z}(R) \supseteq \mathscr{Z}(U)$. The
*relatively free* PI-*algebra of* $R$ is $C\{X\}/\mathscr{Z}(R)$ (terminology to be justified
presently).

Of course any free object also requires a designated set. Writing $U = C\{X\}/A$ and letting $^{-}$ denote the canonical image from $C\{X\}$ to $U$, we propose $\bar{X}$ for this set. Note that if $f \in C\{X\}$ and $\varphi: C\{X\} \to C\{X\}$ is a polynomial then $\varphi f(X_1, \ldots, X_m) = f(\varphi X_1, \ldots, \varphi X_m)$ since $f$ is formed from multiplications and additions.

**Theorem 6.4.5:**   *The following statements are equivalent for $A \lhd C\{X\}$ and $U = C\{X\}/A$:*

  (i) *$A$ is a T-ideal of $C\{X\}$.*
  (ii) *$(U, \bar{X})$ is relatively free.*
  (iii) *$A = \mathscr{L}(U)$.*

**Proof:**   (i) $\Rightarrow$ (iii) Clearly $\mathscr{L}(U) \subseteq A$, since $f(\bar{X}_1, \ldots, \bar{X}_m) = 0$ implies $f \in A$. It remains to show for every $f \in A$ that $f(\bar{h}_1, \ldots, \bar{h}_m) = 0$ for all $\bar{h}_i$ in $U$. But there is a homomorphism $\varphi: C\{X\} \to C\{X\}$ given by $\varphi X_i = h_i$ for $1 \le i \le m$; then $f(h_1, \ldots, h_m) = f(\varphi X_1, \ldots, \varphi X_n) = \varphi f(X_1, \ldots, X_n) \in A$ since $A$ is a *T*-ideal.

(iii) $\Rightarrow$ (ii) We need to show that every map $\sigma: \bar{X} \to R$ can be extended to a unique homomorphism $\psi: U \to R$, for any *C*-algebra $R$ satisfying the identities of $U$. Uniqueness is clear since the $\bar{X}_i$ generate $U$. It remains to verify existence. Define $\varphi: C\{X\} \to R$ such that $\varphi X_i = \sigma \bar{X}_i$ for each $i$. Then $\ker \varphi \supseteq \mathscr{L}(R) \supseteq \mathscr{L}(U) = A$ so we have a homomorphism $\bar{\varphi}: C\{X\}/A \to R$ satisfying $\bar{\varphi} \bar{X}_i = \varphi X$, as desired.

(ii) $\Rightarrow$ (i) Given a surjection $\varphi: C\{X\} \to C\{X\}$ we need to prove $\varphi A \subseteq A$. By hypothesis we have a homomorphism $\psi: U \to U$ such that $\psi \bar{X}_i = \overline{\varphi X_i}$ for each $i$, so for any $f$ in $A$ we have

$$0 = \psi \bar{f} = \bar{f}(\psi \bar{X}_1, \ldots, \psi \bar{X}_m) = \bar{f}(\overline{\varphi X_1}, \ldots, \overline{\varphi X_m}) = \overline{\varphi f}$$

implying $\varphi f \in A$ as desired.      Q.E.D.

**Remark 6.4.6:**   $\mathscr{L}(R)$ is a *T*-ideal of $C\{X\}$, for any *C*-algebra $R$. Thus the study of identities is equivalent to the study of *T*-ideals, which is equivalent to the study of relatively free algebras.

By corollary 6.1.46 $\{\mathscr{L}(M_n(F)): n \in \mathbb{N}\}$ are the only *T*-ideals which are semiprime as ideals when $F$ is an infinite field. We shall study their relatively free PI-algebras, which turn out to be domains, as focal examples in Chapter 7.

## Digression: Relatively Free PI(*)-Rings

The results obtained so far can be obtained analogously for identities of rings with involution by means of the structure theory of §2.13. We discussed identities of rings with involution at the end of §6.1 but did not bother with the free ring in this category. The easiest way of describing it is to start by noting that if $X, Y$ are sets of indeterminates in 1:1 correspondence then the free algebra $C\{X; Y\}$ has the *reversal involution* (*) given by $X_i^* = Y_i$ and $Y_i^* = X_i$, e.g., $(X_1 X_2 Y_3 + X_3 Y_1^2 X_1)^* = Y_1 X_1^2 Y_3 + X_3 Y_2 Y_1$. Rewriting $Y_i$ as $X_i^*$ permits us to denote this ring as $(C\{X\}, *)$.

Clearly $(C\{X\}, *)$ is the free algebra with involution, in the sense that given any $(R, *)$ and any map $f: X \to R$ there is a unique homomorphism $f: (C\{X\}, *) \to (R, *)$ extending $f$. (Indeed, take a homomorphism $C\{X; Y\} \to R$ given by $X_i \mapsto fX_i$ and $Y_i \mapsto (fX_i)^*$.)

An identity of $(R, *)$ can now be viewed as an element of $(C\{X\}, *)$ which is in the kernel of every homomorphism to $(R, *)$. Write $\mathscr{L}(R, *)$ for {identities of $(R, *)$}. A *T-ideal* of $(R, *)$ is an ideal of $(R, *)$ invariant under every surjection $\varphi: (R, *) \to (R, *)$. The (*)-version of theorem 6.4.5 is now

**Theorem 6.4.7:** *The following statements are equivalent for $A \lhd (C\{X\}, *)$ and $(U, *) = (C\{X\}/A, *)$:*

   (i) *$A$ is a T-ideal of $(C\{X\}, *)$.*
   (ii) *$((U, *); \bar{X})$ is relatively free.*
   (iii) *$A = (\mathscr{L}(U), *)$.*

**Proof:** As in theorem 6.4.5.    Q.E.D.

**Theorem 6.4.8:** *If $(R, *)$ is an algebra over an infinite field $F$ then $\mathscr{L}(R \otimes_F H, *) = \mathscr{L}(R, *)$ for any commutative F-algebra H.*

**Proof:** As in theorem 6.1.44.    Q.E.D.

A complication arises when attempting to lift corollary 6.1.46.

**Corollary 6.4.9:** *If R is prime PI and $(R, *)$ is an algebra over an infinite field F then one of the following three cases holds:*

   (i) *$\mathscr{L}(R, *) = \mathscr{L}(M_n(F) \oplus M_n(F)^{op}, \circ)$.*
   (ii) *$\mathscr{L}(R, *) = \mathscr{L}(M_n(F), t)$ where (t) is the transpose.*
   (iii) *$\mathscr{L}(R, *) = \mathscr{L}(M_n(F), s)$ where (s) is the canonical symplectic involution.*

**Proof:**   By theorem 6.1.59 and theorem 6.4.8 we may pass to the ring of central fractions and thus assume $(R, *)$ is simple; tensoring by the algebraic closure $K$ of $Z(R, *)$ we may assume $K$ is algebraically closed. If $R$ is not simple then proposition 2.13.24 yields $(R, *) \approx (M_n(K) \oplus M_n(K)^{\mathrm{op}}, \circ) \approx K \otimes (M_n(F) \oplus M_n(F)^{\mathrm{op}}, \circ)$ and we have (i).

Thus we may assume $R$ is simple; then $R = M_n(K)$ so corollary 2.13.30 says $(R, *) \approx (M_n(K), t) \approx K \otimes (M_n(F), t)$   or   $(R, *) \approx (M_n(K), s) \approx K \otimes (M_n(F), s)$, thereby yielding (ii) and (iii) respectively.         Q.E.D.

(i) implies every identity of $(R, *)$ is special and thus reduces to the non-involuntary case. However (ii) and (iii) yield genuine new examples.

There are involutory analogs to the algebra of generic matrices, which give the explicit constructions of relatively free PI-algebras with involution, cf., exercise 3.

## The Structure of T-Ideals

Various interesting questions belong properly to the realm of $T$-ideals. To start with, when does an identity $f_1$ "imply" an identity $f_2$, i.e., when is $f_2$ in the $T$-ideal generated by $f_1$? In this sense the Capelli polynomial $C_{2t}$ implies the standard polynomial $S_t$ when we substitute 1 at the appropriate positions.

**Example 6.4.10:**   (P. M. Cohn) The identity $f = [[X_1, X_2], X_3]$ does not imply any standard identity. Indeed $f$ is an identity of the exterior algebra $\bigwedge(V)$ on an infinite dimensional vector space V over $\mathbb{Q}$, cf., example 6.1.5, but $S_k(x_1, \ldots, x_k) = k! x_1 \ldots x_k \neq 0$ when $x_1, \ldots, x_k$ are independent elements of V. In fact the exterior algebra plays an important role in the theory of $T$-ideals, as we shall see.

Recently Kemer [84] developed a structure theory for $T$-ideals *as T-ideals* rather than as ideals. He showed that any relatively free ideal has a maximum nilpotent $T$-ideal, and also characterized those $T$-*semiprime* relatively free rings (i.e., those which have no nilpotent $T$-ideals), as either $F\{X\}$, $F_n\{X\}$ (the algebra of generic matrices), $F_n\{X\} \otimes \bigwedge(V)$, or a certain kind of algebra of matrices whose entries are in the $(\mathbb{Z}/2\mathbb{Z})$-homogeneous components of the exterior algebra $\bigwedge(V)$ (his theorem 5). In this sense Kemer has determined all the $T$-ideals, although computing in $T$-ideals remains very complicated.

One obvious question in a $T$-ideal theory would be to find a minimal set of generators of a $T$-ideal (as $T$-ideal). For example, Latyshev proved that any $T$-ideal of $\mathbb{Q}\{X\}$ properly containing $f = [[X_1, X_2], X_3]$ is generated by $f$ and a polynomial of the form $[X_1, X_2] \cdots [X_{2n-1}, X_n]$, cf., Rowen [80B, exercise 6.3.4]. In particular there is

**Specht's Problem 6.4.11:** Is every $T$-ideal of $\mathbb{Q}\{X\}$ finitely generated (as $T$-ideal)?

Three comments: (1) Specht's problem is false in general if one can show there are uncountably many distinct $T$-ideals of $\mathbb{Q}\{X\}$.

(2) On the other hand, Kemer [85, corollary 1] reduces Specht's problem to $T$-ideals of the $T$-prime relatively free algebras $\mathbb{Q}_n\{X\} \otimes \bigwedge(V)$; he refines this result further using $(\mathbb{Z}/2\mathbb{Z})$-grading in his theorem 2.

(3) One might try further to show any maximally non-finitely generated $T$-ideal is prime as an ideal, for this would reduce the conjecture to $T$-ideals of matrices.

Even for $2 \times 2$ matrices Specht's problem is difficult and was solved in this case by Razmyslov [73a]; Drensky [82] has proved that $S_4$ and $[X_1, X_2]^2, X_1]$ in fact generate. This still leaves open the question of how to determine effectively (without substituting) whether a given polynomial is an identity of $2 \times 2$ matrices.

Another motivation for information about $T$-ideals was in Razmyslov [74] in considering the *nilpotence question* (of the nilpotence of the nilradical of an affine PI-algebra); in characteristic 0 he reduced it to, "Does every affine PI-algebra over $\mathbb{Q}$ satisfy a Capelli identity of suitable size?" Note "affine" is necessary, in view of example 6.4.10. Kemer proved this in the affirmative and thus has priority over Braun in answering the nilpotence question in characteristic 0. However, Braun's proof is more understandable than the Razmyslov-Kemer proof. We shall give a related result in exercise 11.

## Homogeneous Identities

The reader may have noted that we have shifted perspective to algebras over $\mathbb{Q}$. There are several reasons for this, which we preface with some general observations.

**Remark 6.4.12:** The free algebra $C\{X_1, \ldots, X_m\}$ is canonically isomorphic to the tensor algebra $T(M)$ where $M = C^{(m)}$ is given the base $X_1, \ldots, X_m$; the isomorphism $T(M) \to C\{X_1, \ldots, X_m\}$ is given by

$$X_{i_1} \otimes \cdots \otimes X_{i_t} \mapsto X_i \cdots X_{i_t}.$$

Thus we can transfer the graded structure of $T(M)$, and, in particular, view $C\{X_1,\ldots,X_m\}$ as an $\mathbb{N}^{(m)}$-graded algebra, graded by the degrees in $X_1,\ldots,X_m$.

Using this point of view we call a polynomial $f$ *homogeneous* if it is homogeneous under this $\mathbb{N}^{(m)}$-grading, i.e., for every $i$ each of the monomials of $f$ has the same degree in $X_i$. We shall now see there are "enough" homogeneous identities by means of a very useful matrix.

**Remark 6.4.13:** "The Vandermonde matrix" If $c_1,\ldots,c_n$ are elements of a commutative ring $C$ then the *Vandermonde matrix*

$$\begin{pmatrix} 1 & c_1 & \cdots & c_1^{n-1} \\ 1 & c_2 & \cdots & c_2^{n-1} \\ \vdots & \vdots & & \vdots \\ 1 & c_n & \cdots & c_n^{n-1} \end{pmatrix}$$

has determinant $\prod_{1 \le i \le k \le n}(c_k - c_i)$, which is nonzero if $C$ is a domain and the $c_i$ are distinct. (Indeed, it is enough to prove this formula in $\mathbb{Z}[\xi_1,\ldots,\xi_n]$ for the Vandermonde matrix $(\zeta_i^{j-1})$ since then we can specialize $\zeta_i \mapsto c_i$ for each $i$. Let $p(\zeta_1,\ldots,\zeta_n)$ be the determinant of $(\zeta_i^{j-1})$. Specializing $\zeta_k \mapsto \zeta_i$ makes two rows of the matrix equal, and thus sends $p \mapsto 0$; hence $(\zeta_k - \zeta_i)|p$ for all $k > i$, implying $\prod_{1 \le i \le k \le n}(\zeta_k - \zeta_i)$ divides $p$. But the total degrees are equal, as are the coefficients of $1\zeta_2 \cdots \zeta_n^{n-1}$ (the contribution of the diagonal), so we conclude $p = \prod_{1 \le i < k \le n}(\zeta_k - \zeta_i)$ as desired.)

The Vandermonde matrix is used in field theory to compute discriminants, cf., Jacobson [85B, p. 258]. Our use here is different.

**Proposition 6.4.14:** *Suppose $R$ is a torsion-free algebra over an integral domain $C$ having $> n$ elements. Then every identity of $R$ having total degree $n$ can be written naturally as a sum of homogeneous identities of degree $n$.*

**Proof:** Suppose $f(X_1,\ldots,X_m)$ is an identity of $R$. We are done unless $f$ is *not* homogeneous in some indeterminate, say $X_1$. Write $f = \sum_{j=0}^{d} f_j$ where each $f_j$ is the sum of all monomials of $f$ having degree $j$ in $X_1$. If we can show each $f_j$ is an identity then we can conclude by repeating this procedure for each $X_i$.

Take distinct $c_1, \ldots, c_{d+1}$ in $C$ and let $c$ be the determinant of the $(d+1) \times$ $(d+1)$ Vandermonde matrix $(c_j^{k-1})$. Then

$$0 = f(c_j r_1, r_2, \ldots, r_m) = \sum_{k=0}^{d} c_j^k f_j(r_1, \ldots, r_m)$$

for any $r_1, \ldots, r_m$ in $R$. Letting $v$ be the column vector whose entries are $f_0(r_1, \ldots r_m), \ldots, f_d(r_1, \ldots, r_m)$, we have the matrix equation $(c_j^{k-1})v = 0$. Left multiplication by the adjoint of $(c_j^{k-1})$ yields $0 = cv$ so $v = 0$ since $R$ is torsion free.          Q.E.D.

**Corollary 6.4.15:**   *Over an infinite field every T-ideal is homogeneous.*

**Proof:**   Every identity is a sum of homogeneous identities.          Q.E.D.

**Remark 6.4.16:**   If $f$ is homogeneous and $h$ is a multilinearization of $f$ (cf., §6.1) then there is $k$ in $\mathbb{N} - \{0\}$ for which $kf$ is obtained by suitable substitution into $h$. (Indeed if $f$ has degree $d$ in $X_1$ then $\Delta_1 f(X_1, \ldots, X_m, X_1) = (2^d - 2)f$, so the result follows by induction on the number of times we apply $\Delta$.)

Combining corollary 6.4.15 and remark 6.4.16 we see for a field of characteristic 0 that the $T$-ideals are generated by multilinear polynomials, which are the most amenable to computation.

Another reason for studying $T$-ideals of $\mathbb{Q}$-algebras is the surprising Nagata-Higman theorem that any nil $\mathbb{Q}$-algebra (without 1) of bounded index is nilpotent, cf., exercise 2. This result pops up in many proofs and was improved by Razmyslov [74] using trace identities. Kemer [84, theorem 6] proves a broad generalization of Nagata-Higman using his theory of $T$-ideals.

## The Quantitative PI-Theory

The "quantitative" PI-theory may be defined as the computations of various invariants of $T$-ideals. Although we have already seen several significant attributes of $T$-ideals, the major impetus to the quantitative theory came from a different quarter. One of the early unsolved questions in PI-theory was the *tensor product question*: Need the tensor product of PI-algebras be a PI-algebra? The tensor product question has the following easy reduction to relatively free algebras:

***Remark 6.4.17:*** Let $\mathcal{U}(R_i)$ denote the relatively free PI-algebras of $R_i$. Then $R_1 \otimes_C R_2$ is a homomorphic image of $\mathcal{U}(R_1) \otimes_C \mathcal{U}(R_2)$, and in particular satisfies all of its identities.

Applying this observation to the structure theoretic results of §6.1 one can obtain rather easily an affirmative answer to the tensor product question whenever $R_2$ satisfies all the identities of $n \times n$ matrices, cf., exercise 4.

On the other hand, Regev found a quantitative approach which still is the only method known for solving the tensor product question in full generality. Let $V_k$ denote the *C*-subspace of $C\{X\}$ spanned by $\{X_{\pi 1} \cdots X_{\pi k} : \pi \in \operatorname{Sym}(k)\}$, a free *C*-module of rank $k!$ Given a *T*-ideal $I$ of $C\{X\}$, let $I_k = I \cap V_k$ and let $c_k(V)$ denote the smallest number of elements which can span $V_k/I_k$ as *C*-module; $c_k(V)$ is called the *k-codimension of* $I$.

**Lemma 6.4.18:** *(Regev) If $I_1, I_2$ are T-ideals of $\mathbb{Z}\{X\}$ such that $c_n(I_1)c_n(I_2) < n!$ then $R = (Z\{X\}/I_1) \otimes_{\mathbb{Z}} (\mathbb{Z}\{X\}/I_2)$ satisfies a multilinear identity of degree n whose coefficients are relatively prime, so R is a PI-ring.*

Regev's lemma is proved in exercise 5. In view of remark 6.4.17 an affirmative solution to the tensor product question lies in proving the "codimension sequence" is exponentially bounded in the following sense:

If $I$ is a *T*-ideal containing a multilinear polynomial of degree $d$, one of whose coefficients is $\pm 1$, then there is $m = m(d)$ such that $c_n(I) \le m^n$ for all $n$.

Regev showed that the codimension sequence indeed is exponentially bounded, and Latyshev [72] produced the result $c_n(I) < (d - 1)^{2n}$, cf., exercises 7 and 6; this is called the *Latyshev bound*. Putting everything together yields

**Theorem 6.4.19:** *(Regev) The tensor product of any two PI-algebras is a PI-algebra.*

These results indicate that any polynomial identity requires a lot of room in $V_k$, and one of the major goals of the quantitative theory was to obtain explicit asymptotic formulas for the codimensions; in particular, cf., Regev [80] and Amitsur [84] and Drensky [84].

The codimension is a rather crude invariant. More precise information can be obtained by giving $V_k$ extra structure, where $C = F$ is any field of characteristic 0. First note there is a vector space isomorphism from the group algebra $F[\mathrm{Sym}(k)] \to V_k$, given by $\sum \alpha_\pi \pi \to \sum \alpha_\pi X_{\pi 1} \cdots X_{\pi k}$. This identification makes $V_k$ a left and right $\mathrm{Sym}(k)$-module, under the action

$$\sigma f(X_1, \ldots, X_m) = f(X_{\sigma 1}, \ldots, X_{\sigma m}).$$

Given a $T$-ideal $I$ we see $I_k = I \cap V_k$ is a left submodule of $V_k$ and thus can be studied in terms of the character theory of representations of the symmetric group. The focus is on the character of $V_k/I_k$, called the *cocharacter* of $I_k$.

An alternate method is by means of $GL(n, F)$. Let $W = \sum_{i=1}^n FX_i \approx F^{(n)}$. Any $T$ in $GL(n, F)$ acts as a transformation on $W$, and thus extends "diagonally" to a transformation on $W^{\otimes k}$ given by $T(w_1 \otimes \cdots \otimes w_k) = Tw_1 \otimes \cdots \otimes Tw_k$. Now identify $F\{X\}$ with the tensor algebra $F \oplus W \oplus W^{\otimes 2} \oplus \cdots$. Clearly $I \cap W^{\otimes k}$ is invariant under the action of $GL(n, K)$ and so can be studied via the representation theory of $G(n, F)$. As before, attention focuses on $W^{\otimes k}/(I \cap W^{\otimes k})$. This approach has become increasingly popular in recent years and does not require the restriction to multilinear identities; also work of Drensky [84] and Berele [82] show this theory includes the $\mathrm{Sym}(k) =$ cocharacter theory.

Suppose $F$ is infinite. By corollary 6.4.15 corresponding relatively free PI-algebra inherits the $\mathbb{N}$-grading. On the other hand, theorem 6.3.25 shows the dimensions of $\{W^{\otimes k}/I \cap W^{\otimes k} : k \in \mathbb{N}\}$ have polynomially bounded growth, so there is an attached Hilbert series. Computing this series is possibly the most pressing problem in the quantitative theory but is very hard to accomplish.

## Trace Identities

Let us bring the good old Hamilton-Cayley theorem into this discussion. Any generic $n \times n$ matrix $Y_1$ satisfies $f(Y_1) = 0$ where $f(Y_1) = Y_1^n + \sum_{i=0}^{n-1} \alpha_i Y_1^i$, and the $\alpha_i$ are the suitable elementary symmetric functions of the eigenvalues of $Y_1$. Thus $M_n(F)$ satisfies a "characteristic coefficient identity" of degree $n$. These are usually called "trace identities" because if $F$ is a $\mathbb{Q}$-algebra Newton's identities permit one to express the characteristic coefficients of $Y_1$ as polynomial expressions in traces of powers of $Y_1$. For example if $n = 2$ then $\det(Y_1) = \frac{1}{2}(\mathrm{tr}(Y_1)^2 - \mathrm{tr}(Y_1^2))$. Razmyslov [74b] and Procesi [76] (also, cf., Helling [74]) proved every identity $f$ of $M_n(\mathbb{Q})$ is a "consequence" of this Hamilton-Cayley trace identity, by showing that $f(X_1, \ldots, X_m)$ is a consequence of the trace identity $\mathrm{tr}(X_{m+1} f)$, and this is a consequence of the Hamilton-Cayley trace identity.

There is a "generic" ring for trace identities, namely the "trace ring" for the generic matrix algebra, and Amitsur-Small [80] showed thereby that any trace identity for $M_n(F)$ is also a trace identity for $M_{n-1}(F)$. Nevertheless, there still does not seem to be an effective procedure for verifying identities, as witnessed by the difficulty of Formanek's verification of Regev's central polynomial

$$f = \sum_{\pi,\,\sigma\,\in\,\text{Sym}(n^2)} (\text{sg}\,\pi)(\text{sg}\,\sigma) X_{\pi 1} X'_{\sigma 1} X_{\pi 2} X_{\pi 3} X_{\pi 4} X'_{\sigma 2} X'_{\sigma 3} X'_{\sigma 4} \cdots$$

where $X_1,\ldots,X_{n^2}, X'_1,\ldots,X'_{n^2}$ are indeterminates, cf., Formanek [86]. It is relatively easy to show $f$ takes on only scalar values, but the hard part is showing $f$ is not an identity; $f$ is very useful in the quantitative theory.

## Invariant Theory

Procesi [76] tied the trace identity theory to the classical invariant theory and thereby extended the roots of PI-theory. The thrust of modern research in the quantitative PI-theory is to explore this relationship and thus learn more about both subjects.

**Definition 6.4.20:**   Letting $F_n\{Y\}$ denote the algebra of generic $n \times n$ matrices (i.e., $Y_k = (\zeta_{ij}^{(k)})$ for $k = 1, 2, \ldots$, and the $\zeta_{ij}^{(k)}$ are each commuting indeterminates over $F$), write $F[\zeta]$ for the commutative polynomial ring $F[\zeta_{ij}^{(k)}: 1 \le i, j \le n, k = 1, 2, \ldots]$. Given $a \in \text{GL}(n, F)$ write $aY_k a^{-1} = (\mu_{ij}^{(k)})$ for suitable $\mu_{ij}^{(k)}$ in $F[\zeta]$; define $a \cdot \zeta_{ij}^{(k)}$ to be $\mu_{ij}(k)$. This action extends naturally to an automorphism $\psi_a: F[\zeta] \to F[\zeta]$ for each $a$ in $\text{GL}(n, F)$.

A (commutative) polynomial $p(\zeta) \in F[\zeta]$ is an *invariant* of $n \times n$ matrices if $a \cdot p = p$ for all $a$ in $\text{GL}(n, F)$, i.e., $p$ is invariant under all elements of $\text{GL}(n, F)$. The set of invariants of $n \times n$ matrices is called *the ring of invariants* (of $n \times n$ matrices).

**Remark 6.4.21:**   Any central polynomial $f(X_1, \ldots, X_m)$ is an invariant, for noting $f$ is a suitable scalar matrix $p(\zeta)1$ we have

$$(a \cdot p)1 = f(aY_1 a^{-1}, \ldots, aY_m a^{-1}) = af(Y_1, \ldots, Y_m)a^{-1} = f(Y_1, \ldots, Y_m) = p1.$$

Thus the ring of invariants contains $Z(F_n\{Y\})$. Procesi [76] showed that the ring of invariants is generated by traces of monomials of generic matrices of degree $\le 2^n - 1$; later this was seen to follow from a classic theorem from invariant theory first proved in full by Gurevich. Using this theorem we see the ring of invariants is merely the center of the $n$-trace ring of $F_n\{Y\}$. Artin and Schofield have shown the $n$-trace ring is in fact a "maximal order" in the sense of Auslander and Goldman, cf., Le Bruyn [86], who also derives consequences of this fact.

Thus writing $R = F_n\{Y\}$, $Z = Z(R)$, $T = T(R)$, and $Z' = Z(T)$, we see $Z'$ is the ring of invariants, and $Z$ and $Z'$ have the same field of fractions. One of the classic questions in invariant theory is to present $Z'$ efficiently in terms of generators and relations: in particular one has

**Question 6.4.22:**   Is the field of fractions of $Z$ purely transcendental?

There is very explicit information for $n = 2$. First of all the ring of invariants $Z'$ is generated over $F$ by tr $Y_1$, tr $Y_2$, det $Y_1$, det $Y_2$, and tr$(Y_1 Y_2)$, and thus is a polynomial ring over $F$, cf., exercise 9. Furthermore, letting

$$C_1 = F[Y_1, \operatorname{tr} Y_1, \operatorname{tr} Y_2, \det Y_2]$$

$$C_2 = F[Y_2, \operatorname{tr} Y_1, \operatorname{tr} Y_2, \det Y_1]$$

$$C = F[\operatorname{tr} Y_1, \operatorname{tr} Y_2, \det Y_1, \det Y_2],$$

Formanek-Schofield [85] prove $T$ is the free product of $C_1$ and $C_2$ over $C$, cf., exercise 10, and conclude $T$ has global dimension 5. (This latter fact had been proved by Small-Stafford [85], who also display $T$ as an iterated Ore extension.)

Formanek showed the answer to question 6.4.22 also is "yes" for $n = 3$ and $n = 4$ (cf., Rowen [80B, p. 197ff] and Formanek [82]). However, Formanek's method increases in complexity with $n$ and is not likely to be applicable for $n = 5$. A positive solution to question 6.4.22 for arbitrary $n$ would have far-reaching applications in the theory of division algebras, as we shall see in the next chapter.

There is yet another method of studying $T$-ideals, using the ring of symmetric functions, defined as the fixed ring of $\mathbb{Z}[\lambda_1, \ldots, \lambda_m]$ under the action of Sym$(m)$. This brings in the theory of fixed subrings under group actions. Formanek [84, theorem 7] shows this method is equivalent to studying the GL$(n, F)$-modules and is also equivalent to studying the Hilbert series of the relatively free PI-algebras of $T$. Formanek [84] is a "must" for anyone interested in entering the quantitative theory; a shorter survey is given in Formanek [85].

# Exercises

## §6.1

1. Every identity is a sum of blended identities. (Hint: Induction on the number of indeterminates in which $f$ is *not* blended, applied to the identities $f(X_1 \mapsto 0)$ and $f - f(X_1 \mapsto 0)$.)
2. If $h$ is a monomial of degree $d > 1$ in $X_i$ then $(2^d - 2)h = \Delta_i h(X_1, \ldots, X_n, X_i)$.

Hence some multiple of $h$ can be recovered from a multilinearization of $h$. Conclude from this that in characteristic 0 any central polynomial can be multilinearized to a multilinear central polynomial, and any identity can be recovered from multilinear identities.

3. Write $f_{(ij)}$ for the polynomial in which $X_i$ and $X_j$ are interchanged. Show that $f$ is $t$-alternating iff $f_{(ij)} = -f$ for all $1 \le i < j \le t$, iff $f_{(i,i+1)} = -f$ for all $1 \le i < t$.

4. Write $\bar{f}$ for the sum of those polynomials of $f$ for which $X_i, \dots, X_t$ occur in ascending order. $f$ is $t$-alternating iff $f = \sum_{\pi \in \text{Sym}(t)} (\text{sg } \pi) \bar{f}(X_{\pi 1}, \dots, X_{\pi t})$, by exercise 3. Conclude that every $t$-normal polynomial can be written in the form $\sum_i h_{i0} C_{2t}(X_1, \dots, X_t, h_{i1}, \dots, h_{it})$ for suitable monomials $h_{ij}$. Thus Capelli identities are "initial" in the sense that if $C_{2t}$ is an identity of $R$ then every $t$-normal polynomial is an identity of $R$.

5. For any commutative ring $C$ every identity $f$ of $M_n(\mathbb{Z})$ is an identity of $M_n(C)$. (Hint: $f$ is an identity of $M_n(\mathbb{Z}[\lambda_1, \lambda_2, \dots])$ by corollary 6.1.45, and $M_n(C)$ is a homomorphic image.)

6. Arguing as in remark 6.1.18 show the only possible multilinear identity of $M_n(C)$ having degree $n$ is $S_{2n}$.

7. The polynomial $\sum_{\pi, \sigma \in \text{Sym}(t)} (\text{sg } \pi)(\text{sg } \sigma)[X_{\pi 1}, X_{t+\sigma 1}] \cdots [X_{\pi t}, X_{t+\sigma t}]$ is a nonzero multiple of the standard polynomial $S_{2t}$. (Hint: It is $t$-normal, with no cancellation of terms.)

8. $S_t = \sum_{i=1}^{t} (-1)^{i-1} X_i S_{t-1}(X_1, \dots, X_{i-1}, X_{i+1}, \dots, X_t) = \sum_{i=1}^{t} (-1)^{t-i} S_{t-1}(X_1, \dots, X_{i-1}, X_{i+1}, \dots, X_t) X_i$. Consequently if $S_{t-1}$ is an identity then so is $S_t$.

9. $2 \text{ tr } S_{2t}(x_1, \dots, x_{2t}) = 0$ for all $x_i$ in $M_n(C)$. (Hint: Exercise 8 shows $2 \text{ tr } S_{2t}(x_1, \dots, x_{2t}) = \sum (-1)^{i-1} \text{tr}[x_i, S_{2t-1}(x_1, \dots, x_{i-1}, x_{i+1}, \dots, x_{2t})] = 0$.)

10. (Amitsur-Levitzki theorem.) $S_{2n}$ is an identity of $M_n(C)$ for every commutative ring $C$. (Hint (Razmyslov): By exercise 5 take $C = \mathbb{Z}$. Any matrix $x$ satisfies the Hamilton-Cayley equation $x^n + \sum_{i=0}^{n-1} \alpha_i x^i = 0$ where $\alpha_i$ can be written in terms of traces by Newton's formulae (cf., Rowen [80B, p. 18]). The multilinearization technique can be applied to this equation to yield $0 = \sum_{\pi \in \text{Sym}(n)} x_{\pi 1} \cdots x_{\pi n} + \sum$ terms involving traces, for all $x_i$ in $M_n(\mathbb{Z})$. Now substituting commutators $[x_i, x_{n+\pi i}]$ in place of the $x_i$ and summing over suitable permutations $\sigma$, one obtains standard polynomials by exercise 7, and the terms involving traces drop out by exercise 9. Thus $0 = \sum_{\pi, \sigma}[x_{\pi 1}, x_{n+\sigma 1}] \cdots [x_{\pi n}, x_{n+\sigma n}] = m S_{2n}(x_1, \dots, x_{2n})$ for some $m$, so $S_{2n}$ is an identity.)

11. $S_{n-1}([X_1^{n-1}, X_2], \dots, [X_1, X_2])$ is not an identity of $M_n(C)$ for any commutative ring $C$. (Hint: Passing to homomorphic images one may assume $C$ is a field. If $|C| \ge n$ take a diagonal matrix $x_1$ having distinct eigenvalues and $x_2 = \sum_{i=1}^{n-1} e_{i,i+1}$. Even when $|C| < n$ one can maneuver into a position to use this substitution by means of "companion matrices.")

12. (Amitsur) A PI-ring $R$ satisfying the identities of $n \times n$ matrices with $\text{Nil}(R)^{n-1} \not\subseteq N(R)$. Let $C$ be a commutative ring containing a nil ideal $I$ which is not of bounded index, e.g., $C = H/\sum_{i \in \mathbb{N}} H\lambda_i^i$ where $H = F[\lambda_1, \lambda_2, \dots]$. Let $R = \{(c_{ij}) \in M_n(C): c_{ij} \in I \text{ for all } i > j\}$, a subring of $M_n(C)$ containing the nil ideal $A = \{(c_{ij}) \in M_n(C): c_{ij} \in I \text{ for all } i \ge j\}$. Then $e_{1n} = e_{12}e_{23} \cdots e_{n-1,n} \in A^{n-1}$ but $e_{1n} \notin N(R)$ since $I \approx Ie_{nn} = (Ie_{n1})e_{1n} \subseteq Re_{1n}$. Thus $A^{n-1} \not\subseteq N(R)$.)

13. Quicker proof of (iii) $\Rightarrow$ (iv) in theorem 6.1.35. Write $g_n = \sum h_{u1} X_1 h_{u2}$ and apply exercise 5.3.11, letting $a_{iju} = h_{u1}(r_2, \dots, r_m)$ and $b_{iju} = r_1 h_{u2}(r_2, \dots, r_m)$.

## Applications of the Embedding Procedure

**14.** If $R/\mathrm{Nil}(R)$ is PI and $S \subseteq R$ is a multiplicatively closed set each of whose elements can be written as a sum of nilpotent elements from $S$ then $S$ is nil.

**15.** If $R/\mathrm{Jac}(R)$ is PI then $R$ is weakly finite. (Hint: See proof of theorem 3.2.37; $M_n(R/\mathrm{Jac}(R))$ is embeddible in matrices for all $n$.)

**16.** If $R$ is prime and satisfies an identity, one of whose coefficients does not annihilate $R$, then $R$ is a PI-ring. (Hint: Assume the identity $f$ is multilinear; then using corollary 3.2.46 assume $R$ is primitive and reprove Kaplansky's theorem.)

## PI-Theory Via Semiprime Rngs

In the following exercises $R_0$ is a rng, i.e., without 1.

**17.** Suppose $R_0$ is prime and satisfies a multilinear identity $f$; write $f(X_1, \ldots, X_n) = f'(X_1, \ldots, X_{n-1})X_n + f''$ where no monomial of $f''$ ends in $X_n$. Then $f'$ is a PI of $\bar{L} = L/L \cap \mathrm{Ann}'\, L$ for any $L < R$. (Hint: $0 = f(a_1, \ldots, a_{n-1}, r)a' = f'(a_1, \ldots, a_{n-1})ra'$ for $a_i \in L$, $a' \in \mathrm{Ann}'\, L$.)

**17′.** Applying PI-induction to $\bar{L}$ of exercise 17 (also, cf., exercise 2.1.3) prove Kaplansky's theorem directly; also prove Posner's original theorem, that all prime PI-rings are Goldie, by verifying the Goldie condition.

**18.** If $L$ is a left ideal of $R_0$ then $[Z(L), R_0]L = 0$. Hint: $[z, r]a = [z, ra] + r[a, z] = 0$ for $a$ in $L$ and $z$ in $Z(L)$.)

**19.** If $R_0$ is semiprime and $L$ is a left ideal of $R_0$ then $L$ is semiprime as a rng and $Z(L) \subseteq Z(R_0)$. (Hint: $[z, r]R[z, r] = 0$ for $z$ in $Z(L)$ by exercise 18.)

**20.** If $R_0$ is semiprime PI and $A \lhd R_0$ then $0 \neq Z(A) \subseteq A \cap Z(R)$. (Hint: Apply a central polynomial to exercise 19.) This alternate proof of theorem 6.1.28 is due to Martindale.

**21.** A left ideal $L$ of a semiprime ring $R$ is large iff $L \cap Z(R)$ is large in $Z(R)$. (Hint: $(\Rightarrow)$ $Z(L) \subseteq Z(R)$ by exercise 19, so apply a central polynomial.)

**22.** (Braun, Schofield, and others?) A short, nonconstructive proof of the existence of central polynomials. (This requires Posner's theorem, cf., exercise 17′.) Let $D$ be the central simple ring of fractions of the generic matrix algebra $\mathbb{Z}_n\{Y\}$. Taking a base of $D$ over $Z(D)$ containing 1, define the projection $\pi: D \to Z(D)$. Viewing $\pi \in \mathrm{End}_{Z(D)} D \subseteq D \otimes D^{\mathrm{op}}$, write $\pi$ as $\sum_i f_i f^{-1} \otimes h^{-1} h_i$ for $f_i, f, h, h_i \in \mathbb{Z}_n\{Y\}$. Then $\sum f_i Y_k h_i = \pi(f Y_k h) \in Z(D)$ so $\mathbb{Z}_n\{Y\}$ has a nontrivial central element!

## PI-Rings Satisfying ACC

**23.** If $R$ is semiprime PI with the ACC on annihilators of ideals then $R$ is semiprime Goldie, and its semisimple Artinian ring of fractions can be had by localizing at the regular elements of $Z(R)$.) (Hint: A finite intersection of primes is 0, so use the results of §3.2.)

**24.** (Cauchon) If $R$ is a semiprime PI-ring satisfying ACC(ideals) then $R$ is Noetherian. (Hint: Reduce to the prime case. Pick $z = g_n(r_1, \ldots, r_m) \neq 0$ and let $t = n^2$. Given $L < R$ and $a$ in $L$ let $z(a)$ denote the vector $(z_1, \ldots, z_t) \in Z^{(t)}$ where $za = \sum_{i=1}^t z_i a_i$. Then $RM \leq R^{(t)}$ is generated by $\{z(a_u): 1 \leq u \leq k\}$ for suitable $a_u \in L$ and suitable $k$; show $L = \sum_{u=1}^k Ra_u$. Here $M$ is as in proposition 6.1.51.)

**24'.** (Cauchon) If $R$ is a PI-ring satisfying ACC(ideals) then $\bigcap_{i \in \mathbb{N}} \text{Jac}(R)^n$ is nilpotent. (Hint: One may assume $R$ is semiprime by lemma 2.6.22; then apply exercise 24.) Of course, this is sharp in view of Herstein's counterexample.

**25.** If $R$ is prime of PI-degree $n$ satisfying ACC(ideals) then $\bigcap_{i \in \mathbb{N}} P^i = 0$ for all $P$ in $\text{Spec}_n(R)$. (Hint: Pass to the Azumaya algebra $R[s^{-1}]$.) Using induction, conclude that there are only a finite number of idempotent prime ideals in any PI-ring satisfying ACC(ideals), where we say $A \lhd R$ is *idempotent* if $A^2 = A$.

**26.** (Robson) Let $C$ be a local Noetherian integral domain with $J = \text{Jac}(C) \neq 0$. Then $\begin{pmatrix} C & C \\ J & C \end{pmatrix}$ is prime Noetherian but $\begin{pmatrix} J & C \\ J & C \end{pmatrix}$ is an idempotent maximal ideal. Thus exercise 25 is sharp.

**27.** (Schelter) Working in $C = \mathbb{Q}(\sqrt{2}, \sqrt{3})[\lambda_1, \lambda_2]$ let $C_1 = \mathbb{Q}[\sqrt{6}, \sqrt{2} + \lambda_1, \lambda_2, \lambda_2\sqrt{2}]$ and $C_2 = \mathbb{Q}[\sqrt{6}, \sqrt{3} + \lambda_1, \lambda_2, \lambda_2\sqrt{2}]$, and $A = C_1\lambda_2 + C_1\lambda_2\sqrt{2}$. Then $A = \lambda_2 C = C_2\lambda_2 + C_2\lambda_2\sqrt{3}$, and $C_1 \cap C_2 = \mathbb{Q}(\sqrt{6}) + A$, which is not Noetherian. Let $R = \begin{pmatrix} C_1 & A \\ A & C_2 \end{pmatrix} \subset M_2(C)$. $R$ is f.g. as $C_1 \times C_2$-module so is Noetherian, and every prime ideal of $R$ has height $\leq 2$. But the "principal ideal theorem" fails for $Z(R)$ since $A$ is $\lambda_2$-minimal but not minimal. In particular $Z(R)$ is not Noetherian.

**28.** (Amitsur) Say a polynomial $f$ is $R$-correct if for each $0 \neq r \in R$ there is a coefficient of $f$ which does not annihilate $R$. Any ring $R$ satisfying an $R$-correct identity is a PI-ring. (Hint: By theorem 6.1.53 one may assume $R$ is semiprime. Then $\bigcap \{P \in \text{Spec}(R): f \text{ is } (R/P)\text{-correct}\} = 0$, so pass to the prime case and apply exercise 16.)

**29.** If $R$ satisfies an $R$-correct identity $f$ of degree $d$ then $S^k_{2[n/2]}$ is a PI of $R$ for some $k$ dependent only on $f$, and $k = 1$ if $R$ is semiprime. (Hint: This comes from the proofs of exercises 28 and 16, in view of the Amitsur-Levitzki theorem.)

## Theory of Generalized Identities

**30.** If $R$ satisfies a proper multilinear GI then $R$ satisfies a proper multilinear GI in two indeterminates. (Hint: Take $f(X_1, \ldots, X_m)$ a proper GI of $R$, such that $f$ can be written as a sum of $k$ monomials with $k$ minimal possible. Let $f_1$ be the sum of those generalized monomials of $f$ in which $X_1$ precedes $X_2$; one may assume $f_1$ is *not* a GI of $R$. But then taking $r_1, \ldots, r_m$ for which $f_1(r_1, \ldots, r_m) \neq 0$ one has $f(X_1, X_2, r_3, \ldots, r_m)$ is proper.)

**31.** Prove Amitsur's theorem for GI's. (Hint: ($\Leftarrow$) By exercise 30 one may assume $R$ satisfies a GI $f = f_1 + f_2$ where $f_1 = \sum r_{1i} X_1 r_{2i} X_2 r_{3i}$ and $f_2 = \sum r'_{1j} X_2 r'_{2j} X_1 r'_{3j}$ and $f_1$ is *not* a GI of $R$. Write $R$ as a dense subring of $\text{End } M_D$. Taking a maximal subfield $F$ of $D$ let $R' = RF$ as in construction 3.2.46. Then $R'$ is dense in $\text{End } M_F$ and $F = Z(R')$. It suffices to show some $0 \neq r \in \text{soc}(R')$ for then $[rM:F] = [rM:D][D:F]$. But $f$ is a proper GI of $R'$, so one may write $f_1$ over $R'$ such that the $r_{1i}$ are linearly independent over $F$. Fix any $x$ in $M$ and let $V = \sum_j Fr'_{3j}x$, a finite dimensional subspace of $M$. For all $r$ in $R'$ and any $r_1$ in $R'$ such that $r_1 V = 0$ one has $0 = f(r_1, r)x = \sum r_{1i} r_1 r_{2i} r r_{3i} x$. By the density theorem there are $\alpha_i$ in $F$ such that for any $y$ in $M$ there is $r$ in $R'$ for which $r r_{3i} x = \alpha_i y$; hence $\sum \alpha_i r_{1i} r_1 r_{2i} = 0$ for all $r_1$ such that $r_1 V = 0$. If some $r_{2i} M \nsubseteq V$ then repeating the application of the density theorem one can find $\alpha_i'$ in $F$ for which $\sum \alpha_i \alpha_i' r_{1i} = 0$, contrary to the independence of the $r_{1i}$.) The proof actually shows that some coefficient of each monomial of $f$ lies in soc($R'$).

32. A (*)-GI is *multilinear* if $X_i$ or $X_i^*$ (but not both) appear exactly once in each monomial. Any *-GI can be multilinearized. If $(R, *)$ is prime of the second kind then every multilinear GI is special. (Hint: Otherwise take $f(X_1, X_1^*, \ldots, X_m, X_m^*)$ a multilinear GI which is *not* special, with $m$ minimal. Write $f = f_1 + f_2$ where $X_1^*$ does not appear in $f_2$, and $X_1$ does not appear in $f_1$, and take $z \neq z^* \in Z(R)$. Then

$$(z - z^*)f_1 = zf(X_1, X_1^*, \ldots, X_m X_m^*) - f(zX_1, (zX_1)^*, \ldots, X_m, X_m^*)$$

is a GI of $(R, *)$ so $f_1 = f_1(X_1^*, X_2, X_2^*, \ldots)$ is a GI of $(R, *)$. But $f_1(r, X_2, X_2^*, \ldots)$ is a special GI for any $r$ in $R$, so $f_1$ special GI; likewise, $f_2$ is special.)

33. Suppose $(R, *)$ is primitive and satisfies a proper GI. Then either (i) every GI of $(R, *)$ is special, or (ii) $R$ is strongly primitive. (Hint: If $R$ is not primitive then take a primitive ideal $P \neq 0$ with $P \cap P^* = 0$ as in proposition 2.13.32. If $f(X_1, X_1^*, \ldots, X_m, X_m^*)$ is a GI of $(R, *)$ then $f(X_1, \ldots, X_{2m})$ is a GI of the rng $P \oplus P^*$ as in example 6.1.58(iii), so is a GI of $P^* \approx (P \oplus P^*)/P$, a nonzero ideal of $R/P$, so $f(X_1, \ldots, X_{2m})$ is a GI of $R/P$ by identity; analogously $f(X_1, \ldots, X_{2m})$ is a GI of $R/P^*$ so is a GI of $R$. Thus one has (i) unless $R$ is primitive. Also one may assume $f = f_1 + f_2$ where $X_1^*$ does not appear in $f_1$ and $X_1$ does not appear in $f_2$, and $f_1$ and $f_2$ are *not* GI's of $(R, *)$. Thus there are $r_2, r_3, \ldots$ in $R$ for which $f_1(X_1, r_2, r_2^*, \ldots)$ is not a GI of $R$. Thus one may assume $f = \sum r_{i1} X_1 r_{i2} + \sum r_{j1}' X_1^* r_{j2}'$. Replace $X_1, X_1^*$, respectively, by $X_2^* X_1, X_1^* X_2$, and repeat the argument of exercise 31 to show soc($R$) $\neq 0$. But now theorem 2.13.21 is applicable and permits us to pass to $RF$ where $F$ is a maximal (*)-symmetric subfield of $D$; thus one concludes as in exercise 31 that $D$ is PI.)

34. (Amitsur) Suppose $(R, *)$ satisfies an identity $f$ of degree $m$, which is $R$-*correct* in the sense that for any $r$ in $R$ there is $\alpha$ in $f$ for which $\alpha r \neq 0$. Then $R$ is a PI-ring, and has PI-degree $\leq m$ if $R$ is semiprime. (Hint: In analogy to exercise 33, one can reduce to the case $(R, *)$ is primitive by means of the (*)-structure theory of §2.13. By exercise 33 one may assume $R$ is strongly primitive, so conclude using exercise 2.13.4 as in the proof of Kaplansky's theorem.)

35. Hypotheses as in exercise 34, $S_{2m}^k$ is an identity of $R$ by Amitsur-Levitzki.

Amitsur's (*)-identity theorem can be applied to Herstein's theory (exercises of §2.13), as exemplified in the next exercise.

36. If $(R, *)$ is semiprime with $(R, *)^+ \subseteq Z(R)$ or $(R, *)^- \subseteq Z(R)$ then $R$ has PI-degree $\leq 2$. (Hint: $(R, *)$ satisfies $[X_1 + X_1^*, X_2]$ or $[X_1 - X_1^*, X_2]$ so apply exercise 34.)

## §6.2

Some results of Montgomery-Small [81] when a ring is affine

1. Suppose $R$ is affine over a commutative Noetherian ring $C$. Then $eRe$ is affine

over $eCe$ under either of the following conditions: (i) $eR$ is f.g. as $eRe$-module; (ii) $ReR$ is f.g. as $R$-module. (Hint: (i) Write $R = C\{r_1,\ldots,r_t\}$ let $eR = \sum_{i=1}^m eRer_i'$ with $r_1' = 1$, and $er_i'er_j = \sum ea_{ijk}er_k'$. Then the $\{er_k'e, ea_{ijk}e\}$ generate $eRe$. (ii) Write $ReR = \sum Rx_i$ and $x_i = \sum_{ij} r_{ij}''er_{ij}'$ for $r_{ij}', r_{ij}''$ in $R$. Then the $er_{ij}'$ generate $eR$ as $ReR$-module, so concude by (i)).

2. Suppose $R$ is left Noetherian and affine over a commutative Noetherian ring $C$. If $G$ is a finite group of automorphisms on $R$ with $|G|^{-1} \in R$ then the fixed ring $R^G$ is affine. (Hint: The skew group ring $R*G$ is affine Noetherian, so $R^G$ is affine by exercise 1 and exercise 2.6.14.)

3. If $T$ is an affine algebra over a commutative Noetherian ring $C$ and $L < T$ satisfies $LT = T$ then $C + L$ is affine over $C$. (Hint: Apply example 6.2.3, noting $Re_{11}R = R$ so that exercise 1(ii) applies.) In particular, this provides a host of affine algebras when $T$ is simple, such as the Weyl algebra $\mathbb{A}_1$. It is an open question if every affine subalgebra of $\mathbb{A}_1$ is Noetherian.

4. (Resco) An affine domain which is left and right Ore, and left but not right Noetherian. Let $\mathbb{A}_1$ be the Weyl algebra $F\{\lambda, \mu\}$ satisfying $\mu\lambda - \lambda\mu = 1$; then $R_0 = F + \mathbb{A}_1\mu$ is left and right Noetherian and affine by exercise 3. Let $t$ denote a commuting indeterminate over $\mathbb{A}_1$, and let $R = \{p \in \mathbb{A}_1[t]$: the constant coefficient of $p$ is in $R_0\}$. Then $R$ is affine and left Noetherian since $R$ is a submodule of $\mathbb{A}_1$ which is f.g. as $R_0$-module; however, $R$ is not right Noetherian since $\mathbb{A}_1$ is not f.g. as right $R_0$-module.)

5. (Small) Suppose $R$ is affine over $C$ but not f.g. over $C$. Then there is $P$ in $\mathrm{Spec}(R)$ for which $R/P$ is not f.g. over $C/(P \cap C)$. (Hint: Write $R = C\{r_1,\ldots,r_n\}$. $\{A \lhd R$: $R/A$ is not f.g. over $C/(A \cap P)\}$ is Zorn, arguing as in Artin-Tate. Thus there is $P \lhd R$ maximal with respect to $R/P$ not f.g. over $C/(P \cap C)$. To show $P$ is prime suppose $P \subset A \lhd R$ and $Ar \subseteq P$. Then $RrR$ is cyclic over $(R/A) \otimes (R/A)^{\mathrm{op}}$ which is f.g. over $C$; hence $RrR$ is f.g. over $C$ implying $R/(P + RrR)$ is not f.g. over the image of $C$, and therefore $r \in P$.)

6. (Martindale) Applying example 2.1.30 to Golod-Shafarevich yields a primitive algebraic algebra which is not locally finite.

7. (Irving-Small [83]) A primitive, affine algebra $R$ with GK-dim 2, having socle $\neq 0$ and failing ACC(Ann). Let $R = F\{r_1, r_2\}$ where $r_1, r_2$ satisfy the relations $r_1^2 = 0$, $r_1r_2^mr_1 = 0$ for $m$ not a power of 2, and $r_1r_2^mr_1 = r_1$ for $m$ a power of 2. Let $M$ be the $F$-vector space with base $\{x_0, x_1, \ldots\}$, made into $R$-module by the actions $r_2x_i = x_{i+1}, r_1x_i = 0$ for $i$ not a power of 2, and $r_1x_i = x_0$ for $i$ a power of 2. Then $M$ is faithful and simple since any submodule contains $x_0$ and thus $M$. Hence $R$ is a primitive ring. Also $\{\mathrm{Ann}\, r_2^mr_1 : m$ is a power of $2\}$ is an infinite ascending chain of annihilators. Furthermore $Rr_1$ is a minimal left ideal since $Rr_1 \approx M$ under the correspondence $r_2^ir_1 \mapsto x_i$.)

8. Prove theorem 6.2.17 without the hypothesis ACC(Ann). (Hint: Continue the given proof. By exercise 3.3.14 it is enough to show $\mathrm{Sing}(R) = 0$. But if $z \in \mathrm{Sing}(R)$ then $0 \neq L \cap \mathrm{Ann}\, z \subseteq L$ so $L \cap \mathrm{Ann}\, z = L$ and thus $L\,\mathrm{Sing}(R) = 0$, implying $\mathrm{Sing}(R) = 0$.)

9. An affine $F$-algebra of GK dim 2 and classical Krull dimension 1, but not having K-dim. Let $R = F\{X_1, X_2\}/\langle X_2\rangle^2$. Then taking $S = \{\bar{X}_1, \bar{X}_2\}$ we have $G_S(n) = (n+1)(n+2)/2$ so GK dim$(R) = 2$, but $R$ has infinite uniform dimension and thus lacks K-dim. Also note GK dim $R/N = 1$ where $N = \langle X_2 \rangle$ is nilpotent.

## Gelfand-Kirillov Dimension

10. If $T$ is the ring of differential polynomials $R[\lambda;\delta]$ with $R$ affine then $\operatorname{GK}\dim(T) = 1 + \operatorname{GK}\dim(R)$. (Hint: ($\geq$) as in remark 6.2.22'). ($\leq$) Note that if $\delta S \subseteq V_i(S)$ then $\delta V_n(S) \subseteq V_{n+i}(S)$.)

11. If $\operatorname{GK}\dim(R_1) \leq 2$ then $\operatorname{GK}\dim(R_1 \otimes R_2) = \operatorname{GK}\dim(R_1) + \operatorname{GK}\dim(R_2)$. If $\operatorname{GK}\dim(R_1) \geq 2$ then $\operatorname{GK}\dim(R_1 \otimes R_2) \geq 2 + \operatorname{GK}\dim(R_2)$. (Hint: Enough to show ($\geq$). For $\operatorname{GK}\dim(R_1) = 1$ then $G_S(n) \geq n$; for $\operatorname{GK}\dim(R_1) > 1$ then by Bergman's theorem $G_S(n) > n^2/2$.)

12. Suppose $S$ is a commutative, multiplicatively closed, left and right denominator set of regular elements of $R$, generated by elements $s$ for which the derivation $[s, ]$ is nilpotent on every finite dimensional $F$-subspace of $R$. Then $\operatorname{GK}\dim(S^{-1}R) = \operatorname{GK}\dim(R)$. (Hint: Let $V$ be an f.d. subspace of $S^{-1}R$, and take $s_1 \cdots s_t V \subseteq R$ for which each $[s_i, ]$ is nilpotent on $V$. Inverting the $s_i$ one at a time reduces to the case where $S$ is generated by a single element $s$. Take an f.d. subspace $V'$ of $R$ such that $1, s \in V'$, $Vs^t \subseteq V'$, and $[s, V'] \subseteq V'$. Letting $\delta = [s, ]$ one has $\delta^m V' = 0$ for some $m$, and so $V^n \subseteq (V's^{-t})^n \subseteq s^{-n(t+m)}(V')^{mn+n}$ yielding $\operatorname{GK}\dim(S^{-1}R) \geq \operatorname{GK}\dim(R)$.) Incidentally the hypothesis that $S$ is a left and right denominator set is superfluous since $S$ is locally triangularizable, but this requires a careful argument, cf., Krause-Lenagan [85B, proposition 4.6].

13. (Bergman [81]) An affine PI-algebra $R$ with an f.g. module $M$ of GK-dim 2, having submodule $N$ for which $\operatorname{GK}\dim(N) = \operatorname{GK}\dim(M/N) = 1$. Let $R = F\{r_1, r_2\}$ where $r_2 r_1 = 0$ and let $M = Rx + Ry$ satisfy the relations $r_1^{n+1} r_2^n x = 0$ and $r_1 r_2^n y = 0$ unless $n$ is a square $m^2$ in which case $r_1 r_2^n y = r_1 r_2^m x$. Let $N = Ry$. Taking $S = \{1, r_1, r_2\}$ note $S^n x + S^n y$ has base $\{r_1^i r_2^j x$ and $r_2^j y: i \leq j$ and $i + j \leq n\}$ so $\operatorname{GK}\dim(M) = 2$. But every $r_1 r_2^j x \in N$ so $\operatorname{GK}\dim(M/N) = 1$. On the other hand, $\operatorname{GK}\dim(N) = 1$ since to get $r_1^i r_2^j x$ in $S^n y$ one needs only $i + j^2 \leq n$. $R$ is PI since $[R, R]^2 = 0$. Bergman also shows how to modify this example to get $M$ arbitrary.

14. (Lenagan) If $R$ is a fully bounded Noetherian ring then $\operatorname{GK}\dim(M) = \operatorname{GK}\dim(R/\operatorname{Ann}_R M)$. (Hint: By exercise 3.5.13 $R$ is an $H$-ring. Let $A = \operatorname{Ann}_R M = \operatorname{Ann}_R\{x_1, \ldots, x_t\}$. Then $R/A \hookrightarrow \bigoplus Rx_i \subseteq M^{(t)}$ so $\operatorname{GK}\dim(R/A) \leq \operatorname{GK}\dim M$.)

15. $\lim_{n \to \infty} G_S(n)^{1/n}$ always exists. (Hint: If $n = pq + r$ then $G_S(n)^{1/n} \leq G_S(p)^{q/n} G_S(r)^{1/n} \leq G_S(p)^{1/p} G_S(p)^{1/n}$; take limits.)

16. If $\gamma(R)$ denotes the Hilbert series of a graded algebra $R$ then $\gamma(R \otimes T)_k = \gamma(R)\gamma(T)$ since $(R \otimes T)_k = \sum R_j \otimes T_{k-j}$. In particular $\gamma(F[\lambda_1, \ldots, \lambda_t]) = (1 - \lambda)^{-t}$.

17. If $R$ is prime Noetherian then $\operatorname{GK}\dim R = \operatorname{GK}\dim L$ for each $0 \neq L \lhd R$. (Hint: Analogous to the proof of proposition 3.5.46.)

# §6.3

## The Bergman-Sarraille Counterexample

1. (Bergman) Let $R = \mathbb{Q}\{x, y\}/\langle[x, z], [y, z], z^2\rangle$ where $z = [x, y]$. $R$ is an affine Noetherian PI-ring not f.g. over any commutative subring. (Hint: $z \in Z(R)$, $(Rz)^2 = 0$, and $R/Rz$ is commutative Noetherian. Hence $[X_1, X_2]^2$ is a PI of $R$, and $R$ is Noetherian since $Rz$ is an f.g. module over $R/Rz$.)

To get the negative properties let $T$ be the subring of $M_3(\mathbb{Q}[\lambda_1, \lambda_2])$ generated by $x = e_{12} + \lambda_1 \cdot 1$ and $y = e_{23} + \lambda_2 \cdot 1$. Then $[x, y] = e_{13} \in Z(T)$ and $[x, y]^2 = 0$, so there is an obvious surjection $\varphi: R \to T$ which is shown to be an isomorphism by describing $T$ explicitly as $\{p \cdot 1 + p_1 e_{12} + p_2 e_{23} + s e_{13} : p, s \in \mathbb{Q}[\lambda_1, \lambda_2]\}$ where $p_i$ denotes $\partial p/\partial \lambda_i$.

One can now show that $R$ cannot be f.g. over a commutative subring $H$. For $\bar{R} = R/Rz$ would be f.g. over $\bar{H} = (H + Rz)/Rz$, so $\bar{H}$ is affine by Artin-Tate; then $\bar{H}$ is f.g. over a subalgebra $\mathbb{Q}[\bar{f}, \bar{g}]$ where $\bar{f}, \bar{g}$ are algebraically independent over $\mathbb{Q}$. But taking $f = p \cdot 1 + \cdots$ and $g = q \cdot 1 + \cdots$ in $H \subset T$ one sees the derivation $\delta = p_1 \partial/\partial_2 - p_2 \partial/\partial_1$ is 0 on $\mathbb{Q}[\bar{f}, \bar{g}]$ and thus on $\bar{R}$, since $\bar{R}$ is integral over $\mathbb{Q}[\bar{f}, \bar{g}]$. Then $0 = \delta\lambda_1 = \delta\lambda_2$ implying $p_1 = p_2 = 0$ so $\bar{f} \in \mathbb{Q}$, contradiction.) Actually this argument shows $R$ is not integral over any commutative subring.

2. (Sarraille) A prime affine Noetherian PI-ring not f.g. (or even integral) over any commutative subring. Let $R' = R[\lambda_3]$ where $\lambda_3$ is a new commuting indeterminate, and $R$ is as in exercise 1, and let $R'' = R' + \lambda_3 M_3(\mathbb{Q}[\lambda_1, \lambda_2, \lambda_3])$ using the description of $R'$ as a subring of $M_3(\mathbb{Q}[\lambda_1, \lambda_2, \lambda_3])$ of example 1. Then $R''$ is a prime PI-ring since localization gives $M_3(\mathbb{Q}(\lambda_1, \lambda_2, \lambda_3))$; however $R$ is a homomorphic image of $R''$, so $R''$ cannot be integral over a commutative subring. To prove $R''$ is affine and Noetherian it suffices to show $\lambda_3 M_3(\mathbb{Q}[\lambda_1, \lambda_2, \lambda_3])$ is an f.g. $R'$-module, as in example 6.2.3. This is done by showing the $\lambda_3 e_{ij}$ span, by a straightforward computation.) Sarraille's example has K-dim 3. This is the lowest dimension for a counterexample, in light of Braun [81].

## Embeddibility into Matrices

3. If $R$ is affine over a field $F$ and f.g. over $C \subseteq Z(R)$ then $R$ is embeddible into matrices. (Hint: $C$ is affine. Furthermore, one may assume $R$ is irreducible, so every element of $C$ is regular or nilpotent. But Noether normalization says $R$ is f.g. over a polynomial ring $C_0$; localizing at $C_0 - \{0\}$ embeds $R$ into a finite dimensional algebra over the field of fractions of $C_0$ and thus into matrices.) The hypothesis "affine" can also be removed from the hypothesis, using a theorem of Cohen from commutative algebra instead of Noether normalization.

4. (Malcev) An affine $F$-algebra $R$ is embeddible in matrices iff $R$ is a subdirect product of finite dimensional algebras over fields, each of which is embeddible in $n \times n$ matrices for some fixed $n$. (Hint: ($\Leftarrow$) Clear. ($\Rightarrow$) One may assume $R \subseteq M_n(C)$ with $C$ affine over $F$ and then assume $R$ irreducible. By Noether normalization $C$ is f.g. over a polynomial ring $C_0$. If $P_i$ is a maximal ideal of $C_0$ then $P_i C \neq C$ by Nakayama so $P_i R \neq R$; pass to $R/P_i R$ and $C/P_i C$ which is f.d. over the field $C_0/P_i$.)

5. (Lewin's original example) Let $C = \mathbb{Q}[\lambda, \mu]$ for commuting indeterminates $\lambda, \mu$. $\begin{pmatrix} \mathbb{Q} + \langle\lambda\rangle & C \\ \langle\lambda\rangle & C \end{pmatrix}$ is affine by example 6.2.3 but has an uncountable number of ideals. Thus there are an uncountable number of affine algebras satisfying the identities of $2 \times 2$ matrices, which are *not* embeddible in matrices.

6. Details of the Irving-Small counterexample. Let $R = F\{X_1, X_2\}/\langle X_1^2, X_2 X_1 X_2\rangle$. Write $\bar{X}_i$ for the image of $X_i$ in $R$. Taking a base $v_1, v_2$ of $V = F[\lambda]^{(2)}$ one can identity $\bar{X}_i$ with $f_i: V \to V$ given by $f_1 v_2 = f_2 v_2 = 0$, $f_1 v_1 = v_2$ and $f_2 v_1 = \lambda v_1$. Thus $R \subseteq M_2(F[\lambda])$ by the regular representation and satisfies the identities of

$2 \times 2$ matrices. Given any subset $I = \{i_1, i_2, \ldots\}$ of $\mathbb{N}$ written in ascending order, let $I' = \{i_1, i_1 + i_2, 2i_1 + i_2 + i_3, 4i_1 + 2i_2 + i_3 + i_4, \ldots\}$ and $A_I = \langle \bar{X}_1 \bar{X}_2^i \bar{X}_1 : i \in I' \rangle \lhd R$. It remains to show each $R/A_I$ has no chains of annihilators of length $> 4$.

Write $T$ for $R/A_I$, and $x_i$ for the image of $\bar{X}_i$ in $T$. Any left annihilator in $T$ is contained in $Tx_1 + Tx_2$; since $(Tx_1 + Tx_2)x_1x_2 = 0$ we see $Tx_1 + Tx_2$ is the unique maximal left annihilator. Likewise $x_1 T + x_2 T$ is the unique maximal right annihilator, implying $Tx_2x_1 = \text{Ann}(x_1 T + x_2 T)$ is the unique minimal left annihilator. Now suppose $L$ is a left annihilator with $Tx_2x_1 < L < Tx_1 + Tx_2$. Then $L$ has some element $a = x_1 p_1(x_2) + p_2(x_2)x_1 + x_1 p_3(x_2)x_1$ where the $p_i$ are polynomials in $x_2$ and $p_1 \neq 0$. Then $\text{Ann}' a = x_1 x_2 T + V(a)$ where $V(a) = \{q(x_2)x_1 : x_1 p_1(x_2)q(x_2)x_1 = 0\}$. It suffices to prove that if $[V(a):F] > 1$ then $L = Tx_2x_1 + Fa$. So assume $[V(a):F] > 1$. In the above notation if $\deg q \leq \deg p_1$ then one can solve for the coefficients of $q$ in terms of the coefficients of $p_1$, so by assumption there is $q$ for which $\deg q > \deg p_1$. But fixing $q$ and performing the same argument on $\text{Ann}'(x_1 x_2 T + q(x_2)x_1 T)$ we have $L = Tx_2x_1 + Fa$ unless there is some $a'$ in $L$ for which the corresponding $p_1'$ has $\deg p_1' > \deg q$. Now again one has some $q'(x_2)x_1$ in $V(a)$ for which $\deg q' > \deg p_1'$. $I'$ must contain $m_1 = \deg p_1 + \deg q$, $m_2 = \deg p_1' + \deg q$, $m_3 = \deg p_1 + \deg q'$, and $m_4 = \deg p_1' + \deg q'$, and $m_4 - m_3 = m_2 - m_1$. By construction of $I'$ one has $m_1 = m_3$ and $m_2 = m_4$, contradiction.

7. The tensor product of affine Noetherian PI-algebras over a field is affine Noetherian (and PI by theorem 6.4.19 or the easier exercise 6.4.4). (Hint: Let $N_i = \text{Nil}(R_i)$ which are nilpotent, and let $R = R_1 \otimes R_2$ and $N = N_1 \otimes R_2 + N_2 \otimes R_1$, a nilpotent ideal. Then each of the factors of the chain $R > N > N^2 > \cdots$ is an f.g. $R/N$-module, so replacing $R_i$ by $R_i/N_i$ one may assume the $R_i$ are semiprime. Then one may assume the $R_i$ are prime. Pass to the "trace rings" and work there.)

8. (Amitsur-Small) There are only a countable number of Noetherian $\mathbb{Q}$-affine PI-algebras. (Hint: There are only a countable number of semiprime ones, since these are embeddible. Thus fix $T = R/N$ where $N = \text{Nil}(R)$ and argue by induction on the index $t$ of nilpotence of $N$. Letting $W = R/N^{t-1}$ one can embed $R$ in $\begin{pmatrix} T & M \\ 0 & W \end{pmatrix}$ where $M$ is an f.g. $R - W$ bimodule; applying induction to $W$ it suffices to prove there are a countable number of f.g. $T - W$ bimodules, i.e., f.g. $T \otimes W^{\text{op}}$-modules. But $T \otimes W^{\text{op}}$ is Noetherian (and countable).)

## Lewin's Embedding

9. Write $M(R; A, B)$ for the universal $R/A - R/B$ bimodule, cf., after remark 6.3.10. This is unique up to isomorphism by "abstract nonsense," and can be defined explicitly as the free bimodule with formal generators $\{\delta r : r \in R\}$ modulo the sub-bimodule generated by all $\delta(c_1 r_1 + c_2 r_2) - c_1 \delta r_1 - c_2 \delta r_2$ and $\delta(r_1 r_2) - (\delta r_1)\bar{r}_2 - \bar{r}_1 \delta r_2$ for $r_i$ in $R$. Consequently $M(R; A, B) \approx R/A \otimes_R M(R; 0, 0) \otimes_R R/B$.

10. (Bergman-Dicks [75]) Using exercise 9 show $M(R; A, B) \approx R/A \otimes K \otimes R/B$ where $K = \ker(R \otimes_C R \to R)$. (Hint: One may assume $A = B = 0$. Given any

derivation $d: R \to N$ define $K \to N$ by $\sum r_i \otimes r_i' \mapsto \sum (dr_i)r_i' = -\sum r_i dr_i'$, a bimodule map.)

11. Applying $\text{Tor}_1^C(\_\_, R/B)$ to $0 \to A \to R \to R/A \to 0$ and comparing to (5) shows $\text{Tor}_1(R/A, R \otimes_C R/B)$ is an image of $\text{Tor}_1^C(R/A, R/B)$. In view of digression 6.3.12 conclude that $\ker \varphi = AB$ whenever $\text{Tor}_1^C(R/A, R/B) = 0$.

## Kurosch's Problem and the Trace Ring

12. Fast PI-solution to Kurosch's problem (Small). By exercise 6.2.5 one may assume $R$ is prime. Then $Z(R)$ is affine algebraic so is a field, and thus $R$ is simple. Conclude with corollary 6.3.2.

13. (Schelter) Suppose $W$ is a given subring of $R$. We say $r \in R$ is *integral over $W$ of degree $u$* if there are generalized $W$-monomials $f_i(X_1)$ of degree $i$ for $0 \le i < u$ such that $r^u = \sum_{i=0}^{u-1} f_i(r)$. $R$ is *$W$-integral* if each element of $R$ is integral over $W$. Prove the more general version of theorem 6.3.23 where $R$ is a centralizing $W$-integral extension of $W$.

14. (Schelter) If $R = W\{r_1, \ldots, r_t\}$ is a prime centralizing extension of $W$ then modifying construction 6.3.28 build a "trace ring" $T(R)$ which is f.g. over $WC'$. (Hint: Use exercise 13.) Thus $R$ and $T(R)$ have the common ideal $Rg_n(R)$, and $T(R)$ inherits many of the properties of $W$ via $WC'$, including the analogue of theorem 6.3.31.

15. Notation as in construction 6.3.28, suppose $\lambda_1, \ldots, \lambda_t$ are commuting indeterminates over $R$ and $a_1, \ldots, a_t \in R$ are arbitrary. Then each characteristic coefficient of $\sum_{i=1}^t a_i \lambda_i$ (as an element of $R[\lambda_1, \ldots, \lambda_t]$) can be expressed as a polynomial in $\lambda_1, \ldots, \lambda_t$ and the characteristic coefficients of the $a_i$. (Hint: This can be read off as a special case of Amitsur [80], but seems to have an easier proof in this special case by multiplying by any $z$ in $g_n(R)$ and applying proposition 6.1.19 and the theory of elementary symmetric functions.)

16. If $R$ is a domain and $W \subseteq R$ then PI-deg($W$) divides PI-deg($R$). (Hint: Passing to ring of fractions of $R$ one may assume $R$ is a division ring; replacing $W$ by $Z(R)W$ one may assume $W$ is a division ring. Take a maximal subfield $E$ of $W$ and a maximal subfield $F$ of $R$ containing $E$. Then PI-deg($R$)/PI-deg($W$) = $[F:E][Z(W):Z(R)]$.)

17. The reduction of the Bergman-Small theorem to $T(R)$.

*Step I.* It suffices to prove the weaker assertion

(*) $\quad n - m = \sum k_i n_i \qquad$ where $k_i \in \mathbb{N}$ and $n_i = \text{PI-deg}(R/P_i)$ for $P_i \in \text{Spec}(R)$.

(Hint: Take this such that the subsum over the maximal ideals $P_i$ is as large as possible. If there is some nonmaximal $P_i$ then taking a maximal ideal $Q_i \supset P_i$ and letting $m' = \text{PI-deg}(R/Q) = \text{PI-deg}(\bar{R}/\bar{Q})$ where $\bar{R} = R/P_i$, one has $n_i - m' = \sum k_i' n_i'$ so substitute back $n_i = m' + \sum k_i' n_i'$ to get a contradiction.)

*Step II.* Take $P$ minimal with respect to PI-deg($R/P$) = $m$. Arguing inductively on $n - m$ one may assume PI-deg($R/Q$) = $n$ for all prime ideals $Q \subset P$. But then LO($P$) holds from $R$ to $T(R)$ by theorem 6.3.31(i), so the passage to $T(R)$ can be made.

18. (Going Down) Suppose $R$ is a prime PI-ring and is integral over an integral domain $C \subseteq Z(R)$, with $C$ "normal" in the sense that $C$ is integrally closed in

its field of fractions. If $P_0 \subset P_1$ in $\text{Spec}(C)$ and $P_1' \in \text{Spec}(R)$ lies over $P_1$ then there is some $P_0' \in \text{Spec}(R)$ lying over $P_0$, with $P_0' \subseteq P_1'$. (Hint, cf., Rowen [80B, theorem 4.4.24]: Let $R_1 = \{r \in R: 0 \neq r + P_1' \in Z(R/P_1')\}$ and $S = C - P_0$, and let $S_1 = \{rs: r \in R_1, s \in S\}$, a submonoid of $R$. It is enough to show $S_1 \cap P_0 R = \varnothing$. Otherwise, $rs \in P_0 R$ for some $r$ in $R_1$, $s$ in $S$. The minimal monic $p(\lambda)$ for $rs$ over $C$ is irreducible by Gauss' lemma. On the other hand, write $rs = \sum a_i r_i$ for $a_i$ in $P_0$ and $r_i$ in $R$. $C\{r_1, \ldots, r_t\}$ is f.g., so using the regular representation and taking determinants show $rs$ satisfies a monic polynomial $q(\lambda)$ all of whose non-leading coefficients are in $P_0$. Hence $p \mid q$ by Gauss' lemma. Write $p = \lambda^d + \sum_{i=0}^{d-1} c_i \lambda^i$. Passing to the domain $\bar{C} = C/P_0$ note $\bar{q}$ is a power of $\lambda$, so $\bar{p} = \lambda^d$ and hence each $c_i \in P_0$.

Now pass to $S^{-1}C$, which is also normal. $\lambda^d + \sum c_i s^{i-d} \lambda^i$ is the minimal polynomial of $r$; since $r$ is integral over $C$ each $c_i s^{i-d} \in C$, and thus $\in P_0$. Thus $r^d \in P_0 R \subseteq P_1'$ so $r + P_1'$ is nilpotent, contradiction.)

19. (Gordon-Small) Every affine PI-algebra over a field has Gabriel dimension. (Hint: By 6.3.36' and 6.3.39 there are prime ideals $P_1, \ldots, P_t$ with $P_1 \ldots P_t = 0$. Considering each $P_{i+1} \cdots P_t / P_i \cdots P_t$ one may pass to $R/P_i$ and assume $R$ is prime. By induction on cl. K-dim one may assume G-dim $R/L$ exists for every large left ideal $L$ of $R$. Thus it suffices to show G-dim $U$ exists for each cyclic uniform left ideal $U$. For any $0 \neq U' < U$ note $U/U'$ is torsion so has the form $R/L'$ for $L'$ large, implying G-dim $U/U'$ exists; conclude G-dim $U$ exists.

## Noetherian PI-rings (following Resco-Small-Stafford [82])

20. If $R$ is left Noetherian PI and $P \in \text{Spec}(R)$ has height $> 1$ then $P$ contains infinitely many height 1 primes. (Hint: Pass to $T(R)$ using proposition 6.3.45. Thus $R$ is integral over $C$. Localizing at $P \cap C$ assume $C$ is local. For each $c \in P \cap C$ there is a prime $P(c) \subset P$ minimal over $c$ and thus of height 1 by the principal ideal theorem. Hence $P(c) \cap C \neq P \cap C$, so by prime avoidance there are an infinite number of these $P(c)$.)

21. If $R$ is prime Noetherian Jacobson of PI-degree $n$ then for every $t \leq \text{K-dim } R$ there is $P$ in $\text{Spec}_n(R)$ of height $t$. (Hint: Induction on $t$; clear for $t = 0, 1$. Using exercise 20 show $\bigcap\{\text{height 1 primes contained in } P\} = 0$. Thus there is $Q$ in $\text{Spec}_n(R)$ of height 1 contained in $P$. Pass to $R/Q$.)

22. If $R$ is prime Noetherian of PI-degree $n$ and $P \in \text{Spec}_n(R)$ then height $P \leq \text{pd } R/P$. (Hint: Let $S = Z(R) - P$. Then $S^{-1}R$ is local so gl. dim $S^{-1}R = \text{pd}(S^{-1}R/S^{-1}P) \leq \text{pd}_R R/P$. But height$(P) = \text{K-dim } S^{-1}R = \text{gl. dim } S^{-1}R$ since $S^{-1}R$ is Azumaya.)

23. If $R$ is prime Noetherian Jacobson PI of finite gl. dim then K-dim $R \leq$ gl. dim $R$. (Hint: exercises 21 and 22.) Resco-Small-Stafford [82] extend this result to the case of an arbitrary semiprime Noetherian ring, by passing to $R[\lambda]$ and localizing at the monic polynomials.

In the following exercises assume the Lorentz-Small theorem (c.f., digression 6.3.46(3)) and refer to assertions (ii)–(v) following it.

24. Prove (ii). (Hint: One needs only show $(\geq)$. Take $P \in \text{Spec}(R)$ with GK dim $R = $ GK dim $R/P$, and let $\bar{R} = R/P$ and $\bar{Z} = Z(R/P)$. Then GK dim $R = \text{tr deg } \bar{Z}/F$ so taking a transcendence base one can show $(\geq)$ explicitly.

25. Prove (iii). (Hint: Take $P$ as in Lorenz-Small and note $A_1 \subseteq P$ or $A_2 \subseteq P$.)
26. Prove (iv). Hint: ($\leq$) clear. ($\geq$) By exercise 3.5.13 there are $x_1, \ldots, x_t$ in $M$ for which Ann $M = \text{Ann}\{x_1, \ldots, x_t\}$, so $R/\text{Ann } M \hookrightarrow \bigoplus Rx_i \leq M^{(t)}$.)
27. Prove (v). (Hint: By (iv) assume $M$ is faithful and GK dim $M$ = GK dim $R$. Let $A_1 = \text{Ann } N$ and $A_2 = \text{Ann } M/N$. Then $A_1 A_2 = 0$ so conclude by (iii), (iv).)
28. If $R$ is Noetherian PI and the reduced rank $\rho(M) = 0$ then GK dim $M <$ GK dim $R$. (Hint: Noetherian induction and exercise 27, as with K dim.)
29. (Small-Stafford [81]) A prime Noetherian ring $R$ having precisely two maximal $P_1, P_2$ such that $P_1^2 = P_1$, $\bigcap_{i \in \mathbb{N}} P_2^i = 0$, such that $P_2$ is neither left nor right localizable but the completion of $R$ at $P_2$ is Noetherian. (Moreover $Z(R)$ is Noetherian.) As with several noncommutative affine counterexamples, this relies heavily on a famous example from commutative ring theory. Nagata found an example of a commutative ring $H$ f.g. over a local ring $C$ such that $H$ has two maximal ideals $M'$ and $M''$ of respective heights 1 and 2, with $M' \cap M'' = \text{Jac}(C)$, cf., Zariski-Samuel [60B, p. 327]. Let $M = \text{Jac}(C) = M' \cap M'' \lhd H$. Let $R = \left\{ \begin{pmatrix} h & b \\ a & c \end{pmatrix} : h \in H, a \in M, c \in C, b \in M', \text{ and } h - c \in M' \right\}$. $R$ is an order in $M_2(H)$, for $\begin{pmatrix} M & M \\ M & M \end{pmatrix} \subset R$. Hence $R$ is a prime. Also $Z(R) \approx C$ is Noetherian, so $R$ is Noetherian. $J = \begin{pmatrix} M & M' \\ M & M \end{pmatrix}$ is a quasi-invertible ideal of $R$, seen by taking determinants. Also $R/J \approx H/M$ is a direct product of two simple rings, by the Chinese Remainder Theorem. Hence $J = \text{Jac}(R)$. Writing $1 = a' + a''$ where $a' \in M'$ and $a'' \in M''$, let $s = \begin{pmatrix} a'' & 0 \\ 0 & 1 \end{pmatrix}$.

Define $P_1 = J + sR$ and $P_2 = \begin{pmatrix} M' & M' \\ M & M \end{pmatrix}$. $P_1$ and $P_2$ are distinct ideals properly containing $J$; conclude $P_1, P_2$ are the unique maximal ideals of $R$ and $P_1 \cap P_2 = J$. To see $P_1^2 = P_1$ show $P_1^2 \supseteq J^2 + (sJ + Js) = J$ since $M' + Ca'' = H$; but $s^2 \in P_1^2 - J$ so $J \subset P_1^2$ implying $P_1^2 = P_1$. On the other hand, $\bigcap P_2^i = 0$ since $\bigcap (M')^i = 0$ by Krull's intersection theorem, and $P_2 \subseteq M_2(M')$.

Let $S = \{$elements regular modulo $P\}$. Then $s \in S$. Take $x$ in $M' - a''M'$. If $S$ were right Ore then there would be $s'$ in $S$ for which $(xe_{12})s' = sr'$ for suitable $r' = (r_{ij})$ and $s' = (s_{ij})$. Now $s' \in S$ requires $s_{22} \notin M$ and thus $s_{22}$ is a unit (in $C$), but $xs_{22} = sa''r_{12} \in a''M'$, contradiction. This proves $S$ is not right Ore, and similarly $S$ is not left Ore.

It remains to show the completion of $R$ at $P_2$ is Noetherian. By exercise 3.5.14 it suffices to show there is an ideal $A$ polycentral modulo $A^2$, for which $P_2^2 \subseteq A \subseteq P_2$, for then the completions of $R$ at $P_2$ and at $A$ are isomorphic. Take $A = P_2^2 + RM$ and show $A^2 + RM = A$ by computation.

30. In the above notation if $R' = \{(r_{ij}) \in R : r_{21} = 0\}$ then $P_2 \cap R'$ is a prime ideal of $R'$ the intersection of whose powers is 0, and is not right localizable, as above. However, $P_2 \cap R'$ *is* left localizable by direct computation. Note $R'$ is f.g. as $C'$-module.

These examples have been extended further by Beachy-Blair [86], to provide pathological examples of completions of Noetherian PI-rings.

## §6.4

1. $S_{n-1}([(X_1 - X_1^*)^n, X_2], [(X_1 - X_1^*)^{n-2}, X_2], [(X_1 - X_1^*)^{n-3}, X_2], \ldots, [X_1 - X_1^*, X_2])$ is a nonspecial identity of $(M_n(F), *)$ for any $n$, because the trace of any skew-symmetric matrix is 0.

2. (Nagata-Higman) Suppose $R_0$ is a $\mathbb{Q}$-algebra without 1 satisfying the identity $X_1^n$. Then $R_0^m = 0$ for some $m$. The *Nagata-Higman* bound is $m = 2^n - 1$; the *Razmyslov* bound is $m = n^2$. (Hint for Nagata-Higman bound: Let $f = \sum_{i=0}^{n-1} X_1^i X_2 X_1^{n-i-1}$, an identity of $R_0$ since $(r_1 + \alpha r_2)^n = 0$ for all $r_i$ in $R_0$ and $\alpha$ in $\mathbb{Q}$. But then $0 = \sum_{j=0}^{n-1} f(r_1, rr_2^j) r_2^{n-j-1} = nr_1^{n-1} rr_2^{n-1} + \sum_{i=0}^{n-2} r_1^i rf(r_2, r_1^{n-i-1})$ so $r_1^{n-1} rr_2^{n-1} = 0$ and thus $AR_0 A = 0$ where $A_0$ is the ideal generated by $\{r^{n-1} : r \in R\}$. But $R_0/A$ is nilpotent of index $2^{n-1} - 1$ by induction.)

3. (Generic matrices with involution.) Let $\{\zeta_{ij}^{(k)} : 1 \le i, j \le n\}$ be a set of commuting indeterminates. Define $(F_n\{Y\}, t)$ to be the subring of $M_n(F[\zeta])$ generated by the generic matrices $Y_k = (\zeta_{ij}^{(k)})$ and their transposes $Y_k^t = (\zeta_{ji}^{(k)})$. The transpose restricts to an involution $(t)$ on $(F_n\{Y\}, t)$, which also is a prime ring. Show it is the relatively free ring with involution corresponding to $\mathscr{L}(M_n(F), t)$ for any infinite field $F$. Perform a similar construction for the canonical symplectic involution when $n$ is even.

4. (Easy special case of Regev's theorem) If $R_2$ satisfies all identities of $M_n(C[\Lambda])$ and $R_1$ is a PI-algebra over $C$ then $R_1 \otimes_C R_2$ is a PI-algebra. (Hint: One may replace $R_2$ by the algebra of generic matrices and thus by $M_n(C[\Lambda])$, so apply corollary 6.1.54.)

5. Prove Regev's lemma 6.4.18. (Hint: One wants to find suitable relatively prime $t_\pi$ in $\mathbb{Z}$ for which $f(X_1, \ldots, X_n) = \sum_{\pi \in \text{Sym}(n)} t_\pi X_{\pi 1} \cdots X_{\pi n}$ is an identity. Formally evaluate $f(r_{11} \otimes r_{12}, \ldots, r_{n1} \otimes r_{n2})$ and obtain a system of $c_n(I_1)c_n(I_2)$ equations in $n!$ indeterminates.)

6. Given any permutation $\pi$ in $\text{Sym}(n)$, construct tables $T_1(\pi) = (t_{ij})$ and $T_2(\pi) = (u_{ij})$ as follows: $t_{11} = 1$ and $u_{11} = \pi 1$. Inductively $t_{1j}$ is the smallest $m$ (if there is one) such that $t_{1,j-1} < m \le n$ and $\pi m > u_{1,j-1}$; then $u_{1j} = \pi t_{1j}$. When it is impossible to continue on the first row start on the second row. Then $\pi$ can be reconstructed uniquely from $T_1(\pi)$ and $T_2(\pi)$. Defining $d(\pi) =$ number of rows in $T_1(\pi)$ prove $d(\pi)$ is the largest number $d$ for which there are $1 \le i_1 < \cdots < i_d \le k$ such that $\pi i_1 > \pi i_2 > \cdots > \pi i_d$.

7. Prove $c_k(I) \le (d - 1)^{2k}$ if $I$ is a $T$-ideal containing a multilinear PI $f$ of degree $d$. (Hint: Notation as in exercise 6 and preceding lemma 6.4.18. Let $V_k'$ be the subspace of $V_k$ spanned by those $X_{\pi 1} \cdots X_{\pi k}$ for which $d(\pi) < d$. Suppose there is $\pi$ such that $h = X_{\pi 1} \cdots X_{\pi k} \notin V_k' + I_k$; choose $\pi$ such that the word $\pi 1 \cdots \pi k$ is minimal possible and obtain a contradiction by writing $h = h_1 X_{\pi i_1} h_2 X_{\pi i_2} \cdots h_d X_{\pi i_d} h_{d+1}$ with $i_1 < i_2 < \cdots < i_d$ and $\pi i_1 > \pi i_2 > \cdots > \pi i_d$ and applying $f$. Thus $V_k' + I_k = V_k$ so $c_k(I)$ is bounded by the number of tables $T_1(\pi)$ having $< d$ rows. But $\pi$ is determined by $T_1(\pi)$ and $T_2(\pi)$, each of which are functions $\{1, \ldots, k\} \to \{1, \ldots, d - 1\}$, so there are most $(d - 1)^{2k}$ such tables.)

8. Multilinearize the Hamilton-Cayley polynomial for $2 \times 2$ generic matrices to get the $2 \times 2$ trace identity

$$Y_1 Y_2 + Y_2 Y_1 - (\text{tr } Y_1) Y_2 - (\text{tr } Y_2) Y_1 + \text{tr } Y_1 \text{ tr } Y_2 - \text{tr}(Y_1 Y_2) = 0.$$

**9.** Using exercise 8 and the Hamilton-Cayley polynomial show the ring of invariants $Z'$ of $2 \times 2$ matrices (using two generic matrices) is generated by tr $Y_1$, tr $Y_2$, det $Y_1$, det $Y_2$, and tr$(Y_1 Y_2)$, and, furthermore, the trace ring is a free $Z'$-module with base $\{1, Y_1, Y_2, Y_1 Y_2\}$.

**10.** Prove the Formanek-Schofield decomposition of the trace ring $T$ of $F_2\{Y_1, Y_2\}$ as the free product $C_1 \coprod C_2$ of $C_1$ and $C_2$ over $C$, where the notation is described in the text. (Hint: First note $C_1$ and $C_2$ generate $T$ since one gets tr$(Y_1 Y_2)$ using exercise 8. Next let $z = Y_1 Y_2 + Y_2 Y_1 - (\text{tr } Y_1)Y_2 - (\text{tr } Y_2)Y_1 \in Z(T)$ by exercise 8, and note $C_1 \coprod C_2 = C[z] + C[z]Y_1 + C[z]Y_2 + C[z]Y_1 Y_2 \approx T$ by exercise 9.

**11.** (Kemer) If $R$ is a $\mathbb{Q}$-algebra satisfying a standard identity $S_n$ then $R$ satisfies some Capelli identity $C_k$. (Hint: Replacing $n$ by $n + 1$ if necessary one may assume $n$ is even. Define the associative algebra $T$ by generators $e_1, \ldots, e_n$ and $x_1, x_2, \ldots$ and relations $e_i w e_j = -e_j w e_i$ for all $i, j$ and all $w$ in $T$; thus $T$ is a generalization of the exterior algebra. Let $M$ be the subspace of $F\{X\}$ spanned by all $C_{2t}(X_1, h_1, \ldots, X_t, h_t)$ for $h_j$ in $F\{X\}$ and let $N$ be the subspace of $T$ spanned by $e_1 h_1(x) e_2 h_2(x) \cdots e_t h_t(x)$ where $h_j(x)$ denotes the evaluation of $h_j$ under the substitution $X_i \mapsto x_i$. There is a vector space isomorphism $\Phi: M \to N$ by $\Phi C_{2t}(X_1, h_1, \ldots, X_t, h_t) = e_1 h_1(x) \cdots e_t h_t(x)$. Let $A$ be a $T$-ideal of $F\{X\}$, and write $A(T) = \{f(T): f \in A\}$. Then $\Phi(A \cap M) = A(T) \cap N$, for if $f(X_1, \ldots, X_{2t})$ is multilinear with $f(e_1, h_1(x), \ldots, e_t, h_t(x)) = w \in N$ then

$$w = (t!)^{-1} \sum_{\pi \in \text{Sym}(t)} (\text{sg } \pi) f(e_{\pi 1}, h_1(x), \ldots, e_{\pi t}, h_t(x)) \in \Phi(A \cap M).$$

Consequently $A$ contains the Capelli polynomial $C_{2m}$ iff $A(T)$ contains all words in which the $e_i$ occur $m$ times, iff $\left(\sum_{i=1}^n Te_i T\right)^m \subseteq A(T)$.

Let $I_u = \{e_{i_1} \cdots e_{i_u}: 1 \leq i_1, \ldots, i_u \leq n\}$. One needs to show that if $I_n \subseteq A(T)$ then $(I_1 T)^m \subseteq A(T)$ for some $m$. Thus it suffices to show for each $k \leq n$ that $TI_k T$ is nilpotent modulo $TI_{k+1}T + A(T)$. Write $n = 2j$ and let $f = \sum_{\pi \in \text{Sym}(j)} X_{\pi 1} \cdots X_{\pi j}$, the multilinearization of $X_1^j$. An easy manipulation in $T$ shows

$$f(a_1 w_1, \ldots, a_j w_j)a_{j+1} \in TI_{k+1}T + A(T) \qquad \text{for all } a_i \text{ in } I_k \text{ and } w_i \text{ in } T;$$

hence $TI_k T$ is nil of bounded index modulo $TI_{k+1}T + A(T)$, so one is done by exercise 2.)

# 7    Central Simple Algebras

This chapter deals with the theory of (finite dimensional) central simple algebras and the Brauer group, in particular, with division algebras. There is some question among experts as to whether this theory belongs more properly to ring theory, field theory, cohomology theory, or algebraic $K$-theory. Accordingly we give an abbreviated account of classical parts of the subject readily found elsewhere; a very thorough treatment of the subject is to be found in the forthcoming book by Jacobson and Saltman, referred to in the sequel as JAC-SAL; certain key parts of this chapter (such as Brauer factor sets) draw on Jacobson's notes for that book. There also has been growing interest in infinite dimensional division algebras, which we shall discuss briefly in 7.1.46ff.

## §7.1 Structure of Central Simple Algebras

In this section we develop the tools for studying the basic structure theory of central simple algebras.

Let us recall some of the properties already proved about central simple algebras. Any simple ring $R$ can be viewed as an algebra over the field $F = Z(R)$, and we consider the case when $[R:F] < \infty$, in which case we say $R$ is a *central simple F-algebra*. Then $[R:F] = n^2$ for some $n$ which we call the *degree* of $R$ (cf., corollary 2.3.25), written deg($R$). By Wedderburn's theorem we can write $R = M_t(D)$ for a division algebra $D$ and suitable $t$. Note that $D \approx \text{End}_R D^{(t)}$ from which we see by means of proposition 2.1.15 that $D$ is unique up isomorphism, and $t$ is uniquely determined since $D$ has IBN. In particular, deg $D$ is a uniquely determined integer ($= n/t$) called the *index* of $R$ and is a very important invariant of $R$. We call $D$ the *underlying division algebra* of $R$. A central simple division algebra is called a *central division algebra*, for short.

## Centralizers and Splitting Fields

We start off with some fundamental structural theorems describing $R$ in terms of its subalgebras. If $R$ is a central simple $F$-algebra and $R'$ is a finite dimensional simple $F$-algebra with center $F'$ then by corollary 1.7.24 and theorem 1.7.27 $R \otimes_F R'$ is a central simple $F'$-algebra, whose dimension over $F'$ is clearly $[R:F][R':F']$. In particular, theorem 2.3.27 says $R \otimes_F R^{\text{op}} \approx M_{n^2}(F) \approx \text{End}_F R$.

This result has an immediate consequence. Recall $C_R(A)$ denotes the centralizer of $A$ in $R$.

**Proposition 7.1.1:**   *Suppose $A$ is a central simple $F$-subalgebra of $R$.*

(i) $R \approx A \otimes_F C_R(A)$; *in particular,* $\deg(A) \mid \deg(R)$.
(ii) *If $A'$ centralizes $A$ with $[A':F][A:F] \geq [R:F]$ then $R \approx A \otimes A'$ and $A' = C_R(A)$.*

*Proof:*

(i) Let $m = [A:F]$. Then $A^{\text{op}} \otimes_F R$ is central simple and contains $A^{\text{op}} \otimes_F A \approx M_m(F)$, and thus has a set of $m \times m$ matric units, implying by proposition 1.1.3 there is an $F$-algebra $R_1$ for which $A^{\text{op}} \otimes R \approx M_m(R_1) \approx M_m(F) \otimes_F R_1 \approx A^{\text{op}} \otimes A \otimes R_1$. By corollary 1.7.29 we see $R$ and $A \otimes R_1$ are each the centralizer of $A^{\text{op}} \otimes 1$ in $A^{\text{op}} \otimes R$; hence $R \approx A \otimes R_1$; implying $R_1 \approx C_R(A)$ by corollary 1.7.29 again. Thus $R \approx A \otimes_F C_R(A)$ and so $\deg R = (\deg A)(\deg C_R(A))$.

(ii) $R \approx A \otimes C_R(A)$ and $A' \subseteq C_R(A)$; counting dimensions shows $A' = C_R(A)$.
                                                                                                                 Q.E.D.

Of course the isomorphism $R \otimes R^{\text{op}} \approx \text{End}_F R$ pertains to the theory of separable subalgebras, which would permit an even faster proof of proposition 7.1.1. We turn now to the key fact that every maximal commutative separable subalgebra $K$ of $R$ splits $R$ and $[K:F] = \deg(R)$. There are several ways of approaching this result: (i) the generalized density theorem (exercise 2.4.1), as utilized in Jacobson [80B, theorem 4.8]; (ii) basic properties of tensor products, cf., exercise 4; (iii) properties of separable algebras, which we present here, following Knus-Ojanguren [74B].

**Remark 7.1.2:** Suppose $T$ is any ring containing $R$. Viewing $T$ as $R - R$ bimodule we have $T \approx R \otimes_F C_T(R)$, by theorem 5.3.24(iii) and remark 5.3.24'. If $T \subseteq R^e = R \otimes R^{op}$ then identifying $R$ with $R \otimes 1$ yields $T = R \otimes C_T(R)$.

**Proposition 7.1.3:** $C_R(K) \otimes_F R^{op} \approx \operatorname{End}_K R$ *for any subalgebra* $K$ *of* $R$.

**Proof:** Let $T$ be the centralizer of $K \otimes 1$ in $R^e \approx \operatorname{End}_F R$; thus $T \approx \operatorname{End}_K R$. Let $R' = C_T(1 \otimes R^{op})$. Then $T \approx R' \otimes R^{op}$ by the opposite of remark 7.1.2. It remains to show $R' = C_R(K) \otimes 1$. ($\supseteq$) holds by inspection. On the other hand, corollary 1.7.29 yields $R \otimes 1 = C_{R^e}(1 \otimes R^{op}) \supseteq R'$; since $R' \subseteq T$ we have $R' \subseteq C_R(K) \otimes 1$. Q.E.D.

**Corollary 7.1.4:** *If* $K$ *is a subfield of* $R$ *then* $C_R(K)$ *is a central simple* $K$-*algebra of degree* $n/[K:F]$, *where* $n = \deg R$.

**Proof:** $C_R(K)$ is simple since $\operatorname{End}_K(R)$ is simple, and likewise $Z(C_R(K)) = K$. Finally if $m = \deg C_R(K)$ then $m^2 n^2[K:F] = [C_R(K) \otimes_F R^{op}:F] = [\operatorname{End}_K R:F] = [R:K]^2[K:F] = n^4/[K:F]$, implying $m = n/[K:F]$. Q.E.D.

**Corollary 7.1.5:** *If a given maximal commutative subalgebra* $K$ *of* $R$ *is a field then* $K$ *splits* $R$ *and* $[K:F] = \deg R = [R:K]$.

**Proof:** $C_R(K) = K$ by remark 0.0.6, so $K$ splits $R^{op}$, and thus $K^{op} = K$ splits $R$. Furthermore $1 = \deg C_R(K) = n/[K:F]$, so $[K:F] = n$ and $[R:K] = [R:F]/[K:F] = n^2/n = n$. Q.E.D.

What if $K$ is not a field? If $K$ is separable over $F$ we have essentially the same outcome, cf., exercise 1. However if $K$ is not separable one may have $[K:F] > \deg R$.

**Example 7.1.6:** Let $R = \operatorname{End}_F V \approx M_n(F)$ where $[V:F] = n$, and let $V'$ be any subspace of $V$ having dimension $[n/2]$ over $F$. Taking a complement $V''$ of $V'$ we define

$$N = \{f \in R: fV' \subseteq V'' \text{ and } fV'' = 0\}$$

Then $N^2 = 0$ so $F \cdot 1 + N$ is a commutative subalgebra of $R$ whose dimension is $1 + [n/2](n - [n/2]) = 1 + [n^2/4]$. Schur proved this is indeed the maximal possible dimension, and Gustafson [76] found an elegant structure-theoretic proof which we give in exercise 3.

Of course it may be impossible to find a maximal commutative subring which is a field; if $R = M_2(\mathbb{C})$ then $R$ has no subfields properly containing $\mathbb{C}$. However, we can bypass this difficulty by passing to the underlying division algebra $D$, for we have

**Remark 7.1.7:** Any commutative subalgebra of a central division algebra $D$ is a field, by remark 2.3.23(ii); consequently, if $K$ is a maximal subfield of $D$ then $C_D(K) = K$ and $K$ splits $D$.

**Remark 7.1.8:** Any splitting field of the underlying division algebra of $R$ splits $R$ (for if $D \otimes_F K \approx M_n(K)$ then $M_t(D) \otimes_F K \approx M_t(F) \otimes D \otimes K \approx M_{tn}(K)$).

Our final result about centralizers is

**Theorem 7.1.9:** (*Double centralizer theorem*) *If $A$ is any simple subalgebra of $R$ then*

   (i) $C_R(A)$ *is simple and $Z(A)$-central;*
   (ii) $C_R(C_R(A)) = A$;
   (iii) $[A:F][C_R(A):F] = [R:F]$.

**Proof:**

   (i) Let $n = \deg(R)$, $K = Z(A)$, and $t = [K:F]$. Let $S = C_R(K)$, which is simple and $K$-central of degree $n/t$ by corollary 7.1.4. Clearly $A \subseteq S$ so $S \approx A \otimes_K C_S(A)$ by proposition 7.1.1, implying $C_S(A)$ is simple and $Z(C_S(A)) = K$. But $C_R(A) \subseteq C_R(K) = S$ so $C_S(A) = C_R(A)$ proving (i).
   (ii) First assume $K = F$. Write $A' = C_R(A)$. Then $R \approx A \otimes A'$ so $A = C_R(A')$ by proposition 7.1.1(ii). In general note $C_R(A') \subseteq C_R(K) = S$ and $A' = C_R(A) \subseteq S$ so $C_R(A') = C_S(A') = C_S(C_S(A)) = A$ as just shown.
   (iii) $(n/t)^2 = [S:K] = [A:K][A':K] = [A:F][A':F]/t^2$.                    Q.E.D.

Another key tool was presaged in corollary 2.9.2:

**Theorem 7.1.10:** (*Skolem-Noether theorem*) *Any isomorphism of simple F-algebras of $R$ can be extended to an inner automorphism of $R$.*

The proof, which can be found in any standard reference such as Jacobson [80B, p. 222], is sketched in exercise 5.

**Digression 7.1.10′:** Chase [84] proves in fact that any isomorphism of separable, maximal commutative subalgebras of $R$ can be extended to an inner automorphism. The proof involves a modification of the argument of exercise 5, using invertible projective modules.

One useful consequence of the Skolem-Noether theorem is

**Theorem 7.1.11:** (*Wedderburn's theorem*) *Every finite simple* ring $R$ *has the form* $M_n(F)$ *for a suitable field* $F$.

**Proof:** It suffices to show every finite division ring $D$ is a field, since we could take $D$ to be the underlying division ring of $R$. Let $K$ be a maximal subfield of $D$ and assume $|K| = m < |D|$. The number of subgroups of $D - \{0\}$ (as multiplicative group) conjugate to $K - \{0\}$ is at most $|D - \{0\}|/|K - \{0\}| = (|D| - 1)/(m - 1)$. But each subgroup contains the element 1 so the number of conjugates of elements of $K - \{0\}$ is at most $(m - 2)(|D| - 1)/(m - 1) + 1 < |D| - 1$, so some element $d$ of $D$ is not conjugate to any element of $K$.

Let $F = Z(D)$, and let $K'$ be a maximal subfield of $D$ containing $F[d]$. Then $|K'| = |F|^{[K':F]} = |F|^{\deg D} = |F|^{[K:F]} = |K|$ so $K' \approx K$ since finite fields of the same order are isomorphic. Hence $K' - \{0\}$ and $K - \{0\}$ are conjugate by the Skolem-Noether theorem, contrary to choice of $d$.      Q.E.D.

This theorem reduces the structure theory of central simple $F$-algebras to the case $F$ is infinite. This enables us to bring in the Zariski topology and to prove the following important result which ties up some loose ends in our previous discussion.

**Theorem 7.1.12:** *Suppose* $\deg(R) = n$ *and* $F$ *is infinite then* $\{a \in R: F[a]$ *is separable of dimension n over* $F\}$ *is dense in the Zariski topology.*

**Proof:** Take a splitting field $K$ of $R$. The generic reduced characteristic polynomials of $R$ and $R \otimes_F K \approx M_n(K)$ are formally the same (since we use the same base); let us call them both $m_x$. But $M_n(K)$ has elements of degree $n$, so $n$ is the degree of $m_x$. It follows from proposition 2.3.35 that $\{$elements of degree $n\}$ is a Zariski open subset of $R$. For these elements the minimal and reduced characteristic polynomials are the same.

On the other hand, $F(r)$ is a separable extension of $R$ iff the roots $r_1,\dots,r_t$ of the minimal polynomial are distinct, iff the discriminant

$$\prod_{1\le i<j\le t} (r_i - r_j)^2 \neq 0.$$

There is a way of calculating the discriminant from the minimal polynomial (cf., Jacobson [85B, p. 258]. Since $M_n(K)$ has elements of degree $n$ of discriminant $\neq 0$, we see the discriminant of the generic element is nonzero. Hence $\{r \in R\colon \text{discriminant}(r) \neq 0\}$ is Zariski open and nonempty and thus dense. Thus

$$\{r \in R\colon F(r) \text{ is separable of degree } n\}$$

$$= \{r \in R\colon \text{discriminant}(r) \neq 0\} \cap \{r \in R\colon r \text{ has degree } n\}$$

is Zariski dense, as desired.        Q.E.D.

**Corollary 7.1.12′ (Koethe's theorem):** *Suppose $K$ is a subfield of a central division algebra $D$ and is separable over $F$. Then $K$ is a subfield of a maximal subfield of $D$ separable over $F$.*

*Proof:* $K$ is infinite by Wedderburn's theorem, so the division algebra $C_D(K)$ has a maximal subfield $L$ separable over $F$, by theorem 7.1.11. Clearly $L$ is separable over $F$ and $[L\colon F] = [L\colon K][K\colon F] = (\deg C_D(K))[K\colon F] = \deg D$ by corollary 7.1.4.        Q.E.D.

*Note:* The proof relied on the fact every central division algebra $D$ contains a separable field extension of $F$. Another proof was already given in exercise 2.5.10. A very quick proof due to Serre runs as follows: Otherwise $F(d)$ is purely inseparable over $F$ for all $d$ in $D$ so $d^n \in F$ for $n = \deg D$. But then $X^n$ is a central polynomial of $D$, and thus of $M_n(F)$ (seen by splitting $D$), which is absurd.

We shall also need criteria for $R$ to be split.

**Lemma 7.1.13:** *Suppose $\deg(R) = n$.*

(i) *$R$ is split if $R$ has $n$ orthogonal idempotents.*

(ii) *If there is some $r \in R$ whose minimal polynomial over $F$ splits into $n$ distinct linear factors, then $F[r] \approx F^{(n)}$ and $R$ is split.*

*Proof:*

(i) $R$ is a direct sum of $n$ left ideals; writing $R = M_t(D)$ we see $t \geq n$, so $t = n$ and thus $D = F$

(ii) $F[r] \approx F[\lambda]/\langle f(\lambda)\rangle \approx \prod_{i=1}^{n} F$ by the Chinese Remainder Theorem; hence we can apply (i).    Q.E.D.

## Examples of Central Simple Algebras

The first known noncommutative division algebra was Hamilton's *quaternions* $\mathbb{H} = \mathbb{R} \oplus \mathbb{R}i \oplus \mathbb{R}j \oplus \mathbb{R}k$ where multiplication is given by $i^2 = j^2 = k^2 = -1$ and $ij = -ji = k$. $\mathbb{H}$ is 4-dimensional over $\mathbb{R}$ and is a ring because we can identify $\mathbb{H}$ as the subalgebra of $M_2(\mathbb{C})$ spanned by $\begin{pmatrix} i & 0 \\ 0 & -i \end{pmatrix}$, $\begin{pmatrix} 0 & 1 \\ -1 & 0 \end{pmatrix}$, and $\begin{pmatrix} 0 & i \\ i & 0 \end{pmatrix}$ (where $i$ is $\sqrt{-1}$ in $\mathbb{C}$), which take on the respective roles of $i, j$, and $k$; $\mathbb{H}$ is a division ring since every nonzero element is invertible by means of the "norm" formula

$$(a_1 + a_2 i + a_3 j + a_4 k)(a_1 - a_2 i - a_3 j - a_4 k) = a_1^2 + a_2^2 + a_3^2 + a_4^2 \in \mathbb{R}^+$$

for $a_1, \ldots, a_4$ in $\mathbb{R}$. Note $\deg \mathbb{H} = 2$, and $\mathbb{C} = \mathbb{R} \oplus \mathbb{R}i$ is a maximal subfield of $\mathbb{H}$.

Inspired by this example we say $R$ is a *quaternion algebra* if $R$ is central simple of degree 2. Quaternion algebras are often studied in the context of the theory of quadratic forms, cf., exercises 1.9.9ff.

Quaternions were later generalized to the *cyclic algebras* $(K, \sigma, \alpha)$ of example 1.6.28, for any field $K$ with automorphism $\sigma$ of order $n$, and any $\alpha$ in $F = K^\sigma$. We recall $(K, \sigma, \alpha)$ is formally $\sum_{i=0}^{n-1} Kz^i$, satisfying the multiplication rules

$$a_1 z^i a_2 z^j = \begin{cases} (a_1 \sigma^i a_2) z^{i+j} & \text{if } i + j < n \\ (\alpha a_1 \sigma^i a_2) z^{i+j-n} & \text{if } i + j \geq n \end{cases}$$

for $a_1, a_2$ in $K$. We shall see that this very simple construction can be obtained under quite general circumstances.

*Note:* Under this definition $M_2(\mathbb{C})$ would not be cyclic because it has no maximal subfields! To resolve this difficulty we could study maximal commutative separable subalgebras in place of maximal subfields, and we shall carry out this idea below. However, it is easier (and customary) to pass instead to the underlying division algebra, since each of its commutative subalgebras is a subfield.

**Remark 7.1.14:**   By definition the cyclic algebra $R = (K, \sigma, \alpha)$ contains

(i) a cyclic field extension $K$ of $F$ of degree $n$, and
(ii) an element $z \notin F$ for which $z^n \in F$.

Perhaps surprisingly, it turns out that

(i) already ensures $R$ is cyclic, and
(ii) by itself ensures $R$ is cyclic when $n$ is prime. The first of these results is easy and will be given now; the second is a theorem of Albert, cf., [JAC-SAL, theorem 2.19]. (Note (ii) implies (i) if $F$ contains a primitive $n$-th root $\zeta$ of 1; the proof of Albert's theorem in characteristic $\neq n$ entails adjoining $\zeta$ and then applying field theory to pass back. When char$(F) = n$ the proof is easy, cf., exercise 8.)

**Proposition 7.1.15:**   *If $R$ is central simple and has a maximal subfield cyclic over $F$ then $R$ is cyclic.*

**Proof:**   Let $\mathrm{Gal}(K/F) = \langle \sigma \rangle$. By the Skolem-Noether theorem there is $z$ in $R$ for which $zaz^{-1} = \sigma a$ for all $a$ in $K$. Let $\alpha = z^n$ where $n = [K:F]$. Then $\alpha a \alpha^{-1} = \sigma^n a = a$ so $\alpha \in C_R(K) = K$; in fact, $\alpha \in F$ since $\sigma \alpha = zz^n z^{-1} = r^n = \alpha$. We claim $R \approx (K, \sigma, \alpha)$. Indeed, there is a homomorphism from the skew polynomial ring $K[\lambda; \sigma] \to R$ given by $\lambda \mapsto z$; the kernel contains $\lambda^n - \alpha$ so we get a homomorphism $(K, \sigma, \alpha) \to R$ which is an injection since $(K, \sigma, \alpha)$ is simple. But $[R:F] = [K:F]^2 = n^2 = [(K, \sigma, \alpha): F]$ so $(K, \sigma, \alpha) \approx R$.     Q.E.D.

**Corollary 7.1.15′:**   *Every quaternion algebra is cyclic (by Koethe's theorem, since every separable quadratic field extension is cyclic).*

## Symbols

The cyclic algebra $R = (K, \sigma, \alpha)$ has a better form when $F$ has a primitive $n$-th root $\zeta = \zeta_n$ of 1. Indeed, then $K = F(y)$ for some $y$ such that $\beta = y^n \in F$, and taking $z$ for which $zyz^{-1} = \sigma y = \zeta y$ we see multiplication in $R$ is now given by the exceedingly simple rules

$$y^n = \alpha, \qquad z^n = \beta, \qquad \text{and} \qquad zy = \zeta yz \tag{0}$$

**Definition 7.1.16:**   The *symbol* $(\alpha, \beta)_n$ denotes the cyclic algebra described by the equations (0), where $\alpha, \beta \in F - \{0\}$. Write $(\alpha, \beta)_n \sim 1$ if $(\alpha, \beta)_n$ is split. (Strictly speaking, the definition depends on the specific choice of $\zeta$, but we shall not worry about this matter.)

Every quaternion algebra of characteristic $\neq 2$ is a symbol by corollary 7.1.15′, since $-1 \in F$. When $\text{char}(F) = n$ clearly $F$ cannot have a primitive $n$-th root of 1, but in this case there is also a version of the symbol, cf., exercise 8. However, we shall assume $\text{char}(F)$ is prime to $n$.

Jumping the gun a bit, let us say two central simple algebras $R_1$ and $R_2$ are *similar*, written $R_1 \sim R_2$, if their underlying division algebras are isomorphic. If $F$ has "enough" roots of 1 then the very deep Merkurjev-Suslin theorem says any central simple $F$-algebra is similar to a tensor product of symbols; thus symbols are the building blocks of the theory, and we would like to see just how easily we can compute with them.

To start off, given $y, z$ as in (0) let us compute $(y + z)^n$. Certainly we can write $(y + z)^n = \sum_{i=0}^{n} f_{i,n}(y, z)$ where the $f_{i,n}$ are polynomials in $y$ and $z$, homogeneous of degree $i$ in $y$ and degree $(n - i)$ in $z$. We shall show $f_{i,n} = 0$ whenever $1 \leq i < n$. Indeed let us write a typical monomial of $f_{i,n}$ as $v_1 \cdots v_n$, where $i$ of the $v_i$ are $y$ and $n - i$ of the $v_i$ are $z$. Then $v_n v_1 \cdots v_{n-1}$ is also a monomial of $f_{i,n}$, and $v_n v_1 \cdots v_{n-1} = \zeta^i v_1 \cdots v_n$. (Indeed if $v_n = z$ then in moving $z$ to the right we must pass over $y$ $i$ times, each time replacing $zy$ by $\zeta yz$ and thereby multiplying altogether by $\zeta^i$; if instead $v_n = y$ then we replace $yz$ by $\zeta^{-1} zy$ a total of $(n - i)$ times, thereby multiplying altogether by $(\zeta^{-1})^{n-i} = \zeta^i$.) Continuing the process shows

$$v_1 \cdots v_n + v_n v_1 \cdots v_{n-1} + \cdots + v_2 \cdots v_n v_1 = (1 + \zeta^i + \cdots + \zeta^{(n-1)i}) v_1 \cdots v_n = 0$$

so partitioning $f_{i,n}$ in this manner shows $f_{i,n}$ is a sum of zeroes and is thus 0! (This lovely argument of G. Bergman was communicated by P. M. Cohn.) Consequently we have

$$(y + z)^n = y^n + z^n = \alpha + \beta.$$

**Proposition 7.1.17:** *(Basic properties of symbols)*:

(i) $(\alpha, \beta)_n \sim 1$ *if there is $r$ in $(\alpha, \beta)_n$ such that $r^n$ is an $n$-th power in $F$ and $1, r, \ldots, r^{n-1}$ are independent.*

(ii) *If $\alpha$ is a norm of the extension $F(\beta^{1/n})$ over $F$ then $(\alpha, \beta)_n \sim 1$. In particular $(1, \beta)_n \sim 1$*

(iii) *If $\alpha + \beta$ is an $n$-th power in $F$ then $(\alpha, \beta)_n \sim 1$*

(iv) $(\alpha, 1 - \alpha)_n \sim 1$

(v) $(\alpha, \beta)_n \otimes (\alpha, \gamma)_n \sim (\alpha, \beta\gamma)_n$

(v') $(\beta, \alpha)_n \otimes (\gamma, \alpha)_n \sim (\beta\gamma, \alpha)_n$

(vi) $(\alpha, -\alpha)_n \sim 1$

(vii) $(\alpha, \beta)_n \sim (\beta^{-1}, \alpha)_n$

(viii) $(\alpha, \beta)_n \otimes (\beta, \alpha^t)_{nt} \sim 1$ *for any* $t$; thus $(\alpha, \beta)_n \sim (\alpha, \beta)_{nt}^{\otimes t}$

*Proof:*

(i) The minimal polynomial of $r$ is $\lambda^n - \gamma^n$ where $\gamma^n = r^n$, so we are done by lemma 7.1.13(ii).

(ii) Special case of (i) for if $\alpha = N(a)$ then $(a^{-1}z)^n = N(a)^{-1}z^n = \alpha^{-1}\alpha = 1 = 1^n$.

(iii) Take $y, z$ as in (0). As computed above $(y + z)^n = \alpha + \beta$ is an $n$-th power in $F$ so we are done by (i).

(iv) $\alpha + (1 - \alpha) = 1$, so done by (iii).

(v) Take $y, z$ in $(\alpha, \beta)_n$ satisfying (0), and take $y', z'$ in $(\alpha, \gamma)_n$ satisfying $z'y' = \zeta y'z'$ and $(y')^n = \alpha$ and $(z')^n = \gamma$. Working in $R = (\alpha, \beta)_n \otimes (\alpha, \gamma)_n$ let $R_1$ be the subalgebra generated by $y \otimes 1$ and $z \otimes z'$, and $R_2$ be the subalgebra generated by $y^{-1} \otimes y'$ and $1 \otimes z'$. Visibly $R_1$ and $R_2$ centralize each other, and $R_1 \approx (\alpha, \beta\gamma)$ and $R_2 \approx (1, \gamma)_n \sim 1$, so $R \approx R_1 \otimes R_2$ by proposition 7.1.1.

(v') As in (v).

(vi) Taking $y, z$ as in (0) where $\beta = -\alpha$ we have $(yz^{-1})^n = \zeta^{-n(n+1)}y^nz^{-n} = -\zeta^{-n(n+1)/2} = (-1)^n$ as seen by considering $n$ even and $n$ odd separately. Hence $(\alpha, -\alpha)$ is split by (i).

(vii) If $y, z$ are as in (0) then $yz^{-1} = \zeta z^{-1}y$ so we could replace $y, z$, respectively, by $z^{-1}, y$ which yield $(\beta^{-1}, \alpha)_n$.

(viii) Take $y, z$ in $(\alpha, \beta)_n$ as in (0), and $y', z'$ such that $(y')^{nt} = \beta$, $(z')^{nt} = \alpha^t$, and $z'y' = \zeta_{nt}y'z'$. Then $R = (\alpha, \beta)_n \otimes (\beta, \alpha^t)_{nt}$ has the centralizing subalgebras $R_1$ generated by $y^{-1} \otimes z'$ and $1 \otimes y'$, and $R_2$ generated by $y \otimes 1$ and $z^{-1} \otimes (z')^t$. Then $R_1 \sim (1, \beta)_{nt}$ and $R_2 \sim (\alpha^{-1}, 1)_n$ so $R \approx R_1 \otimes R_2$ is split. The second assertion follows from (vii).     Q.E.D.

Certain of these properties follow formally from others, e.g., the reader might try deriving (vii) from (v), (v'), and (vi). This becomes significant in the $K_2$-theory in §7.2.

In fact the converse of (ii) also holds. For convenience we assume $F(\beta^{1/n})$ is a field, although this assumption could be removed if we apply digression 7.1.10' in the proof below instead of the Skolem-Noether theorem.

**Proposition 7.1.18:**  (*Suppose* $F(\beta^{1/n})$ *is a field.*) $(\alpha, \beta)_n$ *is split iff* $\alpha$ *is a norm from* $F(\beta^{1/n})$ *to* $F$. (*This is called Wedderburn's criterion.*)

*Proof:* ($\Leftarrow$) is (ii) above. ($\Rightarrow$) Let $R = (\alpha, \beta)_n$. By assumption $R \sim 1 \sim (1, \beta)_n$, so in addition to elements $y, z$ satisfying (0) we also have $y', z'$ in $R$ satisfying $(y')^n = 1$, $(z')^n = \beta$, and $z'y' = \zeta y'z'$. But $F(z')$ and $F(z)$ are each isomorphic to $F(\beta^{1/n})$ so by the Skolem-Noether theorem we may apply a suitable inner automorphism of $R$ and assume $z' = z$. Now let $x = y'y^{-1} \in C_R(z) = F(z)$. Then $1 = (y')^n = (xy)^n = N(x)y^n = N(x)\alpha$ so $\alpha = N(x^{-1})$ as desired.   Q.E.D.

## Overview of the Theory of Central Simple Algebras

In view of major recent advances by Merkurjev and Suslin the theory of central simple algebras can now be presented in a remarkably clear picture. Let us sketch the picture here and spend the rest of the chapter in elaborating the sketch. The underlying idea is to describe central simple algebras in terms of cyclic algebras.

Using "generic" methods one can readily construct a central division algebra which is not cyclic (cf., after remark 7.1.20), leading to the question of whether central division algebras need have maximal subfields Galois over the center. This also has a counterexample, which we shall present shortly.

These examples lead us back to matrices over a division algebra, for perhaps some $M_t(D)$ will be cyclic even when $D$ is not. Accordingly we define $\text{Br}(F)$ as {similarity classes of central simple $F$-algebras}, where we recall $R_1 \sim R_2$ if they have isomorphic underlying division algebras. $\text{Br}(F)$ has multiplication given by $[R_1][R_2] = [R_1 \otimes R_2]$, and is a group since $[R^{\text{op}}] = [R]^{-1}$ by theorem 2.3.27. Thus $[M_n(F)] = [F] = 1$ in $\text{Br}(F)$. $\text{Br}(F)$ is called the *Brauer group*.

The Brauer group turns out to be torsion, i.e., for every $[R]$ in $\text{Br}(F)$ there is $m$ such that $[R]^m = 1$ or, equivalently, $R^{\otimes m} \sim 1$; $m$ is called the *exponent* of $R$. If $D$ is a division algebra of degree $n$ then $m \mid n$ and, furthermore, every prime factor of $n$ divides $m$. Conversely, if $m$ and $n$ satisfy these two requirements then Brauer constructed a cyclic division algebra $D$ of degree $n$ and exponent $m$ to be studied in §7.3.

A fundamental property asked of division algebras is *decomposability*. Namely, if $\deg(D) = n = n_1 n_2$, are there division subalgebras $D_i$ of respective degree $n_i \neq 1$ for which $D \approx D_1 \otimes D_2$? If $n_1$ and $n_2$ are relatively prime the answer is, "Yes," and we thereby can reduce the theory to the case where $n$ is a power of a prime number $p$. But now if $D \approx D_1 \otimes D_2$ and each $n_i \neq 1$ then $n_i$ divides $n/p$ and thus $m$ divides $n/p$. Hence any decomposable division algebra of degree $n$ must have exponent $< n$. In fact there are indecomposable division algebras of exponent $p$ and degree $p^2$ for every odd prime $p$; for exponent 2 there are indecomposable division algebras of degree 8, but every division algebra of degree 4 and exponent 2 is decomposable.

Instead of asking for a division algebra itself to be indecomposable we might instead ask whether it can be decomposed *in the Brauer group*. This leads us to

**Merkurjev-Suslin Theorem:** *If F has a primitive m-th root $\zeta$ of 1 then every division algebra of exponent m is similar to a tensor product of cyclic algebras of degree m. (It follows that* $\mathrm{Br}(F)$ *is a division torsion abelian group when F has "enough" roots of* 1.)

Thus $M_t(D)$ is decomposable as a tensor product of cyclics for suitable $t$. When $\deg(D) = p$ for $p = 3$ then $D$ is already cyclic; for larger primes $p$ this is unknown, but $t \leq (p - 1)!/2$.

Many of the basic results of $\mathrm{Br}(F)$ are proved here by means of "Brauer factor sets," which also serve to identify $\mathrm{Br}(F)$ with a certain second cohomology group.

## The First Generic Constructions

There are two main ways of building and studying central simple algebras. The more classical method is the "arithmetic method" using tools of algebraic number theory such as local field theory. The second method is the "generic method," which is to attach indeterminates to key quantities and then study leading coefficients of polynomials. Both methods are of the utmost importance. In the text we shall focus on the generic method, which gives quick results. The arithmetic method will be discussed all too briefly in Appendix A.

The usual way of employing the generic method is to build first a domain of PI-degree $n$ resembling a skew polynomial ring and the passing to the ring of fractions, as illustrated in the following basic example.

*Example 7.1.19:* Let $\lambda_1, \lambda_2$ be commuting indeterminates over a field $F_0$ which contains a primitive $n$-th root $\zeta$ of 1. Then the symbol $D = (\lambda_1, \lambda_2)_n$ is a division algebra, where $F = F_0(\lambda_1, \lambda_2)$. Indeed, let $D = \sum_{i,j=0}^{n-1} F y^i z^j$ where $y^n = \lambda_1, z^n = \lambda_2$, and $zyz^{-1} = \zeta y$. Letting $C = F_0[y]$ a polynomial ring in $y$ one sees easily that $D$ is the ring of central fractions of the skew polynomial ring $C[z; \sigma]$ where $\sigma$ is given on $C$ by $\sigma(\sum \alpha_i y^i) = \sum \zeta^i \alpha_i y^i$. But $C[z; \sigma]$ is a domain by proposition 1.6.15 so $D$ is a division algebra by the following observation:

**Remark 7.1.20:** If a central simple algebra $D$ is the ring of central fractions of a domain then $D$ is a division algebra. (Indeed, obviously $D$ is a domain, but then $D$ cannot have any nontrivial idempotent so the Wedderburn-Artin theorem says $D$ must be its own underlying division ring.)

Generic methods enable one to readily produce the noncyclic algebra $(\lambda_1, \lambda_2)_m \otimes (\lambda_3, \lambda_4)_n$ of degree $mn$ over the rational fraction field $F = F_0(\lambda_1, \lambda_2, \lambda_3, \lambda_4)$; this is a special case of theorem 7.1.29(ii).

**Digression 7.1.21:** (Tignol) Let us modify this example by taking $F$ to be the field of Laurent series in $\lambda_1, \ldots, \lambda_4$ over $F_0$. Amitsur showed $D = (\lambda_1, \lambda_2)_m \otimes (\lambda_3, \lambda_4)_n$ is a noncyclic division ring, cf., Jacobson [75B, p. 102]. But Tignol-Amitsur [84] implies $D$ cannot have a cyclic splitting field, for this would have a subfield isomorphic to a subfield of $D$ (since $F$ is Henselian), contrary to $D$ not cyclic. Thus $D$ is *not* similar to a cyclic algebra.

Historically (in view of Koethe's theorem), the next question was whether any central simple algebra has a maximal subfield Galois over the center; such an algebra is called a *crossed product*. This has a negative answer, but first we look for a sufficiently general counterexample. We can find a clue by tidying up a loose end from the Artin-Procesi theorem.

**Proposition 7.1.22:** *Suppose* $\deg R = n$. *Then there are elements* $a, b_1, \ldots, b_n$ *in $R$ for which* $\{a^{i-1} b_j : 1 \le i, j \le n\}$ *are a base for $R$ over $F$.*

**Proof:** *Case I.* $R$ is split, i.e., $R = M_n(F)$. Taking $b_i = e_{ii}$ and $a = e_{n1} + \sum_{i=1}^{n-1} e_{i,i+1}$ we see $a^n = 1$ and each $e_{ij} = a^{j-i} b_j$.

*Case II.* $R$ is *not* split. In particular, $F$ is infinite by theorem 7.1.11. We shall translate the assertion to one on PI's. We want to show the $a^{i-1} b_j$ are independent, so, in other words, the Capelli polynomial evaluation

$$C_{2t}(b_1, ab_1, \ldots, a^{n-1} b_1, b_2, \ldots, a^{n-1} b_n, r_1, \ldots, r_t) \ne 0$$

for suitable $r_1, \ldots, r_t$ in $R$ where $t = n^2$. This means we want to show

$$C_{2t}(X_2, X_1 X_2, \ldots, X_1^{n-1} X_2, X_3, \ldots, X_1^{n-1} X_{n+1}, X_{t+1}, \ldots, X_{2t})$$

is a nonidentity of $R$, or, equivalent, by corollary 6.1.45 that this is a nonidentity of $M_n(F)$. But this is clear by case I and corollary 6.1.45. Q.E.D.

We shall now introduce the "most general" division algebra of degree $n$. Let $F_n\{Y\}$ denote the $F$-algebra of generic $n \times n$ matrices $Y_1, Y_2, \ldots$. By proposition 6.1.41 $F_n\{Y\}$ is prime of PI-degree $n$, so its algebra of central fractions is a simple algebra which we denote as $UD(F, n)$. $UD(F, n)$ is central simple of degree $n$ over its center, whatever that may be, but even without knowing the center we can study $UD(F, n)$ effectively by means of PI-theory.

**Proposition 7.1.23:** *$F_n\{Y\}$ satisfies the same identities as any division F-algebra of degree n (over its center).*

**Proof:** Let $D$ be a division $F$-algebra and $F_1 = Z(D) \supseteq F$. Then $F_1$ is infinite by theorem 7.1.11, so by corollary 6.1.46′, $D$ satisfies the same identities as $F_n\{Y\}$.     Q.E.D.

**Theorem 7.1.24:** *$UD(F, n)$ is a division algebra of degree n.*

**Proof:** Otherwise, by the Wedderburn-Artin theorem, $UD(F, n)$ would have some matric unit $e_{12}$; "clearing denominators" we would have $f(Y_1, \ldots, Y_m) \neq 0$ in $F_n\{Y\}$ for which $f^2 = 0$. In other words, $f(X_1, \ldots, X_m)$ is not an identity of $F_n\{Y\}$ but $f^2$ is an identity, cf., proposition 6.1.42; transferring this information to any $F$-division algebra $D$ of degree $n$ (say example 7.1.19) we would have $d_1, \ldots, d_m$ in $D$ for which $f(d_1, \ldots, d_m) \neq 0$ but $f(d_1, \ldots, d_m)^2 = 0$, an absurdity.     Q.E.D.

**Corollary 7.1.25:** *The product of nonidentities of $F_n\{Y\}$ is a nonidentity.*

**Proof:** $f(Y_1, \ldots, Y_m) \neq 0$ iff $f(X_1, \ldots, X_m)$ is a nonidentity, so apply theorem 7.1.24 which says $F_n\{Y\}$ is a domain.     Q.E.D.

We can push this idea still further.

**Corollary 7.1.26:** *Suppose $f_i(Y_1, \ldots, Y_m)g_i(Y_1, \ldots, Y_m^{-1})$ are nonzero elements of $UD(F, n)$ for $1 \leq i \leq t$ with $g_i$ central. Then for any simple F-algebra R of degree n there are $r_1, \ldots, r_m$ in R for which $f_i(r_1, \ldots, r_m)g_i(r_1, \ldots, r_m)^{-1} \neq 0$ for $1 \leq i \leq t$. (Indeed, let $f = f_1(X_1, \ldots, X_m) \cdots f_t(X_1, \ldots, X_m)g_1(X_1, \ldots, X_m) \cdots g_t(X_1, \ldots, X_m)$. Each $f_i$ and $g_i$ is not an identity of $UD(F, n)$ so $f$ is a nonidentity of $F_n\{Y\}$ and thus of R, i.e., there are $r_1, \ldots, r_m$ in R for which $f(r_1, \ldots, r_m) \neq 0$. Hence each $f_i(r_1, \ldots, r_m) \neq 0$ and $g_i(r_1, \ldots, r_m) \neq 0$ so $f_i(r_1, \ldots, r_m)g_i(r_1, \ldots, r_m)^{-1} \neq 0$, noting $g_i(r_1, \ldots, r_m) \in Z(R)$ and is thus invertible.*

This result is applied by encoding structural information into elements.

**Comparison Lemma 7.1.27:** *Suppose* $UD(F, n)$ *has a subfield* $K$ *Galois over* $Z(UD(F, n))$ *with Galois group* $G$. *Then every division* $F$-*algebra* $D$ *of degree* $n$ *has a subfield Galois over* $Z(D)$ *with Galois group* $G$.

The proof is a rather routine translation of information about $UD(F, n)$ into identities and nonidentities, by means of "clearing denominators" and passing to $F_n\{Y\}$, cf., Rowen [80B, proposition 3.2.10].

Having the comparison lemma in our hands we need something to compare. Namely, we want different examples of division algebras whose maximal subfields do *not* have a common Galois group. This is not difficult if we are willing to manipulate indeterminates; we set up the example here and send the reader to Rowen [80B, pp. 189–191] for the straightforward computational details verifying 7.1.28 and 7.1.29.

**Example 7.1.28:** Suppose $C$ is an integral domain with commuting automorphisms $\sigma_1, \ldots, \sigma_t$. Define inductively $T_0 = F_0$ and $T_k = T_{k-1}[\lambda_k; \sigma_k]$ where $\sigma_k$ is extended to an automorphism of $T_{k-1}$ by the rule $\sigma_k \lambda_j = \lambda_j$ for all $j < k$. Then $T_t$ is a domain, and each $\sigma_k$ extends naturally to an automorphism of $T_t$ satisfying $\sigma_k \lambda_j = \lambda_j$ for all $j \neq k$.

**Theorem 7.1.29:** *Given* $n_1, \ldots, n_t$ *in* $\mathbb{N}$ *let* $F_0$ *be a field having a primitive* $n$-*th root* $\zeta$ *of* 1, *where* $n = n_1 \cdots n_t$. *Define* $\sigma_k$ *on the commutative polynomial ring* $C = F_0[\mu_1, \ldots, \mu_t]$ *by* $\sigma_k \mu_j = \mu_j$ *for* $j \neq k$ *and* $\sigma_k \mu_k = \zeta^{n/n_k} \mu_k$. *Building* $T_t$ *as in example* 7.1.28, *let* $D$ *be the ring of central fractions of* $T_t$.

(i) *$D$ is a division algebra of degree $n$ with center*

$$F = F_0(\mu_1^{n_1}, \ldots, \mu_t^{n_t}, \lambda_1^{n_1}, \ldots, \lambda_t^{n_t}).$$

*Furthermore, $D$ has a maximal subfield $K = F_0(\mu_1, \ldots, \mu_t, \lambda_1^{n_1}, \ldots, \lambda_t^{n_t})$ and $K$ is Galois over $F$ with Galois group $G = \langle \sigma_1 \rangle \times \cdots \times \langle \sigma_t \rangle$.*

(ii) *Suppose $L$ is an arbitrary subfield of $D$ Galois over $F$. Then $\mathrm{Gal}(L/F)$ is abelian of exponent dividing the least common multiple of $n_1, \ldots, n_t$.*

(iii) *On the other hand, if $t = 1$ and $L$ is Galois over $F$ then $\mathrm{Gal}(L/F)$ does not have a subgroup isomorphic to $\mathbb{Z}_p \times \mathbb{Z}_p \times \mathbb{Z}_p$.*

Applying the comparison lemma to theorem 7.1.29 gives the following surprising result:

**Theorem 7.1.30:** *(Amitsur) Suppose $n$ is divisible by $p^3$ and $p \nmid \mathrm{char}(F)$. Then $UD(F, n)$ does not have a subfield Galois over $Z(UD(F, n))$ of dimension $p^3$; in particular, $UD(F, n)$ does not have a maximal subfield Galois over the center.*

*Proof:* If $\mathrm{UD}(F, n)$ had a subfield Galois over the center with Galois group $G$ of order $p^3$ then $G$ is abelian of exponent $p$ by theorem 7.1.29(ii) applied to lemma 7.1.27 (taking $n_1 = n_2 = \cdots = p$); but then $G \approx \mathbb{Z}_p \times \mathbb{Z}_p \times \mathbb{Z}_p$, which is impossible by theorem 7.1.29(iii) (taking $n_1 = n$) applied to lemma 7.1.27.

$\hfill$ Q.E.D.

A direct non-crossed product construction has been found recently by Jacob-Wadsworth [86].

Note that these counterexamples all have composite degree, leaving open

*Question 7.1.31:* (The major open question in division algebras) (i) Is every division algebra of prime degree cyclic? (ii) Failing this, is it similar to a cyclic algebra?

## Brauer Factor Sets

So far we know by Koethe's theorem that every finite dimensional division algebra $D$ has a maximal subfield separable over the center, but there may fail to be maximal subfields Galois over the center. This leads us to try to describe the structure of an arbitrary central simple algebra $R$ in terms of a maximal separable subfield, but this is complicated by the fact that the maximal subfields might have the "wrong" dimension, e.g., $R = M_n(\mathbb{C})$, where $\mathbb{C} = Z(R)$ is a maximal subfield! However $R$ always has a suitable commutative separable subring, and we recapture $R$ using *Brauer factor sets.* Brauer's construction, treated in van der Waerden [48B], had been overshadowed by Noether's "crossed product construction" for the case $K$ a Galois field extension of $F$, but has enjoyed a renaissance, largely because of the excellent exposition in Jacobson [83], which is the basis of our treatment here.

We start with a given central simple algebra $R = M_t(D)$, of degree $n$; thus $\deg D = n/t$. By Koethe's theorem $D$ has a maximal separable subfield $K_0$, and we take $K = K_0^{(t)} \subseteq R$ (viewing $K$ as $\begin{pmatrix} K_0 & & \\ & \ddots & \\ & & K_0 \end{pmatrix} \subset R$). Note $[K:F] = n$.

Let $E$ be the normal closure of $K_0$ over $F$. Since $K_0$ splits $R$ we can view

$$R \subseteq R \otimes_F K_0 \approx M_n(K_0) \subseteq M_n(E).$$

A Zariski topology argument analogous to theorem 7.1.12 shows $\{a \in K_0 : K_0 = F(a)\}$ is dense in $K_0$. Thus $\{a \in K : K = F(a)\}$ is dense in $K$, so $K = F(a)$ for some $a$ in $K$, and the conjugates of $a$ under $G$ are therefore distinct. Furthermore, $K \otimes_F K_0$ is a direct product of fields, cf., lemma 7.1.13(ii), so $a$ is diagonalizable in $M_n(E)$ by standard arguments from linear algebra.

**Proposition 7.1.32:** $\{r \in R : r = KrK\}$ is Zariski dense in R.

**Proof:** Let $B = \{a^u r a^v : 0 \le u, v \le n - 1\}$. $KrK = R$ iff $B$ spans $R$; iff $B$ is linearly independent, since $B$ has $n^2$ elements. Writing $B = \{b_1, \ldots, b_t\}$ where $t = n^2$ we see $B$ is independent iff the Capelli polynomial evaluation $C_{2t}(b_1, \ldots, b_t, r_1, \ldots, r_t) \ne 0$ for suitable $r_i$ in $R$ by remark 6.1.16'. But this is clearly a Zariski-open condition, so passing to $R \otimes E \approx M_n(E)$ we may assume by remark 6.1.23' that $a$ is diagonal with distinct entries. Thus the proposition has been reduced to the following assertion:

**Claim 7.1.32':** If $R = M_n(F)$ and $a$ is diagonal with distinct entries $a_1, \ldots, a_n$ then there is $r \in R$ such that $R = \sum_{u,v=0}^{n-1} F a^u r a^v$.

**Proof of Claim:** Take $r = \sum e_{ij}$, i.e., each entry of $r$ is 1. Then $a^u r a^v = (a_i^u a_j^v)$, so we want to show for any matrix $y$ in $R$ there are $\alpha_{uv}$ in $F$ for $0 \le u, v \le n - 1$ such that $y = \sum_{u,v=0}^{n-1} \alpha_{uv}(a_i^u a_j^v) = (\sum_{u,v} \alpha_{uv} a_i^u a_j^v)$. Changing point of view let $w$ be the Vandermonde matrix $(a_i^{j-1}) = \sum_{i,j=1}^{n} a_i^{j-1} e_{ij}$, cf., remark 6.4.13. Then letting $\alpha$ denote the matrix $\sum_{u,v=0}^{n-1} \alpha_{uv} e_{u+1,v+1}$ and $w^t$ denote the transpose of $w$ we would have

$$w \alpha w^t = \sum_{u,v} \alpha_{uv} a_i^u a_j^v = y$$

But $w$ is invertible since the $a_i$ are distinct, so $\alpha = w^{-1} y (w^t)^{-1}$ is the desired solution. Q.E.D.

**Notational Hypothesis 7.1.33:** Having proven proposition 7.1.32 we fix the following notation (when relevant): View $R \subseteq R \otimes_F K_0 \approx M_n(K_0) \subseteq M(E)$ and take a set of matric units $e_{ij}$ in $E$ with respect to which $a = \sum_{i=1}^{n} a_i e_{ii}$ is diagonal; the $a_i$ are then distinct. Choosing $r$ in $R$ for which $R = KrK$ we write $r = (r_{ij})$ in $M_n(E)$. By choice of $r$ we see $M_n(E)$ is spanned by the $a^u r a^v = (r_{ij} a_i^u a_j^v)$ for $0 \le u, v \le n - 1$; in particular, all components $r_{ij}$ must be nonzero.

Since $R = KrK = \sum_{u,v=0}^{n-1} F a^u r a^v$ we can write any arbitrary element $x$ of $R$ as $\sum_{u,v=0}^{n-1} \alpha_{uv} a^u r a^v$ where $\alpha_{uv} \in F$ are uniquely determined. Viewing $x$ as a matrix $(x_{ij})$ in $M_n(E)$ we have $x_{ij} = \sum_{u,v=0}^{n-1} \alpha_{uv} r_{ij} a_i^u a_j^v = \beta_{ij} r_{ij}$ where

$$\beta_{ij} = \sum_{u,v=0}^{n-1} \alpha_{uv} a_i^u a_j^v \in E. \tag{4}$$

The $\beta_{ij}$ are the key to the whole theory.

$E$ is a Galois extension of $F$, whose Galois group $G$ permutes the $a_i$. In this way we can view $G \subseteq \text{Sym}(n)$ by means of the notation $\sigma a_i = a_{\sigma i}$ for every $\sigma$ in $G$. (In other words, we use $\sigma$ to denote both the automorphism and its corresponding permutation of the indices.)

**Proposition 7.1.34:**

(i) *The $\beta_{ij}$ satisfy the conjugacy conditions $\sigma\beta_{ij} = \beta_{\sigma i, \sigma j}$ for all $\sigma$ in $G$.*

(ii) *Conversely, suppose we have a separable field extension $K_0$ of $F$, and $K = K_0^{(t)} = F(a)$, where a has degree n. Let $E$ be the normal closure of $K_0$ over $F$. Given arbitrary $\beta_{ij}$ in $E$ satisfying the conjugacy conditions with respect to $G = \mathrm{Gal}(E/F)$ one can find $\alpha_{uv}$ in $F$, $0 \le u, v \le n - 1$, for which (4) holds.*

**Proof:**

(i) $\sigma\beta_{ij} = \sum \alpha_{uv}\sigma a_i^u \sigma a_j^v = \sum \alpha_{uv} a_{\sigma i}^u a_{\sigma j}^v = \beta_{\sigma i, \sigma j}$

(ii) Let $a_1, \ldots, a_n$ be the conjugates of $a$, and $(a_i^{j-1})$ be the corresponding Vandermonde matrix (cf., proof of claim 7.1.32'). Defining $(\alpha_{uv})$ by the matrix equation

$$(\alpha_{uv}) = (a_i^{u-1})^{-1}(\beta_{ij})(a_j^{v-1})^{-1},$$

we obviously have (4) satisfied, so it remains to show each $\alpha_{uv} \in F$, i.e., $\sigma\alpha_{uv} = \alpha_{uv}$ for all $\sigma$ in $G$. But (4) yields

$$\sum_{u,v} \alpha_{uv} a_{\sigma i}^u a_{\sigma j}^u = \beta_{\sigma i, \sigma j} = \sigma\beta_{ij} = \sum(\sigma\alpha_{uv})(\sigma a_i)^u(\sigma a_j)^v$$

$$= \sum(\sigma\alpha_{uv})a_{\sigma i}^u a_{\sigma j}^v,$$

so matching coefficients yields $a_{uv} = \sigma a_{uv}$ as desired.        Q.E.D.

**Definition 7.1.35:** A set $\{c_{ijk}: 1 \le i, j, k \le n\} \subset E$ is a *Brauer factor set* of $E$ over $F$ if it satisfies the following *Brauer factor set conditions* for all $i, j, k, m$ and all $\sigma$ in $G = \mathrm{Gal}(E/F)$:

BFS1:    $\sigma c_{ijk} = c_{\sigma i, \sigma j, \sigma k}$

BFS2: $c_{ijk}c_{ikm} = c_{ijm}c_{jkm}$.        (Often we view this as $c_{ijk} = c_{ijm}c_{jkm}c_{ikm}^{-1}$.)

**Remark 7.1.36:** Setting $j = i$ and $m = 1$ in BFS2 shows $c_{iik} = c_{ii1}$ for all $i, k$. Likewise, setting $k = j$ yields $c_{ijj} = c_{jjm}$; thus $c_{iij} = c_{iii} = c_{jii}$ for all $i, j$.

**Theorem 7.1.37:**

(i) *(Notation as in 7.1.33) R gives rise to a Brauer factor set $\{c_{ijk}\}$ defined by*

$$c_{ijk} = r_{ij}r_{jk}r_{ik}^{-1} \tag{5}$$

(ii) *Conversely, suppose* $K = K_0^{(t)} = F[a]$ *where* $K_0$ *is a separable field extension of* $F$, *and* $E$ *is the normal closure of* $K_0$, *and let* $n = [K:F]$. *Given a Brauer factor set* $\{c_{ijk}\}$ *define* $R = \{(\beta_{ij}) \in M_n(E): \sigma\beta_{ij} = \beta_{\sigma i, \sigma j} \text{ for all } \sigma \text{ in } G\}$. *Endowing* $R$ *with multiplication*

$$(\beta_{ij})(\beta'_{ij}) = (\beta''_{ik}) \qquad \text{where } \beta''_{ik} = \sum_{j=1}^{n} \beta_{ij} c_{ijk} \beta'_{jk} \tag{6}$$

*we have* $R$ *is a central simple* $F$-*algebra, and there is a canonical injection* $\Phi: R \to M_n(E)$ *given by* $\varphi(\beta_{ij}) = (\beta_{ij} c_{ij1})$. $K$ *is isomorphic to a maximal separable subalgebra of* $R$.

(iii) *Constructions* (i) *and* (ii) *are inverse constructions, as explained in the proof.*

### Proof:

(i) Any element of $R$ can be written in the form $(\beta_{ij} r_{ij})$ as seen preceding (4). Then the product of two elements $(\beta_{ij} r_{ij})$ and $(\beta'_{ij} r_{ij})$ is $(\beta''_{ij} r_{ij})$ where by (5)

$$\beta''_{ik} r_{ik} = \sum_{j=1}^{n} \beta_{ij} r_{ij} \beta'_{jk} r_{jk} = \sum_{j=1}^{n} \beta_{ij} c_{ijk} \beta'_{jk} r_{ik}, \text{ so}$$

$$\beta''_{ik} = \sum_{j=1}^{n} \beta_{ij} c_{ijk} \beta'_{jk}, \tag{7}$$

seen by canceling $r_{ik}$. Now let us take the particular $(\beta_{ij} r_{ij})$ correspond to 1 in $R$. Then $1 = 1 \cdot 1$ implies $\beta_{ik} = \sum_{j=1}^{n} \beta_{ij} c_{ijk} \beta_{jk}$. Applying the conjugancy conditions on $\beta$ we have

$$\sum_{j=1}^{n} \beta_{ij} c_{ijk} \beta_{jk} = \beta_{ik} = \sigma^{-1}\beta_{\sigma i, \sigma k} = \sum_{j} \sigma^{-1}\beta_{\sigma i, \sigma j} \sigma^{-1} c_{\sigma i, \sigma j, \sigma k} \sigma^{-1}\beta_{\sigma j, \sigma k}$$

$$= \sum_{j} \beta_{ij} \sigma^{-1} c_{\sigma i, \sigma j, \sigma k} \beta_{jk}.$$

Thus $0 = \sum_{j=1}^{n} \beta_{ij}(c_{ijk} - \sigma^{-1} c_{\sigma i, \sigma j, \sigma k})\beta_{jk}$. Consider the matrix

$$(\gamma(i)_{jk}) = (r_{ij}^{-1}(c_{ijk} - \sigma^{-1} c_{\sigma i, \sigma j, \sigma k})\beta_{jk})$$

where $i$ is fixed. Then

$$0 = (\beta_{ij} r_{ij})(\gamma(i)_{jk}) = (\gamma(i)_{jk}) \qquad \text{since } (\beta_{ij} r_{ij}) = 1.$$

Hence $(c_{ijk} - \sigma^{-1} c_{\sigma i, \sigma j, \sigma k})\beta_{jk} = 0$ for all $i, j, k$. The same argument applied on the right shows each $c_{ijk} - \sigma^{-1} c_{\sigma i, \sigma j, \sigma k} = 0$, so applying $\sigma$ gives $\sigma c_{ijk} = c_{\sigma i, \sigma j, \sigma k}$, i.e., BFS1. To get BFS2 note

$$c_{ijk} c_{ikm} = c_{ij} c_{jk} c_{ik}^{-1} c_{ik} c_{km} c_{im}^{-1} = c_{ij} c_{jk} c_{km} c_{im}^{-1} = c_{ijm} c_{jkm}.$$

(ii) Clearly, $R$ is an $F$-subspace of $M_n(E)$ and is closed under the given multiplication since

$$\sigma\beta''_{ik} = \sum_{j=1}^{n} \sigma\beta_{ij}\sigma c_{ijk}\sigma\beta'_{jk} = \sum_{j} \beta_{\sigma i, \sigma j}c_{\sigma i, \sigma j, \sigma k}\beta'_{\sigma j, \sigma k} = \beta''_{\sigma i, \sigma k}.$$

$\Phi$ is a homomorphism since

$$c_{ik1}\beta''_{ik} = c_{ij1}c_{jk1}c_{ijk}^{-1}\beta''_{ik} = \sum_{j} c_{ij1}c_{jk1}c_{ijk}^{-1}\beta_{ij}c_{ijk}\beta'_{jk}$$

$$= \sum_{j}(c_{ij1}\beta_{ij})(c_{jk1}\beta'_{jk}).$$

Obviously $\Phi$ is an injection since each $c_{ij1}$ is invertible, so we can identify $R$ with $\bar{R} = \Phi R \subseteq M_n(E)$. To see $K \subseteq \bar{R}$ we note the diagonal matrix $\sum_{i=1}^{n} a_i e_{ii}$ is the image of $\sum_{i=1}^{n} a_i c_{ii1}^{-1} e_{ii}$ which is in $R$ since $\sigma(a_i c_{ii1}^{-1}) = a_{\sigma i}c_{\sigma i, \sigma i, \sigma 1}^{-1} = a_{\sigma i}c_{\sigma i, \sigma i, 1}^{-1}$ by remark 7.1.36. On the other hand, the matrix whose entries are all 1 satisfies the conjugacy conditions and thus yields an element of $R$ whose image under $\Phi$ is $r = (c_{ij1})$ in $M_n(E)$. Since $K \otimes E$ splits we see $\bar{R}E$ contains each $e_{ii}re_{jj} = c_{ij1}e_{ij}$, so each $e_{ij} \in \bar{R}E$. Consequently $M_n(E)$ is a central extension of $\bar{R}$, so $\bar{R}$ is prime and clearly finite dimensional over $F$, implying $\bar{R}$ is simple. It remains to show $F = Z(\bar{R})$. Suppose $z \in Z(\bar{R})$. Then $z \in Z(\bar{R}E) = Z(M_n(E)) = E$. Hence $z = \Phi(\sum_{i=1}^{n} c_{ii1}^{-1}ze_{ii})$ so by the conjugacy condition and remark 7.1.36

$$\sigma c_{ii1}^{-1}\sigma z = \sigma(c_{ii1}^{-1}z) = c_{\sigma i, \sigma i, 1}^{-1}z = c_{\sigma i, \sigma i, \sigma 1}^{-1}z = (\sigma c_{ii1}^{-1})z,$$

implying $z = \sigma z$ for each $\sigma$ in $G$ and thus $z \in F$. Hence $\bar{R}$ (and thus $R$) is central simple, of degree $n$ since PI deg$(\bar{R}) = n$.

(iii) Starting with $R$ we can apply (i) to obtain the Brauer factor set $(c_{ijk})$ and then applying (ii) we get an algebra isomorphic to $R$, seen by confronting (6) with (7). On the other hand, if we are given $(c_{ijk})$ and build $R$ and $\bar{R}$ as in (ii) then taking $r = (c_{ij1}) \in \bar{R}$ we see from the proof of (ii) that $M_n(E) = KrK$; using $r$ as in the discussion preceding (4) we use the procedure of (i) to obtain a Brauer factor set and find we have recaptured (i).     Q.E.D.

It is useful to remember from the proof of (iii) that the $r_{ij}$ and $c_{ij1}$ play the same role.

***Example 7.1.38:*** If $R = M_n(F)$ then taking $r$ to be the matrix each of whose entries is 1 we get $c_{ijk} = 1$ for each $i, j, k$; this is called the *trivial Brauer factor set*. More generally, if $R$ corresponds to the Brauer factor set $(c_{ijk})$ over $K$ then $M_t(R)$ corresponds to the Brauer factor set $(c_{i'j'k'})$ over $K^{(t)}$, where we define $c_{i'j'k'}$ to be $c_{ijk}$ if $i' \equiv i(\text{mod }n)$, $j' \equiv j(\text{mod }n)$ and $k' \equiv k(\text{mod }n)$.

Indeed, write $R = KrK$ in the manner which yields $(c_{ijk})$ and let $r'$ be the $tn \times tn$ matrix formed by $t^2$ blocks each of which is a copy of $r$. Obviously $M_t(R) = K^{(t)}r'K^{(t)}$, viewing $K^{(t)}$ as $t$ copies of $K$ along the diagonal, and this yields the desired $(c_{i'j'k'})$.

**Remark 7.1.39:**

(i) BFS1 implies $\sigma c_{ijk} = c_{ijk}$ for all $\sigma$ fixing $F(a_i, a_j, a_k)$ and thus by Galois theory $c_{ijk} \in F(a_i, a_j, a_k)$.

(ii) Taking $m = 1$ in BFS2 we see $c_{ijk} = c_{ij1}c_{jk1}c_{ik1}^{-1}$ for all $j, k, i$, so the $(c_{ijk})$ are actually determined by the $(c_{ij1})$ in this manner.

Let us now see when two Brauer factor sets yield isomorphic central simple algebras (with respect to the same $K$).

**Theorem 7.1.40:**  *Suppose $(c_{ijk})$ and $(\tilde{c}_{ijk})$ are Brauer factor sets of $E$ over $F$, notation as in 7.1.33, corresponding respectively to central simple algebras $R$ and $\tilde{R}$ as in theorem 7.1.37. There is an isomorphism $R \to \tilde{R}$ fixing $K$ iff there is nonzero $(\gamma_{ij})$ in $M_n(E)$ satisfying the conjugacy conditions for which*

$$\tilde{c}_{ijk} = c_{ijk}\gamma_{ij}\gamma_{jk}\gamma_{ik}^{-1} \qquad \text{for each } i, j, k. \tag{8}$$

(In such a case we say $(c_{ijk})$ and $(\tilde{c}_{ijk})$ are *associate*.)

**Proof:**  $(\Leftarrow)$ As in theorem 7.1.37 we write an element of $R$ as $(\beta_{ij}r_{ij})$ where $r_{ij} = c_{ij1}$ and where the $\beta_{ij}$ satisfy the conjugacy conditons, with multiplication as in (7). Then define $\psi: R \to R'$ by $\psi(\beta_{ij}r_{ij}) = (\beta_{ij}\gamma_{ij}^{-1}\tilde{r}_{ij})$ where $\tilde{r}_{ij} = \tilde{c}_{ij1}$. Clearly, $\psi$ has inverse given by $(\beta_{ij}\tilde{r}_{ij}) \to (\beta_{ij}\gamma_{ij}r_{ij})$ so it remains to show $\psi$ is a homomorphism. Letting $\beta_{ik}'' = \sum_{j=1}^n \beta_{ij}c_{ijk}\beta_{jk}'$ we see $\beta_{ik}''\gamma_{ik}^{-1} = \sum_{j=1}^n \beta_{ij}\gamma_{ij}^{-1}\tilde{c}_{ijk}\beta_{jk}'\gamma_{jk}^{-1}$ implying

$$\psi((\beta_{ij}r_{ij})(\beta_{jk}'r_{jk})) = \psi(\beta_{ik}''r_{ik}) = (\beta_{ik}''\gamma_{ik}^{-1}\tilde{r}_{ik}) = (\beta_{ij}\gamma_{ij}^{-1}\tilde{r}_{ij})(\beta_{jk}'\gamma_k^{-1}\tilde{r}_{jk})$$

$$= \psi(\beta_{ij}r_{ij})\psi(\beta_{jk}'r_{jk}),$$

so $\psi$ is indeed a homomorphism.

$(\Rightarrow)$ Given an isomorphism $\psi: R \to R'$ we may assume $\psi$ fixes $K_0$ by the Skolem-Noether theorem, and thus $\psi$ fixes $E$ in the notation 7.1.33. Writing $\psi(r_{ij}) = (\tilde{r}_{ij})$ in $R'$ for suitable $\tilde{r}_{ij}$ in $E$, let $\gamma_{ij} = r_{ij}\tilde{r}_{ij}^{-1}$, so that $\tilde{r}_{ij} = r_{ij}\gamma_{ij}^{-1}$. We claim for any $(\beta_{ij}r_{ij})$ in $R$ that $\psi(\beta_{ij}r_{ij}) = (\beta_{ij}\gamma_{ij}^{-1}\tilde{r}_{ij})$; indeed $(\beta_{ij}r_{ij})$ can be written as $\sum_{u,v} a_i^u r_{ij}a_j^v$ so

$$\psi(\beta_{ij}r_{ij}) = \left(\sum \psi a_i^u \psi r_{ij} \psi a_j^v\right) = \left(\sum a_i^u r_{ij}\gamma_{ij}^{-1}a_j^v\right) = (\beta_{ij}\gamma_{ij}^{-1}\tilde{r}_{ij})$$

as claimed. In particular, writing 1 as $(\beta_{ij}r_{ij})$ for suitable $\beta_{ij}$, and matching terms in $\psi 1 = \psi 1 \psi 1$ yields

$$\sum_{j=1}^{n} \beta_{ij} c_{ijk} \beta_{jk} \gamma_{ik}^{-1} = \sum_{j=1}^{n} \beta_{ij} \gamma_{ij}^{-1} \tilde{c}_{ijk} \beta_{jk} \gamma_{jk}^{-1},$$

so $\sum_j \beta_{ij} (c_{ijk} \gamma_{ik}^{-1} - \tilde{c}_{ijk} \gamma_{ij}^{-1} \gamma_{jk}^{-1}) \beta_{jk} = 0$. Repeating the last paragraph of the proof of theorem 7.1.37(i) we can peel off the $\beta_{ij}$ and then $\beta_{jk}$ to get $c_{ijk} \gamma_{ik}^{-1} - \tilde{c}_{ijk} \gamma_{ij}^{-1} \gamma_{jk}^{-1} = 0$, i.e., $\tilde{c}_{ijk} = c_{ijk} \gamma_{ij} \gamma_{jk} \gamma_{ik}^{-1}$.          Q.E.D.

**Corollary 7.1.41:**  $R \approx M_n(F)$ *iff there is a Brauer factor set satisfying* $c_{ijk}$ $= \gamma_{ij} \gamma_{jk} \gamma_{ik}^{-1}$ *for suitable* $(\gamma_{ij})$ *in* $M_n(E)$ *satisfying the conjugacy conditions.*

**Proof:**  Apply (8) to the trivial Brauer factor set of example 7.1.38.     Q.E.D.

## Noether Factor Sets

When the maximal subfield $K$ of $R$ is already Galois over $F$ the previous construction becomes considerably more explicit and yields the more familiar "crossed product" construction of E. Noether. The theory is a natural generalization of that of cyclic algebras. Let $G = \text{Gal}(K/F)$. For each $\sigma$ in $G$ the Skolem-Noether theorem gives an invertible $u_\sigma$ in $R$ for which $u_\sigma a u_\sigma^{-1} = \sigma a$ where $K = F(a)$. Let $c_{\sigma,\tau} = u_\sigma u_\tau u_{\sigma\tau}^{-1}$. Then $c_{\sigma,\tau} \in C_R(K)$ $= K$, and obviously

$$u_\sigma u_\tau = c_{\sigma,\tau} u_{\sigma\tau}. \qquad (9)$$

Applying (9) to the associative law $(u_\sigma u_\tau) u_\rho = u_\sigma (u_\tau u_\rho)$ yields the *Noether factor set conditions*

$$c_{\sigma,\tau} c_{\sigma\tau,\rho} = c_{\sigma,\tau\rho} \sigma c_{\tau,\rho} \qquad \text{for all } \sigma, \tau, \rho \text{ in } G. \qquad (10)$$

Using a Dedekind independence argument one sees readily that $\sum_{\sigma \in G} K u_\sigma$ is a subalgebra of $R$ which is $|G|$-dimensional over $K$, implying $R = \sum_{\sigma \in G} K u_\sigma$.

Conversely, any set $\{c_{\sigma,\tau} : \sigma, \tau \in G\}$ satisfying (10) is called a (*Noether*) *factor set* and gives rise to an algebra $R = \sum_{\sigma \in G} K u_\sigma$ under multiplication

$$\left( \sum_\sigma \beta_\sigma u_\sigma \right) \left( \sum_\tau \beta'_\tau u_\tau \right) = \sum_\rho \left( \sum_{\sigma\tau = \rho} c_{\sigma,\tau} \beta_\sigma \sigma \beta'_\tau \right) u_\rho \qquad \text{where the } \beta_\sigma \in K. \quad (11)$$

Customarily one proves by direct verification that the multiplication of (11) is associative and furthermore defines a central simple algebra; however, in exercise 10 we shall see this as a special case of theorem 7.1.37.

**Example 7.1.42:**  Suppose $(K, \sigma, \alpha)$ is a cyclic algebra. Taking $z$ such that

$z^n = \alpha$ we could take $u_{\sigma^i} = z^i$ for each $0 \le i < n$. Now (9) yields $c_{\sigma^i, \sigma^j} = 1$ if $i + j < n$ and $\alpha$ otherwise. Thus cyclic algebras have particularly nice Noether factor sets and permits us to generalize some of the results on symbols, cf., exercise 12.

## Algebras of Degree 2,3,4

**Proposition 7.1.43:** (*Brauer*) *Any central division algebra D of degree n has an element d for which* $\text{tr}(d) = \text{tr}(d^{-1}) = 0$.

**Proof:** Notation as in theorem 7.1.37 we want to find $d = (\beta_{ij} c_{ij1})$ with $\text{tr}(d) = \text{tr}(d^{-1}) = 0$. We pick a special sort of $\beta_{ij}$; let $\beta_{ii} = 0$ for all $i$ and $\beta_{ij} = f(a_i)^{-1}$ where $f = \sum_{i=0}^{n-1} \alpha_i \lambda^i$ with the $\alpha_i$ to be determined. Clearly the conjugacy conditions hold since $\beta_{\sigma i, \sigma i} = 0 = \beta_{ii}$ and $\beta_{\sigma i, \sigma j} = f(a_{\sigma i})^{-1} = \sigma f(a_i)^{-1} = \sigma \beta_{ij}$ for $i \ne j$. Furthermore, $\text{tr}(d) = \sum \beta_{ii} c_{ii1} = 0$. Thus far we have made no restriction on the $\alpha_i$. We get $\text{tr}(d^{-1}) = 0$ iff $\sum \beta'_{ii} = 0$ where the $\beta'_{ii}$ are the diagonal entries of the adjoint matrix to $(\beta_{ij} c_{ij1})$. But $f(a_1) \cdots f(a_n) \sum \beta'_{ii} = 0$ unravels to a linear condition on the $\alpha_i$, so we can solve for $\alpha_0$ in terms of the other $\alpha_i$, and this solution yields the desired $d$. Q.E.D.

In other words if $\sum_{i=0}^{n} \alpha_i d^i = 0$ we have $\alpha_1 = \alpha_{n-1} = 0$.

**Theorem 7.1.44:** (*Wedderburn*) *Every division algebra D of degree n = 2 or 3 is cyclic.*

**Proof:** For $n = 2$ the assertion is corollary 7.1.15′.

For $n = 3$ we note that by proposition 7.1.43 there is some $d \ne 0$ in $D$ for which $\text{tr}(d) = \text{tr}(d^{-1}) = 0$, implying $d^3 \in F$, and thus $D$ is cyclic. Q.E.D.

A similar argument also handles degrees 4 and 12.

**Theorem 7.1.45:** (*Albert*) *If* $\text{index}(R) = 4$ *then R has a splitting field with Galois group* $(\mathbb{Z}/2\mathbb{Z})^{(2)}$ *over F.*

**Proof:** We may assume $R$ is a division algebra; we shall treat only the char $\ne 2$ case. By proposition 7.1.43 there is $d$ in $R$ for which $d^4 + \alpha_2 d^2 + \alpha_0 = 0$ for suitable $\alpha_0, \alpha_2$ in $F$. Hence $(d^2 + \alpha_2/2)^2 \in F$, proving $R$ has an element $d_1 \notin F$ whose square is in $F$. Pick $0 \ne r$ in $R$ for which $r d_1 r^{-1} = -d_1$, by Skolem-Noether. Then $r^2$ centralizes $d_1$ but $r$ does not, so $F(r^2) \subset F(r)$.

If $r^2 \in F$ then $r$ and $d_1$ generate a central quaternion subalgebra $Q$ of $R$, so $R \approx Q \otimes Q_1$ where $Q_1 = C_R(Q)$. $Q_1$ is also quaternion and thus cyclic; taking a quadratic subfield $F(d_2)$ over $F$ we see $R$ contains the subfield $F(d_1, d_2)$ which has Galois group $(\mathbb{Z}/2\mathbb{Z})^{(2)}$ over $R$.

If $r^2 \notin F$ then $[F(r^2):F] = 2$ so $F(d_1, r^2)$ is the desired subfield. Thus in either case we proved the result for degree 4.    Q.E.D.

An alternate proof of these results is sketched in exercises 15–20.

## Infinite Dimensional Division Algebras

The theory of an arbitrary division ring $D$ is much less developed than the finite dimensional theory, but there are certain decisive trends of considerable interest. First of all there is the surprisingly useful

**Cartan-Brauer-Hua Theorem:**   (*exercise 23*). *Any proper subring of $D$ which is invariant with respect to all inner automorphisms of $D$ is central.*

This theorem is used to describe the structure of $D$ in terms of sets invariant under inner automorphisms, such as $[D, D]$, and also has been generalized to rings with involution, cf., Herstein [76B]. Cartan-Brauer-Hua was applied already in, cf., exercises 2.5.31ff. Jacobson [64B] incorporates the Cartan-Brauer-Hua theorem into a Galois theory for division rings. Certainly any division ring $D$ can be viewed as a left vector space over a division subring $D_1$. When $D_1$ is finite dimensional over its center then $[D:D_1]$ is the same as what one would get by viewing $D$ as *right* vector space over $D_1$. However, this is not necessarily the case in general as first proved by P. M. Cohn, and Schofield [85] shows for any $1 < m, n \le \infty$ there are division rings $D_1 \subset D$ such that the left dimension (of $D$ over $D_1$) is $m$, whereas the *right* dimension is $n$; thus the Galois theory meets a serious problem with its most fundamental invariant. Schofield has produced a rather easy example of this pathelogical behavior, which we give in exercise 41.

This leaves us to search for an alternate way to measure the "size" of a division algebra. The most natural candidate may be the GK-dimension. However the GK-dimension is often "too small," as evidenced by

*Example 7.1.46:*  (Makar-Limanov [83]) The division ring of fractions of the Weyl algebra $\mathbf{A}_1(\mathbb{Q})$ has a free subalgebra generated by $(\mu\lambda)^{-1}$ and $(\mu\lambda)^{-1}(1 - \mu)^{-1}$, where $\mu, \lambda$ are the generators in example 1.6.32. We do not go into the details here, which involve working in the subring $\sum_{i \in \mathbb{Z}} \mu^i \mathbb{Q}((t))$ where $t = \mu\lambda$ and showing the monomials which are strings in $t^{-1}$ and $(1 - \mu)^{-1}$ are independent.

Incidentally, Makar-Limanov [84] proved *every* noncommutative division ring with uncountable center contains a free group. Kurosch's problem remains unsolved for division rings. One prerequisite for an effective theory is

**Question 7.1.47:**  If $D$ is finite dimensional as left vector space over a division subring $D_1$ then does $GK \dim D = GK \dim D_1$? (i.e., if $GK \dim D_1 < \infty$ then is $GK \dim D < \infty$ and does equality hold?)

Another useful measure of the size might be by means of maximal subfields, cf., exercises 32, 33 and work of Resco on the transcendence degree. Stafford [83] recently showed the Krull dimension of $D \otimes D^{op}$ can be an important invariant, and in an interesting survey Schofield [84] suggests concentrating on $D$ such that $D \otimes D^{op}$ is Noetherian. This includes the division rings of fractions of the various Noetherian domains studied in Chapter 8, specifically of group algebras and enveloping algebras, and these are explored in depth by Lorenz [84], cf., exercise 34.

The particular case of group algebras is discussed at the end of §8.2. Lichtman [84a] has studied the structure of subrings of matrix rings over such division rings.

Algebraic elements of arbitrary division rings have certain very nice properties, the foremost of which may well be Wedderburn's factorization theorem, exercises 15–18.

## §7.2 The Brauer Group and the Merkurjev-Suslin Theorem

In the overview in §7.1 we defined the Brauer group of a field, denoted $Br(F)$, as the set of similarity classes of central simple algebras (two algebras are *similar* if they have the same underlying division algebra); the group operation is tensor product. Our goal in this section is to obtain some of the properties of the Brauer group and to examine the Merkurjev-Suslin theorem, proving some easy special cases. Actually, we shall be interested mostly in the $p$-periodic part of $Br(F)$; special attention is paid to the case $p = 2$, for this ties in with the theory of rings with involution.

One obvious question to ask is, "What is the group structure of $Br(F)$, for a given field $F$?" If $F$ is algebraically closed then $Br(F) = (1)$ since there are no nontrivial $F$-central division algebras. The Tsen-Lang theorem (Jacobson [80B, p. 649]) implies more generally that $Br(F) = (1)$ whenever $F$ is an algebraic extension of $F_0(\lambda)$ where $F_0$ is algebraically closed. Another instance of $Br(F) = (1)$ is when $F$ is a finite field, by theorem 7.1.11. It is also easy to check $Br(\mathbb{R}) \approx \mathbb{Z}/2\mathbb{Z}$, cf., exercise 1. Usually $Br(F)$ is infinite and difficult to ascertain, as one sees when investigating $Br(\mathbb{Q})$, cf., appendix A.

## General Properties of the Brauer Group

There are three major approaches to the Brauer group. The first is by looking at each algebra individually; the second is by means of crossed products, and the third is via cyclic algebras. So far we have favored the first route, gleaning as much as we can from the Brauer factor set construction. Often one passes to the underlying division algebra, also, cf., exercise 2. The crossed product method relies on the fact that every class in the Brauer group is represented by a crossed product (corollary 7.2.5 below) and ties in with the powerful techniques of cohomology theory since the product in the Brauer group turns out to be compatible with the group operation in $H^2$. Finally the approach using cyclic algebras is perhaps both the oldest and the freshest, bringing in algebraic $K$-theory.

**Remark 7.2.1:** For any fields $F \subseteq K$ there is a group homomorphism $\text{res}_{K/F} \, \text{Br}(F) \to \text{Br}(K)$ called the *restriction*, given by $[R] \to [R \otimes_F K]$. Thus Br is a functor from {fields} to {abelian groups}. $\text{Ker}(\text{res}_{K/F}) = \{[R] \in \text{Br}(F): R \text{ is split by } K\}$. Thus the functor Br contains the theory of splitting fields.

One fundamental result needed later to work with restriction is

**Proposition 7.2.2:** *Suppose $K \supseteq F$ is a subfield of $R$. Then $R \otimes_F K \sim C_R(K)$.*

**Proof:** By proposition 7.1.3 $C_R(K) \otimes_K K \otimes_F R^{\text{op}} \approx \text{End}_K R$ is split as $K$-algebra, so $C_R(K) \sim R \otimes K$.     Q.E.D.

## First Invariant of $\text{Br}(F)$: The Index

Recall index($R$) is defined as the degree of the underlying division algebra $D$ of $R$. Note for any field $K \supset F$ that $\text{index}(R \otimes K) = \text{index}(D \otimes K)$ divides $\text{degree}(D \otimes K) = \text{degree}(D) = \text{index}(R)$.

**Theorem 7.2.3:** *Let $t = \text{index}(R)/\text{index}(R \otimes_F K)$. Then $t \mid [K:F]$ and $K$ is isomorphic to a subfield of $M_{[K:F]/t}(D)$, where $D$ is the underlying division algebra of $R$.*

**Proof:** Let $k = [K:F]$. By the regular representation $K \subseteq M_k(F)$, so $M_k(D) \supseteq D \otimes K = M_t(D_1) \supseteq M_t(F)$ for some division algebra $D_1$. By proposition 7.1.1 there is a division algebra $D'$ for which $M_k(D) \approx M_t(F) \otimes M_u(D') \approx M_{tu}(D')$; hence $u = k/t$ and $D' \approx D$. But $K = Z(D \otimes K)$ centralizes $M_t(F)$ so $K \subseteq M_u(D') \approx M_u(D)$.     Q.E.D.

**Corollary 7.2.4:** *If* $[K:F]$ *and* index($R$) *are relatively prime then* index($R \otimes K$) = index($R$).

## The Identification with $H^2$

The modern way of thinking about the Brauer group is via cohomology of groups; the identification is based on the following result.

**Corollary 7.2.5:** *Every central simple algebra $R$ is similar to a crossed product.*

**Proof:** We may assume $R$ is a division algebra of degree $t$. Take a maximal separable subfield $L$ of $R$ (over $F$) and let $K$ be the normal closure of $L$. Then $K$ is a Galois extension of $F$. Let $R' = M_{[K:F]/t}(R)$. $K$ is a subfield of $R'$ by theorem 7.2.3, and deg $R' = [K:F]$ so $K$ is maximal.     Q.E.D.

**Theorem 7.2.6:** *If $K$ is Galois over $F$ with Galois $G$ then $H^2(G, K - \{0\}) \approx$* Br($K/F$).

Since we want to push off in a different direction we shall not give full details of this proof, which can be found in many standard texts including Jacobson [80B, theorem 8.11]. However, the general idea of the proof is very instructive. $K - \{0\}$ is a $G$-module by means of the action of $G$ as automorphisms on $K$, where the operation of $K - \{0\}$ is multiplication. Rewriting formula (2) in example 5.2.13 taking $M = K - \{0\}$ and using multiplication instead of addition we write $c_{\sigma,\tau}$ for $f(\sigma, \tau)$ and observe any $f$ in $Z^2(G, K - \{0\})$ satisfies

$$1 = \delta f(\sigma, \tau, \rho) = \sigma c_{\tau,\rho} \, c_{\sigma,\tau}^{-1} c_{\sigma\tau,\rho}^{-1} c_{\sigma,\tau\rho} \qquad \text{for all } \sigma, \tau, \rho \text{ in } G.$$

Thus $c_{\sigma,\tau} c_{\sigma\tau,\rho} = c_{\sigma,\tau\rho} \sigma c_{\tau,\rho}$, which is the Noether factor set condition. In this way $f$ gives rise to a central simple algebra, yielding a map $\psi \colon Z^2(G, K - \{0\}) \to$ Br($K/F$).

$\psi$ is onto by corollary 7.2.5, and $f \in \ker \psi$ iff the $c_{\sigma,\tau}$ have $\gamma_\sigma$ for which $c_{\sigma,\tau} = \gamma_\sigma \sigma \gamma_\tau \gamma_{\sigma\tau}^{-1}$ by exercise 7.1.11, so $\ker \psi = B^2(G, K - \{0\})$. Thus the proof boils down to showing $\psi$ is a group homomorphism or, equivalently,

**Lemma 7.2.7:** *If $R, R'$ are crossed products with respective Noether factor sets $(c_{\sigma,\tau})$ and $(c'_{\sigma,\tau})$ over $K$, then $R \otimes_F R'$ is similar to a crossed product $R''$ having Noether factor set $(c_{\sigma\tau} c'_{\sigma\tau})$ over $K$.*

The crux of the proof of lemma 7.2.7 is to locate $R''$ such that we can compute the factor set. Certainly $K \otimes_F K \subseteq R \otimes_F R'$, and it turns out $R'' = e(R \otimes R')e$ where $e$ is as in exercise 6.

The above approach relied on the delicate connection to crossed products; one would like a version of lemma 7.2.7 which held more generally for Brauer factor sets, i.e.,

**Lemma 7.2.7′:** *If $R, R'$ have respective Brauer factor sets $(c_{ijk})$ and $(c'_{ijk})$ with respect to a maximal separable subfield $K$ then $R \otimes R'$ is similar to an algebra having Brauer factor set $(c_{ijk} c'_{ijk})$.*

This lemma is proved in Jacobson [83] analogously to lemma 7.2.7, also, cf., van der Waerden [49B, pp. 207–215]. However, there is a rather direct proof once we free ourselves from fields. Namely, as in notational hypothesis 7.1.33 write $R = KrK$ and $R' = Kr'K$. Then $R \otimes R' = (K \otimes K)(r \otimes r')(K \otimes K)$ and the $n^2 \times n^2$ matrix $r \otimes r'$ can be computed to yield the Brauer factor set of $R \otimes R'$ over $K \otimes K$; this is easily seen to be associated to the Brauer factor set obtained by repeating $(c_{ijk} c'_{ijk})$ in blocks, completing the proof of lemma 7.2.7′. The reader is strongly encouraged to fill in the details of this proof before referring to the literature.

One can also use lemma 7.2.7′ to compute the Brauer factor set of $R^{op}$.

**Proposition 7.2.8:** *If the Brauer factor set $(c_{ijk})$ corresponds to $R$ the $(c_{kji})$ corresponds to $R^{op}$.*

**Proof:** In view of lemma 7.2.7′ we want $(c_{ijk} c_{kji})$ to be associated to the trivial Brauer factor set. Let $\gamma_{ij} = c_{iji} c_{ii1}$. Then by remark 7.1.36

$$\sigma \gamma_{ij} = c_{\sigma i, \sigma j, \sigma i} c_{\sigma i, \sigma i, \sigma 1} = c_{\sigma i, \sigma j, \sigma i} c_{\sigma i, \sigma i, 1} = \gamma_{\sigma i, \sigma j},$$

yielding the conjugacy condition. Furthermore, BFS2 yields

$$\gamma_{ij} = c_{iji} c_{iim} = c_{ijm} c_{jim} \text{ (taking } k = i\text{) and, likewise,}$$

$$\gamma_{ij} = c_{iji} c_{mii} = c_{mij} c_{mji} \text{ by left-right symmetry.}$$

Hence $\gamma_{ij} \gamma_{jk} \gamma_{ik}^{-1} = (c_{ijk} c_{jik})(c_{jki} c_{kji})(c_{jik} c_{jki})^{-1} = c_{ijk} c_{kji}$ as desired.     Q.E.D.

## Second Invariant of $\mathrm{Br}(F)$: The Exponent

One of the major features of the Brauer group is that it is torsion. This can be seen at once by combining exercise 5.2.2 with theorem 7.2.6, but better results can be obtained directly. To wit, we define the *exponent* of a central simple algebra $R$, written $\exp(R)$, to be the order of $[R]$ in the Brauer group; i.e., $\exp(R)$ is the smallest number $m$ for which $R^{\otimes m} \sim 1$, where we recall $R^{\otimes m}$ denotes $R \otimes_F \cdots \otimes_F R$ taken over $m$ copies of $R$.

**Example 7.2.9:** If $R = (\alpha, \beta)_n$ and $F(\beta^{1/n})$ is a field then $\exp(R)$ is the smallest $m$ for which $\alpha^m$ is a norm from $F(\beta^{1/n})$ to $F$, by Wedderburn's criterion (proposition 7.1.18).

**Theorem 7.2.10:** $\exp(R)$ *divides* $\text{index}(R)$.

**Proof:** Let $n$ be the degree of $R$. We need to show $R^{\otimes n} \sim 1$, or, equivalently, (by lemma 7.2.7') $(c_{ijk})^n$ is associated to the trivial Brauer factor set, if $(c_{ijk})$ is a Brauer factor set of $R$. Putting $\gamma_{ij} = \prod_{m=1}^n c_{ijm}$ yields (by condition BFS2 of definition 7.1.35)

$$\gamma_{ij}\gamma_{jk}\gamma_{ik}^{-1} = \prod_{m=1}^n c_{ijm}c_{jkm}c_{ikm}^{-1} = \prod_m c_{ijk} = (c_{ijk})^n.$$

To proceed further we need the following elementary way of cutting out the "unwanted" part of the index.

**Proposition 7.2.11:** *Suppose* $\text{index}(R) = p^j t$ *where* $p$ *is a prime number not dividing* $t$. *Then there is a field extension* $L$ *over* $F$ *whose dimension is relatively prime to* $p$, *for which* $\text{index}(R \otimes_F L) = p^j$.

**Proof:** Let $E$ be the normal closure of a separable splitting subfield $K$ of $R$; $[E:F] = [E:K][K:F] = p^{j'}t'$ for suitable $j' \geq j$ and $t'$ prime to $p$. Let $H$ be a Sylow $p$-subgroup of $\text{Gal}(E/F)$ and let $L$ be the fixed subfield of $E$ under $H$. By Galois theory $[L:F] = t'$ is prime to $p$, so by theorem 7.2.3 $\text{index}(R \otimes_F L) = p^j t''$ for some $t''$ dividing $t$. On the other hand, $(R \otimes_F L) \otimes_L E \approx R \otimes_F E$ is split so $p^j t''$ divides $[E:L] = p^{j'}$, implying $t'' = 1$, proving $\text{index}(R \otimes_F L) = p^j$.          Q.E.D.

**Theorem 7.2.12:** *If* $n, m$ *are the respective index and exponent of a suitable central simple algebra* $R$ *then the following two conditions hold*:

(i) $m \mid n$
(ii) *every prime divisor of* $n$ *also divides* $m$.

**Proof:**

(i) Holds by Theorem 7.2.10.
(ii) Suppose $p \mid n$. By proposition 7.2.11 there is a field $L$ with $[L:F]$ prime $p$, for which $R \otimes L$ has index $p^j$ with $j \geq 1$. But then $\exp(R \otimes L)$ is a nontrivial power of $p$ which obviously divides $m$.          Q.E.D.

We shall see in §7.3 that (i) and (ii) are the only numerical connections between the index and exponent.

**Theorem 7.2.13:** *Suppose $D$ is a central division algebra of degree $n = p_1^{\alpha(1)} \cdots p_k^{\alpha(k)}$ for suitable distinct $p_1, \ldots, p_k$ primes and $\alpha(1), \ldots, \alpha(k)$ in $\mathbb{N} - \{0\}$. Then $D \approx D_1 \otimes \cdots \otimes D_k$ for central division subalgebras $D_i$ of degree $p_i^{\alpha(i)}$, $1 \leq i \leq k$.*

*Proof:* It suffices to prove that if $\deg(D) = n = n'n''$ where $n'$ and $n''$ are relatively prime then $D \approx D_1 \otimes D_2$ where $\deg(D_1) = n'$ and $\deg(D_2) = n''$, for then we can continue the decomposition by induction. Take $u, v$ in $\mathbb{N}$ for which $un' + vn'' \equiv 1 \pmod{n}$ and take $D_1 \sim D^{\otimes vn''}$ and $D_2 \sim D^{\otimes un'}$. Then $D_1 \otimes D_2 \sim D^{\otimes (un' + vn'')} \sim D$. On the other hand, $D_1^{\otimes n'} \sim D^{\otimes vn} \sim 1$ so $\exp(D_1) | n'$, implying by theorem 7.2.12 $\deg(D_1)$ is prime to $n''$; but $\deg(D_1)$ divides $\deg(D) = n'n''$ so $\deg(D_1)$ divides $n'$. Likewise, $\deg(D_2)$ divides $n''$. Thus

$$\deg(D_1 \otimes D_2) \leq n'n'' = n = \text{index}(D) = \text{index}(D_1 \otimes D_2) \leq \deg(D_1 \otimes D_2)$$

so equality holds; in particular, $D_1 \otimes D_2$ is a division algebra so $D_1 \otimes D_2 \approx D$.
                                                                                    Q.E.D.

In view of theorem 7.2.13 the structure theory of division algebras reduces to the case where the degree (and exponent) is a prime power.

*Digression 7.2.13':* The zero divisor question for tensor products (also see exercise 7.1.42). Conversely to theorem 7.2.3, we take a brief look at the question: If $D_1$ and $D_2$ are central division algebras not having a common subfield over $F$ then is $D_1 \otimes D_2$ a division algebra? Answer: "Yes" if $\deg(D_1)$ and $\deg(D_2)$ are relatively prime (exercise 22) or if $\deg(D_1) = \deg(D_2) = 2$ (exercise 23). However, there is a counterexample, in general.

*Example 7.2.13':* (Tignol-Wadsworth [86]) Cyclic division algebras $D_1$ and $D_2$ of degree $p$ for $p$ an arbitrary odd prime, not having any common maximal subfield, such that $D_1 \otimes D_2$ is not a division algebra. Suppose $F_0$ is a field containing a primitive $p$-th root $\zeta$ of 1 and $F = F_0(\lambda_1, \lambda_2, \lambda_3)$ for commutative indeterminates $\lambda_i$ over $F_0$. Let $D_i = (\alpha_i, \beta_i)_p$ where $\alpha_1 = \lambda_1$, $\beta_1 = \lambda_2, \alpha_2 = \lambda_1(\lambda_3 - 1)\lambda_2^{-1}$, and $\beta_2 = \lambda_1 \lambda_3$. Taking $y_i, z_i$ such that $y_i^p = \alpha_i$, $z_i^p = \beta_i$, and $z_i y_i = \zeta y_i z_i$ one sees the element $a = (y_1 \otimes 1 + z_1 \otimes y_2)(1 \otimes z_2^{-1})$ of $D_1 \otimes D_2$ is a $p$-th root of 1. Indeed, letting $a_1 = z_1 y_1^{-1} \otimes y_2$ and writing 1 for $1 \otimes 1$ we have

$$a^p = ((1 + a_1)(y_1 \otimes z_2^{-1}))^p = (1 + a_1)(1 + \zeta^2 a_1) \cdots (1 + \zeta^{2(p-1)} a_1)(y_1 \otimes z_2^{-1})^p.$$

But $\lambda^p + 1 = \prod_{i=0}^{p-1} (\zeta^i \lambda + 1)$ since $\zeta^{-i}$ is a zero of $\lambda^p + 1$ for each $i$. Substituting $a_1$ for $\lambda$ yields $(1 + a_1)(1 + \zeta^2 a_1) \cdots (1 + \zeta^{2(p-1)} a_1) = a_1^p + 1 = \beta_1 \alpha_1^{-1} \alpha_2 + 1$, so

$$a^p = (\beta_1 \alpha_1^{-1} \alpha_2 + 1)(\alpha_1 \beta_2^{-1}) = (\beta_1 \alpha_2 + \alpha_1)\beta_2^{-1} = (\lambda_1(\lambda_3 - 1) + \lambda_1)(\lambda_1 \lambda_3)^{-1} = 1$$

as desired.

Thus $D_1 \otimes D_2$ cannot be a division algebra. Tignol-Wadsworth verify using valuation theory that $D_1$ and $D_2$ have no common maximal subfield. However, this follows at once from the following observation of Saltman, based on a leading monomial argument which foreshadows §7.3:

**Claim:** $D_1 \otimes D_2^{\mathrm{op}}$ is a division algebra.

**Proof of Claim:** Let $D = D_1 \otimes D_2^{\mathrm{op}}$. $D_2^{\mathrm{op}} \approx (\beta_2, \alpha_2)_p \sim (\lambda_1, \alpha_2)_p \otimes (\lambda_3, \alpha_2)_p$ since $\beta_2 = \lambda_1 \lambda_3$. Thus $D = D_1 \otimes D_2^{\mathrm{op}} \sim (\lambda_1, \beta_1 \alpha_2)_p \otimes (\lambda_3, \alpha_2)_p = (\lambda_1, \lambda_1(\lambda_3 - 1))_p \otimes (\lambda_3, \lambda_1(\lambda_3 - 1)\lambda_2^{-1})$. But now $\lambda_2$ appears only in the last position. Letting $D_3 = (\lambda_1, \lambda_1(\lambda_3 - 1))_p$ and $R = D_3 \otimes_F F(\lambda_3^{1/p})$, we have identified $D$ as the ring of central fractions of the skew polynomial ring $R[\mu; \sigma]$ where $\sigma$ fixes $D_3$ and $\sigma \lambda_3^{1/p} = \zeta \lambda_3^{1/p}$. (We view $\lambda_2$ as $\lambda_1(\lambda_3 - 1)\mu^p$.) If $D$ were not a division ring then neither would $R[\mu; \sigma]$; taking leading coefficient of zero-divisors would yield zero-divisors in $R$. But

$$D_3 \sim (\lambda_1, -\lambda_1)_p \otimes (\lambda_1, 1 - \lambda_3)_p \sim (\lambda_1, 1 - \lambda_3)_p$$

and the indeterminate $\lambda_1$ is obviously not a norm from $F(\lambda_3^{1/p}, (1 - \lambda_3)^{1/p})$ to $F(\lambda_3^{1/p})$, so $R$ cannot be split, by proposition 7.1.18.     Q.E.D.

Saltman's observation raises the following new tensor product question:

**Question 7.2.13′′′:** If $\mathrm{index}(D_1 \otimes D_2^{\otimes u}) < \mathrm{index}(D_1)\mathrm{index}(D_2)$ for all $1 \le u < \exp(D_2)$ then do $D_1$ and $D_2$ necessarily have a common subfield?

Although this question remains open, Schofield has pointed out a lovely way of attacking such problems generically and has some very nice counterexamples, cf., appendix C.

## Outer Automorphisms

We have seen that for fields $F \subseteq K$ there is the very natural restriction map $\mathrm{Br}(F) \to \mathrm{Br}(K)$ sending $[R] \mapsto [C_R(K)]$. Thus we are led to ask whether there is an inverse map from $\mathrm{Br}(K)$ to $\mathrm{Br}(F)$. This is too much to hope for, but there is a useful map, the *corestriction*, which remained undiscovered until the Brauer group was examined from a cohomological point of view. Our approach, however, is ring-theoretical, following Tignol [85], which modifies Draxl's presentation of work of Riehm [70] and Tamagawa. The start by examining the following set-up:

$G$ is a finite group of automorphisms of an arbitrary algebra $R$ over a field $E$, and we assume that each $\sigma$ in $G$ restricts to a nontrivial automorphism of $E$. (Such automorphisms are often called "outer.") Write $\operatorname{tr} r$ for $\sum_{\sigma \in G} \sigma r$, and write $F = E^G$ (the fixed subfield of $E$ under $G$) and $R_0 = \{\operatorname{tr} r : r \in R\}$. Note $R_0 \subseteq R^G$ are vector spaces over $F$.

**Proposition 7.2.14:**   $R^G = R_0$ and $R \approx R_0 \otimes_F E$.

**Proof:**   First we claim $R_0 E = R$. Otherwise, taking $r$ in $R \backslash R_0 E$ we can find a map $f: R \to E$ (as vector spaces over $E$) such that $fr = 1$ and $f(R_0 E) = 0$. For each $\sigma$ in $G$ write $\alpha_\sigma$ for $f(\sigma r)$ in $E$. Then for each $\gamma$ in $E$ we have

$$0 = f(\operatorname{tr}(\gamma r)) = f\left( \sum_{\sigma \in G} \sigma \gamma \sigma r \right) = \sum (\sigma \gamma) f(\sigma r) = \sum \alpha_\sigma \sigma \gamma,$$

contrary to the Dedekind independence theorem (Jacobson [85B, p. 291]) since the $\sigma: E \to E$ are presumed distinct. This proves the claim.

Now we have a canonical map $\varphi: R^G \otimes_F E \to R^G E = R$ whose restriction to $R_0 \otimes E$ is epic. It remains to prove $\varphi$ is monic, since then necessarily $R^G = R_0$, showing $\varphi$ is an isomorphism. So assume, on the contrary, $0 \neq r = \sum_i r_i \otimes \alpha_i \in \ker \varphi$ for suitable $\alpha_i$ in $E$, and for suitable $r_i$ in $R^G$ which are independent over $F$. Then $\sum r_i \alpha_i = 0$ so for every $\gamma$ in $E$ we have

$$0 = \operatorname{tr}\left( \gamma \sum_i r_i \alpha_i \right) = \operatorname{tr}\left( \sum_i \alpha_i \gamma r_i \right) = \sum_i \operatorname{tr}(\alpha_i \gamma) r_i.$$

Hence for each $i$ we have $0 = \operatorname{tr}(\alpha_i \gamma) = \sum_\sigma (\sigma \alpha_i) \sigma \gamma$; again this contradicts the Dedekind independence theorem unless each $\alpha_i = 0$.     Q.E.D.

(The "outer automorphism" hypothesis can be applied much more generally, cf., Montgomery [80B].)

**Corollary 7.2.15 :**   *Suppose $R$ is a simple (not necessarily central) $E$-algebra of degree $n$. Then $R^G$ is a simple $F$-algebra of degree $n$, $Z(R^G) = Z(R)^G$, and $[R:E] = [R^G:F]$.*

**Proof:**   From $R \approx R^G \otimes_F E$ we see $R^G$ is simple, $Z(R^G) \otimes E \subseteq Z(R)$, and

$[R:E] = [R^G:F]$. But $Z(R)$ is invariant under $G$, so using $Z(R)$ instead of $R$ in the proposition shows $Z(R) \approx Z(R)^G \otimes E$; since, obviously, $Z(R)^G \subseteq Z(R^G)$, we conclude $Z(R^G) = Z(R)^G$ and thus $[R:Z(R)] = [R^G:Z(R^G)]$. Q.E.D.

**Corollary 7.2.16:** *Suppose $R_1$ and $R_2$ are simple $E$-algebras with $Z(R_1) = E$, and $G = \mathrm{Gal}(E/F)$ extends naturally to a group of automorphisms on each of $R_1$ and $R_2$. Letting $G$ act diagonally on $R_1 \otimes R_2$ (i.e., $\sigma(r_1 \otimes r_2) = \sigma r_1 \otimes \sigma r_2$) we have*

$$(R_1 \otimes_E R_2)^G = R_1^G \otimes_F R_2^G.$$

*Proof:* ($\supseteq$) is clear. ($\subseteq$) Let $n_i = [R_i^G:F]$. By corollary 7.2.15

$$[R_1:E][R_2:E] = n_1 n_2 = [(R_1 \otimes R_2)^G:F],$$

so we are done by proposition 7.1.1(ii).     Q.E.D.

## The Corestriction

We now have the prerequisites to define the important corestriction map from $\mathrm{Br}(L)$ to $\mathrm{Br}(F)$, for any separable extension $L$ of $F$. This is also called the *transfer map* or *trace*, and when defined in cohomology the groups are written in additive notation, instead of the multiplicative notation used here.

*Definition 7.2.17:* Given $\sigma \in \mathrm{Gal}(E/F)$ and an $E$-algebra $R$ define the algebra $\sigma^{-1}R$ to have the same ring structure as $R$ but with the new scalar product of $\alpha$ in $E$ and $r$ in $\sigma^{-1}R$ to be defined as $(\sigma\alpha)r$ (the product taken in $R$.) The intuition here is to think of an element of $\sigma^{-1}R$ in the form $\sigma^{-1}r$, and then $\alpha\sigma^{-1}r = \sigma^{-1}(\sigma\alpha)\sigma^{-1}r = \sigma^{-1}((\sigma\alpha)r)$.

*Remark 7.2.18:*

(i) The $\sigma$-semilinear homomorphisms $\psi: R_1 \to R_2$ (i.e., $\psi(\alpha r) = (\sigma\alpha)\psi r$), are precisely the algebra homomorphisms $\psi: R_1 \to \sigma^{-1}R_2$.

(ii) If $\sigma$ fixes a subfield $L$ of $E$ and $A \in L\text{-}\mathscr{A}\mathit{lg}$ and $R \in E\text{-}\mathscr{A}\mathit{lg}$ then $\sigma^{-1}(A \otimes_L R) \approx A \otimes_L \sigma^{-1}R$ for any $L$-algebra $A$ (identifying $E$ with $L \otimes E$).

(iii) $(\sigma_1 \sigma_2)^{-1}R \approx \sigma_2^{-1}\sigma_1^{-1}R$.

(iv) $\sigma^{-1}E \approx E$. (Indeed (i) is clear, and consequently we get (ii) by taking $\psi: A \otimes \sigma^{-1}R \to A \otimes R$ to be the identity. (iii) and (iv) are also immediate from (i).)

To define the corestriction for an arbitrary separable extension $L$ of $F$, let $E$ be a Galois extension of $F$ containing $L$, let $G = \mathrm{Gal}(E/F)$ and $H = \mathrm{Gal}(E/L)$, and let $\sigma_1, \ldots, \sigma_t$ be a left transversal of $H$ in $G$, i.e., $G = \bigcup H\sigma_i$. (Thus $t = [L\!:\!F]$.)

**Definition 7.2.19:** The corestriction $\mathrm{cor}_{L/F} R$ of the central simple $L$-algebra $R$, notation as above. Define $\tilde{R} = \bigotimes_{i=1}^{t} \sigma_i^{-1}(R \otimes_L E)$, where the undecorated tensor is taken over $E$, and use the following rule for extending any $\sigma$ in $G$ to an automorphism *of* $\tilde{R}$:

First note there is $\pi$ in $\mathrm{Sym}(t)$ for which $\sigma_i \sigma \in H\sigma_{\pi i}$. Indeed, $\pi$ is 1:1 for if $\sigma_i \sigma$ and $\sigma_j \sigma$ are in the same coset then $\sigma_i \sigma_j^{-1} = (\sigma_i\sigma)(\sigma_j\sigma)^{-1} \in H$ so $H\sigma_i = H\sigma_j$ and $i = j$.

Now put $\sigma_i \sigma = \tau_i \sigma_{\pi i}$ for suitable $\tau_i$ in $H$ and define

$$\sigma\left( \bigotimes_{i=1}^{t}(r_i \otimes \beta_i) \right) = \bigotimes_{i=1}^{t}(r_{\pi i} \otimes \tau_i \beta_{\pi i}) \tag{1}$$

for $r_i$ in $R$ and $\beta_i$ in $E$. To see that this definition makes sense, let us note that we would expect $\sigma\sigma_{\pi i}^{-1}(R \otimes_L E) = \sigma_i^{-1}\tau_i(R \otimes E) = \sigma_i^{-1}(R \otimes \tau_i E)$ by remark 7.2.18(ii), i.e., $\sigma$ sends the $\pi i$ component to the $i$ component, "twisted" by $\tau_i$. It is easily checked that $\sigma$, indeed, is an automorphism of $\tilde{R}$, and so we finally can define

$$\mathrm{cor}_{L/F} R = \tilde{R}^G.$$

Note $\tilde{R} \in \mathrm{Br}(E)$ so $\mathrm{cor}_{L/F} R \in \mathrm{Br}(F)$ by corollary 7.2.15; thus the corestriction is a map from $\mathrm{Br}(L)$ to $\mathrm{Br}(F)$, which we want to show is a group homomorphism. To do this we must first free the construction from the various choices which were made along the way.

**Proposition 7.2.20:** *The construction* $\mathrm{cor}_{L/F} R$ *depends (up to isomorphism) only on* $L, F,$ *and* $R$.

**Proof:** First we show that if we started with a different transversal $\sigma_1', \ldots, \sigma_n'$ we end up with an isomorphic result. Write $\sigma_i' = \rho_i \sigma_i$ for $\rho_i$ in $H$, and note $\sigma_i'\sigma = \rho_i \tau_i \sigma_{\pi i} = \tau_i'\sigma_{\pi i}'$ where $\tau_i' = \rho_i \tau_i \rho_{\pi i}^{-1}$. Also by remark 7.2.18(iii) (where $R, E$, respectively, replace $A, R$),

$$\sigma_i'^{-1}(R \otimes_L E) = \sigma_i^{-1}\rho_i^{-1}(R \otimes_L E) \approx \sigma_i^{-1}(R \otimes_L \rho_i^{-1} E) \approx \sigma_i^{-1}(R \otimes_L E),$$

so $\tilde{R} \approx \bigotimes_{i=1}^{n} \sigma_i'^{-1}(R \otimes_L E)$ (which we denote as $\tilde{R}'$) under the isomorphism $\Phi$ which sends $\bigotimes_{i=1}^{n}(r_i \otimes \beta_i) \mapsto \bigotimes_{i=1}^{n}(r_i \otimes \rho_i^{-1}\beta_i)$. Extending any $\sigma$ in $G$ to

$\tilde{R}$ under the rule $\sigma\big(\bigotimes_{i=1}^{n}(r_i \otimes \rho_i^{-1}\beta_i)\big) = \bigotimes_{i=1}^{n}(r_{\pi i} \otimes \tau_i \rho_{\pi i}^{-1}\beta_{\pi i})$, we have the commutative square

$$
\begin{array}{ccc}
\tilde{R} & \xrightarrow{\ \Phi\ } & \tilde{R}' \\
{\scriptstyle \sigma}\downarrow & & \downarrow{\scriptstyle \sigma} \\
\tilde{R} & \xrightarrow{\ \Phi\ } & \tilde{R}'
\end{array}
$$

and thus $\Phi$ restricts to an isomorphism of the fixed rings of $\tilde{R}$ and $\tilde{R}'$.

It remains to show that if $\bar{E}$ is another Galois extension of $L$ containing $F$ then we could have used $\bar{E}$ instead of $E$ in defining $\mathrm{cor}_{L/F} R$. Replacing $\bar{E}$ by a field composition $E\bar{E}$ of $E$ and $\bar{E}$ we may assume $E \subset \bar{E}$. Let $\bar{G} = \mathrm{Gal}(\bar{E}/F)$, $\bar{H} = \mathrm{Gal}(\bar{E}/L)$, and $N = \mathrm{Gal}(\bar{E}/E)$. By Galois theory $N \lhd \bar{G}$, and $G \approx \bar{G}/N$ and $H \approx \bar{H}/N$; $\sigma_1, \ldots, \sigma_n$ is also a transversal for $\bar{G}$ over $\bar{H}$. Let $\bar{R} = \bigotimes_{i=1}^{n} \sigma_i^{-1}(R \otimes_L \bar{E})$, and extend $\bar{G}$ to act on $\bar{R}$ in analogy to (1). We want to show $\bar{R}^N \approx \tilde{R}$ canonically so that $\bar{R}^{\bar{G}} = (\bar{R}^N)^G \approx \tilde{R}^G$. To this end note

$$
\sigma_i^{-1}(R \otimes_L \bar{E}) \approx \sigma_i^{-1}(R \otimes_L E) \otimes_E \bar{E}) \approx \sigma_i^{-1}(R \otimes_L E) \otimes_E \sigma_i^{-1}\bar{E}
$$

$$
\approx \sigma_i^{-1}(R \otimes_L E) \otimes_E \bar{E}
$$

so $\bar{R} \approx \tilde{R} \otimes_E \bar{E}$. If $\sigma \in N$ then $\sigma_i \sigma = \sigma'\sigma_i$ for suitable $\sigma'$ in $N$ so in applying (1) we see $\pi = 1$ and $\sigma$ acts like $1 \otimes \sigma'$ on $\tilde{R} \otimes_E \bar{E}$. Thus $\bar{R}^N = \tilde{R} \otimes_E \bar{E}^N \approx \tilde{R} \otimes_E E \approx \tilde{R}$ canonically, as desired.     Q.E.D.

**Corollary 7.2.21:**   $\mathrm{cor}_{L/F}$ *is a homomorphism from* $\mathrm{Br}(L)$ *to* $\mathrm{Br}(F)$.

*Proof:*   By using the same Galois extension $E$ over $F$ and the same transversal, we need only show that if $\tilde{R}_1$ and $\tilde{R}_2$ are central simple $E$-algebras then $(\tilde{R}_1 \otimes_E \tilde{R}_2)^G = \tilde{R}_1^G \otimes_F \tilde{R}_2^G$ where $G$ is a group of automorphisms of $\tilde{R}_i$ for $i = 1, 2$ which restricts to $\mathrm{Gal}(E/F)$, and $G$ acts diagonally on $\tilde{R}_1 \otimes \tilde{R}_2$; but this is corollary 7.2.16.     Q.E.D.

Of course we would like to relate the corestriction to the restriction map $\mathrm{Br}(F) \to \mathrm{Br}(L)$. To this end we shall recall the *skew group algebra* $R*G$ from definition 1.6.40; this is a free $R$-module of rank $|G|$.

**Proposition 7.2.22:**   *Notation of proposition* 7.2.14. *If* $R$ *is a central simple* $E$-*algebra of degree* $n$ *then* $R^G \otimes_F (R*G)^{\mathrm{op}} \approx \mathrm{End}_F R$. *Thus* $R*G$ *is a central simple* $F$-*algebra of degree* $|G|n$, *and* $R*G \sim R^G$.

**Proof:** Define $\varphi: R^G \to \mathrm{End}_F R$ by the right regular representation, and $\varphi: (R*G)^{\mathrm{op}} \to \mathrm{End}_F R$ by $\psi(\sum r_\sigma \sigma) r = \sum r_\sigma(\sigma r)$. Clearly $\varphi$ and $\psi$ are injections whose images centralize each other, and $[R^G : F][R*G : F] = n^2[R : F]|G| = n^4|G|^2 = [\mathrm{End}_F R : F]$. Since $R^G$ is central simple we see by proposition 7.1.1(ii) that $\varphi(R*G)^{\mathrm{op}}$ is the centralizer of $\psi R^G$; hence $\psi(R*G)^{\mathrm{op}}$ is central simple over $F$, and $R^G \otimes (R*G)^{\mathrm{op}} \approx \mathrm{End}_F R \sim 1$, implying $R^G \sim R*G$.

Q.E.D.

**Digression 7.2.23:** Before proceeding further we would like to refer the reader to P. Schneider's dissertation (1977), which also treats the corestriction algebraically, deriving the properties obtained here plus others. For example, given a group $G$ of automorphism on $K$ define $\mathrm{Br}(K)^G = \{[R] \in \mathrm{Br}(K): \sigma^{-1} R = R$ for all $\sigma$ in $G\}$. For $F = K^G$ he obtains the exact sequence

$$0 \to \ker(\mathrm{res}_{K/F}) \hookrightarrow \mathrm{Br}(F) \xrightarrow{\text{res}} \mathrm{Br}(K)^G \to H_3(G, K - \{0\}).$$

If $K/F$ is cyclic then $1 = H^1(G, K - \{0\}) = H^3(G, K - \{0\})$, so in this case the sequence is exact.

## Goldman's Element

To proceed further we require some general results about central simple $F$-algebras. Let us view the reduced trace $\mathrm{tr}: R \to F$ as a map from $R$ to $R$ (since $F \subseteq R$). Since $\mathrm{End}_F R \approx R \otimes_F R^{\mathrm{op}} \approx R \otimes R$ as vector spaces over $F$, we have a unique element $a = \sum x_i \otimes y_i$ in $R \otimes R$ such that $\mathrm{tr}(r) = \sum x_i r y_i$ for all $r$ in $R$. We call this *Goldman's element*.

**Proposition 7.2.24:** *Goldman's element $a \in R \otimes R$ has the following properties:*

(i) $a^2 = 1$.
(ii) $a(r_1 \otimes r_2) = (r_2 \otimes r_1)a$ for all $r_1, r_2$ in $R$.
(iii) *If $K$ splits $R$ then viewing $a \otimes 1 \in (R \otimes R) \otimes K \approx M_n(K) \otimes_K M_n(K)$, we have $a \otimes 1 = \sum_{i,j} e_{ij} \otimes e_{ji}$.*

**Proof:**

(ii) $a(r_1 \otimes r_2) = \sum x_i r_1 \otimes y_i r_2$ corresponds to the map

$$r \mapsto \sum x_i r_1 r y_i r_2 = \mathrm{tr}(r_1 r) r_2,$$

and $(r_2 \otimes r_1)a$ corresponds to $r \mapsto \sum r_2 x_i r r_1 y_i = r_2 \mathrm{tr}(r r_1) = \mathrm{tr}(r_1 r) r_2$; hence $a(r_1 \otimes r_2) = (r_2 \otimes r_1)a$.

(iii) $a \otimes 1$ is Goldman's element for $R \otimes K \approx M_n(K)$. But $\sum_{i,j} e_{ij} \otimes e_{ji}$ clearly yields the trace in $M_n(K)$ so is $a \otimes 1$ by uniqueness.

(i) By (iii) we see $1 \otimes 1 = (a \otimes 1)^2 = a^2 \otimes 1$, so $a^2 = 1$.     Q.E.D.

Goldman's element arose in the study of Azumaya algebras, cf., Knus-Ojanguren [74B, p. 112] and Saltman [81], which extend many of the notions of central simple algebras to Azumaya algebras. Let us apply proposition 7.2.24 to higher tensor powers.

**Corollary 7.2.25:** *Consider the action of* $\text{Sym}(n)$ *on* $R^{\otimes t}$, *by* $\pi(r_1 \otimes \cdots \otimes r_t)$ $= r_{\pi 1} \otimes \cdots \otimes r_{\pi t}$. *There is a group homomorphism* $\psi: \text{Sym}(n) \to \text{Unit}(R^{\otimes t})$ *given by* $\pi \mapsto a_\pi$ *such that* $a_\pi^{-1} x a_\pi = \pi x$ *for all* $x$ *in* $R^{\otimes t}$.

*Proof:* Write $R^{\otimes t}$ as $R_1 \otimes \cdots \otimes R_t$ where each $R_i = R$. If $\pi$ is the transposition $(ij)$ then take $a_\pi$ to be the Goldman's element $a$ of proposition 7.2.24 calculated in $1 \otimes \cdots \otimes 1 \otimes R_i \otimes 1 \cdots \otimes R_j \otimes 1 \otimes \cdots \otimes 1$. Since every permutation is a product of transpositions, we can extend the definition of $a_\pi$ to all permutations, and well-definedness follows from the uniqueness of Goldman's element.    Q.E.D.

## Basic Properties of the Corestriction

We are ready for the most important property of the corestriction.

**Theorem 7.2.26:** $\text{cor} \circ \text{res} = [L:F]$. *More precisely, if* $A \in \text{Br}(F)$ *and* $L$ *is a separable extension field of* $F$ *with* $[L:F] = t$ *then* $\text{cor}_{L/F}(A \otimes_F L) \approx A^{\otimes t}$.

*Proof:* Let $R = A \otimes_F L$, and let $E$ be the normal closure of $L$ over $F$. Under notation of definition 7.2.19 there is a map $\varphi: \tilde{R} \to A^{\otimes t} \otimes_F E$ given by $\bigotimes_{i=1}^t ((a_i \otimes \gamma_i) \otimes \beta_i) \mapsto (\bigotimes_{i=1}^t a_i) \otimes \prod \sigma_i^{-1}(\gamma_i \beta_i)$ where $a_i \in A$, $\gamma_i \in L$, and $\beta_i \in E$, noting $R$ is spanned by elements of the form $a_i \otimes \gamma_i$. $\varphi$ is visibly a surjection of $F$-algebras and is thus an isomorphism because $[\tilde{R}:F] = [A^{\otimes t} \otimes_F E:F]$. Thus $\varphi^{-1}((\bigotimes a_i) \otimes \beta) = \bigoplus (a_i \otimes \beta_i)$ where $\beta_1 = \sigma_1 \beta$ and all other $\beta_i = 1$. By (1) we get $\sigma(\bigotimes(a_i \otimes \beta_i)) = \bigotimes(a_{\pi i} \otimes \tau_i \beta_{\pi i})$. Let $j = \pi^{-1} 1$. For all $i \neq j$ we have $\tau_i \beta_{\pi i} = \tau_i 1 = 1$. Applying $\varphi$ to each side defines an action of $G$ on $A^{\otimes t} \otimes_F E$, by

$$\sigma((\bigotimes a_j) \otimes \beta) = (\bigotimes a_{\pi i}) \otimes \sigma_j^{-1} \tau_j \beta_{\pi j} = (\bigotimes a_{\pi i}) \otimes \sigma \beta, \tag{2}$$

since $\sigma_j^{-1} \tau_j \beta_{\pi j} = (\sigma_j^{-1} \tau_j)\sigma_1 \beta = \sigma \beta$. Thus $G$ acts on $A^{\otimes t} \otimes E$.

In other words, let $\hat{\sigma}$ denote the action of $\sigma$ on $A^{\otimes t}$ by permuting the components according to the action of $\pi$ on the indices; the action (2) is really the diagonal action $\sigma((\bigotimes a_i) \otimes \beta) = \hat{\sigma}(\bigotimes a_i) \otimes \sigma \beta$, and under this action

$$\text{cor}_{L/F}(A \otimes_F L) \approx ((A^{\otimes t}) \otimes_F E)^G.$$

On the other hand, we define $\bar{G} = \{1 \otimes \sigma : \sigma \in G\}$, a group isomorphic to $G$ which also acts on $A^{\otimes t} \otimes_F E$, i.e., $\bar{\sigma}((\otimes a_i) \otimes \beta) = (\otimes a_i) \otimes \sigma\beta$. Obviously $(A^{\otimes t} \otimes_F E)^{\bar{G}} \approx A^{\otimes t} \otimes_F E^{\bar{G}} \approx A^{\otimes t}$, so we want to show $(A^{\otimes t} \otimes_F E)^G \approx (A^{\otimes t} \otimes_F E)^{\bar{G}}$ or, equivalently, by proposition 7.2.22 we need find an isomorphism $\psi : (A^{\otimes t} \otimes E)*G \rightarrow (A^{\otimes t} \otimes E)*\bar{G}$.

Now corollary 7.2.25 provides us a multiplicative subgroup $\{w_\sigma : \sigma \in G\}$ of $A^{\otimes t}$ such that $w_\sigma^{-1} x w_\sigma = \hat{\sigma}x$ for all $x$ in $A^{\otimes t}$. Thus for any $\sigma$ in $G$ and any $x$ in $A^{\otimes t}$, $\beta$ in $E$, we have

$$\sigma(x \otimes \beta) = w_\sigma^{-1} x w_\sigma \otimes \sigma\beta = (w_\sigma \otimes 1)^{-1} \bar{\sigma}(x \otimes \beta)(w_\sigma \otimes 1).$$

Define $\psi$ by

$$\sum s_\sigma \sigma \mapsto \sum s_\sigma (w_\sigma \otimes 1)^{-1} \bar{\sigma}$$

for $s_\sigma$ in $A^{\otimes n} \otimes E$. $\psi$ is a homomorphism which is invertible by inspection and thus is the desired isomorphism.     Q.E.D.

**Corollary 7.2.27:**   *If $m$ is relatively prime to $[L:F]$ then* cor: $\mathrm{Br}_m(L) \rightarrow \mathrm{Br}_m(F)$ *is a retraction and, in particular, is onto.*

**Proof:**  Let $\varphi : \mathrm{Br}_m(F) \rightarrow \mathrm{Br}_m(F)$ be given by $A \rightarrow A^{\otimes t}$ where $t = [L:F]$. Then $\varphi$ is an automorphism of $\mathrm{Br}_m(F)$, and cor $\circ$ res $\circ$ $\varphi^{-1}$ is the identity.
                                                                               Q.E.D.

**Example 7.2.28:**   Let $L = \mathbb{Q}(\mu, \mu_1, \ldots, \mu_p)$ where $\mu, \mu_1, \ldots, \mu_p$ are commuting indeterminates over $\mathbb{Q}$ and $\sigma$ is defined on $L$ by $\sigma\mu = \mu$ and $\sigma\mu_i = \mu_{i+1}$ subscripts mod $p$. Let $R = (\mu_1^{-1}\mu_2, \mu)_n$. Then $\mathrm{cor}_{L/F} R \sim 1$, seen either by direct computation or by proposition 7.2.32 below, but $R$ is a division algebra. This shows cor: $\mathrm{Br}_m(L) \rightarrow \mathrm{Br}_m(F)$ may not be 1:1 even when $[L:F]$ is prime to $m$.

**Remark 7.2.29:**   Suppose $E/F$ is Galois with Galois group $G$, and $R \in \mathrm{Br}(E)$.

(i) $\mathrm{cor}_{E/F} \sigma R \approx \mathrm{cor}_{E/F} R$ for every $\sigma$ in $G$.

(ii) res $\circ$ cor $R \sim \tilde{R} = \otimes_{\sigma \in G} \sigma R$. (Indeed, (i) is by definition 7.2.19, and (ii) translates to $\tilde{R}^G \otimes_F E \approx \tilde{R}$ which is proposition 7.2.14.)

One easy application of the corestriction enables us to throw out the "prime-to $p$" part.

**Proposition 7.2.30:**   *Suppose $R$ is a central simple $F$-algebra of degree $p^t$. Then $R$ is similar to a corestriction of a crossed product of degree a power of $p$.*

**Proof:** Let $K$ be a separable splitting field of $R$ and $E$ the normal closure of $K$ over $F$. Let $L$ be the fixed subfield of a Sylow $p$-subgroup $H$ of $\text{Gal}(E/F)$ and let $m = [L:F]$ which is prime to $p$. $E$ is a splitting field of $R \otimes_F L$, which thus is similar to a crossed product of degree $p^u$, and taking $m'$ such that $m'm \equiv 1$ (modulo $p^u$) we have $\text{cor}(R^{\otimes m'} \otimes_F L) \sim R$ by theorem 7.2.26.

$$Q.E.D.$$

**Digression:** There are more cohomological definitions of the corestriction, based on example 5.2.13. Conceptually one defines the *restriction map* $\text{Res}: H^n(G, M) \to H^n(H, M)$ for any subgroup $H$ of $G$ by viewing every complex over $H$ in the natural way as a complex over $G$. At the 0 level this injects $M^G$ into $M^H$. The map $N_{G/H}$ in the opposite direction is given by $Nx = \sum_{i=1}^{t} \sigma_i x$ where $\{\sigma_1, \ldots, \sigma_t\}$ is a given transversal of $G$; one extends this to a map $\text{cor}: H^n(H, M) \to H^n(G, M)$ by proceeding along an injective resolution. This formulation of the corestriction arose long before the concrete description we have used here. A computational cohomological definition is given in exercise 15.

## The Corestriction of a Symbol

These results become extremely powerful when confronted by the computation of the corestriction of a symbol, which we now perform.

**Remark 7.2.31:** If $R = (\alpha, \beta)_n$ is a symbol over $E$ and $\sigma \in \text{Gal}(E/F)$ then $\sigma R \approx (\sigma\alpha, \sigma\beta)$, by remark 7.2.18(i) (sending generator to respective generator).

We return to the set-up of definition 7.2.19. Namely $L$ is a separable extension of $F$, and $E$ is the normal closure of $L$ over $F$, so that $E$ is Galois over $F$ and $G = \text{Gal}(E/F)$. We take a transversal $\sigma_1, \ldots, \sigma_t$ of $\text{Gal}(E/L)$ in $G$. Given $\beta$ in $L$ we define the *norm* $N_{L/F}\beta = \sigma_1 \beta \cdots \sigma_t \beta$; this is clearly independent of the choice of transversal. Also we now write $(\alpha, \beta; L)_n$ instead of the symbol $(\alpha, \beta)_n$ to specify that the center of the simple algebra is $L$, thereby removing possible ambiguity.

**Proposition 7.2.32:** ("*Projection formula*") Suppose $\alpha \in F$, $\beta \in L$, and $F$ has a primitive $n$-th root $\zeta$ of 1. Then $\text{cor}_{L/F}(\alpha, \beta; L)_n \sim (\alpha, N_{L/F}\beta; F)_n$.

**Proof:** (Schneider [77]; Tignol [85]) Let $R = (\alpha, \beta; L)_n$. Then $R \otimes_L E \approx (\alpha, \beta; E)_n$, so under the notation of 7.2.19 we have $\tilde{R} = \bigotimes_{i=1}^{t} \sigma_i^{-1}(\alpha, \beta; E)_n \approx \bigotimes_{i=1}^{t}(\alpha, \sigma_i^{-1}\beta; E)_n$. Take $y, z$ in $R$ such that $zy = \zeta yz$, $y^n = \alpha$, and $z^n = \beta$.

Letting $y_i, z_i$, respectively, be $y \otimes 1$, $z \otimes 1$ in the $i$-th tensor component $(\alpha, \sigma_i^{-1}\beta; E)_n$ we see that $\sigma y_i = y_{\pi i}$ and $\sigma z_i = z_{\pi i}$ in the notation of (1), where $\pi$ is the permutation of the cosets of $\text{Gal}(E/L)$ corresponding to $\sigma$. It remains to show under this action that $(\alpha, N_{L/F}\beta; F)_n \sim \tilde{R}^*G$, in view of proposition 7.2.22. To this end, letting $\tilde{z} = z_1 \cdots z_t = (z \otimes 1)^{\otimes t}$ we claim $\tilde{R}$ contains a set of invertible elements $\{\tilde{r}_\sigma : \sigma \in G\}$ satisfying the following properties for all $\sigma, \tau \in G$ and all $1 \le i \le t$:

$$\tilde{r}_\sigma y_i = y_{\pi i}\tilde{r}_\sigma, \qquad \tilde{r}_\sigma \tilde{z} = \tilde{z}\tilde{r}_\sigma \quad \text{and} \quad \tilde{r}_\sigma(\sigma\tilde{r}_\tau) = \tilde{r}_{\sigma\tau}. \tag{3}$$

We shall call such a set a *Tignol set*. Given a Tignol set we define

$R_1 = F$-subalgebra of $\tilde{R}^*G$ generated by $E$ and the $\tilde{r}_\sigma \sigma$.

$R_2 = F$-subalgebra of $\tilde{R}^*G$ generated by $y_1$ and $\tilde{z}$.

$R_3 = C_{\tilde{R}^*G}(R_1 R_2)$.

Let us compute $R_1, R_2$, and $R_3$.

$R_1 \sim 1$; indeed $\tilde{r}_\sigma \sigma \mapsto \sigma$ shows $R_1 \approx E^*G$ which by proposition 7.2.22 is similar to $E^G = F$.

$R_2 \approx (\alpha, N_{L/K}\beta)_n$ since $\tilde{z}y_1 = \zeta y_1 \tilde{z}$, $y_1^n = \alpha$, and $\tilde{z}^n = \prod_{i=1}^t \sigma_i^{-1}\beta = N_{L/F}\beta$.

$R_3$ is split. Indeed, $R_3$ has degree $n^{t-1}$ since $\deg(\tilde{R}^*G) = |G|n^t$, $\deg R_1 = |G|$, and $\deg R_2 = n$. On the other hand, let $C = F(y_2 y_1^{-1}, \ldots, y_t y_1^{-1}) \subseteq R_3$. Each $F(y_i y_1^{-1}) \approx F^{(n)}$ by lemma 7.1.13(ii) since $(y_i y_1^{-1})^n = \alpha\alpha^{-1} = 1$. Hence $C \approx F^{(n^{t-1})}$ so we are done by lemma 7.1.13(i).

But $R_1, R_2, R_3$ centralize each other, so $\tilde{R}^*G \approx R_1 \otimes R_2 \otimes R_3 \sim R_2 \sim (\alpha, N_{L/K}\beta)_n$ as desired.

It remains to find the desired Tignol set. First note each $y_1^{-1}y_i$ acts on $C$ (defined above) by left multiplication, and this action has order $n$ since $(y_1^{-1}y_i)^n = 1$. By iteration one sees the fixed subring (without 1) under all these actions contains a nonzero idempotent which we call $e$. Thus $y_i e = y_1 e$ for each $i$. Now for any $t$-tuple $\mathbf{j} = (j(1), \ldots, j(t))$ where $0 \le j(i) < n$ for each $i$, write $z^{\mathbf{j}}$ for $z_1^{j(1)} \cdots z_t^{j(t)}$. Thus $z^{\mathbf{j}}y_i y_1^{-1} = \zeta^{j(i)-j(1)}y_i y_1^{-1}z^{\mathbf{j}}$ for each $i \ge 2$. Define

$$e_{\mathbf{j}} = z^{-\mathbf{j}}e z^{\mathbf{j}}.$$

Each $e_{\mathbf{j}}$ is clearly a primitive idempotent, and we claim the $\{e_{\mathbf{j}} : j(1) = 0\}$ are all distinct. Indeed, otherwise, there is $\mathbf{j}$ for which $z^{\mathbf{j}}$ commutes with $e$ but $j(i) \ne 0$ for suitable $i \ge 2$; then $y_1 e z^{\mathbf{j}} = z^{\mathbf{j}}y_1 e = z^{\mathbf{j}}y_i e = \zeta^{j(i)}y_i e z^{\mathbf{j}} = \zeta^{j(i)}y_1 e z^{\mathbf{j}}$, contradiction.

Since the primitive idempotents of $C \approx F^{(n^{t-1})}$ are orthogonal and $n^{t-1}$ in number, we see $\{e_{\mathbf{j}} : j(1) = 0\}$ is a complete set of primitive idempotents of $C$. In particular, $e = e_0$ is unique, and thus $\sigma e = e$ for all $\sigma$ in $G$ (since $\sigma e$ plays the same role as $e$). Now we let

$$\tilde{r}_\sigma = \sum_{j(\pi 1)=0} z^{-\mathbf{j}} e z^{\pi \mathbf{j}},$$

where $\pi$ is the permutation on the cosets of $H$ defined by $\sigma$, and $\pi \mathbf{j}$ is defined as $(j(\pi 1), j(\pi 2), \ldots, j(\pi t))$. Note $\sigma(z_i) = z_{\pi i}$ implies $\sigma(z^{\pi \mathbf{j}}) = z^{\mathbf{j}}$. Let us verify (3).

$$\tilde{r}_\sigma y_i = \sum_{\mathbf{j}} z^{-\mathbf{j}} e z^{\pi \mathbf{j}} y_i = \sum \zeta^{j(\pi i)} z^{-\mathbf{j}} y_i e z^{\pi \mathbf{j}} = \sum \zeta^{j(\pi i)} z^{-\mathbf{j}} y_{\pi i} e z^{\pi \mathbf{j}}$$

$$= y_{\pi i} \sum z^{-\mathbf{j}} e z^{\pi \mathbf{j}} = y_{\pi i} \tilde{r}_\sigma.$$

$r_\sigma \tilde{z} = \tilde{z} \tilde{r}_\sigma$ since $\tilde{z} = z_1 \cdots z_t$ commutes with each $z_i$ and with each $y_i y_1^{-1}$ (and thus with $C$).

The last equation is trickier. First note $e z^{\mathbf{j}} e = z^{\mathbf{j}} e_{\mathbf{j}} e = 0$ unless $\mathbf{j} = 0$. Also recall $e$ is fixed by $G$. Now taking any $h$ in $G$ and its corresponding permutation $\rho$ on the cosets of $H$ we have

$$\tilde{r}_\sigma \sigma(\tilde{r}_\tau) = \sum_{j(\pi 1)=0,\, j'(\rho 1)=0} z^{-\mathbf{j}} e z^{\pi \mathbf{j}} \sigma(z^{-\mathbf{j}'} e z^{\rho \mathbf{j}'})$$

$$= \sum_{\mathbf{j},\mathbf{j}'} z^{-\mathbf{j}} e z^{\pi \mathbf{j}} z^{-\pi^{-1}\mathbf{j}'} e z^{\rho \pi^{-1}\mathbf{j}'} = \sum_{\mathbf{j}} z^{-\mathbf{j}} e z^{\rho \pi \mathbf{j}},$$

since we just saw that $e z^{\pi \mathbf{j}} z^{-\pi^{-1}\mathbf{j}'} e = 0$ unless $\mathbf{j}' = \pi^2 \mathbf{j}$. But this shows $\tilde{r}_\sigma \sigma(\tilde{r}_\tau) = \tilde{r}_{\sigma\tau}$ as desired, verifying (3). Furthermore, $\tilde{r}_1 = \sum e_{\mathbf{j}} = 1$; taking $\tau = \sigma^{-1}$ we see $\tilde{r}_\sigma \sigma(\tilde{r}_\tau) = \tilde{r}_1 = 1$, proving each $\tilde{r}_\sigma$ is invertible.    Q.E.D.

This is enough to prove a result of Tignol on the corestriction for cyclic algebras, cf., exercise 17; to proceed further we need to calculate $\text{cor}(\alpha, \beta)_n$ when $\alpha \notin F$.

**Definition 7.2.33:** Given relatively prime polynomials $f, g$ in $F[\lambda]$ define the *Rosset-Tate* symbol $\left(\dfrac{f}{g}\right)$ in $\text{Br}(F)$ by induction on the length of factorization of $g$ into irreducibles in $F[\lambda]$ as follows, where $n$ is fixed arbitrarily:

(i) If $g = \lambda$ or $g$ is constant then $\left(\dfrac{f}{g}\right) = 1$

(ii) If $g \neq \lambda$ is monic irreducible then let $L = F(a)$ where $a$ is a root of $g$, and

$$\left(\frac{f}{g}\right) = \text{cor}_{L/F}(a, f(a))_n.$$

(iii) $\left(\dfrac{f}{gh}\right) = \left(\dfrac{f}{g}\right)\left(\dfrac{f}{h}\right).$

**Remark 7.2.34:** $\left(\dfrac{f + gh}{g}\right) = \left(\dfrac{f}{g}\right)$, in view of (ii).

**Notation:** If $g = \sum_{i=u}^{v} \alpha_i \lambda^i$ with $\alpha_u, \alpha_v \neq 0$ write $\bar{g}$ for $\alpha_u^{-1} \sum_{i=u}^{v} \alpha_i \lambda^{i-u}$ (i.e., the lowest coefficient is 1), and $c(g)$ for $(-1)^v \alpha_v \in F$. Note $\overline{gh} = \bar{g}\bar{h}$ and $c(gh) = c(g)c(h)$.

**Remark 7.2.35:** In (ii) above we have $\bar{g} = g(0)^{-1}g$ so $c(\bar{g}) = (-1)^{\deg(g)}g(0)^{-1} = N_{L/F}a^{-1}$.

**Lemma 7.2.36:** (*Rosset-Tate reciprocity law*) $\left(\dfrac{f}{g}\right) \sim \left(\dfrac{\bar{g}}{f}\right)(c(\bar{g}), c(f))_n^{-1}$.

**Proof:** The idea is to reduce quickly to (ii) and then apply remark 7.2.35 to manipulate symbols according to proposition 7.1.17. Explicitly, assume first $g$ is constant or $g = \lambda$. Then $\bar{g} = 1$ so $c(\bar{g}) = 1$ and every term is 1. Thus we may assume $g \neq \lambda$ is nonconstant; in view of (iii) we may assume $g$ is monic irreducible, and so the constant term $g(0) \neq 0$. Then $c(\bar{g}) = N_{L/F}a^{-1}$ by remark 7.2.35.

For any constant $\alpha$ in $F$ we have by (ii) and the projection formula

$$\left(\frac{\alpha}{g}\right) = \operatorname{cor}_{L/F}(a, \alpha)_n \sim (N_{L/F}a, \alpha)_n \sim (c(\bar{g}), \alpha)_n^{-1} \sim \left(\frac{\bar{g}}{\alpha}\right)(c(\bar{g}), c(\alpha))_n^{-1},$$

since $c(\alpha) = \alpha$ and $\left(\dfrac{\bar{g}}{\alpha}\right) = 1$ by (i). Likewise, using the fact $(a, -a) \sim 1$ we have

$$\left(\frac{\lambda}{g}\right) = \operatorname{cor}_{L/F}(a, a)_n \sim \operatorname{cor}_{L/F}(a, -1)_n \sim (N_{L/F}a, -1)_n$$

$$= (c(\bar{g}), -1)_n^{-1} \sim \left(\frac{\bar{g}}{\lambda}\right)(c(\bar{g}), c(\lambda))_n^{-1}$$

since $c(\lambda) = -1$ and $\left(\dfrac{\bar{g}}{\lambda}\right) = 1$.

Thus we are done unless $f \neq \lambda$ is nonconstant, and it clearly suffices to prove the case $f$ (as well as $g$) is monic irreducible. Let $b$ be a root of $f$ and let $K$ be a field containing $L = F(a)$ and $L' = F(b)$. Note $N_{K/L}(a - b) = f(a)$. We have $(a, a - b)_n \sim (a, 1 - a^{-1}b)_n(a, -a)_n(a, -1)_n$. But $(a, -a)_n \sim 1 \sim (a^{-1}b, 1 - a^{-1}b)_n$ so

$$(a, a-b)_n \sim (a, 1 - a^{-1}b)_n(a^{-1}b, 1 - a^{-1}b)_n(a, -1)_n \sim (b, 1 - a^{-1}b)_n(a, -1)_n. \quad (4)$$

We shall conclude by computing the corestriction of each side of (4).

$$\mathrm{cor}_{K/F}(a, a - b)_n \sim \mathrm{cor}_{L/F}(\mathrm{cor}_{K/L}(a, a - b)_n)$$

$$\sim \mathrm{cor}_{L/F}(a, N_{K/L}(a - b))_n$$

$$\sim \mathrm{cor}_{L/F}(a, f(a))_n = \left(\frac{f}{g}\right). \tag{5}$$

$$\mathrm{cor}_{K/F}(b, 1 - a^{-1}b)_n \sim \mathrm{cor}_{L'/F}(\mathrm{cor}_{K/L'}(b, 1 - a^{-1}b)_n))$$

$$\sim \mathrm{cor}_{L'/F}(b, N_{K/L'}(1 - a^{-1}b))_n.$$

But $N_{K/L'}(1 - a^{-1}b) = N_{K/L'}((a - b)a^{-1}) = N_{K/L'}(b - a)N_{K/L'}(-a)^{-1} = g(b)g(0)^{-1} = \bar{g}(b)$ so

$$\mathrm{cor}_{K/F}(b, 1 - a^{-1}b)_n \sim \mathrm{cor}_{L'/F}(b, \bar{g}(b))_n = \left(\frac{\bar{g}}{f}\right). \tag{6}$$

Finally $N_{K/L'}a = N_{L/F}a = c(\bar{g})^{-1}$ so

$$\mathrm{cor}_{K/F}(a, -1)_n \sim \mathrm{cor}_{L'/F}(\mathrm{cor}_{K/L'}(a, -1)_n) \sim \mathrm{cor}_{L'/F}(N_{K/L'}a, -1)_n$$

$$\sim \mathrm{cor}_{L'/F}(c(\bar{g})^{-1}, -1)_n \sim (c(\bar{g})^{-1}, (-1)^{\deg f})_n = (c(\bar{g}), c(f))_n^{-1}, \tag{7}$$

so equating (5) with (6), (7) yields the reciprocity law.    Q.E.D.

## Interpretation of the Rosset-Tate Symbol

Suppose $[L:F] = t$ and $0 \neq a, b \in L$. Then $N_{L/F(a)}b \in F(a)$, so $N_{L/F(a)}b = f(a)$ for some $f \in F[\lambda]$ relatively prime to the monic minimal polynomial $g$ of $a$. Definition 7.2.33(ii) then says $\mathrm{cor}_{L/F}(a, b)_n = \left(\frac{f}{g}\right)$.

**Theorem 7.2.37:**    (Rosset-Tate) $\mathrm{cor}_{L/F}(a, b) \sim \bigotimes_{i=1}^{k}(c(g_i), c(\bar{g}_{i-1}))_n$ where the $\bar{g}_i$ are defined as follows: $g_0 = g$ and $g_1 = f$ chosen as immediately above; inductively if $g_i \nmid \bar{g}_{i-1}$ then using Euclid's algorithm write $\bar{g}_{i-1} = q_i g_i + g_{i+1}$ with $\deg g_{i+1} < \deg g_i$. In particular, $k \leq \deg g \leq [L:F]$.

**Proof:** $\left(\frac{g_{i+1}}{g_i}\right) = \left(\frac{\bar{g}_{i-1}}{g_i}\right)$ by remark 7.2.34, so the reciprocity law shows

$$\left(\frac{g_i}{g_{i-1}}\right) \sim \left(\frac{\bar{g}_{i-1}}{g_i}\right)(c(\bar{g}_{i-1}), c(g_i))_n^{-1} \sim \left(\frac{g_{i+1}}{g_i}\right)(c(g_i), c(\bar{g}_{i-1}))_n. \tag{8}$$

The last nonzero $g_k$ is constant by Euclid's algorithm, so $\left(\dfrac{\bar{g}_{k-1}}{g_k}\right) = 1$, and

we get the desired formula by starting with $\left(\dfrac{g_1}{g_0}\right) = \left(\dfrac{f}{g}\right)$ and plugging into (8) inductively.     Q.E.D.

The first term $(c(f), c(\bar{g}))_n \approx ((c(f), N_{L/F}a^{-1})_n \approx (N_{L/F}a, c(f))_n$, so we can rephrase the Rosset-Tate theorem as

**Corollary 7.2.38:**  *Suppose $L$ is a separable extension of $F$ and $[L{:}F] = t$, $F$ containing a primitive $t$-th root of 1. Then for any symbol $R = (a, b)_n$ in $\mathrm{Br}(L)$, $\mathrm{cor}_{L/F} R$ is similar to a tensor product of at most $t$ symbols of degree $n$ in $\mathrm{Br}(F)$, the first of which has the form $(N_{L/F}a, \alpha)_n$ for suitable $\alpha$.*

For $t = 2$ we have a stronger result.

**Corollary 7.2.39:**  *If $[L{:}F] = 2$, $a, b \in L$, and $\mathrm{char}(F) \neq 2$ then $\mathrm{cor}(a, b)_2 \sim (N_{L/F}a, \alpha) \otimes_F (N_{L/F}a', \alpha')$ for suitable $a' \in L$, and $\alpha', \alpha \in F$ (which are all given explicitly in the proof).*

**Proof:**  If $a \in F$ or $b \in F$ then we are done by the projection formula so assume $a, b \notin F$. Let $\sigma$ be the nontrivial automorphism of $L$ over $F$, and let $g$ be the minimal polynomial of $a$; then $g = (\lambda - a)(\lambda - \sigma a)$. Now writing $b = \gamma_0 a + \gamma_1$ for $\gamma_0, \gamma_1$ in $F$ we have $f = \gamma_0 \lambda + \gamma_1$ as preceding theorem 7.2.37 so by elementary theory of polynomials we see the remainder $g_1$ obtained from dividing $\bar{g}$ by $f$ is

$$g_2 = \bar{g}(-\gamma_1\gamma_0^{-1}) = (N_{L/F}a)^{-1}g(-\gamma_1\gamma_0^{-1}) = (N_{L/F}a)^{-1}N_{L/F}(-\gamma_1\gamma_0^{-1} - a)$$
$$= N_{L/F}(\gamma_1\gamma_0^{-1}a^{-1} + 1) = N_{L/F}(\gamma_0^{-1}a^{-1}b),$$

so theorem 7.2.37 shows

$$\mathrm{cor}(a, b)_2 \sim (c(f), N_{L/F}a)_2 \otimes (N_{L/F}\gamma_0^{-1}a^{-1}b, c(\bar{f}))_2,$$

yielding the desired result since $Q \sim Q^{\mathrm{op}}$ for any quaternion algebra.     Q.E.D.

It would be extremely interesting to have a version of Rosset-Tate calculating the corestriction of a cyclic algebra when $F$ does *not* contain primitive $n$-th roots of 1.

*Example 7.2.40:*  To see the sharpness of the Rosset-Tate theorem let $F_0$ be a field containing a primitive $n$-th root of 1, let $L = F_0[\lambda_1, \ldots, \lambda_t, \mu_1, \ldots, \mu_t]$

where the $\lambda_i$ and $\mu_i$ are commuting indeterminates over $F_0$, and let $F = L^\sigma$ where $\sigma\lambda_i = \lambda_{i+1}$ and $\sigma\mu_i = \mu_{i+1}$ (subscripts mod $t$). Let $R = (\lambda_1, \mu_1)_n$. By remark 7.2.29 res cor $R \sim (\lambda_1, \mu_1)_n \otimes \cdots \otimes (\lambda_t, \mu_t)_n$ which is a division algebra. Thus cor $R$ cannot be written as a tensor product of fewer than $t$ symbols of degree $n$.

Other interesting immediate applications include the following cute result based on work of Albert, Perlis, Rowen-Saltman, and Mammone-Tignol.

**Proposition 7.2.41:** *Let $R$ be a central simple algebra of degree $n$ over a field $F$ containing a primitive $n$-th root of 1. Suppose $R$ is split by a Galois extension $K$ over $F$ whose Galois group $G$ has a cyclic normal subgroup $H$ of order $n$, and suppose that $|G| = pn$ where $p$ and $n$ are relatively prime. Then $R$ is similar to a product of $\leq (p - 1)$ cyclics of degree $n$.*

**Proof:** Let $L = K^H$; then $[K:L] = n$ and $[L:F] = p$. $R \otimes_F L$ is split by $K$, which is cyclic over $L$; hence $R \otimes_F L \sim (K, \sigma, b)$ for suitable $b$. Writing $K = L(y)$ with $a = y^n \in L$ we have $R \otimes_F L \sim (a, b)_n$. By theorem 7.2.26 we have $R^{\otimes p} \sim \text{cor}_{L/F}(a, b)_n$ so $R \sim \text{cor}_{L/F}(a, b^{p'})_n$ where $p'p \equiv 1(\text{mod } n)$. Then corollary 7.2.38 shows $R$ is similar to a product of at most $p$ symbols, the first of which has the form $(N_{L/F}a, \alpha)_n$. Thus we are done if we can show $N_{L/F}a$ is an $n$-th power in $F$, for then $(N_{L/F}a, \alpha) \sim 1$. Write $H = \langle \sigma \rangle$ and take $\tau \in G$ of order $p$. Note $\tau^{-1}\sigma\tau = \sigma^u$ for $u$ satisfying $u^p \equiv 1 \pmod{n}$.

We may write $\sigma y = \zeta y$ where $\zeta$ is a primitive $n$-th root of 1; then $\sigma\tau y = \tau\sigma^u y = \zeta^u\tau y$ so $\tau y = cy^u$ for suitable $c$ in $L$. Hence $y\tau y \cdots \tau^{p-1}y \in Lyy^u \cdots y^{u^{p-1}} = Ly^t$ where $t = 1 + u + \cdots + u^{p-1} = (u^p - 1)/(p - 1) \equiv 0 \pmod{n}$, implying $y\tau y \cdots \tau^{p-1}y$ is some element $\gamma$ of $L$. But $N_{L/F}a = a\tau a \cdots \tau^{p-1}a = \gamma^n \in F$, so $\gamma \in F$ as desired since $n$ is prime to $p = [L:F]$. Q.E.D.

**Corollary 7.2.42:** *When $p = 2$ and $n$ is odd we see any central simple algebra of degree $n$ split by the dihedral group of order $2n$ is cyclic, cf., exercise 19. In fact, by exercise 19 one can choose $b$ such that $L(b^{1/n})$ is Galois (dihedral) over $F$; then $N_{L/F}b$ is an $n^{th}$ power so now in the notation of corollary 7.2.40 we have* $\text{cor}(a, b)_n \sim (N_{L/F}\gamma_0, \gamma_1^{-1}\gamma_0) \sim (\gamma_1, \gamma_0^2)_n$.

**Theorem 7.2.43:** *(Mostly Rosset) Suppose $p = \deg(R)$ is prime $> 2$ and $F$ has a primitive $p$-th root of 1. Then $R$ is similar to a product of $(p - 1)!/2$ cyclic algebras of degree $p$.*

*Proof:* We start as in proposition 7.2.30. Let $E$ be the normal closure of a separable splitting field $K$ of $F$. We assume the "worst" case, i.e., $\text{Gal}(E/F) = \text{Sym}(p)$, which contains a dihedral group $H$ of order $2p$; let $L = E^H$. Then $R \otimes_F L$ is cyclic by corollary 7.2.42. But $[L:F] = (p-1)!/2$, which is prime to $p$, so $R$ is the corestriction of a symbol (from $L$ to $F$). We conclude by applying theorem 7.2.37.     Q.E.D.

## Central Simple Algebras with Involution

A major application of the corestriction is a description of central simple algebras with involution, thereby explaining a lovely body of results of Albert whose previous proofs had been very computational in nature. We shall describe the "modern" aspects of the theory, leaving some of the older computational results to the literature (or to the exercises). The key to this subject is the following connection with the exponent.

**Theorem 7.2.44:** *A central simple algebra $R$ has exponent 2 iff $R$ has an involution of the first kind.*

*Proof:* ($\Leftarrow$) Any involution is an anti-automorphism, so clearly $R \approx R^{\text{op}}$ as $F$-algebras, implying $R \otimes R \sim R \otimes R^{\text{op}} \sim F$.

($\Rightarrow$) Let $K_0$ be a maximal subfield of the underlying division algebra of $R$, and let $(c_{ijk})$ be a Brauer factor set of $R$ with respect to $K = K_0^{(t)}$. We use the notational hypothesis 7.1.33, i.e., $R = KrK$, $E$ is the normal closure of $K_0$, and $r = (r_{ij})$. Since $R \sim R^{\text{op}}$ we see by proposition 7.2.8 there are $\gamma_{ij}$ in $E$ satisfying the conjugacy conditions such that

$$c_{kji} = c_{ijk}\gamma_{ij}\gamma_{jk}\gamma_{ik}^{-1}.$$

Note, in particular, $c_{iii} = c_{iii}\gamma_{ii}$ so $\gamma_{ii} = 1$, and $c_{iji} = c_{iji}\gamma_{ij}\gamma_{ji}\gamma_{ii}^{-1}$ yielding

$$\gamma_{ij}\gamma_{ji} = \gamma_{ii} = 1. \tag{9}$$

By theorem 7.1.37 we can view $R$ as $\{(\beta_{ij}r_{ij}) \in M_n(E)$: the $\beta_{ij}$ satisfy the conjugacy conditions$\}$, where $r_{ij} = c_{ij1}$. Define (*) on $R$ by

$$(\beta_{ij}r_{ij})^* = (\bar{\beta}_{ij}r_{ij}) \quad \text{where } \bar{\beta}_{ij} = \beta_{ji}\gamma_{ij}.$$

Then $(\beta_{ij}r_{ij})^{**} = (\bar{\beta}_{ij}r_{ij})^* = (\beta_{ji}\gamma_{ij}r_{ij})^* = (\beta_{ij}\gamma_{ji}\gamma_{ij}r_{ij}) = (\beta_{ij}r_{ij})$ by (7).
Furthermore, if $(\beta''_{ij}r_{ij}) = (\beta_{ij}r_{ij})(\beta'_{ij}r_{ij})$ so that $\beta''_{ij} = \sum_{k=1}^n \beta_{ij}c_{ijk}\beta'_{kj}$ then

$$(\beta'_{ij}r_{ij})^*(\beta_{ij}r_{ij})^* = \left(\sum_k \bar{\beta}'_{ik}c_{ikj}\bar{\beta}_{kj}r_{ij}\right) = \left(\sum_k \beta'_{ki}\gamma_{ik}c_{ikj}\beta_{jk}\gamma_{kj}r_{ij}\right)$$

$$= \left( \sum_k \beta_{jk} c_{ikj} \gamma_{ik} \gamma_{kj} \beta'_{ki} r_{ij} \right) = \left( \sum_k \beta_{jk} c_{jki} \beta'_{ki} \gamma_{ij} r_{ij} \right)$$

$$= (\beta''_{ji} \gamma_{ij} r_{ij}) = (\beta''_{ij} r_{ij})^*$$

proving (*) is indeed an anti-automorphism of degree 2.      Q.E.D.

The next result enables us to focus on a particular subfield when dealing with involutions.

**Proposition 7.2.45:** *Suppose* $(D, *)$ *is a central division algebra with involution and K is a subfield of D having an automorphism* $\sigma$ *of order* $\leq 2$, *such that* (*) *and* $\sigma$ *agree on F. Then D has an involution whose restriction to K is* $\sigma$.

**Proof:**  By Skolem-Noether the isomorphism $K \to (\sigma K)^*$ given by $a \mapsto (\sigma a)^*$ extends to an inner automorphism of $D$, i.e., for suitable $d$ in $D$ we have $(\sigma a)^* = d^{-1}ad$ for all $a$ in $K$. Then $d^*d^{-1}a(d^*d^{-1})^{-1} = d^*(\sigma a)^*(d^{-1})^* = (d^{-1}(\sigma a)d)^* = (\sigma(\sigma a))^{**} = a$ for all $a$ in $K$, implying $d^*d^{-1} \in C_D(K)$. Let $A = C_D(K)$. Thus $d^* \in Ad$ so $d^* \pm d \in Ad$. Since $d$ could be replaced by any element of $Ad$ we may use $d + d^*$ or $d - d^*$ in place of $d$, and certainly one of these is nonzero. Hence we may assume $d^* = \pm d$, in which case $D$ has the involution $(J)$ given by $x \mapsto dx^*d^{-1}$. But now $a^J = da^*d^{-1} = d(d^{-1}\sigma ad)d^{-1} = \sigma a$, for all $a$ in $K$.      Q.E.D.

A further refinement is given in exercise 26. We shall use the following observation to pass to involutions of the *second* kind.

**Remark 7.2.46:**  (i) If $R$ has an involution (*) of the first kind, and if $K$ is a quadratic extension of $F$ with nontrivial automorphism $\sigma$ over $F$, then (*) $\otimes \sigma$ is an involution of the second kind on $R \otimes_F K$.
(ii) Suppose, moreover, $K$ is a subfield of $R$ and the restriction of (*) to $K$ is $\sigma$. Then (*) restricts to an involution of $C_R(K)$ of the second kind over $F$.

The tie to the corestriction actually comes with involutions of the second kind, via corestricting along a quadratic extension.

**Remark 7.2.47:**  Suppose $[L:F] = 2$ with $1/2 \in F$, and write $\text{Gal}(L/F) = \{1, \sigma\}$. In the notation of 7.2.19 we have $\tilde{R} = R \otimes \sigma R$, and $\sigma$ acts on $\tilde{R}$ by $\sigma(r_1 \otimes r_2) = r_2 \otimes r_1$. Thus $\text{cor}_{L/F}$ is spanned by $\{r \otimes r : r \in R\}$, for if $x = \sum r_{i1} \otimes r_{i2} \in \text{cor } R$ then

$$2x = \sum(r_{i1} \otimes r_{i2} + r_{i2} \otimes r_{i1}) = \sum((r_{i1} + r_{i2}) \otimes (r_{i1} + r_{i2}) - r_{i1} \otimes r_{i1} - r_{i2} \otimes r_{i2}).$$

**Theorem 7.2.48:** (*Albert-Riehm-Scharlau*) *Suppose* $[L:F] = 2$ *and* $R$ *is a central simple $L$-algebra with an anti-automorphism* (\*) *which restricts to the nontrivial automorphism* $\sigma$ *of $L$ over $F$. Then there is $r_0$ in $R$ for which* $r^{**} = r_0 r r_0^{-1}$ *for all $r$ in $R$. Any such $r_0$ satisfies $r_0^* r_0 \in F$, and writing $\alpha = r_0 r_0^*$ we have* $\mathrm{cor}_{L/F} R \sim (L, \sigma, \alpha)$, *a quaternion algebra.*

**Proof:** The existence of $r_0$ is by the Skolem-Noether theorem applied to (\*)$^2$. Then

$$(r_0^*)^{-1} r^* r_0^* = (r_0 r r_0^{-1})^* = r^{***} = r_0 r^* r_0^{-1}$$

for all $r$ in $R$, implying $r_0^* r_0 \in Z(R) = L$. But $r_0^{**} = r_0^{-1} r_0 r_0 = r_0$ so letting $\alpha = r_0^* r_0$ we see $\alpha = \alpha^*$ and thus $\alpha \in F$.

Let $\varphi$ denote the composition $R \otimes_\sigma R \xrightarrow{1 \otimes *} R \otimes R^{\mathrm{op}} \approx \mathrm{End}\, R_L \subseteq \mathrm{End}\, R_F$, and let $Q$ be the centralizer of $\varphi(\mathrm{cor}_{L/F} R)$ in $\mathrm{End}\, R_F$. Then $\varphi(\mathrm{cor}_{L/F} R) \otimes Q \approx \mathrm{End}\, R_F$: counting dimensions we see $Q$ is a quaternion algebra over $F$, and

$$\varphi(\mathrm{cor}_{L/F} R) \sim Q^{\mathrm{op}} \sim Q.$$

We want to display $Q$ explicitly. Note $\varphi$ is given by $\varphi(r_1 \otimes r_2)v = r_1 v r_2^*$ where $r_i, v \in R$; thus by remark 7.2.47 we see $f \in Q$ iff $f$ commutes with all $\varphi(r \otimes r)$, i.e., $f(r v r^*) = r(fv)r^*$.

In particular, $Q$ contains the left regular representation of $L$, since $L = Z(R)$. Hence $L$ is a maximal subfield of $Q$. Also $Q$ contains the map $z: R \to R$ given by $v \mapsto (v r_0)^*$, since

$$(r v r^* r_0)^* = r_0^* r^{**} v^* r^* = r_0^* r_0 r r_0^{-1} v^* r^* = \alpha r r_0^{-1} v^* r^* = r \alpha r_0^{-1} v^* r^*$$

$$= r r_0^* v^* r^* = r(v r_0)^* r^*.$$

Since $z^2 v = ((v r_0)^* r_0)^* = r_0^* r_0 (v r_0) r_0^{-1} = \alpha v$ we see (taking $v = 1$) that $z^2 = \alpha$; furthermore, $z(\beta v) = \beta^* z v$, so conjugation by $z$ induces $\sigma$ on $L$, and thus we have displayed $Q$ as $(L, \sigma, \alpha)$.    Q.E.D.

Before applying this theorem let us list two facts whose easy verifications are left to the reader.

**Remark 7.2.49:**

(i) Any involution (\*) of $R$ extends to $M_n(R)$ by $(r_{ij})^* = (r_{ji}^*)$.

(ii) If a division ring $D$ has an antiautomorphism (\*) satisfying $d^{**} = ada^{-1}$ for all $d$ in $D$ and if $a^* = a^{-1}$ then $D$ has an involution given by $d \mapsto (1 + a)^{-1} d^* (1 + a)$.

The converse to remark 7.2.49(i) relies on the structure theory of central simple algebras (cf., exercise 24) and actually is false for Azumaya algebras.

**Theorem 7.2.49′:** *Suppose* $[L:F] = 2$. *A central simple L-algebra R has an involution (*) of second kind over F iff* $\mathrm{cor}_{L/F} R \sim 1$.

**Proof:** $(\Rightarrow)$ $(*)^2 = 1$ so take $r_0 = 1 = \alpha$; then $\mathrm{cor}_{L/F} R \sim (L, \sigma, 1) \sim 1$.
$(\Leftarrow)$ (Scharlau) In view of 7.2.49(i) we may replace $R$ by its underlying division algebra. By hypothesis $\tilde{R}$ has a set of matric units and is thus split, so $\sigma R \sim R^{\mathrm{op}}$, yielding an anti-automorphism (*) of $R$ over $F$. Taking $r_0, \alpha$ as in the theorem we have $(L, \sigma, \alpha)$ split so $\alpha = N_{L/F}\beta$ for some $\beta$ in $L$. Replacing $r_0$ by $\beta^{-1} r_0$ yields $r_0^* r_0 = 1$ so we get the desired involution by remark 7.2.49(ii). Q.E.D.

**Theorem 7.2.50:** *Suppose R is a central simple F-algebra which contains a quadratic field extension K of F. The following assertions are equivalent:*

(i) $\exp(R) = 2$.
(ii) $\exp C_R(K) = 2$, *and* $C_R(K)$ *has an involution of second kind over F.*
(iii) $C_R(K)$ *has an involution of second kind over F.*

**Proof:** (i) $\Rightarrow$ (ii). By proposition 7.2.45 there are involutions $(*)_1$ and $(*)_2$ of $R$ of first kind, whose respective restrictions to $K$ are the identity and the nontrivial automorphism over $F$. The respective restrictions of $(*)_1$ and $(*)_2$ to $C_R(K)$ are involutions of the first and second kinds, and $\exp C_R(K) = 2$ by theorem 7.2.44.
(ii) $\Rightarrow$ (iii) *a fortiori.*
(iii) $\Rightarrow$ (i) $C_R(K) \sim R \otimes_F K$ so $R^{\otimes 2} \sim \mathrm{cor}\, C_R(K) \sim 1$. Q.E.D.

Passing to the centralizer of a quadratic extension of the center will be a very useful inductive procedure. The last ingredient to be analyzed here is the *type* of involution. Suppose $R$ has an involution (*) of the first kind. By theorem 2.13.21 and definition 2.13.27 (*) has either orthogonal or symplectic type. However, these results do not lend themselves readily to computation. It is more convenient to tensor by the algebraic closure $\bar{F}$ of $F$ and pass to $(R \otimes_F \bar{F}, *) \approx (M_n(\bar{F}), *)$ since the type is preserved under tensor extension; now using theorem 2.13.29 and example 2.13.22 we may replace (*) by the transpose ($t$) or the canonical symplectic involution ($s$), cf., example 2.13.6, depending on whether (*) has orthogonal or symplectic type, respectively.

According to exercise 26 one may also choose the type of (\*). Although it is tempting to choose orthogonal type since the transpose is so familiar, symplectic type has a significant advantage.

**Proposition 7.2.51:** *Suppose the division algebra D has a symplectic type involution* (s). *Then for every* (s)-*symmetric element d the characteristic values of d occur in pairs, and d has degree* $\leq n/2$, *where* $n = \deg D$.

*Proof:* Splitting by the algebraic closure we may assume $R = M_n(F)$ and (s) is the canonical symplectic involution. Extend (s) further to $M_n(F[\lambda])$ by putting $\lambda^s = \lambda$, and view $d$ in $M_n(F[\lambda])$. Let $a = \begin{pmatrix} 0 & I \\ -I & 0 \end{pmatrix}$. Then $a^s = -a$ and $\det a = 1$. But $(\lambda - d)^s = \lambda - d$ so $(a(\lambda - d))^t = -a(\lambda - d)$ by remark 2.13.5. Hence the characteristic polynomial

$$\det(\lambda - d) = \det a \det(\lambda - d) = \det a(\lambda - d) = f_d^2$$

by Jacobson [85B, theorem 6.4], where $f_d$ is called the *Pfaffian* of $a(\lambda - d)$. Hence $f_d(d)^2 = 0$ in $D$ so $f_d(d) = 0$; since $\det(\lambda - d) = f_d^2$ we conclude its roots occur in pairs and are the characteristic values of $d$.                Q.E.D.

This result holds more generally for arbitrary central simple algebras, as can be seen using exercise 28; a direct computational approach is given in Rowen [80B, p. 143].

The Pfaffian leads one to believe many properties of $M_n(F)$ will be inherited by $(M_{2n}(F), s)$ where $s$ is a symplectic involution. In fact, Rowen proved in analogy to theorem 7.1.45 that every simple algebra of index 8 and exponent 2 has a splitting field with Galois group $(\mathbb{Z}/2\mathbb{Z})^{(3)}$ over $F$ and is thus a crossed product. The proof given in Rowen [80B, theorem 3.2.40] is rather computational; a proof using Brauer factor sets is given in exercises 29–32.

## The Merkurjev-Suslin Number

We turn now to the most important theorem in the theory of central simple algebras. The Merkurjev-Suslin theorem is really a result linking the Brauer group to algebraic $K$-theory, as is explained in appendix B. We are interested here in its algebraic content, which is

**Merkurjev-Suslin "onto part".** *If* $F$ *has a primitive m-th root of* 1 *where*

$m = \exp(R)$ *then $R$ is similar to a tensor product of cyclic algebras each of degree dividing $m$.* (*This is easy for $R$ cyclic, cf., exercise 18.*)

**Digression.** The assumption that $F$ has a primitive $m$-th root $\zeta$ of 1 is crucial because it enables one to deal with symbols. This raises the question of what happens when $\zeta \notin F$. The case $m = 3$ is easy, cf., exercise 34, and Merkurjev [86] has shown that if $F(\zeta)/F$ is cyclic (which is often the case, e.g., when $m$ is not divisible by 8) then $R$ is similar to a tensor product of corestrictions (from $F(\zeta)$ to $F$) of symbols. Thus the major remaining question here is whether the corestriction of a symbol from $F(\zeta)$ to $F$ is similar to a tensor product of cyclic algebras.

The following elementary computation is quite relevant to this discussion, showing us how to "push" elements of a symbol into a smaller field.

**Remark 7.2.52:** Suppose $[K:F] = 2$. For any $a, b$ in $K$ there are $\alpha, \beta$ in $F$ for which $(a, b)_m \sim (a, \beta)_m \otimes (\alpha, \beta^{-1}b)_m$. (Indeed, this is obvious unless $a, b$ are independent over $F$. But $[K:F] = 2$ so $\alpha^{-1}a + \beta^{-1}b = 1$ for suitable $\alpha, \beta$ in $F$. Then $1 \sim (\alpha^{-1}a, \beta^{-1}b)_m \sim (a, b)_m \otimes (a, \beta^{-1})_m \otimes (\alpha^{-1}, \beta^{-1}b)_m$ as desired.)

**Important.** One of the chief applications of the Merkurjev-Suslin theorem is that the Brauer group is divisible in the presence of enough roots of 1. Indeed, the problem of finding the $p$-th root of $[R]$ is reduced to the case when $R$ is a symbol $(\alpha, \beta)_m$, whose $p$-th root was computed in proposition 7.1.17(viii) to be $(\alpha, \beta)_{pm}$ when $F$ has a primitive $pm$-root of 1. Further results concerning divisibility are in exercises 35–38'.

The remainder of this discussion is spent on the Merkurjev-Suslin theorem; we shall prove several special cases (also, cf., exercise 18) but *not* the theorem itself (although we give parts of the proof). We start with a related notion.

**Definition 7.2.53:** $R$ is *decomposable* if $R \approx R_1 \otimes R_2$ with $R_1, R_2 \neq F$. $R$ is *exp-decomposable into cyclics* if $R$ is a tensor product of cyclics each of degree dividing $\exp(R)$.

The Merkurjev-Suslin theorem says for any $R$ we can find $t$ such that $M_t(R)$ is exp-decomposable into cyclics; we shall call the smallest such $t$ the *Merkurjev-Suslin* number $\mu(R)$. It is useful to consider the degree, and we define $\mu(n)$ to be $\max\{\mu(R): \deg R = n\}$. In appendix C we shall see $\mu(n) = \mu(UD(F, n))$ and thus exists.

Our objective is to compute upper bounds for $\mu(n)$ for various $n$ prime and $n = 4$, thereby proving the Merkurjev-Suslin theorem in these cases. *Lower* bounds for $\mu(n)$ are obtained in Tignol-Amitsur [85] by the following approach: If $\mu(R) = t$ and $\deg(R) = n$ then the direct product of cyclic subfields of the cyclic components provides an abelian splitting field of dimension $tn$ over $F$; they prove this is impossible for certain $R$ if $t$ is small enough. In particular, $\mu(p^2) > 1$, cf., digression 7.1.21.

**Remark 7.2.54:** $\mu(2) = \mu(3) = 1$ by theorem 7.1.44. Question 7.1.31(ii) can be rephrased as, "Is $\mu(p) = 1$ for all primes $p$?" $\mu(p) \le p^{(p-1)!/2}$ by theorem 7.2.43.

**Reduction 7.2.55:**

(i) *If $n = n_1 n_2$ for $n_1, n_2$ relatively prime, then*

$$\mu(n) = \mu(n_1)\mu(n_2).$$

(ii) *To prove the Merkurjev-Suslin theorem (and find an upper bound for $\mu(R)$) we may assume $\exp(R)$ is prime.*

*Proof:*

(i) by theorem 7.2.13.

(ii) by (i) we may assume $\text{index}(R) = p^j$ for some prime $p$, so $\exp(R) = p^u$ for some $u$. We argue by induction on $u$; we are given the case $u = 1$. $\exp(R^{\otimes p}) = p^{u-1}$ so by induction is similar to a central simple algebra $R_1$ exp-decomposable into cyclics. $R_1$ has a $p$-th root $R_2$ by remark 7.2.52. Thus $(R \otimes R_2)^{\otimes p} \sim 1$ so by hypothesis $R \otimes R_2 \sim R_3$ which is exp-decomposable into cyclics, as is $R_2^{\text{op}} \otimes R_3 \sim R$.          Q.E.D.

**Digression 7.2.55':** We may assume $R$ is split by a field $K$ Galois over $F$ of dimension a prime power (and thus solvable). This is seen easily by applying the corestriction to a Galois splitting field to remove the prime-to-$p$ part, via theorem 7.2.37.

**Digression 7.2.55":** We may assume $R$ is split by a field whose Galois group over $F$ is a direct product of copies of $\mathbb{Z}/p\mathbb{Z}$. This reduces the theorem to the study of abelian crossed products, cf., exercise 7.1.14, and it is hoped that their explicit description in Amitsur-Saltman [78] may soon lead to an elementary proof of the Merkurjev-Suslin theorem. See exercise 33 for the proof.

Define $\mu(p, p^j) = \max\{\mu(R): \deg(R) = p^j$ and $\exp(R) = p\}$. We have an upper bound for $\mu(p, p) = \mu(p)$, but do not know how to arrive at arbitrary $p^j$. However, when $p = 2$ one has the theory of algebras with involution at our disposal, and algebraic proofs of the Merkurjev-Suslin theorem (onto part) are available, cf., Wadsworth [82]. (In fact, this case was Merkurjev's original theorem.) We shall carry out the theory for low values of $j$, both as a relatively easy illustration, and also because the results themselves have some merit. Assume char$(F) \neq 2$. Clearly $\mu(2) = \mu(2, 2) = 1$.

**Lemma 7.2.56:** (*Albert's descent*) *Suppose $Q$ is a quaternion $K$-algebra with involution* (*) *of second kind such that $F = Z(Q, *)$. Then there is a quaternion $F$-algebra $Q_0$ such that $Q \approx Q_0 \otimes_F K$. Furthermore, given any $d = d^* \in Q$ we may choose $Q_0$ such that $d \in Q_0$.*

**Proof:** If $Q$ is split then take $Q_0 = M_2(F)$; hence we may assume $Q$ is a division algebra. Take $d_1 = d_1^* \neq 0$ in $Q$; then $d_1$ is quadratic over $F$. There is $d_2$ for which $d_2 d_1 d_2^{-1} = -d_1$; replacing $d_2$ by $d_2 \pm d_2^*$ we may assume $d_2^* = \pm d_2$. Then $d_2^2 \in Z(Q, *) = F$ so $d_1, d_2$ generate the desired quaternion subalgebra $Q_0$.     Q.E.D.

**Theorem 7.2.57:** (*Albert*) $\mu(2, 4) = 1$, *i.e., every division algebra $D$ of degree 4 and exponent 2 is a tensor product of quaternion subalgebras.*

**Proof:** $D$ has a maximal subfield $K_1 K_2$ where each $[K_i : F] = 2$, by theorem 7.1.45. By proposition 7.2.45 we may assume (*) induces the nontrivial automorphism of $K_2$ over $F$. Let $Q = C_D(K_2)$, and apply the lemma to get $Q_0$; then $Q_1 = C_D(Q_0)$ and $D \approx Q_0 \otimes Q_1$.     Q.E.D.

To push on to degree 8 we need some results about symbols, again assuming characteristic $\neq 2$.

**Lemma 7.2.58:** *Suppose the symbols $(\alpha_1, \beta_1)_2 \approx (\alpha_2, \beta_2)_2$ are not split. Then there is $\alpha \in F - \{0\}$ for which $(\alpha_1, \beta_1)_2 \approx (\alpha, \beta_1)_2 \approx (\alpha, \beta_2)_2 \approx (\alpha_2, \beta_2)_2$.*

**Proof:** Take $y_i, z_i$ in $(\alpha_i, \beta_i)_2$ for which $y_i^2 = \alpha_i, z_i^2 = \beta_i$, and $z_i y_i = -y_i z_i$. Since $(\alpha_1, \beta_1)_2 \approx (\alpha_2, \beta_2)_2$ we can work in the same algebra $(\alpha_1, \beta_1)_2$. Let $y = [z_1, z_2]$. If $y = 0$ then $F(z_1) = F(z_2)$ so $z_1 \in F(z_2)$ and the assertion is trivial. (Just take $\alpha = \alpha_1$.) So assume $y \neq 0$. Then $z_i y = -y z_i$ by inspection for $i = 1, 2$; furthermore, $y$ has reduced trace 0, being a commutator, implying $y^2 = \alpha \in F$ for some $\alpha$. Thus we have exhibited $(\alpha_1, \beta_1)$ also as $(\alpha, \beta_1)$ and as $(\alpha, \beta_2)$.     Q.E.D.

**Lemma 7.2.59:** (*Tignol*) *Suppose R is simple of exponent* 2 *and possesses an involution* (*) *of second kind with* $F = Z(R, *)$; *furthermore, assume there are quadratic field extensions* $K_1, K_2$ *of F such that* $K = K_1 K_2 Z$ *is a splitting field of R where* $Z = Z(R)$. *Then there are quaternion F-algebras* $Q_1, Q_2, Q_3$ *such that* $M_2(R) \approx Q_1 \otimes_F Q_2 \otimes Q_3 \otimes_F K$.

**Proof:** Write $(\alpha, \beta)$ for $(\alpha, \beta)_2$. By theorem 7.2.57 $R$ is a tensor product of two quaternion algebras (over $Z$) split respectively by $K_1$ and $K_2$; writing $K_i = F(\sqrt{\alpha_i})$ for $\alpha_i \in F$ we thus can write $R$ as $(\alpha_1, c_1) \otimes (\alpha_2, c_2)$ where $c_1, c_2 \in Z$. On the other hand, $R$ has an involution of the second kind, and letting $\sigma$ be the nontrivial automorphism of $Z$ over $F$ we have $(\alpha_1, N_\sigma c_1) \otimes (\alpha_2, N_\sigma c_2) \sim \text{cor}_{Z/F} R \sim 1$ by theorem 7.2.49'. But these are quaternion algebras so letting $\beta_i = N_\sigma c_i$ we have $(\alpha_1, \beta_1) \approx (\alpha_2, \beta_2)$, and there is $\alpha$ such that $(\alpha_1, \beta_1) \approx (\alpha, \beta_1) \approx (\alpha, \beta_2) = (\alpha_2, \beta_2)$ by lemma 7.2.58. Hence

$$M_2(F) \approx (\alpha_1 \alpha, \beta_1) \approx (\alpha_2 \alpha, \beta_2) \approx (\alpha, \beta_1 \beta_2),$$

implying $(\alpha_1 \alpha, c_1)$, $(\alpha_2 \alpha, c_2)$, and $(\alpha, c_1 c_2)$ each has involution of second kind over $F$, again by theorem 7.2.49'.

Let $A_1 = (\alpha_1 \alpha, c_1)$, $A_2 = (\alpha_2 \alpha, c_2)$, and $A_3 = (\alpha, c_1 c_2)$. Then

$$A_1 \otimes A_2 \otimes A_3 \sim ((\alpha_1, c_1) \otimes (\alpha, c_1)) \otimes ((\alpha_2, c_2) \otimes (\alpha, c_2)) \otimes ((\alpha, c_1) \otimes (\alpha, c_2))$$

$$\sim (\alpha_1, c_1) \otimes (\alpha_2, c_2) \otimes (\alpha^2, c_1) \otimes (\alpha^2, c_2)$$

$$\sim (\alpha_1, c_1) \otimes (\alpha_2, c_2) \otimes F \otimes F \sim R.$$

On the other hand, we can write $A_i = Q_i \otimes_F Z$ by lemma 7.2.56 where $Q_i$ is a quaternion algebra over $F$, so $R \sim Q_1 \otimes Q_2 \otimes_F Q_3 \otimes_F Z$, as desired.
                                                                                                         Q.E.D.

A little extra care in the proof would have enabled us to choose $Q_i$ to be split by $K_i$ for $i = 1, 2$.

**Theorem 7.2.60:** (*Tignol*) *If R has degree* 8 *and exponent* 2 *then* $M_2(R)$ *is decomposable into quaternions. In other words* $\mu(2, 8) \leq 2$.

**Proof:** We may assume $R$ is a division algebra. By Rowen's theorem (exercise 32) there is a maximal subfield $K = K_1 K_2 K_3$ where each $[K_i : F] = 2$. Let $R_1 = C_R(K_3)$. Then $R_1$ has an involution of second kind over $F$ and $\exp(R_1) = 2$, so by lemma 7.2.59 $M_2(R_1) \approx Q_1 \otimes Q_2 \otimes Q_3 \otimes_F K_3$ for suitable quaternion algebras $Q_i$ over $F$. Let $Q_4$ be the centralizer of $Q_1 \otimes Q_2 \otimes Q_3$ in $M_2(R)$. Then $M_2(R) \approx Q_1 \otimes Q_2 \otimes Q_3 \otimes Q_4$.     Q.E.D.

Amitsur-Rowen-Tignol [79] proved $\mu(2,8) > 1$, so by Tignol's theorem $\mu(2,8) = 2$. We can apply the above theory in a lovely result of Snider.

**Theorem 7.2.61:** *If $\sqrt{-1} \in F$ then $\mu(4) \leq 16$; in fact, any division algebra of degree 4 is similar to a tensor product of a cyclic algebra of degree 4 and four quaternion algebras.*

**Proof:** If $\exp(D) = 2$ then $\mu(D) = 1$ by theorem 7.2.57. Thus we may assume $\exp(D) = 4$. Then $D \otimes D$ has exponent 2. Furthermore, we claim $\text{index}(D \otimes D) = 2$. To see this note $D$ has a subfield $K$ quadratic over $F$ by theorem 7.1.45, so $\text{index}(D \otimes_F K) = 2$ and thus $(D \otimes D) \otimes_F K \approx (D \otimes_F K)^{\otimes 2}$ is split. Hence we can write $D \otimes D \approx M_2(Q)$ where $Q = (\alpha, \beta)_2$ for $\alpha, \beta$ in $F$, so $Q$ has a square root $R = (\alpha, \beta)_4$. Thus $(D \otimes R)^{\otimes 2} \sim (Q \otimes Q) \sim 1$. But $D$ and $R$ each contain $K$ so $D \otimes R$ has zero-divisors, i.e., we can write $D \otimes R \approx M_2(D_1)$ where $D_1$ has degree 8. By Tignol's theorem $M_2(D_1) \approx Q_1 \otimes Q_2 \otimes Q_3 \otimes Q_4$ for quaternion algebras $Q_i$. Hence $D \sim R^{\text{op}} \otimes Q_1 \otimes Q_2 \otimes Q_3 \otimes Q_4$.          Q.E.D.

This proof contained an exponent reduction argument which is peculiar insofar as the index goes up (to 8 here).

## *"Proof" of the Merkurjev-Suslin Theorem (onto part)*

Fix a prime $p$. Assume $F$ has a primitive $p$-th root of 1, and write $\text{Br}_{\text{cyc}}(F)$ for the subgroup of $\text{Br}(F)$ generated by symbols of degree $p$, i.e., the classes of algebras which are tensor products of cyclics of degree $p$. Also given $L \supseteq F$ write $R_L$ for $R \otimes_F L$. We shall use the exponential notation $A^\sigma$ for $\sigma A$ and $A^{\sigma-1}$ for $A^{\text{op}} \otimes \sigma A$. Let us prove the Merkurjev-Suslin theorem (onto part), modulo

**Missing Link 7.2.62:** Suppose $L/F$ is cyclic of degree $p$, and $[A] \in \text{Br}_{\text{cyc}}(L)$. If $\text{cor}_{L/F} A \sim 1$ then $A \sim (A')^{\sigma-1} \otimes C_L$ where $[A'] \in \text{Br}_{\text{cyc}}(L)$ and $[C] \in \text{Br}_{\text{cyc}}(F)$.

*Note:* The "missing link" is true since it follows readily from the Merkurjev-Suslin theorem (1:1 part) applied to Hilbert's theorem 90 in $K$-theory (cf., appendix B). However, Hilbert's theorem 90 in K-theory is very difficult, as is the Merkurjev-Suslin theorem (1:1 part). Thus an algebraic proof of the "missing link" would certainly revolutionize the subject. Note that for $p = 2$ the hypothesis says $A$ has an involution of second kind over $F$, but the result still seems very hard to prove directly; we have obtained the special cases $\deg(A) = 2$ or 4 (in which case $A' \sim 1$).

Since $1 \otimes \sigma$ acts on $R_L$ we have $R_L^\sigma \approx R_L$ for any $R$ in $\mathrm{Br}(F)$, implying $R_L^{\sigma-1} \sim 1$.

We turn to the proof now that any $R$ in $\mathrm{Br}(F)$ is exp-decomposable into cyclics. By reduction 7.2.55(ii) we may assume $\exp(R)$ is prime; in fact, by digression 7.2.55' we may assume $R$ is split by a Galois extension $K$ of $F$ of dimension $p^t$ for some $t$. We induct on $t$. Since $\mathrm{Gal}(K/F)$ is a $p$-group there is a subgroup of index $p$; let $L$ be its fixed subfield, which is Galois of dimension $p$ over $F$, and let $\sigma$ generate $\mathrm{Gal}(L/F)$.

Now let $A = R_L$. $A$ is split by $K$ and $[K:L] = p^{t-1}$, so by induction on $t$ we have $[A] \in \mathrm{Br}_{\mathrm{cyc}}(L)$. Furthermore, $\mathrm{cor}\, A \sim \mathrm{cor}\,\mathrm{res}\, R \sim R^{\otimes p} \sim 1$ so by the "missing link" we have $A \sim (A')^{\mathrm{op}} \otimes \sigma A' \otimes C_L = (A')^{\sigma-1} \otimes C_L$ for suitable $A', C$ as above.

**Claim 1:** *If $B \in \mathrm{Br}_p(L)$ then $B^{(\sigma-1)^{p-1}} \sim (\mathrm{cor}\, B)_L$.*

**Proof of Claim 1:** $B^{(\sigma-1)p} \sim B^{\sigma^p - 1p} \sim 1$ since $\mathrm{Br}_p(L)$ is $p$-torsion. But $(\sigma - 1)(\sigma^{p-1} + \cdots + 1) = 1$ so $B^{(\sigma-1)^{p-1}} \sim B^{\sigma^{p-1} + \cdots + 1} \sim \mathrm{res}\,\mathrm{cor}\, B \sim (\mathrm{cor}\, B)_L$ by remark 7.2.29.

**Claim 2:** *For each $i \le p - 1$ there is $B$ in $\mathrm{Br}_{\mathrm{cyc}}(L)$ and $C$ in $\mathrm{Br}_{\mathrm{cyc}}(F)$ such that $A \sim C_L \otimes_L B^{(\sigma-1)^i}$. Furthermore, if $i = 1$ we can take $C \sim 1$.*

**Proof of Claim 2:** For $i = 1$ we are done taking $B = A'$. In general we proceed by induction on $i$. Suppose we have found $B$ and $C$ such that $A \sim C_L \otimes_L B^{(\sigma-1)^i}$, for $i < p - 1$. Then

$$1 \sim R_L^{\sigma-1} = A^{\sigma-1} \sim B^{(\sigma-1)^{i+1}} \otimes C_L^{\sigma-1} \sim B^{(\sigma-1)^{i+1}},$$

so $1 \sim B^{(\sigma-1)^{p-1}} \sim (\mathrm{cor}\, B)_L$ by claim 1, and thus $\mathrm{cor}\, B$ is split by $L$. Write $L = F(a)$ where $\sigma a = \zeta a$ and $\alpha = a^p \in F$. Then $\mathrm{cor}\, B = (\alpha, \beta)_p$ for suitable $\beta$ in $F$, and by the projection formula $(\alpha, \beta)_p = \mathrm{cor}(a, \beta)_p$. Hence $\mathrm{cor}(B \otimes (a, \beta)_p^{\mathrm{op}}) \sim 1$ so the "missing link" yields

$$B \sim (a, \beta)_p \otimes (B')^{\sigma-1} \otimes C_L' \qquad \text{where } [B'] \in \mathrm{Br}_{\mathrm{cyc}}(L) \text{ and } [C'] \in \mathrm{Br}_{\mathrm{cyc}}(F).$$

Substituting back yields

$$A \sim C_L \otimes_L (C_L' \otimes (a, \beta)_p \otimes (B')^{\sigma-1})^{(\sigma-1)^i} \sim C_L \otimes (a, \beta)_p^{(\sigma-1)^i} \otimes B'^{(\sigma-1)^{i+1}}.$$

Now $(a, \beta)_p^{\sigma-1} \sim (a^{-1}\sigma a, \beta)_p \sim (\zeta, \beta)_p$ which is a restriction since $\zeta, \beta \in F$; the claim is proved.

**Claim 3:**   *There is C in* $\mathrm{Br}_{\mathrm{cyc}}(F)$, *and B in* $\mathrm{Br}_{\mathrm{cyc}}(L)$ *such that* $A \sim (C \otimes \mathrm{cor}\, B)_L$.

**Proof of Claim 3:**   Apply claim 1 to claim 2 for $i = p - 1$.

To conclude the proof we note $(R^{\mathrm{op}} \otimes C \otimes \mathrm{cor}\, B)_L \sim A^{\mathrm{op}} \otimes (C \otimes \mathrm{cor}\, B)_L \sim 1$ so $R^{\mathrm{op}} \otimes C \otimes \mathrm{cor}\, B$ is split by $L$ and thus similar to some cyclic $C'$. Hence $R \sim C \otimes C' \otimes \mathrm{cor}\, B$. But $B \in \mathrm{Br}_{\mathrm{cyc}}(L)$, so $R \in \mathrm{Br}_{p,\,\mathrm{cyc}}(F)$ by theorem 7.2.37.

                                                                   Q.E.D.

## Digression: p-Algebras

A *p-algebra* is a central simple algebra of degree a power of $p$, having characteristic $p$; $p$-algebras fill the gap in the theory developed here so far. Ironically the theory of $p$-algebras is easier than the characteristic 0 theory, largely because of the use of derivations in key places, and one might hope for a proof of the Merkurjev-Suslin theorem by passing to modulo $p$. (This is yet to be done.) Many of the characteristic 0 techniques can be modified for characteristic $D$. There is a $p$-symbol (cf., exercise 7.1.8) and a corresponding version of the Rosset-Tate theorem, cf., Mammone [86]. Some of the highlights of the $p$-algebra theory are outline below; the reader should consult JAC-SAL for details.

**Teichmuller's theorem:**   *[JAC-SAL, theorem 4.2.22] Suppose K is a purely inseparable splitting field of the p-algebra R, and $K = K_0(a)$ for some a such that $a^p \in K_0$. If $\lambda^{p^n} - \alpha$ is the minimal polynomial of a over F and if $K_0$ is not a splitting field of R then there is a cyclic algebra $(L, \sigma, \alpha)$ of degree $p^n$ for which $R \otimes (L, \sigma, \alpha^{-1})$ is split by $K_0$. (The case $n = 1$ is exercise 7.1.8.)*

On the other hand, purely inseparable splitting fields of $p$-algebras exist [cf., JAC-SAL, theorem 4.1.15], and, in fact, the minimal dimension of such a splitting field is the exponent of the algebra. Thus an induction argument applied to Teichmuller's theorem shows that every $p$-algebra is decomposable into cyclics. This was known by Albert, who proved a stronger result [cf., JAC-SAL, theorem 5.5.7]:

**Albert's theorem:**   *The tensor product of cyclic p-algebras is cyclic. Consequently every p-algebra is similar to a cyclic p-algebra!*

A good bound for the degree of the cyclic algebra is found in Mammone [86], using the corestriction. (This theorem is false in characteristic 0, cf., digression 7.1.21.)

Thus we are left with the old crossed product question. The first noncyclic division $p$-algebra was only constructed in 1978 by Amitsur-Saltman [78] and is a rather delicate matter, using abelian crossed products. Soon thereafter Saltman [78] extended their result and found a noncrossed product division $p$-algebra of degree $p^3$. It is still unknown whether or not any $p$-algebra of degree $p$ is cyclic. Saltman has recently devised a theory which shows the theory of crossed products is characteristic-free, but his methods do not seem to extend to the Brauer group.

## §7.3 Special Results

In this section we obtain examples of indecomposable algebras and noncrossed products of low exponent, based on an example of Brauer. These method should lend some intuition into the workings of division algebras. In order to get the noncrossed product results we need a "generic" division algebra of given degree and exponent, which is developed in appendix C.

**Lemma 7.3.1:** *Suppose $K$ has an automorphism $\sigma$ whose order $n$ is a power of $p$, and $F = K^\sigma$ has a primitive $m$-th root $\zeta$ of $1$ where $m \mid n$. Let $K_0$ be the fixed field of $\sigma^{n/m}$. Then $K = K_0(a)$ for some $a$ in $K$ satisfying $\sigma^{n/m}a = \zeta a$ and $\sigma a = aa_0$ for suitable $a_0$ in $K_0$. Furthermore, $N_0(a_0) = \zeta$ for any such $a_0$, where $N_0$ denotes the norm from $K_0$ to $F$.*

**Proof:** Let $u = n/m$. $K = K_0(a)$ with $\sigma^u a = \zeta a$ by Hilbert's theorem 90 (cf., Jacobson [85B, theorem 4.35]). Then $\sigma^u(a^{-1}\sigma a) = (\sigma^u a)^{-1}\sigma^{u+1}a = (\zeta a)^{-1}\sigma(\zeta a) = a^{-1}\sigma a$, proving $a^{-1}\sigma a \in K_0$. Thus $\sigma a = aa_0$ for some $a_0$ in $K_0$. But then $\zeta a = \sigma^u a = \sigma^{u-1}(aa_0) = \sigma^{u-2}(aa_0\sigma a_0) = \cdots = aN_0(a_0)$, proving $N_0(a_0) = \zeta$.     Q.E.D.

### Brauer's Algebra

The underlying example is a cyclotomic algebra discovered by Brauer. Fix a prime number $p$, and in general write $\zeta_n$ for a primitive $n$-th root of $1$. *All numbers $q, n, t$ to be considered are powers of $p$ greater than 1.*

**Example 7.3.2:** Given $q, n, t$ with $n \le t \le qn$ write $E_{q,t} = \mathbb{Q}(\zeta_q)(\mu_1, \ldots, \mu_t)$

where the $\mu_i$ are commuting indeterminates over $\mathbb{Q}$. $E_{q,t}$ has an automorphism $\sigma$ of order $t$, which permutes the $\mu_i$ cyclically and fixes $\zeta_q$; let $K_{q,n,t}$ be the fixed subfield of $E_{q,t}$ under $\sigma^n$. Thus $\sigma$ restricts to an automorphism of order $n$ on $K_{q,n,t}$, so we have the cyclic algebra $R_{q,n,t} = (K_{q,n,t}, \sigma, \zeta_q)$. When $n = t$ then $K_{q,n,t} = E_{q,t}$, so $R_{q,n,n} = (E_{q,t}, \sigma, \zeta_q)$. Brauer proved $R_{q,n,n}$ is a division algebra, as shown in JAC-SAL; we shall repeat his proof in slightly greater generality in order to handle other examples, following Rowen [82]. We retain the above notation in what follows.

**Remark 7.3.3:** $\zeta_q$ is not an $n^{\text{th}}$ power in $E_{q,t}$. Otherwise $(fg^{-1})^n = \zeta_q$ for suitable $f, g$ in $\mathbb{Q}(\zeta_q)[\mu_1, \ldots, \mu_t]$. Then $f^n = \zeta_q g^n$ so letting $\alpha, \beta$ be the leading coefficients of $f, g$, respectively, (under the lexicographic order) we see $\alpha^n = \zeta_q \beta^n$; thus $\zeta_q = (\alpha\beta^{-1})^n$ in $\mathbb{Q}(\zeta_q)$, which cannot be, since $\mathbb{Q}(\zeta_q)$ does not contain a primitive $pq$-root of 1 by Jacobson [85B, theorem 4.21].) It follows at once that if $\zeta$ is a root of $\lambda^n - \zeta_q$ in an extension field of $E_{q,n}$ then $\zeta$ is a primitive $nq$-root of 1.

**Lemma 7.3.4:** *Suppose $t \mid qk$ and $k/t$.*

(i) *For any $\mathbf{j}$ there is a suitable power $\zeta$ of $\zeta_{t/k}$ for which $\sigma^k a = \zeta a$ for all $a$ in $H_{\mathbf{j}}$.*

(ii) *For any $\mathbf{j}$ there is some homogeneous $b$ in $H$ such that $bH_{\mathbf{j}} \subseteq K_{q,k,t}$ and $H_{\mathbf{j}} b \subseteq K_{q,k,t}$.*

**Proof:**

(i) $(\sigma^k)^{t/k} = 1$ so $\sigma^k$ induces a transformation on each $V_i$ whose order divides $t/k$ and thus divides $q$; hence $V_i$ has some eigenvector $v_i$, i.e., $\sigma^k v_i = \zeta(i) v_i$ for some power $\zeta(i)$ of $\zeta_{n/k}$. But $\{v \in V_i : \sigma^k v = \zeta(i)v\}$ is a $\sigma$-submodule of $V_i$, for if $\sigma^k v = \zeta v$ then $\sigma^k(\sigma v) = \sigma^{k+1}v = \sigma(\zeta v) = \zeta(\sigma v)$; since $V_i$ is simple as $\sigma$-module we see $\sigma^k v = \zeta(i)v$ for all $v$ in $V_i$. But now taking $\zeta = \zeta(1)^{j(1)} \cdots \zeta(u)^{j(u)}$ we see $\sigma^k a = \zeta a$ for all $a$ in $H_{\mathbf{j}} = V_1^{j(1)} \cdots V_u^{j(u)}$.

(ii) Continuing the proof of (i) we see that if $\zeta(i)$ is a characteristic root of $H$ then so is its complex conjugate $\overline{\zeta(i)} = \zeta(i)^{-1}$; thus $\zeta(i)^{-1} = \zeta(i')$ for suitable $i'$, giving rise to $b_i$ in $V_{i'}$ with $\sigma^k b_i = \zeta(i')b_i$. Letting $b = b_1^{j(1)} \cdots b_u^{j(u)}$ we see $b$ is homogeneous and $\sigma^k b = \zeta^{-1}b$; hence $\sigma^k(ba) = \zeta^{-1}\zeta ba = ba$ for all $a$ in $H_{\mathbf{j}}$, implying $bH_{\mathbf{j}} \in K_{q,k,t}$. Likewise $ab \in K_{q,k,t}$.     Q.E.D.

**Remark 7.3.5:** Consequences of lemma 7.3.4(i):

(i) Each $H_{\mathbf{j}}$ is a $\sigma$-module.
(ii) Either $H_{\mathbf{j}} \cap K_{q,n,t} = 0$ or $H_j \subset K_{q,n,t}$.

(iii) If $a \in H$ and $\sigma^k a = \zeta a$ then $\sigma^k a_j = \zeta a_j$ for each component $a_j$ of $a$.

(iv) $H \cap K_{q,n,t}$ inherits the grade of $H$ (seen by taking $k = n$ and $\zeta = 1$ in (iii).)

We want to use this set-up in $R_{q,n,t} = (K_{q,n,t}, \sigma, \zeta_q)$. To this end write $R_{q,n,t} = \sum_{i=0}^{n-1} K_{q,n,t} z^i$ where $z a z^{-1} = \sigma a$ for each $a$ in $K_{q,n,t}$ and $z^n = \zeta_q$. Let $W = \sum (H \cap K_{q,n,t}) z^i$. For any $r$ in $R_{q,n,t}$ there is $c$ in $W$-$\{0\}$ for which $cr = \sum_{i=0}^{n-1} a_i z^i$ for suitable $a_i$ in $H \cap K_{q,n,t}$, i.e., $cr \in W$. Thus $W$ is an order in $R_{q,n,t}$, which we shall call the *Brauer order* of $R_{q,n,t}$. We "grade" $W$ according to the grade on $H$, i.e., put $W_j = \{\sum a_i z^i \in W: \text{each } a_i \in H_j\}$. We shall call this the *Brauer grade* on $W$; given $w \in W$ we say the *leading term of $w$* is the nonzero component in $W_j$, for $j$ maximal in $\mathbb{N}^{(u)}$ according to the lexicographic order.

**Lemma 7.3.6:**

(i) *If $d \mid n$ the centralizer of $z^d$ in $R_{q,n,t}$ is isomorphic to $R_{qn/d,d,t}$.*

(ii) *If $d \mid q$ then $R_{q,n,t}^{\otimes d} \sim R_{q/d,n,t} \otimes_F \mathbb{Q}(\zeta_q)$ where $F = \mathbb{Q}(\zeta_{q/d})$*

**Proof:**

(i) Let $R'$ be the centralizer of $z^d$. $R_{q,n,t} = \bigoplus_{i=0}^{n-1} K_{q,n,t} z^i$ as a vector space over $F$, so conjugating by $z^d$ we see $R' = \bigoplus_{i=0}^{n-1} K_{q,d,t} z^i$ since $K_{q,d,t}$ is the subfield fixed by $\sigma^d$. Furthermore, $(z^d)^{n/d} = z^n = \zeta_q$, so $K_{q,d,t}(z^d) \approx K_{q,d,t}(\zeta_{qn/d}) = K_{qn/d,d,t}$, seen by counting dimensions over $K_{q,d,t}$. Hence

$$R' = \bigoplus_{i=0}^{d-1} K_{q,d,t}(z^d) z^i \approx \bigoplus_{i=0}^{d-1} K_{qn/d,d,t} z^i = R_{qn/d,d,t}.$$

(ii) $R_{q,n,t}^{\otimes d} \sim (K_{q,n,t}, \sigma, \zeta_q^d) = (K_{q,n,t}, \sigma, \zeta_{q/d}) \approx R_{q/d,n,t} \otimes_F \mathbb{Q}(\zeta_q)$.          Q.E.D.

The notion of passing to the leading coefficient lies at the heart of the matter and often enables us to pass from $E_{q,t}$ to $F$. However, we want to work in $R_{q,n,t}$, and therefore need more structure. Let $F_0 = \mathbb{Q}(\zeta_q)$ and $H = F_0[\mu_1, \ldots, \mu_t]$, which is $\mathbb{N}$-graded with respect to total degree. Then the 1-component $H_1 = \sum_{i=1}^{t} F_0 \mu_i$ is a $\sigma$-module, by which we mean $\sigma H_1 = H_1$; thus we can view $H_1$ as $F_0[\lambda]$-module by $(\sum \alpha_i \lambda^i) x = \sum \alpha_i \sigma^i x$ for $\alpha_i$ in $F_0$ and $x$ in $H_1$. Since $\sigma^t = 1$ we see $\lambda^t - 1 \in \text{Ann } H_1$, so $H_1$ is a module over $F_0[\lambda]/\langle \lambda^t - 1 \rangle$; writing $F_0[\lambda]/\langle \lambda^t - 1 \rangle$ as a direct product of fields $L_1 \times \cdots \times L_u$ by the Chinese Remainder Theorem and letting $V_i = L_i H_1$ we see $H_1 \approx V_1 \oplus \cdots \oplus V_u$ and the $V_i$ are simple as $\sigma$-modules. Since the $V_i$ are vector spaces over $F_0$ we can grade $H$ lexicographically by the $V_i$, i.e., given $j = (j(1), \ldots, j(u))$ in $\mathbb{N}^{(u)}$ we put $H_j = V_1^{j(1)} \cdots V_u^{j(u)}$.

**Remark 7.3.7:** Remark 7.3.5(i) shows $zH_j = H_j z$ for all $j$, implying $W_j W_{j'} \subseteq W_{j+j'}$; i.e., the Brauer grade on $W$ is indeed a grade.

We shall usually analyze $R_{q,n,t}$ by passing to $W$, relying heavily on the fact $R_{q,n,t}$ is the ring of central fractions of $W$. Lemma 7.3.6 can be used to set up an induction via "leading terms," provided $k$ (in lemma 7.3.4) is less than $n$. Taking $k = n/p$ the hypothesis of lemma 7.3.4 now says $pt \mid qn$, so we put $m = qn/t \geq p$ and prove

**Theorem 7.3.8:** Let $m = qn/t$. $R_{q,n,t}$ is a division algebra of degree $n$ and exponent $m$ whenever $m \geq p$. (In particular $\exp(R_{q,n,n}) = q$.)

**Proof:** First we prove $R_{q,n,t}$ is a division algebra, by induction on $n$. Take $R$ as above. Obviously $R_{q,1,t}$ is a field, so assume $n > 1$. Suppose $ab = 0$ in $W$. Passing to leading terms we may assume $a, b$ are homogeneous. Thus $a \in \sum_i H_j z^i$ and $b \in \sum_i H_{j'} z^i = \sum_i z^i H_{j'}$ for suitable $j, j'$, cf., remark 7.3.7. Taking $k = n/p$ in lemma 7.3.6(ii) we have homogeneous $a', b'$ such that $a'a \in \sum K_{q,n/p,t} z^i$ and $bb' \in \sum z^i K_{q,n/p,t}$. But these centralize $z^{n/p}$ so in $C_R(z^{n/p}) \approx R_{pq,n/p,t}$ we have $(a'a)(bb') = 0$; by induction $a'a = 0$ or $bb' = 0$, implying $a = 0$ or $b = 0$. This proves $W$ is a domain and thus $R_{q,n,t}$ is a division algebra of degree $n$.

It remains to show $\exp(R_{q,n,t}) = m$. Let $K = K_{q,n,t}$ and let $K'$ be the fixed subfield of $E_{q,t}$ under $\sigma^{t/q}$. Since $t/q \leq n$ we see $K' \subseteq K$, and, furthermore, $\zeta_q = N_{K'/F}(a)$ for suitable $a$ in $K'$, by lemma 7.3.1. Thus $\zeta_q^m = N_{K'/F}(a)^m = N_{K/F}(a)$, proving $\exp(R_{q,n,t}) \mid m$ by Wedderburn's criterion. To conclude the proof it suffices to show $R_{q,n,t}^{\otimes(m/p)}$ is not split. By lemma 7.3.6 (noting $m/p = q(n/pt)$)

$$R_{q,n,t}^{\otimes(m/p)} \approx R_{pt/n,n,t} \otimes_{\mathbb{Q}(\zeta_{pt/n})} \mathbb{Q}(\zeta_q).$$

$[\mathbb{Q}(\zeta_q) : \mathbb{Q}(\zeta_{pt/n})] = m/p < n$, and $R_{pt/n,n,t}$ is a division algebra of degree $n$, so we see the right hand side is not split, as desired.     Q.E.D.

*Note:* The proof of theorem 7.3.8 shows $\text{index}(R_{q,n,t}^{\otimes(m/p)}) = n/(m/p) = pn/m = pt/q$, and the same argument (cf., corollary 7.3.10 below) shows that each $p$-tensor power reduces the index by the factor $p$, until finally the index plunges to 1 at the last stage. A little-asked but interesting question is "What is $\text{index}(R)/\text{index}(R^{\otimes p})$ for a given central simple algebra $R$?" This question has more than passing interest, because of

**Remark 7.3.9:** Suppose $D$ is a division algebra with $\deg(D)$ a $p$-power, and $\deg(D) \leq p \, \text{index}(D^{\otimes p})$. Then $D$ is *indecomposable* in the sense that $D$ cannot be written as $D_1 \otimes_F D_2$ for $D_1, D_2 \neq F$. (Otherwise $D^{\otimes p} \approx D_1^{\otimes p} \otimes D_2^{\otimes p}$ so index$(D^{\otimes p}) \leq$ index$(D_1^{\otimes p})$index$(D_2^{\otimes p}) \leq (\deg(D_1)/p)(\deg(D_2)/p) \leq \deg(D)/p^2$, contrary to hypothesis.)

**Corollary 7.3.10:** $R_{q,n,t}$ *is an indecomposable division algebra of exponent* $m = qn/t$ *whenever* $m > p$.

**Proof:** Write $R_q$ for $R_{q,n,t}$, fixing $n$ and $t$. We shall verify remark 7.3.9. Indeed by lemma 7.3.6 we have $R_{q,n,t}^{\otimes p} \sim R_{q/p,n,t} \otimes \mathbb{Q}(\zeta_q)$. But $[\mathbb{Q}(\zeta_q): \mathbb{Q}(\zeta_{q/p})] = p$ and thus index$(R_{q,n,t}^{\otimes p}) \geq$ index$(R_{q/p,n,t})/p = n/p$ since $R_{q/p,n,t}$ is a division algebra (by theorem 7.3.8).        Q.E.D.

The Brauer division rings can also be used to construct indecomposables of exponent $p$, but the situation is rather more intricate, as reflected in the fact that whereas every algebra of degree 4 and exponent 2 is decomposable, we shall find indecomposable division algebras of degree $p^2$ and exponent $p$ for odd primes $p$. Thus we need to differentiate between 2 and the odd primes. This is hardly surprising in view of the fact the Galois group of $[\mathbb{Q}(\zeta_n): \mathbb{Q}]$ is cyclic for $n$ odd but is the direct product of $\mathbb{Z}/2\mathbb{Z}$ and a cyclic for $n$ a power of 2 (cf., Jacobson [85B, p. 276]); we shall ultimately rely on this fact.

The sibling question to indecomposability is the crossed product question. As a byproduct of this analysis, we shall construct a division algebra of exponent $p$ which is *not* a crossed product with respect to the group $(\mathbb{Z}/p\mathbb{Z})^{(3)}$ for $p$ odd; this leads to a noncrossed product of exponent $p$ and degree $p^3$, cf., appendix C.

## p-Central Sets

We say an element $r$ in a central simple algebra $R$ is $p$-central if $r^p \in F - \{0\}$ but $r \notin F$. The use of $p$-central elements has already pervaded our discussion of crossed products and of decomposability into cyclics, and we now provide a uniform treatment which will be useful in discussing examples. *Assume throughout that $F$ has a primitive $p$-th root $\zeta$ of 1.*

**Definition 7.3.11:** We say elements $r, r'$ of $R$ are $\zeta$-*commuting* if $rr' = \zeta^i r' r$ for some $i$. A subset $S$ of $R$ is $p$-*central* if

(1) Each $s$ in $S$ is $p$-central.
(2) Any two elements of $S$ are $\zeta$-commuting.

(3) For any $s, s'$ in $S$ there is some invertible element $r$ in $R-\{0\}$ for which $rsr^{-1} = s$ and $rs'r^{-1} = \zeta s'$.

We say $S$ is *strongly p-central* if, furthermore, we have the condition

(4) If $s_1, \ldots, s_t \in S$ and $\gamma_i = s_i^p \in F$ then $\gamma_1^{u(1)} \cdots \gamma_t^{u(t)}$ is not a $p$-th power in $F$ unless each $u(i)$ is divisible by $p$.

**Example 7.3.12:**

(i) The symbol $(\alpha, \beta)_n$ has the $p$-central set $\{s_1, s_2\}$ where $s_1^p = \alpha$, $s_2^p = \beta$, and $s_2 s_1 = \zeta s_1 s_2$.

(ii) If $R_1 \otimes R_2$ is a division algebra and $S_i$ is a strongly $p$-central set of $R_i$ for $i = 1, 2$ then $S_1 \cup S_2$ is a strongly $p$-central set of $R_1 \otimes R_2$. In particular, any division algebra which is a tensor product of $t$ cyclics of degree $p$ has a $p$-central set of order $2t$.

(iii) If a subfield $L$ of $R$ is Galois over $F$ with Galois group $(\mathbb{Z}/p\mathbb{Z})^{(t)}$ then $L$ has a commutative strongly $p$-central set of order $t$. (Indeed $L = L_1 \cdots L_t$ where $L_i = F(a_i)$ and $a_i^p \in F$; then $\{a_1, \ldots, a_t\}$ is the strongly $p$-central set.)

Condition (3) of definition 7.3.11 is included to ensure $s \notin Fs'$ for $s, s'$ in $S$ and usually follows automatically from the Skolem-Noether theorem. Condition (4) is included to provide the following converse of example 7.3.12(iii).

**Remark 7.3.13:** If $S$ is strongly $p$-central then each $F(s_1^{u(1)} \cdots s_t^{u(t)})$ is a field extension of dimension $p$ over $F$. (Indeed, we want to show the polynomial $f = \lambda^p - \gamma_1^{u(1)} \cdots \gamma_t^{u(t)}$ is irreducible over $F$. Let $a$ be a root of an irreducible factor of $f$, and let $K = F(a)$. Then $f$ has the $p$ distinct roots $\zeta^u a, 0 \le u < p$, each in $K$, proving $[K:F] = p$ so $f$ is irreducible.)

Our object is to show any strongly $p$-central set $S$ gives rise to something like example 7.3.12. To this end write $S = \{s_1, s_2, \ldots\}$; given $\mathbf{j} = (j(1), j(2), \ldots)$ with $0 \le j(u) < p$ and almost all $j(u) = 0$, write $\mathbf{s}^{\mathbf{j}}$ for the product $s_1^{j(1)} s_2^{j(2)} \ldots$, and $\tilde{S}$ for the set of all these $\mathbf{s}^{\mathbf{j}}$.

**Proposition 7.3.14:** $\tilde{S}$ *is independent over F. In particular* $|\tilde{S}| \le [R:F]$.

**Proof:** Otherwise take a nontrivial dependence relation $\sum \alpha_j \mathbf{s}^{\mathbf{j}} = 0$ with a minimal number of terms, which we rewrite more suggestively as $\sum \alpha_i \mathbf{s}_i = 0$ with $\alpha_1, \alpha_2 \ne 0$. Multiplying through by $\mathbf{s}_1^{p-1} s_1$ we may assume $\mathbf{s}_1 = s_1$. If each $\mathbf{s}_i$ is a power of $s_1$ then $[F(s_1):F] < p$, contradicting remark 7.3.13. Thus we may assume $\mathbf{s}_2$ is not a power of $s_1$, implying by (3) there is $r$ in $R$ for which $rs_1r^{-1} = s_1$ and $rs_2r^{-1} = \zeta s_2$. Now

$$0 = \sum \alpha_i s_i - r \sum \alpha_i s_i r^{-1} = \sum_{i>1} (1 - \zeta^{u(i)}) \alpha_i s_i,$$

a dependence relation of shorter length, contradiction.        Q.E.D.

**Corollary 7.3.15:**  $|\tilde{S}| = p^{|S|}$.

Now let us write a strongly $p$-central set $S$ as $\{s_1, \ldots, s_t\}$, and let $\tilde{R} = \sum\{Fs : s \in \tilde{S}\}$. Obviously $\tilde{R}$ is an $F$-subalgebra of $R$ with $[\tilde{R}:F] = p^t$, so any $p$-central set in $\tilde{R}$ having $t$ elements is maximal. One easy way to produce a new strongly $p$-central set is to take $\{s_1 s_2^i, s_2, \ldots, s_t\}$ for any $i$; rearranging the order of the elements of $S$ and iterating this procedure enables us to modify $S$ by taking suitable products of its elements. Let us define the *center* $Z(S)$ to be $\{z \in S : zs = sz \text{ for all } s \text{ in } S\}$. By the manipulation just described we might possibly increase $|Z(S)|$; we say $S$ has *best possible form* if $|Z(S)|$ is largest possible.

**Proposition 7.3.16:**  *Suppose $S$ is a strongly $p$-central set of best possible form, with $|Z(S)| = k$. Then $\tilde{R}$ as defined above is decomposable as a tensor product*

$$R_1 \otimes_F \cdots \otimes_F R_m \otimes_F \tilde{F},$$

*where each $R_i$ is cyclic of degree $p$ over $F$, $\tilde{F}$ is the field generated over $F$ by $Z(S)$, and $m = (t - k)/2$.*

**Proof:**  Write $S = \{s_1, \ldots, s_t\}$ such that $Z(S) = \{s_{t-k+1}, \ldots, s_t\}$. Note by remark 7.3.13 that $\tilde{F}$ is a field. We are done if $k = t$ so we proceed by induction on $t - k$, assuming $t > k$. By hypothesis we may assume $s_1$ does not commute with $s_2$; replacing $s_2$ by a suitable power we may assume $s_2 s_1 = \zeta s_1 s_2$, so the subalgebra $R_1$ of $\tilde{R}$ generated by $s_1$ and $s_2$ is visibly isomorphic to the symbol $(\gamma_1, \gamma_2)_p$ (where $\gamma_i = s_i^p$) and is thereby cyclic.

Now write $s_3 s_1 = \zeta^u s_1 s_3$ and $s_3 s_2 = \zeta^v s_2 s_3$, and let $s_3' = s_1^v s_2^{p-u} s_3$. Then $s_3'$ commutes with $s_1$ and $s_2$; replacing $s_3$ by $s_3'$ and repeating this procedure for each $i$ we may assume $s_i$ commutes with $s_1$ and $s_2$ for all $i > 3$. Let $\tilde{R}_1$ be the subalgebra of $\tilde{R}$ generated by $S - \{s_1, s_2\}$. Then $\tilde{R}_1$ centralizes $R_1$, and $\{s_3, \ldots, s_t\}$ is a strongly $p$-central set for $\tilde{R}_1$ implying $[\tilde{R}_1 : F] = p^{t-2}$. Hence $[\tilde{R}:F] = [\tilde{R}_1 : F][R_1 : F]$ so $\tilde{R} = R_1 \otimes_F \tilde{R}_1$. By induction we can write $\tilde{R}_1 \approx R_2 \otimes \cdots \otimes R_m \otimes_F \tilde{F}$, and the assertion follows at once.        Q.E.D.

**Digression 7.3.17:**  If $S$ is only required to be $p$-central then in proposition 7.3.16 we can only conclude $\tilde{F}$ is a direct product of fields over $F$ and thus is Azumaya. The only reservation with proposition 7.3.16 as a converse to example 7.3.13 is condition (4), which, of course, is apparent if $R$ has no zero divisors, cf., digression 7.2.13′.

## The Leading Monomial Technique

Structural questions require considerable information about the subfields. There are two available methods: One way is to obtain precise information about *all* the maximal subfields, using cohomological and arithmetic techniques, cf., Tignol-Amitsur [85], [86]. The other method is to obtain precise information about *certain* maximal subfields and to show how this information can be passed on to all maximal subfields of a related algebra. More explicitly, we can start with an algebra $R$ in hand, modify it slightly, and then using "leading monomials" land in a specific subalgebra of $R$; thus we exploit what is already known about $R$. We shall develop this method for $R$ cyclic, following Rowen [82] (which unfortunately is marred by an error, leaving a gap when $p = 2$), and thereby obtain the desired indecomposability and crossed product results.

The general setup: Given $R = (K, \sigma, \alpha)$ of degree $n$ and exponent $m$, where $F = K^{\sigma}$ contains a primitive root $\zeta_p$ of 1, we assume $p^2 \mid n$. Let $K_0$ be the fixed subfield of $K$ under $\sigma^{n/p}$. Then $[K : K_0] = p$ so $K = K_0(k)$ for suitable $k$ in $K$ with $\sigma^{n/p}k = \zeta_p k$. Thus $\sigma k = a_0 k$ for some $a_0$ in $K_0$, by lemma 7.3.1.

Write $R = \sum_{i=0}^{n-1} Kz^i$ where $z^n = \alpha$ and $zaz^{-1} = \sigma a$ for each $a$ in $K$; then $zkz^{-1} = \sigma k = a_0 k$ so $zk = a_0 kz$. Letting $R_0$ be the subring of $R$ generated by $K_0$ and $z^p$, we can identify $z^{n/p}$ with $\alpha^{1/p}$ in $R_0$ since $\sigma^{n/p} = 1$ on $K_0$; and thus $R_0 \approx (K_0(\alpha^{1/p}), \sigma^p, \alpha^{1/p})$, a cyclic algebra of degree $n/p^2$ over its center $F(\alpha^{1/p})$. If $n = p^2$ then $R_0$ is isomorphic to the ring $K_0(\alpha^{1/p})$.

*We assume* throughout $R_0$ *is a division ring.* This is a rather mild assumption, which for $n = p^2$ merely says $\alpha$ is not a $p$-power in $K_0$.

**Remark 7.3.18:** $(R_0 k^i z^j)(R_0 k^{i'} z^{j'}) = R_0 k^{i+i'} z^{j+j'}$ since $zk = a_0 kz$. Thus $R$ is $(\mathbb{Z}/p\mathbb{Z})^{(2)}$-graded, by $R_{ij} = R_0 k^i z^j$. The idea is to reduce questions to homogeneous elements under this grading, to which end we introduce indeterminates.

Working in $R(\lambda_1, \lambda_2) = R \otimes_F F(\lambda_1, \lambda_2)$ where $\lambda_1, \lambda_2$ are commuting indeterminates over $F$ we extend $\sigma$ to $K(\lambda_1, \lambda_2)$ by putting $\sigma \lambda_i = \lambda_i$ for $i = 1, 2$ and define

$$k' = k\lambda_1, \, \alpha' = \alpha\lambda_2^n, \, z' = z\lambda_2.$$

$$F' = F(\lambda_1^p, \lambda_2^p), \, K_0' = K_0(\lambda_1^p, \lambda_2^p), \quad \text{and} \quad K' = K_0'(k').$$

Then $\sigma k' = (\sigma k)\lambda_1 = a_0 k\lambda_1 = a_0 k'$ so $\sigma$ restricts to an automorphism of $K'$, whose fixed field is $F'$. Our object of interest is

$$R' = (K', \sigma, \alpha'),$$

which can easily be identified with the subring of $R(\lambda_1, \lambda_2)$ generated by $K'$ and $z'$, since $(z')^n = \alpha'$ and conjugation by $z'$ induces the automorphism $\sigma$ on $K'$. Note $R_0$ is a subalgebra both of $R$ and $R'$.

Let $A$ be the subring of $R'$ generated by $K_0[\lambda_1^p, \lambda_2^p]$, $k'$, and $z'$. $R'$ is the central localization of $A$ at $F[\lambda_1^p, \lambda_2^p] - \{0\}$, so we can study the structure of $R'$ in terms of the structure of $A$, which is $\mathbb{N}^{(2)}$-graded according to the lexicographic order in degree of $\lambda_1$ and $\lambda_2$. For any $f$ in $A$ write $v(f)$ for the "leading" monomial under this $\mathbb{N}^{(2)}$-grade; since $\lambda_1$ is "attached" to $k$ and $\lambda_2$ is "attached" to $z$ we see $v(f)$ must have the form $rk^i z^j \lambda_1^i \lambda_2^j$ for suitable $r$ in $R_0$ where $1 \le i, j \le p$. Thus we have

**Remark 7.3.19:** If $f \ne 0$ then $v(f)$ is invertible; also $v(f_1 f_2) = v(f_1)v(f_2)$. If $f \in Z(A)$ then $v(f) \in F'$. (Indeed, the first assertion is obvious. To check the last assertion it suffices to show $v(f)$ commutes with every monomial $g$ in $A$. But then $g = v(g)$ so

$$v(f)g = v(f)v(g) = v(fg) = v(gf) = v(g)v(f) = gv(f).)$$

**Remark 7.3.20:** Let tr denote the reduced trace. Then $\text{tr}(v(f)) = v(\text{tr}(f))$. In particular, if $\text{tr}(f) = 0$ then $\text{tr}(v(f)) = 0$.

**Proposition 7.3.21:** $R'$ is a division algebra of degree $n$ and exponent dividing $m$.

**Proof:** Suppose $f_1 f_2 = 0$ in $R'$ with $f_1, f_2 \ne 0$. Clearing denominators we may assume $f_1, f_2 \in A$. But then $0 = v(f_1 f_2) = v(f_1)v(f_2)$, contrary to $v(f_1)$, $v(f_2)$ invertible (by remark 7.3.19). To check $\exp(R') | m$ we write $\alpha^m = N_{K/F}(a)$ for some $a$ in $K$, and thus $(\alpha')^m = N_{K'/F'}(a\lambda_2^m)$, noting $\lambda_2^m = (\lambda_2^p)^{m/p} \in F'$.

Q.E.D.

We shall want to obtain precise information about leading monomials of elements of degree $p$. Given $f$ in $A$ write $\hat{f}$ for the coefficient of $v(f)$ (in $R$), i.e., the "leading coefficient" of $f$.

**Remark 7.3.22:** Suppose $f$ is a monomial and $[F'(f):F'] \leq p$. Either $\hat{f} \in R_0$ or $f^p \in F'$. (Indeed, write $f = rk^i z^j \lambda_1^i \lambda_2^j$. If $p \mid j$ then 1 and $f^p$ are $F'$-dependent, seen by matching coefficients of $\lambda_2$, so $f^p \in F'$; likewise, if $p \mid i$ then $f^p \in F'$. But if $p \mid i$ and $p \mid j$ then $\hat{f} \in R_0$.)

**Proposition 7.3.23:** Suppose $\mathrm{tr}(f) = 0$ and $[F'(f):F'] = p$. Then $[F'(v(f)):F'] = p$; furthermore, if $v(f)^p \notin F'$ then $f \in R_0$. The minimal polynomial of $v(f)$ can be obtained as follows:

Write $\sum_{i=0}^{p} c_i f^i = 0$ for suitable $c_i$ in $A \cap F'$, and take the leading monomials among the terms to get $\sum_{i=0}^{p} c_{i0} v(f)^i = 0$. (When nonzero $c_{i0} = v(c_i)$, but various $c_{i0}$ can be 0).

**Proof:** The last assertion is clear, and some $c_{i0} \neq 0$. If $c_{p0} = 0$ then $[F'(v(f)):F'] < p$; but $R'$ is a division algebra so $[F'(v(f)):F']$ divides $n$, a power of $p$, implying $v(f) \in F'$, contrary to $\mathrm{tr}(f) = 0$. Thus $v(f)$ has degree $p$, and in fact $\hat{f} \in R_0$ or $v(f)^p \in F'$, by remark 7.3.22.      Q.E.D.

**Corollary 7.3.24:** If $L'$ is a subfield of $R'$ of degree $p$ over $F'$ then there is $f$ in $A \cap L'$ such that $F'(v(f))$ is a field extension of $F'$ of degree $p$.

**Proof:** Take $f$ in $A \cap L'$ having reduced trace 0, noting $\mathrm{char}(F') \neq p$, and apply proposition 7.3.23.      Q.E.D.

Our machinery is to be applied to $p$-central sets. However, we must first know they exist and thereby confront the possible discrepancy between subalgebras of degree $p$ and *cyclic* subalgebras of degree $p$. For $p = 2$ or 3 this is not a problem since every algebra of degree 2 or 3 is cyclic. However, for $p > 3$ we need a method of reducing to the cyclic subalgebra case; this is done in the next proposition, which relies on the following useful result of Wedderburn, proved in exercises 7.1.15–7.1.18:

**Wedderburn's Factorization Theorem:** *For any element $d$ of a division ring $D$ the minimal polynomial of $d$ over $Z(D)$ can be factored as $(\lambda - d_n) \cdots (\lambda - d_1)$ where each $d_i$ is a conjugate of $d$.*

**Proposition 7.3.25:** *Suppose $R'$ is a tensor product of $F'$-subalgebras of degree $p$. Then $R'$ has two commuting cyclic $F'$-subalgebras of degree $p$.*

**Proof:** Let $R'_1, \ldots, R'_u$ be the given central subalgebras of degree $p$ which centralize each other; we shall find cyclic $R''_1, \ldots, R''_u$ in $R'$ which centralize each other. Write $R'_1 = (K_1, \sigma_1, \alpha_1)$ and pick $d_1 \in A \cap K_1$ with $\mathrm{tr}(d_1) = 0$. By Wedderburn's result the minimal polynomial of $d_1$ is of the form $(\lambda - d_p) \cdots (\lambda - d_1)$ where each $d_i = r_i d_1 r_i^{-1}$ for suitable $r_i$ in $R_1$. Thus $r_i$ has degree $p$, and we may assume $r_i \in A$; furthermore, multiplying through by a suitable central element (the product of the "denominators" of the $d_i$) we may assume each $d_i \in A \cap R_1$.

By proposition 7.3.23 the minimal polynomial of $v(d_1)$ is $(\lambda - v(d_p)) \cdots (\lambda - v(d_1))$; since $\mathrm{char}(F') \neq p$ we see $d_1$ cannot appear as a multiple root, so $v(d_1), \ldots, v(d_p)$ are distinct. But $d_2 r_2 = r_2 d_1$ implies $v(d_2)v(r_2) = v(r_2)v(d_1)$ so $v(d_1)$ does not commute with $v(r_2)$, and, consequently, $v(d_1)$ and $v(r_2)$ generate a noncommutative subalgebra $R''_1$ of $R'$. $R''_1$ is a division algebra since $R'$ is. Let $n_1 = [R''_1 : F]$. We have $n_1 \geq p^2$, and we shall now prove equality.

Obviously $R''_1$ is spanned by the monomials in $v(d_1)$ and $v(r_2)$, e.g., $v(d_1)^2 v(r_2)v(d_1)$ and so forth. But $R'_1$ is spanned by the monomials in $d_1$ and $r_2$, and so $p^2 + 1$ of them are dependent; taking the "leading part" of a dependence relation in $R'_1$ yields a dependence relation in $R''_1$, so any $p^2 + 1$ monomials in $v(d_1)$ and $v(r_2)$ are dependent, proving $n_1 \leq p^2$ as desired. Hence $n_1 = p^2$, so $R'_1$ is $F'$-central of degree $p$.

Now applying "leading monomials" to each of the $R'_i$ in turn we have $F'$-central algebras $R''_i$ each of degree $p$, generated by elements which we now call $v(d_{1i})$ and $v(r_{1i})$. By proposition 7.3.23 we see $R'_1$ is cyclic (using remark 7.1.15) unless $\hat{d}_{1i} \in R_0$ and $\hat{r}_{1i} \in R_0$. Thus we are done unless $u - 1$ of the $R''_i$ lie in $R_0(\lambda_1^p, \lambda_2^p)$. But then $\deg R_0(\lambda_1^p, \lambda_2^p) \geq p^{u-1} = n/p$, contrary to $\deg(R_0) = n/p^2$.        Q.E.D.

**Proposition 7.3.26:** *Suppose $n = p^2$. Let $N$ denote the norm from the field $R_0 = K_0(\alpha^{1/p})$ to $F(\alpha^{1/p})$. If $R'$ has a central simple subalgebra of degree $p$ then $N(a)\alpha^{1/p} \in F$ for some $a \neq 0$ in $R_0$.*

**Proof:** Let $R_1$ be a central simple subalgebra of $R'$ of degree $p$ and $R_2 = C_{R'}(R_1)$. Then $R' \approx R_1 \otimes R_2$ and by proposition 7.3.25 we may assume $R_1, R_2$ are cyclic; thus we have a $p$-central set of 4 elements. Taking the leading monomials gives us a strongly $p$-central set $v(S)$; taking their products gives us a set of $p^4$ linearly independent elements $s_u = r_u k^{i(u)} z^{j(u)} \lambda_1^{i(u)} \lambda_2^{j(u)}$, $1 \leq u \leq p^4$, where each $s_u^p \in F'$ and the $s_u$ $\zeta$-commute, cf., proposition 7.3.14. We may assume $0 \leq i(u), j(u) < p$, and we claim for any given $(i, j)$ there are at most $p^2$ elements $s_u$ with $i(u) = i$ and $j(u) = j$; indeed, otherwise, we may assume $i(u) = i$ and $j(u) = j$ for $1 \le u \le p^2 + 1$, in which case $s_1^{-1} s_u \in R_0$ for $1 \le u \le p^2 + 1$, implying $[R_0 : F] > p^2$, contradiction.

Now by counting we have exactly $p^2$ elements $s_u$ for each $(i, j)$ such that $i(u) = i$ and $j(u) = j$; in particular, taking $i = 0$ and $j = 1$, we have some $s_u = r_u z \lambda_2$, so $N(r_u) \alpha^{1/p} \lambda_2^p = s_u^p \in F'$, implying $N(r_u) \alpha^{1/p} \in F$, as desired.     Q.E.D.

This norm condition ties in with the following arithmetic result.

**Proposition 7.3.27:**   *In example 7.3.2 suppose $p$ is an odd prime, $qn/t = p$ and $q \ge p^2$. Write $N$ for the norm with respect to $\sigma^{n/p}$. Then $N(K_{q,n,t}) \zeta_q \cap K_{q/p, n/p, t} = 0$.*

*Proof:*  Straightforward, but rather long with all the gory details: Let $H_q = \mathbb{Q}(\zeta_q)[\mu_1, \ldots, \mu_t]$. We want to prove the impossibility of

$$a_1 N(h_1) = a_2 N(h_2) \zeta_q, \tag{1}$$

where $h_1, h_2 \in H_q$, $h_2 h_1^{-1} \in K_{q,n,t}$, and $a_1, a_2 \in H_{q/p}$.

We may assume $a_1, a_2$ are relatively prime, and writing $h_i$ as a product of $d_i$ irreducibles in $H_q$, that $d_1 \ge d_2$. We shall induct on $d_1$.

First assume $d_1 = 0$, i.e., $h_1, h_2 \in \mathbb{Q}(\zeta_q)$. Then $(h_2 h_1^{-1} \zeta_{pq})^p = a_1 a_2^{-1}$ in $\mathbb{Q}(\zeta_{pq})$ which is cyclic over $\mathbb{Q}(\zeta_{q/p})$. But $a_1 a_2^{-1} \in \mathbb{Q}(\zeta_{q/p})$; thus $h_2 h_1^{-1} \zeta_{pq} \in \mathbb{Q}(\zeta_q)$ (the unique subfield of $\mathbb{Q}(\zeta_{pq})$ of dimension $p$ over $\mathbb{Q}(\zeta_{q/p})$) implying $\zeta_{pq} \in \mathbb{Q}(\zeta_q)$, which is false.

Thus we may assume $d_1 \ne 0$. Write the factorizations $h_1 = \prod f_i$ and $h_2 = \prod g_i$ for $f_i, g_i$ irreducible. Each $f_i$ and $g_j$ are relatively prime, for, otherwise, we could cancel $f_i$ from $h_1$ and $g_j$ from $h_2$ and still have (1), contrary to minimality of $d_1$. But $\prod f_i / \prod g_i = \sigma^n(\prod f_i / \prod g_i) = \prod \sigma^n f_i / \prod \sigma^n g_i$ so that we see $(\prod f_i)(\prod \sigma^n g_i) = (\prod g_i)(\prod \sigma^n f_i)$ and thus $f_1$ divides some $\sigma^n f_i$.

$\mathbb{Q}(\zeta_q)$ is cyclic over $\mathbb{Q}(\zeta_{qn/t})$, of dimension $q/p = t/n$. Let $\tau$ be the generating automorphism, viewed as an automorphism of $H_q$ over $H_p$. Note $\tau$ commutes with $\sigma$. $\tau^{t/pn}$ fixes $a_1$ and $a_2$.

Write $f_{1u}$ for $\tau^{ut/pn} f_1$. We claim $f_{11}$ is associate to some $\sigma^{jn/p} f_i$. If not then $f_{11}$ is not associate to $f_{10} = f_1$ (seen by taking $i = 1$ and $j = 0$), implying the $f_{1u}$ are not associate; furthermore, $f_{11}$ does not divide $\sigma^{jn/p} h_1$ for any $j$. But $f_1 | a_1$ implies $f_{11} | a_2$ so $f_{11} | N(h_1) a_1$ by (1); hence $f_{11} | a_1$ so $f_{10} \cdots f_{1, p-1}$ divides both $a_1$ and $a_2$, contradiction.

Repeating this argument shows for each $i$ there is suitable $i'$ and $j'$ for which $\tau^{t/pn} f_i$ is associate to some $\sigma^{j'n/p} f_{i'}$. But then we could cancel $f_{10} \cdots f_{1, p-1}$ from $h_1$ and $N(f_{10} \cdots f_{1, p-1})$ from $a_2$ unless some repetition occurs, i.e., indeed some $f_{1u}$ is associate to some $\sigma^{jn/p} f_1$.

Let $\tau' = \tau^{ut/pn}$ and $\sigma' = \sigma^{jn/p}$, and write $\tau' f_1 = \alpha\sigma' f_1$ for suitable $\alpha$ in $\mathbb{Q}(\zeta_u)$. Then

$$\tau'^2 f_1 = \tau'(\alpha\sigma' f_1) = (\tau'\alpha)\sigma'\tau' f_1 = (\alpha\tau'\alpha)\sigma'^2 f_1.$$

Iterating the procedure yields $f_1 = (\alpha\tau'\alpha\tau'^2\alpha \cdots (\tau')^{p-1}\alpha) f_1$. Thus $\alpha\tau'\alpha \cdots = 1$, so by Hilbert's theorem 90 there is some $\beta$ in $\mathbb{Q}(\zeta_q)$ for which $\alpha = (\tau'\beta)^{-1}\beta$. Now

$$\tau'(\beta f_1) = (\tau'\beta)\alpha\sigma' f_1 = \beta\sigma' f_1 = \sigma'(\beta f_1),$$

so $N(\beta f_1)$ is fixed under $\tau'$. Thus we can cancel $\beta f_1$ from $h_1$ and $N(\beta f_1)$ from $a_2$. Proceeding in like manner with each $\sigma^{ni} f_1$ enables us to shrink the length of $h_1$, contrary to assumption.     Q.E.D.

We are ready to combine the various techniques to produce the desired examples.

**Theorem 7.3.28:**   *Suppose $R = R_{q,n,pq}$ is Brauer's division algebra, where $p$ is odd prime and $n = p^2$. Then $R'$ is indecomposable.*

**Proof:**   If $R'$ were decomposable then we would have $N(a)\zeta_{pq} \in F$ by proposition 7.3.26; this is impossible by proposition 7.3.27 (replacing $q$ by $pq$, and $n$ by $p$).     Q.E.D.

**Digression 7.3.29:**   For $p = 2$ one would of course need to take $n = 8 = p^3$. An indecomposable example was found in Amitsur-Rowen-Tignol [79]. In Rowen [82] an alternative is sought, relying on an analogue to proposition 7.3.27, but now the argument seems to me to be flawed, although possibly salvageable.

Let us turn to the crossed product result. The methods are largely the same, but the use of the Brauer grade is more intricate for the twofold reason that we are working in higher degrees but with more restricted $p$-central sets.

**Theorem 7.3.30:**   $p$ *is an odd prime. Suppose $qn/t = p$, and $n = p^3$. Let $R = R_{q,n,t}$. Then $R'$ has exponent $p$ but is not a crossed product with respect to $(\mathbb{Z}/p\mathbb{Z})^{(3)}$.*

**Proof:**   On the contrary, assume $R'$ has a commuting $p$-central set $S = \{s_1, s_2, s_3\}$; furthermore, there are $r_u$ in $R$ such that $r_u s_u = \zeta_p s_u r_u$ and $r_u s_v = s_v r_u$ for all $u \neq v$. Replacing $s_u, r_u$ by $v(s_u), v(r_u)$ we may assume $s_u = v(s_u)$ for $1 \leq u \leq 3$. Write $s_u = d_u k^{i(u)} z^{j(u)}$ where $d_u \in R_0$ and $0 \leq i(u), j(u) \leq p - 1$. Rearranging the $s_u$, and if necessary using $s_1$ to cancel out $k$, we may assume $i(u) = 0$ for all $u > 1$. Let $L_0$ be the subfield of $R$ spanned by all $s_u, u > 1$. If

$z^{n/p} \notin L_0$ then we could replace $s_1$ by $z^{n/p}$. On the other hand, if $z^{n/p} \in L_0$ then (by Galois theory) $z^{n/p}$ is a product of powers of the $s_u, u > 1$, so we could assume some other $s_u = z^{n/p}$; then $i(1) = 0$ since $s_1$ commutes with each $s_u$, and we could interchange $s_1$ and $s_u$. In this way we may always assume $s_1 = z^{n/p}$.

$K_{q,n,t}$ is cyclic over $F$ and thus has some element $y \notin F$ with $y^p \in F$. Repeating the argument of the previous paragraph we may assume $s_2 = y$. Note $y$ commutes with $z^p$. Thus we have attained

***Reduction 1:*** $R'$ has a commuting $p$-central set $\{s_1, s_2, s_3\}$ with $s_1 = z^{n/p}$ and $s_2 = y$.

Let $K_i$ be the fixed field of $K_{q,n,t}$ under $\sigma^{n/p^{i+1}}$, e.g., $K_0$ is as previously defined, and let $T_i$ be the subring of $R$ generated by $K_i$ and $z$. Thus $T_0 \supset T_1 \supset T_2 \cdots$. Now view $T_0$ as the cyclic algebra $(K_0, \sigma, \zeta_{pq}) = R_{pq, n/p, t}$, giving rise to a new Brauer order and Brauer grade. Since $s_3 \in R_0 \subseteq T_0$ we may multiply by a suitable element of $F$ to assume $s_3$ is in the Brauer order of $T_0$, and take its leading component $s_3'$. Note $(s_3')^p \in Z(T_0) = F(z^{p^2})$. Clearly $s_3' \in T_1 k_0^i$ for some $i$, where $k_0 \in K_0$ with $k_0^p \in K_1$, but $s_3' \in R_0$ so matching powers of $z$ shows $s_3' \in R_1 k_0^i$ where $R_1$ is the subring of $R$ generated by $K_1$ and $z^p$.

***Reduction 2:*** $s_3' \notin Z(R_0)$. For suppose $s_3' \in Z(R_0)$ and write $s_3 = s_3' + b$ and let $b'$ be the leading term of $b$. Then $(s_3')^p + p(s_3')^{p-1} b + \cdots = s_3^p \in Z(T_0)$ so subtracting $(s_3')^p$ and taking leading terms shows $p(s_3')^{p-1}b' \in Z(T_0)$ and thus $b' \in Z(T_0)s_3'$. Continuing down the line shows $b \in Z(T_0)s_3'$ so $s_3 \in Z(T_0)s_3' \subseteq Z(R_0)$. But $Z(R_0)$ is spanned by the products of powers of $s_1$ and $s_2$, contradicting proposition 7.3.14.

$R_1$ is the field $K_1(z^p)$ which is Galois over $F(z^p)$. If $s_3' \in R_1$ then $s_3' \in Fy^v z^{pj}$ for suitable $v, j$; multiplying $s_3$ by a suitable power of $s_2 = y$ we may assume $v = 0$. On the other hand, if $s_3' \notin R_1$ then $1 \leq i < p$. In these two respective cases we see $s_3'$ and $k_0$, or $s_3'$ and $z^p$, generate an $F(z^{n/p})$-central subalgebra of $T_0$ of degree $p$, so its centralizer $W$ has degree $p$. Furthermore, $y \in W$ so there is some $w$ in $W$ for which $wy = \zeta_p yw$. But $w$ commutes with $z^p$ so is in $T_1$; writing $w = \sum_{i=0}^{p-1} w_i z^i$ for $w_i$ in $K_1(z^p)$ we see $\sum w_i yz^i = yw = \zeta_p^{-1} wy = \zeta_p^{-1} \sum \zeta_p^i w_i yz^i = \sum \zeta_p^{i-1} w_i yz^i$, so matching components yields $w = w_1 z$. But then $N(w_1)z^p = w^p \in F(z^{n/p})$, contrary to proposition 7.3.27.     Q.E.D.

## Noncrossed Products of Prime Exponent

We turn to the question of determining noncrossed product division algebras of arbitrary exponent. Inspired by Amitsur's noncompatibility results we should look for two division algebras of exponent $p$ and degree $p^t$, one of which is a crossed product only with respect to the group $(\mathbb{Z}/p\mathbb{Z})^{(t)}$ and the latter of which is *not* a crossed product with respect to $(\mathbb{Z}/p\mathbb{Z})^{(t)}$. The former is provided by theorem 7.1.29, and the latter by theorem 7.3.30 when $t \geq 3$ (for $p$ odd).

To finish the argument one needs some sort of "generic" of exponent $p$ and degree $p^t$, which will fill the role $UD(F, n)$ played in the "comparison" lemma 7.1.27. There is a more general construction (cf., appendix C) which works for any exponent, thereby yielding

**Theorem 7.3.31:** *There is a noncrossed product division algebra of degree $p^3$ and exponent $p$, for any odd prime $p$.*

# Appendix A: The Arithmetic Theory and the Brauer Group

One of the most important areas of recent investigation in finite dimensional division algebras has been the generalization of valuation theory from fields to division algebras, thereby enabling one to apply standard arithmetic techniques to division rings. Although the scope of this book does not permit a close look at this subject, let us glance at its highlights.

An (exponential) *valuation* on a division ring is a function $v: D\text{-}\{0\} \to G$ where $G$ is an ordered group called the *valuation group* such that

$$v(d_1 d_2) = v(d_1) + v(d_2)$$

$$v(d_1 + d_2) \geq \min\{v(d_1), v(d_2)\}.$$

The *valuation ring* $R$ of $D$ is $\{r \in D : v(r) \geq 0\}$ and is clearly seen to be local with maximal ideal $M = \{r \in D : v(r) > 0\}$. $R/M$ is called the residue division ring. Although $G$ is not required to be abelian in this generality, we still have used $+$ to denote its group operation.

*Example A1:* Suppose $D_0$ is a division ring with valuation $v: D_0 \to G$, and $\sigma$ is an automorphism on $D_0$ such that $v(\sigma d) = v(d)$ for all $d$ in $D_0$. Let $D = D_0((\lambda; \sigma))$, i.e., $D$ has the additive structure of the skew Laurent series ring $D_0((\lambda))$, but additive structure given by $\lambda d = (\sigma d)\lambda$, cf., definition 1.6.39. Ordering $\mathbb{Z} \times G$ lexicographically, we define the valuation $\tilde{v}$ on $D$ by $\tilde{v}\left(\sum_{i \geq t} d_i \lambda^i\right) = (t, v(d_t))$.

In case $D_0$ is a field and $\sigma$ is an automorphism of order $n$ one sees without difficulty that $Z(D) = Z_0((\lambda^n))$ where $Z_0$ is the fixed field of $D_0$ under $\sigma$, so $[D:Z(D)] = n^2$ and $D$ has degree $n$. In particular, if $F_0$ is a field containing a primitive $n$-th root $\zeta$ of 1 we could take $D_0$ to be the field of Laurent series $F_0((\mu))$, where $\sigma$ is given by $\sigma\mu = \zeta\mu$. Note $D_0$ has the valuation given by $v(\sum_{t \geq j} \alpha_j \mu^j) = j$, so $D$ has valuation group $\mathbb{Z} \times \mathbb{Z}$.

More generally one could iterate this construction for any abelian group $G$ and form an *iterated Laurent series division* ring, in analogy to example 7.1.16. The advantage of this construction is that valuation theory is now available. Amitsur [72] exploited this fact; for a detailed exposition see Jacobson [75B, pp. 94–103]. Recently Tignol-Amitsur [85] have found a cohomological description of this division ring, cf., exercise 7.2.11.

Suppose now $D$ is a central division algebra of degree $n$ over $F$. A valuation $v$ on $F$ extends to $D$ iff $v$ extends uniquely to each maximal subfield of $D$, cf., Wadsworth [86], so in this case the structure of $D$ is likely to be closely bound to those of the maximal subfields. A classic instance is when $F$ is complete with respect to a discrete valuation $v$, since then $v$ extends uniquely to all finite dimensional field extensions of $F$, cf., Jacobson [80B, theorems 9.5 and 9.12]; then Hensel's lemma applies, cf., Jacobson [80B, p. 573].

We say a field $F$ as above is a *local field* if the residue field is finite; examples include the field of $p$-adic numbers $\mathbb{Q}_p$ or a field of Laurent series over a finite field. In this case $D$ has a maximal subfield $W$ which is an unramified extension of $F$, cf., Jacobson [80B, theorem 9.21] or Reiner [75B, p. 145]. Extending $v$ to a valuation of $W$ and letting $\bar{W}$, $\bar{F}$ be the respective residue fields of $W$ and $F$, let $q = |\bar{F}|$; then $|\bar{W}| = q^n$ for some $n$, so by Hensel's lemma $W$ is the splitting field of the polynomial $\lambda^{q^n} - \lambda$ over $F$. In particular, $W$ is uniquely determined up to isomorphism (and thus up to conjugation in $D$ by Skolem-Noether), and one shows easily $\mathrm{Gal}(W/F) \approx \mathrm{Gal}(\bar{W}/\bar{F})$. Since the latter group is generated by the Frobenius automorphism $\sigma: a \mapsto a^q$ we see $D \approx (W, \sigma, \alpha)$ for suitable $\alpha$.

Note that $W$ and $\sigma$ are uniquely determined, so let us examine $\alpha$ further. Pick $\pi \in F$ such that $v(\pi)$ generates the value group $G$. Then $\alpha = \alpha_0 \pi^k$ for suitable $k \in \mathbb{Z}$ and suitable $\alpha_0$ such that $v(\alpha_0) = 1$. Another application of Hensel's lemma shows $\alpha_0$ is a norm from $W$ (cf., Jacobson [80B, proposition 9.8]) so $D \approx (W, \sigma, \pi^k)$ by exercise 7.1.12. Similarly $\pi^n$ is a norm so we may take $1 \leq k < n$. In fact, $i$ must be relatively prime to $n$ for $D$ to be a division algebra, and, conversely, each such $(W, \sigma, \pi^k)$ is a division algebra by Wedderburn's criterion, and these are nonisomorphic. The classic work on local fields is Serre [79B].

## Particular Brauer Groups

These results show $\operatorname{Br}(F) \approx \mathbb{Q}/\mathbb{Z}$ for any local field $F$. Indeed, we saw any division algebra $D$ of degree $n$ over $F$ has the form $(W, \sigma, \pi^k)$ where $v(\pi)$ generates the value group of $F$ and $1 \le k < n$ with $k$ relatively prime to $n$. Fixing $\pi$ we define the *Hasse invariant map* $\operatorname{Br}(F) \to \mathbb{Q}/\mathbb{Z}$ by $[D] \mapsto k/n$, and it is now straightforward to show this is a group isomorphism. Furthermore, one sees that the exponent and index are equal for central simple algebras over local fields.

Local fields are useful in studying the Brauer group of an arbitrary field, for one can take the restriction map $\varphi_v \colon \operatorname{Br}(F) \to \operatorname{Br}(F_v)$ where $F_v$ is the completion of $F$ at an absolute value $v$. This procedure becomes meaningful for the following kinds of fields:

(1) *algebraic number field*, defined as a finite extension of $\mathbb{Q}$; the valuations arise from the maximal ideals of the corresponding ring of algebraic integers, and the archimedian absolute values arise from the embeddings in $\mathbb{C}$.

(2) *function field*, defined as a finite extension of a field of rational fractions $F_0(\lambda)$ where $F_0$ is finite; the valuations arise from the discrete valuation subrings.

Fields $F$ of the form (1) or (2) are called *global fields* and using theorem 7.2.13 to reduce to the prime power case one sees that any $[R] \in \operatorname{Br}(F_v)$ lies in the kernel of almost all $\varphi_v$. Thus there is a map $\varphi \colon \operatorname{Br}(F) \to \bigoplus \operatorname{Br}(F_v)$. The celebrated Hasse-Brauer-Noether-Albert theorem says this map is monic, and one can conclude that any central simple algebra over the global field $F$ is cyclic (cf., Reiner [75B, theorem 32.20]).

Hasse actually pushed the monic $\varphi$ further. Noting as above that each $\operatorname{Br}(F_v) \approx \mathbb{Q}/\mathbb{Z}$ for each nonarchimedean absolute value $v$, and $\operatorname{Br}(F_v) \approx \mathbb{Z}/2\mathbb{Z}$ for each archimedean absolute value, he defined $\psi \colon (\mathbb{Z}/2\mathbb{Z})^{(t)} \oplus (\mathbb{Q}/\mathbb{Z})^{(\mathbb{N})} \to \mathbb{Q}/\mathbb{Z}$ by taking the sum of all nonzero components in $\mathbb{Q}/\mathbb{Z}$ and proved there is an exact sequence

$$1 \to \operatorname{Br}(F) \to \bigoplus \operatorname{Br}(D_v) \to \mathbb{Q}/\mathbb{Z} \to 0$$

for any global field $F$. Hence $\operatorname{Br}(F) \approx (\mathbb{Z}/2\mathbb{Z})^{(t)} \oplus (\mathbb{Q}/\mathbb{Z})^{(\mathbb{N})}$ but in a tricky way. For example $\operatorname{Br}(\mathbb{Q}) \approx (\mathbb{Z}/2\mathbb{Z}) \oplus (\mathbb{Q}/\mathbb{Z})^{(\mathbb{N})}$.

An important tool in Brauer group theory is the *Auslander-Brumer-Faddeev Theorem*: if $F$ is an infinite field and $\operatorname{char}(F) \mid p$ then

$$p\text{-}\operatorname{Br}(F((\lambda))) \approx p\text{-}\operatorname{Br}(F) \oplus \left( \bigoplus_{K/F \text{ finite}} \operatorname{Hom}(\operatorname{Gal} F_s/F, \mathbb{Q}/\mathbb{Z}) \right),$$

where $F_s$ is the separable closure of $F$. Using this result Fein, Schacher and Sohn have computed the Brauer group of a purely transcendental extension of a global field, cf., Fein-Schacher [81] and Fein-Schacher-Sonn [85]. This team continues to produce interesting new results about Brauer groups. In view of the Merkujev-Suslin theorem, one basic question in Brauer group theory is

*Question A2:* Is every abelian, divisible torsion group the Brauer group of a suitable field?

## The Brauer Group of a Commutative Ring

Since we have had occasion to consider maximal commutative separable sub-algebras of $R$ in place of maximal subfields (e.g., when $F = \mathbb{C}$), remark 7.2.1 leads us to study the Brauer group of an arbitrary commutative ring $C$, defined as the set of equivalence classes of Azumaya algebras, modulo the relation $\text{End}_C P \sim 1$ where $P$ is a $C$-progenerator (which by remark 4.1.8' means $P$ is faithful and projective). Then $\text{Br}(C)$ is indeed a group, since the "opposite" is the inverse. This leads us to study the "restriction" for standard ring constructions involving fields, which was carried out in the pioneering paper of Auslander-Goldman [60]. For example, suppose $\lambda$ is a commuting indeterminate over $C$. The composition $C \xrightarrow{i} C[\lambda] \xrightarrow{j} C$ (where the latter map sends $\lambda \mapsto 0$) yields a composition $\text{Br}(C) \xrightarrow{\text{Br}\,i} \text{Br}(C[\lambda]) \to \text{Br}(C)$ which is the identity, implying $\text{Br}(C[\lambda]) \approx \text{Br}(C) \oplus \ker \text{Br}\,j$; in particular, $\text{Br}\,i$ is monic. When $C$ is a field Auslander-Goldman [60, 7.5] showed $\text{Br}\,i$ is an isomorphism iff $C$ is perfect. However, the canonical monic $\text{Br}(F) \to \text{Br}(F[\lambda_1, \lambda_2])$ will never be epic since we can use the "generic" ideas of §7.1 to construct "cyclic" Azumaya algebras over $F[\lambda_1, \lambda_2]$. When $C$ is not a field there are general instances when $\text{Br}(C) \approx \text{Br}(C[\lambda])$.

On the other hand, one can return to fields by means of the fields of fractions; if $F$ is the field of fractions of $C$ then $\text{Br}(C) \to \text{Br}(F)$ often is monic by Auslander-Goldman [60, 7.2] but is rarely epic, e.g., $\text{Br}(\mathbb{Z}) = 0$ but $\text{Br}(\mathbb{Q})$ is not! These facts (and much more) as described in the excellent survey article of Zelinsky [77].

## Appendix B: The Merkurjev-Suslin Theorem and *K*-Theory

So far we have discussed only half of the Merkurjev-Suslin theorem! To state the other half we must bring in the $K_2$-theory of a field, which can be done by means of the elementary matrices. Let $\{e_{ij}: 1 \le i, j \le m\}$ be a given set of matric units and $\mathscr{E}(m, F) = \{1 + \alpha e_{ij} : \alpha \in F \text{ and } i \neq j\} \subset M_m(F)$. (Here "1" denotes the identity matrix.) Analogously to exercise 7.1.38 one can show easily by means of elementary transformations that $\mathscr{E}(m, F)$ generates $GL(m, F)$ as a group.

**Remark B1:** If $m \geq 3$ then $\mathscr{E}(m, F)$ satisfy the following relations (where $[\ ,\ ]$ is the *group* commutator).

(i) $(1 + \alpha e_{ij})(1 + \beta e_{ij}) = 1 + (\alpha + \beta)e_{ij}$.

(ii) $[1 + \alpha e_{ij}, 1 + \beta_{jk}] = 1 + \alpha\beta e_{ik}$    *for $i \neq k$.*

(iii) $[1 + \alpha e_{ij}, 1 + \beta e_{kl}] = 1$    for $i \neq l$ and $j \neq k$.

Formalizing these relations we now define the *Steinberg group* $St(m, F)$ to be generated by elements designed $\{x_{ij}^\alpha : 1 \leq i, j \leq m, i \neq j, \text{ and } \alpha \in F - \{0\}\}$ satisfying the relations for all $\alpha, \beta$ in $F - \{0\}$:

$$x_{ij}^\alpha x_{ij}^\beta = x_{ij}^{\alpha + \beta}, \ [x_{ij}^\alpha, x_{jk}^\beta] = x_{ik}^{\alpha\beta} \text{ for } i \neq k, \text{ and } [x_{ij}^\alpha, x_{kl}^\beta] = 1$$

whenever $i \neq \ell$ and $j \neq k$. Since these relations are precisely those of remark B1 we have an epic $\Phi_n : St(m, F) \to GL(m, F)$ given by $x_{ij}^\alpha \mapsto 1 + \alpha e_{ij}$. We define $K_2(m, F) = \ker \Phi_n$. There are natural embeddings of $St(m, F) \to St(m + 1, F)$ and $\mathscr{E}(m, F) \to \mathscr{E}(m + 1, F)$, so we have $K_2(m, F) \subseteq K_2(m + 1, F) \subseteq \cdots$. We define $K_2 F = \bigcup_{n \in \mathbb{N}} K_2(m, F)$, or, more precisely, the direct limit.

Define $w_{ij}(\alpha) = x_{ij}^\alpha x_{ji}^{\alpha^{-1}} x_{ij}^\alpha$ and $h_{ij}(\alpha) = w_{ij}(\alpha)w_{ij}(-1)$, and define the *Steinberg symbol* $\{\alpha, \beta\} = h_{ij}(\alpha\beta)h_{ij}(\alpha)^{-1}h_{ij}(\beta)^{-1}$. The reason we do not include $ij$ in the notation is that $\{\alpha, \beta\}$ does not depend on the choice of $i, j$ and turns out to be in the center of the Steinberg group; we leave this as an exercise for the reader (or else check Milnor [71B]), as well as the following identities:

(1) $\{\alpha, 1 - \alpha\} = 1$

(2) $\{\alpha, \beta\gamma\} = \{\alpha, \beta\}\{\alpha, \gamma\}$

(3) $\{\beta\gamma, \alpha\} = \{\beta, \alpha\}\{\gamma, \alpha\}$

(4) $\{\alpha, -\alpha\} = 1$

(5) $\{\alpha, \beta\} = \{\beta, \alpha\}^{-1}$

Also one sees easily that each Steinberg symbol lies in $K_2(F)$.

Now comes the surprise. By a deep theorem of Matsumoto (cf., Milnor [71B, §11.1 and §12]) $K_2(F)$ is generated by the Steinberg symbols, and all relations among the symbols are consequences of (1), (2), and (3). In particular, $K_2(F)$ is an abelian group, and its operation is usually written as addition.

On the other hand, the symbol algebras $(\alpha, \beta)_m$ satisfy conditions (1)–(5), assuming $F$ contains a primitive $m$-th root of 1. Thus there is a group homomorphism $K_2(F) \to \mathrm{Br}_m(F)$ given by $\{\alpha, \beta\} \mapsto (\alpha, \beta)_m$. (Here $\mathrm{Br}_m(F)$ denoted $\{[R] : \exp(R) \text{ divides } m\}$.) Obviously $mK_2(F)$ lies in the kernel, so we have a map

$$R_{m,F} : K_2(F)/mK_2(F) \to \mathrm{Br}_m(F).$$

This map is dependent on the particular choice of the $m$-th root $\zeta$ of 1, since the structure of the algebra $(\alpha, \beta)_m$ depends on the choice of $\zeta$. To free ourselves of this choice, we could let $\mu_m$ denote the group of $m$-th roots of 1 and define

$$K_2(F)/mK_2(F) \to \mathrm{Br}_m(F) \otimes \mu_m$$

by $\{\alpha, \beta\} \mapsto (\alpha, \beta)_m \otimes \zeta$. We are ready to state the Merkurjev-Suslin theorem properly.

**Theorem B2:**  (*The real Merkurjev-Suslin Theorem*) $R_{m,F}$ *is an isomorphism.*

Before continuing let us interpret this result. $R_{m,F}$ epic implies $\mathrm{Br}_m(F)$ is generated by cyclics of degree $m$; this is what we have been calling the Merkurjev-Suslin theorem. On the other hand, $R_{m,F}$ monic means the whole theory of central simple algebras can be reread in $K_2$-theory. Suslin [82] also obtained characteristic $p$ results, e.g., $K_2(F)$ has no $p$-torsion. An important special case of the Merkurjev-Suslin theorem is

**Bloch's Theorem:**  *If $E$ is a purely transcendental field extension of $F$ then $R_{m,F}$ and $R_{m,E}$ have isomorphic kernels and cokernels.*

In particular, if $R_{m,F}$ is epic then $R_{m,E}$ is also epic; since every central simple algebra over an algebraic number field is cyclic, one concludes $\mathrm{Br}_m(F)$ is generated by cyclics whenever $F$ is a purely transcendental extension of an algebraic number field. In presenting the Auslander-Brumer theorem, Fein-Schacher [80] give a relatively straightforward proof that every central simple $F(\lambda)$-algebra is similar to the tensor product of a restriction of a central simple $F$-algebra with a finite number of corestrictions of symbols (the corestriction taken from finite dimensional extensions of $F(\lambda)$. Hence the epic part of Bloch's theorem follows easily from Rosset-Tate.

On the other hand, one could prove the "epic" part of Merkurjev-Suslin if one could show the center of the generic division algebra $\mathrm{UD}(F, n)$ is purely transcendental over $\mathbb{Q}$. This observation (due to Rosset) led to a spate of attempts to build various "generic" division algebras whose centers were purely transcendental, and, although some success was achieved, the theorems thereby obtained have been proved more succinctly by the methods we have already presented. Until now the Bloch-Rosset approach has led to various open cases of Noether's famous problem of whether the fixed subfield under a group of automorphisms of a purely transcendental field extension is purely transcendental. An elementary reduction of the Merkurjev-Suslin theorem to Bloch's theorem would be most welcome.

Let us say a few words about the proof of the Merkurjev-Suslin theorem. The Rosset-Tate reciprocity law only relies on formal properties of symbols, as the reader can easily verify, and thus holds in $K_2$ (where the corestriction is replaced by the "transfer" Tr, defined in Milnor [71B]; in fact, Rosset and Tate framed their results more generally for "Milnor functors" (taking values in an abelian group) which possess the formal symbols $(\alpha, \beta)$ and the transfer maps, satisfying the following properties, written now in additive notation, where we always assume $\alpha, \beta \neq 0$:

$$(\alpha, -\alpha) = 0 = (\alpha, 1 - \alpha) \qquad \text{where here } \alpha \neq 1$$

$$\text{Tr}_{L/F} \, \text{Tr}_{K/L} = \text{Tr}_{K/F}$$

$$\text{Tr}_{L/F}(\alpha, \beta) = (\alpha, N_{L/F} \beta) \qquad \text{for } \alpha \in F, \beta \in L.$$

Note that the $K_2$-version gives an explicit formula for Tr of a symbol and thus for Tr on all of $K_2$, since the symbols generate. Consequently, we have the following commutative diagram, fixing an $m$-th root $\zeta$ of 1:

$$
\begin{array}{ccc}
K_2(L)/mK_2(L) & \longrightarrow & \text{Br}_m(F) \\
\downarrow{\scriptstyle \text{Tr}_{L/F}} & & \downarrow{\scriptstyle \text{cor}_{L/F}} \\
K_2(F)/mK_2(F) & \longrightarrow & \text{Br}_m(F)
\end{array}
$$

This diagram permits us to pass back and forth from $K_2$ to $\text{Br}_m(F)$, thereby permitting us to bring in the powerful tool from $K$-theory.

**Hilbert's Theorem 90 for $K$-Theory:**   *If $L/F$ is cyclic of prime degree $p$ and $\text{Tr}\, x = 0$ for some $x$ in $K_2(F)$ then there is $y$ in $K_2(F)$ for which $x = \sigma y - y$, where $\text{Gal}(L/F) = \langle \sigma \rangle$.*

This result is the key obstacle to an "elementary" proof of the Merkurjev-Suslin theorem, for its current proof relies on results from "higher" $K$-theory.

# Appendix C: Generics

In this discussion we sketch a method for obtaining generic division algebras of different classes and also see what they are good for.

**Definition C1:**   A *pseudo-place* $\varphi: W \to R$ of algebra is a map $\varphi: W \to R \cap \{?\}$ where ? is a formally undefined element, such that if $W_1 = \varphi^{-1}R$ then the restriction of $\varphi$ to $W_1$ is a surjection $\varphi_1: W_1 \to R$.

An algebra $U$ in a class $\mathscr{C}$ is *generic* if for every $R$ in $\mathscr{C}$ and every *finite* subset $S$ of $U - \{0\}$ there is a pseudo-place $\varphi: U \to R$ for which $0 \neq \varphi u \in R$ for each $u$ in $S$.

**Note:**   We require $\varphi_1$ to be a surjection so that the image of central elements is central. The reader should compare "pseudo-place" to the classical definition of a "place" of fields, cf., Jacobson [80B].

**Remark C2:**   We shall always take $\mathscr{C}$ to be a class of central simple algebras which includes division algebras. Then the proof of theorem 7.1.24 shows any generic is a division algebra.

The motivation behind the definition is that any finite $p$-central set of $U$ would be mapped via suitable $\varphi$ to a $p$-central set of $R$, so decomposability and crossed product properties are thereby preserved. More precisely, let us say $R$ is *split* by a group $G$ if $R$ has a splitting field Galois over $Z(R)$ with Galois group $G$; in particular, if $|G| = \deg R$ then $R$ is a *G-crossed product*.

**Proposition C3:**   *Suppose $U$ is generic for $\mathscr{C}$.*

(i) *If $U$ is a G-crossed product then every division algebra in $\mathscr{C}$ is a G-crossed product; if $U$ is split by $G$ then every central simple algebra in $\mathscr{C}$ is split by a subgroup of $G$.*

(ii) *If $U$ is a tensor product of cyclics of exponents $m_1, \ldots, m_j$ then every algebra in $\mathscr{C}$ is a tensor product of cyclics of exponents dividing $m_1, \ldots, m_j$.*

**Proof:**   Left to the reader, since the idea is exactly as in the lemma 7.1.27; the reason we must pass to subgroups of $G$ in the general case is that we have direct products of subfields. For (ii) one writes $U = (K_1, \sigma_1, \alpha_1) \otimes \cdots \otimes (K_j, \sigma_j, \alpha_j)$, encodes all the information into elements, and takes $u$ in $U$ to be the product of all relevant numerators and denominators, stipulating the denominators are central.     Q.E.D.

***Example C4:***   Let $\mathscr{C} = \{$simple $F$-algebras of cardinality $\leq \alpha$ and degree $n\}$, where $\alpha$ is an arbitrary infinite cardinal $\geq |F|$. Then $U = \mathrm{UD}(F, n)$ obtained from $\alpha$ generic matrices; indeed given any $R$ in $\mathscr{C}$ and

$$u = f(Y_1, \ldots, Y_m)g(Y_1, \ldots, Y_m)^{-1}$$

in $U$ we take $r_1, \ldots, r_m$ such that $f(r_1, \ldots, r_m)g(r_1, \ldots, r_m)^{-1} \neq 0$, and listing the elements of $R$ as $\{r_i : i \in I\}$ with $|I| = \alpha$ one defines the pseudo-place $U \to R$ to be the specialization $Y_i \mapsto r_i$ for each $i$ in $I$.

It remains to show that generics exist. This was done for arbitrary exponent by Saltman and for other classes of interest by Rosset and Snider. Now we shall sketch a geometric approach of unifying these constructions, and soon shall consider a very useful technique of Schofield.

Let $F$ be an infinite field and let $\mathscr{C} = \{$prime PI-algebras over $F$ having PI-degree $n\}$. To make things simpler, conceptually we consider each object in $\mathscr{C}$ as a pair $(R, Z)$ where $Z = Z(R)$. Let $\mathscr{C}_\varphi$ denote the subclass defined by one additional sentence

$$\varphi = \exists \mathbf{x} \forall \mathbf{y} (f(\mathbf{x}) \neq 0 \wedge \psi(\mathbf{x}, \mathbf{y}))$$

where $\mathbf{x} = (x_1, \ldots, x_k)$, $\mathbf{y} = (y_1, \ldots, y_{k'})$, the various $x_i, y_j$ taken from $R$ or $Z$, according to specification, and where $f$ is a given polynomial and $\psi$ is a conjunction of atomic formulas (i.e., polynomials whose values in $(\mathbf{x}, \mathbf{y}, \ldots)$ are 0). We want a generic for $\mathscr{C}_\varphi$. To accomplish this we first observe that if $(R, Z) \in \mathscr{C}_\varphi$ then $(R[\lambda], Z[\lambda]) \in \mathscr{C}_\varphi$ as in corollary 6.1.45. (One uses the same $\mathbf{x}$ and looks at homogeneous components in $\mathbf{y}$.)

Now consider the algebra $F[\Lambda]_n\{Y\}$ of generic $n \times n$ matrices over the commutative ring $F[\Lambda]$ where $\Lambda$ is a set of commuting indeterminates $\{\lambda_{iu} : 1 \leq u \leq n^2, i \in \mathbb{N}\}$. Let $t = n^2$. Noting the generic matrices $Y_1, \ldots, Y_t$ are a base of the ring of fractions of $F[\Lambda]_n\{Y\}$ over its center, we let $x_i = \sum_{u=1}^t \lambda_{iu} Y_u$ for $1 \leq i \leq k$ and $y_j = \sum_{u=1}^t \lambda_{j+t, u} Y_u$. The conditions on $\varphi$ translate to a set of polynomial conditions in the $\lambda_{iu}$ and the $\xi$ (viewing the $Y_u$ as matrices whose entries are from another set $\xi$ of commuting indeterminates); so letting $A$ be the ideal of $F[\Lambda; \xi]$ generated by these polynomials, we take $U$ to be the image of $F[\Lambda]_n\{Y\}$ in $M_n(F[\Lambda, \xi]/A)$.

By construction $U$ is generic if it is prime. But $U$ has a central extension which is $M_n(F[\Lambda; \xi]/A)$ and so will be prime when $A$ is a prime ideal in $F[\Lambda; \xi]$, i.e., when the corresponding algebraic variety $V$ is irreducible, which will usually be the case. In fact, the only situation where $V$ will *not* be irreducible is if every specialization of the $\lambda_{iu}$ to $Z$ gives an algebraic equation in $Z$, e.g., if $\varphi = \{\exists x : x^2 = 1\}$. But in this case we name a solution for each such equation and introduce corresponding new constant symbols; rewriting $\varphi$ in terms of these new symbols then produces a variety $V'$, and we get a generic by taking an irreducible component.

In applications we are usually interested in a class of simple algebras of degree $n$, which of course is obtained by taking the rings of central fractions.

*Example C5:*  Let us describe {simple algebras of degree $n$ and exponent $\leq m$} in terms of elementary sentences. Let $t = n^2$ and $v = t^m$. Then $\exp(R) \leq m$ iff $R^{\otimes m}$ has a set of $v \times v$ matric units. In other words, given $1 \leq k(1), \ldots, k(m) \leq t$ write $\mathbf{k} = (k(1), \ldots, k(m))$ and $b_{\mathbf{k}}$ for $b_{k(1)} \otimes \cdots \otimes b_{k(m)}$ in $R^{\otimes m}$; existence of a set of $v \times v$ matric units now translates to the existence of $\alpha_{ij\mathbf{k}}$ in $Z(R)$, $1 \leq i, j \leq v$ such that $e_{ij} = \sum_{\mathbf{k}} \alpha_{ij\mathbf{k}} b_{\mathbf{k}}$ satisfies the matric unit conditions of definition 1.1.1.

This is clearly in the desired form and, therefore, yields the generic of degree $n$ and exponent $m$ that we have looking for.

*Remark C6:*  (Amitsur) If $U$ is generic for $\mathscr{C}_\varphi$ then $M_n(U)$ is generic for $\{M_n(R) : R \in \mathscr{C}_\varphi\}$. Note that the existence of a set of $n \times n$ matric units is of the desired form, so this defines a class $\mathscr{C}_{\varphi'}$. Thus by passing to $\mathscr{C}_{\varphi'}$ we can obtain the following instant consequences of proposition C3:

**Proposition C7:**

(i) *If $M_n(U)$ is split by $G$ then $M_n(R)$ is split by a subgroup of $G$, for every central simple algebra $R$ of $\mathscr{C}_\varphi$.*

(ii) *If $M_n(U)$ is a tensor product of cyclics then $M_n(R)$ is a tensor product of cyclics.*

**Corollary C8:**  *The Merkurjev-Su.lin number $\mu(n) = \mu(\mathrm{UD}(F, n))$.*

*Proof:*  ($\geq$) is obvious; ($\leq$) by (ii).     Q.E.D.

## Generic Splitting Fields and Schofield's Method

*Definition C9:*  A *generic splitting field* for $R$ is a splitting field $E$ of $R$ such that for each splitting field $K$ of $R$ there is a place $E \to K$.

Intuitively a generic splitting field is the "most general" splitting field for $R$ and can be found if we can determine a varietal criterion for $R$ to be split. The most famous criterion is by means of Grassmannians and their Plucker coordinates, giving rise to the *Brauer-Severi* variety, which is instrumental in the proof of the Merkurjev-Suslin Theorem. A more direct criterion, found independently by Peterson and Jacobson, comes from noting that $R$ is split iff there is an element $r$ in $R$ such that $[rRr: Z(R)] = 1$; this criterion translates readily to equations and thus to varieties. For details the reader can consult JAC-SAL.

Given a generic splitting field construction, we can easily build generics for various classes by tensoring by a suitable generic splitting field. For example, to build a generic division algebra of degree $n$ and exponent $m$ we start with the generic $D = \mathrm{UD}(F, n)$ of degree $n$ and let $E$ be the generic splitting field for $D^{\otimes m}$. Then $D \otimes_F E$ has exponent $m$ and degree $n$, and is generic such as seen by an easy specialization argument.

The advantage of this approach is that one can bring in the following result.

**Theorem:** *(van den Bergh [83]-Schofield). If $E$ is the generic splitting field of $R_2$ then* $\mathrm{index}(R_1 \otimes E) = \min\{\mathrm{index}(R_1 \otimes_F R_2^{\otimes i}: 1 \le i \le \exp(R_2)\}$.

This theorem is proved using Quillen's $K$-theory and obviously bears on the tensor product question; indeed, Schofield [87] has found very natural counterexamples (i.e., $D_1 \otimes D_2$ has zero divisors although $D_1$ and $D_2$ have no subfields in common) for wide ranges of values of $\deg(D_1)$ and $\deg(D_2)$. Thus he "explains" and extends the Tignol-Wadsworth example.

# Exercises

## §7.1

### Maximal Commutative Subalgebras

1. Suppose $K$ is a maximal commutative subalgebra of a central simple $F$-algebra $R$. If $K$ is separable over $F$ then $R$ is projective as $K$-module, and $[K:F] = \deg R$. (Hint: $K$ is projective as $K^e$-module so $R \approx R \otimes_K K$ is projective over $R \otimes_K K^e \approx R \otimes_K (K \otimes_F K) \approx R \otimes_F K$, which is free over $K$. Now using exercise 2.11.15 note $[R:F] = [R \otimes_F K:K] = [\mathrm{End}_K R:K] = [\mathrm{End}_{K_P} R_P:K_P] = [R_P:K_P]^2$ for any $P$ in $\mathrm{Spec}(K)$; but $[R \otimes_F K:K] = [R \otimes_K K^e:K] = [R_P \otimes (K^e)_P:K_P] = [R_P:K_P][K:F]$.)

2. Any central simple algebra has a maximal commutative subalgebra which is separable. (Hint: Write $R = M_t(D)$ and take $K^{(t)}$ where $K$ is a maximal separable subfield of $D$. Appeal to the counting argument of exercise 1 to show $C_R(K^{(t)}) = K^{(t)}$.

3. Prove *Schur's inequality* $[C:F] \le [n^2/4] + 1$ for any commutative subalgebra $C$ of $M_n(F)$. (Hint: (Gustafson) Writing $M_n(F) \approx \operatorname{End}_F V$ where $[V:F] = n$ note $V$ is a faithful $C$-module. Switching points of view, one need only show if $C$ is a commutative $F$-algebra and $V$ is a faithful $C$-module with $[V:F] \le n$ then $[C:F] \le [n^2/4] + 1$. $C$ is Artinian and thus semiperfect commutative, implying $C$ is a direct product of local rings. It suffices to assume $C$ is local. Let $J = \operatorname{Jac}(C)$. Let $\bar{b}_1, \ldots, \bar{b}_m$ be a base for $V/JV$ as $C/J$-module; choosing $b_i$ such that $\bar{b}_i = b_i + JV$ define $f: J \to \operatorname{Hom}(V/JV, JV)$ sending $a$ to the map $f_a$ given by $f_a \bar{b}_i = ab_i$. Then $f$ is monic by Nakayama's lemma so $[J:F] \le m(n - m) \le [(n/2)^2]$. But $[C:F] = 1 + [J:F]$.)

4. (Sweedler) An alternate approach to showing $D \otimes_F K \approx \operatorname{End}_K D$ for a maximal subfield $K$ of a central division algebra $D$. First show if $L$ is a subfield with $C_D(C_D(L)) = L$ and $L \subseteq T \subseteq D$ then there is a monic $\psi: T \otimes_L D^{\mathrm{op}} \to \operatorname{Hom}_{C_{D(T)}}(C_D(L), D)$ given by $\psi(a \otimes d)x = axd$. (To prove $\psi$ is monic take $0 \ne r = \sum_{i=1}^n a_i \otimes d_i \in \ker \psi$ with $n$ minimal. One may assume $d_n = 1$, so $\sum_{i=1}^{n-1} a_i \otimes [x, d_i] \in \ker \psi$ for any $x$ in $C_D(L)$, implying each $d_i \in L$; hence $r = \left(\sum a_i d_i\right) \otimes 1 = 0$, contradiction.

   Apply this fact twice. First $T = K$ and $L = F$ yields $\psi_1: K \otimes_F D \hookrightarrow \operatorname{End}_K D$. But taking $T \ne D$ and $L = K$ yields $[D:K]^2[K:F] \le [K:F][D:F]$ so $[D:K] \le [K:F]$ and $\psi_1$ is thus onto.)

## Skolem-Noether Theorem

5. Prove the Skolem Noether theorem. (Hint: Suppose $\psi: A \to A'$ is an isomorphism of simple $F$-subalgebras and form $T = A \otimes R^{\mathrm{op}}$. $R$ is an $A - R$ bimodule and thus $T$-module in two ways: the obvious way and by the action $(a \otimes r)x = (\psi a)rx$. These module actions are isomorphic since $T$ has a unique simple module whose multiplicity in $R$ is determined by $[R:F]$. Hence there is $f: R \to R$ satisfying $f(axr) = (\psi a)(fx)r$. Take $a = 1$ to show $f$ is given by left multiplication by an invertible element $u$ of $R$; then take $r = 1$.)

6. Weaken the hypothesis in the Skolem-Noether theorem to "finite dimensional simple subalgebras of an arbitrary simple Artinian aglebra $R$." (Hint: Use the underlying division ring of $R$ in place of $F$.)

7. Counterexample to generalizing Skolem-Noether to semisimple subalgebras. Let $A = F[a]$ and $A' = F[a']$ where $R = M_3(F)$, $a = \begin{bmatrix} 1 & 0 & 0 \\ 0 & 2 & 0 \\ 0 & 0 & 2 \end{bmatrix}$ and $a' = \begin{bmatrix} 1 & 0 & 0 \\ 0 & 1 & 0 \\ 0 & 0 & 2 \end{bmatrix}$. Then $a$ and $a'$ have the same minimal polynomial so $A \approx A'$ but $ar \ne ra'$ for all invertible $r$ in $R$.

8. Suppose $\operatorname{char}(F) \ne p$. Define a *p-symbol* $[\alpha, \beta]_n$ (where $n$ is a power of $p$) to be the cyclic algebra $(K, \sigma, \beta)$ where $K = F(y)$, $y^p = y + \alpha$, $zyz^{-1} = y + 1$, and $z^p = \beta$. If $\deg(R) = p$ and $R$ has an element $r \notin F$ with $r^p \in F$ then $R$ is a $p$-symbol. (Hint: exercise 2.5.10.) Also check $[\alpha_1, \beta] \otimes [\alpha_2, \beta] \sim [\alpha_1 + \alpha_2, \beta]$ and $[\alpha, \beta_1] \otimes [\alpha, \beta_2] \sim [\alpha, \beta_1 \beta_2]$.

## Brauer and Noether Factor Sets

**9.** A Brauer factor set is *normalized* if $c_{iik} = 1 = c_{ikk}$ for all $i, k$. This can be achieved via theorem 7.1.40 by taking $\gamma_{ij} = 1$ if $i \neq j$ and $\gamma_{ii} = c_{iii}$. A Noether factor set $(c_{\sigma\tau})$ is *normalized* if $c_{1\sigma} = c_{\sigma 1} = 1$ for all $\sigma$ in $G$; this is achieved by taking $u_1 = 1$.

**10.** Suppose $K = F(a)$ is Galois over $F$ and $a_1 = a, a_2, \ldots, a_n$ are the conjugates of $a$. Then any $\sigma \in \mathrm{Gal}(K/F)$ is defined by its action on $a_1$. There is a 1:1 correspondence $\Psi : \{\text{Brauer factors sets}\} \to \{\text{Noether factor sets}\}$ given by defining $c_{\sigma\tau} = c_{1,\sigma 1,\sigma\tau 1}$; $\Psi^{-1}$ is given by defining $c_{1jk} = c_{\sigma,\tau}$ where $\sigma 1 = j$ and $\tau 1 = \sigma^{-1}k$, and then putting $c_{ijk} = c_{1ij}c_{1jk}c_{1ik}^{-1}$. Furthermore, prove the corresponding factor sets arise from isomorphic central simple algebras. In particular, a central simple algebra $R$ is a crossed product iff $R$ has a maximal subfield Galois over $F$. (Hint: The 1:1 correspondence is a computation and the last assertion is a trick. First, given $(c_{ijk})$ show $\Psi$ is defined by verifying (10), applying BFS2 (where $i = 1$, $j = \sigma 1$, $k = \sigma\tau 1$, and $m = \sigma\tau\rho 1$).

Conversely, given a Noether factor set $(c_{\sigma\tau})$ note $c_{1\tau} = c_{1,\tau\rho}$ (taking $\sigma = 1$ in (10)) so $c_{11j} = c_{11k}$ for all $j, k$. The definition $c_{ijk} = c_{1ij}c_{1jk}c_{1ik}^{-1}$ is well-defined (since for $i = 1$ one has $c_{11j}c_{11k}^{-1} = 1$), and this condition yields BFS2 at once. Taking $\sigma, \tau, \rho$ in $G$ for which $\sigma 1 = i$, $\tau 1 = \sigma^{-1}j$ and $\rho 1 = (\sigma\tau)^{-1}k$ (so that $\sigma\tau\rho 1 = k$) one has

$$c_{ijk} = c_{1ij}c_{1jk}c_{1ik}^{-1} = c_{\sigma\tau}c_{\sigma\tau,\rho}c_{\sigma,\tau\rho}^{-1} = \sigma c_{\tau,\rho}.$$

Thus $\mu c_{ijk} = c_{\mu i,\mu j,\mu k}$ for all $\mu$ in $G$.

By inspection the two correspondences $\Psi$ and $\Psi^{-1}$ are indeed inverses. Finally given $\Psi(c_{ijk}) = (c_{\sigma\tau})$ and letting $R$ be the central simple algebra giving rise to $(c_{ijk})$, one wants to identify $R$ with the algebra giving rise to $(c_{\sigma\tau})$. Taking $u_\sigma$ in $R$ for which $u_\sigma a u_\sigma^{-1} = \sigma a$ and building a Noether factor set $(\tilde{c}_{\sigma\tau})$ as in (9) in the text one gets a corresponding Brauer factor set $(\tilde{c}_{ijk}) = \Psi^{-1}(\tilde{c}_{\sigma\tau})$, which thus must be associated to $(c_{ijk})$, i.e, there are $(\gamma_{ij})$ for which (8) holds. Replacing $u_\sigma$ by $\gamma_{1,\sigma 1}^{-1}u_\sigma$ yields a Noether factor set $(c_{\sigma\tau}')$ corresponding to $(c_{ijk})$.

**11.** A Noether factor set $(c_{\sigma\tau})$ corresponds to a matrix algebra iff there are $\gamma_\sigma$ in $K$ for which $c_{\sigma,\tau} = \gamma_\sigma\sigma\gamma_\tau\gamma_{\sigma\tau}^{-1}$. (Hint: The corresponding Brauer factor set $(c_{ijk})$ is associated to (1) iff there are $\gamma_{ij}$ satisfying the conjugacy conditions, for which $c_{ijk} = \gamma_{ij}\gamma_{jk}\gamma_{ik}^{-1}$. To prove ($\Rightarrow$) let $\gamma_\sigma = \gamma_{1,\sigma 1}$ and note $c_{\sigma,\tau} = \gamma_\sigma\sigma\gamma_\tau\gamma_{\sigma\tau}^{-1}$.)

**12.** (Wedderburn's criterion) The cyclic algebra $(K, \sigma, \alpha)$ is split iff $\alpha$ is a norm from $K$ to $F$. (Hint: If $\alpha = N_\sigma(\gamma)$ then taking $\gamma_{\sigma^i} = \gamma\delta\gamma\cdots\sigma^{i-1}\gamma$ one sees the Noether factor set of example 7.1.42 corresponds to $M_n(F)$. Conversely, if there are $\gamma_{\sigma^i}$ as in exercise 11 then taking $\gamma = \gamma_\sigma$ one has $\gamma_{\sigma^i} = \gamma\sigma\gamma\cdots\sigma^{i-1}\gamma$ for each $i < n$, by induction on $i$, so $\alpha = c_{\sigma,\sigma^{n-1}} = \gamma\sigma(\gamma\cdots\sigma^{n-2}\gamma) = N_\sigma\gamma$.)

**13.** A cyclic algebra $R = (K, \sigma, \alpha)$ has Brauer factor set $(c_{ijk})$ given by $c_{ijk} = 1$ if $i \leq j \leq k$ or $i \geq k \geq j$, and $c_{ijk} = \alpha$, otherwise. (Hint: Translate example 7.1.42 using exercise 10.)

**14.** (Abelian crossed products) Suppose $K$ is Galois over $F$ with abelian Galois group $G = \langle\sigma_1\rangle \times \cdots \times \langle\sigma_t\rangle$, with $\sigma_i$ of order $n_i$. If $K$ is a maximal subfield of $R$ then there are invertible $u_1, \ldots, u_t$ in $R$ such that $u_i a u_i^{-1} = \sigma a$ for all $a$ in $K$. Write $c_{ij} = u_i u_j u_i^{-1} u_j^{-1} \in K$ and $b_i = u_i^{n_i}$, an "abelian factor set" corresponding to $R$, and derive the ensuing abelian factor set conditions: $c_{ii} = 1$, $c_{ij}^{-1} = c_{ji}$, $\sigma_j(b_i) =$

$N_i(c_{ji})b_i$, $N_iN_j(c_{ij}) = 1$, and $\sigma_ic_{jk}\sigma_jc_{ki}\sigma_kc_{ij} = c_{jk}c_{ki}c_{ij}$. Here $N_i$ denotes the norm with respect to $\sigma_i$. Conversely, any abelian factor set determines a crossed product with respect to the corresponding abelian group. The algebra corresponding to the abelian factor set is split iff there are $a_i$ in $K$ satisfying $b_i = N(a_i)$ and $c_{ij} = (\sigma_ia_j)a_j^{-1}a_i(\sigma_ja_i)^{-1}$. This important construction due to Dickson was rediscovered by Amitsur-Saltman [78], who used it to get a concrete description of crossed products when the underlying group is abelian, with striking consequences for $p$-algebras; Amitsur-Rowen-Tignol [79] used this construction to build an indecomposable division algebra of degree 8 with involution.

## Wedderburn's Division Algorithm (see Rowen [80B, pp. 178–179]

**15.** We say $f$ *divides* $g$ in $D[\lambda]$ if $g = qf$ for some $q$ in $D[\lambda]$. Using the Euclidean algorithm (remark 1.6.20) show $\lambda - d$ divides $g(\lambda) - g(d)$, and thus $(\lambda - d)$ divides $g$ iff $g(d) = 0$.

**16.** If $\lambda - d$ divides $gh$ and *not* $h$ then $\lambda - h(d)dh(d)^{-1}$ divides $g$. (Hint: $\lambda - d$ divides $g(\lambda)h(d)$ by exercise 15.)

**17.** If $h(d) = 0$ for all conjugates $d$ of $d_1$ then the minimal polynomial $p$ of $d_1$ over $Z(D)$ divides $h$. (Hint: Take a counterexample $h$ of minimal degree; writing $h = qp + r$ for $\deg r < \deg p$ show the coefficients of $r$ lie in $Z(D)$.)

**18.** (Wedderburn [21]) If $g(\lambda)$ is irreducible in $F[\lambda]$ and $g$ has a root $d_1$ in some $F$-central division algebra $D$ then $g$ can be written as a product $(\lambda - d_n)\cdots(\lambda - d_1)$ where the $d_i$ are each conjugates of $d_1$. (Hint: Write $g = q(\lambda)(\lambda - d_k)\cdots(\lambda - d_1)$ for $k$ as large as possible, and let $h = (\lambda - d_k)\cdots(\lambda - d_1)$. Then $h$ satisfies the criterion of exercise 17, so the minimal polynomial $p$ of $d_1$ over $Z(D)$ divides $h$ and thus $g$. Hence $p = g$ implying $h = g$.) Actually, the proof does not require $[D:F] < \infty$.

**19.** Give Wedderburn's proof that every central division algebra of degree 3 is cyclic, using exercise 18. (Hint: The minimal polynomial of $d_1$ can be written $(\lambda - d_3)(\lambda - d_2)(\lambda - d_1)$ where $d_2$ can be taken to be $[d, d_1]d_1[d, d_1]^{-1}$ for arbitrary $d$, cf., exercise 16. Let $x = [d_1, d_2]$. $d_1 + d_2 + d_3 \in F$, so commuting with $d_i$ shows $[d_{\pi 1}, d_{\pi 2}] = \pm x$ for every $\pi$ in Sym(3). But then $d_3 = xd_2x^{-1}$ by exercise 16, and, likewise, $d_{i+1} = xd_ix^{-1}$ for each $i$. Hence $x^3 \in F$. By Zariski topology there is $d_1$ such that $x \notin F$.)

**20.** Using Wedderburn's method reprove the fact that every $D$ of degree 4 is a crossed product. (Hint: Take $d \in D$ of degree 4. The minimal polynomial is of the form $(\lambda^2 + a'\lambda + b')(\lambda^2 + a\lambda + b)$, and matching parts one can show $a^2$ is quadratic over $F$, cf., Rowen [80B, p. 181]. One concludes easily $D$ is a crossed product.)

**Division Rings** The next exercises contain a collection of results about arbitrary division rings. We start with a very easy proof of the theorem of Cartan-Brauer-Hua.

**21.** Suppose $V$ is a vector space over a division ring $D$, and $A$ is an additive subgroup of $V$ containing two $D$-independent elements. If $f: A \to V$ is a group homomorphism satisfying $fa \in Da$ for every $a$ in $A$ then there is $d$ in $D$ such that $fa = da$

for all $a$ in $A$. (Hint: If $a_1, a_2$ are $D$-independent and $fa_i = d_i a_i$ then $f(a_1 + a_2) = d_1 a_1 + d_2 a_2$ implies $d_1 = d_2$.)

22. (Treur [77]) Suppose $D \subseteq E$ are division rings, such that $D$ is invariant with respect to all inner automorphisms determined by an additive subgroup $A$ of $E$. Assume $1 \in A$. Then either $A \subseteq D$ or $D \subseteq C_E(A)$. (Hint: Fixing $d$ in $D$, let $f_d: A \to D$ be right multiplication by $d$. Exercise 21 implies either $ad = da$ for all $a$ in $A$, or else every element of $A$ is $D$-dependent on 1 so that $A \subseteq D$.)

23. (Cartan-Brauer-Hua) Suppose $D$ is a division subring of $E$, invariant with respect to all inner automorphisms of $E$. Then $D = E$ or $D \subseteq Z(E)$.

24. If $\frac{1}{2} \in E$ prove the analogue of exercise 23, using derivations instead of automorphisms.

25. (Faith [58]) If $E$ is a division ring which is not a finite field and if $D$ is a proper division subring then the multiplicative group $D - \{0\}$ is of infinite index in $E - \{0\}$. (Hint: $E$ is infinite by Wedderburn's theorem, so one may assume $D$ is infinite. If $[E - \{0\}: D - \{0\}] < \infty$ then for any $x$ in $E$ one can find $d_1 \neq d_2$ in $D$ with $x + d_1 \in D(x + d_2)$; solve to get $x \in D$.)

26. (Herstein) Any noncentral element of a division ring has infinitely many conjugates. (Hint (Faith): In exercise 25 take $D$ to be the centralizer.)

The next few exercises describe and illuminate Bergman's example of a division ring $D$ f.g. as a module over a subring $R$ which is *not* a division ring, answering a question of P. F. Smith.

27. (P. F. Smith) No such example exists if $R$ is left Noetherian. (Hint: $D$ would be a Noetherian $R$-module so for any $r$ in $R$ the ascending chain $Rr^{-1} \leq Rr^{-2} \leq \cdots$ implies $r$ is invertible in $R$.)

28. (P. F. Smith) No such example exists if $R$ is right Ore. (Hint: One could take $D$ to be the ring of fractions, but then by common denominators $D = Rr^{-1}$ for some $r$ in $R$ so $D = R$.)

29. (Bergman) Let $G$ be the free group on three elements $x, y, z$ ordered such that $z$ is minimal among $\{x, y, z, x^{-1}, y^{-1}, z^{-1}\}$. (Just reverse the usual order.) Let $E = \mathbb{Q}((G))$ and let $D$ be the division subalgebra generated by $x, y, z$. Define $v: E \to G$ by taking $v(f)$ to be the least element of $\mathrm{supp}(f)$. Since $D$ is countable there is a monic function $\varphi: D \to \mathbb{N} - \{0\}$ such that $yz^{\varphi d}x < v(d)$ for each $d$ in $D$. Note $\{y, z, yz^i x : i \in \mathbb{N}\}$ generate a free subsemigroup; thus the $\mathbb{Q}$-subalgebra $R$ of $D$ generated by $y, z$, and $\{d + yz^{\varphi d}x : d \in D\}$ is free and so is not a division ring. But $D = R + Rx$ since $d = (d + yz^{\varphi d}x) - (yz^{\varphi d})x$.

30. (Bergman [83]) If in exercise 29 one takes $G$ generated only by two elements $y, z$ and now stipulates $\varphi$ satisfies $yz^{2\varphi d + 1}y < v(d)$, then the $\mathbb{Q}$-subalgebra $R$ generated by $y, z^2$ and $d + yz^{2\varphi d + 1}y$ is still a free $\mathbb{Q}$-algebra, and now $D = R + (yz)R = (yz)R + R$. Bergman also modifies the construction to provide an instance where $R$ is *left* Ore.

31. (Vamos) The following are equivalent for fields $F \subseteq K$: $(i) K \otimes_F K$ is Noetherian; (ii) {fields between $F$ and $K$} satisfies ACC; (iii) $K$ is an f.g. field extension of $F$. (Hint: (i) $\Rightarrow$ (ii) If $F_1 \subset F_2 \subset \cdots$ are fields define $\varphi_i: K \otimes_F K \to K \otimes_{F_i} K$ canonically and take $A_i = \ker \varphi_i$; then $A_i \subset A_{i+1}$ since $a \otimes 1 - 1 \otimes a \in \ker \varphi_{i+1} - \ker \varphi_i$ for any $a$ in $F_{i+1} - F_i$. (ii) $\Rightarrow$ (iii) Take $K_1$ to be a maximal f.g. field extension of $F$; then $K_1 = K$. (iii) $\Rightarrow$ (i). See exercise 32 ($\Leftarrow$).)

32. Suppose $K$ is a subfield of $D$. $D \otimes_{Z(D)} K$ is Noetherian iff $K$ is an f.g. field extension of $Z(D)$. (Hint: $(\Rightarrow)$ $D \otimes K$ is a free $K \otimes K$-module so $K \otimes K$ is Noetherian by sublemma 2.5.32'. Apply (i) $\Rightarrow$ (iii) of exercise 21. $(\Leftarrow)$ $D \otimes K$ is f.g. as module over $D(a_1, \ldots, a_t)$ where $(a_1, \ldots, a_t)$ is a transcendence base of $K$ over $Z(D)$.) This easy result of Resco-Small-Wadsworth [79] is surprisingly useful, as we see in the next exercise.

33. Let $D$ be the division ring of fractions of the Weyl algebra $A_n(F)$ where $\text{char}(F) = 0$. Then every subfield $K$ of $D$ is finitely generated over $F$. (Hint: $D \otimes K$ is a localization of $A_n(F) \otimes K \approx A_n(K)$ which is Noetherian, so apply exercise 32.)

34. (Lorenz [84]) Generalizing exercise 33, suppose $R$ is an $F$-algebra such that $R \otimes_F K$ is left Noetherian for each field extension $K$ of $F$. If $S$ is a left denominator set for $R$ then every subfield $L$ of $S^{-1}R$ is f.g. over $F$ and, furthermore, satisfies $\text{tr deg } L/F \leq K\text{-dim } R \otimes_F L$. (Hint: $L$ is f.g. as in exercise 33. $\text{tr deg } L/F = K\text{-dim } L = K\text{-dim } L \otimes L \leq K\text{-dim } S^{-1}R \otimes L \leq K\text{-dim } R \otimes L$, cf., exercise 3.5.26.)

## The Dieudonne Determinant (cf., Draxl [83B])

Assume $D$ is a division ring and $n \geq 3$.

35. Suppose $r \in GL(n, D)$, the invertible $n \times n$ matrices over $D$. Then $r$ can be written in the form $a_1 dp a_2$ where $a_1$ is lower triangular with 1's on the diagonal, $a_2$ is upper triangular with 1's on the diagonal, $d$ is diagonal, and $p$ is a permutation matrix. This is called the *Bruhat decomposition*. (Hint: The standard linear algebra argument using elementary transformations, noting left resp. right multiplication by an elementary matrix corresponds to an elementary row (resp. column) transformation.)

36. In the Bruhat decomposition $d$ and $p$ are uniquely determined. (Hint: If $a_1 dp a_2 = a_1' d' p' a_2'$ then $d^{-1}a_1^{-1}a_1'd'$ is lower triangular but equals $pa_2(a_2')^{-1}(p')^{-1}$ so $p = p'$ and hence the diagonal $d^{-1}d' = 1$.) Consequently, writing $d$ as the diagonal $(d_1, \ldots, d_t)$ and $p$ as the permutation matrix $\sum_{i=1}^n e_{i, \pi i}$ in the Bruhat decomposition of $r$ we can define $\tilde{d}(r) = (sg \, \pi)d_1 \cdots d_t$.

37. (Dieudonne determinant) If $r, r_1 \in GL(n, D)$ then in the above notation $\tilde{d}(r_1)\tilde{d}(r_2)$ $\tilde{d}(r_1 r_2)^{-1}$ is a product of group commutators in $D - \{0\}$. Thus writing $G^{ab}$ for $G/G'$ where $G'$ is the group commutator of $G$ and writing $R^{ab}$ for $\{$units of $R\}^{ab}$ for any ring $R$, we get a group homomorphism $\det: M_n(D) \to D^{ab}$ which is the composite of $\tilde{d}$ with the canonical map $D \to D^{ab}$. This is called the *Dieudonne determinant*.

The Dieudonne determinant can be tied in with algebraic $K$-theory, as we shall now see.

38. Define $SL(n, D) = \{r \in GL(n, D): \det r = 1\}$. Then $SL(n, D)$ is generated by elementary matrices (i.e., of the form $1 + e_{ij}$ for $i \neq j$) and is its own commutator subgroup for all $n \geq 3$. (Hint: The elementary matrix $1 + e_{12}$ can be written as the commutator of $1 + e_{13}$ and $1 + e_{32}$, so it suffices to prove the first assertion. For any $u, v$ in $D$ write $\begin{pmatrix} u^{-1} & 0 \\ 0 & u \end{pmatrix}$ and $\begin{pmatrix} uvu^{-1}v^{-1} & 0 \\ 0 & 1 \end{pmatrix}$ as products of elementary matrices by transforming them to the identity up elementary row and column transformations.)

**39.** The conclusion of exercise 38 also holds for $n = 2$ if $|D| \geq 4$.

**40.** The Dieudonne determinant gives an isomorphism $M_n(D)^{ab} \to D^{ab}$ for all $n$, except for $n = 2$ and $D = \mathbb{Z}/2\mathbb{Z}$ or $\mathbb{Z}/3\mathbb{Z}$. (Hint: exercises 37–39.) This is a useful piece of information since it relates to the Whitehead group $K_1(D)$, cf., Milnor [71B, §3].

**41.** (Cohn-Schofield) Division rings $D \subset E$ such that the left dimension $[E:D] = 2$ and the right dimension is infinite. Let $E$ be a division ring generated by the free algebra $F\{X_1, X_2\}$, cf., corollary 1.3.41, and let $D$ be the division subring generated by $X_1, X_2^2$, and $X_1^2 X_2 - X_2 X_1$. (Hint: Clearly $E = D + DX_2$, since the last generator enables one to pull $X_1$ to the left. It remains to show the right dimension is infinite. First observe that nowhere in exercises 6.3.9 or 6.3.10 was $C$ required to be commutative! Thus taking $C = D$ and $R = E$ in exercise 6.3.10 it suffices to show $M(E; 0, 0) \approx \ker(E \otimes_D E \to E)$ is infinite dimensional as right module over $D$. But $M(E; 0, 0)$ has bimodule generators $\delta X_1, \delta X_2$ with relations $\delta X_1 = 0$, $0 = \delta X_2^2 = X_2 \delta X_2 + (\delta X_2)X_2$, and $0 = \delta(X_1^2 X_2 - X_2 X_1) = X_1^2 \delta X_2 - (\delta X_2)X_1$, i.e., $M(E; 0, 0)$ can be viewed as the $E - E$ bimodule $E(\delta X_2)E$ where $(\delta X_2)X_2 = -X_2 \delta X_2$ and $(\delta X_2)X_1 = X_1^2 \delta X_2$. The right dimension over $D$ is infinite because $\{(X_1 X_2^n)\delta X_2 : n \in \mathbb{N}\}$ are independent.)

**The Zero Divisor Question** The following exercises are addressed to the following question, which we call the *zero divisor question for tensor products*:

If $R_1$ and $R_2$ are domains over an algebraically closed field $F$ then is $R_1 \otimes R_2$ a domain? The reason the zero divisor question is posed here is that it is usually asked more specifically for division algebras and is still open in this case. Bergman [75] indeed found domains $R_1, R_2$ over $F$ for which $R_1 \otimes_F R_2$ is not a domain, but in this example $R_1 \otimes_F D$ and $D \otimes_F R_2$ *are* domains for every division algebra $D$. (In particular, neither $R_1$ nor $R_2$ can be a subring of a division algebra, and, in fact, he shows there is no nontrivial homomorphism from $R_1$ or $R_2$ to a division algebra!) The following exercises present his example. Incidentally, when $F$ is not algebraically closed a related zero divisor question can be raised for finite dimensional division algebras, and its solution is given in digression 7.2.13'ff.; also see appendix C.

**42.** Define a *bilinear system* over $F$ to be a system $A = (A_1, A_2, A_3; \varphi_A)$ where $\varphi: A_1 \times A_2 \to A_3$ is a bilinear map of vector spaces over $F$. Let $\mathscr{B}\mathit{ilin}$ denote the category of bilinear systems, where morphisms are defined in the natural way (so that the appropriate diagram commutes.) There is a forgetful functor $G: \mathscr{F}\text{-}\mathscr{A}\mathit{lg} \to \mathscr{B}\mathit{ilin}$ given by $GR = (R, R, R; \varphi_R)$; in other words, any algebra $R$ provides a bilinear system by taking $\varphi_R$ to be the ring multiplication and then forgetting the ring structure. Also note that if $A$ and $B$ are bilinear systems one can define their tensor product $A \otimes B = (A_1 \otimes B_1, A_2 \otimes B_2, A_3 \otimes B_3; \varphi_A \otimes \varphi_B)$, where $(\varphi_A \otimes \varphi_B)(a_1 \otimes b_1, a_2 \otimes b_2) = \varphi_A(a_1, a_2) \otimes \varphi_B(b_1, b_2)$ for $a_i$ in $A_i$ and $b_i$ in $B_i$.

**43.** The universal from a bilinear system $A = (A_1, A_2, A_3; \varphi_A)$ to the functor $G$ of exercise 42 can be constructed as follows: Define $F\langle A \rangle$ to be the tensor algebra $T(A_1 \oplus A_2 \oplus A_3)$ modulo the ideal generated by all $a_1 a_2 - \varphi_A(a_1, a_2)$ for $a_i$ in $A_i$. Show there is a natural vector space embedding $A \to F\langle A \rangle$ (which, more precisely, is an injection $A \to G(F\langle A \rangle)$ in $\mathscr{B}\mathit{ilin}$).

**44.** $F\langle A \rangle$ is not a domain iff $A$ *has zero divisors* in the sense that there are nonzero

$a_1, a_2$ in $A_1, A_2$, respectively, such that $\varphi(a_1, a_2) = 0$. (Hint: Take two distinct prime integers $p_1, p_2$ and grade $T(A_1 \oplus A_2 \oplus A_3)$ by putting $\deg(A_i) = p_i$ for $i = 1, 2$ and $\deg(A_3) = p_1 p_2$. This induces a grading on $F\langle A \rangle$, which has zero divisors iff there are homogeneous zero divisors.) Thus the zero divisor question has been transferred to $\mathscr{B}i\ell in$.

**45.** Show the zero divisor question in $\mathscr{B}i\ell in$ depends on finite dimensional bilinear systems.

**46.** Given a bilinear system $A = (A_1, A_2, A_3; \varphi_A)$ one has an induced map $\bar{\varphi}_A : A_1 \otimes A_2 \to A_3$. Let $A_0 = \ker \bar{\varphi}_A$. Replacing $A_3$ by $\varphi_A(A_1 \times A_2)$ one may assume $\varphi_A$ (and thus $\bar{\varphi}_A$) is onto, providing an exact sequence $0 \to A_0 \to A_1 \otimes A_2 \to A_3 \to 0$. Let $A_i^*$ denote the dual space $\operatorname{Hom}_F(A_i, F)$. Then $(A_1 \otimes A_2)^* \approx A_1^* \otimes A_2^*$ via the dual base (by a dimension count), so exercise 1.5.5 provides an exact sequence $0 \to A_3^* \to A_1^* \otimes A_2^* \to A_0^* \to 0$. Define $A^* = (A_1^*, A_2^*, A_0^*, \psi)$ to be the corresponding bilinear system where $\psi$ is obtained from the dual of the map $A_0 \to A_1 \otimes A_2$. The bilinear system $A \otimes A^*$ then has the "canonical" zero divisors $s_1 = \sum b_{1j} \otimes f_{1j}$ and $s_2 = \sum b_{2j} \otimes f_{2j}$ where for $i = 1, 2$ the $b_{ij}$ (resp. $f_{ij}$) are a base of $A_i$ (resp. $A_i^*$).

**47.** Notation as above. Let $H = \{a_1 \otimes a_2 : a_1 \in A_1 \text{ and } a_2 \in A_2\}$. If $[A_i:F] = n_i$ for $1 \le i \le 3$ then $[A_0:F] = n_1 n_2 - n_3$ and as an algebraic set $H$ has dimension $n_1 + n_2 - 1$ (since $\alpha a_1 \otimes a_2 = a_1 \otimes \alpha a_2$). Hence $A_0 \cap H \ne 0$ if $(n_1 n_2 - n_3) + (n_1 + n_2 - 1) > n_1 n_2$, i.e., $n_3 < n_1 + n_2 - 1$. This is a sufficient condition for $A$ to have zero-divisors. Likewise, $A^*$ has zero-divisors if $n_3 + [H:F] > n_1 n_2$, i.e., if $n_3 > n_1 n_2 + 1 - n_1 - n_2$. Thus for $A$ and $A^*$ each *not* to have zero divisors we need $n_1 + n_2 - 1 \le n_3 \le n_1 n_2 + 1 - n_1 - n_2$, implying $(n_1 - 2)(n_2 - 2) \ge 2$. Conversely, if $(n_1 - 2)(n_2 - 2) \ge 2$ then $n_3$ can be found as desired; any nondegenerate $A$ (under certain geometric stipulations) satisfying $[A_i:F] = n_i$ would have $A$ and $A^*$ without zero divisors.

**48.** Bergman's concrete example. By symmetry one may assume $n_1 \ge n_2$. By exercise 47 the minimal counterexample would be $n_1 = 4$, $n_2 = 3$, $n_3 = 4 + 3 - 1 = 6$. Let $\{b_1, \ldots, b_4\}$ be a base of $A_1$ and $\{c_1, c_2, c_3\}$ be a base of $A_2$. Choosing $\alpha \in F$ arbitrarily $\ne 0, 1$ define $A(\alpha) = (A_1, A_2, A_3; \varphi)$ where $A_3 = (A_1 \otimes A_2)/A_0$ and $A_0$ is spanned by the following elements (taking $j$ modulo 3):

$$b_1 \otimes c_j - b_3 \otimes c_{j+1} \quad \text{for } 1 \le j \le 3$$

$$b_2 \otimes c_j - b_4 \otimes c_{j+1} \quad \text{for } 1 \le j \le 2 \quad \text{and} \quad b_2 \otimes c_3 - \alpha b_4 \otimes c_1.$$

Define $\varphi_A$ by $\varphi(a_1, a_2) = a_1 \otimes a_2 + A_0$ for all $a_i$ in $A_i$. Then $A(\alpha^{-1}) = (A(\alpha))^*$. Moreover, if $f: F\langle A(\alpha) \rangle \to D$ is a ring homomorphism to a division algebra $D$ for $\alpha \ne 0, 1$ then $A(\alpha) \subseteq \ker f$. (Otherwise, writing $d_j, e_j$ for the respective images in $D$ of $b_j, c_j$ one would have $d_2 e_3 = \alpha d_4 e_1$ but solving the various relations yields $d_2 e_3 = d_4 e_1$ so $d_4 = 0$ or $e_1 = 0$ from which it follows all $d_j = 0 = e_j$.)

**49.** $F\langle A(\alpha) \rangle \otimes D$ is a domain for every division algebra $D$, so in particular $F\langle A(\alpha) \rangle$ is a domain. (Hint: Otherwise, $A(\alpha) \otimes U$ has nontrivial zero-divisors, i.e., $\varphi(y_1, y_2) = 0$, where $U$ is the underlying bilinear system of $D$. The universality of the construction gives us a morphism $f: A(\alpha^{-1}) \to U$ such that $\varphi s_i = y_i$ for $i = 1, 2$, where $s_i$ is as in exercise 46, thereby contradicting exercise 48.)

# §7.2

1. (Frobenius) The only central division algebras over $\mathbb{R}$ are $\mathbb{R}$ and the quaternion algebra $\mathbb{H}$. (Hint: If $D \neq \mathbb{R}$ then $\mathbb{C}$ is a maximal subfield of $D$, so $\deg D = 2$ and thus $D$ is isomorphic to the symbol $(\alpha, -1)_2 \approx (\pm 1, -1)_2$.) Thus $|\mathrm{Br}(\mathbb{R})| = 2$.

2. $eRe \sim R$ for any idempotent $e$ of $R$. (Hint: Let $L = Re \approx L_0^{(t)}$ where $L_0$ is a minimal left ideal of $R$; $eRe \approx \mathrm{End}_R Re \approx \mathrm{End}_R L_0^{(t)} \approx M_t(\mathrm{End}_R L_0)$.

3. (Cancellation) If $[K:F]$ is prime to $\deg(R)$ and $R \otimes K \approx R' \otimes K$ then $R \approx R'$. (Hint: $R^{\mathrm{op}} \otimes R' \otimes K$ is split.)

4. Suppose $\deg(R) = n = m_1 \cdots m_t$ where $m_1, \ldots, m_t$ are distinct prime powers. If there are fields $K_i$ with $[K_i:F] = n/m_i$ such that each $C_R(K_i)$ is a division ring, then $R$ is also a division ring. (Hint: Otherwise, let $m = n/\mathrm{index}(R)$; take some $m_i$ not prime to $m$ and reach a contradiction.)

5. Suppose $R$ has a maximal subfield $K$ Galois over $F$ with Galois group $G$. Letting $K_p$ be the fixed subfield of $K$ under the Sylow $p$-subgroup of $G$, show that if each $C_R(K_p)$ is a division ring then $R$ is also a division ring. (Hint: exercise 4.)

## Brauer Factors Sets and Cohomology

6. Suppose $K$ is a separable field extension of $F$. $K \otimes K$ has a primitive idempotent $e$ such that $(a \otimes a')e = (a' \otimes a)e$ for all $a, a'$ in $K$; if $K$ is Galois over $F$ then $(\sigma \otimes \sigma)e = e$ for all $\sigma$ in $\mathrm{Gal}(K/F)$. (Hint: $K \otimes_F K$ is a direct product of copies of $F$. Define $\psi: K \otimes K \to K$ by $a \otimes a' \mapsto aa'$, and write $\ker \psi = (K \otimes K)e_0$ for $e_0$ idempotent. Let $e = 1 - e_0$. Then $Ke \approx K$ so $e$ is primitive. $\psi e = 1$ implies $(a \otimes a')e = (a' \otimes a)e$. If $K$ is Galois over $F$ then $\ker \psi$ is invariant under $\sigma \otimes \sigma$ for any $\sigma$ in $\mathrm{Gal}(K/F)$. Hence $(\sigma \otimes \sigma)e_0$ is an idempotent in $\ker \psi$, so $(\sigma \otimes \sigma)e_0 = e_0$.)

7. Prove lemma 7.2.7 by taking $e$ from exercise 6 and showing $e(R \otimes R')e$ is a crossed product with respect to $Ke$ having Noether factor set $(c_{\sigma\tau} c'_{\sigma\tau})$.

8. If $\mathrm{index}(R)$ is odd then $R$ has a Brauer factor set $(c'_{ijk})$ with $c_{iij} = 1$ and $c_{ijk} = c_{jki} = c_{kij} = c_{kji}^{-1} = c_{jik}^{-1} = c_{ikj}^{-1}$ for all $i, j, k$. (Hint: Take $(c_{ijk})$ for $R$; then $(c_{ijk} c_{kji}^{-1})$ is associated to $R \otimes (R^{\mathrm{op}})^{\mathrm{op}} \approx R \otimes R$; take $c'_{ijk} = (c_{ijk} c_{kji}^{-1})^{(n+1)/2}$.)

9. (Inflation) Any group homomorphism $\varphi: G_1 \to G$ induces a map $H^2(G, K - \{0\}) \to H^2(G_1, K - \{0\})$ given by $f \mapsto \bar{f}$ where $\bar{f}(\sigma, \tau) = f(\varphi\sigma, \varphi\tau)$ for all $\sigma, \tau$ in $G$. This is called the *inflation* of $\varphi$. In particular, when $G$ and $G_1$ are groups of automorphisms on $K$ the inflation of $\varphi$ yields a map $\mathrm{Br}(F) \to \mathrm{Br}(K_1)$ where $F, K_1$ are the fixed subfields of $K$ under $G, G_1$, respectively. Show the inflation corresponds to the restriction if $\varphi$ is monic.

10. (Neumann) Suppose $G$ is an arbitrary ordered group (not necessarily finite) acting as automorphisms on a field $K$. Any 2-cocycle $f: G^{(2)} \to K - \{0\}$ gives rise to a division ring $\mathscr{D}(K, \bar{f})$ whose elements are those formal sums $\{\sum_g a_g \lambda_g : a_g \in G\}$ for which the support $\{g: a_g \neq 0\}$ is well-ordered, where addition is componentwise (in terms of coefficients of each $\lambda_g$) and multiplication is given by the rules $\lambda_g a = (\sigma a)\lambda_g$ for all $a$ in $K$, and $\lambda_g \lambda_h = f(g, h)\lambda_{gh}$. (Hint: Use the argument of propositions 1.2.22 and 1.2.24.) If $G = \mathbb{Z}^{(n)}$ the additive structure of $D$ is that of Laurent series $K((\lambda_1, \ldots, \lambda_n))$ with multiplication given by $\lambda_i a = (\sigma_i a)\lambda_i$ and $\lambda_j \lambda_i = f(\sigma_j, \sigma_i) f(\sigma_i, \sigma_j)^{-1} \lambda_i \lambda_j$ where $\sigma_1, \ldots, \sigma_n$ are the generators of $\mathbb{Z}^{(n)}$.

11. (Tignol-Amitsur [85]) If $G$ is an arbitrary finite abelian group and $f: G^{(2)} \to K$ is

a Noether factor set then taking a group surjection $\varphi\colon \mathbb{Z}^{(n)} \to G$ yields a Neumann division algebra $D = \mathscr{D}(K, \bar{f})$ where $\bar{f}$ is the inflation of $f$ under $\varphi$ to a 2-cocycle $\bar{f}\colon \mathbb{Z}^{(n)} \times \mathbb{Z}^{(n)} \to K$. Show $D$ is naturally isomorphic to the iterated Laurent power series ring. Given $\mathbf{i} = (i(1), \ldots, i(n))$ in $\mathbb{Z}^{(n)}$ write $\lambda^{\mathbf{i}}$ for $\lambda_1^{i(1)} \cdots \lambda_n^{i(n)}$ and $\sigma^{\mathbf{i}}$ for $\sigma_1^{i(1)} \cdots \sigma_n^{i(n)}$. Using the notation at the end of exercise 10, say an element $a$ of $K$ is $\mathbf{i}$-good if for all $\mathbf{j}$ in $\mathbb{Z}^{(n)}$ we have $f(\sigma^{\mathbf{i}}, \sigma^{\mathbf{j}}) f(\sigma^{\mathbf{j}}, \sigma^{\mathbf{i}})^{-1} = a(\sigma^{\mathbf{j}} a)^{-1}$. Let $\Gamma = \{\mathbf{i} \in \mathbb{Z}^{(n)}\colon \sigma^{\mathbf{i}}$ is the identity on $K$ and there is an $\mathbf{i}$-good element$\}$. Then $Z(D) = \sum_{\mathbf{i} \in \Gamma}\{a_{\mathbf{i}}\lambda^{\mathbf{i}}\colon a_{\mathbf{i}}$ is $\mathbf{i}$-good$\}$. Hence $(\deg D)^2 = [\mathbb{Z}^{(n)}\colon \Gamma][K\colon F]$ where $F = K^G$; since each of the right-hand terms divides $|G|$ show $\deg D$ divides $|G|$. Amitsur-Tignol continue by describing all the subfields of $D$ which are Kummer extensions of the center.

## Factor Sets and Roots of 1

12. (Rowen [84]) If $R$ has odd degree then there is an element $d$ in $R$ for which $\operatorname{tr}(d) = \operatorname{tr}(d^2) = 0$ where tr is the trace. (Hint: Take the Brauer factor set $(c_{ijk})$ of exercise 8. One wants $(\beta_{ij})$ such that $d = (\beta_{ij}c_{ij1})$ satisfies $0 = \operatorname{tr}(d) = \sum \beta_{ii}c_{ii1} = \sum \beta_{ii}$ and $0 = \operatorname{tr}(d^2) = \sum \beta_{ij}\beta_{ji}$. Thus one may assume $(c_{ijk})$ is trivial, in which case $R$ is split and the result is obvious.)

13. (Brauer) If $\exp(R) = m$ and $F$ has a primitive $m$-th root $\zeta$ of 1 then $R$ is similar to an algebra having a Brauer factor set all of whose elements are roots of 1. (Hint: Take a Brauer factor set $(c_{ijk})$ for a maximal separable subfield $K$, and take $\gamma_{ij}$ such that $1 = c_{ijk}^m \gamma_{ij}\gamma_{jk}\gamma_{ik}^{-1}$. Let $L = K[\gamma_{ij}^{1/m}\colon 1 \le i, j \le n]$, also separable over $F$. If $[L\colon K] = t$ then $M_t(R)$ has a Brauer factor set $(c'_{ijk})$ with respect to $L$, comprised of blocks of $(c_{ijk})$. Since $\operatorname{Gal}(L/K)$ sends $\gamma_{ij}^{1/m}$ to $\zeta^u \gamma_{ij}^{1/m}$ for $u = u(i, j)$ one can use the conjugacy conditions to expand the $n \times n$ matrix $(\gamma_{ij})$ to a $tn \times tn$ matrix $(\gamma'_{ij})$ with each $\gamma_{ij}'^m = \gamma_{ij}$. Now let $c''_{ijk} = c'_{ijk}\gamma'_{ij}\gamma'_{jk}\gamma_{ik}'^{-1}$, an $m$-th root of 1. The $(c''_{ijk})$ is a Brauer factor set associated to $(c'_{ijk})$.)

14. (Benard-Schacher) $R$ is *cyclotomic* if there is a root $\zeta$ of 1 for which $F(\zeta)$ is a maximal subfield of $R$ and the corresponding Noether factor set consists only of roots of 1. Show if $R$ is cyclotomic of exponent $m$ then $F$ contains a primitive $m$-th root of 1. (Hint (Janusz [72]): Choose $\zeta'$ in the Noether factor set $(c_{\sigma\tau})$ such that each $c_{\sigma\tau}$ is a power of $\zeta'$. If $(\zeta')^t = 1$ then each $c_{\sigma\tau}^t = 1$ so $m \mid t$. It suffices to show $\zeta' \in F$. Write $R = \sum_\sigma F(\zeta)u_\sigma$ as in (11). For any $\rho$ in $G = \operatorname{Gal}(F(\zeta)/F)$ build the new crossed product $R_\rho = \sum_\sigma F(\zeta)u_\sigma$ with respect to $(\rho c_{\sigma\tau})$. One has an isomorphism $R \to R_\rho$ sending $au_\sigma$ to $(\rho a)u_\sigma$. But $\rho\zeta = \zeta^i$ for some $i$, so $R^{\otimes i} \sim R_\rho \sim R$. Thus $m \mid (i - 1)$ and hence $\rho\zeta' = \zeta'$ for each $\rho$ in $G$, proving $\zeta' \in F$.)

## Corestriction

15. (Tate [76]) A cohomological description of the corestriction. Defining the co-complex $\mathbb{A} = \{A^n(G, M)\colon n \in \mathbb{N}\}$ and the cohomology groups $H^n(G, M)$ as in example 5.2.13. If $H$ is of finite index in $G$ and $\mathbf{b} = \{b_1, \ldots, b_t\}$ is a left transversal of $H$ define $\operatorname{tr}_{\mathbf{b}}\colon A^n(H, M) \to A^n(G, M)$ by

$$(\operatorname{tr}_{\mathbf{b}} f)(g_1, \ldots, g_n) = \sum_{i=1}^{t} b_{0i}^{-1} f(b_{0i}g_1 b_{1i}^{-1}, b_{1i}g_2 b_{2i}^{-1}, \ldots, b_{n-1, i}g_n b_{ni}^{-1}),$$

where $b_{0i} = b_i$ and, inductively, $b_{ui} \in Hb_{u-1,i}g_i$ for $1 \leq u \leq n$. Then $\text{tr}_\mathbf{b}$ is independent of the transversal $\mathbf{b}$ and induces the desired corestriction map from $H^n(H, M) \to H^n(G, M)$. Verify the formula of theorem 7.2.26.

16. (Tate) Suppose $B: M_1 \times M_2 \to M$ is a bilinear map of $G$-modules satisfying $B(gx_1, gx_2) = gB(x_1, x_2)$. $B$ induces the *cup product* maps $A^m(G, M_1) \times A^n(G, M_2) \to A^{m+n}(G, M)$ defined by $(f_1, f_2) \mapsto f_1 \cup f_2$ where

$$(f_1 \cup f_2)(g_1, \ldots, g_{m+n}) = B(f_1(g_1, \ldots, g_m), g_1 \cdots g_n f_2(g_{m+1}, \ldots, g_n)).$$

Show $d(f_1 \cup f_2) = (df_1) \cup f_2 + (-1)^m f_1 \cup df_2$ where $d$ is as in example 5.2.13, and $\text{cor}(f_1 \cup \text{res} f_2) = (\text{cor} f_1) \cup f_2$ in the cohomology groups.

17. (Tignol [83]) Suppose $K$ is a cyclic field extension of $F$ of dimension $n$, i.e., $\text{Gal}(K/F) = \langle \sigma \rangle$ with $\sigma^n = 1$, and $F$ has a primitive $m$-th root of 1. Let $L$ be the fixed field of $K$ under $\sigma^m$. Then the image of $\text{cor}_{L/F}: \text{Br}(K/L) \to \text{Br}(K/F)$ is $\text{Br}_m(K/F)$. (Hint: "$\subseteq$" is clear. Suppose $(\alpha, \beta)_n \in \text{Br}_m(K/F)$, where $K \approx F(\beta^{1/n})$. Then $(\alpha, \beta^m)_n \sim 1$ implying $\alpha$ is a norm from $L$, so use the projection formula.)

18. (Tignol [83]) Every cyclic $F$-algebra of degree $n$ and exponent $m$ is similar to a tensor product of $n/m$ symbols of degree $m$, provided $F$ has a primitive $m$-th root of 1. (Hint: Apply Rosset-Tate to exercise 17.)

19. (Rowen-Saltman [82]) Short proof of corollary 7.2.42. (Hint: Write $G = \langle \sigma, \tau:$ $\sigma^n = 1 = \tau^2$ and $\sigma\tau = \tau\sigma^{-1} \rangle$. Let $L$ be a splitting field of $R$ with $\text{Gal}(L/F) = G$. Let $K$ be the fixed field $L^\tau$, and $L_0 = L^\sigma$. $K$ splits $R$ since $[L:K]$ is prime to $n$ so $K \subseteq R$, and also one can view $L$ in the cyclic algebra $R \otimes L_0$. Take $a \in L$ such that $\sigma a = \zeta a$, and take $b$ in $R \otimes L_0$ such that $ab = \zeta ba$. Then $\tau a \in Fa$. Let $\tau' = 1 \otimes \tau$. Replacing $b$ by $b^{(n+1)/2} \tau' b^{-(n+1)/2}$ one may assume $\tau' b = b^{-1}$. But $L_0(b)$ also has Galois group $G$ over $F$, so reversing the roles of $a$ and $b$ one may assume $\tau a = a^{-1}$. Let $d = (b + b^{-1})(a + a^{-1})^{-1}$. Using Newton's identities one can show the odd characteristic coefficients of $d$ are 0 by showing $\text{tr}(d^{2k+1}) = 0$ for $k < n/2$. But the same holds for $d^{-1}$, so the minimal polynomial of $d$ has the form $\lambda^n + \alpha$, i.e., $d^n \in F$.)

20. Reprove every division algebra of degree 3 is cyclic by noting it is split by the dihedral group of order 6. (This assumes characteristic 0.)

21. Do exercise 20 in char. $\neq 0$.

## The Zero-Divisor Question

22. If $D_1, D_2$ are division algebras having relatively prime degrees $n_1, n_2$ then $D_1 \otimes D_2$ is a division algebra. (Hint: Tensoring by $D_2^{op}$ shows $n_1 \mid \text{index}(D_1 \otimes D_2)$.)

23. (Albert) If $Q = (\alpha, \beta)_2$ is a quaternion division algebra and $D$ is a division algebra such that no subfield of $D$ splits $Q$ then $D \otimes Q$ is a division algebra. (Hint: Take $y, z$ in $Q$ with $y^2 = \alpha$, $z^2 = \beta$, and $yz = -zy$. Suppose $D \otimes Q$ has a zero-divisor $r_1 + r_2 z$ where $r_1, r_2 \in D \otimes F(y)$. Then $r_2 \neq 0$, so one may assume $r_2 = 1$. Writing $r_1 = d_1 + d_2 y$ yields the zero-divisor $(d_1 - d_2 y - z)(r_1 + z) = d_1^2 - d_2^2 \alpha - \beta + [d_1, d_2]y$ in $D \otimes F(y)$, so this is 0. Hence $[d_1, d_2] = 0$ so $F(d_1, d_2)$ is a field; thus $F(d_1, d_2) \otimes Q$ is a division algebra in which $(r_1 z)^2 = \beta^2$, so $r_1 z = \pm \beta$, contradiction.)

## Central Simple Algebras with (*)

**24.** If $M_n(R)$ has an involution then $R$ has an involution of the same kind. (Hint: For first kind use theorem 7.2.44; for second kind translate to the corestriction.)

**25.** Any involution (*) of first kind on $R$ has either symplectic or orthogonal type. Furthermore, for any other involution $(J)$ of the first kind on $R$ there is some $r_0$ in $R$; with $r_0^* = \pm r_0$, such that $r^J = r_0^{-1} r^* r_0$ for all $r$ in $R$; in characteristic $\neq 2$, (*) and $(J)$ have the same type iff $r_0^* = +r_0$. (Hint: Skolem-Noether shows the automorphism $r \mapsto (r^J)^*$ is given by conjugation by some $r_0$. But then $r_0^* r_0^{-1} \in Z(R) = F$, so $r_0^* \in Fr_0$; $r_0^* = \pm r_0$ since $r_0^{**} = r_0$. To prove the last assertion one may tensor by the algebraic closure $\bar F$ of $F$ and thereby assume $R = M_n(\bar F)$; now apply theorem 2.13.29 and remark 2.13.5.)

**26.** In proposition 7.2.45 we may take (*) to be of symplectic or orthogonal type at our discretion and may assume $d^* = d$. (Hint: Notation as in that proof, one is done if $Ad$ has both symmetric and antisymmetric elements. But, otherwise, there is $\zeta = \pm 1$ such that $(ad)^* = \zeta ad$ for all $a$ in $A$, yielding $da^* = \zeta(ad)^* = ad$, and hence $A$ is commutative; conclude $K$ is maximal and $\sigma = 1$.)

**27.** Suppose $n$ is a power of 2. The generic matrix ring with involution $(F_n\{Y\}, t)$ of exercise 6.4.3 is a domain whose ring of central fractions is a division algebra of exponent 2 and degree $n$. The same holds for $(F_n\{Y\}, s)$.

**28.** Suppose $(a_{ij})^t = -(a_{ij})$. Define the *Pfaffian adjoint matrix* to have its $ij$-entry be $(-1)^{i+j-1}$ times the Pfaffian of the $(n-2) \times (n-2)$ matrix obtained by striking out both the $i$ and $j$ rows *and* columns. Using the Pfaffian adjoint prove by induction $Pf(a_{ij}) = \sum (\text{sg } \pi) a_{\pi 1, \pi 2} \cdots a_{\pi(n-1), \pi n}$, summed over all permutations $\pi$ for which $\pi(2k-1) < \pi(2k)$ for each $k$. For example, $Pf(a_{ij}) = a_{12}$ for $n = 2$, and $Pf(a_{ij}) = a_{12}a_{34} - a_{13}a_{24} + a_{14}a_{23}$ for $n = 4$.

**29.** (Jacobson [83]) Suppose $R$ has orthogonal type involution. Then $R$ has a maximal commutative separable subalgebra $K = F[a]$ with $a^* = a$, and there is $r = r^*$ in $R$ with $R = KrK$. Under notational hypothesis 7.1.33 $(\beta_{ij}r_{ij})^* = (\beta_{ji}r_{ij})$ and the Brauer factor set $(c_{ijk})$ satisfies $c_{ijk} = c_{kji}$ for all $i, j, k$. (Hint: Using Zariski topology one may assume $R = M_n(E)$; take $a$ diagonal and $r = \sum_{i,j} e_{ij}$. A typical element is $b = \sum_{u,v} \alpha_{uv} a^{u-1} r a^{v-1}$ so $b^* = \sum \alpha_{uv} a^{v-1} r a^{u-1}$; writing $b = (\beta_{ij}r_{ij})$ note $\beta_{ij} = \sum \alpha_{uv} a_i^{u-1} a_j^{v-1}$.)

**30.** Suppose $R$ has symplectic type involution. $R$ has a maximal commutative separable subalgebra $K = F[a]$ with $a^* = -a$, and there is $r = -r^*$ in $R$ with $R = KrK$. (*) is now given by $(\beta_{ij}r_{ij})^* = -(\beta_{j+m, i+m}r_{ij})$, where $m = n/2$ and subscripts are taken modulo $m$. (Hint: Take $a_{i+m} = -a_i$ for $1 \le i \le m$. Note if $\sigma a_i = a_j$ then $\sigma a_{i+m} = \sigma a_{j+m}$. Apply exercise 29 to the involution $x \mapsto a^{-1}x^*a$ to get $r_0$ such that $r_0 = a^{-1}r_0^*a$ and $R = Kr_0K = Kar_0K$; take $r = ar_0$.)

**31.** If $\deg(R) = n = 2m$ and $R$ has an involution (*) of symplectic type then there is $b = b^*$ for which $\text{tr}(b) = \text{tr}(b^{-1}) = 0$. (Hint: Notation as in exercise 30, take $\beta_{ij} = f(a_i) - f(a_{j+m})$ subscripts mod $n$, where $f = \sum_{i=0}^{m-1} \alpha_i \lambda^i$ is to be determined. Then $b = (\beta_{ij}r_{ij})$ is symmetric, and $\text{tr } b = 0$. Arguing as in proposition 7.1.43, using the Pfaffian polynomial, one needs to solve a linear condition on the $\alpha_i$ to ensure $\text{tr}(b^{-1}) = 0$.)

**32.** Any division algebra $D$ of degree 8 and exponent 2 contains a maximal subfield

$K$ Galois over $F = Z(D)$ with Galois group $(\mathbb{Z}/2\mathbb{Z})^{(3)}$. (Hint: If $D$ has a central quaternion subalgebra $Q$ then $D \approx Q \otimes C_D(Q)$, so one is done by theorem 7.1.45. In general, by exercise 31 there is $b^* = b$ satisfying $b^4 + a_1 b^2 + \alpha_2 = 0$ for suitable $\alpha_i$ in $F$; hence $D$ does contain a quadratic field extension $K_1$ of $F$. By exercise 25, $D$ has a symplectic-type involution which restricts to the nontrivial automorphism $\sigma$ of $K_1$ over $F$; take $d = d^*$ for which $dad^{-1} = \sigma a$ for all $a$ in $K$. If $d^2 \in F$ then $K_1$ and $d$ generate a quaternion subalgebra, so assume $d^2 \notin F$ and let $K_2 = F(d^2)$. $[K_2 : F] = 2$. Now one may assume (*) restricts to $\sigma \otimes 1$ on $K_1 \otimes K_2 = K_1 K_2$ and repeat the argument.)

## Merkurjev-Suslin Theorem

**33.** In proving the Merkurjev-Suslin theorem one may assume that $R$ has prime exponent $p$ and has a splitting field Galois over the center with Galois group an abelian $p$-group. (Hint: By digression 7.2.55' one may assume that $R$ has a splitting field $L$ whose Galois group is a $p$-group. Thus $L$ contains a cyclic extension $F_1$ over $F$ of degree $p$. Using the corestriction it is enough to show that $R \otimes_F K$ is similar to a tensor product of cyclics of degree $p$, for some field $K$ with $[K : F]$ prime to $p$. By induction $C_R(F_1)$ is similar to a tensor product of cyclics over $F_1$. Consider such a cyclic algebra $(a, b; F_1)_p$. Then $b = f(a)$ for some $f \in F[\lambda]$ of degree $< p$; taking $K$ to be the splitting field of $f$ over $F$, one sees $[K : F]$ is prime to $p$ and thus may replace $F$ by $K$; thus one may assume $f$ splits over $F$. Write $f = (\lambda - \alpha_1) \cdots (\lambda - \alpha_t)$. Then $b = (a - \alpha_1) \cdots (a - \alpha_t)$, so $(a, b)_p$ is similar to $(a, a - \alpha_1)_p \otimes \cdots \otimes (a, a - \alpha_t)_p$. But $(a, a - \alpha_i)_p \sim (\pm \alpha_i, 1 - a^{-1} \alpha_i)_p$ by an easy computation, and this contains the cyclic field extension $F(\alpha_i^{1/p})$ of $F$; do this for each $i$ and put them together.)

**34.** If $\zeta^3 = 1$ then any symbol of degree 3 over $F(\zeta)$ is similar to the restriction of a tensor product of two cyclic algebras over $F$, by remark 7.2.52. (Hint: Take $K = F(\zeta)$.) Conclude the Merkurjev-Suslin theorem can be extended to arbitrary base fields in case $m = 3$.

## Divisibility in the Brauer Group

**35.** Suppose $\zeta$ is a primitive $m$-th root of 1. Then $F(\zeta)$ is cyclic over $F$ provided $m$ is odd or $\sqrt{-1} \in F$. (Hint: Use Jacobson [85B, theorem 7.19 and 4.20].)

**36.** Suppose $K_0$ is a cyclic field extension of $F$ with $[K_0 : F] = p^t$, and $F$ has a primitive $p$-th root $\zeta$ of 1. Then $K_0$ is contained in a cyclic field extension $K$ of $F$ with $[K : F] = p^{t+1}$, iff $\zeta = N_0(a_0)$ for some $a_0$ in $K_0$, where $N_0$ is the norm from $K_0$ to $F$. (Hint: ($\Rightarrow$) by lemma 7.3.1. ($\Leftarrow$) Let $\sigma$ be the generating automorphism of $\text{Gal}(K_0/F)$. By Hilbert's theorem 90 there is $b_0$ in $K_0$ for which $b_0^{-1} \sigma b_0 = a_0^p$. Let $K = K_0(b)$ where $b$ is a root of the polynomial $f(\lambda) = \lambda^p - b_0$. Then $\sigma f = \lambda^p - \sigma b_0$ so $\sigma$ extends to an automorphism of $K$ given by $\sigma b = \zeta^j a_0 b$. Applying $N_0$ shows $b \notin K_0$ so $[K : K_0] = p$.)

**37.** Suppose $K$ is a cyclic field extension of $F$ having a subfield $K_0$ with $[K : F] = n$ and $[K_0 : F] = m$. Then $[(K, \sigma, \alpha)]^{n/m} = [(K_0, \sigma, \alpha)]$ in $\text{Br}(F)$ where $\sigma$ generates $\text{Gal}(K/F)$. (Hint: Let $k = n/m$. It suffices to show $R = (K_0, \sigma, \alpha) \otimes (K, \sigma, \alpha^{-k})$ is split. Take $z_0$ in $(K_0, \sigma, \alpha)$ with $z_0^m = \alpha$ and $z_0 a z_0^{-1} = \sigma a$ for all $a$ in $K_0$; take $z$

in $(K, \sigma, \alpha^{-k})$ with $z^n = \alpha^{-k}$ and $zaz^{-1} = \sigma a$ for all $a$ in $K$. Let $R_1$ be the subalgebra of $R$ generated by $z_0 \otimes z$ and $1 \otimes K$; $R_1$ is split since $(z_0 \otimes z)^n = 1$. Let $K' = (K_0 \otimes K_0)^{\sigma \otimes \sigma}$ and let $R_2$ be the subalgebra of $R$ generated by $K'$ and $z_0 \otimes 1$. It suffices to prove $K'$ has $m$ orthogonal idempotents, for then $[R_2 : F] \geq m^2$ and thus $R_2 = C_R(R_1)$ and is also split, proving $R$ is split. Take the primitive idempotent $e \in K'$ as in exercise 6; $\{(\sigma^i \otimes 1)e : 1 \leq i \leq n\}$ are distinct primitive idempotents, as desired.)

38. Suppose $F$ has a primitive $p$-th root $\zeta$ of 1. $[R]$ has a cyclic $p$-th root in $\mathrm{Br}(F)$ if $R$ has a cyclic splitting field from which $\zeta$ is a norm. (Hint: One may assume $\deg(R) = q^t$ for $q$ prime, and the result is trivial unless $q = p$. If $R \sim (K_0, \sigma, \alpha)$ take $K$ as in exercise 36 and apply exercise 37.)

39. If $\zeta$ is a reduced norm from a splitting field of $R$ then $[R]$ has a $p$-th root in $\mathrm{Br}(F)$. In particular, if $F$ has a primitive $p^{t+1}$-root of 1 then the $p^t$-torsion subgroup of $\mathrm{Br}(F)$ is $p$-divisible. (Hint: Apply Merkurjev-Suslin to exercise 36.)

### $p$-Algebras

40. Suppose $L$ is cyclic over $F$ of dimension $n = p^t$. Then $L$ is contained in a field $L(y)$ cyclic over $F$ of dimension $p^{t+1}$; $y^p = y + a$ where $a \in L$ satisfies $\sigma a - a = b^p - b$ for suitable $b$ in $L$. (Hint: Parallel to exercise 36. $\mathrm{Tr}_{L/F} b = 1$ for some $b$ in $L$ by the Dedekind independence of characters, cf., Jacobson [85B, p. 297]. $\mathrm{Tr}(b^p - b) = 0$ so Hilbert's theorem 90 (additive form) yields $a$ in $L$ for which $\sigma a - a = b^p - b$. Let $y$ be a root of $\lambda^p - \lambda - a$; $\sigma$ extends to an automorphism $\tilde{\sigma}$ of $L(y)$ given by $\tilde{\sigma} y = y + b + u$ for some $u$ in $\mathbb{N}$. $\tilde{\sigma}^n y = y + 1$, so $\tilde{\sigma}$ has order $np = p^{t+1}$.)

41. Any cyclic $p$-algebra is a $p$-power of a cyclic $p$-algebra. (Hint: by exercises 37, 40.

42. If $\mathrm{char}(F) = p$ then the $p$-Sylow subgroup of $\mathrm{Br}(F)$ is $p$-divisible. (Hint: exercise 41 and Albert's theorem.)

43. In characteristic 2 prove the following: Every algebra of degree 4 is a crossed product; every algebra of degree 4 and exponent 2 is decomposable into cyclics; every algebra $R$ of degree 8 and exponent 2 is a crossed product, and $M_2(R)$ is decomposable into cyclics. (Hint: Use the computation that if $y^2 + \gamma y \in F$ then for all $r$ in $R$ one has $[y, r]y = -(y + \gamma)[y, r]$.)

### Appendix

1. Analogously to example C5 find a generic crossed product with respect to a specified Galois group $G$. (This can also be done using group rings, cf., exercises 8.2.7ff.)

2. Find a generic crossed product of exponent $\leq m$ with respect to a specified Galois group $G$.

# 8 Rings from Representation Theory

## §8.1 General Structure Theory of Group Algebras

In Chapter 1 we defined monoid algebras as a tool for studying polynomial rings. In this section we turn to the case where the monoid is a group $G \neq (1)$; $C[G]$ is then called a *group algebra* and is a $C$-algebra with base consisting of the elements of $G$. We shall concentrate on the situation for which the base ring $C$ is a field $F$ and shall study group algebras in terms of their ring-theoretic structure.

$F[G]$ is never simple since there is a surjection $F[G] \to F$ sending each $g$ in $G$ to 1. On the other hand, when $G$ is finite $F[G]$ is finite dimensional and thus Artinian and, in fact, is semisimple Artinian when $\operatorname{char}(F) = 0$ by Maschke's theorem (8.1.8' below).

Thus many interesting structural questions pertain only to infinite groups. On the other hand, the study of group algebras of finite groups is enhanced by the rich theory of characters of group representations, and we would like that theory to be available. We shall try to circumvent this paradox by outlining some of the basic character theory, emphasizing that part which will yield intuition for some of the later results. However, we shall focus on general structure theory, which does not rely much on the character theory. Structural results on Noetherian group algebras are to be had in §8.2. Most of this material follows Passman [77B], which is the standard text (as well as reference for results published through 1976). A few newer results are given in the exercises.

Throughout this section we assume that $C$ is a commutative ring, $F$ is a field, and $G$ is a group, written multiplicatively. We identify $G$ canonically as a subgroup of $C[G]$ or $F[G]$, by the injection $g \mapsto 1g$. In this sense the group algebra contains all of the information about the group. On the other hand, sometimes $\mathbb{C}[G] \approx \mathbb{C}[H]$ even when $G \not\approx H$.

**Remark 8.1.1:** If $R$ is a $C$-algebra and $H$ is a multiplicative subgroup of $R$ then any group homomorphism $f: G \to H$ extends uniquely to an algebra homomorphism $C[G] \to R$ given by $\sum c_g g \mapsto \sum c_g fg$. (This is a special case of proposition 1.3.16′.)

In particular, any group homomorphism $G \to H$ extends to an algebra homomorphism $C[G] \to C[H]$, thereby yielding a functor $\mathcal{G}\imath\rho \to C\text{-}\mathcal{A}\ell g$ given by $G \to C[G]$. On the other hand, the functor $Unit: C\text{-}\mathcal{A}\ell g \to \mathcal{G}\imath\rho$, sending $R$ to its multiplicative group of invertible elements (and restricting homomorphisms), also enters into the picture.

**Remark 8.1.2:** If $G$ is a group then $C[G]$, together with the canonical injection $G \to C[G]$, is a universal from $G$ to the functor $Unit$. (Indeed, any group homomorphism $f: G \to Unit(R)$ extends uniquely to an algebra homomorphism $\hat{f}: C[G] \to R$, which restricts, in turn, to a group homomorphism $Unit(C[G]) \to Unit(R)$; this completes the appropriate diagram and is unique.)

In view of these observations it makes sense to transfer some terminology from algebras to groups, and in definition 5.2.12 we introduced $G$-modules; we see at once that any $G$-module becomes a $\mathbb{Z}[G]$-module under the action $\left(\sum_g n_g g\right)x = \sum n_g(gx)$, and thus $G\text{-}\mathcal{M}od$ and $\mathbb{Z}[G]\text{-}\mathcal{M}od$ are isomorphic categories.

To proceed further we shall consider some of the basic structural properties of groups. Some minor ambiguities arise since the same terminology has a different connotation for groups than for rings and will depend on the context. In particular,

$N \lhd G$ denotes $N$ is a normal subgroup of $G$.

$[g_1, g_2]$ will denote the *group* commutator $g_1 g_2 g_1^{-1} g_2^{-1}$.

The *center* $Z(G) = \{z \in G : [z, G] = (1)\}$.

The subgroup of $G$ generated by a subset $S$ is denoted $\langle S \rangle$ or as $\langle g_1, \ldots, g_m \rangle$ if $S = \{g_1, \ldots, g_m\}$.

$C_G(A)$ denotes the *centralizer* of a subset $A$ of $G$, which is $\{g \in G: [g, A] = (1)\}$; recall for any $g$ in $G$ that $[G: C_G(g)]$ is the number of conjugates of $g$.

Recall $\mathrm{supp}(\sum c_g g) = \{g : c_g \neq 0\}$.

The order of an element $g$ in $G$ is also called the *period* of $g$; $G$ is called a $p'$-*group* if no element has period $p$. It turns out that many of the structural results hold for $F[G]$ when $\mathrm{char}(F) = 0$ or when $p = \mathrm{char}(F) > 0$ and $G$ is a $p'$-group. We shall unify these two situations by means of the terminology, "$G$ is a $\mathrm{char}(F)'$-group".

## An Introduction to the Zero-Divisor Question

One of the most elementary questions one can ask about $F[G]$ is whether or not it has zero-divisors. If $g \in G$ has period $n$ then $0 = 1 - g^n = (1 - g) \cdot (1 + g + \cdots + g^{n-1})$, so $C[G]$ certainly is *not* a domain. What about the torsion-free case?

*Questions for Torsion-Free Groups:*

(i) *Zero-divisor question.* Is $F[G]$ necessarily a domain?

(ii) Failing (i), can $F[G]$ have nontrivial idempotents?

Amazingly, these questions are still open, although $F[G]$ is now known to be a domain in many cases, to be discussed in §8.2. When $G$ is abelian the answer is easy.

**Proposition 8.1.3:** *If $G$ is torsion-free abelian and $C$ is an integral domain then $C[G]$ is an integral domain.*

*Proof:* Suppose $r_1 r_2 = 0$ for $r_1, r_2$ in $C[G]$. Replacing $G$ by the subgroup generated by $\mathrm{supp}(r_1) \cup \mathrm{supp}(r_2)$, we may assume $G$ is finitely generated abelian, so $G \approx \langle g_1 \rangle \times \cdots \times \langle g_t \rangle$ for suitable $g_i$ in $G$. But now $C[G]$ can be identified with a subring of the Laurent series ring $C((G))$, which is a domain.
$$\text{Q.E.D.}$$

An alternate conclusion to the previous proof is by the following convenient observation.

**Remark 8.1.4:** $C[G_1 \times G_2] \approx C[G_1] \otimes C[G_2]$. (Indeed, define $\varphi: C[G_1] \times C[G_2] \to C[G_1 \times G_2]$ by $\varphi(\sum c_g g, \sum c'_h h) = \sum c_g c'_h (g, h)$. Clearly $\varphi$ is a balanced map which induces a surjection $C[G_1] \otimes C[G_2] \to C[G_1 \times G_2]$, whose inverse is given by $\sum c_{(g,h)}(g, h) \mapsto \sum c_{(g,h)} g \otimes h$.

## *Maschke's Theorem and the Regular Representation*

Group algebras have an extra piece of structure.

**Remark 8.1.5:** Suppose $C$ has a given automorphism $\sigma$ of degree 2. (In particular, one could take $\sigma = 1$). Then $C[G]$ has an involution given by $\left(\sum_{g \in G} c_g g\right)^* = \sum (\sigma c_g) g^{-1}$.

**Proposition 8.1.6:** *Suppose $C$ is a subring of $\mathbb{C}$ closed under complex conjugation, and take* (∗) *as in remark 8.1.5. If $a \in C[G]$ and $a^*a = 0$ then $a = 0$. In particular, $C[G]$ is semiprime.*

**Proof:** Write $a = \sum a_g g$. The coefficient of 1 in $a^*a$ is $0 = \sum_g \bar{a}_g a_g = \sum |a_g|^2$ so each $a_g = 0$. Hence $a = 0$. It follows that if $A \lhd (C[G], *)$ with $0 = A^2 = A^*A$ then $A = 0$, so $C[G]$ is semiprime by the easy proposition 2.13.34.     Q.E.D.

**Corollary 8.1.7:** *(Maschke's theorem for $\mathbb{C}$) $\mathbb{C}[G]$ is semisimple Artinian for any finite group G.*

**Proof:** $\mathbb{C}[G]$ is semiprime and Artinian.     Q.E.D.

Maschke's theorem is the fundamental theorem in the subject of group algebras. Let us try to generalize it in various directions. Using field theory we shall first show that the main hypothesis in proposition 8.1.6 is superfluous. For that proof to work, the only property required of $\mathbb{C}$ was that $\mathbb{C} = K[\sqrt{-1}]$ for a suitable field $K$ in which 0 cannot be written as a sum of nonzero squares; in the proof we took $K = \mathbb{R}$. We shall now see that *all* algebraically closed fields of characteristic 0 have this property.

**Theorem from the Artin-Schreier Theory of Real-Closed Fields:** *Any algebraically closed field $E$ of characteristic 0 has a subfield $K$ satisfying the following properties:*

(1) *Any sum of nonzero squares is a nonzero square.*
(2) $E = K[i]$ *where $i^2 = -1$.*

**Proof:** Obviously $-1$ is not a sum of squares in $\mathbb{Q}$; so by Zorn's lemma, $E$ has a subfield $K$ maximal with respect to $-1$ is not a sum of squares. By Lang [65B, theorem 11.2.1] $K$ is real closed; in particular, every sum of nonzero squares in $K$ is a nonzero square in $K$, and $K[i]$ is algebraically closed. It remains to show $E$ is an algebraic extension of $K$, for then we shall have $K[i] = E$.

Suppose $a \in E - K$. Then $K(a) \supset K$ so

$$-1 = \sum (p_i(a)/q(a))^2$$

for $p_i, q$ in $K[\lambda]$ taken with $\deg q$ minimal. Thus $-q(a)^2 = \sum p_i(a)^2$, so $q(a)^2 + \sum p_i(a)^2 = 0$.

We claim the polynomial $q(\lambda)^2 + \sum p_i(\lambda)^2$ is nonzero. Otherwise, taking the constant terms $p_{i0}, q_0$ of $p_i, q$ we get $0 = q_0^2 + \sum p_{i0}^2$ in $K$, implying each $p_{i0} = q_0 = 0$. But then we can cancel $\lambda$ from $q$ and from each $p_i$, contrary to supposition, so the claim is established. It follows at once that $a$ is algebraic over $K$, as desired.     Q.E.D.

**Lemma 8.1.8:**  *If $F$ is algebraically closed of characteristic 0 then $F[G]$ has no nil left ideal $L \neq 0$.*

**Proof:**  Otherwise take $0 \neq a \in L$. As seen above we can write $F = F_0[i]$ where $F_0$ is real closed. Then the proof of proposition 8.1.6 is available (for $C = F_0$), implying $b = a^*a \neq 0$, so $b^2 = b^*b \neq 0$, and by iteration each power of $b \neq 0$, contrary to $L$ being nil.     Q.E.D.

**Theorem 8.1.8':**  *(Maschke's theorem in characteristic 0 for arbitrary groups). If $\mathrm{char}(F) = 0$ then $F[G]$ is semiprime. In particular, if $G$ is finite then $F$ is semisimple Artinian.*

**Proof:**  Let $\bar{F}$ be the algebraic closure of $F$. Then $\bar{F}[G] \approx F[G] \otimes \bar{F}$ has no nonzero nilpotent ideals, so $F[G]$ has no nonzero nilpotent ideals.
                                                                                Q.E.D.

Next we aim for an alternate proof of Maschke's theorem, which also deals with characteristic $p$, by looking at conjugacy classes.

**Proposition 8.1.9:**  *If $h \in G$ has only a finite number of conjugates $h_1, \ldots, h_t$ then $\sum_{i=1}^{t} h_i \in Z(C[G])$, and conjugation by any $g$ in $G$ permutes the conjugates.*

**Proof:**  Let $h_{\pi i} = g h_i g^{-1}$, a conjugate of $h$. Clearly $\pi i \neq \pi j$ for $i \neq j$, so $\pi$ permutes the subscripts. Write $a = \sum_{i=1}^{n} h_i$. Then $gag^{-1} = \sum_{i=1}^{n} h_{\pi i} = \sum h_i = a$, showing $a$ commutes with each $g$ in $G$; thus $a \in Z(C[G])$.
                                                                                Q.E.D.

This leads us to the following important definition.

**Definition 8.1.10:**  $\Delta(G) = \{g \in G : g$ has only a finite number of conjugates$\} = \{g \in G : C_G(g)$ has finite index in $G\}$.

**Remark 8.1.11:**  $\Delta(G)$ is a characteristic subgroup of $G$, and $\Delta(\Delta(G)) = \Delta(G)$. More generally, if $g_i$ has $n_i$ conjugates for $i = 1, 2$ then $g_1^{-1}$ has $n_1$ conjugates and $g_1 g_2$ has at most $n_1 n_2$ conjugates (seen by observing $g(g_1 g_2)g^{-1} = (g g_1^{-1} g)(g g_2^{-1} g))$.

**Proposition 8.1.12:**  (*Converse to proposition 8.1.9*) *Suppose* $z \in Z(C[G])$). *If* $h \in \operatorname{supp}(z)$ *then every conjugate of $h$ is in* $\operatorname{supp}(z)$, *so* $h \in \Delta(G)$. *Consequently,* $Z(C[G])$ *is a free C-module whose base consists of all elements of the form* $\sum\{all\ conjugates\ of\ h\}$ *where* $h \in \Delta(G)$; *in particular,* $Z(C[G]) \subseteq C[\Delta(G)]$.

**Proof:**  Write $z = \sum c_h h$. For any $g$ in $G$ we have $\sum c_h h = z = gzg^{-1} = \sum c_h g h g^{-1}$, implying every conjugate of $h$ appears in $\operatorname{supp}(z)$ with the same coefficient. Since $\operatorname{supp}(z)$ is finite this proves the first assertion. Moreover, writing $z_h$ for $\sum\{$conjugates of $h\}$ we can rewrite $z = \sum c_h z_h$, summed over representatives of the conjugacy classes in $\operatorname{supp}(z)$. Thus all the $z_h$ span $Z(C[G])$ and are independent because distinct conjugacy classes are disjoint.    Q.E.D.

**Corollary 8.1.13:**  *Suppose $H$ is a finite subgroup of $G$, and $r = \sum_{h \in H} h$. Then $r^2 = |H|r$; moreover, $r \in Z(C[G])$ iff $H \lhd G$.*

**Proof:**  For each $h$ in $H$ we have $rh = \sum_{h' \in H} h' h = \sum_{h' \in H} h' = r$, so $r^2 = \sum_{h \in H} rh = \sum_{h \in H} r = |H|r$. The other assertion follows at once from propositions 8.1.9 and 8.1.12.    Q.E.D.

The element $\sum_{h \in H} h$ used in this proof has fundamental significance in the theory of group algebras over finite groups. To apply this result, suppose $|G| = n$. Recall that there is a group injection $\operatorname{Sym}(n) \to \operatorname{GL}(n, F)$ sending a permutation $\pi$ to the *permutation matrix* $\sum_{i=1}^{n} e_{\pi i, i}$; composing this with the group injection $G \to \operatorname{Sym}(n)$ given by Cayley's theorem yields a (faithful) representation $G \to \operatorname{GL}(n, F)$ called the *regular representation*, which is the restriction to $G$ of the regular representation of $F[G]$ as $F$-algebra. Writing $G = \{g_1, \ldots, g_n\}$ we see that the matrix $\hat{g}$ corresponding to $g$ is $\sum e_{\pi i, i}$ where $g g_i = g_{\pi i}$. Note $\operatorname{tr}(\hat{g}) = 0$ if $g \neq 1$, since then $g g_i \neq g_i$ for all $i$; $\operatorname{tr}(1) = n$.

**Theorem 8.1.14:** (*Maschke's theorem for arbitrary fields*) *Suppose* $|G| = n$. $\mathrm{Nil}(F[G]) = 0$ *iff* $\mathrm{char}(F) \nmid n$, *in which case* $F[G]$ *is semisimple Artinian.*

**Proof:** ($\Rightarrow$) Let $z = \sum_{g \in G} g$. Then $0 \neq z \in Z(F[G])$, so $0 \neq z^2 = |G|z = nz$, implying $\mathrm{char}(F) \nmid n$.

($\Leftarrow$) Take $r = \sum_{g \in G} \alpha_g g$ in $\mathrm{Nil}(F[G])$. For any $h$ in $\mathrm{supp}(r)$ we have $h^{-1}r = \alpha_h 1 + \sum_{g \neq h} \alpha_g h^{-1} g$, so (taking traces via the regular representation) we get

$$\mathrm{tr}(h^{-1}r) = \mathrm{tr}(\alpha_h 1) + \sum_{g \neq h} \alpha_g \mathrm{tr}(h^{-1}g) = n\alpha_h + 0 = n\alpha_h.$$

But $h^{-1}r$ is nilpotent so $\mathrm{tr}(h^{-1}r) = 0$. Thus $n\alpha_h = 0$ implying $\alpha_h = 0$. This proves $r = 0$, as desired.

The last assertion is clear since $F[G]$ is Artinian.          Q.E.D.

## Group Algebras of Subgroups

Later on we shall consider the question of when a group algebra satisfies a polynomial identity. For the present we record one easy result which places much of the PI-theory at our disposal. Recall a *transversal* of a subgroup $H$ in $G$ is a set of representatives of the left cosets of $H$ in $G$.

**Remark 8.1.15:** Any transversal $B$ of $H$ is a $C[H]$-base for $C[G]$; if $H \lhd G$ then $B$ is $C[H]$-normalizing. (Indeed $G = \bigcup_{b \in B} Hb$ implying $C[G] = C[H]B$, proving $B$ spans. To prove $B$ is a base suppose $\sum r_i b_i = 0$ where $r_i = \sum c_{ih} h \in C[H]$ and $b_i \in B$; then $0 = \sum c_{ih} h b_i$ so each $c_{ih} = 0$ implying $r_i = 0$. The last assertion is clear.)

This has an immediate application.

**Proposition 8.1.16:** *If* $[G:H] = n$ *then there is a canonical injection* $C[G] \to M_n(C[H])$ *since* $C[G]$ *is a free* $C[H]$-*module of dimension* $n$; *in particular, if* $G$ *has an abelian subgroup* $H$ *of index* $n$ *then* $C[G]$ *satisfies all polynomial identities of* $n \times n$ *matrices.*

**Proof:** The first assertion follows from remark 8.1.15 by means of corollary 1.5.14; the second assertion is then immediate since $C[H]$ is abelian.
                                                                        Q.E.D.

## Augmentation Ideals and the Characteristic $p$ Case

It remains to consider the case where $\operatorname{char}(F) = p$ and $G$ has $p$-torsion. The situation becomes much more complicated, although the answer is very neat in one particular case. If $H \lhd G$ the group homomorphism $G \to G/H$ extends to a unique $C$-algebra surjection $\rho_H : C[G] \to C[G/H]$. In particular, the surjection $\rho_G : C[G] \to C$ is given by $\rho(\sum c_g g) = \sum c_g$.

**Definition 8.1.17:**   The *augmentation ideal* $\omega(F[G])$ is

$$\ker \rho_G = \{\textstyle\sum c_g g : \sum c_g = 0\}.$$

**Remark 8.1.18:**   If $F$ is a field then $\omega(F[G])$ is a maximal ideal since $F[G]/\omega(F[G]) \approx F$. More generally, if $\operatorname{char}(F) = p$ and $G/H$ is a finite $p'$-group then $\ker \rho_H$ is an intersection of a finite number of maximal ideals, by Maschke's theorem.

**Remark 8.1.19:**   $\omega(C[G]) = \sum_{g \in G} C(g - 1)$. (Indeed $(\supseteq)$ is clear; conversely, if $r = \sum c_g g \in \omega(C[G])$ then $r = \sum c_g g - \sum c_g = \sum c_g(g - 1)$.)

**Theorem 8.1.20:**   *Suppose $G$ is a finite $p$-group and $\operatorname{char}(F) = p$. Then $F[G]$ is a local ring whose Jacobson radical is nilpotent.*

**Proof:**   For any $g$ in $G$ we have $g^q = 1$ for a suitable power $q$ of $p$, so $(g - 1)^q = g^q - 1 = 0$. Thus $\omega(F[G]) = \sum F(g - 1)$ is spanned by nilpotent elements and thus is nilpotent by proposition 2.6.32. We are done since $\omega(F[G])$ is a maximal ideal.     Q.E.D.

One can actually use the augmentation ideal to compute $\ker \rho_H$ for $H \lhd G$.

**Proposition 8.1.21:**   $\ker \rho_H = \omega(C[H])C[G] = C[G]\omega(C[H])$.

**Proof:**   By symmetry we need only show the first equality. Let $\{g_i : i \in I\}$ be a transversal and write $\bar{G} = G/H$. By remark 8.1.15 any $r$ of $C[G]$ can then be written as $\sum r_i g_i$ where $r_i \in C[H]$, and we have $\rho_H r = \sum (\rho_H r_i)\bar{g}_i$. Thus $r \in \ker \rho_H$ iff each $\rho_H r_i = 0$ in $C$, i.e., iff each $r_i \in \omega(C[H])$. This proves $\ker \rho_H \subseteq \omega(C[H])C[G]$, and the reverse inclusion is clear.     Q.E.D.

## Major Digression: The Structure of Group Algebras of Finite Groups

Starting with Maschke's theorem one can branch off to a detailed study of the structure of $F[G]$ for $G$ finite. This subject is of utmost significance in the study of finite groups as well as other branches of algebra, but is so extensively treated in the literature that any presentation here would be inadequate and superfluous. Nevertheless, since we need the intuition to be had from that theory, we shall recall certain aspects of the theory, our goal being to indicate how the character theory can be used to study the structure of the group ring. The reader should consult Jacobson [80B] for a beautiful summary or Curtis-Reiner [62B, 82B] for a thorough account of the representation theory. The best source of results on characters which apply to the group algebra structure may be Isaacs [76B]. For the remainder of this digression $G$ is finite and $\operatorname{char}(F) = 0$. We already have the following information about the semisimple Artinian ring $R = F[G]$:

(i) The number of nonisomorphic simple $R$-modules equals the number of simple components of $R$, by theorem 2.3.13.

(ii) The number of conjugacy classes of $G$ is $[Z(R):F]$, by proposition 8.1.12.

(iii) These numbers are all the same iff the center of each simple component of $R$ is canonically isomorphic to $F$.

Of course, if $R$ is split, i.e., a direct product of matrix algebras over $F$, then (iii) holds; in this important case we call $F$ a *splitting field* for $G$. We shall return to splitting fields shortly.

The key to investigating $F[G]$ is the *group representation of degree n*, by which we mean a group homomorphism $\rho: G \to \operatorname{GL}(n, F)$; sometimes it is convenient to extend $\rho$ to an $F$-algebra homomorphism $F[G] \to M_n(F)$. There is a fundamental correspondence between representations and $F[G]$-modules which are finite dimensional over $F$. Suppose $V$ is an $F[G]$-module and $[V:F] = n$. Left multiplication by any $g$ in $G$ yields a linear transformation $\rho_g: V \to V$. The map $g \mapsto \rho_g$ yields a homomorphism $G \to \operatorname{Aut}_F V \approx \operatorname{GL}(n, F)$, which is a representation of degree $n$. Conversely, viewing $M_n(F)$ as $\operatorname{End}_F V$ for a vector space $V$ over $F$, we can use any representation $\rho$ to study $V$ naturally as a $G$-module or thus as an $F[G]$-module. So the representation theory is part of the module theory of $F[G]$; in particular, we say $\rho$ is *irreducible* if the corresponding $F[G]$-module $V$ is simple. In this case $D = \operatorname{End}_{F[G]} V$ is a division algebra by Schur's lemma.

If $\rho$ is a representation of degree $n$ then $\rho 1$ is the $n \times n$ identity matrix so $\mathrm{tr}(\rho 1) = n$. This leads us to define the *character* $\chi$ *afforded by* $\rho$ to be the function $\chi: G \to F$ given by $\chi g = \mathrm{tr}(\rho g)$. Characters are much easier to handle than the representations which afford them, but still carry much of the information contained by the representation; consequently they are the main tool of representation theory.

*Crucial Remark:* If $g \in G$ has order $m$ then the minimal polynomial of $g$ divides $\lambda^m - 1$, so the characteristic values of $\rho g$ must be $m$-th roots of 1. Thus $\chi g$ is a sum of $m$-th roots of 1.

*Examples:*

(i) If $\chi 1 = 1$ we say $\chi$ is a *linear character*; in this case $\chi = \rho: G \to F$ is a group homomorphism,

(ii) The *unit character* is the linear character given by $\chi g = 1$ for all $g$ in $G$.

(iii) The regular representation has degree $|G|$ and affords the character $\chi$ given by $\chi 1 = |G|$ and $\chi g = 0$ for all $g \neq 1$.

Until further notice assume $F$ is a splitting field for $G$, and let $t$ be the number of conjugacy classes of $G$. Thus $F[G] \approx \prod_{i=1}^{t} M_{n_i}(F)$ for suitable $n_i$, and a dimension count shows $\sum_{i=1}^{t} n_i^2 = |G|$. *Note that the ring structure of $F[G]$ is completely determined by the $n_i$.*

Let us describe the commutative part of $F[G]$. If $H \lhd G$ then the map $G \to G/H$ yields a surjection $F[G] \to F[G/H]$, so $F[G/H]$ is isomorphic to the direct product of some of the simple components of $F[G]$. Let $G'$ be the commutator subgroup. Then $G/G'$ is abelian so $F[G/G']$ is commutative. We claim $F[G/G']$ is isomorphic to the direct product of *all* of the commutative components of $F[G]$. Indeed, we may pass to a splitting field and assume $F$ is a splitting field. But then each commutative component of $F[G]$ corresponds to a linear character $\chi: G \to F$; since $\ker \chi \supseteq G'$ we see $\chi$ factors through $G \to G/G' \to F$, yielding a linear character of $G/G'$ as desired.

Let $\rho_1, \ldots, \rho_t$ correspond to a maximal set of nonisomorphic simple $F[G]$-modules, which thus are a maximal set of inequivalent irreducible representations. Let $\chi_i$ denote the character afforded by $\rho_i$. Then $\deg(\chi_i) = n_i$, so the characters yield the ring structure of $F[G]$. Furthermore, the $\chi_i$ are distinct. Indeed letting $e_i = \rho_i 1$ we have

$$\chi_j(e_i g) = \mathrm{tr}\, \rho_j(e_i g) = \mathrm{tr}\, \delta_{ij} e_j \rho_j g = \delta_{ij} \chi_j g,$$

so taking $g_i$ such that $\chi_i g_i \neq 0$ we see $\chi_i(e_i g_i) = \chi_i g_i \neq 0$ but $\chi_j(e_i g_i) = 0$ for $i \neq j$.

A fairly straightforward computation enables one to prove that when $F = \mathbb{C}$ there is a Hermitian form on characters, given by

$$\langle \chi', \chi'' \rangle = \frac{\left( \sum_{g \in G} \chi' g \overline{\chi'' g} \right)}{|G|},$$

cf., Jacobson [80B, pp. 272–275].

**Brauer's Theorem on Splitting Fields [80B, p. 314]:** *Suppose G has exponent m. Then any field F in which the polynomial $\lambda^m - 1$ splits is a splitting field for G.*

*Note:* The proof in characteristic $p$ is actually quite trivial: Let $F_1$ be the splitting field of $\lambda^m - 1$ over the characteristic subfield of $F$, a finite field. Then every simple component $R_i$ of $F_1[G]$ has the form $M_{n_i}(K_i)$ for a suitable field $K_i$, by Wedderburn's theorem that finite division rings are fields. But for each $a$ in $R_i$, $\operatorname{tr} a_i$ is a sum of roots of 1 and thus belongs to $F_1$, proving $K_i = F_1$, as desired.

Note $\chi g = n$ iff $\rho g = 1$, and $|\chi g| = n$ iff $\rho g$ is a scalar. This is seen by passing to the algebraic closure of $F$, over which $\rho g$ is conjugate to a diagonal matrix whose entries clearly are roots of 1. We have the following two normal subgroups of $G$: $\ker \chi = \{ g \in G : \chi g = n \} = \ker \rho$ and $Z(\chi) = \{ g \in G : |\chi g| = n \}$.

**Proposition:** *Each $n_i^2 \leq [G : Z(\chi_i)]$, equality holding iff $\chi_i g = 0$ for all $g \notin Z(\chi_i)$.*

*Proof:* $|G| = \sum_{g \in G} |\chi_i g|^2 = n_i^2 |Z(\chi_i)| + \sum \{ |\chi_i g|^2 : g \notin Z(\chi_i) \}$.    Q.E.D.

**Corollary:** *If $G/Z(\chi_i)$ is abelian then $|G : Z(\chi_i)| = n_i^2$.*

*Proof:* One must show $\chi_i g = 0$ for all $g \notin Z(\chi_i)$. But then $\rho_i g$ is not a scalar matrix so there is $h$ in $G$ such that the group commutator $[\rho_i g^{-1}, \rho_i h] \neq 1$ so $z = [g^{-1}, h] \notin \ker \chi_i$. But $z \in Z(\chi_i)$ by hypothesis so $\rho_i z = \zeta 1$ for some $\zeta \neq 1$. Thus $\chi_i g = \chi_i (hgh^{-1}) = \chi_i gz = \zeta \chi_i g$ implying $\chi_i g = 0$.    Q.E.D.

Finally, we would like to quote a theorem of Schur (Jacobson [80B, p. 283]), that the degree of each irreducible character divides $[G : Z(G)]$. With these results in hand we can determine quite a few group algebras.

*Examples (letting $Z = Z(G)$):* Recall from group theory that if $G$ is a $p$-group $\neq 1$ then $Z \neq (1)$ and $[G : Z] \neq p$. In particular, if $|G| = p^2$ then $G$ is abelian. Assume $\operatorname{char}(F) = 0$ for simplicity.

(i) $G$ is abelian. Remark 8.1.4 reduces the structure of $F[G]$ to the case $G$ is cyclic. Write $G = \langle g \rangle$ where $g^n = 1$. If $F$ has a primitive $n^{\text{th}}$ root $\zeta$ of 1 then $F[G] \approx F^{(n)}$ where the $j^{\text{th}}$ component is generated by the idempotent $\sum_{i=1}^{n} \zeta^{ij} g^i$. At the other extreme let us determine $\mathbb{Q}[G]$. Write $F_m = \mathbb{Q}[\zeta^{n/m}]$ for all $m \mid n$. Then $[F_m : \mathbb{Q}] = \varphi(m)$ where $\varphi$ is Euler's $\varphi$ function. Moreover, for each $m \mid n$ there is a surjection $F[G] \to F_m$ given by the group map $g \to \zeta^{n/m}$, so $F_m$ is a component of $F[G]$. Counting dimensions shows $\mathbb{Q}[G] \approx \prod_{m \mid n} F_m$. In particular, if $n$ is prime then $\mathbb{Q}[G] \approx \mathbb{Q} \times \mathbb{Q}[\zeta]$.

(ii) Suppose $|G| = p^3$ for $p$ prime and $G$ is *not* abelian. Then $|G/Z| = p^2$, so $G/Z$ is abelian. Thus $G' = Z$ so $G$ has $|G/G'| = p^2$ linear characters. Any other irreducible character $\chi_i$ has $n_i^2 = [G : Z(\chi_i)] \mid p^3$ implying $n_i = p$. Now $p^3 = |G| = \sum n_i^2 = p^2 \cdot 1 + (t - p^2)p^2$, whence $p = 1 + t - p^2$, implying $t = p^2 + p - 1$. This implies $F[G] \approx F^{(p^2)} \times M_p(F)^{(p-1)}$ for any splitting field $F$ of $G$.

For $p = 2$ there are two nonisomorphic nonabelian groups of order 8, namely the dihedral group $D_4$ and the quaternion group $Q_8$. Each has exponent 4 and so by Brauer's theorem is split by $\mathbb{Q}[i]$; their group algebras over $\mathbb{Q}[i]$ are isomorphic by the last paragraph. Let us compare their group algebras over $\mathbb{Q}$. In each case $G/G' \approx \mathbb{Z}_2 \times \mathbb{Z}_2$, and $\mathbb{Q}[\mathbb{Z}_2] \approx \mathbb{Q}^{(2)}$ by (i), so $\mathbb{Q}[G/G'] \approx \mathbb{Q}^{(2)} \otimes_{\mathbb{Q}} \mathbb{Q}^{(2)} \approx \mathbb{Q}^{(4)}$, which is the commutative part of $\mathbb{Q}[G]$. There is room only for one more component, of dimension 4 over $\mathbb{Q}$.

For $G = D_4 = \langle a, b : a^4 = 1 = b^2$ and $b^{-1}ab = a^3 \rangle$, we have a surjection $\mathbb{Q}[G] \to M_2(\mathbb{Q})$ given by $a \mapsto \begin{pmatrix} 0 & 1 \\ -1 & 0 \end{pmatrix}$ and $b \mapsto \begin{pmatrix} 1 & 0 \\ 0 & -1 \end{pmatrix}$; hence $M_2(\mathbb{Q})$ is one of the components of $\mathbb{Q}[D_4]$, which thus is $M_2(\mathbb{Q}) \times \mathbb{Q}^{(4)}$. Thus $\mathbb{Q}$ is a splitting field for $D_4$.

For $G = Q_8 = \langle a, b : a^4 = 1 = b^4, a^2 = b^2$, and $b^{-1}ab = a^3 \rangle$, we take Hamilton's quaternion algebra $\mathbb{H} = \mathbb{Q} \oplus \mathbb{Q}i \oplus \mathbb{Q}j \oplus \mathbb{Q}k$ where $ij = -ji = k$ and $i^2 = j^2 = k^2 = -1$; there is a surjection $\mathbb{Q}[G] \to \mathbb{H}$ given by $a \mapsto i$ and $b \mapsto j$, so $\mathbb{Q}[Q_8] \approx \mathbb{H} \times \mathbb{Q}^{(4)}$.

(iii) $G$ is the Dihedral group $D_m = \langle a, b : a^m = 1 = b^2$ and $b^{-1}ab = a^{m-1} \rangle$. (We treated $m = 4$ in (ii).) Then $[b, a^j] = a^{-2j}$ so $\langle a^2 \rangle \subseteq G'$; on the other hand, $G/\langle a^2 \rangle$ is abelian so $G' = \langle a^2 \rangle$ has index $\leq 4$ in $G$. $\langle a \rangle$ has index 2 in $G$ so by proposition 8.1.16 we see $F[G]$ has PI degree $\leq 2$, implying each component $R_i$ has degree $\leq 2$. Passing to the splitting field and counting linear characters, we see each $R_i$ is $F$-central, so $[R_i : F] = 1$ or 4.

If $m$ is even then $\langle a^2 \rangle \neq \langle a \rangle$ so $[G : G'] = 4$. This implies $4(t - 4) + 4 = 2m$, i.e., $t = m/2 + 3$, which is usually proved by describing the conjugacy classes explicitly. Thus $F[G] \approx F^{(4)} \times \prod_{i=1}^{m/2-1} R_i$ for quaternion $F$-algebras $R_i$.

If $m$ is odd then $\langle a^2 \rangle = \langle a \rangle$ so $G/G'' \approx \mathbb{Z}_2$ and $F[G/G'] \approx F^{(2)}$; as before, we see $F[G] \approx F^{(2)} \times \prod_{i=1}^{(m-1)/2} R_i$ where each $R_i$ is a quaternion $F$-algebra.

One can display the complex representations of $D_m$ explicitly (cf., Jacobson [80B, p. 261]), and show that many of the characters of degree 2 have values not in $\mathbb{Q}$ whenever $m > 3$. Thus $\mathbb{Q}$ is *not* a splitting field for $D_m$ for $m \geq 3$. However, every finite field whose characteristic does not divide $2m$ is a splitting field for $D_m$, by the characteristic $p$ proof of Brauer's theorem given above. Likewise, every finite field of characteristic $\neq 2$ is a splitting field for $Q_8$.

(iv) $G = \text{Sym}(n)$. There is an extensive literature on the representations. $\mathbb{Q}$ is a splitting field, and to each representation $\rho$ there corresponds a *partition of* $n$, by $a_1 \geq a_2 \geq \cdots \geq a_k$ with $\sum a_i = n$. The *Young-Frobenius* theorem states

$$\deg(\rho) = \left( n! \prod_{i>j}(b_i - b_j) \right) \bigg/ \prod_{i=1}^{k} b_i! \qquad \text{where } b_i = a_i + k - i.$$

(v) Suppose $[G:Z(G)] = p^3$ for $p$ prime. By the earlier results, for any irreducible character $\chi_i$ of degree $n_i$, we have $n_i^2 \leq p^3$ and $n_i \mid p^3$. Thus $n_i \in \{1, p\}$, so $F[G] \approx F^{(j)} \times M_p(F)^{t-j}$ for any splitting field $F$, where $j = [G:G']$.

## The Kaplansky Trace Map and Nil Left Ideals

To generalize theorem 8.1.14 to infinite groups we need a trace map which does not depend on matrices.

**Definition 8.1.22:** Define the (Kaplansky) *trace map* $\text{tr}: C[G] \to C$ for any group $G$ and any commutative ring $C$, by $\text{tr} \sum c_g g = c_1$. (For finite groups this differs from the trace of the regular representation by a factor of $|G|$.) For the remainder of this section tr *always* denotes the Kaplansky trace map.

**Remark 8.1.23:** tr is a trace map according to definition 1.3.28. (Indeed

$$\text{tr}\left(\sum c_g g\right)\left(\sum c'_g g\right) = \sum_{gh=1} c_g c'_h = \sum_{hg=1} c'_h c_g = \text{tr}\left(\sum c'_g g\right)\left(\sum c_g g\right).)$$

Consequently, every group algebra has IBN by proposition 1.3.29.

The following computation helps in characteristic $p$.

**Lemma 8.1.24:** *Suppose* $r = \sum c_g g \in C[G]$. *If* $\text{char}(C) = p$ *is prime and* $q$ *is a power of* $p$ *then* $\text{tr}(r^q) = \sum \{c_g^q : g^q = 1\}$.

***Proof:***   It is enough to show $\operatorname{tr}(r^q) = \operatorname{tr}\left(\sum c_g^q g^q\right)$. Expanding $\left(\sum c_g g\right)^q$ yields terms of the form $c_1 \cdots c_q g_1 \cdots g_q$, where $c_i$ is written in place of $c_{g_i}$. When two $g_i$ are distinct we also have $c_1 \cdots c_q g_j \cdots g_q g_1 \cdots g_{j-1}$ for each $j \le q$. Since $\operatorname{tr}(g_j \cdots g_q g_1 \cdots g_{j-1}) = \operatorname{tr}(g_1 \cdots g_q)$ and $\operatorname{char} C = p$ we see that the terms in $\operatorname{tr}(r^q)$ drop out except those from $\operatorname{tr}\left(\sum c_g^q g^q\right)$, as desired.

**Proposition 8.1.25:**   *Suppose C is a domain. For any nilpotent r in C[G] we have* $\operatorname{tr}(r) = 0$ *provided the following condition is satisfied:*

$\operatorname{char}(C) = p \ne 0$ *and* supp(r) *does not have a nonidentity element whose period is a power of* p.

***Proof:***   Take a large enough power $q$ of $p$ such that $r^q = 0$. Writing $r = \sum c_g g$ we have (by hypothesis) $g^q \ne 1$ for every $g \ne 1$ in supp(r), so $0 = \operatorname{tr}(0) = \operatorname{tr}(r^q) = \sum\{c_g^q \cdot g^q = 1\} = c_1^q$, implying $0 = c_1 = \operatorname{tr}(r)$.    Q.E.D.

***Remark 8.1.25′:***   The conclusion of proposition 8.1.25 also holds in characteristic 0. Indeed, suppose, on the contrary, $r \ne 0$ is nilpotent with $\operatorname{tr}(r) \ne 0$. Then $r^m = 0$ for some $m$. Write $r = \sum_{i=1}^{t} c_i g_i$. The fact $r^m = 0$ can be expressed as a polynomial equality $f(c_1, \dots, c_t) = 0$. Taking a language containing a constant symbol for each element of $G$, we can appeal to remark 2.12.37′ as follows:

Let $\varphi$ denote the sentence $f(c_1, \dots, c_t) = 0 \Rightarrow c_1 = 0$. Any maximal ideal $P$ of $C$ yields an ideal $C[G]P$ of $C[G]$; picking $P$ such that $C/P$ has characteristic $> \max\{$periods of $g_1, \dots, g_t\}$ we see by proposition 8.1.25 that $\varphi$ holds in $(C/P)[G]$. But remark 2.12.37′ shows there are "enough" such $P$ to conclude $\varphi$ holds in characteristic 0, i.e., $\operatorname{tr}(r) = 0$, contradiction.)

This argument of "passing to characteristic $p$" is extremely useful in group algebras.

Armed with this "nilpotent implies trace 0" result, we are finally ready for a result which combines and improves both theorem 8.1.8′ and 8.1.14.

**Theorem 8.1.26:**   *F[G] has no nonzero nil left ideals, if G is a* char(F)′-*group.*

***Proof:***   Suppose $L$ is a nil left ideal and $a = \sum \alpha_g g \in L$. For each $g$ in supp(r) we have $g^{-1} r \in L$ so $\alpha_g = \operatorname{tr}(g^{-1}r) = 0$ by proposition 8.1.25, implying $a = 0$. Thus $L = 0$.    Q.E.D.

(Of course, the characteristic 0 part of this result relies on remark 8.1.25′ which is quite deep; in most of our applications we shall fall back on theorem 8.1.8′.)

Having considered finite groups, we turn to infinite groups for the remainder of this section.

## Subgroups of Finite Index

One of the techniques of studying infinite groups is in terms of "nice" subgroups of finite index, especially in terms of transversals; see remark 8.1.15 and proposition 8.1.16, for example. We take the opportunity of collecting several results along this vein, in preparation for the major structure theorems. We are interested in relating the structure of $C[G]$ to $C[H]$ where $[G:H] < \infty$, and start with some easy general observations about cosets of groups.

**Remark 8.1.27:**

(i) If $[G:H] = n$ then for any subgroup $G_1$ of $G$ we have $[G_1:H \cap G_1] \leq n$ (since any transversal of $H \cap G_1$ in $G_1$ lifts to distinct coset representatives of $H$ in $G$.)

(ii) If $[G:H_i] = n_i$ for $1 \leq i \leq t$ then $[G:\bigcap H_i] \leq n_1 \cdots n_t$. (Indeed, any coset $(\bigcap H_i)g = \bigcap H_i g$ is determined by the cosets $H_1 g, \ldots, H_t g$, so there are $\leq n_1 \cdots n_t$ possible choices.)

**Proposition 8.1.28:** *Any subgroup of finite index contains a normal subgroup of finite index. In fact, if $[G:H] = n$ and $H_1 = \bigcap \{g^{-1}Hg : g \in G\}$ then $H_1 \triangleleft G$ and $[G:H_1] \leq n!$*

**Proof:** Right multiplication by any element of $G$ permutes the cosets of $H$, yielding a group homomorphism $\varphi: G \to \text{Sym}(n)$; since $H_1$ is the largest normal subgroup of $G$ contained in $H$ we have $[G:H_1] \leq [G:\ker \varphi] \leq n!$
Q.E.D.

To continue the investigation we recall from proposition 8.1.16 that there is an injection $\psi: C[G] \to M_n(C[H])$ when $[G:H] = n$. Let us describe the action of $\psi$ on $G$. Fixing a transversal $g_1, \ldots, g_n$ we note $g_i g = h_i g_{\sigma i}$ for suitable $h_i$ in $H$ and $1 \leq \sigma i \leq n$. Moreover, $\sigma: \{1, \ldots, n\} \to \{1, \ldots, n\}$ is 1:1 for if $\sigma i = \sigma j$ then $h_i^{-1} g_i g = h_j^{-1} g_j g$ so $h_i^{-1} g_i \in Hg_i \cap Hg_j$, implying $i = j$. Thus $\sigma$ is a permutation and $\psi g = \sum_{i=1}^{n} h_i e_{\sigma i, i}$. In case $H$ is abelian we can take determinants to get $|\psi g| = (sg\ \sigma) h_1 \cdots h_n$; forgetting about $sg\ \sigma$ gives us a group homomorphism $\tilde{\psi}: G \to H$ defined by $\tilde{\psi}g = h_1 \cdots h_n$. Note $G' \subseteq \ker \tilde{\psi}$.

**Proposition 8.1.29:** *If* $[G: Z(G)] = n$ *then* $G' \cap Z(G)$ *has exponent* $n$.

*Proof:* Taking $H = Z(G)$ in the preceding paragraph we see $\sigma = (1)$ and each $h_i = g$ whenever $g \in G' \cap H$, so $1 = \tilde{\psi}g = g^n$. Q.E.D.

*Digression 8.1.30:* We have made a superficial application of a useful tool. Even if $H$ is not abelian we could pass to $C[H/H']$ before taking the determinant, thereby yielding a map $\tilde{\psi}: G \to H/H'$; since $G' \subseteq \ker \tilde{\psi}$ we have a map ver: $G/G' \to H/H'$ which is called the *transfer map* in Passman [77B].

Before utilizing proposition 8.1.29 we need another group-theoretic result.

**Lemma 8.1.31:** *If* $G$ *is a finitely generated group then every subgroup of finite index is finitely generated. More precisely, if* $G = \langle a_1, \ldots, a_t \rangle$ *and* $[G:H] = n$ *then* $H$ *is generated by* $tn$ *elements.*

*Proof:* Take a transversal $\{1 = g_1, g_2, \ldots, g_n\}$. Then $(Hg_i)a_j = Hg_{\sigma i}$ for a suitable permutation $\sigma$, implying $g_i a_j = h_{ij} g_{\sigma i}$ for suitable $h_{ij}$ in $H$. Let

$$H_0 = \langle h_{ij}: 1 \le i \le t, 1 \le j \le n \rangle \quad \text{and} \quad G_0 = \bigcup_{i=1}^{n} H_0 g_i.$$

To show $H_0 = H$ it suffices to show $G_0 = G$. Then $H = H \cap (\bigcup H_0 g_i) = H_0$. But

$$G_0 a_j = \bigcup H_0 g_i a_j = \bigcup H_0 h_{ij} g_{\sigma i} = \bigcup H_0 g_{\sigma i} = G_0,$$

implying $G_0 = G_0 G \supseteq G$. Q.E.D.

The commutator subgroup $G'$ now enters in an interesting fashion.

**Theorem 8.1.32:** *If* $[G: Z(G)] = n$ *then* $|G'| \le n^{n^3 + 1}$.

*Proof:* Take a transversal $\{g_1, \ldots, g_n\}$ of $G$ over $Z = Z(G)$. Any group commutator has the form $[z_1 g_i, z_2 g_j] = z_1 g_i z_2 g_j (z_1 g_i)^{-1} (z_2 g_j)^{-1} = [g_i, g_j]$ where $z_1, z_2 \in Z$; hence there are at most $n^2$ distinct commutators, and these generate $G'$. Let $t = n^3$. The abelian group $G' \cap Z$ has index $\le n$ in $G'$ and thus is generated by $\le t$ elements by lemma 8.1.31, implying $G' \cap Z$ is a direct product of $t$ cyclic groups each of exponent (and thus order) $\leq n$, by proposition 8.1.29. Thus $|G' \cap Z| \leq n^t$ so $|G'| \leq n^{t+1}$ as desired.     Q.E.D.

**Corollary 8.1.33:**   *Suppose $H = \langle h_1, \ldots, h_t \rangle$ is a subgroup of $\Delta(G)$.*

(i) $[H : Z(H)] \leq [G : C_G(H)] \leq \prod [G : C_G(h_i)]$.

(ii) $H'$ *is a finite group.*

(iii) $\operatorname{tor} H = \{$*elements of $H$ of finite order*$\}$ *is a finite normal subgroup of $H$, and $H/\operatorname{tor} H$ is torsion-free abelian.*

**Proof:**

(i) Clear from remark 8.1.27 since $Z(H) = H \cap C_G(H)$ and $C_G(H) = \bigcap_{i=1}^t C_G(h_i)$.

(ii) Apply (i) to theorem 8.1.32, since each $C_G(h_i)$ has finite index.

(iii) Since $H'$ is finite we have $(\operatorname{tor} H)/H' \approx \operatorname{tor}(H/H')$, which is finite and normal in the finitely generated abelian group $H/H'$. Thus $\operatorname{tor}(H)$ is finite and normal in $H$, and $H/\operatorname{tor} H \approx (H/H')/\operatorname{tor}(H/H')$ is torsion-free abelian.

Q.E.D.

**Lemma 8.1.34:**   *Suppose $G$ is a finite union of cosets of subgroups $H_1, \ldots, H_m$. Then some $H_i$ has finite index in $G$.*

**Proof:**   Induction on $m$. Since the assertion is obvious for $m = 1$, we may assume $m > 1$ and $[G : H_m]$ is infinite. Write $G = \bigcup_{i=1}^m \bigcup_{j=1}^{t_i} H_i g_{ij}$. Some coset $H_m g$ does not appear among the $H_m g_{mj}$; since the cosets of $H_m$ are disjoint we have

$$H_m g \subseteq \bigcup_{i=1}^{m-1} \bigcup_{j=1}^{t_i} H_i g_{ij}.$$

But each $H_m g_{mj} = H_m g g^{-1} g_{mj} \subseteq \bigcup_{i=1}^{m-1} \bigcup_{j=1}^{t_i} H_i g_{ij} g^{-1} g_{mj}$ so $G$ is a finite union of cosets of $H_1, \ldots, H_{m-1}$, and we are done by induction.     Q.E.D.

## Prime Group Rings

Let us turn to the question of when $C[G]$ is prime, which turns out to be much easier than the zero-divisor question. Of course, $C$ must be a domain. Otherwise, the answer depends only on $G$ and requires passing to $\Delta(G)$ by means of the following projection map.

**Definition 8.1.35:**   Suppose $H$ is a subgroup of $G$. Define $\pi_H : C[G] \to C[H]$ by $\pi_H \sum c_g g = \sum_{g \in H} c_g g$, i.e., we restrict the support to $H$. Write $\pi_\Delta$ for $\pi_{\Delta(G)}$.

**Remark 8.1.36:** $\pi_H$ is a map both in $C[H]$-$\mathcal{M}od$ and $\mathcal{M}od$-$C[H]$. (Indeed given $h$ in $H$ we have $g \in H$ iff $hg \in H$ so $\pi_H(h\sum_{g \in G} c_g g) = \pi_H(\sum c_g hg) = \sum_{g \in H} c_g hg = h\pi_H(\sum c_g g)$. The other side is symmetric.)

**Lemma 8.1.37:** Suppose $r_{i1}, r_{i2}$ in $C[G]$ satisfy $\sum_{i=1}^{t} r_{i1} g r_{i2} = 0$ for all $g$ in $G$. Then $\sum_{i=1}^{t} r_{i1} \pi_\Delta r_{i2} = 0$ and $\sum_{i=1}^{t} (\pi_\Delta r_{i1})(\pi_\Delta r_{i2}) = 0$.

**Proof:**   It is enough to prove the first assertion (for then

$$0 = \pi_\Delta \left( \sum_{i=1}^{t} r_{i1} \pi_\Delta r_{i2} \right) = \sum \pi_\Delta(r_{i1} \pi_\Delta r_{i2}) = \sum (\pi_\Delta r_{i1})(\pi_\Delta r_{i2}).$$

Let $A = (\bigcup_{i=1}^{t} \text{supp}(r_{i2})) - \Delta(G)$, and $B = \bigcup_{i=1}^{t} \text{supp}(r_{i1})$, both finite subsets of $G$. Thus there are $c_{ai}$ in $C$ such that $r_{i2} = \pi_\Delta r_{i2} + \sum_{a \in A} c_{ai} a$.

For any $a$ in $A$ and any $b$ in $B$ such that $b^{-1} g_0$ is a conjugate of $a$ we pick a specific $g_{ab}$ in $G$ such that $b^{-1} g_0 = g_{ab}^{-1} a g_{ab}$. Since $A$ and $B$ are finite we have a finite number of the $g_{ab}$.

Let $H = \bigcap \{C_G(g) : g \in \text{supp } \pi_\Delta r_{i2} \text{ for some } i\}$. Then $H$ centralizes all $\pi_\Delta r_{i2}$; by hypothesis each $h$ in $H$ satisfies

$$0 = \left( \sum_{i=1}^{t} r_{i1} h^{-1} r_{i2} \right) h = \sum_{i=1}^{t} r_{i1}(h^{-1} r_{i2} h)$$

$$= \sum_{i=1}^{t} r_{i1} \pi_\Delta r_{i2} + \sum_{i=1}^{t} \sum_{a \in A} r_{i1} c_{ai} h^{-1} a h.$$

Suppose $\sum r_{i1} \pi_\Delta r_{i2} \neq 0$; we aim for a contradiction. There is some $g_0 \neq 0$ in its support, so $g_0 \in \text{supp}(r_{i1} c_{ai} h^{-1} a h)$ for some $i$ and some $a$. Then $g_0 = bh^{-1} a h$ for some $b$ in $B$, implying $g_{ab}^{-1} a g_{ab} = b^{-1} g_0 = h^{-1} a h$. Hence $hg_{ab}^{-1} \in C_G(a)$, i.e., $h \in C_G(a) g_{ab}$. Thus $H$ is a finite union of cosets of the $C_G(a)$. But $H$ has finite index in $G$ by remark 8.1.27(ii), so by lemma 8.1.34 some $C_G(a)$ has finite index, i.e., $a \in \Delta(G)$ contrary to $a \in A$.     Q.E.D.

**Corollary 8.1.38:**   If $A_1, A_2 \lhd C[G]$ with $A_1 A_2 = 0$ then $\pi_\Delta A_1 \pi_\Delta A_2 = 0$.

**Theorem 8.1.39:**   (Connell's theorem) Suppose $C$ is an integral domain. The following assertions are equivalent:

   (i) $C[G]$ is prime.
   (ii) $Z(C[G])$ is an integral domain.
   (iii) $G$ has no finite normal subgroup $\neq (1)$.
   (iv) $\Delta(G)$ is torsion-free abelian (or (1)).
   (v) $C[\Delta(G)]$ is an integral domain.

**Proof:**   (i) $\Rightarrow$ (ii) is clear; (ii) $\Rightarrow$ (iii) follows from corollary 8.1.13 since in its notation if $H \lhd G$ is finite then $r(r - |H|1) = 0$ implying $r = |H|1$ and thus $r = 1 = |H|$.

(iii) $\Rightarrow$ (iv) Suppose $g_1, g_2 \in \Delta(G)$. The conjugates of $g_1$ and $g_2$ generate a normal subgroup $H$ of $G$. But corollary 8.1.33(iii) shows tor $H$ is a finite normal subgroup of $G$, so tor $H = (1)$ and $H \approx H/\text{tor } H$ is torsion-free abelian. This shows $g_1 g_2 = g_2 g_1$ and $g_1$ is torsion-free; since $g_i$ were arbitrary in $\Delta(G)$ we conclude $\Delta(G)$ is torsion-free abelian.

(iv) $\Rightarrow$ (v) by proposition 8.1.3.

(v) $\Rightarrow$ (i) by corollary 8.1.38.      Q.E.D.

We saw in theorem 8.1.8′ that $F[G]$ is semiprime if $\text{char}(F) = 0$; an analysis of the characteristic $p$ case is given in exercise 14.

## Primitive Group Rings

Having settled the question of which group rings are prime, we might wonder when $F[G]$ is primitive, for $G \neq (1)$. $G$ can neither be finite nor abelian since in either case $F[G]$ would be simple, which is impossible. On the other hand, $\Delta(G)$ is torsion-free abelian or (1), by theorem 8.1.39. For some time no primitive group rings were known, and later there were no known examples for $\Delta(G) \neq (1)$. In the 1970s the construction of a primitive group ring was the initiation to the elite club of group ring theorists. Using free products Formanek constructed the first primitive group algebras satisfying $\Delta(G) \neq 1$, cf., exercise 24. In the main text we examine the structural relation between $F[G]$ and $K[G]$ for $F \subseteq K$, mainly for the case $\Delta(G) = (1)$. Write $\Delta$ for $\Delta(G)$ in what follows.

**Lemma 8.1.40:**   (*M. Smith*) *Suppose* $F[G]$ *is prime and $K$ is a field extension of $F$. If $0 \neq A \lhd K[G]$ then $A \cap zF[G] \neq 0$ for some $z$ in $Z(K[G])$.*

**Proof:**   Note $K[G]$ is prime by theorem 8.1.39. Let $\{\alpha_i : i \in I\}$ be a base for $K$ over $F$. Any $a$ in $A$ can be written in the form $\sum \alpha_i r_i$ for $r_i$ in $F[G]$; we choose $a \neq 0$ such that the number $t$ of nonzero terms is minimal. If $t = 1$ then we are done by taking $z = \alpha_1$, so we may assume $t \geq 2$.

Reordering the $\alpha_i$ we may write $a = \sum_{i=1}^{t} \alpha_i r_i$. For any $g$ in $G$ we have

$$\sum_{i=2}^{t} \alpha_i (r_1 g r_i - r_i g r_1) = r_1 g a - a g r_1 \in A,$$

so by minimality of $t$ we see $r_1 g r_i = r_i g r_1$ for each $i$; by lemma 8.1.37 we conclude $r_1 \pi_\Delta r_i = r_i \pi_\Delta r_1$.

Pick some $g$ in supp$(r_1)$; replacing $a$ by $g^{-1}a$ we may assume $1 \in$ supp$(r_1)$. In particular $\pi_\Delta r_1 \neq 0$. Let $b = \pi_\Delta a \neq 0$. Let $b_1 = b, b_2, \ldots, b_u$ be the distinct conjugates of $b$ with respect to elements of $G$, and $z = b_1 \cdots b_u \in K[\Delta]$. By theorem 8.1.39 we see $K[\Delta]$ is an integral domain, from which it follows that $0 \neq z \in Z(K[G])$; hence $zr_1 \neq 0$. It remains to show $zr_1 \in A$.

$$r_1 b = r_1 \pi_\Delta a = \sum \alpha_i r_1 \pi_\Delta r_i = \sum \alpha_i r_i \pi_\Delta r_1 = a\pi_\Delta r_1 \in A,$$

so $zr_1 = r_1 z = (r_1 b)b_2 \cdots b_u \in A$ as desired.        Q.E.D.

**Theorem 8.1.41:** *(Passman) Suppose $F[G]$ is primitive and $K \supseteq F$. If either $\Delta = (1)$ or $K$ is an algebraic extension of $F$ then $K[G]$ also is primitive.*

**Proof:** Suppose $L$ is a maximal left ideal of $F[G]$ having core 0. By sublemma 2.5.32' $KL \approx K \otimes_F L$ is a proper left ideal of $K[G]$ so is contained in a maximal left ideal $L'$. Let $A = $ core$(L')$, a primitive ideal. Then $A \cap F[G] \subseteq$ core$(L) = 0$. We want to show $A = 0$. If $\Delta = (1)$ then $Z(K[G]) = K$ by proposition 8.1.12, so $A = 0$ by the contrapositive of lemma 8.1.40.

Thus we may assume $\Delta \neq (1)$. By hypothesis $K$ is an algebraic extension of $F$, so the integral domain $K[\Delta]$ is algebraic over $F[\Delta]$; one sees $A \cap K[\Delta] = 0$ (for if $a \neq 0$ in $A \cap K[\Delta]$ satisfied $\sum_{i=0}^t c_i a^i = 0$ for $c_i \in F[\Delta]$ and $t$ minimal then $0 \neq c_0 = -\sum_{i=1}^t c_i a^i \in F[\Delta] \cap A \subseteq A \cap F[G] = 0$, a contradiction). Hence $A \cap Z(K[G]) = 0$ by proposition 8.1.12.

If $A \neq 0$ then by lemma 8.1.40 we have $z$ in $Z(K[G])$ and $r$ in $F[G]$ such that $0 \neq zr \in A$; but $z \notin A$ so $r \in A$, contrary to $A \cap F[G] = 0$. We conclude $A = 0$ as desired.        Q.E.D.

**Corollary 8.1.42:** *If $F[G]$ is primitive with $|F| > |G|$ then $\Delta = (1)$.*

**Proof:** We may assume $F$ is algebraically closed by the theorem. $Z(F[G]) \approx F$ by example 2.12.28, so $\Delta = (1)$ by proposition 8.1.9.        Q.E.D.

## F Supplement: The Jacobson Radical of Group Algebras

Other than the "zero divisor problem," the leading question in group algebras is probably whether the Jacobson radical is necessarily nil. A positive answer would show that all the group algebras of theorem 8.1.26 were semiprimitive. Although there is no known counterexample, a positive solution seems to be out of reach. We shall present here some of the basic positive results. Some of these rely on the structure theory of rings, using only the most superficial properties of groups.

**Theorem 8.1.43:** *If* $H < G$ *then* $C[H] \cap \text{Jac}(C[G]) \subseteq \text{Jac}(C[H])$ *(by lemma 2.5.32 applied to lemma 2.5.17(i), for* $C[H]$ *is a summand of* $C[G]$.)

**Remark 8.1.44:** If $\text{Jac}(C[H])$ is nil for every finitely generated subgroup $H$ of $G$ then $\text{Jac}(C[G])$ is nil. (Indeed, take any $r \in \text{Jac}(C[G])$ and let $H = \langle \text{supp}(r) \rangle$. Then $r \in C[H] \cap \text{Jac}(C[G]) \subseteq \text{Jac}(C[H])$ is nilpotent.

**Theorem 8.1.45:** *(Amitsur, Herstein) If* $F$ *is uncountable then* $\text{Jac}(F[G])$ *is nil.*

**Proof:** By remark 8.1.44 we may assume $G$ is finitely generated and thus countable, so $\text{Jac}(F[G])$ is nil by theorem 2.5.22.     Q.E.D.

**Corollary 8.1.46:** *If* $F$ *is uncountable and* $G$ *is a* $\text{char}(F)'$*-group then* $F[G]$ *is semiprimitive.*

**Proof:** Apply theorem 8.1.45 to theorem 8.1.26.     Q.E.D.

The characteristic 0 case can be improved to

**Theorem 8.1.47:** *(Amitsur) Suppose* $\text{char}(F) = 0$ *and* $F$ *is not algebraic over its characteristic subfield* $F_0 \approx \mathbb{Q}$. *Then* $F[G]$ *is semiprimitive.*

**Proof:** Let $\Lambda = \{\lambda : i \in I\}$ be a transcendence base of $F$ over $F_0$ and let $F_1 = F_0[\Lambda]$. Then $F_1[G] \approx (F_0[G])(\Lambda)$ is semiprimitive by corollary 2.5.42 applied to theorem 8.1.10'; thus $F[G]$ is semiprimitive by theorem 2.5.36.
                                                                              Q.E.D.

## Polynomial Identities

Our next objective is to consider when a group ring satisfies a polynomial identity (PI). This condition is very natural because of the following link to representation theory:

**Remark 8.1.48:** Suppose $G$ is finite and $\text{char}(F) = 0$. The PI-degree of the semisimple Artinian ring $F[G]$ equals the maximal degree of the irreducible complex representations. (Indeed, this is true in case $F = \mathbb{C}$ by the Amitsur-Levitzki theorem since $F[G]$ is then split; in general, pass to $K \otimes_F F[G] \approx K[G] \approx K \otimes_{\mathbb{C}} \mathbb{C}[G]$ where $K$ is a field compositum of $F$ and $\mathbb{C}$, observing that $F[G]$ and $\mathbb{C}[G]$ have the same PI-degree.)

Thus the PI-degree contains important information; let us recall that if $G$ has an abelian group of index $n$ then $F[G]$ has PI-degree $\leq n$, by proposition 8.1.16.

The theorem we are interested in is the *Isaacs-Passman theorem* (cf., Passman [77B, theorem 3.8]): If $F[G]$ is semiprime of PI-degree $n/2$ then $G$ has an abelian subgroup of index $\leq [n^2/3]$! By theorem 8.1.8', we can conclude for char$(F) = 0$ that $F[G]$ is PI iff $G$ has an abelian subgroup of finite index bounded by a function of the PI-degree. Although this assertion is false for char$(F) > 0$ (cf., exercise 30) there also is a characteristic $\neq 0$ version due to Passman, involving $p$-abelian subgroups. The Isaacs-Passman theorem is treated very carefully in Passman [77B], so we shall not repeat the proof here. However, there is some room for streamlining parts of the proofs using PI-theory. First some examples.

### Examples 8.1.49:

(i) If $G_1$ and $G_2$ have abelian subgroups $H_1$, $H_2$ of index $n_1$, $n_2$, respectively, then $H_1 \times H_2$ has index $n_1 n_2$ in $G_1 \times G_2$. If $F[G_i]$ has PI-degree $n_i'$ then $F[G_1 \times G_2] \approx F[G_1] \otimes F[G_2]$ has PI-degree $n_1' n_2'$ by exercise 6.4.4.

(ii) For an example where the PI-degree is less than the index of a maximal abelian subgroup, let us turn to Sym$(n)$. The smallest index of an abelian subgroup is computed in exercise 31 and 32, and for $n \geq 4$ is larger than the PI-degree (exercise 33).

The aim of this discussion is to reduce the Isaacs-Passman theorem to the finite group case, which is accessible to the powerful methods of character theory by remark 8.1.48. We shall not handle this case since the results needed from character theory are beyond the scope of this book; the reader can consult Passman [77B, pp. 234–243] for a self-contained discussion, or Isaacs [78B, Chapter 12] for a more detailed treatment in the context of character theory.

## *Digression: Related Results from Character Theory*

It might be of interest to the reader to see a brief sketch of Isaac's treatment, which yields a sharper bound on the index of the abelian subgroup and also contains other interesting results of Isaacs-Passman, such as

**Theorem:** (*Isaacs [76B, theorem 12.11]*) *All the complex characters of a finite nonabelian group $G$ have degree 1 or $p$ (for $p$ prime), iff $G$ satisfies one*

*of the properties:*

(i) $[G:Z(G)] = p^3$.
(ii) *G has a normal abelian subgroup A of index p.*

We saw in the earlier digression that (i) is sufficient; the sufficiency of (ii) follows from a well-known theorem of Ito that the degree of any irre- ducible character divides $[G:A]$. Ito's theorem is proved in Isaacs [76B, theorem 6.15] by means of an easy argument on induced characters. Thus the hard direction is the converse.

The result from Isaacs' book most relevant to our discussion is

**Theorem:** *(Isaacs [76B, theorem 12.23]) If $F[G]$ has PI-degree $n$ for* $\text{char}(F) = 0$ *and $G$ finite then $G$ has an abelian subgroup of index $\leq (n!)^2$.*

The proof starts by taking $H \lhd G$ maximal with respect to $G/H$ non-abelian. Then $F[H]$ has PI-degree $\leq n/2$ by a result of Gallagher on induced characters, so induction applies, and it remains to study $\bar{G} = G/H$. Note that any (nontrivial) normal subgroup of $\bar{G}$ contains $\bar{G}'$, so $\bar{G}'$ is the unique minimal normal subgroup of $G$. Thus the following lemma becomes crucial.

**Lemma 8.1.50:** *(Isaacs [77B, lemma 12.3]) Suppose $G'$ is the unique minimal normal subgroup of $G$. Then all nonlinear complex characters of $G$ have the same degree m, and $G$ itself has one of the following forms:*

(i) *G is nonsolvable.*
(ii) *G is a p-group, $Z(G)$ is cyclic, and $G/Z(G)$ is a direct product of cyclic groups of order p.*
(iii) *G is a Frobenius group with Frobenius kernel $G'$ (i.e., $C_G(g) \subseteq G'$ for all g in $G' - \{1\}$) and its Frobenius complement is an abelian group A of order m (i.e., $A \cap G' = (1)$ and $AG' = G$).*

Let us outline the proof of the lemma. We may assume $G$ is solvable, so $G'$ is an abelian $p$-group for some $p$. If $Z(G) \neq (1)$ then $G' \subseteq Z(G)$ so $|G'| = p$ and $Z(G)$ is cyclic. If $a, b \in G$ then $a^p b = a^{p-1}[a,b]ba = [a,b]a^{p-1}ba = \cdots = [a,b]^p ba^p = ba^p$ proving $a^p \in Z(G)$, i.e., $G/Z(G)$ has exponent $p$, yielding (ii). If $Z(G) = (1)$ then $G$ is *not* a $p$-group. Taking a Sylow $q$-subgroup $Q$ for some $q \neq p$ one shows the normalizer of $Q$ is a Frobenius complement of $G'$, since $G' \cap C_G(a) = (1)$ for all $a$ in $A$.

When using lemma 8.1.50 in proving the above theorems, one can use the rich theories of $p$-groups and of Frobenius groups to dispose of cases (ii) and (iii) quite easily; in fact, the bound for the index of the abelian subgroup could be vastly improved. The weak link of the theory is case (i), in which the same abelian subgroup is used for $H$ and for $G$.

One final word concerning the finite case—the shortest current proof of the existence of a bound $\varphi(n)$ on the index of an abelian subgroup is found in Passman [77B, pp. 194–196], but this $\varphi(n)$ is much worse than the Isaacs-Passman bound quoted above.

## The Reduction Techniques

Having disposed so cavalierly with the finite group case of the Isaacs-Passman theorem we want to reduce the proof of the theorem to the finite case. Here the techniques are quite different, and we shall see the reduction is quite amenable to ring-theoretic methods.

**Lemma 8.1.51:** *Suppose $f$ is a multilinear central polynomial of $R = C[G]$ and is a sum of $t$ monomials.*

(i) *Any element in* supp $f(R)$ *has at most $t$ conjugates.*

(ii) *Suppose further $R$ has PI-degree $n$. Then we may assume $f$ is $n^2$-normal, and $[G: \langle \operatorname{supp} f(R) \rangle] \le n^2$; thus $[G: \Delta(G)] \le n^2$. If $[G: Z(G)] > n^2$ then some noncentral element of $G$ has at most $t$ conjugates.*

*Proof:*

(i) $f(G)$ spans $f(R)$ since $G$ is a base. Thus any element $g$ of supp $f(R)$ lies in supp$(z)$ for suitable $z$ in $f(G)$. By proposition 8.1.12 each conjugate of $g$ appears in supp$(z)$, so the number of conjugates is at most $|\operatorname{supp}(z)| \le t$.

(ii) By lemma 6.1.32 we cannot have $n^2 + 1$ elements of $R$ independent over $f(R)$, so $[G: \langle \operatorname{supp} f(R) \rangle] \le n^2$. Since supp $f(R) \subseteq \Delta(G)$ by (i) we see $[G: \Delta(G)] \le n^2$. Finally if $[G: Z(G)] > n^2$ then supp $f(R) \nsubseteq Z(G)$ so again we are done by (i).    Q.E.D.

**Theorem 8.1.52:** *Suppose $C[G]$ is semiprime of PI-degree $n > 1$. Then $[G: \Delta(G)] \le n^2$.*

*Proof:* By lemma 8.1.51(ii).    Q.E.D.

We focus on a special case, in order to tighten the bound on the index of $\Delta(G)$.

**Lemma 8.1.53:** (*M. Smith*) *Suppose* $C[G]$ *is prime of PI-degree* $n$. *Then* $[G:\Delta(G)] = n$.

**Proof:** Let $\Delta = \Delta(G)$. $\Delta$ has finite index by theorem 8.1.52, so $C_G(\Delta) \subseteq \Delta$; hence $C_G(\Delta) = \Delta$ noting that $C[\Delta]$ is an integral domain by Connell's theorem. Let $k = [G:\Delta(G)]$; $k \geq n$ by proposition 8.1.16. Let $Z = Z(C[G])$, $S = Z - \{0\}$, $R = S^{-1}C[G]$, $F = S^{-1}Z$, and $L = S^{-1}C[\Delta] \supset F$. Then $[R:L] = [C[G]:C[\Delta]] = k$ and, furthermore, $G/\Delta$ acts as a group of distinct automorphisms on $C[\Delta]$ (and thus extends to $L$) since $C_G(\Delta) = \Delta$. Hence $k \leq [L:F]$ so $k^2 \leq [R:L][L:F] = [R:F] = n^2$, proving $k \leq n$.

Q.E.D.

The final reduction to the finite case requires adapting an idea we encountered in §2.2. for rings.

**Definition 8.1.54:** A group $G$ is *residually finite* if there is a set $\{N_i : i \in \mathbb{N}\}$ of normal subgroups such that each $G/N_i$ is finite and $\bigcap_{i \in \mathbb{N}} N_i = (1)$; we say, moreover, $G$ is *strongly residually finite* if the intersection of every infinite collection of the $N_i$ is (1).

**Remark 8.1.55:** In the above notation write $G_i = G/N_i$. The canonical map $G \to \prod_{i \in \mathbb{N}} G_i$ given by $g \mapsto (N_i g)$ is monic iff $G$ is residually finite. If $G$ is strongly residually finite and $\mathscr{F}$ is any ultrafilter on $\mathbb{N}$ containing the cofinite filter then the canonical map $G \to (\prod G_i)/\mathscr{F}$ is monic.

**Theorem 8.1.56:** *Suppose there is a function* $\varphi : \mathbb{N} \to \mathbb{N}$ *such that every finite group* $G$ *whose group algebra* $\mathbb{Q}[G]$ *has PI-degree* $n$ *possesses an abelian subgroup of index* $\leq \varphi(n)$. *Then every group whose group algebra* $\mathbb{Q}[G]$ *has PI-degree* $n$ *possesses an abelian subgroup of index* $\leq \psi(n) = n\varphi(n)$.

**Proof:** By proposition 1.4.21 we can embed $G$ in an ultraproduct $\hat{G}$ of finitely generated subgroups $\{G_i : i \in I\}$. If each $G_i$ has an abelian subgroup of finite index $\leq \psi(n)$ with transversal $g_{i1}, \ldots, g_{i\psi(n)}$ then the corresponding ultraproduct $\hat{H}$ of the $H_i$ is abelian and $(g_{i1}), \ldots, (g_{i\psi(n)})$ is a transversal of $\hat{H}$ in $\hat{G}$; hence $[\hat{G}:\hat{H}] \leq \psi(n)$ so $H = G \cap \hat{H}$ is abelian with $[G:H] \leq \psi(n)$ by remark 8.1.27.

So we have reduced the proof to the case $G$ is finitely generated. Note $[G:\Delta(G)] < \infty$ by lemma 8.1.51; thus $\Delta(G)$ is finitely generated by lemma 8.1.31. By corollary 8.1.33 we have a finite subgroup $\mathrm{tor}(\Delta(G)) \lhd G$ such that $\Delta(G)/\mathrm{tor}(\Delta(G))$ is finitely generated abelian. We claim $[G:\Delta(G)] \leq n$. To see this let us pass to $\bar{G} = G/\mathrm{tor}(\Delta(G))$. Since $\mathrm{tor}(\Delta(G))$ is finite one sees easily $\Delta(\bar{G}) = \overline{\Delta(G)}$ which is torsion-free abelian, implying by theorem 8.1.39 that $F[\bar{G}]$ is prime. But $F[\bar{G}]$ has PI-degree $\leq n$, so lemma 8.1.53 shows $n \geq [\bar{G}:\Delta(\bar{G})]$, proving the claim.

Clearly it suffices to show $\Delta(G)$ has an abelian subgroup of index $\leq$ $\psi(n)/n = \varphi(n)$; we may replace $G$ by $\Delta(G)$, and thereby assume $G = \Delta(G)$. Since $G$ is finitely generated we see by corollary 8.1.33(i) that $[G:Z(G)] < \infty$. Thus $Z(G)$ is finitely generated; let $Z$ be its torsion-free part. Then $[Z(G):Z] = |Z(G)/Z| = |\text{tor } Z(G)| < \infty$. Letting $N_i = \{z^i : z \in Z\}$ for each $i \in \mathbb{N}$ we see any infinite intersection of the $N_i$ is (1). But each $G/N_i$ is finite since $Z/N_i$ is torsion abelian and thus finite. By remark 8.1.55 we can embed $G$ into an ultraproduct of the $G/N_i$; as in the first paragraph it suffices to prove the theorem for the finite groups $G/N_i$, and this was the desired reduction.       Q.E.D.

## Q Supplement: A Prime but Nonprimitive Regular Group Algebra

As mentioned in §2.11, until recently the major question in the theory of (von Neumann) regular rings was whether a prime regular ring need be primitive. Domanov [77] has constructed a counterexample which, although intricate, is rather easy to describe; moreover, it is a group algebra, which simplifies the verification. Since we have seen when a group algebra is prime, we shall now determine when a group algebra is regular.

**Proposition 8.1.57:** *Suppose $G$ is a* char$(F)'$-*group. If $G$ is locally finite then $F[G]$ is regular.*

*Proof:* Suppose $r \in F[G]$. Then $H = \langle \text{supp}(r) \rangle$ is a finite group so $F[H]$ is semisimple Artinian by Maschke's Theorem. Since $r \in F[H]$ we have $a$ in $F[H]$ such that $rar = r$, and this holds in $F[G]$, proving $F[G]$ is regular.       Q.E.D.

The converse is also true, cf., exercise 18, but is not needed here. We are ready for Domanov's example, following the exposition Passman [84].

**Lemma 8.1.58:** *Suppose $G$ has a normal abelian subgroup $A$ such that $G/A$ is countable but $F[A]$ has more than $2^{|G/A|}$ idempotents. Then $F[G]$ is not right (or left) primitive.*

*Proof:* Otherwise $R = F[G]$ possesses a maximal right ideal $L$ having core

0. Since $F[A]$ is commutative, $L \cap F[A] \lhd F[A]$ and is thus contained in a maximal ideal $B$ of $F[A]$. Suppose $e$ is idempotent in $F[A]$. For any $g$ in $G$ note that $geg^{-1}$ is idempotent and thus $geg^{-1} + B$ is 0 or 1 in the field $F[A]/B$.

Let $\bar{G} = G/A$, and define $\hat{e}$ to be the $\bar{G}$-tuple $(geg^{-1} + B)$, indexed over a set of representatives $\{\bar{g}: \bar{g} \in \bar{G}\}$. Since each entry is 0 or 1 there are at most $2^{|\bar{G}|}$ such tuples, so by hypothesis there are idempotents $e_1 \neq e_2$ with $\hat{e}_1 = \hat{e}_2$. Letting $e = e_1 - e_1 e_2$ we thus have $\hat{e} = 0$. Note that if $\bar{g}_1 = \bar{g}_2$ then their action by conjugation on $A$ is the same. In particular, $geg^{-1} \in B$ for all $g$ in $G$.

We claim $e = 0$. Indeed let $I = \sum_{g \in G} Rgeg^{-1} \lhd R$. If $e \neq 0$ then $L + I = R$, i.e.,

$$1 = x + \sum_{j=1}^{t} \alpha_j g_j e g_j^{-1} \qquad \text{for suitable } x \text{ in } L, \alpha_j \text{ in } R, \text{ and } g_j \text{ in } G.$$

Multiplying on the right by $\prod_j (1 - g_j e g_j^{-1})$, which annihilates each $g_j e g_j^{-1}$, yields

$$\prod_j (1 - g_j e g_j^{-1}) = x \prod_j (1 - g_j e g_j^{-1}) + 0 \in F[A] \cap L = B.$$

But each $g_j e g_j^{-1} \in B$; hence $1 \in B$, contradiction.

Having proved the claim we see $e_1 = e_1 e_2$ and, likewise, $e_2 = e_1 e_2$ proving $e_1 = e_2$, contradiction. Thus $\text{core}(L) \neq 0$. (In fact, the proof shows $I \subseteq \text{core}(L)$.) Q.E.D.

Given groups $A$ and $H$ let $\tilde{A} = \prod_{h \in H} A_h$, the direct product of copies of $A$ indexed by the elements of $H$. Then $H$ acts on $\tilde{A}$ by permuting its components, and so we can define the semidirect product of $\tilde{A}$ by $H$, which is called the *wreath product* of $A$ and $H$. Thus $(\tilde{a}, h)(\tilde{a}', h') = (\tilde{a}(h\tilde{a}'), hh')$.

**Example 8.1.59:** (Domanov) A group $G$ such that $F[G]$ is a prime regular group algebra which is not primitive, for *every* field $F$ of characteristic $\neq 2$. Let $c$ denote the cardinality of the continuum. For any prime $p$ let $A$ and $H$ be elementary abelian $p$-groups (i.e., direct sums of copies of $\mathbb{Z}/p\mathbb{Z}$), where $H$ is countable and $|A| > c$. Let $G$ be the wreath product of $A$ and $H$. Then $\Delta(G) = 1$ and for any field $F$ of characteristic $p$, $F[G]$ is prime regular but not primitive.

**Proof:** Any element of $G$ is a pair $(\tilde{a}, h)$ where $h \in H$ and $\tilde{a} \in \tilde{A}$. It is easy to see that any nonidentity element of $G$ has an infinite number of conjugates, by examining separately the cases when $h = 1$ or $h \neq 1$; we leave this to the reader. Hence $\Delta(G) = 1$ so $F[G]$ is prime by Connell's theorem. It is also easy to see $G$ is locally finite since $H$ and $A$ each are locally finite; by proposition 8.1.57 $F[G]$ is regular. But $F[G]$ is not primitive, by lemma 8.1.58.    Q.E.D.

## §8.2 Noetherian Group Rings

This section is concerned with group algebras of polycyclic-by-finite groups, to be defined presently. These are the only known Noetherian group rings and as such have been the subject of intensive research. The main result presented here is the Farkas-Snider solution of the zero-divisor question for torsion-free polycyclic-by-finite groups.

**Remark 8.2.0:**   Since $F[G]$ has an involution, $F[G]$ is left Noetherian iff it is right Noetherian. We shall use this observation implicitly.

### Polycyclic-by-Finite Groups

**Definition 8.2.1:**   A *subnormal series* of subgroups of $G$ is a chain $G = G_m \triangleright G_{m-1} \triangleright \cdots \triangleright G_0 = (1)$, i.e., each $G_{i-1} \triangleleft G_i$. $G$ is *polycyclic* if $G$ has a subnormal series with each factor $G_i/G_{i-1}$ cyclic; $G$ is poly-{infinite cyclic} if each factor in the series is infinite cyclic. $G$ is *polycyclic-by-finite*, or *virtually polycyclic*, if $G$ has a polycyclic group of finite index.

$G$ is *solvable* if there is a subnormal series with each factor abelian. $G$ is *nilpotent* if the lower central series terminates at $G$. Thus every nilpotent group is polycyclic, and every polycyclic group is solvable. On the other hand, any finitely generated abelian group is a finite direct product of cyclic groups and thus nilpotent. An easy but enlightening example of a non-nilpotent polycyclic group is the infinite dihedral group $G = \langle ab: b^{-1}ab = a^{-1}$ and $b^2 = 1 \rangle$.

**Theorem 8.2.2:**   *If $G$ is polycyclic-by-finite then $F[G]$ is Noetherian.*

**Proof:**   Let $H$ be polycyclic of finite index. Then $F[G]$ is an f.g. left and right $F[H]$-module, so it suffices to prove $F[H]$ is Noetherian. Taking a subnormal series $H = H_m \triangleright \cdots \triangleright H_0 = (1)$ with each $H_i/H_{i-1}$ cyclic, we shall prove by induction that each $F[H_i]$ is Noetherian. This is obvious for $i=0$, so suppose inductively that $R = F[H_{i-1}]$ is Noetherian. If $H_i/H_{i-1}$ is finite we are done as before, so assume $H_i/H_{i-1}$ is infinite cyclic, generated by some element $g$. Since $H_{i-1} \triangleleft H_i$ we see conjugation by $g$ produces an automorphism $\sigma$ of $R$. Let $T = R[\lambda; \sigma]$. Then $T$ is left Noetherian by proposition 3.5.2;

but $S = [\lambda^i : i \in \mathbb{N}\}$ is clearly a left denominator set so $S^{-1}T$ also is left Noetherian by proposition 3.1.13.

There is a ring homomorphism $\varphi: T \to F[H_i]$ given by $\varphi(\sum r_i \lambda^i) = \sum r_i g^i$ for $r_i$ in $R$; extending this naturally to a surjection $S^{-1}T \to F[H_i]$ shows $F[H_i]$ is left (and thus right) Noetherian.     Q.E.D.

The idea behind this proof was to obtain a polycyclic-by-finite group ring from an iterated skew Laurent extension; we return to this idea in §8.4. Conversely, let us see what can be said of $G$ when $F[G]$ is Noetherian.

**Lemma 8.2.3:**  *If $R = F[G]$ is Noetherian then $G$ has no infinite chain $H_1 < H_2 < \cdots$ of subgroups. In particular $G$ is finitely generated.*

**Proof:**  $R\omega F[H_1] \subseteq R\omega F[H_2] \subseteq \cdots$ must terminate so suppose

$$R\omega(F[H_i]) = R\omega(F[H_{i+1}]).$$

Write $H = H_i$. For any $h$ in $H_{i+1}$ we have $\pi_H(h - 1) \in \pi_H(\omega(F[H_{i+1}])) \subseteq \pi_H(R\omega(F[H])) = \omega(F[H])$ so $\pi_H(h - 1) \neq -1$, thereby implying $h \in H$. Thus $H_i = H_{i+1}$.     Q.E.D.

**Proposition 8.2.4:**  *If $F[G]$ is Noetherian and $G$ is solvable then $G$ is polycyclic.*

**Proof:**  Take a subnormal chain with each factor $G_i/G_{i+1}$ abelian. $F[G_i]$ is Noetherian by sublemma 2.5.32′ since $F[G]$ is free over $F[G_i]$. Hence $F[G_i/G_{i+1}]$ is Noetherian. By lemma 8.2.3 each $G_i/G_{i+1}$ is finitely generated abelian and thus polycyclic, implying $G$ is polycyclic.     Q.E.D.

**Remark 8.2.5:**  The familiar proofs about finite solvable groups show that any subgroup or homomorphic image of a polycyclic group is polycyclic. Explicitly, if $G$ has a subnormal series $G = G_m \triangleright \cdots \triangleright G_0 = (1)$ and $H < G$ then taking $H_i = H \cap G_i$ we see $H = H_m \triangleright \cdots \triangleright H_0 = (1)$ is a subnormal series of $H$, and the factor $H_i/H_{i+1}$ is isomorphic to $H_i G_{i+1}/G_{i+1}$, a subgroup of $G_i/G_{i+1}$.

**Proposition 8.2.6:**  *If $G$ has a subnormal series each of whose factors is finite or cyclic then $G$ is polycyclic-by-finite. Moreover, any polycyclic-by-finite group has a characteristic poly-{infinite cyclic} subgroup $H$ of finite index.*

**Proof:**  Let  $G = G_m \triangleright \cdots \triangleright G_0 = (1)$. We proceed by induction on $m$;

assuming $G_{i-1}$ has a characteristic poly-{infinite cyclic} subgroup $H_{i-1}$ of finite index, we want to prove this for $G_i$. If $G_i/G_{i-1}$ is finite we merely take $H_i = H_{i-1}$, so assume $G_i/G_{i-1}$ is cyclic. Pick $g \in G_i$ such that the coset $G_{i-1}g$ generates $G_i/G_{i-1}$. Note $H_{i-1} \lhd G_i$.

Let $\tilde{H} = H_{i-1}\langle g \rangle$, a subgroup of $G_i$ of finite index $\leq [G_{i-1} : H_{i-1}]$. By induction $\tilde{H}_i$ has a subnormal series each of whose factors is infinite cyclic, but $\tilde{H}_i$ need not be characteristic in $G$. We finish at once by means of the following two lemmas, the first of which improves on proposition 8.1.28.

**Lemma 8.2.7:**   *Suppose $G$ is finitely generated. For any $n$ there are only finitely many subgroups of index $n$ in $G$. In particular, any subgroup $H$ of finite index contains a characteristic subgroup of finite index.*

**Proof:**   Write $G = \langle g_1, \ldots, g_t \rangle$, and suppose $[G:H] = n$. As observed in proposition 8.1.28 right multiplication by elements of $G$ permutes the cosets of $H$, thereby yielding a group homomorphism $\varphi: G \to \mathrm{Sym}(n)$ with $\ker \varphi \subseteq H$. But $\varphi$ is determined by its action on $g_1, \ldots, g_t$ so there are $\leq t^{n!}$ possible homomorphisms. Moreover, $H$ must be the preimage of a subset of $\mathrm{Sym}(n)$, of which there are $2^{n!}$, so there are $\leq (2t)^{n!}$ possibilities for $H$.

Thus $\bigcap\{$subgroups of $G$ having index $n\}$ is a characteristic subgroup (contained in $H$) of finite index in $G$.     Q.E.D.

**Lemma 8.2.8:**   *Suppose $H$ is poly-{infinite cyclic}. Then the same holds for every subgroup of $H$.*

**Proof:**   By remark 8.2.5, noting every subgroup of an infinite cyclic group is infinite cyclic.     Q.E.D.

The infinite factors actually give us an important invariant.

**Definition 8.2.9:**   The *Hirsch number* $h(G)$ of a polycyclic-by-finite group $G$ is the number of infinite cyclic factors in a subnormal series each of whose factors is infinite cyclic or finite.

**Proposition 8.2.10:**   *Suppose $G$ is polycyclic-by-finite. Then $h(G)$ is well-defined, and $h(G) = 0$ iff $G$ is finite. If $N \lhd G$ then $h(G) = h(N) + h(G/N)$.*

**Proof:**   We borrow the idea of proof of the Jordan-Holder theorem, dealing now with subnormal series whose factors are either infinite cyclic or finite.

First suppose we have

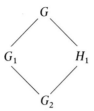

i.e., $G_2 = G_1 \cap H_1$. Recall from remark 8.1.27 that $[G:G_1] < \infty$ iff $[H_1:G_2] < \infty$; likewise, $[G:H_1] < \infty$ iff $[G_1:G_2] < \infty$. Thus we see that the number of infinite cyclic factors is the same for the left side as for the right side.

In general, suppose we have two subnormal series $G \triangleright G_1 \triangleright \cdots \triangleright G_m = (1)$ and $G \triangleright H_1 \triangleright \cdots > H_n = (1)$. Then $G_1 \cap H_1$ is polycyclic-by-finite by proposition 8.2.6 and has a subnormal series whose factors are all infinite cyclic except the first, which is finite. Using the above observation plus induction applied to $G_1$ and $H_1$ show $h(G)$ is well-defined; the finishing touches are left to the reader.

Clearly $h(G) = 0$ iff there are no infinite factors, iff $G$ is finite. To prove the last assertion note $N$ and $G/N$ each are polycyclic-by-finite; putting together their subnormal chains shows by inspection $h(N) + h(G/N) = h(G)$.

Q.E.D.

To illustrate the use of the Hirsch number we present the next proposition.

**Remark 8.2.11:** Any infinite polycyclic-by-finite group $G$ has a characteristic torsion-free abelian subgroup $\neq (1)$. (Indeed, take $H$ as in proposition 8.2.6. The next-to-last subgroup in the derived series of $H$ is clearly characteristic abelian and is torsion-free since $H$ clearly is torsion-free.)

**Proposition 8.2.12:** *Any polycyclic-by-finite group $G$ is residually finite.*

**Proof:** Remark 8.2.11 shows $G$ has a torsion-free characteristic abelian subgroup $A \neq (1)$. Let $A_i = \{a^i : a \in A\}$. Then $h(G/A_i) = h(G) - h(A_i) \leq h(G) - 1$ so by induction each $G/A_i$ is residually finite. Since $\bigcap A_i = (1)$ we conclude $G$ is residually finite. Q.E.D.

## Digression: Growth of Groups

An obvious question to ask at this juncture is whether polycyclic-by-finite groups have finite GK-dimension, for then we could apply the theory of §6.2

(which, in fact, will be very useful for enveloping algebras). Let us back-track a bit.

**Definition 8.2.13:** Suppose $G$ is a finitely generated group generated by $g_1, \ldots, g_t$. The *growth function* $\varphi_G$ is defined by taking $\varphi_G(n)$ to be the number of products of length $n$ of the $g_i$ and the $g_i^{-1}$ for $1 \le i \le t$.

**Remark 8.2.14:** We can define polynomial, subexponential, and exponential growth just as in definition 6.2.10. In fact, the growth of $G$ precisely matches the growth of $F[G]$ (as an affine algebra) since we can use the generating set of $g_1, \ldots, g_n$ and their inverses.

Although growth of groups is an older subject than growth of algebras, we shall frame our results for group algebras since then we can draw on the results from §6.2. In particular we have

**Remark 8.2.15:** If $[G:H] < \infty$ then $F[G]$ and $F[H]$ have the same growth, by remark 6.2.14(iv).

One might have expected polycyclic-by-finite group algebras to have polynomial growth, by a skew version of proposition 6.2.22. However, a completely different story emerges.

**Theorem 8.2.16:** (*Milnor-Wolf-Gromov*) A finitely generated group has polynomial growth iff it is nilpotent-by-finite.

Although we shall not prove this theorem, we shall point to some readily accessible results in the literature. Bass [72, theorem 2] shows that any f.g. nilpotent group $G$ with lower central series $G \supset G_1 \supset \cdots \supset G_t = (1)$ has polynomial growth of degree $d = \sum_{n=1}^{t} nd_n$ where $d_n$ is the rank of the f.g. abelian group $G_{n-1}/G_n$.

On the other hand, Bass [72,§3] gave a rather simple proof that any polycyclic-by-finite group $G$ having subexponential growth is nilpotent-by-finite. Sketch of proof: Using remark 8.2.15 one may assume $G$ is poly-{infinite cyclic}, so there is a normal polycyclic subgroup $H$ with $G/H$ infinite cyclic. By induction on the Hirsch number $H$ is nilpotent-by-finite, so by lemma 8.2.7 contains a nilpotent subgroup $N$ normal in $G$ of finite index. Let $g$ be the generator of $G/H$, and refine the lower central series of $N$ to a series $N > N_1 > N_2 > \cdots$ for which each factor is cyclic. The subexponential growth implies $g^n$ acts trivially on each factor for some $n > 1$, cf., Bass [72, lemma 2].

Letting $K$ be the subgroup of $G$ generated by $N$ and $g^n$ we see $K$ is also nilpotent. But $K$ has the same Hirsch number as $G$ so $[G:K] < \infty$.

**Example 8.2.17:** Let $G$ be the group with generators $g_1, g_2$, and $z$ and relations $[g_1, g_2] = z$ and $[g_i, z] = 1$. Then $G' = \langle z \rangle$ and $G/G'$ is abelian, so $G$ is nilpotent. It is instructive to work out Bass' proof for this example, and to calculate GK-dim $\mathbb{Q}[G]$.

Based on work of Milnor, Rosset [76] proved that if $G$ is a finitely generated group of subexponential growth and $N$ is a normal subgroup with $G/N$ solvable then $N$ is finitely generated. Applying this to $N = G'$ and applying induction on the length of the derived series, one sees easily that if $G$ is solvable of subexponential growth then $G$ is polycyclic and thus nilpotent-by-finite by the last paragraph.

All of the work described thus far is "elementary" albeit rather ingenious. Gromov's proof that any finitely generated group of polynomial growth is solvable and thus nilpotent-by-finite is much harder, cf., Tits [ ]. Incidentally there are groups of subexponential but not polynomial growth, but of course these are not solvable.

## Homological Dimension of Group Algebras

**Proposition 8.2.18:** *If $G$ possesses a subnormal series each of whose factors is infinite cyclic then* $\mathrm{gl.\,dim}\, F[G] \le h(G)$.

**Proof:** Let $G = G_m \triangleright \cdots \triangleright G_0 = (1)$ be the given series; clearly $m = h(G)$. By induction on $h(G)$ we have $\mathrm{gl.\,dim}\, F[G_{m-1}] \le m - 1$. But there is $g$ in $G$ such that $G/G_{m-1} \approx \langle \bar{g} \rangle$. Conjugating by $g$ induces an automorphism $\sigma$ on $F[G_{m-1}]$. The subring $R$ of $F[G]$ generated by $F[G_{m-1}]$ and $g$ is isomorphic to $(F[G_{m-1}])[\lambda; \sigma]$ and thus has $\mathrm{gl.\,dim} \le m$ by corollary 5.1.25. But $F[G] = S^{-1}R$ where $S = \{g^i : i \in \mathbb{N}\}$, so $\mathrm{gl.\,dim}\, F[G] \le m$ by proposition 5.1.26.
                                                                               Q.E.D.

We want to improve this to polycyclic-by-finite groups. First let us show it is enough to prove $pd_{F[G]} F < \infty$. To see this we shall make some general observations about modules over arbitrary group rings.

**Proposition 8.2.19:** $\mathrm{gl.\,dim}\, F[G] = pd_{F[G]} F$ *for any group $G$, where $F$ has the trivial $G$-action* $g\alpha = \alpha$.

***Proof:*** $\geq$ is true by definition. On the other hand, let

$$0 \to P_n \to \cdots P_0 \to F \to 0$$

be a projective resolution of $F$ as $F[G]$-module. Consider the functor $\underline{\quad} \otimes_F M$ from $F[G]$-$\mathcal{M}od$ to $F[G]$-$\mathcal{M}od$ where any $N \otimes M$ is viewed as $F[G]$-module via the diagonal action $g(x \otimes y) = gx \otimes gy$. This functor is exact since it is obviously exact as a functor from $F$-$\mathcal{M}od$ to $F$-$\mathcal{M}od$, since $F$ is a field. In particular, each $P_i \otimes_F M$ is a projective $F[G]$-module, and $M \approx F \otimes_F M$ has the projective resolution

$$0 \to P_n \otimes_F M \to \cdots \to P_0 \otimes_F M \to M \to 0.$$

Thus $pd\, M \leq n$, proving gl. dim $F[G] \leq n = pd_{F[G]} F$.      Q.E.D.

**Theorem 8.2.20:**  *If $G$ is polycyclic-by-finite and is a* char$(F)'$-*group then* gl. dim $F[G] < \infty$.

***Proof:*** Letting $H$ be as in proposition 8.2.6 we have $[G:H] < \infty$ and gl. dim $F[H] < \infty$ by proposition 8.2.18. Thus it suffices to prove

**Theorem 8.2.21:**  *(Serre) Suppose $G$ has a subgroup $H$ of finite index such that* gl. dim $F[H] < \infty$. *If $G$ is a* char$(F)'$-*group then* gl. dim $F[G] < \infty$.

***Proof:*** By proposition 8.2.19 it suffices to find a projective resolution of finite length of $F$ as $F[G]$-module. We are given a projective resolution $0 \to P_n \overset{d_n}{\to} \cdots \to P_0 \overset{d_0}{\to} F \to 0$ as $F[H]$ module. View this is an acyclic complex $(P,d)$ where $P = \left(\bigoplus_{u=0}^n P_u\right) \oplus F$, and $d = \bigoplus_{i=1}^n d_u$. Let $m = [G:H]$. Then the acyclic complex $\tilde{P} = (\bigotimes P, \tilde{d})$ can be constructed as in corollary 5.2.57. Write $\tilde{P} = \tilde{P}' \oplus F$ (identifying $F \otimes_F \cdots \otimes_F F$ with $F$). We shall give $\tilde{P}$ a natural $F[G]$-module structure with respect to which $\tilde{P}'$ is projective, and this provides a projective resolution of length $\leq mn$ over $F$, by remark 5.2.55.

Write $G = \bigcup_{i=1}^m g_i H$; then for any $g$ in $G$ we have $g^{-1}g_i = g_{\sigma i}h_{\sigma i}^{-1}$ where $\sigma \in \mathrm{Sym}(m)$ depends on $g$. $G$ acts on $\tilde{P}$ by

$$g(a_1 \otimes \cdots \otimes a_m) = \pm h_{\sigma 1}a_{\sigma 1} \otimes \cdots \otimes h_{\sigma m}a_{\sigma m} \tag{1}$$

for $a_i$ homogeneous in $P_i$ where the sign arises because of the sign switch in graded tensor products (see after 5.2.57) and can be computed to be $(-1)$ to the power $\sum \deg a_i \deg a_j$ summed over all $i < j$ for which $\sigma i > \sigma j$. Then $\tilde{P}$ is a $G$-module and thus an $F[G]$-module, and $\tilde{d}$ is an $F[G]$-module map. (We skip the straightforward but tedious computation that the sign works out; complete details are given in Passman [77B, pp. 443–448].)

Write $P = P' \oplus F$ where $P' = \bigoplus_{i=1}^{n} P_i$ is projective as $F[H]$-module, and take $P''$ such that $M_0 = P' \oplus P''$ is a free $F[H]$-module. Let $M = M_0 \oplus F$, and form $\tilde{M} = \bigotimes_{i=1}^{m} M$ which can be written as $\tilde{M}' \oplus F$. $\tilde{M}'$ is an $F[G]$-module by (1), and $\tilde{P}'$ is a summand of $\tilde{M}'$ since the canonical epic $\tilde{M}' \to \tilde{P}'$ splits. Hence it suffices to show $\tilde{M}'$ is projective.

Explicitly $\tilde{M}'$ is the direct sum of all tensor products involving $j$ copies of $M_0$ and $(m - j)$ copies of $F$, for each $j \geq 1$. Thus taking a base $B$ for $M_0$ over $F[H]$ we see $\tilde{M}'$ has a base over $F$ consisting of all $w = w_1 \otimes \cdots \otimes w_m$ where $w_i$ has the form $hb$ for $h \in H$ and $b \in B \cup \{1\}$, with not all $w_i = 1$. The given action of $G$ on $\tilde{P}'$ extends naturally to $\tilde{M}'$. Let $K_w = \{g \in G : gw \in Fw\}$. The $G$-orbit of $w$ spans an $F[G]$-module isomorphic to $F[G] \otimes_{F[K_w]} Fw$; since $\tilde{M}'$ decomposes as the direct sum of these it suffices to prove each $F[G] \otimes_{F[K_w]} Fw$ is projective. Hence it suffices to show each $Fw$ is projective as $F[K_w]$-module.

We claim $K_w$ is finite. Indeed $H$ contains a normal subgroup $N$ of $G$ of finite index by proposition 8.1.28 and it suffices to prove $N \cap K_w = (1)$. But if $g \in N \cap K_w$ then $g^{-1}g_i = g_i h'_i$ where $h'_i = g_i^{-1} g^{-1} g_i \in N$, so the corresponding permutation $\sigma$ is (1); writing $w = w_1 \otimes \cdots \otimes w_m$ with $w_i = h_i b_i$ we get $\alpha w = gw = h'_1 h_1 b_1 \otimes \cdots \otimes h'_m h_m b_m$ for some $\alpha$ in $F$. Matching components for $b_i \neq 1$ shows some $h'_i = 1$; hence $g^{-1} = 1$, proving the claim.

By hypothesis $|K_w| \neq 0$ in $F$. By Maschke's theorem $F[K_w]$ is semisimple Artinian, so each of its modules (including $Fw$) is projective. Q.E.D.

The fact that polycyclic-by-finite group algebras have finite global dimension is one of the keys to their structure. The proof given above, due to Serre, was taken from Swan [69].

## Traces and Projectives

We shall now begin to focus on the zero-divisor question. Formanek [73] recognized that it might first be wise to consider idempotents, since a domain cannot have nontrivial idempotents; he proved in characteristic 0 that every Noetherian group algebra over a torsion-free group has no nontrivial idempotents. The proof (exercise 2) is an ingenious application of the (Kaplansky) trace map. Actually a slight generalization of Formanek's theorem is needed. Let $\operatorname{tr} : M_n(F[G]) \to F[G]$ be the usual matrix trace, i.e.,

$$\operatorname{tr}(r_{ij}) = \sum_{i=1}^{n} r_{ii} \in F[G].$$

**Remark 8.2.22:**   Any trace map $t: F[G] \to F$ yields a trace map

$$\text{Tr}: M_n(F[G]) \to F$$

given by $\text{Tr}(a) = t(\text{tr } a)$ for $a \in M_n(F)$.

Our first use of remark 8.2.27 is somewhat technical.

**Lemma 8.2.23:**   *(Formanek [73]). Define* $t_h: F[G] \to F$ *by* $t_h(\sum \alpha_g g) = \sum \{\alpha_g: g$ *is conjugate to* $h\}$, *where* $h \in G$. *If for all* $n \neq \pm 1$ $h$ *is not conjugate to* $h^n$ *then* $\text{Tr}_h e = 0$ *for every idempotent* $e$ *of* $M_n(F[G])$.

**Proof:**   First assume $\text{char}(F) = p$. Write $e = (r_{ij})$ and $\text{tr}(e) = \sum \alpha_g g = \sum_{i=1}^n r_{ii}$. We claim there is some $q = p^m$ such that no $g^q$ is conjugate to $h$, for all $g$ in $\text{supp}(\text{tr } e)$. This surely follows if we can show for any $g$ in $G$ there is at most one positive power $q'$ of $p$ such that $g^{q'}$ is conjugate to $h$ (for then take $q$ greater than all of the $q'$ pertaining to $\text{supp}(\text{tr } e)$.) So assume $g^{q'}$ and $g^{q''}$ each are conjugate to $h$, with $q' > 0$ minimal possible. Then $q' | q''$; letting $n = q''/q'$ we see $h^n$ is conjugate to $(g^{q'})^n = g^{q''}$ so by hypothesis $n = \pm 1$. But then $q'' = q'$ as desired, proving the claim.

Note $e_{ij} = [e_{ii}, e_{ij}]$ for any $i \neq j$. Thus 8.1.5(i) yields

$$e = e^q \in \sum r_{ii}^q e_{ii} + [M_n(F[G]), M_n(F[G])],$$

so $\text{Tr}_h e = \sum t_h(r_{ii}^q) = 0$, proved as in lemma 8.1.24.

Having proved the result for characteristic $p$, we get the result for characteristic 0 by a "Nullstellensatz" argument, as in remark 8.1.25′.    Q.E.D.

**Proposition 8.2.24:**   *If* $G$ *is torsion-free and* $F[G]$ *is Noetherian then* $\text{Tr}_h e = 0$ *for every idempotent* $e$ *of* $M_n(F[G])$, *and all* $h \neq 1$ *in* $G$.

**Proof:**   We need to verify $h$ is not conjugate to $h^n$. Indeed, if $ghg^{-1} = h^n$ let $h_i = g^{-i} h g^i$ and $H_i = \langle h_i \rangle$. Then

$$h_i^n = g^{-i} h^n g^i = g^{-i} g h g^{-1} g^i = g^{-(i-1)} h g^{i-1} = h_{i-1}$$

implying $H_{i-1} \leq H_i$, and $H_{i-1} < H_i$ since $G$ is torsion-free. But this gives an infinite ascending chain of subgroups of $G$ contrary to lemma 8.2.3.

Q.E.D.

Now let $t_G$ denote the Kaplansky trace map, so that if $r = \sum \alpha_g g$ then $\alpha_g = t_G(rg^{-1})$. (Note in §8.1 we used the notation "tr" instead.)

**Corollary 8.2.25:** *If* $F[G]$ *is Noetherian and* $G$ *is torsion-free then* $\sum_{g \neq 1} t_G((\text{tr } e)g^{-1}) = 0$ *for any idempotent* $e$ *of* $M_n(F[G])$.

Suppose $P$ is a projective $F[G]$-module. Then $P = eM_n(F[G])$ for a suitable idempotent $e$ of $M_n(F[G])$. We are now ready for an important invariant for projective modules.

**Definition 8.2.26:** If $P$ is $F[G]$-projective define $\text{Tr}_G P = \text{Tr}_G e = t_G(\text{tr } e)$.

**Theorem 8.2.27:** *Suppose* $G$ *is torsion-free and* $F[G]$ *is Noetherian. Then* $\text{Tr}_G P = [F \otimes_{F[G]} P : F]$ *and thus is a well-defined integer* (*independent of* $n$ *or the choice of* $e$).

**Proof:** Let $\pi : F[G] \to F$ be the augmentation homomorphism given by $\pi g = 1$. Using corollary 8.2.25 we have

$$\text{Tr}_G P = t_G(\text{tr } e) = \sum_{g \in G} t_G((\text{tr } e)g^{-1}) = \text{tr}(\pi e) = [F \otimes_{F[G]} P : F],$$

the last equality holding since the trace of an idempotent in $M_n(F)$ is its rank, and $\pi e \in M_n(F)$ is idempotent.    Q.E.D.

**Corollary 8.2.28:** $\text{Tr}_G(P_1 \oplus P_2) = \text{Tr}_G P_1 + \text{Tr}_G P_2$ *for any two projective modules* $P_1$ *and* $P_2$, *provided* $G$ *is torsion free.*

In this manner we have a powerful tool for studying the projective modules over Noetherian group algebras. Recent research has pushed the theory of $\text{Tr}_G$ to the non-torsion-free case, cf., Cliff [85]. Formanek's original theorem is given in exercise 2.

## The Zero Divisor Question

We are finally ready to turn to the zero divisor question. First some interesting preliminary results. Recall from proposition 1.2.17 that $F[G]$ is a domain if $G$ is an ordered group. We can weaken the hypothesis slightly.

**Definition 8.2.29:** $G$ is a *left-ordered* group if $G$ has a total order $\leq$ such that whenever $g_1 < g_2$ we have $gg_1 < gg_2$ for all $g$ in $G$.

As in lemma 1.2.32 one can characterize a right order in terms of the *positive cone* $P(G, \leq) = \{g \in G : g \geq 1\}$.

**Proposition 8.2.30:** *If $G$ is left ordered then $P = P(G, \leq)$ is a submonoid satisfying $P \cap P^{-1} = \{1\}$ and $P \cup P^{-1} = G$. Conversely, given any such submonoid $P$ we can make $G$ left ordered such that $P = P(G, \leq)$.*

**Proof:** The first assertion is straightforward. The second assertion can be extracted from the last sentence of the proof of lemma 1.2.32.     Q.E.D.

Note that any left ordered group is right ordered (exercise 3), and we now have a result decisively stronger than proposition 1.2.33.

**Proposition 8.2.31:** *If $G$ has a subnormal series whose factors are left ordered, then $G$ is left ordered.*

**Proof:** By induction it suffices to show if $G$ has a left ordered normal subgroup $N$ and $G/N$ is left ordered then so is $G$. But $P = \{g \in G : g \in P(N, \leq)$ or $Ng \in P(G/N, \leq)\}$ satisfies the conditions of proposition 8.2.30.     Q.E.D.

**Corollary 8.2.32:** *If $G$ has a subnormal series each of whose factors is finitely generated torsion-free abelian then $G$ is left ordered.*

**Proof:** By the proposition we may assume $G$ is torsion-free abelian, so we are done by remark 1.2.31.     Q.E.D.

We are interested in left-ordered groups because they provide an instance for which there is an easy solution to the zero-divisor question.

**Proposition 8.2.33:** *If $G$ is torsion-free and left-ordered then $F[G]$ is a domain.*

**Proof:** Given $r_1, r_2 \neq 0$ in $F[G]$ choose $h$ maximal in $\mathrm{supp}(r_2)$, and then choose $g$ in $\mathrm{supp}(r_1)$ such that $gh$ is maximal possible. We claim $gh \neq g_1 h_1$ for any other $g_1$ in $\mathrm{supp}(r_1)$ and $h_1$ in $\mathrm{supp}(r_2)$. This will certainly imply $r_1 r_2 \neq 0$. Indeed, suppose $gh = g_1 h_1$; then $g_1 h \leq gh = g_1 h_1$ so multiplying by $g_1^{-1}$ yields $h \leq h_1$, and thus $h = h_1$. But now $gh = g_1 h$ so $g = g_1$, as desired.     Q.E.D.

A related property of left-ordered groups is given in exercise 5.

**Corollary 8.2.34:** *If $G$ is as in corollary 8.2.32 then $F[G]$ is a domain.*

**Proof:** $G$ is torsion-free since any torsion passes down the series.     Q.E.D.

These results pretty well summarize what was known about the zero-divisor question in 1973 when Formanek affirmed it for "supersolvable" groups $G$, i.e., $G$ has a series $G \geq G_1 \geq \cdots \geq G_t = 0$ with each $G_i \lhd G$ and $G_i/G_{i+1}$ cyclic. This motivated work on the important abelian-by-finite case, which has applications to the second cohomology group and thus to division algebras (cf., exercise 7ff; actually Farkas [75] managed to obtain the key applications by means of a clever reduction to corollary 8.2.34). K. A. Brown settled the abelian-by-finite case in characteristic 0, and building on this, Farkas-Snider [76] proved their famous theorem:

**Theorem 8.2.35:** *(Farkas-Snider) If $G$ is torsion-free polycyclic-by-finite and* $\operatorname{char}(F) = 0$ *then $F[G]$ is a domain.*

*Proof:* The idea behind the Farkas-Snider proof is quite intuitive. $R = F[G]$ is a prime Noetherian ring by Connell's theorem and theorem 8.2.2 and thus has a classical ring of fractions $S^{-1}R = M_n(D)$. Suppose $R$ is not a domain. There is a left ideal $L$ of $R$ with $0 < [S^{-1}L:D] \leq n$. We would like to conclude as in the proof of theorem 5.1.45 with the contradiction that $n^2$ divides $[S^{-1}L:D]$, but we do not know if every projective $F[G]$-module is stably free.

Nevertheless, by proposition 8.2.6 there is a normal polycyclic subgroup $H$ of finite index $t$, each of whose factors is infinite cyclic. Thus $F[H]$ is a Noetherian domain and so $S_1^{-1}F[H]$ is a division ring $D_1$, where $S_1 = \{$regular elements of $F[H]\}$. But $S_1 \subset S$ in view of remark 8.1.36. Hence $0 < [S_1^{-1}L:D_1] < [S_1^{-1}R:D_1]=t$ (for if $S_1^{-1}L=S_1^{-1}R$ then localizing further at $S$ would yield $S^{-1}L = S^{-1}R$, contrary to assumption). Thus to arrive at the desired contradiction it suffices to prove $t \mid [S_1^{-1}L:D_1]$.

By theorem 8.2.20 there is an f.g. projective resolution $0 \to P_m \to \cdots \to P_0 \to L \to 0$ of $L$ in $R\text{-}\mathcal{M}od$. Each $P_i$ is f.g. projective as $F[H]$-module, so $0 \to S_1^{-1}P_m \to \cdots \to S_1^{-1}P_0 \to S_1^{-1}L \to 0$ is a projective resolution in $D_1\text{-}\mathcal{M}od$. By remark 5.1.43 we need prove the following.

*Claim 1.* If $P$ is a projective $R$-module then $t$ divides $[S_1^{-1}P:D_1]$.

To prove claim 1 we use $\operatorname{Tr}_G$ of definition 8.2.26. Note $P$ is also $F[H]$-projective so we can also define $\operatorname{Tr}_H P$. Coupling theorem 5.1.34 with proposition 5.1.32 one sees every projective $F[H]$-module is stably free. Writing $P \oplus F[H]^{(u)} \approx F[H]^{(v)}$ we tensor on the left by $D_1 = S_1^{-1}F[H]$ to get

$$S_1^{-1}P \oplus D_1^{(u)} \approx D_1^{(v)}.$$

Thus $[S_1^{-1}P:D_1] = v - u$. On the other hand,

$$v = \mathrm{Tr}_H(F[H]^{(v)}) = \mathrm{Tr}_H(P \oplus F[H]^{(u)}) = \mathrm{Tr}_H P + u,$$

so $\mathrm{Tr}_H P = [S_1^{-1}P:D_1]$. Thus to prove claim 1 we need to show $t$ divides $\mathrm{Tr}_H P$. Since $\mathrm{Tr}_G P$ is an integer (theorem 8.2.27) it suffices to prove

*Claim 2.* $\mathrm{Tr}_H P = (\mathrm{Tr}_G P)t$.

*Proof of Claim 2.* Let $e_1, \ldots, e_n$ be a canonical right base for $R^{(n)}$ over $R$. Writing $P = eM_n(R)$ and $ee_i = \sum_{j=1}^n e_j r_{ji}$ we can identify $e$ with the matrix $(r_{ji})$, and thus $\mathrm{Tr}_G P = \sum t_G r_{ii}$. Now let $g_1, \ldots, g_t$ be a left transversal of $G$ over $H$. Then $\{e_i g_k : 1 \le k \le t, 1 \le i \le n\}$ is a right base of $R^{(n)}$ over $F[H]$ and

$$e(e_i g_k) = (ee_i)g_k = \left(\sum e_j r_{ji}\right)g_k = \sum_j e_j g_k(g_k^{-1} r_{ji} g_k),$$

so we can identify $e$ with the $nt \times nt$ matrix whose diagonal has the entries $g_k^{-1} r_{ii} g_k$ for $1 \le k \le t$ and $1 \le i \le n$. Thus

$$\mathrm{Tr}_H P = \sum_{k=1}^t \sum_{i=1}^n t_G(g_k^{-1} r_{ii} g_k) = \sum_{k=1}^t \sum_{i=1}^n t_G r_{ii} = \sum_{k=1}^t \mathrm{Tr}_G P = (\mathrm{Tr}_G P)t,$$

as desired. This proves claim 2 and thus concludes the proof of the Farkas-Snider theorem.          Q.E.D.

Cliff [80] has verified the zero-divisor conjecture in characteristic $\ne 0$ for torsion-free polycyclic-by-finite groups.

These results greatly enhanced the status of group algebras of polycyclic-by-finite groups, especially since they are the only known examples of Noetherian group algebras, and below we shall describe the flurry of activity in this area. Special attention should be paid to the following situation. Suppose $\mathrm{char}(F) = 0$ and $G$ is polycyclic-by-finite. $F[G]$ is a semiprime Goldie ring and thus has a semisimple Artinian ring of fractions $Q$. If $G$ has no finite normal subgroups $\ne (1)$ then $F[G]$ is prime so $Q \approx M_n(D)$ for some division ring $D$ and suitable $n$. J. Moody [86] has proved the following result (cf., Farkas [80, #18]):

**Goldie Rank Theorem:**  *$n$ is the least common multiple of the orders of the finite subgroups of $G$. (Equivalently, $\mathrm{Tr}_G P \in \mathbb{Z}$ for every projective module $P$ over $F[G]$.)*

Further progress also has been made on the zero-divisor question for solvable groups of finite rank, cf., Crawley-Boerey-Kropholler-Linnell [87].

Strojnowski [86] found a proof of claim 1 which appeals more directly to dependence of elements and thus applies in more general settings, cf., exercises 12–15; he also obtains results concerning when idempotents are trivial, cf., exercise 16.

## Digression: The Structure of Polycyclic-by-Finite Group Algebras

Spurred by Roseblade [78], there has been a spurt of activity concerning the polycyclic-by-finite group algebras. Following Snider [80] we shall sketch Roseblade's method in bold strokes and also indicate recent further research. Two excellent recent references are Passman [84] (which contains proofs of many of the results given here) and Farkas [86];

**Definition 8.2.36:** A group $G$ is *orbitally sound* if for any subgroup $H$ whose normalizer is of finite index, $H$ contains a normal subgroup of $G$ of finite index. An ideal $A$ of $F[G]$ is *faithful* if $1 - g \notin A$ for all $g \neq 1$ in $G$. A field $F$ is *absolute* if $F$ is an algebraic extension of a finite field (so, in particular, $\operatorname{char}(F) \neq 0$).

**Roseblade's Theorem:** *Let* $\Delta = \Delta(G)$. *If* $G$ *is orbitally sound and* $P$ *is a faithful prime ideal of* $F[G]$ *then* $P = (P \cap F[\Delta])F[G]$. *Furthermore, for* $F$ *nonabsolute, an ideal* $P$ *is primitive iff* $F[\Delta]/(P \cap F[\Delta])$ *is finite dimensional over* $F$.

Assume throughout the remainder of this discussion that $G$ is polycyclic-by-finite. Then one has the following information:

(1) $G$ has an orbitally sound subgroup of finite index; this is proved by Roseblade [78], and an improved version (with easier proof) due to Wehrfritz is presented in Farkas-Snider [79, theorem 2.1]. Thus Roseblade's theorem is available.

(2) Suppose $F$ is a nonabsolute field. $F[G]$ is primitive iff $\Delta(G) = 1$. (Compare with the discussion in §8.1.) This is due to Domanov [78] and Farkas-Passman [78] in characteristic 0 and to Rosenblade [78] in general.

(3) Every faithful primitive ideal has the form $\operatorname{Ann}(F[G] \otimes_{F[H]} M)$ for a suitable subgroup $H$ and a suitable simple $F[H]$-module $M$ of dimension 1 over $F$.

(4) Lewin [82] proved that if $G$ is non-nilpotent then $G$ has non-free projective modules. On the other hand, Stafford found a rather easy to construct non-free projectives in case $G$ is nilpotent and nonabelian, cf., §8.4. This was used by Farkas to provide a counterexample to a "folklore" conjecture about idempotents, cf., exercise 18.

(5) Roseblade [73] proved $F[G]$ is Jacobson and satisfies the Nullstellensatz. These results will be seen to be a consequence of more general ring theoretic results in §8.4.

(6) When $F$ is absolute, all simple $F[G]$-modules are finite dimensional over $F$; thus all primitive ideals are maximal, and, in particular, $F[G]$ is not primitive. (Compare with #2 and Roseblade's theorem above.) This is used to show that certain finitely generated groups are residually finite, cf., Passman [77B, theorem 12.3.13].

(7) Lorenz-Passman [81] have determined the prime ideals of $F[G]$; this work is described in Passman [84, §11].

(8) What is the largest possible transcendence degree $d$ of a maximal subfield of the division ring of fractions $D$ of $F[G]$? Resco showed $d \le h(G)$ the Hirsch number of $G$, cf., exercises 7.1.32–7.1.34; Stafford [83] proved more generally that $h(G) = \text{gl. dim } D \otimes_F D = K\text{-dim } D \otimes_F D$.

(9) Farkas-Schofield-Snider-Stafford [82] proved that the division ring of fractions of the group algebra (over $\mathbb{Q}$) of a torsion-free finitely generated nilpotent group is uniquely determined in the sense that if $\mathbb{Q}[G_1]$ and $\mathbb{Q}[G_2]$ have isomorphic division rings of fractions $D_1, D_2$ then $G_1 \approx G_2$. The idea is to order $G_i$ for $i = 1, 2$ and embed $\mathbb{Q}[G_i]$ into the Mal'cev-Neumann ring of "crossed product" Laurent series, cf., exercise 7.2.10, the map back to $G_i$ taking the term of lowest order enables one to restrict the given isomorphism $D_1 \to D_2$ to a group isomorphism $G_1 \to G_2$. (This sketch glosses over some tricky points.) This theorem cannot be extended to arbitrary torsion-free polycyclic groups, as evidenced by the following counterexample:

Let $G_1 = \langle x, y, z : [z, x] = 1 = [x, y]$ and $zyz^{-1} = y^{-1} \rangle$ and $G_2 = \langle a, b, c : ab, cac^{-1} = b$, and $cbc^{-1} = a \rangle$. Then $G_1 \not\approx G_2$ since $G_1/G_1'$ is not torsion-free (for $y^2$ is a commutator) but $G_2/G_2'$ is torsion-free (for $G_2' = \langle ab^{-1} \rangle$). On the other hand, $D_1 \approx D_2$; indeed, $F[G_2]$ contains the subalgebra $R$ generated by $x_1 = a + b$, $y_1 = ab^{-1}$, and $z_1 = c$, so $D_2$ contains the ring of fractions $Q$ of $R$ which is an isomorphic copy of $D_1$, and, in fact, $D_2 = Q$ since $a = (a + b)(ab^{-1})(ab^{-1} + 1)^{-1} \in Q$.

(10) Passman [82] showed the division ring of fractions of the group algebra of a poly-infinite cyclic group is the "universal field of fractions" in Cohn's sense and so is amenable to arithmetic techniques.

(11) Utilizing a result of Markar-Limanov, Lorenz [84] showed that a division $F$-algebra $D$ generated as division algebra by a polycyclic-by-finite group $G$ contains a free subalgebra iff $G$ is not abelian-by-finite. In particular, GK dim $D < \infty$ iff $G$ is abelian-by-finite, cf., proposition 6.2.14.

Many of these results were inspired by earlier results for enveloping algebras of finite dimensional Lie algebras, and now there is considerable interplay between the two theories. We shall explore some general ring-theoretic approaches in §8.4, as well as other properties of prime ideals which are relevant in this situation.

## §8.3 Enveloping Algebras

The object of this section is to describe some of the theory of enveloping algebras of Lie algebras using methods of ring theory; the next section continues this project from an even more ring-theoretic point of view. Unfortunately, we do not have the space to deal adequately with semisimple Lie algebras; a brief overview of the subject is given at the end of the section.

We shall prove only the basic general facts about enveloping algebras and then focus on nilpotent and solvable Lie algebras, which are more amenable to ring theoretic techniques. We draw heavily from the classic Dixmier [74B] and the excellent reference Borho-Gabriel-Rentschler [73B], henceforth referred to as respectively as DIX and BGR. These books have a large overlap, and we usually refer to BGR. The classic text on Lie algebras is still Jacobson [62B]. Recall the basic definitions from 1.6.10ff, and suppose $L$ is a Lie algebra over a commutative ring $C$. The following concept also will be useful.

**Definition 8.3.0:** A *Lie L-module* is a $C$-module $M$ endowed with scalar multiplication $L \times M \to M$ satisfying $[ab]x = a(bx) - b(ax)$ for all $a, b$ in $L$ and all $x$ in $M$.

**Remark 8.3.1:** There is a functor $F: C\text{-}\mathcal{A}\ell g \to C\text{-}\mathcal{L}ie$ given by $FR = R^-$, where given a morphism $f: R_1 \to R_2$ we define $Ff: R_1^- \to R_2^-$ to be the same map viewed as a Lie homomorphism.

We are led to consider the universal from a given Lie algebra $L$ to the functor of remark 8.3.1; in other words, we want an algebra $U(L)$ together with a morphism $u: L \to U(L)^-$, such that for any associative algebra $R$ and any Lie homomorphism $f: L \to R^-$ there is a unique homomorphism $g: U(L) \to R$ such that $f = gu$.

**Proposition 8.3.2:** $U(L)$ *exists and can be taken to be* $T(L)/A$ *where* $T(L) = T_C(L)$ *is the tensor algebra for* $L$ *(as C-module) and* $A$ *is the ideal generated by all* $a \otimes b - b \otimes a - [ab]$ *for all* $a,b$ *in* $L$; $u: L \to T(L)/A$ *is the map* $a \mapsto a + A$.

**Proof:** Suppose $f: L \to R^-$ is given. By proposition 1.9.14 there is an algebra homomorphism $\tilde{f}: T(L) \to R$ given by $\tilde{f}(a_1 \otimes \cdots \otimes a_t) = fa_1 \cdots fa_t$. In particular, $\tilde{f}(a \otimes b - b \otimes a - [ab]) = [fa, fb] - f[ab] = 0$ for all $a,b$ in $L$, so $A \subseteq \ker \tilde{f}$. This gives us the desired homomorphism $g: T(L)/A \to R$, and uniqueness of $g$ follows from the fact that $uL$ generates $T(L)/A$ as an algebra.    Q.E.D.

By abstract nonsense $U(L)$ is unique up to isomorphism; the key additional fact is

**Theorem 8.3.3:** *(Poincare-Birkhoff-Witt = PBW theorem) Suppose $L$ is free in $C$-$\mathcal{M}od$ (e.g., if $C$ is a field). Then $u: L \to U(L)$ is monic. Explicitly if $X = \{x_i : i \in I\}$ is a base for $L$ over $C$ then well-ordering $I$ and writing $y_i$ for $ux_i$ in $U(L)$ we have the base $B = \{y_{i_1} \cdots y_{i_t} : t \in \mathbb{N}$ and $i_1 \leq \cdots \leq i_t\} \cup \{1\}$ of $U(L)$.*

**Proof:** Certainly the second assertion implies the first, for if $\sum c_i x_i \in \ker u$ then $\sum c_i y_i = 0$ and each $c_i = 0$. To prove $B$ is a base grade $T = T(L)$ by $\mathbb{N}$ in the usual way and let $\tilde{T}_m = \sum_{i=0}^{m} T_i$, $U_m = \tilde{T}_m/(\tilde{T}_m \cap A) \approx (\tilde{T}_m + A)/A \subset U(L)$, and $B_m = \{y_{i_1} \cdots y_{i_t} : t \leq m$ and $i_1 \leq \cdots \leq i_t\}$. Since $B = \bigcup B_m$ and $U(L) = \bigcup U_m$, it suffices to prove

**Claim:**  *Each $B_m$ is a base of $U_m$.*

The claim will be proved by induction on $m$; it is clear for $m = 0, 1$ since $B_0 = \{1\}$ is a base of $U_0 \approx C$, and $B_1 = $ is a base of $U_1 \approx \tilde{T}_1 \approx C \oplus L$ (because $\tilde{T}_1 \cap A = 0$).

Note that if $\pi \in \text{Sym}(m)$ and $a_1, \ldots, a_m \in L$ then

$$(ua_1) \cdots (ua_m) - (ua_{\pi 1}) \cdots (ua_{\pi m}) \in U_{m-1}.$$

(Indeed, $\pi$ is a product of transpositions of the form $(i, i + 1)$, so we may assume $\pi = (i, i + 1)$. But then

$$(ua_1) \cdots (ua_m) - (ua_{\pi 1}) \cdots (ua_{\pi m}) = ua_1 \cdots ua_{i-1} u[a_i a_{i+1}] ua_{i2} \cdots ua_{im} \in U_{m-1}$$

as desired.

Choosing the permutation $\pi$ to rearrange the indices in ascending order, we see any element $b_{i_1} \cdots b_{i_m}$ is spanned by $B_m \cup U_{m-1}$, so by induction $U_m$ is spanned by $B_m \cup B_{m-1} = B_m$.

It remains to show $B$ is independent, which we shall do by means of the regular representation. Write $\mathcal{L} = \{$finite ascending sequences in $I\}$, and let $V$ be the free $C$-module with base $\mathcal{L}$. Given $J = (i_1, \ldots, i_m)$ in $\mathcal{L}$ we write $y_J$ for $y_{i_1} \cdots y_{i_m}$ in $U(L)$. We say $i \le J$ if $i \le i_1$, in which case we denote $(i, i_1, \ldots, i_m)$ by $iJ$. For convenience write $x_i x_j v_J$ for $x_i(x_j v_J)$. Consider the scalar multiplication $L \times V \to V$ defined inductively on the length of $J$, as follows:

$$
\begin{aligned}
x_i v_J &= v_{iJ} && \text{if } i \le J \\
[x_i x_j] v_{J'} + x_j x_i v_{J'} && \text{if } J = jJ' \text{ and } i > j
\end{aligned}
\tag{1}
$$

**Claim:** *V is a Lie L-module under this multiplication.*

This would ensure that the regular representation $f: L \to \text{End}_C V$ given by $(fx_i)v_J = x_i v_J$ is a Lie map and thus lifts to a homomorphism $\tilde{f}: U(L) \to \text{End}_C V$. But now for any $\alpha_J$ in $C$ not all 0 we must have $\sum \alpha_J y_J \ne 0$ since taking $i$ greater than any index appearing in the $J$ would yield $\tilde{f}(\sum \alpha_J y_J)v_i = \sum \alpha_J v_{Ji} \ne 0$ in $V$.

Thus it remains to verify the claim, which would follows immediately from

$$
x_k x_j v_{J'} = [x_i x_k] v_J + x_k x_i v_J \qquad \text{for all } i, k.
\tag{2}
$$

If $k \le J$ this is immediate from (1). Thus we may assume $k \not\le J$. Using double induction we may assume (2) holds if $J$ is replaced by a sequence of smaller length or if $k$ is replaced by a smaller index. Thus, writing $J = jJ'$ with $k > j$ we may assume

$$
x_k x_j v_{J'} = [x_k x_j] v_{J'} + x_j x_k v_{J'} \qquad \text{and}
$$

$$
\begin{aligned}
[x_i x_k] x_j v_{J'} &= [[x_i x_k] x_j] v_{J'} + x_j [x_i x_k] v_{J'} \\
&= [[x_i x_k] x_j] v_{J'} + x_j (x_i x_k v_{J'} - x_k x_i v_{J'})
\end{aligned}
$$

and (since $j < k$)

$$
x_i x_j (x_k v_{J'}) = [x_i x_j] x_k v_{J'} + x_j x_i (x_k v_{J'}).
$$

Plugging in yields

$$
\begin{aligned}
x_i x_k v_J &= x_i (x_k x_j v_{J'}) = x_i ([x_k, x_j] v_{J'} + x_j x_k v_{J'}) \\
&= x_i [x_k x_j] v_{J'} + x_i x_j x_k v_{J'} \\
&= ([x_i [x_k x_j]] v_{J'} + [x_k x_j] x_i v_{J'}) + ([x_i x_j] x_k v_{J'} + x_j x_i x_k x_{J'}).
\end{aligned}
$$

On the other hand, we are done if we can prove (2) with $i$ and $k$ interchanged, so the same analysis yields

$$x_k x_i v_J = [x_k[x_i x_j]]v_{J'} + [x_i x_j]x_k v_{J'} + [x_k x_j]x_i v_{J'} + x_j x_k x_i v_{J'}.$$

Taking the difference and using the Jacobi identity we get

$$x_i x_k v_J - x_k x_i v_J = ([x_i[x_k x_j]] - [x_k[x_i x_j]])v_{J'} + x_j(x_i x_k v_{J'} - x_k x_i v_{J'})$$

$$= [x_j[x_k x_i]]v_{J'} + x_j[x_i x_k]v_{J'}$$

$$= [x_i x_k]x_j v_{J'}$$

$$= [x_i x_k]v_J$$

proving (2) as desired.     Q.E.D.

We say a *Lie representation* is a Lie map $L \to \text{End}_C V$ where $V$ is an arbitrary $C$-module.

**Corollary 8.3.4:**   *Any Lie representation $L \to \text{End}_C V$ extends uniquely to an algebra homomorphism $U(L) \to \text{End}_C V$.*

**Corollary 8.3.5:**   *Any Lie map $f: L_1 \to L_2$ extends uniquely to an algebra homomorphism $Uf: U(L_1) \to U(L_2)$, thereby yielding a functor $U: C\text{-}\mathcal{L}ie \to C\text{-}\mathcal{A}lg$. If $L_1, L_2$ are free $C$-modules and $f$ is monic (resp. epic) then $Uf$ is an injection (resp. surjection).*

***Proof:***   The composition $L_1 \to L_2 \hookrightarrow U(L_2)$ yields the desired homomorphism $U(L_1) \to U(L_2)$. The other assertions are straightforward.     Q.E.D.

## Basic Properties of Enveloping Algebras

The sequel will be limited to a Lie algebra $L$ over a field $F$, and we shall view $L \subset U(L)$ by means of the PBW theorem. As in group algebras, we strike the dilemma that the structure of $U(L)$ must be intricate enough so as to include the representation theory of $L$; thus the study of $U(L)$ is both desirable and forbidding. Certain properties however are easy consequences of the PBW theorem.

**Proposition 8.3.6:**   $U(L_1 \times \cdots \times L_t) \approx U(L_1) \otimes \cdots \otimes U(L_t).$

***Proof:***   It suffices to prove this for $t = 2$. Then the map $f: L_1 \times L_2 \to$

$U(L_1) \otimes U(L_2)$ given by $f(a_1, a_2) = ua_1 \otimes ua_2$ extends to a homomorphism $U(L_1 \times L_2) \to U(L_1) \otimes U(L_2)$. Conversely $U(L_i) \subseteq U(L_1 \times L_2)$ for $i = 1, 2$ by corollary 8.3.5, and the balanced map $g: U(L_1) \times U(L_2) \to U(L_1 \times L_2)$ given by $g(r_1, r_2) = r_1 r_2$ yields a homomorphism $g: U(L_1) \otimes U(L_2) \to U(L_1 \times L_2)$ which is the inverse of $f$ on $L_1 \times L_2$ and thus on $U(L_1 \times L_2)$. Q.E.D.

**Remark 8.3.7:** If $K$ is a field extension of $F$ then $L \otimes_F K$ is a Lie algebra under "extension of scalars," and $U(L) \otimes_F K$ is isomorphic to the enveloping algebra $U(L \otimes_F K)$ of $L \otimes_F K$ over $K$. Indeed, there is a monic $L \otimes K \to U(L) \otimes K$; by corollary 8.3.5 this extends to an injection $U(L \otimes K) \to U(L) \otimes K$ which is onto by the PBW.)

**Remark 8.3.8:** If $L_1$ is a Lie subalgebra of $L$ then expanding a base $B_1$ of $L_1$ to a base $B$ of $L$ we see by the PBW theorem that $U(L)$ is free as $U(L_1)$-module. (Indeed, write $B' = B - B_1 = \{x_i : i \in I\}$, and well-order $I$. Then $\{x_{i_1} \cdots x_{i_m} : i_1 \leq \cdots \leq i_m\}$ is a base, since elements of $B_1$ can be pushed to the left by means of the Lie multiplication.)

**Proposition 8.3.9:** *If $A$ is a Lie ideal of $L$ then $U(L)A = AU(L) \triangleleft U(L)$ and $U(L/A) \approx U(L)/U(L)A$.*

**Proof:** The Lie map $L \to L/A$ gives a surjection $\varphi: U(L) \to U(L/A)$ as in corollary 8.3.5; viewing $U(L)$ as a free $U(A)$-module we see the base of remark 8.3.8 is sent to a base of $U(L/A)$, so $U(L)A = \ker \varphi \triangleleft U(L)$; symmetrically $AU(L) = \ker \varphi$. Q.E.D.

**Definition 8.3.10:** Let $U_m(L)$ be the subspace of $U(L)$ generated by all products $\{a_1 \cdots a_n : n \leq m$ and $a_i \in L\}$. $\{U_m : m \in \mathbb{N}\}$ will be called the *standard filtration* of $U(L)$. We say the *order* of $r \in U(L)$ is the smallest $m$ for which $r \in U_m$.

**Note:** This is a filtration according to definition 1.2.13, if we change the indices to negative number as in remark 1.2.13'; thus $U(L)$ is filtered according to definition 5.1.36.

**Remark 8.3.11:** Each $U_m(L)$ is an $L$-module under Lie multiplication, since $[a, a_1 \cdots a_m] = \sum a_1 \cdots a_{i-1}[aa_i]a_{i+1} \cdots a_m$. Of course, if $L$ is finite dimensional as $F$-algebra then $U_m(L)$ also is f.d.

We shall now use the associated graded algebra of $U(L)$, denoted as $G(L)$ (cf., definition 3.5.30) to study $U(L)$. Recall $G(L) = \bigoplus G_i$ where $G_i = U_i(L)/U_{i-1}(L)$, in particular, $G_0 = F$ and $G_1 = L$; multiplication is transferred from $U(L)$.

**Remark 8.3.12:** $G(L)$ is commutative, since $[r_i, r_j] \in U_{i+j-1}(L)$ for all $r_i$ in $U_{i-1}(L)$ and $r_j$ in $U_{j-1}(L)$ (as in remark 8.3.11). Moreover, $G_n = (G_1)^n$.

**Proposition 8.3.13:** $G(L)$ *is canonically isomorphic* (*as graded algebra*) *to the symmetric algebra* $S(L)$ *defined in example 1.9.17.*

**Proof:** Let $\{x_i : i \in I\}$ be a base of $L$. Define $\varphi: T_F(L) \to G(L)$ by sending a monomial $x_{i_1} \otimes \cdots \otimes x_{i_m}$ to its image in $G_m$. Note each $x_i \otimes x_j - x_j \otimes x_i \in \ker \varphi$ since, viewed in $U(L)$, $x_i x_j + U_1(L) = x_j x_i + U_1(L)$ (because $[x_i, x_j] \in L$). Thus we have a homomorphism $\bar{\varphi}: S(L) \to G(L)$ which sends base to base (since $\{x_{i_1} \cdots x_{i_m} : i_1 \leq \cdots \leq i_m\}$ can be viewed as a base for the $m$-th component in each). Thus $\bar{\varphi}$ is an isomorphism.          Q.E.D.

(We shall show below this implies $U(L)$ has GK-dim when $L$ is finite dimensional.) Since $S(L)$ is a domain we see $G(L)$ is a domain, providing the following consequence.

**Theorem 8.3.14:** $U(L)$ *is a domain.* $\text{Jac}(R) = 0$ *for every* subring $R$ *of* $U(L)$ *not contained in* $F$. *In fact, if* $r_i \in U(L)$ *has order* $m_i$ *for* $i = 1, 2$ *then* $r_1 r_2$ *has order* $m_1 + m_2$.

**Proof:** The last assertion is clear by passing to "leading terms" in the domain $G(L)$. Hence $U(L)$ is a domain. Furthermore, if $0 \neq r \in \text{Jac}(R)$ then taking $r_1 \in R - F$ we see $r_1 r$ has order $> 0$ so $1 - r_1 r$ also has order $> 0$ and cannot be invertible, contradiction; thus $\text{Jac}(R) = 0$.          Q.E.D.

**Corollary 8.3.14′ (Irving):** *Suppose* $L$ *is a Lie algebra over an integral domain* $C$ *and is faithful* (*as* $C$-*module*). *Then* $\text{Jac}(U(L)) = 0$.

This result generalizes the commutative version of Amitsur's theorem (by taking $L$ to be 1-dimensional).

**Proof:** Let $S = C - \{0\}$ and $F = S^{-1}C$. Then $S^{-1}U(L)$ satisfies the universal property of the universal enveloping algebra of the Lie algebra $L \otimes_C F$ over $F$; making this identification we see $\text{Jac}(U(L)) = 0$ by theorem 8.3.14. Furthermore, $L \to L \otimes F \to U(L \otimes F)$ is monic so $L \to U(L)$ is monic.          Q.E.D.

**Proposition 8.3.15:**   *If* $[L:F] < \infty$ *then* $U(L)$ *is left and right Noetherian.*

**Proof:**   $G(L)$ is Noetherian by the Hilbert basis theorem so apply corollary 3.5.32(ii).    Q.E.D.

See exercise 15–18 for instances of $U(L)$ Ore, even when $L$ is infinite dimensional.

**Remark 8.3.16:**   There is an anti-automorphism of $L$ given by $a \to -a$, which can be viewed as an isomorphism of $L$ to the opposite algebra $L^{\mathrm{op}}$. This extends to an automorphism $U(L) \to U(L)^{\mathrm{op}}$, i.e., an involution of $U(L)$, called the *principal involution*.

## Addition Theory for Characteristic 0

Let us recapitulate the discussion from remark 8.3.12 on, to gain more precise information. We assume char$(F) = 0$. There is a commutative diagram

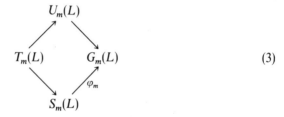

$$(3)$$

where the subscript denotes the $m$-th component. When unambiguous the $L$ shall be dropped from the notation. Diagram (3) can be restricted to a diagram of isomorphisms. Indeed $\varphi_m : S_m \to G_m$ is already an isomorphism by proposition 8.3.13. To get isomorphisms in the other directions, we take the subset $T'_m$ of $T_m$ consisting of *homogeneous* tensors, defined to be elements of the form $\sum_{\pi \in \mathrm{Sym}(m)} a_{\pi 1} \otimes \cdots \otimes a_{\pi m}$ for $a_i$ in $L$, and let $U^m$ denote the canonical image of $T'_m$ in $U_m$. Thus (3) restricts to the diagram

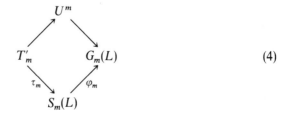

$$(4)$$

**Proposition 8.3.17:**   *Each map in diagram (4) is bijective. Thus there is a group isomorphism* $\psi : S(L) \to U(L)$.

**Proof:**  $\tau_m$ has the inverse map given by

$$x_1 \otimes \cdots \otimes x_m \mapsto (m!)^{-1} \sum x_{\pi 1} \otimes \cdots \otimes x_{\pi m}$$

and is thus bijective; hence the map $T'_m \to G_m$ is bijective, so each upper arrow is bijective.     Q.E.D.

**Proposition 8.3.18:**  $U_m = U_{m-1} \oplus U^m$.

**Proof:**  $U_m / U_{m-1} = G_m \approx U^m$, so $U_m = U_{m-1} + U^m$, and clearly $U_{m-1} \cap U^m = 0$ by comparing bases.     Q.E.D.

**Corollary 8.3.19:**  $U(L) = \bigoplus_{m \in \mathbb{N}} U^m(L)$, *and this is functorial (in the sense that any Lie map $L_1 \to L_2$ extends to a graded map $U(L_1) \to U(L_2)$).*

Let us turn to derivations, cf., definition 1.6.5. Any $a_0$ in $L$ induces the *adjoint map* $\mathrm{ad}\,a_0 : L \to L$ given by $(\mathrm{ad}\,a_0)a = [a_0, a]$; this is a derivation of $L$ (by the Jacobi identity), and all such derivations are called *inner*.

**Proposition 8.3.20:**  *Suppose $\delta$ is a derivation on $L$. Then $\delta$ extends to a unique derivation $\delta'$ on $U(L)$ and to a unique derivation $\delta''$ on $S(L)$, and $\delta'\psi = \psi\delta''$ where $\psi : S(L) \to U(L)$ is the canonical group isomorphism of proposition 8.3.17. If $\delta = \mathrm{ad}\,x$ for some $x$ in $L$ then $\delta' = [x, \ ]$ is inner.*

**Proof:**  Define  $\tilde{\delta} : T(L) \to T(L)$  by  $\tilde{\delta}(a_1 \otimes \cdots \otimes a_t) = \sum_{i=1}^{t} a_1 \otimes \cdots \otimes a_{i-1} \otimes \delta a_i \otimes a_{i+1} \otimes \cdots \otimes a_t$. Then  $\tilde{\delta}((a_1 \otimes \cdots a_t) \otimes (a_{t+1} \otimes \cdots \otimes a_m)) = \tilde{\delta}(a_1 \otimes \cdots \otimes a_t) \otimes a_{t+1} \otimes \cdots \otimes a_m + a_1 \otimes \cdots \otimes a_t \otimes \tilde{\delta}(a_{t+1} \otimes \cdots \otimes a_m)$, from which one sees at once that $\tilde{\delta}$ is a derivation on $T(L)$. But $\tilde{\delta}(a \otimes b - b \otimes a - [ab]) = (\delta a \otimes b - b \otimes \delta a - (\delta a)b) + (a \otimes \delta b - \delta b \otimes a - a\delta b)$. Thus we see $\tilde{\delta}A \subseteq A$ where $A$ is as in proposition 8.3.2, so $\tilde{\delta}$ induces a derivation $\delta' : U(L) \to U(L)$. On the other hand, $\tilde{\delta}(a \otimes b - b \otimes a) = (\delta a \otimes b - b \otimes \delta a) + (a \otimes \delta b - \delta b \otimes a)$, so $\tilde{\delta}$ induces a derivation $\delta'' : S(L) \to S(L)$, and one sees easily that $\delta'\psi = \psi\delta''$. Finally, if $\delta = \mathrm{ad}\,x$ then in $U(L)$ we have $\delta'(a_1 \cdots a_t) = \sum_{i=1}^{t} a_1 \cdots a_{i-1}[x, a_i]a_{i-1} \cdots a_t = [x, a_1 \cdots a_t]$, proving the last assertion. The uniqueness of $\delta'$ and $\delta''$ follows from the fact that $L$ generates $U(L)$ and $S(L)$.     Q.E.D.

**Corollary 8.3.21:**  $\psi : S(L) \to U(L)$ *is an $L$-module isomorphism.*

**Corollary 8.3.22:**  $\psi(\mathrm{Ann}'_{S(L)} L) = Z(U(L))$.

**Proof:** $z \in Z(U(L))$ iff $[a, z] = 0$ for each $a$ in $L$. Q.E.D.

This map is not necessarily a homomorphism since it may not preserve multiplication (exercise 3). However, for $F$ algebraically closed one always has $\text{Ann}'_{S(L)} L \approx Z(U(L))$, cf., [DIX, theorem 10.4.5]. One more general result of interest is

**Theorem 8.3.23:** *(Quillen) If $L$ is a finite dimensional algebra over a field then $U(L)$ is Jacobson.*

The proof involves certain ideas of Duflo [73] which have then been extended by McConnell to much more general classes of rings and will be discussed in §8.4, cf., corollary 8.4.14. This result enhances interest in the primitive ideals (since they are now seen to be dense in the prime spectrum).

## Nilpotent and Solvable Lie Algebras

Before continuing, we should quote some of the standard Lie algebra theory. First some definitions in analogy to group theory. *Assume that $L$ is finite dimensional over* a field $F$ of characteristic 0.

**Definition 8.3.24:** Writing $L^2$ for $[LL]$ and inductively $L^n$ for $[L^{n-1}L]$, we say $L$ is *nilpotent of index $n$* if $L^n = 0$ for $n$ minimal such. Likewise, writing $L^{(1)} = L$ and inductively $L^{(n)}$ for $[L^{(n-1)}L^{(n-1)}]$, we say $L$ is *solvable of length $n$* if $L^{(n)} = 0$ for some $n$. In particular, every nilpotent algebra is solvable.

In analogy to the associative radical theory, every finite dimensional Lie algebra has a unique largest solvable ideal, called the *radical* or $\text{rad}(L)$; $\text{rad}(L/\text{rad } L) = 0$.

Consider a Lie algebra $L$ of endomorphisms of a finite dimensional vector space $V$ over $F$. We say $L$ is *triangularizable* (resp. *strictly triangularizable*) if there is a change of base of $V$ with respect to which each element of $L$ is upper triangular (resp. strictly upper triangular). If every element of $L$ is nilpotent as a matrix then $L^n = 0$ for some $n$, by proposition 2.6.30. It follows easily that $L$ is strictly triangularizable. Indeed, take $k$ maximal for which $L^k \neq 0$, and take $0 \neq v_1 \in L^k V$; noting $Lv_1 = 0$ we see $[LV:F] < [V:F]$, so proceeding by induction on $[V:F]$ we can expand $v_1$ to a base with respect to which $L$ is strictly upper triangular. Since the hypothesis is satisfied by the adjoint representation of a nilpotent Lie algebra, we have proved the following:

L is nilpotent iff $\operatorname{Ad} L$ is nilpotent as a set of matrices, in which case $\operatorname{Ad} L$ is strictly triangularizable.

For solvable Lie algebras the situation is almost as nice. Suppose $\rho\colon L \to \operatorname{End}_F V$ is a representation. Given a subalgebra $L_1$ of $L$ we write $L_1^*$ for the dual space $\operatorname{Hom}(L_1, F)$ of linear functionals. If $\alpha \in L_1^*$ we define the *eigenspace* $V_\alpha = \{v \in V\colon (\rho a)v = \alpha(a)v$ for all $a$ in $L_1\}$, clearly a subspace invariant under the action of $L_1$.

**Lie's Lemma:** $[DIX, lemma\ 1.3.11]$ *If* $[V\colon F] < \infty$ *and* $L_1 \lhd L$ *then in the above notation* $V_\alpha$ *is invariant under the action of* $L$.

**Lie's Theorem:** $[DIX, theorem\ 1.3.12]$ *Any solvable Lie algebra of matrices over an algebraically closed field is triangularizable.*

The proof is a straightforward induction based on Lie's lemma. Let us now call a Lie algebra $L$ *completely solvable* if $\operatorname{Ad} L$ is triangularizable. Every completely solvable Lie algebra is solvable; conversely, every solvable Lie algebra over an algebraically closed field is completely solvable. For this reason the study of enveloping algebras over solvable Lie algebras usually is based on the hypothesis $F$ is algebraically closed. Lie's theorem provides $a \in L$ for which $Fa \lhd L$.

**Remark 8.3.25:** If $Fa \lhd L$ then $U(L)a \lhd U(L)$ and $U(L/Fa) \approx U(L)/U(L)a$. (By proposition 8.3.9.)

This remark is exceedingly useful in the study of enveloping algebras of solvable Lie algebras. One application ties in with McConnell's theory of completions discussed in §3.5. We say an ideal $A$ of $R$ is *polynormal* if $A = \sum_{i=1}^{t} Ra_i$ where $a_1$ is $R$-normalizing and inductively the image of $a_k$ in $\bar{R} = R/\sum_{i=1}^{k-1} Ra_i$ is $\bar{R}$-normalizing for each $1 < k \leq t$. (Compare to "polycentral," definition 3.5.34.) The $a_i$ are called the *polynormal generating set* of $A$.

**Theorem 8.3.26:** *(McConnell [68]) (i) If* $L$ *is a nilpotent Lie algebra then every ideal of* $U(L)$ *is polycentral. (ii) If* $L$ *is solvable and* $F$ *is algebraically closed then every ideal of* $U(L)$ *is polynormal.*

**Proof:** Suppose $A \lhd U(L)$. $L$ acts on $A$ by the adjoint action. Let us prove (ii), for (i) is analogous. As noted above there is some $a_1$ in $A$ for which

$Fa_1 \lhd L$ and $U(L/Fa_1) \approx U(L)/U(L)a_1$. Proceeding by induction on $[L:F]$ we see $A/Fa_1$ is polynormal with a suitable polynormal base $a_2 + Fa_1, \ldots,$ $a_t + Fa_1$, so $a_1, \ldots, a_t$ is a polynormal generating set of $A$.    Q.E.D.

**Corollary 8.3.27:**  *The completion of $U(L)$ with respect to any ideal is Noetherian if $L$ is nilpotent.*

**Proof:**  Apply theorem 3.5.36 to theorem 8.3.26.    Q.E.D.

We shall continue this train of thought in §8.4.

Nilpotent Lie algebras are "better behaved" in Dixmier's theory than solvable Lie algebras. For example, $\psi$ of corollary 8.3.22 *is* an associative homomorphism for $L$ nilpotent, cf., [DIX, proposition 4.8.12].

## Solvable Lie Algebras as Iterated Differential Polynomial Rings

Differential polynomial rings were introduced in §1.6, under the notation $R[\lambda; 1, \delta]$ where $\delta$ is a derivation of $R$. Since "1" is always understood here we shall now use the abbreviated notation $R[\lambda; \delta]$. In studying enveloping algebras of solvable Lie algebras the following remark is important. We say a ring $R$ is an *iterated differential polynomial ring* if there is a chain of subrings $R_\alpha$ of $R$ such that $R_{\alpha+1} \approx R_\alpha[\lambda_\alpha; \delta_\alpha]$ is a differential polynomial ring for every $\alpha$.

**Proposition 8.3.28:**  *If $L$ is a solvable Lie algebra over a field $F$ then $U(L)$ is an iterated differential polynomial ring.*

**Proof:**  $L^{(t)} = 0$ for some $t$. By induction on $t$ we have $U(L')$ is an iterated differential polynomial ring. Take any $a$ in $L - L'$. Then $L_1 = L' + Fa$ is a Lie subalgebra of $L$, and ad $a$ induces a derivation of $L_1$ which thus extends to a derivation $\delta$ of $U(L)$. Let $R = U(L')[\lambda; \delta]$. There is an epimorphism $\varphi: R \to U(L_1)$ given by $\lambda \mapsto a$; but $\ker \varphi = 0$ for if $0 = \varphi(\sum r_i \lambda^i) = \sum r_i a^i$ for $r_i$ in $U(L')$ then each $r_i = 0$ by remark 8.3.8. Continuing inductively, given $L_i \supset L'$ we can take $a_{i+1}$ in $L - L_i$ and let $L_{i+1} = L + Fa_i$; then $U(L_{i+1}) \approx U(L_i)[\lambda_i; \text{ad } a_{i+1}]$ and eventually we reach $L$.    Q.E.D.

This result is the cornerstone of our study of enveloping algebras of solvable Lie algebras and leads us to look a bit closer at properties of a derivation $\delta$ on a ring $R$.

**Remark 8.3.29:**   If $A \lhd R$ then $\delta(A^2) \subseteq A$. (Indeed

$$\delta(a_1 a_2) = a_1 \delta a_2 + (\delta a_1)a_2 \in A.)$$

**Lemma 8.3.30:**   *If S is a denominator set on R then $\delta$ extends uniquely to a derivation on $S^{-1}R$.*

**Proof:**   We start with uniqueness. Clearly $0 = \delta(s^{-1}s) = s^{-1}\delta s + (\delta s^{-1})s$ so we must have

$$\delta s^{-1} = -s^{-1}(\delta s)s^{-1} \text{ and thus}$$

$$\delta(s^{-1}r) = s^{-1}\delta r - s^{-1}(\delta s)s^{-1}r. \tag{1}$$

To verify that (1) indeed defines a derivation, first we show it is well-defined. In view of theorem 3.1.4(iii) we need show that $\delta((r's)^{-1}r'r) = \delta(s^{-1}r)$ for every $r'$ in $R$ such that $r's \in S$. Indeed

$$(r's)^{-1}\delta(r'r) - (r's)^{-1}\delta(r's)(r's)^{-1}r'r$$

$$= (r's)^{-1}(r'\delta r + (\delta r')r) - (r's)^{-1}(r'\delta s + (\delta r')s)(r's)^{-1}r'r$$

$$= s^{-1}\delta r + (r's)^{-1}(\delta r')r - s^{-1}(\delta s)s^{-1}r - (r's)^{-1}(\delta r')r$$

$$= s^{-1}\delta r - s^{-1}(\delta s)s^{-1}r$$

as desired.       Q.E.D.

This lemma also holds for more general quotient ring constructions, as one may expect.

**Proposition 8.3.31:**   *Suppose R is semiprime Goldie and $\delta$ is a derivation on R. Then $\delta P \subseteq P$ for every minimal prime ideal P of R.*

**Proof:**   Let $Q$ be the semisimple Artinian ring of fractions of $R$. By theorem 3.2.27 we know some simple component $Q \approx Q/A$ is the ring of fractions of $R/P$, where $A$ is a maximal ideal of $Q$ containing $P$. But $\delta$ extends to $Q$ by lemma 8.3.30, and $A = Qe$ for some central idempotent $e$ of $Q$. Hence $A^2 = A$ so $\delta A \subseteq A$ by remark 8.3.29; hence $\delta P \subseteq R \cap \delta A \subseteq R \cap A = P$.       Q.E.D.

We shall say $P \in \text{Spec}(R)$ is *$\delta$-invariant* if $\delta P \subseteq P$ and write $\delta\text{-Spec}(R)$ for $\{\delta\text{-invariant primes}\}$. We just proved $\delta\text{-Spec}(R)$ contains the minimal primes for $R$ semiprime Goldie.

**Proposition 8.3.32:** *Suppose* $\text{char}(F) = 0$ *and* $R$ *is a Noetherian* $F$-*algebra with a derivation* $\delta$. *Let* $R' = R[\lambda; \delta]$.

(i) *If* $A \lhd R$ *and* $\delta A \subseteq A$ *then* $R'A = AR' \lhd R'$.

(ii) *If* $P \in \delta\text{-Spec}(R)$ *then* $PR' \in \text{Spec}(R')$.

(iii) *There is a surjection* $\text{Spec}(R') \to \delta\text{-Spec}(R)$ *given by* $P' \mapsto P' \cap R$.

*Proof:*

(i) $\lambda a = a\lambda + \delta a$ for all $a$ in $A$, implying $\lambda(AR') \subseteq AR'$ and thus $AR' \lhd R'$. Hence $R'A \subseteq AR'$, and by symmetry $AR' \subseteq R'A$.

(ii) $PR' \lhd R'$ by (i). It remains to show the ideal $PR'$ is prime. Passing to $R/P$ we may assume $R$ is prime and need to show $R'$ is prime. But this follows from a leading term argument analogous to that preceding definition 1.6.14; if $\left(\sum_{i=0}^{m} r_i \lambda^i\right) R \left(\sum_{j=0}^{n} r'_j \lambda^j\right) = 0$ with $r_m, r'_n \neq 0$ then $0 = r_m R r'_n \lambda^{m+n}$ + terms of lower degree, so $r_m R r'_n = 0$, contrary to $R$ prime.

(iii) $P = PR' \cap R$ is seen by checking the constant term of $PR'$, so LO holds from $\delta\text{-Spec}(R)$ to $\text{Spec}(R')$. It remains to show for each $P'$ in $\text{Spec}(R')$ that $P' \cap R \in \delta\text{-Spec}(R)$. Let $A = P' \cap R$. Clearly $\delta A = [\lambda, A] \subseteq P' \cap R = A$, so we need to show $A$ is prime. Passing to $R/A$ and $R'/AR'$ we may assume $A = 0$. Let $N = \text{Nil}(R)$. $N^t = 0$ for some $t$, and $N$ is $\delta$-invariant by proposition 2.6.28. By (i) we have $(NR')^t \subseteq N^t R' = 0$ so $NR' \subseteq P'$ and thus $N \subseteq P' \cap R = 0$. Let $P_1, \ldots, P_m$ be the minimal primes of $R$, which are $\delta$-invariant by proposition 8.3.31. The same argument as above shows $(P_1 R') \cdots (P_m R') \subseteq P_1 \cdots P_m R' = 0$, so some $P_i R' = P'$ and thus $P_i = 0$, proving $R$ is indeed prime.     Q.E.D.

We say a prime ideal $P$ of $R$ is *principal* if $P = Ra = aR$ for some $a$ in $P$.

**Corollary 8.3.33:** *Suppose* $R$ *is prime Noetherian. If* $P' \in \text{Spec}(R[\lambda; \delta])$ *has height 1 and* $P' \cap R$ *is nonzero principal then* $P'$ *is principal.*

*Proof:* Let $P = P' \cap R$, which by hypothesis has the form $aR = Ra$. Let $R' = R[\lambda; \delta]$. Then $aR' = PR' \in \text{Spec}(R')$ so $aR' = P'$; analogously $P' = R'a$.     Q.E.D.

One can frame these results in a setting more appropriate to arbitrary enveloping algebras, cf., exercise 9. Now we shall couple these results with proposition 1.6.36.

## Completely Prime Ideals (Following BGR)

**Definition 8.3.34:** An ideal $P$ of $R$ is *completely prime* if $R/P$ is a domain.

Although we studied this idea in §2.6, we did not have many examples of completely prime ideals of domains. We shall see now that *every* prime ideal of an enveloping algebra of a solvable Lie algebra is completely prime, thereby showing how the solvable theory begins to diverge from the semi-simple theory, in view of the next example.

**Example 8.3.35:** Suppose $F = \mathbb{C}$ and $R = U(L)$ where $L$ is a Lie algebra which is *not* solvable. Then $L$ has a simple module $M$ with $n = [M:\mathbb{C}] > 1$, and we view $M$ as $R$-module. Let $P = \operatorname{Ann} M$, a primitive ideal. By the density theorem $R/P$ is dense in $\operatorname{End}_{\mathbb{C}} M \approx M_n(\mathbb{C})$, and so $R/P \approx M_n(\mathbb{C})$ which surely is not a domain.

**Theorem 8.3.36:** *If $R = U(L)$ with $L$ solvable then every prime ideal of $R$ is completely prime.*

**Proof:** Induction on $n = [L:F]$; the assertion is trivial for $n = 1$. Let $F = L_0 \lhd L_1 \lhd \cdots \lhd L_n = L$, where each $[L_i:F] = i$, and let $R_i = U(L_i)$. We shall show by induction on $i$ that each prime ideal of $R_i$ is completely prime. This assertion is trivial for $i = 0$; assuming it holds for $i$ we shall prove it for $i + 1$. By propsition 8.3.28 we have $R_{i+1} \approx R_i[\lambda; \delta]$ for a suitable derivation $\delta$ of $R_i$. Suppose $P \in \operatorname{Spec}(R_{i+1})$, and let $P_0 = P \cap R_i \in \delta\text{-Spec}(R)$ by proposition 8.3.32(iii).

Let $\bar{R}_i = R_i/P_0$ and $\bar{R}_{i+1} = R_{i+1}/P_0 R_{i+1} \approx \bar{R}_i[\lambda; \delta]$. By induction $\bar{R}_i$ is a domain, so $\bar{R}_{i+1}$ is a domain, and we are done unless $\bar{P} \neq 0$. Let $S = \bar{R}_i - \{0\}$ and let $D = S^{-1}\bar{R}_i$, the division ring of fractions of $\bar{R}_i$. By lemma 8.3.30 $\delta$ extends to $D$, so we can view $\bar{R}_{i+1} \subseteq D[\lambda; \delta]$. Furthermore, $0 \neq S^{-1}\bar{P} \lhd D[\lambda; \delta]$. Hence proposition 1.6.36 shows $\delta$ is inner on $D$, and $D[\lambda; \delta]$ is a polynomial ring $D[\mu]$ where $\mu = \lambda - d$, and also $S^{-1}\bar{P}$ generated by some central element $f \neq 0$. Let $Z = Z(D)$ and $K = Z[\mu]/Z[\mu]f$, a field.

$D[\lambda; \delta]/S^{-1}\bar{P} \approx D \otimes_Z K$ is simple by theorem 1.7.27. On the other hand, there is a composition

$$\psi: U(L_i \otimes_F K) \approx U(L_i) \otimes_F K \subseteq D \otimes_F K \to D \otimes_Z K.$$

$\psi U(L_i \otimes_F K)$ is prime since $S^{-1}\psi U(L_i \otimes_F K) \approx D \otimes_Z K$ is simple, and therefore $\ker \psi \in \operatorname{Spec}(U(L_i \otimes_F K))$. But $[L_i \otimes_F K:K] \leq i$, implying $\ker \psi$ is completely prime by the induction hypothesis. Hence $D \otimes_Z K$ is a domain, and consequently $S^{-1}\bar{P}$ (and thus $\bar{P}$) is completely prime.    Q.E.D.

One strategy of BGR is beginning to shape up. Given a finite dimensional solvable Lie algebra $L$ we can write $U(L)$ in the form $U(L_1)[\lambda; \delta]$ *where $L_1$ is any subspace of codimension 1 containing $L'$*. But localizing at suitable elements inside $U(L_1)$ will give this differential polynomial ring better form, thereby enabling us to break $\text{Spec}(U(L))$ into simpler and better-known spectra.

## Dimension Theory on $U(L)$

The two dimensions used for enveloping algebras are Krull dimension and Gelfand-Kirillov dimension. Some of the basic results can be obtained using K-dim, cf., exercise 10, but we shall focus on GK dim. *We assume throughout that $L$ is finite dimensional over $F$.*

**Remark 8.3.37:**   GK dim $U(L) = [L:F]$. (Indeed, by proposition 8.3.13 the associated graded algebra $\approx F[\lambda_1, \ldots, \lambda_n]$ where $n = [L:F]$, and GK dim $F[\lambda_1, \ldots, \lambda_n] = n$ by proposition 6.2.22. Hence GK dim $U(L) = n$ by remark 6.2.14(vi).)

When$[L:F]$ is infinite dimensional $U(L)$ does not have GK dim, cf., exercise 13. Nonetheless, there are instances when $U(L)$ has subexponential growth, cf., exercises 12, 14, 15.

**Lemma 8.3.38:**   *Suppose $L$ is solvable and let $0 = L_0 \lhd L_1 \lhd \cdots \lhd L_n = L$ where each $[L_i:F] = i$. Suppose $P \in \text{Spec}(U(L))$ and let $P_i = (P \cap U(L_i))U(L)$. Then the chain $P_0 \subseteq P_1 \subseteq \cdots \subseteq P_n = P$ in $U(L)$ has length $n - \text{GK dim}(U(L)/P)$ and is in $\text{Spec}(U(L))$.*

**Proof:**   Induction on $n$. Let $R = U(L)$, $R_i = U(L_i)$, and $Q_i = (P \cap U(L_i))R_{n-1}$. Note $P_{n-1} = Q_{n-1}R$. By induction $Q_0 \subseteq Q_1 \subseteq \cdots \subseteq Q_{n-1}$ has length $m = (n-1) - \text{GK dim}(R_{n-1}/Q_{n-1})$ in $\text{Spec}(R_{n-1})$. But $R \approx R_{n-1}[\lambda; \delta]$ and each $Q_i$ is $\delta$-invariant; hence $P_i = Q_i R \in \text{Spec}(R)$. Clearly $P_0 \subseteq \cdots \subseteq P_{n-1}$ also has length $m$. Let $d = \text{GK dim}(R_{n-1}/Q_{n-1})$; thus $m = n - 1 - d$.
$R/P_{n-1} = R/Q_{n-1}R \approx (R_{n-1}/Q_{n-1})[\lambda; \delta]$. Hence GK dim $R/P_{n-1} = d + 1$ by exercise 6.2.10 (a minor modification of proposition 6.2.22). Therefore $m = n - \text{GK dim}(R/P_{n-1})$, and we are done if $P = P_{n-1}$. Thus we may assume $P_n \supset P_{n-1}$. Now the chain $P_0 \subseteq \cdots \subseteq P_n$ has length $m + 1$, so it remains to prove GK dim $R/P = d$. But "$\geq$" is clear since $Q_{n-1} = P \cap R_{n-1}$ (so $R_{n-1}/Q_{n-1} \hookrightarrow R/P$), and "$\leq$" follows from theorem 6.2.25 which shows

$$\text{GK dim } R/P \leq \text{GK dim } R/P_{n-1} - 1 = d. \qquad \text{Q.E.D.}$$

**Theorem 8.3.39:** (*Tauvel's height formula*) *If* $L$ *is a solvable Lie algebra over a field* $F$ *of characteristic* 0 *then*

$$[L:F] = \text{height}(P) + \text{GK dim } U(L)/P$$

*for every* $P$ *in* Spec($R$).

*Proof:* First assume $F$ is algebraically closed. $[L:F] = \text{GK dim } U(L)$ by remark 8.3.37. Thus we have ($\geq$) by theorem 6.2.25, and ($\leq$) follows from the chain constructed in lemma 8.3.38 (in view of Lie's theorem).

In general we tensor by the algebraic closure of $F$, noting this affects neither GK dim (by remark 6.2.26') nor height (by theorem 3.4.13'). Q.E.D.

This result leads us to wonder next whether or not $U(L)$ is catenary. The catenarity of $U(L)$ for $L$ nilpotent was proved by Lorenz-Rentschler using the properties of the Dixmier Map (to be described below) and by Malliavin using Smith's theory of localization at local homologically regular rings. Recently Gabber has proved catenarity for enveloping algebras of solvable Lie algebras over $\mathbb{C}$. The reader can consult Krause-Lenagan [85B, pp. 129–135] for a proof of Gabber's result and its application to computing GK-dimensions of modules.

## Examples

Often one must localize in order to maneuver into a position to utilize these results. This is best illustrated by means of the low-dimensional examples of Lie algebras. The Weyl algebra $\mathbf{A}_n$ begins to crop up and, in fact, plays a key role in the general theory, as we shall see. Write $L'$ for $[LL]$. Given a normalizing element $x$ of $U(L)$ write $\mathscr{S}(x)$ for those prime ideals containing $x$; hence $\mathscr{S}(x)$ is homeomorphic to Spec($U(L)/U(L)x$). Assume char($F$) = 0.

*Example 8.3.40:* We say $L$ is *abelian* if $L' = 0$. Then $U(L)$ is commutative and isomorphic to a polynomial ring in view of proposition 8.3.6. (If $L$ is infinite dimensional one could apply exercise 1.)

*Example 8.3.41:* Let $L$ be a nonabelian Lie algebra of dimension 2. Then $L$ has elements $x, y$ with $[xy] = x$ and thus is unique up to isomorphism. $L$ is solvable but not nilpotent, and $R = U(L)$ is primitive by exercise 2.1.4. We can determine the nonzero prime ideals by localizing at $x$. Indeed, $yx = x(y - 1)$ implies $x$ is a normalizing element, so $S = \langle x \rangle$ is a left denominator set of the Ore domain $R$, and $S^{-1}R \approx F(x)[\lambda; \delta]$ where $\delta$ is the natural extension of ad $y$ to $F(x)$. (In other words, ad $y$ is given by $x \cdot \partial/\partial x$.) Thus $S^{-1}R$ is simple by proposition 1.6.36, so any prime ideal $P \neq 0$ of $R$ intersects $S$ nontrivially. But this implies $x \in P$ since $x$ is normalizing, so we see $P/Rx$ is a prime ideal of $R/Rx \approx F[y]$. In other words, $\operatorname{Spec}(R) = \{0\} \cup \mathscr{S}(x)$, and $\mathscr{S}(x)$ is homeomorphic to the spectrum of a polynomial ring over a field.

Note $L$ is solvable but not nilpotent. Every derivation of $L$ is inner by exercise 5.

*Example 8.3.42:* We wish to describe the 3-dimensional nonabelian Lie algebras $L$ over a field $F$, and the prime spectra of their enveloping algebras. For more details see Jacobson [62B, pp. 10–14].

First we show that if the non-abelian 2-dimensional Lie algebra is isomorphic to a Lie ideal $L_2$ of $L$ then $L \approx L_1 \times L_2$ for some 1-dimensional Lie subalgebra $L_1$. Indeed, pick $a$ in $L - L_2$; ad $a$ restricts to a derivation on $L_1$ which as noted in example 8.3.41 has the form ad $a'$ for some $a'$ in $L_2$, so take $L_1 = F(a - a')$. In this case $U(L) \approx U(L_1) \otimes_F U(L_2) \approx U(L_2)[\lambda]$, where $\lambda$ is a commuting indeterminate. Now we proceed by a case-by-case analysis based on $L'$.

*Case I.* $[L':F] = 1$.

(i) If $L'$ is central then $L$ is nilpotent and has a base $x$, $y$, $z$ where $[xy] = z$ and $[xz] = [yz] = 0$. This is called the *Heisenberg* Lie algebra. Note $z \in Z(U(L))$. Letting $C = F[y, z]$ we see $U(L) \approx C[\lambda; \delta]$ where $\delta$ is given on $F[y, z] = U(Fy + Fz)$ by extending the action of ad $x$. Let $S = \langle z \rangle = \{z^i : i \in \mathbb{N}\}$. Then $S^{-1}U(L) \approx A_1(S^{-1}F[z])$. To see this we merely note that $x$ and $z^{-1}y$ generate a Weyl subalgebra whose centralizer is $Z(S^{-1}U(L)) = S^{-1}F[z]$. Thus the complement of $\mathscr{S}(z)$ in $\operatorname{Spec}(U(L))$ is homeomorphic to $\operatorname{Spec}(A_1(F[z, z^{-1}]))$ which by corollary 1.6.34 is homeomorphic to $\operatorname{Spec}(F[z, z^{-1}])$, the spectrum of a polynomial ring with one point (corresponding to $z$) removed. On the other hand, $\mathscr{S}(z) = \operatorname{Spec}(U(L)/zU(L)) = \operatorname{Spec}(U(L/Lz)) = \operatorname{Spec}(F[x, y])$ since $L/Lz$ is abelian. In this way $\operatorname{Spec}(U(L))$ has been described as the union of spectra of two well-known commutative rings.

(ii) If $L'$ is not central then there is a 2-dimensional nonabelian ideal $L_2$ containing $L'$, so $L \approx L_1 \times L_2$ as in the first paragraph, and $L \approx U(L_2)[\lambda]$. Note $L$ is solvable but not nilpotent.

*Case II.* $[L':F] = 2$. $L'$ is abelian for, otherwise, we would get thrown into case I by the first paragraph. Now the structure of $L$ is determined by ad $a$ on $L'$ where $a \in L - L'$. Since we could replace $a$ by any multiple, $L$ is determined by an element of PGL$(2, F)$. When $F$ is closed under quadratic extensions one can find a base $x, y, z$ satisfying one of the following multiplication tables:

$$[xy] = 0, [xz] = x \text{ and } [yz] = \alpha y \text{ for some } 0 \neq \alpha \in F;$$

or   $$[xy] = 0, [xz] = x, \text{ and } [yz] = x + y.$$

Then $U(L) \approx C[\lambda; \delta]$ where $C = F[x, y]$ and $\delta = \text{ad } z$. In case $[yz] = -y$ this can be analyzed quite easily. Note $xy \in Z(U(L))$; also since $x$ is a normalizing element we can take $S = \langle x \rangle$, and $S^{-1}U(L)$ has a Weyl subalgebra generated by $x$ and $x^{-1}z$, from which one sees $S^{-1}U(L) \approx A_1(S^{-1}F[\mu])$ where $\mu = xy$. Hence the complement of $\mathscr{S}(x)$ in Spec$(U(L))$ is homeomorphic to Spec$(S^{-1}F[\mu]) \approx$ Spec$(F[\mu])$. On the other hand, $\mathscr{S}(x)$ is homeomorphic to Spec$(U(L/Fx))$, which is given in example 8.3.41.

*Case III.* $L' = L$. It follows that $L$ is simple (i.e., has no proper ideals) since if $A \neq 0$ were a proper ideal then $A$ and $L/A$ would each have dimension 1 or 2 and thus be solvable by example 8.3.41; hence $L$ would be solvable, contrary to $L' = L$. When $F$ is algebraically closed $L$ is unique up to isomorphism, cf., Jacobson [62B, p. 13].

One such example is $L = \text{sl}_2(F)$, the Lie algebra of $2 \times 2$ matrices over $F$ having trace 0, for $\{e_{12}, e_{21}, e_{11} - e_{22}\}$ is a base of $L$ and thus $L = L'$ by direct computation. $U(\text{sl}_2(F))$ is an amazingly rich example of a ring, cf., fact 8.3.49 below.

***Remark 8.3.43:***   If $L$ is the Heisenberg algebra used in example 8.3.42 then $A_1(F)$ is a homomorphic image of $U(L)$ obtained by $z \mapsto 1$. It is easy enough to generalize this definition of Heisenberg algebra and thereby show that every Weyl algebra is a homomorphic image of the enveloping algebra of a solvable Lie algebra, cf., exercise 4.

## *Digression: A Sketch of Some of the Deeper Results*

For the remainder of this section we outline parts of the enveloping algebra theory which would otherwise be outside the scope of this book in order to give the reader an idea of the broader picture. Throughout $L$ is a finite dimensional Lie algebra over $\mathbb{C}$. Reductions from arbitrary base fields of characteristic 0 are discussed in §8.4.

**The Semicenter**

**Definition 8.3.44:** The *eigenspace* of an $L$-module $M$ under a linear functional $f: L \to \mathbb{C}$ is defined to be $\{x \in M : [ax] = (fa)x$ for all $a$ in $L\}$. If $P \in \operatorname{Spec} U(R)$ then the *semicenter* $Sz(R)$ of $R = U(L)/P$ is the direct sum of the eigenspaces of the linear functionals of $L$.

Obviously $Z(R) \subseteq Sz(R)$, seen by taking $f = 0$, and elementary linear algebra shows $R^f R^g \subseteq R^{f+g}$ so $Sz(R)$ is a subalgebra of $R$. $Sz(R) = Z(R)$ iff $0$ is the only eigenvalue. In particular, this is the case when $L$ is nilpotent, although exercises 3 and 8 also provide a non-nilpotent example, cf., [BGR, §6.11].

**Fact 8.3.45:** Key properties of the semicenter. Write $Q(\ )$ for the simple Artinian ring of fractions of a prime Goldie ring. Taking $P \in \operatorname{Spec}(U(L))$, let $R = U(L)/P$ and write $H(P)$ for $Z(Q(R))$.

(1) $Sz(R)$ is an integral domain [BGR, theorem 6.1], whose field of fractions contains $H(P)$, by [BGR, lemma 6.5].

(2) The following three conditions are equivalent: $P$ is primitive; $H(P) = \mathbb{C}$; $P$ is locally closed in $\operatorname{Spec}(R)$, i.e., $P$ is *not* an intersection of prime ideals properly containing $P$. These conditions are called *Dixmier's conditions* and will be studied below and in §8.4. Their equivalence for $L$ solvable follows easily from (1); their equivalence in general relies on Moeglin's work on polarization to be discussed below.

(3) Let $Z = Z(R)$. If $Sz(R) = Z$ then there is $z$ in $Z$ for which $Z[z^{-1}]$ is f.g. over $\mathbb{C}$ and $R[z^{-1}] \approx Z[z^{-1}] \otimes \mathbf{A}_n$ for suitable $n$, where $\mathbf{A}_n = \mathbf{A}_n(\mathbb{C})$, cf., [BGR, proposition 6.8].

(4) In case $L$ is nilpotent, every primitive ideal $P$ is maximal and $R[z^{-1}]$ is a Weyl algebra; this follows from (3). On the other hand, when $L$ is not nilpotent there always will be a non-maximal primitive ideal of $U(L)$. Indeed, any solvable non-nilpotent $L$ has a non-abelian 2-dimensional image, which by example 8.3.41 is primitive; in the non-solvable case one can appeal to the theory of semisimple Lie algebras.

**Polarization** There is an important technique, called *polarization*, for constructing simple modules and thus primitive ideals (which, of course, are their annihilators).

**Remark 8.3.46:** Any $f: L \to F$ yields a bilinear form $B_f: L \times L \to F$ given by $B_f(a_1, a_2) = f[a_1, a_2]$.

We say a Lie subalgebra $L_0$ of $L$ is a *polarization* of $f$ if $L_0$ is a maximal isotropic subspace of $L$ under $B_f$ (i.e., $f[L_0, L_0] = 0$). Thus the Witt theory of maximal isotropic subspaces (Jacobson [85B, Chapter 6]) will play an important role. In particular, $[L_0:F] = ([L:F] + [\text{Rad } B_f:F])/2$.

**Proposition 8.3.47:** *(Vergne)*   *Suppose*   $0 = L_0 \lhd L_1 \lhd \cdots \lhd L_n = L$   *with* $[L_i:F] = n$. *With respect to the bilinear form* $B_f$, $W = \sum_{i=1}^{n} L_i \cap L_i^{\perp}$ *is a polarization of* $f$.

The proof is by induction, using the Jacobi identity, cf., [BGR, §9.2 and 9.4]. If $W$ is a polarization of $f$ we can define a $W$-module action on $Fx$ (a one-dimensional vector space over $F$) by $a \cdot x = (fa)x$ for all $a$ in $W$. (Then $f[a_1, a_2]x = 0 = (fa_1)(fa_2)x - (fa)_2(fa_1)x$, so this is indeed a Lie module action.) We call this module $F_f$.

**Theorem (Conze-Vergne, cf., [BGR, §9.5]):**   *If $L$ is solvable over $\mathbb{C}$ and $W$ is a polarization of $f$ then the "induced" $U(L)$-module*

$$M(f, W) = U(L) \otimes_{U(W)} F_f$$

*is simple.*

We now have a host of simple modules and thus primitive ideals. These primitive ideals depend on $W$ as well as $f$. This is circumvented by modifying the definition a bit.

Given a polarization $W$ of $L$ note that any $w$ in $W$ defines a map $\text{ad } w : L/W \to L/W$, i.e., $\text{ad } w(a + W) = [wa] + W$. Define the linear functional $f_W : W \to F$ by $f_W w = \frac{1}{2} \text{tr}(\overline{\text{ad } w})$. Then $f_W([W, W]) = 0$ so we have a Lie map $W \to U(W)$ given by $a \mapsto a + f_W a$; by universality this extends to an algebra homomorphism $\varphi_W : U(W) \to U(W)$. Thus any $W$-module $M$ obtains a new scalar product given by

$$w \cdot x = (\varphi_W w)x,$$

and we let $\tilde{M}$ denote the corresponding module structure. In particular, we let $\tilde{M}(f, W) = U(L) \otimes_{U(W)} \tilde{F}_f$. When $W$ acts nilpotently on $L/W$ each $\text{tr}(\overline{\text{ad } w}) = 0$ so $\tilde{M}(f, W) = M(f, W)$. However, in general $\tilde{M}(f, W)$ can differ from $M(f, W)$, and, in fact, Ann $\tilde{M}(f, W)$ is independent of $W$. Write $P(f)$ for Ann $\tilde{M}(f, W)$. Thus for $L$ solvable we have the well-defined *Dixmier map* $\text{Dix} : L^* \to \{\text{primitive ideals of } U(L)\}$ given by $f \mapsto P(f)$, cf., [BGR, §10]. The celebrated Dixmier-Duflo theorem [BGR, §11] says the Dixmier map is surjective for solvable Lie algebras. Conze showed $P(f)$ is a completely prime ideal, so we actually have

$$\text{Dix}: L^* \to \text{Prim}_c(U(L)),$$

where $\text{Prim}_c(U(L))$ is the set of completely prime, primitive ideals. Duflo [75] extended these results to arbitrary $L$ by considering the *solvable polarizations* to be those polarizations which are solvable as Lie algebras; then any solvable polarization gives rise as before to a completely prime, primitive ideal, yielding the Dixmier-Duflo map

$$\text{Dix}: L_{SP}^* \to \text{Prim}_c(U(L))$$

where $L_{SP}^* = \{f \in L^*: f$ has a solvable polarization$\}$. The image is called the set of *Duflo ideals*. Obviously $L_{SP}^* = L^*$ when $L$ is solvable; when $L$ is semisimple Duflo [75, lemma 1.1] shows $f$ has a solvable polarization iff the radical of $B_f$ is minimal among all $\{B_h: h \in L^*\}$. Duflo [75] showed the Dixmier-Duflo map is well-defined, continuous with respect to the Zariski topology on $L^*$, and the Duflo ideals are dense in the primitive spectrum. Moeglin [80] proved the stronger result that every minimal primitive ideal is a Duflo ideal. In this way, we have "enough" primitives to prove the following results:

Duflo [75, theorem 5.2]. $Z(U(L))$ is isomorphic to the fixed subalgebra of the symmetric algebra of $S(L)$ under the action of $L$.

Moeglin [80]. Dixmier's conditions (cf., fact 8.3.45(2)) are equivalent for arbitrary $L$.

Moeglin [80a]. Every nonzero ideal of $U(L)$ contains a *semi-invariant*, i.e., a nonzero eigenvector under the action of $\text{Ad}\, L$ on $U(L)$. Furthermore, if $\text{Rad}\, L$ is nilpotent there is some $z$ in $Z = Z(U(L))$ such that each maximal ideal $m$ of $Z$ *not* containing $z$ has only finitely many primitives of $U(L)$ lying over $m$, and there is a primitive Duflo ideal over $m$ which is contained in all the primes lying over $m$. Her methods are continued in Moeglin-Rentschler [81], especially see their theorem 3.7 which generalize further Dixmier's conditions.

**Algebraic Lie Algebras and the Adjoint Group** A *linear algebraic group* is a multiplicative subgroup $G$ of $\text{End}\, F^{(n)}$ defined as the invertible elements of the zero set of a set $\mathscr{S}$ of polynomials in the commutative polynomial ring $F[\Lambda] = F[\lambda_{ij}: 1 \le i, j \le n]$ where the action of $F[\Lambda]$ on $\text{End}\, F^{(n)}$ is obtained by identifying $\text{End}\, F^{(n)}$ with $M_n(F)$, i.e., given $f \in F[\Lambda]$ and $(a_{ij})$ in $M_n(F)$ we substitute $\lambda_{ij} \mapsto a_{ij}$ in $f$. As usual we can define $\mathscr{I} = \mathscr{I}(G) = \{f \in F[\Lambda]: f(g) = 0 \text{ for all } g \text{ in } G\} \lhd F[\Lambda]$ and can define $A = F[\Lambda]/\mathscr{I}$, the *algebra of polynomial functions* on $G$; by definition $G$ is the annihilator of $\mathscr{I}$ in $\text{GL}(n, F)$. $G$ is *connected* if $A$ is an integral domain. Given a subgroup $H$

of $G$ we can define the *algebraic hull* of a subgroup $H$ of $G$ to be the annihilator of $\mathscr{I}(H)$ in $G$, clearly the smallest algebraic subgroup containing $H$.

The action of $F[\Lambda]$ on $G$ induces an action of $A$ on $G$. For any $g \in G$ we get a map $\ell_g : A \to A$ by putting $\ell_g \bar{f}(x) = \bar{f}(g^{-1}x)$ for all $\bar{f}$ in $A$ and all $x$ in $G$. The *Lie algebra* of $G$ is $\{\delta \in \text{Deriv}(A): \delta\ell_g = \ell_g\delta\}$, cf., proposition 1.6.12. Such a Lie algebra is called *algebraic*. The standard references are Chevalley [68B, §2.8] and Borel [69B, pp. 117–118] (which is more geometric in outlook).

***Example 8.3.48:*** The "standard example" is the familiar group $SL(n, F)$, which is an algebraic group whose corresponding Lie algebra is $\text{sl}_n(F)$, the matrices of trace 0. We have already encountered $\text{sl}_2$ in example 8.3.42.

Let us continue Chevalley's theory. A (Lie) *automorphism* of $L$ is a Lie homomorphism from $L$ to $L$ which is 1:1 and onto. Aut $L$ is then an algebraic group whose Lie algebra Deriv $L$ contains the Lie ideal Ad $L$. As explained in [BGR, §12.8] there is a smallest algebraic Lie algebra $L_0$ containing Ad $L$. $L_0$ is called the algebraic *hull* of $L$; the irreducible algebraic group giving rise to $L_0$ is called the *adjoint algebraic group* $\Gamma$. $\Gamma$ acts on $U(L)$ by extending the automorphisms of $L$; likewise, $\Gamma$ acts on the enveloping algebra of any ideal of $L$.

The ideas of the last two paragraphs can be viewed quite generally in purely algebraic terms, by means of Hopf algebras, and this approach has proved very useful in Moeglin-Rentschler [81]. The reader can find a lovely Hopf algebra treatment in Hochschild [71B]; the basic concepts are given in example 8.4.55′.

The Dixmier-Duflo map now induces a continuous map $L_{\text{SP}}^*/\Gamma \to \text{Prim}_c(U(L))$ which is injective by Moeglin [80] and is a homeomorphism when $L$ is nilpotent by Conze [75]. Surjectivity for $L$ solvable was proved by Rentschler, but continuity of the inverse remains open (cf., Tauvel [82] for certain cases). Moeglin-Rentschler [84] have refined these results to pass from a primitive ideal of $U(L)$ to a primitive ideal of $U(L_0)$ where $L_0$ is a Lie ideal of $L$. Borho [77, theorem 6.6] generalized the Dixmier map $L^*/\Gamma \to \text{Prim}(U(L))$ for $L = \text{sl}_n$, and Borho-Jantzen [77, theorem 5.15] proved it is 1:1 but *not* onto.

**Enveloping Algebras of Algebraic Lie Algebras** The Lie algebras of greatest interest to Lie group theorists are, of course, the algebraic Lie algebras defined above, and these have been the subject of much recent research because they are amenable to a wider variety of techniques than abstract Lie algebras.

Suppose $P \in \mathrm{Prim}(U(L))$, the primitive spectrum. Let $R = U(L)/P$, and let $Q(\ )$ denote the classical ring of fractions. We noted in 8.3.45(4) that if $L$ is nilpotent then for suitable $n$, $Q(R) \approx \mathbf{D}_n \otimes K$, where $\mathbf{D}_n = Q(A_n(\mathbb{C}))$, called the *Weyl* (*skew*) *field*. In general by Goldie's theorem $Q(R)$ will have the form $M_t(D)$ for a suitable division algebra $D$, leading to several basic structure questions:

*Generalized Gelfand-Kirillov Conjecture:* $D \approx \mathbf{D}_n \otimes K$ where $n = \frac{1}{2}\mathrm{GK}\text{-dim } R$.

*Goldie-Rank Question:* What is $t$?

More generally, for any prime ideal $P$ one could conjecture $Q(R) \approx M_n(\mathbf{D}_n) \otimes_{\mathbb{C}} K$ for some field $K \supseteq \mathbb{C}$; for $P$ primitive we would necessarily have $K = \mathbb{C}$ by 8.3.45(2). This conjecture was posed for $P = 0$ in Gelfand-Kirillov[66], i.e., is necessarily $Q(U(L)) \approx \mathbf{D}_n \otimes \mathbb{C}(\lambda_1, \ldots, \lambda_k)$ for suitable $n, k$ where $2n + k = [L:\mathbb{C}]$? (Gelfand-Kirillov [66, lemma 6] showed $\mathrm{GK\,tr\,deg\,} Q(U(L)) = [L:\mathbb{C}]$, and $\mathrm{GK\,tr\,deg\,} \mathbf{D}_n = 2n$, (where $\mathrm{GK\,tr\,deg}$ is their version of GK dim; actually, $\mathrm{GK\,dim\,} \mathbf{D}_n$ does not exist!) The reason for the assumption of algebraicity on $L$ is that the conclusion clearly implies $[L:\mathbb{C}] = 2n$ when $U(L)$ is primitive, but there are examples of solvable Lie algebras of odd dimension having primitive enveloping algebras, e.g., example 8.3.42, case II, with $\alpha$ irrational. (Of course, these are not algebraic!)

Gelfand-Kirillov [66] proved their conjecture for $L$ nilpotent and for $L = \mathfrak{sl}_n$. The Gelfand-Kirillov conjecture was verified in the solvable case independently Joseph [74] and McConnell [74] and Borho (cf., [BGR, §8.4], which exploits the semicenter). McConnell also determined the structure for nonalgebraic solvable Lie algebras, cf., McConnell [77a] for a quick proof. Joseph [80] has posed and verified the generalized Gelfand-Kirillov conjecture for $L = \mathfrak{sl}_n$; see Borho [86, §6.3] for a general strategy.

(In view of the Gelfand-Kirillov conjecture one would not expect PI-rings to play much of a role in enveloping algebras. However, Joseph [86] has discovered a serious application of PI-theory to their prime images.)

Suppose $L$ is the Lie algebra of a connected linear algebraic group $G$. Then $G$ acts as automorphisms on $U(L)$, and any ideal of $U(L)$ is $G$-stable. Fix a primitive ideal $J$ of $U(L)$. The fixed subring $Q(U(L)/J)^G$ is $Z(Q(U(L)/J))$ which is $\mathbb{C}$ by fact 8.3.45(2); hence for any Lie ideal $L_0$ of $L$, one sees $Q(U(L_0)/L_0 \cap J)^G \approx \mathbb{C}$. Moeglin-Rentschler [81] proved the following:

(1) There is a primitive ideal $J_0$ of $U(L_0)$ such that $J \cap U(L_0) = \bigcap_{g \in G} gJ_0$.
(2) Any other primitive ideal of $U(L_0)$ as in (1) is conjugate to $J_0$ under $G$.

(3) Now Let $G_1 = \{g \in G : gJ_0 = J_0\}$, an algebraic subgroup of $G$, and let $L_1$ be its Lie algebra, i.e., $L_1 = \{a \in L : [a, J_0] = J_0\}$. Given $I \lhd U(L_0)$ we write $\mathrm{Ind}(I, L_1 \uparrow L) = \bigcap_{g \in G} g(U(L)I) \lhd U(L)$. Then there exists a primitive ideal $J_1$ of $U(L_1)$ lying over $J_0$ such that $J = \mathrm{Ind}(J_1, L_1 \uparrow L)$.

(4) (Moeglin-Rentschler [84]) Any other primitive ideal of $U(L_1)$ as in (3) is conjugate to $J_1$.

**A Few Words About Semisimple Lie Algebras** Enveloping algebras of semisimple Lie algebras have been the subject of much recent research and now possess a lovely theory which we shall summarize along the lines of Borho [86] and Joseph [83]. An excellent overall reference is Jantzen [83B]. Then we shall note how $\mathrm{sl}_2$, the "easiest" nontrivial example, is actually rich enough to provide many interesting counterexamples in ring theory, discovered only in the last few years.

We start with some Lie algebraic preliminaries. These are developed quickly in [DIX, Chapter 1] and explained more fully in Jacobson [62B]. Assume throughout $L$ is a semisimple Lie algebra over $\mathbb{C}$. A *Cartan subalgebra* $H$ is a nilpotent Lie subalgebra which is its own normalizer, i.e., $[xH] \subseteq H$ iff $x \in H$. From [DIX, §1.9] we see $H$ exists and is unique up to an "elementary" automorphism.

Given a Cartan subalgebra $H$ and an $H$-module $M$ define the *weight space* $M^\alpha$ of $\alpha \in H^*$ to be the eigenspace of $\alpha$; $\alpha$ is called a *weight* if $M^\alpha \neq 0$. Now let $M = L$ viewed as $H$-module via the adjoint action; the weights are then called *roots*. Each $[L^\alpha : F] = 1$, and $L = H \oplus (\bigoplus L^\alpha)$, summed over all roots $\alpha$. If $\alpha, \beta$ are roots then $[L^\alpha, L^\beta] \subseteq L^{\alpha + \beta}$; also $-\alpha$ is a root [DIX, theorem 1.10.2(i), (ii)].

Define the *Killing form* on $L$ by $\langle a, b \rangle = \mathrm{tr}(\mathrm{ad}\, a)(\mathrm{ad}\, b)$, which is nondegenerate (by Cartan's criterion). If $\alpha, \beta$ are roots with $\alpha \neq \pm \beta$ then $L^\alpha$ and $L^\beta$ are orthogonal; on the other hand, $L^\alpha + L^{-\alpha} + [L^\alpha, L^{-\alpha}]$ is a three-dimensional Lie subalgebra of $L$ isomorphic to $\mathrm{sl}_2$. Furthermore, $[L^\alpha, L^{-\alpha}] \subseteq H$ and contains a unique element $h_\alpha$ such that $\alpha(h_\alpha) = 2$ [DIX, theorem 1.10.2]. $\beta(h_\alpha) \in \mathbb{Z}$ for every root $\beta$. Let $H_\mathbb{Q} = \sum_\alpha \mathbb{Q} h_\alpha$. We switch over to Jacobson [62B, pp. 117ff]. $H_\mathbb{Q}$ contains all the roots, and the Killing form restricts to a positive definite symmetric bilinear form on $H_\mathbb{Q}$. Labeling the roots $\alpha_1, \ldots, \alpha_t$ where $t = [H : \mathbb{C}] = [H_\mathbb{Q} : \mathbb{Q}]$, we see any root $\beta = \sum_{i=1}^t u_i \alpha_i$ for suitable $u_i \in \mathbb{Q}$, and thus we can order the roots according to the lexicographic order on $n$-tuples $(u_1, \ldots, u_t)$; in particular, we say $\beta$ is *positive* if $u_1 > 0$, and $\beta$ is *negative* if $u_1 < 0$. Let $N_+ = \bigoplus_{\alpha > 0} L^\alpha$ and $N_- = \bigoplus_{\alpha < 0} L^\alpha$, each clearly nilpotent Lie subalgebras. Then we have the *triangular decomposition* $L = H \oplus N_+ \oplus N_-$.

For each root $\alpha$ define the *reflection* $s_\alpha \colon H^* \to H^*$ by

$$x \mapsto x - \frac{2\langle x, \alpha \rangle}{\langle \alpha, \alpha \rangle}\, \alpha.$$

The subgroup $W$ of Aut $H^*$ generated by the reflections is called the *Weyl group*; clearly $W$ is finite.

Of course the Killing form enables us to identify $L$ with $L^*$. For any $I \lhd U(L)$, the associated graded ideal in the symmetric algebra $S(L)$ has a zero set $V$ (viewing $S(L)$ as the ring of polynomial functions on $L^*$), called the *associated variety* of $I$. $V$ is stable under ad $L$ so is a union of $G$-orbits. The study of associated varieties is one of the themes of Borho [86], who explains how the associated variety of a primitive ideal is irreducible and thus a nilpotent orbit.

There is the *Harish-Chandra isomorphism* between $Z = Z(U(L))$ and $S(H)^W$, the fixed subalgebra of $S(H)$ under the natural action of $W$, cf., [DIX, theorem 7.4.5], and $S(H)^W$ is a polynomial ring by a theorem of Chevalley. This leads us to examine the "fibers" of Spec $U(L)$ over max spec $Z$. In other words, given a maximal ideal $m$ of $Z$, what can be said about $\mathscr{F}(m) = \{$prime ideals of $U(L)$ lying over $m\}$?

(Of course the theory is much easier for those primitive ideals $P$ corresponding to finite dimensional representations, for then $R/P$ is PI and thus simple of finite dimension over its center; but $R/P$ is finite dimensional over $\mathbb{C}$ by the Nullstellensatz, so $Z(R/P) \approx \mathbb{C}$ and the Goldie rank of $R/P$ is the dimension of the simple module whose annihilator is $P$.) To describe the general situation we need some more notions.

One main ingredient is the *Harish-Chandra bimodule*, which is an $L \times L$-module $M$ satisfying the following two finiteness conditions, where $M$ is viewed also as $L$-module via the diagonal embedding $a \mapsto (a, a)$ of $L$ into $L \times L$:

(i) $[U(L)x \colon \mathbb{C}] < \infty$    for each $x$ in $M$.

(ii) $[\operatorname{Hom}_L(N, M) \colon \mathbb{C}] < \infty$    for each finite dimensional $L$-module $N$.

Note $M$ is also a module over $U(L \times L) \approx U(L) \otimes U(L) \approx U(L) \otimes U(L)^{\mathrm{op}}$, viewing $U(L) \approx U(L)^{\mathrm{op}}$ via the principal involution; hence the usage "bimodule." It turns out for any maximal ideal $m$ of $Z$ that $U(L)/mU(L)$ is a Harish-Chandra bimodule. Borho [86, §4] defines a *direct link* from $P$ to $Q$ in $\operatorname{Spec}(R)$ to be a bimodule $M = A/B$ with $A, B \lhd R$, such that $\operatorname{Ann}_R M = P$ and $\operatorname{Ann}'_R M = Q$; direct links of linked primitive ideals are Harish-Chandra bimodules. (Note Borho's definition is weaker than the definition given in §3.5, although it coincides in the solvable case.) Some important results:

(1)  $U(L)/mU(L)$ has finite length as $U(L \times L)$-module.
(2)  It follows from (1) that each fiber $\mathscr{F}(m)$ is finite.
(3)  By (2) any prime ideal lying over $m$ is a $G$-ideal and is thus primitive by fact 8.3.45(2). In other words, $\mathscr{F}(m)$ only consists of primitive ideals.
(4)  Each fiber has a unique maximum and unique minimum member.
(5)  The map $P \to Z \cap P$ yields a continuous, clopen map from the primitive spectrum of $U(L)$ to Spec($Z$) which is an affine variety. Consequently, max spec $U(L)$ is homeomorphic to max spec $Z$.

Even more decisive information can be had by studying modules of heighest weight, cf., Jantzen [83, Chapter 4 and 5]. A module of *heighest weight* $\alpha - \beta \in H^*$ is any $U(L)$-module generated by a *heighest weight vector* $x_\alpha$ satisfying $ax_\alpha = 0$ for all $a$ in $N_+$, and $hx_\alpha = \langle \alpha - \beta, h \rangle x_\alpha$ for all $h$ in $H$. Each $\alpha$ in $H^*$ gives rise in a canonical way to a heighest weight module which has a unique simple image whose annihilator is denoted $J_\alpha$. Thus $\alpha \mapsto J_\alpha$ is a function from $H^*$ to {primitive ideals of $U(L)$}, which Duflo [77] proved is onto! This result is especially powerful when coupled with the Harish-Chandra isomorphism $Z \approx S(H)^W$. Picking $J_\alpha$ in the fiber $\mathscr{F}(m)$ yields an onto function $W \to \mathscr{F}(m)$ given by $w \mapsto J_{w\alpha}$.

Even more precise information can be had by focusing on the *lattice of integral weights*, which is $\{\beta \in H^*: 2\langle \alpha, \beta \rangle / \langle \alpha, \alpha \rangle \in \mathbb{Z}$ for all roots $\alpha\}$, but the results become quite intricate and we refer the reader to Joseph [83] for further information; this subject is still developing.

**Ring-Theoretic Properties of $sl_2$** As remarked earlier, $sl_2$ is an incredibly rich source of counterexamples, and we want to collect some of the relevant properties.

*Fact 8.3.49:*  Ring-theoretic properties of $R = U(sl_2(\mathbb{C}))$. Write $e = e_{21}$, $f = e_{12}$, and $h = e_{22} - e_{11}$, the "standard" base of $sl_2$, noting $[ef] = h$, $[fh] = 2f$, and $[eh] = -2e$. Let $z = 4fe + h^2 + 2h$, the "Casimir" element, and write $A = U(L)L$, the "augmentation" ideal. Let $R = U(L)/Az$.

(1)  (S. P. Smith [81a] and Brown [81, example 6.4]). $R$ is a Noetherian ring with precisely two nonzero prime ideals: $P_1 = U(L)z/Az$ and $P_2 = A/Az$. $P_1 \subset P_2$ are, in fact, primitive ideals; $P_2^2 = P_2$ and $P_1P_2 = P_2P_1 = 0$. Thus $P_1 = $ Nil($R$). K-dim $R = $ K-dim $R/P_1 = 1$, but $R/P_2 \approx \mathbb{C}$.

(2)  (Dean-Stafford [86, theorem 3.3]) $R$ is *not* embeddible in an Artinian ring. The proof involves the "easy" direction of a theorem of Schofield con-

cerning "Sylvester rank functions" on the modules of any ring embeddible into an Artinian ring, coupled with computations involving the projective modules arising in Stafford [85a]. Incidentally, Stafford's theory of $\hat{g}(M)$ studied in §3.5 enters here also.

(3) (Stafford [85a, theorem 3.2]) Let $R' = R \otimes_{\mathbb{C}} R \approx U(\mathrm{sl}_2 \times \mathrm{sl}_2)$. $R'$ is Noetherian, but is not "weakly ideal invariant."

(4) Stafford [85a, corollary 4.3]) $M_2(R')$ has a simple module which has infinite length as $R'$-module. (Compare with proposition 2.5.29.)

(5) (Stafford [85a, theorem 2.7]) $R'$ has a simple module $M$ for which $2 \operatorname{GK} \dim M > \operatorname{GK} \dim(R/\operatorname{Ann} M)$. The significance lies in the fact that equality holds for Harish-Chandra modules; Gabber proved "$\geq$" holds in general.

## Digression: The Restricted Enveloping Algebra

The basic structure theorems of Lie algebras fail over fields of characteristic $p > 0$, cf., Jacobson [62B, pp, 52–53]. On the other hand, in characteristic $p$ one often has a richer structure; for example, if $\delta$ is a derivation then $\delta^p$ is also a derivation by Leibniz' formula. Accordingly we have

**Definition 8.3.50:** A *restricted Lie algebra* is a Lie algebra $L$ over a field $F$ of characteristic $p$, together with a map $a \mapsto a^{[p]}$ such that (for all $a, b$ in $L$ and all $\alpha$ in $F$)

$$(\alpha a)^{[p]} = \alpha^p a^{[p]}$$

$$(a + b)^{[p]} = a^{[p]} + b^{[p]} + \sum_{i=1}^{p-1} f_i(a, b),$$

where $f_i(a, b)$ is the coefficient of $\lambda^{i-1}$ in $(\operatorname{ad}(\lambda a + b))a$, and

$$[ab^{[p]}] = (\operatorname{ad} b)^p a.$$

A *restricted Lie map* is a Lie map which preserves these properties, i.e., the morphism in the appropriate category. Let us call this category $F\text{-}\mathscr{L}ie_p$. Summarizing the discussion until now, if $F$ is a field of characteristic $p > 0$ we have a functor $G: F\text{-}\mathscr{A}lg \to F\text{-}\mathscr{L}ie_p$ given by $A \mapsto A^-$; verification of the above properties are in Jacobson [62B, pp. 185–187]. The universal from a restricted Lie algebra $L$ to $G$ is denoted as $u(L)$.

**Proposition 8.3.51:** (*Jacobson* [62B, *theorem* 5.12]) $u(L)$ *exists and is isomorphic to* $U(L)/A$ *where* $A$ *is the ideal of* $U(L)$ *generated by all* $\{x^p - x^{[p]} : x \in L\}$. *There is a canonical monic* $L \to u(L)$, *and* $[u(L): F] = p^{[L:F]}$.

***Sketch of Proof:***   Let $\{x_i : i \in I\}$ be a base of $L$ over $F$. If $x = \sum \alpha_i x_i$ then $x^p - x^{[p]} = \sum \alpha_i (x_i^p - x_i^{[p]})$, so it follows that the $x_i + A$ are linearly independent and there is a monic $L \to u(L)$. The universal property is straightforward, and $\{x_1^{j(1)} \cdots x_t^{j(t)} : 0 \leq j(i) \leq p - 1\}$ is a base of $u(L)$. (This is the trickiest of the calculations.)

Thus $u(L)$ is analogous to $U(L)$ with the notable exception that for $L$ finite dimensional $u(L)$ is also finite dimensional and hence the PI-theory of Chapter 6 is applicable. In fact, the PI-degree of $u(L)$ is obviously a power of $p$. This tempts one to pass to characteristic $p$ when studying enveloping algebras, cf., remark 2.12.37′ and exercise 20.

## §8.4  General Ring Theoretic Methods

In this section we touch on one of the more promising directions of current endeavor in ring theory. Throughout the history of the subject, ring theorists have tried to prove deep theorems in other subjects with economy of effort. As the allied subjects have progressed, this task has become more and more difficult, with the gaping gulf of triviality always dangerously nearby. Nevertheless, several elegant theories have been constructed recently in an effort to unify the results of polycyclic group algebras and enveloping algebras. Since these often involve Noetherian rings, parts of this section could be viewed as an extension of §3.5, although the outlook is completely different. Most of the results also apply to other interesting classes of rings such as affine PI-algebras.

### *Vector Generic Flatness and the Nullstellensatz*

Following McConnell [82a] we shall begin by unifying a large number of verifications of the Nullstellensatz. Note that these results are characteristic-free.

***Definition 8.4.1:***   A ring $T$ is called an *almost normalizing extension* of $R$ if $T$ is generated over $R$ *as a ring* by elements $a_1, \ldots, a_t$ such that

  (i)  $R + Ra_i = R + a_i R$     for $1 \leq i \leq t$

  (ii)  $[a_i, a_j] \in R + \sum Ra_i$     for $1 \leq i, j \leq t$.

***Example 8.4.2:***

  (i) Any Ore extension $R[\lambda; \sigma, \delta]$ is an almost normalizing extension of $R$.

(ii) Any enveloping algebra $U(L)$ of a finite dimensional Lie algebra $L$ is an almost normalizing extension, where we take the $a_i$ to be a base of $L$.

(iii) Any finite normalizing extension (cf., definition 2.5.28ff) is an almost normalizing extension.

(iv) If $S = \{\lambda^i : i \in \mathbb{N}\}$ then $S^{-1}R[\lambda; \sigma] = R[\lambda, \lambda^{-1}; \sigma]$ is an almost normalizing extension of $R$.

(v) Polycyclic group algebras are built as a series of almost normalizing extensions in the form of (iv).

**Remark 8.4.3:** Any almost normalizing extension $T$ of $R$ is spanned by the monomials $a_1^{i(1)} \cdots a_t^{i(t)}$ as left (or right) $R$-module, for $i(1), \ldots, i(t)$ in $\mathbb{N}$, notation as in definition 8.4.1. Write $\mathbf{i}$ for $(i(1), \ldots, i(t))$ and $a^{\mathbf{i}}$ for $a_1^{i(1)} \cdots a_t^{i(t)}$, and write $|\mathbf{i}|$ for $i(1) + \cdots + i(t)$. Then $T$ is a filtered ring, seen by taking $T_n = \{\sum Ra^{\mathbf{i}} : |\mathbf{i}| \leq n\}$, cf., definition 5.1.36 and remark 1.2.13'.

**Theorem 8.4.4:** *Any almost normalizing extension $T$ of a left Noetherian ring $R$ is left Noetherian.*

**Proof:** Using the above filtration we see the associated graded ring $\operatorname{Gr} T$ is generated over $R$ by elements $\bar{a}_1, \ldots, \bar{a}_t$ such that $R\bar{a}_u = \bar{a}_u R$ and $[\bar{a}_u, \bar{a}_v] = 0$ for $1 \leq u, v \leq t$; thus $\operatorname{Gr} T$ is left Noetherian by proposition 3.5.2, implying $T$ is left Noetherian by corollary 3.5.32(i).     Q.E.D.

This gives us a very broad class of Noetherian rings, and we should check them for Nullstellensatz properties. We would like to use generic flatness and theorem 2.12.36, but it is not obvious how to lift generic flatness directly. McConnell's solution was to require a stronger condition, which does lift easily. To this end we examine the filtration on $T$. Using the total order of example 1.2.18, we can speak of the *leading coefficient* of $\sum r_i a^{\mathbf{i}}$ to be that nonzero $r_{\mathbf{i}}$ for $\mathbf{i}$ maximal. Also we define $T_{\mathbf{i}} = \sum_{\mathbf{j} \leq \mathbf{i}} Ra^{\mathbf{j}}$ and $T_{\mathbf{i}}^- = \sum_{\mathbf{j} < \mathbf{i}} Ra^{\mathbf{j}}$. Given $L < T$ let $\operatorname{lead}_{\mathbf{i}}(L) = \{r \in R : r$ is the leading coefficient of some element of $L \cap T_{\mathbf{i}}$.

Let us say a set $\{L(\mathbf{i}) : \mathbf{i} \in \mathbb{N}^{(t)}\}$ of left ideals is *weakly ascending* if $L_{(\mathbf{i})} \subseteq L_{(\mathbf{j})}$ whenever $i(u) \leq j(u)$ for all $1 \leq u \leq t$.

**Remark 8.4.5:** $\{\operatorname{lead}_{\mathbf{i}}(L) : \mathbf{i} \in \mathbb{N}^{(t)}\}$ is a weakly ascending set of left ideals of $R$. If $R$ is left Noetherian then $\{\operatorname{lead}_{\mathbf{i}}(L) : \mathbf{i} \in \mathbb{N}^{(t)}\}$ is finite. (Indeed, the first assertion is immediate, and the second assertion follows at once from the following combinatorial fact:

*Claim 8.4.6:* An infinite sequence of $\mathbb{N}^{(t)}$ has an infinite subsequence which is increasing (or stationary) in each component. (It suffices to prove the claim for $t = 1$, for then we could refine the sequence to an infinite subsequence increasing in the last component and are done by induction. For $t = 1$ suppose $\{n_1, n_2, \ldots\}$ is a sequence in $\mathbb{N}$. Some $n_u$ is smallest in the sequence; cutting off the first $(u - 1)$ terms we may assume $n_1$ is minimal. Now we choose $n_v$ minimal among $n_2, n_3, \ldots$ and throwing away $n_2, \ldots, n_{v-1}$ we may assume $n_v = n_2$. Continuing this selection procedure indefinitely yields $n_1 \leq n_2 \leq n_3 \leq \cdots$.)

**Lemma 8.4.7:** *Suppose $T$ is an almost normalizing extension of $R$. For $L < T$ let $^-$ denote the canonical image in the $R$-module $M = T/L$. Then $\bar{T}_i/\bar{T}_i^- \approx R/\mathrm{lead}_i(L)$.*

*Proof:* Define $T_i \to R/\mathrm{lead}_i(L)$ by taking the leading coefficient of any element of $T_i$ modulo $\mathrm{lead}_i(L)$. Clearly $L \to 0$ so we have a map $\varphi \colon \bar{T}_i \to R/\mathrm{lead}_i(L)$. Suppose $x = \sum_{j \leq i} r_j a^j \in T$. Then $\bar{x} \in \ker \varphi$ iff $\bar{r}_i = 0$; so that $\bar{x} \in \bar{T}_i^-$.          Q.E.D.

*Definition 8.4.8:* An algebra $R$ over an integral domain $C$ satisfies *vector generic flatness* if for any $k \in \mathbb{N}$ and for any weakly ascending set $\{L(\mathbf{i}) \colon \mathbf{i} \in \mathbb{N}^{(k)}\}$ of left ideals of $R$ one can find $s$ in $C - \{0\}$ such that $(R/L(\mathbf{i}))[s^{-1}]$ is a free $C[s^{-1}]$-module for all $\mathbf{i}$.

It turns out that vector generic flatness is satisfied by the rings in which we are interested but is a stronger condition than generic flatness. First note any integral domain $C$ satisfies vector generic flatness over itself. Indeed, by claim 8.4.6 any weakly ascending set of ideals $\{L(\mathbf{i}) \colon \mathbf{i} \in \mathbb{N}^{(k)}\}$ has only finitely many minimal nonzero members, which we call $L_1, \ldots, L_m$; taking $0 \neq s \in \bigcap_{i=1}^m L_u$ yields $(C/L(\mathbf{i}))[s^{-1}] = 0$ whenever $L(\mathbf{i}) \neq 0$, and clearly $C[s^{-1}]$ is free as $C[s^{-1}]$-module. To see vector generic flatness implies generic flatness we need the following observation:

*Remark 8.4.9:* If $M = M_0 > \cdots > M_t = N$ is a chain of modules for which every factor module $M_{i-1}/M_i$ is free then $M/N$ is free. (Indeed, we may assume $N = 0$. Suppose $B_i$ is a set of elements of $M_{i-1}$ which modulo $M_i$ is a base in $M_{i-1}/M_i$. Then $\bigcup B_i$ surely spans $M$, and we claim it is a base of $M$. For, otherwise, $\sum_{i,j} r_{ij} b_{ij} = 0$ for suitable $b_{ij}$ in $B_i$; then $\sum_j r_{1j} b_{1j} \in M_1$ implying each $r_{1j} = 0$; continuing inductively, once we have proved $r_{uj} = 0$ for all $u < i$ we can conclude the $r_{ij} = 0$.)

**Remark 8.4.10:** Vector generic flatness implies generic flatness. (Indeed, suppose $M$ is an f.g. $R$-module. Then there is a chain $M = M_0 > M_1 > \cdots > M_t = 0$ with each factor cyclic. But any cyclic module has the form $R/L$, so we can write each $M_{i-1}/M_i = R/L_i$ for suitable $L_i < R$. Picking $s$ such that each $(R/L_i)[s^{-1}]$ is free over $C[s^{-1}]$, we see $M[s^{-1}] = M_0[s^{-1}] > M_1[s^{-1}] > \cdots > M_t[s^{-1}] = 0$ satisfies the hypothesis of remark 8.4.9, so $M[s^{-1}]$ is free, as desired.)

Having seen that vector generic flatness is merely an infinitistic version of generic flatness, we now feel confident in improving the results of §2.12.

**Theorem 8.4.11:** *Suppose $R \subseteq T$ are algebras over an integral domain $C$, and $R$ satisfies vector generic flatness over $C$. Then $T$ also satisfies vector generic flatness over $C$, given either of the additional hypothesis:*

(i) *$T$ is f.g. as $R$-module.*
(ii) *$T$ is an almost normalizing extension of $R$.*

**Proof:** Let $\{L(\mathbf{i}): \mathbf{i} \in \mathbb{N}^{(k)}\}$ be a weakly ascending set of left ideals of $T$. Our strategy in either case is to build for each $T/L(\mathbf{i})$ a chain of cyclic modules in $R$, apply vector generic flatness to the corresponding factors, and conclude by means of remark 8.4.9.

(i) Write $T = \sum_{i=1}^{k} Rx_i$ and for each $j \le k$ let $L_j(\mathbf{i})$ denote $\{r \in R : rx_j \in L(\mathbf{i}) + \sum_{u=1}^{j-1} Rx_u\}$. For each $j$ the set $\{L_j(\mathbf{i}): \mathbf{i} \in \mathbb{N}^{(k)}\}$ is weakly ascending, so there is $s(j)$ in $C$ such that $(R/L_j(\mathbf{i}))[s(j)^{-1}]$ is free over $C[s(j)^{-1}]$. Taking $s = s(1) \cdots s(t)$ we see each $(R/L_j(\mathbf{i}))[s^{-1}]$ is free over $C[s^{-1}]$. But by definition right multiplication by $x_j$ gives an isomorphism $R/L_j(\mathbf{i}) \approx (L(\mathbf{i}) + \sum_{u=1}^{j} Rx_u)/(L(\mathbf{i}) + \sum_{u=1}^{j-1} Rx_u)$, so we have a chain from $T$ to $L(\mathbf{i})$ whose factors are $R/L_j(\mathbf{i})$, implying by remark 8.4.9 that each $(T/L(\mathbf{i}))[s^{-1}]$ is free.

(ii) Let $a_1, \ldots, a_t$ be as in definition 8.4.1, and let $L(\mathbf{i}, \mathbf{j}) = \text{lead}_{\mathbf{j}}(L(\mathbf{i})) \le R$. Then $\{L(\mathbf{i}, \mathbf{j}): \mathbf{i} \in \mathbb{N}^{(k)}, \mathbf{j} \in \mathbb{N}^{(t)}\}$ can be viewed as a weakly ascending set of left ideals of $R$, indexed by $\mathbb{N}^{(k+t)}$, so there is $0 \ne s \in C$ such that each $(R/L(\mathbf{i}, \mathbf{j}))[s^{-1}]$ is free over $C[s^{-1}]$. But iterating lemma 8.4.7 shows there is a chain from $T$ to $L(\mathbf{i})$ whose factors are isomorphic to the $R/L(\mathbf{i}, \mathbf{j})$; consequently $(T/L(\mathbf{i}))[s^{-1}]$ is free by remark 8.4.9.     Q.E.D.

**Theorem 8.4.12:** *Suppose $R \subseteq T$ are $C$-algebras, and $T$ can be obtained from $R$ by a finite series of extensions as in (i) or (ii) of theorem 8.4.11. Furthermore, assume $R[\lambda_1, \lambda_2]/PR[\lambda_1, \lambda_2]$ satisfies vector generic flatness over $C[\lambda_1]/P$*

*for all P in* $\text{Spec}(C[\lambda_1])$. *If C is a Jacobson ring then T satisfies the weak Nullstellensatz and MN (the maximal Nullstellensatz, cf., definition 2.12.26).*

**Proof:** It is enough to show these vector generic flatness hypotheses lift to $T$, since then we are done by remark 8.4.10 and theorem 2.12.36. By iteration we may assume $T$ satisfies hypothesis (i) or (ii) of theorem 8.4.11, which enables us to obtain the vector generic flatness. Q.E.D.

**Corollary 8.4.13:** *If C is a Jacobson ring and T is built from C by a finite series of extensions as in (i) or (ii) of theorem 8.4.11, then T is Jacobson and satisfies MN over C.*

**Proof:** Passing to prime homomorphic images of $C$, we may assume $C$ is an integral domain. Taking $R = C$ we see $T$ satisfies the weak Nullstellensatz and MN over $C$. It remains to show for every prime ideal $P$ of $T$ that $\text{Nil}(T/P) = 0$ (for then $\text{Jac}(T/P) = 0$ by the weak Nullstellensatz). Let $P_0 = C \cap P \in \text{Spec}(C)$. Passing to $T/P$ and $C/P_0$ we may assume $P = P_0 = 0$, so $C$ is an integral domain. Let $S = C - \{0\}$. Then $S^{-1}C$ is a field so $S^{-1}T$ is Noetherian by theorem 8.4.4. But $S^{-1}\text{Nil}(T)$ is a nil ideal and is thus 0, so $\text{Nil}(T/P) = 0$ as desired. Q.E.D.

**Corollary 8.4.14:**

(i) (*Quillen-Duflo*) *Every enveloping algebra of a finite dimensional Lie algebra is Jacobson and satisfies the Nullstellensatz.*

(ii) (*Roseblade*) *Every group algebra of a polycyclic-by-finite group is Jacobson and satisfies the Nullstellensatz. (Note the base ring can be taken to be arbitrary Jacobson.)*

*Proof:*

(i) By example 8.4.2(ii).

(ii) Each cyclic group corresponds to an almost normalizing extension; each finite group factor corresponds to an f.g. extension. Q.E.D.

The generic flatness techniques have been seen to apply to almost normalizing extensions of commutative Jacobson rings and to affine PI-algebras (theorem 6.3.3). There is a nice theorem which unifies these results.

**Theorem 8.4.15:** (*McConnell* [84a]) *Suppose R is an affine PI-algebra over a commutative Jacobson ring C. Suppose T is built from R by a finite series of* *extensions as in (i) or (ii) of theorem 8.4.11, such that at each stage the new generators centralize R. Then T is Jacobson and satisfies MN.*

**Proof:** Any primitive ideal of $T$ intersects $R$ at a prime ideal by proposition 2.12.39 (since the generators centralize $R$), so we may assume $R$ is prime PI of some PI-degree $n$. By theorem 6.1.33 if we take $s$ in $g_n(R)$ then $R[s^{-1}]$ is f.g. free over $Z(R)[s^{-1}]$. $Z(R)[s^{-1}]$ is affine over $C$ by proposition 6.3.1 and is certainly an almost normalizing extension of $C$; hence $R[s^{-1}]$ and thus $T[s^{-1}]$ satisfies the hypotheses of corollary 8.4.13. Therefore $T[s^{-1}]$ is Jacobson and satisfies MN.

To prove MN for $T$, suppose $M$ is a simple $T$-module. We want to show $\text{End}_T M$ is algebraic. As observed in the first sentence we may assume $M$ is $R$-faithful, and thus $M$ is naturally a $T[s^{-1}]$-module by lemma 2.12.31. Obviously $M$ is simple over $T[s^{-1}]$, so MN for $T[s^{-1}]$ implies $0 = \text{Ann}_C M$ is maximal in $C$. Therefore $C$ is a field, and $\text{End}_T M \approx \text{End}_{T[s^{-1}]} M$ is algebraic over $C$, proving MN for $T$.

$T[\lambda]$ satisfies MN (using $T[\lambda]$ instead of $T$ above), so the weak Nullstellensatz holds by lemma 2.12.35. It remains to prove $\text{Nil}(T/P) = 0$ for every $P$ in $\text{Spec}(T)$. Passing to $T/P$ we may assume $T$ is prime. But then taking $S = Z(R) - \{0\}$ we see from the first paragraph that $S^{-1}T$ is obtained by a series of almost normalizing and f.g. extensions of the field $S^{-1}C$, so is Noetherian by theorem 8.4.4; hence $\text{Nil}(S^{-1}T) = 0$ implying $\text{Nil}(T) = 0$.
Q.E.D.

Let us pause a moment to try to digest these results. As shown in exercise 2.12.5, there is an example of a commutative local Jacobson domain having an Ore extension which is not Jacobson. On the other hand, Goldie-Michler [75] proved that any skew polynomial ring $R[\lambda; \sigma]$ over a Noetherian Jacobson ring $R$ is Jacobson, provided $\sigma$ is an automorphism; their results are given in exercises 2 through 10. McConnell's theory itself can be "skewed" somewhat, cf., Brown [82]. Part of McConnell's theorem can also be freed of generic flatness, cf., exercise 13.

When $\sigma$ is not onto the situation becomes much more complex, and Irving has made a careful study of these rings and their prime and primitive spectra in case $R$ is commutative. In particular, Irving [79b] determines exactly when the Ore extension is a left or right primitive ring, putting the left-but-not-right primitive examples of Bergman and Jategaonkar in a broader context; the Nullstellensatz is also considered there.

In contrast to Roseblade's results about polycyclic group algebras, Irving [78] has produced a Noetherian Jacobson domain which is affine over an absolute field but which is primitive and thus fails the Nullstellensatz and generic flatness, cf., exercise 11. This casts doubt on whether a good theory about the Nullstellensatz could be developed without generic flatness.

## Digression: Subalgebras

In attempting to generalize corollary 8.4.13, we may wonder what can be said of affine *subalgebras* of $T$. In view of proposition 8.3.15 and the Farkas-Snider theorem $T$ is often a Noetherian domain; letting $D$ be the division ring of fractions Lichtman [85] proved the striking result that for any affine $F$-subalgebra of $M_n(D)$ the Jacobson radical is nilpotent of index $\leq n$. The method is totally different from the theory of Duflo-Irving-McConnell described above. Instead Lichtman shows that it is enough to prove the theorem in characteristic $p$ for all $p > 0$ and observes that the PI-theory is applicable by passing to the restricted enveloping algebra. Lichtman's paper also contains interesting observations on removing the finite dimensional hypothesis.

## The Primitive Spectrum

Having gotten past the Nullstellensatz, we can look for even finer information about the primitive spectrum; for example, is it locally closed in $\operatorname{Spec}(R)$? This question can be formulated algebraically. Since $P$ in $\operatorname{Spec}(R)$ is the intersection of an open set and a closed set iff $P$ is a $G$-ideal, we are in effect asking if every primitive ideal is a $G$-ideal. This is patently true if every primitive ideal is maximal but is much more difficult for enveloping algebras of non-nilpotent Lie algebras and can be false for polycyclic group algebras. Thus a very interesting challenge is to find exactly how ring theory can be brought in to bear on the matter. We shall discuss this in some detail, getting at once to the heart of the matter.

**Definition 8.4.16:** The *heart* of a prime ideal $P$, written Heart($P$), is the extended centroid of $R/P$.

Usually the heart is defined only for Noetherian rings, as the center of the classical ring of fractions $Q$ of $R/P$. Definition 8.4.16 is a straightforward generalization and also has the following immediate application.

**Remark 8.4.17:** Suppose $P$ is a primitive ideal. Writing $P = \operatorname{Ann}_R M$ for a simple module $M$ and $D = \operatorname{End}_R M$, we have a canonical injection $\varphi: \operatorname{Heart}(P) \to Z(D)$ given in example 3.4.16.

We are now ready to examine the following possible properties for a prime ideal $P$ in an $F$-algebra $R$;

(1) $P$ is primitive.
(2) Heart($P$) is algebraic over $F$.
(3) $P$ is a $G$-ideal.

These properties are called *Dixmier's conditions*. Dixmier's condition (2) is less intuitive than (1) and (3), but could be thought of as follows: If $F$ is algebraically closed and (2) holds then Heart($P$) = $F$; thus $R/P$ is centrally closed with center $F$. Thus in utilizing (2) we should expect to rely on the theory of centrally closed algebras developed in §3.4, and this will indeed be the case.

**Proposition 8.4.18:** *If* $P \in J\text{-Spec}(R)$ *then* (3) $\Rightarrow$ (1). *If* $R$ *satisfies the Nullstellensatz then* (1) $\Rightarrow$ (2).

**Proof:** (3) $\Rightarrow$ (1). $P$ is an intersection of primitive ideals, one of which must be $P$, by (3). (1) $\Rightarrow$ (2) by remark 8.4.17.     Q.E.D.

Thus (3)$\Rightarrow$(1)$\Rightarrow$(2) whenever $R$ is Jacobson and satisfies the Nullstellensatz. To complete the chain of implications we would like to show (2) $\Rightarrow$ (3). This is a rather touchy issue; fortunately, there is a weakening of (3) which is more amenable.

**Definition 8.4.19:** We say a set $S \subset R - \{0\}$ *separates* Spec($R$) if every nonzero prime ideal contains an element of $S$. We say $P \in R$ *satisfies property* (3') if there is a countable set separating Spec($R/P$).

Obviously, any $G$-ideal satisfies property (3') so we have (3) $\Rightarrow$ (3'); the reason we weaken (3) is the following result of Irving. (The proof is modified according to an idea of M. Smith.)

**Proposition 8.4.20:** *If* $R$ *is an* $F$-algebra with a countable base $B$ then (2) $\Rightarrow$ (3').

**Proof:** Writing $B = \{b_i : i \in \mathbb{N}\}$ and $b_i b_j = \sum_k \alpha_{ijk} b_k$ with $\alpha_{ijk}$ in $F$, let $F_0$ be the subfield of $F$ generated by the $\alpha_{ijk}$, and let $R_0 = F_0 B$. Then $R_0$ is a countable subalgebra of $R$. We shall prove $R_0$ itself separates Spec($\hat{R}$) where $\hat{R}_0, \hat{R}$ are the respective central closures of $R_0, R$. Note every ideal of $\hat{R}$ intersects $\hat{R}_0$ nontrivially by theorem 3.4.11, and obviously every ideal of $\hat{R}_0$ intersects

$R_0$ nontrivially since $\hat{R}_0$ is an essential extension of $R_0$. Hence $R_0$ separates $\text{Spec}(\hat{R})$.

To prove the proposition we may assume $P = 0$. Then (2) shows $Z(\hat{R})$ is algebraic over $F$. Hence LO holds from $R$ to $RZ(\hat{R}) = \hat{R}$, by theorem 2.12.48 and proposition 2.12.49. We conclude $R_0$ separates $\text{Spec}(R)$.          Q.E.D.

Thus far we have proved $(3) \Rightarrow (1) \Rightarrow (2) \Rightarrow (3')$ under quite general circumstances and should note that $(3')\Rightarrow(3)$ can fail, also, cf., Lorenz [77].) Thus we led first to examine $(3')\Rightarrow(1)$, to try to complete a general chain of equivalence. Note that we should require $F$ uncountable, since for $F$ countable $F[\lambda]$ is countable (so that $(3')$ is trivial for $R = F[\lambda]$), whereas $F[\lambda]$ is not primitive. When $F$ is uncountable there is a host of examples of $(3') \Rightarrow (1)$, including enveloping algebras (Dixmier [77]), polycyclic-by-finite group algebras (Farkas [79]), and related constructions (Irving [79]). Unfortunately, there is not yet a sweeping result stating that $(3') \Rightarrow (1)$ for Noetherian algebras over an uncountable field. To stimulate such a result let us pose

**Conjecture 8.4.21:**   *Suppose $R$ is an affine algebra over an uncountable field $F$. Then $(3') \Rightarrow (1)$ for any prime, semiprimitive ideal $P$.*

In the absence of an answer to this conjecture, let us consider a general strategy mapped out by Farkas, which in fact lies behind all the known positive results.

Call a ring $R$ *Kaplansky* if the primitive spectrum of each prime homomorphic image of $R$ is a Baire space. (Recall a topological space is Baire if any countable intersection of open sets is dense.)

**Proposition 8.2.22:**   *If $R$ is Jacobson and Kaplansky over an uncountable field then $(3') \Rightarrow (1)$.*

*Proof:*   Let $P \in \text{Spec}(R)$ such that $\text{Spec}(R/P)$ has a countable separating set $S$ disjoint from $P$. Passing to $R/P$ we may assume $P = 0$, and $0 \notin S$. Given any $s$ in $S$ let $U_s = \{$primitive ideals not containing $s\}$ a dense subset of $\text{Spec}(R)$ by corollary 2.12.5'. By hypothesis there is some $P' \in \bigcap_{s \in S} U_s$, and since each $s \notin P'$ by construction we see $0 = P'$ is primitive.          Q.E.D.

Thus we are reduced to showing that the rings of interest to us are Kaplansky. Extending Farkas [79], Lorenz-Passman [79] proves that polycyclic-by-finite group algebras are Kaplansky. Irving [79] gives many other instances. Also, cf., Dixmier [77]. In summary we have

**Fact 8.4.23:** Over an algebraically closed field of characteristic 0, (2) ⇒ (1) for enveloping algebras and polycyclic-by-finite group algebras.

Let us return now to the question of when (1), (2), and (3) are equivalent. In view of proposition 8.4.18 the "missing link" is (2) ⇒ (3), which by Lorenz [77] fails for various polycyclic group algebras. (2) ⇒ (3) does hold for any group algebra of finitely generated nilpotent by-finite groups, for which, in fact, primitive ideals are maximal.

We shall focus here on the story for enveloping algebras, which involves an interesting blend of Lie techniques with pure ring theory. Assume char($F$) = 0. We just observed (2) ⇒ (1) when $F$ is algebraically closed; more recently Moeglin [80a] proved (1) ⇒ (3) when $F$ is uncountable and algebraically closed, thereby yielding (2) ⇒ (3) in this case. Using these results we can apply some purely ring-theoretic methods of Irving-Small [80] to remove these extra restrictions from $F$.

## Reduction Techniques

**Proposition 8.4.24:** *Suppose $K$ is a separable field extension of $F$, $R$ is an $F$-algebra, and $R' = R \otimes_F K$ satisfies* ACC(ideals).

(i) *If $B$ is a semiprime ideal of $R$ then $BR'$ is a semiprime ideal of $R'$ which can be written as $P'_1 \cap \cdots \cap P'_t$ for suitable $t$, where each $P'_j$ is a minimal prime over $BR'$.*

(ii) *If, furthermore, $B \in \text{Spec}(R)$ then $B = P'_j \cap R$ for some $j$; if $B$ is primitive and $[K:F] < \infty$ then $P'_j$ is primitive.*

**Proof:**

(i) Passing to $R'/BR'$ we may assume $B = 0$, i.e., $R$ is semiprime. Then $R'$ is semiprime by theorem 2.12.52 and has a finite number of minimal prime ideals $P'_1, \ldots, P'_t$ by corollary 3.2.26.

(ii) Continuing the proof of (i) we may assume $R$ is prime and

$$(P'_1 \cap R) \cdots (P'_t \cap R) = 0,$$

implying some $P'_j \cap R = 0 = B$. The last assertion is corollary 3.4.14.

Q.E.D.

Note: The separability assumption was needed only in showing $R'$ is semiprime, but, anyway, this is satisfied when $\text{char}(F) = 0$.

We can improve proposition 8.4.24(ii) if we bring in the Nullstellensatz (cf., definition 2.12.26).

**Proposition 8.4.24':** Let $\bar{F}$ be the algebraic closure of a field $F$ of characteristic 0. Suppose $R$ is an $F$-algebra, $R \otimes_F \bar{F}$ satisfies the Nullstellensatz and is left Noetherian, and $K$ is a field extension of $F$ such that $R' = R \otimes_F K$ is left Noetherian. Let $M$ be any simple $R$-module. Then

(i) $[\text{End}_R M : F] < \infty$,
(ii) $M' = M \otimes_F K$ is cyclic and completely reducible as $R'$-module,
(iii) If $R'$ is prime and $R$ is primitive then $R'$ is primitive.

**Proof:** First we prove (ii) for $K = \bar{F}$. View $M = M \otimes 1 \subset M'$. If $M = Rx$ then $M' = R'x$, so $M'$ is cyclic and thus Noetherian. Suppose $N'$ is a submodule of $M'$. Then $N'$ is f.g. and thus is contained in $M \otimes F'$ for some finitely generated (and thus finite dimensional) field extension $F'$ of $F$. But $M \otimes F'$ is completely reducible (proved in analogy to corollary 3.4.14), and thus contains a complement $N''$ of $N'$, and $N''$ tensors up to a complement of $N'$ in $M'$, thereby proving $M'$ is completely reducible.

To prove (i), now take a simple submodule $N'$ of $M'$ (still taking $K = \bar{F}$). As above we view $N'$ in $M \otimes F'$; passing to the normal closure of $F'$ we may assume $F'$ is Galois over $F$. Let $G = \text{Gal}(F'/F)$. The group $\tilde{G} = \{1 \otimes \sigma : \sigma \in G\}$ acts naturally on $M \otimes F'$; $\sum(1 \otimes \sigma)N'$ is a submodule of $M \otimes F'$ fixed under $\tilde{G}$ and thus equal to $M \otimes F'$ (since $M$ is simple). Hence $M'$ is a finite sum of copies of $N' \otimes \bar{F}$, implying $\text{End}_{R'} M'$ is finite dimensional over $\bar{F}$. But $M'$ is cyclic over $R'$ so $\text{End}_{R'} M' \approx (\text{End}_R M) \otimes \bar{F}$, proving (i).

To prove (ii) for arbitrary $K$, let $D = \text{End}_R M$ and note

$$M' \approx (M \otimes_D D) \otimes_F K \approx M \otimes_D (D \otimes K).$$

One proves in analogy to theorem 3.4.11 that each submodule of $M'$ has the form $M \otimes L$ for some left ideal $L$ of $D \otimes K$. But $D \otimes K$ is semisimple Artinian by (i), so $M'$ is a finite direct sum of simple submodules $M'_1, \ldots, M'_t$ proving (ii). Now let $P'_i = \text{Ann}\, M'_i$. $P'_1 \cdots P'_t = 0$, so some $P'_i = 0$ if $R'$ is prime, proving (iii).     Q.E.D.

**Proposition 8.4.25:** If $R' = R \otimes_F K$ is primitive and left Noetherian then $R \otimes_F K_1$ is primitive for some finitely generated field extension $K_1$ of $F$.

***Proof:*** Take a maximal left ideal $L$ of $R'$ having core 0. If $L = \sum_{i=1}^{t} R'a_i$ then writing $a_i = \sum_{j=1}^{t(i)} r_{ij} \otimes a_{ij}$ for $a_{ij}$ in $K$ let $K_1$ be the subfield of $K$ generated by the $a_{ij}$. $R'$ is a central extension of $R \otimes K_1$. Let $L_0 = L \cap (R \otimes K_1)$, clearly a proper left ideal of $R \otimes K_1$ with core 0 (for if $A = \mathrm{core}(L_0)$ then $L \supseteq AK \lhd R'$). $L_0$ is also maximal (for if $L_1 \supset L_0$ then $R'L_1 = R'$ implying $L_1 = R \otimes K_1$). Hence $R \otimes K_1$ is primitive. Q.E.D.

**Lemma 8.4.26:** *Suppose $R$ is an algebra over a field $F$, such that $R \otimes_F K$ is semiprime, left Noetherian for every algebraic extension $K$ of $F$. Let $Q$ be the classical ring of fractions of $R$. If $Z(Q)$ is algebraic over $F$ then there is a finite dimensional field extension $K_1$ of $F$ such that $Q \otimes_F K_1$ is split, i.e., a direct product of matrix rings over $K_1$.*

***Proof:*** We claim the ring of fractions of the semiprime Goldie ring $R \otimes K$ is $Q \otimes K$. This is clear for $K$ finite over $F$ for then $Q \otimes K$ is semisimple Artinian; the claim is established by passing to direct limits.

Now take $K$ to be the algebraic closure of $F$. By corollary 1.7.24 $Z(Q) \otimes K = Z(Q \otimes K)$, a direct product of fields since $Q \otimes K$ is semisimple Artinian. But by assumption $Z(Q)$ is algebraic over $F$, implying $Z(Q) \otimes K$ is a finite direct product of algebraic extensions of $K$ and thus a direct product of copies of $K$ itself. Writing out the idempotents explicitly one obtains these in a suitable finite dimensional extension of $F$, and this is the desired $K_1$. Counting dimensions shows that if $Z(Q) \otimes K \approx K^{(n)}$ then $Z(Q) \otimes K_1 \approx K_1^{(n)}$. Thus $Q \otimes K_1$ is split. Q.E.D.

We also need some results concerning $G$-ideals. We say a prime ring $R$ is $G$-prime if 0 is a $G$-ideal.

**Lemma 8.4.26':**

(i) *If a $G$-prime ring $T$ is a direct limit of finite centralizing extensions of $R$ then $R$ is also $G$-prime.*

(ii) *If $R \otimes_F K$ is $G$-prime for a field extension $K$ of $F$ then $R$ is also $G$-prime.*

***Proof:***

(i) Let $A = \bigcap\{\text{nonzero primes of } T\}$. Then $R \cap A \neq 0$ by the theorem 3.4.13 and is contained in every nonzero prime $P$ of $R$ by LO. (There is some $P'$ lying over $P$ by theorem 2.12.48 and proposition 2.12.49, and $R \cap A \subset R \cap P' = P$.)

(ii) $K$ is algebraic over a purely transcendental extension $K_0$ of $R$, and $R \otimes K_0$ is $G$-prime by (i). Write $K_0 = F(\Lambda)$ for a suitable set of commuting indeterminates $\Lambda$ over $F$; then $R \otimes K_0 \approx R(\Lambda)$, and some $\sum r_i \lambda_1^{i(1)} \cdots \lambda_t^{i(t)}$ is in every nonzero prime ideal of $R(\Lambda)$. But if $0 \neq P \in \operatorname{Spec}(R)$ then $P(\Lambda) \in \operatorname{Spec} R(\Lambda)$ since $R(\Lambda)/P(\Lambda) \approx (R/P)(\Lambda)$. Hence $\sum r_i \lambda_1^{i(1)} \cdots \lambda_t^{i(t)} \in P(\Lambda)$ implying each $r_i \in P$, i.e., each $r_i \in \bigcap \{\text{nonzero prime ideals of } R\}$, as desired.    Q.E.D.

**Theorem 8.4.27:** (*Irving-Small* [80]) *Assume $R$ is a countable dimensional $F$-algebra and $R \otimes_F K$ is left Noetherian and Jacobson satisfying the Nullstellensatz for every field extension $K$ of $F$. Furthermore, assume* $\operatorname{char}(F) = 0$, *and* $(2) \Rightarrow (1)$ *in* $R \otimes \bar{F}$ *where $\bar{F}$ is the algebraic closure of $F$, and* $(2) \Rightarrow (3)$ *in* $R \otimes \bar{K}$ *for some uncountable algebraically closed field $\bar{K} \supseteq F$. Then Dixmier's conditions* (1), (2), *and* (3) *are equivalent in $R$.*

**Proof:** We shall make use of two lemmas, assuming the hypothesis of the first sentence.

**Lemma 8.4.28:** *If* $(1) \Rightarrow (3)$ *in* $R \otimes_F K$ *for some separable (not necessarily algebraic) field extension $K$ of $F$ then* $(1) \Rightarrow (3)$ *in $R$.*

**Proof of Lemma 8.4.28:** Let $P$ be a primitive ideal of $R$, and let $\bar{R} = R/P$ and $Q$ be the ring of fractions of $\bar{R}$. Then $Z(Q)$ is algebraic over $F$ by proposition 8.4.18, so by lemma 8.4.26 we can split $Q$ by a finite dimensional field extension $K_1$ of $F$. By proposition 8.4.24 $R \otimes K_1$ contains a primitive ideal $P'$ lying over $P$. Cleary heart($P'$) $= K_1$; in view of lemma 8.4.26'(ii) we may replace $R$ by $R \otimes K_1$, and thereby assume $Z(Q) = F$.

Now $\bar{R} \otimes_F K$ is prime and thus primitive by proposition 8.4.24', so by hypothesis is $G$-prime; hence $\bar{R}$ is $G$-prime by lemma 8.4.26'(ii).    Q.E.D.

**Lemma 8.4.29:** *If* $(2) \Rightarrow (1)$ *in* $R \otimes_F K$ *where $K$ is the separable algebraic closure of $F$ then* $(2) \Rightarrow (1)$ *in $R$.*

**Proof of Lemma 8.4.29:** Suppose $P \in \operatorname{Spec}(R)$ and $\bar{R} = R/P$. By proposition 8.4.24 we see $\bar{R} \otimes_F K$ is semiprime and satisfies ACC(ideals) (since it is an image of $R \otimes_F K$). Let $Q$ be the ring of fractions of $\bar{R}$. By lemma 8.4.26 we see $Q \otimes K_1$ is split for some field $K_1$ finite over $F$. By proposition 8.4.24 there is a minimal prime ideal $P'$ of $\bar{R} \otimes K_1$ with $P' \cap \bar{R} = (0)$, so the ring of fractions of $(\bar{R} \otimes K_1)/P'$ is one of the (split) components of $Q$. This implies $K_1 = \operatorname{heart}(P')$ so, by hypothesis, $P'$ is primitive. Hence $\bar{R}$ is primitive by corollary 2.5.30.    Q.E.D.

*Proof of Theorem 8.4.27:* By hypothesis we have $(2) \Rightarrow (3)$ in $R \otimes \bar{K}$, and $(1) \Rightarrow (2)$ by proposition 8.4.18, so $(1) \Rightarrow (3)$ holds in $R$ by lemma 8.4.28. $(3) \Rightarrow (2)$ by proposition 8.4.18, and $(2) \Rightarrow (1)$ by lemma 8.4.29.    Q.E.D.

**Corollary 8.4.30:** (*modulo Moeglin's theorem*). *Dixmier's conditions are equivalent for any enveloping algebra of a finite dimensional Lie algebra over a field of characteristic 0.*

*Proof:* The hypotheses of theorem 8.4.27 are satisfied by fact 8.4.23 and Moeglin's theorem.    Q.E.D.

## An Additivity Principle for Goldie Rank

In §6.3 we discussed the Bergman-Small "additivity principle" which expresses the PI-degree of $R/P$ as a linear combination of PI-degrees of prime homomorphic images of $R$. Bergman-Small [75] also discuss the analogous result concerning the PI-degree of a prime subring $W$ of $R$, and the question arose in the study of Jantzen's conjecture for enveloping algebras, whether a suitable non-PI version of this easier Bergman-Small theorem is available. Although we do not have the PI-degree to work with, we could content ourselves with the *Goldie rank* of a prime ring $R$, which we recall is that $n$ for which the classical ring of fractions $Q(R) \approx M_n(D)$ for a suitable division ring $D$. (This is also the uniform dimension of $R$.) Joseph-Small [78] indeed proved such an additivity principle for enveloping algebras by means of GK dimension, and Borho [82] found a general axiomatic approach. We reproduce an elegant treatment of Warfield [83] which penetrates quickly to the main result, with minimal requirements on the rings.

Let us carry the following notation: $R$ is prime left Goldie, and $Q(\ )$ denotes the classical ring of fractions. Write *rank* for the Goldie rank. Thus $Q(R) \approx M_n(D)$ for a suitable division ring $D$ and for $n = \mathrm{rank}(R)$.

*Remark 8.4.31:* If $R$ contains a simple Artinian ring of rank $t$ then $t \mid \mathrm{rank}(R)$. (Indeed, we have a set of $t \times t$ matric units so proposition 1.1.3 shows $R \approx M_t(R')$ for a suitable subring $R'$ of $R$; then $M_n(D) \approx Q(R) \approx Q(M_t(R')) \approx M_t(Q(R')$ so $\mathrm{rank}(R) = t \cdot \mathrm{rank}(R')$.)

**Theorem 8.4.32:** (*Warfield*) *Suppose $W \subseteq R$ and all prime images of $W$ are left Goldie. Let $Q = Q(R)$, and let $Q = M_0 > M_1 > \cdots > M_t = 0$ be a composition series of $Q$ as $W - Q$ bimodule, and let $P_i = \mathrm{Ann}_w M_i/M_{i+1}$ and $\mathscr{P} = \{P_i : 0 \leq i < t\}$. Then $\mathscr{P} \subseteq \mathrm{Spec}(W)$, every minimal prime of $W$ belongs* *to $\mathscr{P}$, and there are positive integers $m_P$ such that*

$$\text{rank}(R) = \sum_{P \in \mathscr{P}} m_P \text{rank}(W/P)$$

*Moreover, each $M_i/M_{i+1}$ is torsion-free as $W/P_i$-module.*

**Proof:** First note each $P_i$ is prime, for if $AB \subseteq P_i$ with $A, B \supset P_i$ then $M_i \supset M_{i+1} + BM_i \supset M_{i+1}$, contrary to hypothesis. Also $P_t \cdots P_1 \subseteq \text{Ann } Q = 0$ so each minimal prime ideal of $W$ belongs to $\mathscr{P}$. It remains to prove the last assertion. Let $W_i = R/P_i$ and $\bar{M}_i = M_i/M_{i+1}$ a faithful $W_i$-module. $\bar{M}_i$ is simple as a bimodule and is torsion-free as $W_i$-module by proposition 3.5.77. Left multiplication by any regular element of $W_i$ thus yields a bijection from $\bar{M}_i$ to itself, and thus $Q(W_i)$ is contained in the endomorphism ring $E_i$ of $\bar{M}_i$ over $Q$. Hence remark 8.4.31 yields $m_i$ such that $m_i \text{rank}(Q(W_i)) = \text{rank}(E_i)$; since rank $W_i = \text{rank } Q(W_i)$ we conclude

$$\sum m_i \text{rank}(W_i) = \sum \text{rank}(E_i) = \text{rank}(Q) = \text{rank}(R)$$

as desired.      Q.E.D.

The natural question arises to when $\mathscr{P}$ consists *only* of the minimal primes of $W$. Equivalently, we want to show the primes in $\mathscr{P}$ are incomparable. To answer this we require a new notion.

**Definition 8.4.32':**   $R$ is a *restricted extension* of a subring $W$ if $WrW$ is f.g. both as left $W$-module and right $W$-module.

**Example 8.4.33:**   Let $L_1 \subseteq L$ be finite dimensional Lie algebras. Then $U(L)$ is a restricted extension of $U(L_1)$. Indeed, clearly $L_1 U_n(L) = U_n(L)L_1$ where $U_n(L)$ is the component under the standard filtration. Hence $U(L_1)U_n(L) = U_n(L)U(L_1)$ is f.g. and thus Noetherian as left and right $U(L_1)$-module. But any $r$ in $U(L)$ lies in some $U_n(L)$ and thus $U(L_1)rU(L_1) \subseteq U(L_1)U_n(L)$ is f.g., as desired.

The same argument shows that any Ore extension of a Noetherian ring is restricted. On the other hand, Warfield [83] also gives some easy examples of extensions of Noetherian rings which are not restricted. In the next result "Goldie" means left and right Goldie.

**Proposition 8.4.34:**   *Suppose $R$ is prime Goldie and is a restricted extension of $W$, and all prime images of $W$ are Goldie. Then each $P_i$ and $P_j$ of theorem 8.4.32 are bonded primes (cf., definition 3.5.86).*

***Proof:*** Since $Q = Q(R)$ is simple Artinian each $M_{i+1}$ has a complement $I_i$ in $M_i$ as right $Q$-module, i.e., $I_i \oplus M_{i+1} = M_i$. Since $I_i \leq Q$ we have $I_i = e_i Q$ for suitable idempotents $e_i$, and $\{e_0, \ldots, e_{t-1}\}$ are orthogonal idempotents of $Q$ with $\sum e_i = 1$. $R \cap Qe_i$ and $R \cap e_i Q$ are, respectively, nonzero left and right ideals of $R$, since $Q$ is left and right essential over $R$; since $R$ is prime we have $0 \neq (R \cap e_i Q)(R \cap Qe_j) \subseteq R \cap e_i Q e_j$ for all $1 \leq i, j \leq t$; select $0 \neq r_{ij}$ in $R \cap e_i Q e_j$. Also note for $j < i$ that $e_j M_i \subseteq e_j M_{j+1} = 0$.
We now use the fact $M_i$ is a $W$-module to establish the key fact:

$$e_j W e_i = 0 \qquad \text{for any } j < i \qquad \text{since } e_j W e_i Q \subseteq e_j W M_i \subseteq e_j M_i = 0.$$

Letting $E = \bigoplus_{j \geq i} e_j Q e_i$ we have $W \subseteq E$ by applying the Pierce decomposition $Q = \bigoplus_{i,j=1}^{t} e_i Q e_j$ to (1).
Note each $e_i Q e_i$ is simple Artinian, by lemma 2.7.12(iv). Let

$$J = \sum_{j > i} e_j Q e_i \lhd E.$$

Then $J^t = 0$ and $E/J \approx \sum_i e_i Q e_i$ is semisimple Artinian. Fixing $i$ let $P_i' = J + \sum_{u \neq i} e_u Q e_u$. Clearly $E/P_i' \simeq e_i Q e_i$ so $P_i'$ is a maximal ideal of $E$.
$M_i = \sum_{u=i}^{t-1} e_u Q$ and is thus naturally an $E - E$ bimodule, and $P_i' M_i \subseteq \sum_{u=i+1}^{t-1} e_u Q = M_{i+1}$. Hence $\mathrm{Ann}_E(M_i/M_{i+1}) = P_i'$ since $P_i'$ is maximal so

$$W \cap P_i' = W \cap \mathrm{Ann}_E(M_i/M_{i+1}) = \mathrm{Ann}_W(M_i/M_{i+1}) = P_i,$$

and we have $W/P_i \hookrightarrow E/P_i'$. But $M/M_{i+1}$ is torsion-free over $W/P_i$ by theorem 8.4.32, so each regular element of $W/P_i$ remains regular in the simple Artinian ring $E/P_i'$. Hence we can form the simple Artinian ring of fractions $Q_i$ of $W/P_i$ inside $E/P_i'$.
Now define the $E - E$ bimodule $L_i = \sum_{u=1}^{i} Q e_u$; let $N_{ij} = M_{i+1} + L_{j-1}$, $N_{ij}' = N_{ij} + W r_{ij} W$, and $\bar{N}_{ij} = N_{ij}'/N_{ij}$, which are $W - W$ bimodules since $W \subseteq E$. Furthermore, $\bar{N}_{ij}$ is an image of $W r_{ij} W$ so (by restrictedness) is f.g. both as left and right module. To conclude we shall display $\bar{N}_{ij}$ as a submodule of an $E - E$ bimodule annihilated, respectively, on the left and right by $P_i'$ and $P_j'$; then $\bar{N}_{ij}$ is a $Q_i - Q_j$ bimodule and thus a torsion-free $W/P_i - W/P_j$ bimodule.
Let $N_{ij}'' = N_{ij} + e_i Q e_j$, clearly an $E - E$ bimodule containing $N_{ij}'$; we have

$$P_i' e_i Q e_j \subseteq P_i' e_i Q \subseteq \sum_{u=i+1}^{t-1} e_u Q = M_{i+1}$$

$$e_i Q e_j P_j' \subseteq Q e_j P_j' \subseteq \sum_{u=1}^{j-1} Q e_u = L_{j-1}.$$

Hence $N''_{ij}/N_{ij}$ is annihilated respectively on the left and right by $P'_i$ and $P'_j$, as desired.     Q.E.D.

The conclusion of theorem 8.4.34 raises the following

**Question 8.4.35:**   Are bonded prime ideals of a given Noetherian ring $W$ necessarily incomparable?

A positive answer would show that $\mathscr{P}$ (of theorem 8.4.32) is precisely the set of minimal primes of $W$ whenever $R$ is a restricted extension of $W$. In particular, this is the case when GK dim$(W) < \infty$, as we shall see presently.

**Proposition 8.4.36:**   (*Borho, Lenagan*) *Suppose M is an* $R - T$ *bimodule, f.g. as R-module, where R, T are affine algebras over a field F. Then* GK-dim$_R M \geq$ GK-dim$_T M$, *the right-handed GK-dimension of M as right T-module.*

*Proof:*   Choose finite dimensional $F$-subspaces $M_0, T_0$ of $M, T$, respectively. Since $M$ is f.g. it has an f.d. subspace $M'_0 \supseteq M_0$ such that $RM'_0 = M \supseteq M'_0 T_0$. Thus $R$ has an f.d. subspace $R_0$ such that $R_0 M'_0 \supseteq M'_0 T_0$ (since $M'_0 T_0$ is f.d.), implying $R^n_0 M'_0 \supseteq M'_0 T^n_0$ for all $n$. Taking limits yields the desired result.     Q.E.D.

**Corollary 8.4.36′ (GK Symmetry):**   *If M is an* $R - R$ *bimodule f.g. on each side and* GK-dim $R < \infty$ *then the left and right GK module dimensions of M are equal.*

**Corollary 8.4.37:**   *Question 8.4.35 has an affirmative answer for all W of finite* GK dim.

*Proof:*   Write $\ell$-GK and $\imath$-GK for the left and right GK module dimensions, respectively. If $M$ is an f.g. torsion-free $W/P_1 - W/P_2$ bimodule then GK dim $R/P_2 = \imath$-GK $M = \ell$-GK $M =$ GK dim $R/P_1$, so $P_1$ and $P_2$ are incomparable by theorem 6.2.25.     Q.E.D.

When GK dim $R = \infty$ we would hope to make do with K-dim. Question 8.4.35 has a positive answer for almost fully bounded rings, seen via Jategaonkar's theory ($J$ supplement in §3.5), and this yields an additivity principle for group algebras of polycyclic-by-finite groups. However, a general attempt to answer question 8.4.35 using K-dim runs up against the hoary Krull-symmetry problem. (For solvable Lie algebras P. Polo [86] has proved Krull symmetry for f.g. $U(L)$-bimodules.)

## Stably Free Modules and Left Ideals

The study of projective modules in this section is motivated by the following recapitulation of previous results.

**Theorem 8.4.38:** *If $R$ is the enveloping algebra of a finite dimensional Lie algebra or the group algebra of a poly-{infinite cyclic} group then every f.g. projective $R$-module is stably free.*

**Proof:** By theorem 5.1.41 for the enveloping algebra (since the associated graded algebra is a commutative polynomial algebra and thus satisfies the conditions of the theorem); by theorem 5.1.34 and proposition 5.1.32 for the group algebra.   Q.E.D.

This raises the question of when all stably frees are free. Stafford has found a direct, unifying treatment, which we present here and in the exercises. Surprisingly, there is a stably free left *ideal* which is not free. We start off with a result of independent interest. In what follows $\sigma$ is *always an automorphism* of $R$, so $R[\lambda; \sigma, \delta]$ is a left and right Ore extension of $R$.

**Proposition 8.4.39:** *Suppose $T = R[\lambda; \sigma, \delta]$ with $R$ left Noetherian. Then the set $S$ of "monic" polynomials is a left denominator set for $T$.*

**Proof:** Given $a \in T$ and $s \in S$ we note right multiplication by $a$ induces a map $T \to (Ta + Ts)/Ts$ whose kernel is $Tsa^{-1}$. Hence $\bar{T} = T/Tsa^{-1}$ is isomorphic to a submodule of the f.g. $R$-module $T/Ts$ and is thus Noetherian, implying $\bar{T}$ is generated by $\bar{1}, \bar{\lambda}, \dots, \bar{\lambda^n}$ for some $n$. But then there are $r_0, \dots, r_n$ in $R$ for which $\lambda^{n+1} - \sum_{i=0}^{n} r_i \lambda^i \in Tsa^{-1}$, i.e., $(\lambda^{n+1} - \sum r_i \lambda^i)a \in Ts$, proving the desired Ore condition.   Q.E.D.

The ring $S^{-1}T$ is very useful, cf., Resco-Small-Stafford [82]; however, let us return to the matter at hand.

**Theorem 8.4.40:** *Suppose $T = R[\lambda; \sigma, \delta]$ with $R$ a left Noetherian domain. If there are $r_0, r_1$ in $R$, $r_0$ noninvertible such that $T = Tr_0 + T(\lambda + r_1)$ then $L = T(\lambda + r_1)r_0^{-1}$ is a non-free left ideal of $T$ satisfying $L \oplus T \approx T^{(2)}$.*

**Proof:** Since $T$ is a domain we have $Ta \approx T$ for all $a$ in $T$, so the composite map

$$T^{(2)} \approx Tr_0 \oplus T(\lambda + r_1) \to Tr_0 + T(\lambda + r_1) = T$$

is an epic whose kernel $K$ is $\{(a_1, a_2) \in T^{(2)} : a_1 r_0 + a_2(\lambda + r_1) = 0\}$. Since $T$ is free this epic splits, i.e., $T^{(2)} \approx K \oplus T$. On the other hand, $(a_1, a_2) \in K$ iff $a_1 r_0 = -a_2(\lambda + r_1)$ so the map $f : K \to L$ given by $f(a_1, a_2) = a_1$ is an isomorphism. (Indeed, if $a_1 = 0$ then $a_2(\lambda + r_1) = 0$, implying $a_2 = 0$, so $\ker f = 0$; to see $f$ is epic, note by definition of $L$ that if $a \in L$ then $ar_0 = a'(\lambda + r_1)$ for some $a'$ in $T$, so $(a, -a') \in K$.) Thus $T^{(2)} \approx L \oplus T$.

It remains to show $L$ is not free; since $T$ has IBN we need merely show $L$ is not principal. Since $R$ is left Ore, there are $r_0', r_1'$ in $R$ for which $r_0' r_0 = r_1'(\delta r_0 - (\sigma r_0) r_1)$; putting $a = r_1' \lambda - r_0'$ we have

$$ar_0 = (r_1' \lambda - r_0') r_0 = r_1' \lambda r_0 - r_1' \delta r_0 + r_1'(\sigma r_0) r_1 = r_1' \sigma r_0(\lambda + r_1) \in R(\lambda + r_1)$$

implying $a \in L$. Note $\deg a = 1$.

Now suppose $L$ were principal, i.e., $L = Tb$. Then $\deg b \leq \deg a = 1$; in fact $\deg b = 1$ since

$$br_0 = r(\lambda + r_1) \qquad \text{for some } r \text{ in } R \tag{4}$$

In view of proposition 8.4.39 we can find some monic $s$ in $T$ for which $sr_0 \in T(\lambda + r_1)$, i.e., $s \in L = Tb$; hence the leading coefficient of $b$ is a unit.

On the other hand, by hypothesis $1 \in Tr_0 + T(\lambda + r_1)$. Write $1 = d_1 r_0 + d_2(\lambda + r_1)$ for $d_i$ in $T$. Dividing $b$ into $d_1$ yields $d_1 = r' + d_1' b$ for suitable $r'$ in $R$ and thus

$$1 = (r' + d_1' b) r_0 + d_2(\lambda + r_1)$$
$$= r' r_0 + d_1'(br_0) + d_2(\lambda + r_1)$$
$$= r' r_0 + (d_1' r + d_2)(\lambda + r_1)$$

by (4). Comparing degrees shows this is impossible unless $d_1' r + d_2 = 0$, so $1 = r' r_0$, and $r_0$ is invertible, the desired contradiction.    Q.E.D.

There is an analogous result for the skew Laurent extension $R[\lambda, \lambda^{-1}; \sigma]$, which we recall is the localization of $R[\lambda; \sigma]$ at the set $\{\lambda^i : i \in \mathbb{N}\}$.

**Theorem 8.4.41:** *Suppose $T$ is a skew Laurent extension of $R$. If there are $r_0, r_1$ in $R$ with $r_0$ noninvertible such that $T = Tr_0 + T(\lambda + r_1)$ and $(\sigma r_0) r_1 \notin Rr_0$ then $L = T(\lambda + r_1) r_0^{-1}$ is a non-free left ideal satisfying $L \oplus T \approx T^{(2)}$.*

**Proof:** Repeat the previous proof through (4), noting that by multiplying through by a suitable power of $\lambda$ we could assume $b \in R[\lambda; \sigma]$ with non-zero constant term; then, as before, we have $b$ monic of degree 1. Write

$b = \lambda + r''$. Then (4) becomes

$$(\sigma r_0)\lambda + \delta r_0 + r''r_0 = r\lambda + rr_1.$$

Matching components shows $\sigma r_0 = r$ so $\delta r_0 + r''r_0 = (\sigma r_0)r_1$, contrary to hypothesis. Thus $L$ is not principal.     Q.E.D.

**Remark 8.4.42:**   In the two results above we have $L \approx Lr_0 = Tr_0 \cap T(\lambda + r_1)$, which thus could be used in place of $L$.

As a first application we quickly rederive exercise 5.1.9.

**Corollary 8.4.43:**   *If $D$ is a noncommutative division ring then $D[\lambda_1, \ldots, \lambda_t]$ has a stably free, nonfree left ideal for any $t \geq 2$.*

**Proof:**   Take noncommutative elements $a, b$ in $D$. Let $R = D[\lambda_1, \ldots, \lambda_{t-1}]$, $\delta = 0$, $\sigma = 1$, $r_0 = \lambda_{t-1} + a$, $r_1 = b$, and $\lambda = \lambda_t$. Then the unit $[a, b] = [r_0, \lambda + b] \in Rr_0 + R(\lambda + r_1)$ so theorem 8.4.40 is applicable.     Q.E.D.

**Corollary 8.4.44:**   *Suppose $T = R[\lambda; \sigma, \delta]$ with $R$ left Noetherian, and there is a nonunit $r$ in $R$ for which $\sum R\delta^i r = R$. Then $T\lambda \cap Tr$ is a stably free, nonfree left ideal of $T$.*

**Proof:**   $1 \in \sum_{i=0}^{n-1} R\delta^i r$ for some $n$; since $\delta^{i+1} r = \lambda \delta^i r - \sigma(\delta^i r)\lambda \in T\delta^i r + T\lambda$ we see by induction on $n$ that $1 \in Tr + T\lambda$, so we conclude by theorem 8.4.40 together with remark 8.4.42.     Q.E.D.

**Corollary 8.4.45:**   *If $R_0$ is a Noetherian domain then the $n$-th Weyl algebra $A_n(R_0)$ has a stably free, nonfree left ideal.*

**Proof:**   View $A_n(R_0)$ as $R[\lambda; 1, \delta]$ where $R = A_{n-1}[\mu]$ as in example 1.6.32, and take $r = \mu$.     Q.E.D.

Since theorem 8.4.40 "works" best when the structure of $R$ is easy to describe (so that we can locate $r_0$ and $r_1$), it is useful to have an iterative procedure to pass to more manageable rings.

**Proposition 8.4.46:**   *Suppose $R \subseteq R'$ are domains satisfying the hypotheses:*
(S$_1$) $R \cap R'L = L$ *for every* $L < R$.
(S$_2$) $R'$ *is flat as $R$-module.*

($S_3$) *If* $a_i \in R'$ *satisfies* $0 \neq a_2 a_1 \in R$ *then there are* $r_i$ *in* $R$ *and a unit* $u$ *in* $R'$ *for which* $a_1 = ur_1$ *and* $a_2 = r_2 u^{-1}$.

If $L$ is a stably free nonprincipal left ideal of $R$ then $R'L$ is a stably free nonprincipal left ideal of $R'$. In fact, if $L \oplus R^{(n)} \approx R^{(m)}$ then $R'L \oplus R'^{(n)} \approx R'^{(m)}$.

**Proof:** Since $R' \otimes_R R^{(m)} \approx R'^{(m)}$ canonically and $R'$ is flat we see $R' \otimes L \approx R'L$ viewing $L$ in $R'^{(m)}$. Thus $R'L \oplus R'^{(n)} \approx (R' \otimes L) \oplus (R' \otimes R^{(n)}) \approx R' \otimes (L \oplus R^{(n)}) \approx R'^{(m)}$. It remains to show $R'L$ is not principal. Suppose, on the contrary, $R'L = R'a_1$, and pick $0 \neq r \in L$. Then $r = a_2 a_1$ for some $a_2$ in $R'$ so by hypothesis ($S_3$) there is $r_1$ in $R_1$ for which $a_1 = ur_1$; thus we have $R'L = R'r_1$. But ($S_1$) implies $Rr_1 = R \cap R'r_1 = R \cap R'L = L$, contrary to $L$ is not principal.     Q.E.D.

**Remark 8.4.47:**  If $R' = R[\lambda; \sigma, \delta]$ or $R' = R[\lambda, \lambda^{-1}; \sigma]$ then the hypothesis (and conclusion) of proposition 8.4.46 are satisfied since $R'$ is a free $R$-module, cf., sublemma 2.5.32'.

**Theorem 8.4.48:**   *If* $G$ *is nonabelian and poly-{infinite cyclic} then the group algebra* $F[G]$ *has a stably free, non-free left ideal.*

**Proof:**  Induction on the Hirsch number $t$ of $G$. There is a torsion-free normal subgroup $H = \langle h_1, \ldots, h_t \rangle$ of $G$ such that $G = H\langle g \rangle$ and conjugation by $g$ induces a nontrivial automorphism $\sigma$ of $H$. Letting $R = F[H]$ and $T = F[G]$ we have $T \approx R[\lambda, \lambda^{-1}; \sigma]$ where $\lambda$ is identified with $g$. Applying remark 8.4.47 we are done unless each stably free left ideal of $R$ is free; by induction on $t$ we may assume that $H$ is abelian, so $R$ is commutative. Hence for any $r$ in $R$ we have

$$1 = r\lambda^{-2}(\sigma r) + \lambda^{-2}(\lambda - \sigma r)(\lambda + r) \in T\sigma r + T(\lambda + r).$$

Taking $r_0 = \sigma r$ and $r_1 = r$ in theorem 8.4.41 we need to find some nonunit $r$ for which $(\sigma^2 r)r \notin R\sigma r$. This is not difficult since $R$ is the localization of the commutative polynomial ring $F[h_1, \ldots, h_t]$ at the powers of $h_1 \cdots h_t$ and thus is a unique factorization domain. Indeed, since $\sigma \neq 1$ we may assume $\sigma h_1 \neq h_1$. Taking $r = 1 - h_1$ we need

$$(1 - \sigma^2 h_1)(1 - h_1) \notin R(1 - \sigma h_1).$$

This is clear unless $1 - \sigma h_1$ is an associate of $1 - h_1$ or $1 - \sigma^2 h_1$, which implies $\sigma h_1 = h_1^{-1}$. Thus we are done unless $\sigma h_1 = h_1^{-1}$, in which case we can take $r = 1 + h_1 + h_1^3$.     Q.E.D.

Lewin [82] showed that all torsion-free polycyclic-by-finite group algebras have nonfree stably free left ideals; to get his result we would need some analogue of proposition 8.4.46 for $T$ f.g. as $R$-module. I think such a result is not yet known. Stafford also applied his techniques to enveloping algebras of nonabelian Lie algebras, cf., exercises 17ff.

## Lie Superalgebras

A recent flurry of activity has surrounded Lie superalgebras. The techniques of this section enable one readily to extend the results about enveloping algebras of Lie algebras to those of Lie superalgebras. We shall indicate briefly how this is done. Suppose throughout the $F$ is a field of characteristic $\neq 2$. We start with the tensor-product view of the multiplication of a (not necessarily associative) $F$-algebra $A$ as a map $A \otimes A \to F$ given by $a_1 \otimes a_2 \mapsto a_1 a_2$. Suppose $A$ happens to be $(\mathbb{Z}/2\mathbb{Z})$-graded, as vector space. It often is natural to use the graded version of the tensor product to identify $a_i \otimes a_j$ with $(-1)^{ij} a_j \otimes a_i$ for $a_i$ in $A_i$ and $a_j$ in $A_j$, leading to condition (1) in the following definition:

**Definition 8.4.49:**  A *Lie superalgebra* is a nonassociative algebra which is $\mathbb{Z}/2\mathbb{Z}$-graded as a vector space and satisfies the rules

(1) $[a_i a_j] = (-1)^{ij} [a_j a_i]$

(2) (Graded Jacobi identity) $(-1)^{ik} [a_i[a_j a_k]] + (-1)^{ij} [a_j[a_k a_i]]$

$+ (-1)^{jk} [a_k[a_i a_j]] = 0$

for homogeneous $a_i$ in $A_i$, $a_j$ in $A_j$, and $a_k$ in $A_k$, where $[\ ]$ denotes the graded Lie product.

Note $L = L_0 \oplus L_1$ where $L_0$ is a Lie algebra in the usual sense. We say $a \in L$ is *even* (resp. *odd*) if $a \in L_0$ (resp. $a \in L_1$).

**Remark 8.4.50:**  If $R$ is an associative $(\mathbb{Z}/2\mathbb{Z})$-graded algebra then we can define $R^-$ by the graded Lie bracket $[r_i, r_j] = r_i r_j - (-1)^{ij} r_j r_i$. Thus we have a functor from $\{(\mathbb{Z}/2\mathbb{Z})$-graded $F$-algebras$\}$ to $\{$Lie superalgebras$\}$, giving rise to a universal, described explicitly as follows.

Given a graded Lie algebra $L$ let $T(L)$ be the corresponding graded tensor algebra and $U(L) = T(L)/\langle a \otimes b - b \otimes a - [ab]\rangle$ over all $a, b \in L$. If, instead, we used the usual ungraded tensor algebra we would have to mod out elements of the form $a_i \otimes a_j - (-1)^{ij} a_j \otimes a_i - [a_i a_j]$ where $a_i, a_j$ are in $L_i, L_j$, respectively. We have a canonical map $L \to U(L)$ which by a graded

Poincare-Birkhoff-Witt theorem is monic and, in fact, has a base which can be described quite precisely: If $\{y_i : i \in I\}$ is a base of $L_0$ and $\{z_j : j \in J\}$ a base of $L_1$ with $I, J$ well-ordered, then $U(L)$ has the base

$$\{y_{i_1} \cdots y_{i_t} z_{j_1} \cdots z_{j_{t'}} : t, t' \in \mathbb{N}, i_1 \leq \cdots \leq i_t, j_1 < \cdots < j_{t'}\}.$$

Of course, the reason the $z_j$ do not repeat is that $2z_j^2 = z_j^2 - (-1)z_j^2 = [z_j z_j] \in L_0$. For the same reason $U(L)$ need not be a domain.

This description of $U(L)$ shows it is a free module over $U(L_0)$ and is f.g. if $[L_1 : F] < \infty$. In particular, assuming $L$ is finite dimensional we see $U(L)$ is Noetherian and is Jacobson by corollary 8.4.13 (or by exercise 13). Also $U(L)$ has an obvious filtration by degree, and the associated graded algebra is the tensor product of a polynomial ring by an exterior algebra. Armed with these facts one can readily translate much of the enveloping algebra theory to the "super" case, cf., exercise 26.

## *R Supplement: Hopf Algebras*

The material we shall now present is of a completely different flavor than the rest of this section and does not have any obvious connection to Noetherian ring theory. Nevertheless, it is of considerable interest, providing a unified view of group algebras and enveloping algebras and has many important applications. The standard reference is Sweedler [69B], which we shall refer to as SW.

The underlying idea is to dualize the operations of an algebra. For convenience we work in the category of algebras over a field $F$. Multiplication in an associative algebra $A$ defines the map $\mu: A \otimes A \to A$ in which the following diagram is commutative (all tensors over $F$):

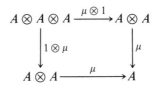

Furthermore, the canonical injection $u: F \to A$ (given by $\alpha \mapsto \alpha \cdot 1$) satisfies the commutative diagram

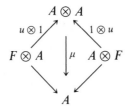

where the bottom arrows are the natural isomorphisms. Actually, we saw in exercise 1.7.4 that these properties characterized an $F$-vector space $A$ being an associative algebra. Dualizing yields the following definition:

**Definition 8.4.51:** A *coalgebra* is a triple $(A, \Delta, \varepsilon)$ where $A$ is a vector space (over $F$), $\Delta: A \to A \otimes A$ is called the *diagonalization* (or *comultiplication*) and $\varepsilon: A \to F$ is called the *augmentation* (or *counit*), satisfying the following commutative diagrams:

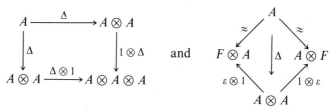

Let as adopt Sweedler's shorthand $\Delta a = \sum a_{(1)} \otimes a_{(2)}$ for $a$ in $A$. We thus have $\sum \Delta a_{(1)} \otimes a_{(2)} = \sum a_{(1)} \otimes \Delta a_{(2)}$ and $a = \sum (\varepsilon a_{(1)}) a_{(2)} = \sum (\varepsilon a_{(2)}) a_1$. An algebra $A$ is a *bialgebra* if $A$ is also a coalgebra such that $\Delta$ and $\varepsilon$ are algebra homomorphisms. (This is equivalent to saying $\mu$ and $u$ are coalgebra morphisms, as is seen by writing these conditions diagrammatically.)

**Example 8.4.52:** If $S$ is a monoid then $F[S]$ is a coalgebra under the maps given by $\Delta s = s \otimes s$ and $\varepsilon s = 1$ for all $s$ in $S$. (Note that $\varepsilon$ corresponds to the usual augmentation map.)

In what follows $A^*$ *always* denotes $\mathrm{Hom}(A, F)$; given $\varphi: A \to B$ we let $\varphi^{\#}$ denote the canonical map $B^* \to A^*$ given by $\varphi^{\#} f = f\varphi$, cf., example 0.1.8. We shall often rely on the natural identification $(A^*)^* \approx A$ when $[A:F] < \infty$.

**Remark 8.4.53:** If $A$ is a coalgebra then $A^*$ is an algebra where multiplication is the composite $A^* \otimes A^* \to (A \otimes A)^* \overset{\Delta^{\#}}{\to} A^*$ and $u = \varepsilon^{\#}$. This relies on the fact there is a monic $\rho: A^* \otimes A^* \to (A \otimes A)^*$ sending $f \otimes g$ to the map $h: A \otimes A \to F$ defined by $h(a_1 \otimes a_2) = f a_1 g a_2$. In case $A$ is finite dimensional over $F$, we see by counting dimensions that $\rho$ is bijective; reversing the arrows we see that if $A$ is an algebra then $A^*$ is a coalgebra, and if $A$ is a bialgebra then $A^*$ is a bialgebra.

**Definition 8.4.54:** An *antipode* (if it exists) of a bialgebra $A$ is a map $S: A \to A$ such that $\mu(S \otimes 1_A)\Delta = u\varepsilon = \mu(1_A \otimes S)\Delta$, i.e., $\sum (Sa_{(1)})a_{(2)} = u\varepsilon a = \sum a_{(1)} Sa_{(2)}$ for all $a$ in $A$. A *Hopf algebra* is a bialgebra with antipode.

***Example 8.4.55:*** If $G$ is a group the involution of $F[G]$ given by $g \to g^{-1}$ is an antipode, so $F[G]$ is a Hopf algebra.

If $L$ is a Lie algebra then $U(L)$ is a Hopf algebra, where $\Delta a = a \otimes 1 + 1 \otimes a$ for $a$ in $L$; $\varepsilon$ is given by $\varepsilon L = 0$, and $S$ is the principal involution.

***Example 8.4.55′:*** In §8.3 we briefly described the theory of algebraic Lie algebras of linear algebraic groups. The general theory can be described concisely in terms of Hopf algebras, as was shown by Hochschild [71B]; more full details are to be found in Abe [77B]. Let $G$ be any group and write $M_F(G)$ for the $F$-algebra of functions from $G$ to $F$. $G$ acts on $M_F(G)$ by translations, i.e., for $x$ in $G$ and $f$ in $M_F(G)$ we define $xf$ by $(xf)g = f(gx^{-1})$ for $g$ in $G$. Define $\mathscr{R}_F(G) = \{ f \in M_F(G) : Gf$ spans a finite dimensional vector space over $F \}$.

We have a homomorphism $\sigma : M_F(G) \to M_F(G \times G)$ given by $\sigma f(g_1, g_2) = f(g_1 g_2)$, and we have a monic $\pi : M_F(G) \otimes M_F(G) \to M_F(G \times G)$ given by $\pi(f_1 \otimes f_2)(g_1, g_2) = f_1 g_1 f_2 g_2$. Then $f \in \mathscr{R}_F(G)$ iff $\sigma f \in \pi(M_F(G) \otimes (M_F(G))$. Now we define the following Hopf algebra structure on $\mathscr{R}_F(G)$:

comultiplication $\Delta = \pi^{-1}\sigma : \mathscr{R}_F(G) \to \mathscr{R}_F(G) \otimes \mathscr{R}_F(G)$;
counit $\varepsilon : \mathscr{R}_F \to F$ given by $\varepsilon f = fe$ where $e$ is the neutral element of $G$;
antipode $S$ given by $(Sf)g = fg^{-1}$.

Given any Hopf subalgebra $A$ of $\mathscr{R}_F(G)$ we let $A^0$ denote the dual space $A^* = \operatorname{Hom}(A, F)$ endowed with the convolution product (exercise 29). Call $h : A \to A$ *proper* if $h$ commutes with each translation by $G$ or equivalently the diagram

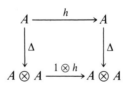

is commutative. Then $\{$proper endomorphisms of $A\}$ is a subalgebra of $\operatorname{End}_F A$, and there is an algebra isomorphism $\Psi : \{$proper endomorphisms of $A\} \to A^0$ given by $h \mapsto \varepsilon \circ h$ ($\Psi^{-1}$ is given by $\varphi \mapsto (1_A \otimes \varphi) \circ \Delta$.) Let $\mathscr{G}(A) = \{$algebra homomorphisms from $A$ to $F\}$, a subgroup of $A^0$ since $\varphi^{-1} = \varphi \circ S$ for $\varphi$ in $\mathscr{G}(A)$; $\Psi^{-1}$ maps $\mathscr{G}(A)$ onto $\{$proper automorphisms of $A\}$. A *differentiation* of $A$ is a map $\partial : A \to F$ satisfying $\partial(a_1 a_2) = \varepsilon a_1 \partial a_2 + (\partial a_1) \varepsilon a_2$. $\delta = \Psi^{-1} \partial$ then satisfies $\delta(a_1 a_2) = a_1 \delta a_2 + (\delta a_1) a_2$, i.e., $\delta \in \operatorname{Deriv}(A)$. $\operatorname{Deriv}(A)$ is called the *Lie algebra* of $\mathscr{G}(A)$ and is isomorphic to the Lie algebra of differentiations of $A$ under the Lie product

$$[\partial_1 \partial_2] = (\partial_1 \otimes \partial_2 - \partial_2 \otimes \partial_1) \circ \Delta.$$

Now we can define "algebraic Lie algebra" in this general setting, and Hochschild [71B, Chapter 13] has rather quick proofs of the basic facts about algebraic Lie algebras.

Let us return to the Hopf algebras of example 8.4.55.

**Definition 8.4.56:** A coalgebra $A$ is *cocommutative* if the following diagram (dualizing commutativity of multiplication) is commutative:

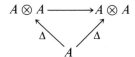

where the top arrow is the isomorphism given by $a_1 \otimes a_2 \rightarrow a_2 \otimes a_1$; in other words, $\sum a_{(1)} \otimes a_{(2)} = \sum a_{(2)} \otimes a_{(1)}$.

**Remark 8.4.57:** It is enough to check cocommutativity on a base of $A$; thus group algebras are cocommutative (checked on $G$), and enveloping algebras are cocommutative.

So far we have generalized the augmentation of a group algebra and its canonical involution. To generalize the trace map we need another definition. Given any Hopf algebra $A$ define the homomorphism $\pi: A^* \rightarrow F$ by $\pi f = f1$. View $A^*$ as algebra via remark 8.4.53.

**Definition 8.4.58:** A *(left) integral of* $A^*$ is an element $f_0$ in $A^*$ satisfying $ff_0 = (\pi f)f_0$ for every $f$ in $A^*$. {left integrals of $A^\#$} is denoted as $\int$, and is an $F$-vector space called the *space of* integrals.

For *any* algebra homomorphism $\varphi: A \rightarrow F$ one has $\ker \varphi = \{a - \varphi a : a \in A\}$. Thus $\int = $ Ann$'\ker \pi$.

**Example 8.4.59:** The trace map tr of $F[G]^*$ is a left integral by direct verification.

There is a monic $\int \otimes A \rightarrow A^*$ sending $f_0 \otimes a$ to the map $f: A \rightarrow F$ given by $fa = f_0Sa$. In case $A$ is finite dimensional this is an isomorphism (cf., [SW, corollary 5.1.6]), from which one concludes $S$ is bijective and dim $\int = 1$. In this case, however, we have a canonical isomorphism $A \approx (A^*)^*$ so $A$ possesses a left integral (arising from $A^*$); noting $\varepsilon$ takes the place of $\pi$

above we have {left integrals of $A$} $\approx \text{Ann}' \ker \varepsilon$, a 1-dimensional vector space. For $|G|$ finite one sees $\sum_{g \in G} g$ is a left (and right) integral of $F[G]$. Since $\varepsilon(\sum g) = |G|$ exercise 32 generalizes Maschke's theorem.

To give the reader a better picture of Hopf algebras, we sketch some of the other important definitions and results, leading to a statement of Kostant's theorem.

**Definition 8.4.60:** A *subcoalgebra* of a coalgebra is a subspace $V$ satisfying $\Delta V \subseteq V \otimes V$.

If $B$ is a subcoalgebra of a coalgebra $A$ then the inclusion map $i: B \to A$ is a morphism yielding an algebra homomorphism $i^{\#}: A^* \to B^*$ whose kernel is $B^{\perp} = \{f \in A^*: fB = 0\}$. In particular $B^{\perp} \lhd A^*$. Conversely, if $B$ is a subspace of $A$ such that $B^{\perp} \lhd A^*$ then $B$ is a subcoalgebra of $A$. (This verification is longer but straightforward, cf., SW, proposition 1.4.3). Thus the ideal structure of $A^*$ corresponds to the subcoalgebra structure of $A$.

One is thus led to dualize the structure theory of algebras; note that the subcoalgebra theory is "weaker" than the ideal theory, since any theorem about ideals (stated in $A^*$) yields a theorem about subcoalgebras, but not vice versa. For *finite dimensional* bialgebras $A$ we have $A \approx (A^*)^*$, so the subcoalgebra theory and the ideal theory are equivalent in this case. This is significant because subcoalgebras turn out to be quite tractable, as we shall see now.

**Definition 8.4.61:** A coalgebra $A$ is *irreducible* if any two nonzero subcoalgebras intersect nontrivially. $A$ is *simple* if $A$ has no proper subcoalgebras $\neq 0$; $A$ is *pointed* if all simple subcoalgebras are 1-dimensional.

The *fundamental theory of coalgebras* (cf., exercise 34 or [SW, theorem 2.2.1]) states that any element of a coalgebra is contained in a finite dimensional subcoalgebra. Consequently, any simple coalgebra $A$ is finite dimensional.

**Corollary 8.4.62:** *Any cocommutative coalgebra $A$ over an algebraically closed field $F$ is pointed.*

**Proof:** Let $B$ be a simple subcoalgebra. Then $[B:F] < \infty$. Hence $B^*$ is a commutative, simple finite dimensional $F$-algebra and thus by the Wedderburn-Artin theorem is a field extension of $F$; since $F$ is algebraically closed $B^* = F$ so $[B:F] = 1$.    Q.E.D.

The last ingredient needed is the smash product, a generalization of the group algebra crossed product (exercise 8.1.35).

**Definition 8.4.63:** Suppose $H$ is a Hopf algebra. An $F$-algebra $R$ is an $H$-module algebra if $R$ is an $H$-module such that for all $h$ in $H$ and $r, s$ in $R$ we have

$$h \cdot (rs) = \sum (h_{(1)} \cdot r)(h_{(2)} \cdot s) \quad \text{and} \quad h \cdot 1 = \varepsilon(h)1$$

where $\cdot$ denotes the scalar multiplication. In this case define the *smash product* $R \# H$ to be the vector space $R \otimes_F H$ provided with multiplication

$$(r \# g)(s \# h) = \sum r(g_{(1)} \cdot s) \# g_{(2)} h$$

for $r, s$ in $R$ and $g, h$ in $H$. It is easy to check this really is an algebra.

**Kostant's Theorem:** ([*SW, theorems 8.1.5 and 13.0.1*] *Any cocommutative pointed Hopf algebra $H$ is isomorphic (as an algebra) to a smash product of an enveloping algebra (of a suitable Lie algebra $L$) and a group algebra $F[G]$. (One obtains $G$ and $L$ via exercise 31.*)

In an effort to study enveloping algebras an $\mathbb{N}$-graded version of Hopf algebras has been introduced, where the graded tensor product is used (i.e., there is a sign switch for the odd components).

In view of Kostant's theorem we should expect the theory of left Noetherian, cocommutative pointed Hopf algebras to be rich enough to yield many of the results of this chapter. I do not know of results in this direction.

Recently Hopf algebras have been seen to be the "correct" setting for many results about group actions on rings, see especially Bergman [85] and work by M. Cohen and Montgomery.

# Appendix to Chapter 8: Moody's Theorem

As noted in the body of the text, Moody has proved the Goldie rank conjecture. We shall outline the proof according to a nice (unpublished) exposition of Farkas and Linnell; the reader should consult Moody's dissertation for a complete proof. The key is the Grothendieck group $\mathscr{G}_0(F[G])$, where $G$ is a polycyclic-by-finite group. To get a hold on $\mathscr{G}_0(F[G])$ we proceed by induction on the Hirsch number $h(G)$. In other words, take $N \lhd G$ with $h(N) = h(G) - 1$. Then $h(G/N) = 1$, so $G/N$ is cyclic-by-finite. One can display $F[G]$ as a "crossed product" of $F[N]$ with respect to $G/N$. (We use the definition and notation in exercise 8.1.35, and shall call $t$ the "twist.")

Thus we are led to a related question: How do we compute $\mathscr{G}_0(R^*\Gamma)$, for a Noetherian ring $R$ and a cyclic-by-finite group $\Gamma$? The answer is found in

**Moody's Fundamental Theorem:** *If $R$ is a Noetherian ring and $\Gamma$ is a f.g. abelian-by-finite group then there is a surjection $\bigoplus \mathscr{G}_0(R^*K) \to \mathscr{G}_0(R^*\Gamma)$, where the direct sum is taken over all finite subgroups $K$ of $\Gamma$.*

The proof is based on a series of results about $\mathscr{G}_0$ and crossed products; let us start with one application of the Grothendieck-Serre-Quillen theorem. We assume throughout that $R$ is Noetherian.

**Theorem 8B.1:** *("GSQ Reduction") Suppose $\Gamma$ is the semidirect product $A \rtimes H$ where $A$ is f.g. torsion free abelian and $H$ is finite, and assume $A$ is a free $\mathbb{Z}[H]$-module. Then $\mathscr{G}_0(R^*H) \to \mathscr{G}_0(R^*\Gamma)$ is surjective.*

*Proof:* Since $A$ is free on $\mathbb{Z}[H]$ we can find a base $B$ of $A$ on which $H$ acts by permutation. Let $\bar{B}$ be the submonoid of $A$ generated by $B$. Since $H\bar{B} = \bar{B}H$ we can form $R^*(H\bar{B})$, which is free over $R^*H$ and graded by total degree in the elements of $B$; the 0-component is $R^*H$. Theorem 5A.2 gives an isomorphism $\mathscr{G}_0(R^*H) \to \mathscr{G}_0(R^*H\bar{B})$, and localizing (cf., remark 5A.3) gives a surjection $\mathscr{G}_0(R^*H\bar{B}) \to \mathscr{G}_0(R^*\Gamma)$, as desired.          Q.E.D.

In the sequel we view $R^{(n)}$ as the ring $R \times \cdots \times R$, whose canonical central idempotents we denote as $e_1, \ldots, e_n$. Two other necessary results for crossed products:

**Theorem 8B.2. Corestriction Theorem for Crossed Products:** *(compare with proposition 7.2.22). If $L < G$ are groups with $[G:L] = n$ then any crossed product $R^*L$ can be extended to a crossed product $R^{(n)}*G$ such that*

(i) $R^{(n)}*G \approx M_n(R^*L)$

(ii) *Conjugation by $G$ permutes the components of $R^{(n)}$ transitively, and the stabilizer of one of these components is $L$.*

*Proof:* (sketch) Write $G = \bigcup_{i=1}^n Lg_i$ with $g_1 = 1$. There is a homomorphism $G \to \mathrm{Sym}(n)$ obtained by permuting the cosets by right multiplication, i.e., we send $a \in G$ to $\hat{a}$ corresponding to $Hg_i \to Hg_i a$. Note $g_i a = h_{i,a} g_{\hat{a}(i)}$ for suitable $h_{i,a}$ in $L$. Define an action of $G$ on $R^{(n)}$ by $a(r_i) = (h_{i,a} r_{\hat{a}(i)})$. If $t$ is the twist of $R^*L$ we now can find a twist $\tilde{t}$ for $R^{(n)}*G$, i.e. $\tilde{t}: G \times G \to \mathrm{Unit}\,(R^{(n)})$ is given by

$$(\tilde{t}(a, b))_i = t(h_{i,a}, h_{\hat{a}(i), b}).$$

(ii) holds since $L$ is the stabilizer of $e_1$. Moreover, $e_1(R^{(n)}*G)e_1 \approx R*L$, and $\{g_i e_1 g_j^{-1}: 1 \leq i, j \leq n\}$ are a set of $n \times n$ matrix units, proving (i).          Q.E.D.

**Remark 8B.3:**   For any finite subgroup $H$ of $G$ one can show under the previous set-up that $R^{(n)}*H \approx \prod M_{n(K)}(R*K)$ for a suitable finite subgroups $K$ of $L$, and suitable $n(K)$ in $\mathbb{N}$; the proof is a computation obtained by partitioning $\{e_1, \ldots, e_n\}$ into orbits under $H$.

**Sketch of the Proof of Moody's Fundamental Theorem:**   Note there is an exact sequence $1 \to A \to \Gamma \to H \to 1$, where $A$ is f.g. free abelian, and $H$ is a finite group. First assume the following condition:

(*) $\cdots \mathbb{Q} \otimes_{\mathbb{Z}} A$ is a free $\mathbb{Q}[H]$-module under the action induced from this exact sequence.

Pick a base $x_1, \ldots, x_n$ for $\mathbb{Q} \otimes_{\mathbb{Z}} A$ over $\mathbb{Q}[H]$, and let $F$ be the free $\mathbb{Z}$-module with formal base $\{hx_i : h \in H, 1 \leq i \leq n\}$.

Localizing at $\mathbb{N} - \{0\}$ we see $A \subseteq F[n^{-1}]$ for some $n$ in $\mathbb{N}$, with $F[n^{-1}]$ free over $\mathbb{Z}[H]$; also $F[n^{-1}]/A$ is finite since $F[n^{-1}]$ and $A$ have the same rank as abelian groups. Thus we can build an exact sequence

$$1 \to F[n^{-1}] \to G \to H \to 1$$

so $\mathscr{G}_0(R^{(n)}*H) \to \mathscr{G}_0(R^{(n)}*G)$ is onto, by the GSQ reduction. Moreover, theorem 8B.2 yields $\mathscr{G}_0(R^{(n)}*G) \approx \mathscr{G}_0(M_n(R*\Gamma)) \approx \mathscr{G}_0(R*\Gamma)$ (since $R*\Gamma$ and $M_n(R*\Gamma)$ are Morita equivalent). Using "Shapiro's lemma" from cohomology theory one can conclude by applying remark 8B.3.

In general, to reduce to the case (*) note $\mathbb{Q} \otimes_{\mathbb{Z}} A$ is a projective $\mathbb{Q}[H]$-module, so take a finite dimensional $\mathbb{Q}[H]$-module $M$ such that $(\mathbb{Q} \otimes_{\mathbb{Z}} A) \oplus M$ is free over $\mathbb{Q}[H]$. It is easy to find an f.g. free $\mathbb{Z}[G]$-module $N < M$ such that $\mathbb{Q} \otimes_{\mathbb{Z}} N \approx M$, so we have an extension $1 \to N \times A \to N \times \Gamma \to H \to 1$, to which case (*) is applicable. The passage from $N \times \Gamma$ to $\Gamma$ is via group homological techniques, as explained in pp. 11–12 of the Farkas-Linnell preprint.

# Exercises

## §8.1

1. (Snider [81]) If char$(F) = 0$ then $F[G]$ is a nonsingular ring. (Hint: Passing to the algebraic closure one may assume $F$ is algebraically closed; Thus $F[G]$ has an involution (*) for which $r^*r = 0$ implies $r = 0$. If $a \in$ Sing$(R)$ then $Ra^* \cap$ Ann $a \neq 0$;

but if $ra*a = 0$ one has $(ra*)(ra*)* = 0$ so $ra* = 0$, contradiction.) His paper also deals with characteristic $p$.

2. If $C$ is a semiprime $\mathbb{Q}$-algebra then $\text{Nil}(C[G]) = 0$. (Hint: Pass to prime images of $C$ and then to their fields of fractions.)

## Group Algebras of Finite Groups When Char($F$) > 0

3. Suppose $\text{char}(F) = p$ and $G_p$ is a Sylow $p$-subgroup of $G$. Then every f.g. projective $F[G_p]$-module $M$ is free so $|G_p|$ divides $[M:F]$. Thus $|G_p|$ divides the $F$-dimension of every projective $F[G]$-module. Conclude finally that $[\text{Jac}(F[G]):F] \geq |G_p| - 1$. (Hint: $F[G_p]$ is local. But $F \approx F[G]e/\text{Jac}(F[G])e$ for an idempotent $e$; count dimensions.)

4. If $g_1, g_2$ are conjugates then $g_2 - g_1$ is a ring commutator in $C[G]$ (seen by computing $[g, g_1 g^{-1}]$). Conversely, if $r \in [C[G], C[G]]$ then the sum of the coefficients over each conjugacy class is 0 (since $[g,h] = gh - g^{-1}(gh)g$.)

5. (i) Suppose $A$ is an $F$-algebra of characteristic $p$. (i) $(a + b)^t - a^t - b^t \in [A, A]$ whenever $t$ is a power of $p$. (Hint: By induction assume $t = 1$.) (ii) Let $T = \{a \in A : a^t \in [A, A]$ for some power $t$ of $p\}$. $T$ is an $F$-subspace of $A$. Moreover, if $F$ is algebraically closed and $[A:F] < \infty$ then the number of simple components of the semisimple Artinian algebra $A/\text{Jac}(A)$ is $[A/T:F]$. (Hint: $\text{Jac}(R)$ is nilpotent so is in $T$; thus one may assume $\text{Jac}(A) = 0$, and then assume $A$ is simple, i.e., $A \approx M_n(F)$. But $[A, A] \subseteq T \subset A$ so a dimension count shows $T = [A, A]$, as desired.)

6. (Brauer) Suppose $G$ is finite and $F$ is algebraically closed. We say $g$ in $G$ is $p$-regular if $g$ has period prime to $p$. Show the number of simple components of $F[G]/\text{Jac}(F[G])$ is the number of conjugacy classes of $\text{char}(F)$-regular elements. (Hint: By exercise 5 it suffices to show a set $B$ of representatives of conjugacy classes of $p$-regular elements is a base of $R/T$ over $F$. $B$ spans all $p$-regular elements by exercise 4. Furthermore, if $g \in G$ then $g = g_1 g_2$ where the $g_i$ are powers of $g$ with $g_1$ $p$-regular and $g_2^t = 1$ for some power $t$ of $p$; then $(g_1 g_2 - g_1)^t = 0$ so $g_1 g_2 - g_1 \in T$ proving $B$ spans $G$ (modulo $T$), hence $B$ spans $R/T$. It remains to show $\sum \alpha_i b_i \in T$ implies all $\alpha_i = 0$. Take a large power $t$ of $p$ such that $b_i^t = b_i$ for each $i$. Then $\sum \alpha_i^t b_i \in [A, A]$ so each $\alpha_i^t = 0$ by exercise 4.)

7. Suppose $G$ is a *doubly transitive* subgroup of $\text{Sym}(n)$, i.e., for each $i_1 \neq i_2$ and $j_1 \neq j_2$ in $\{1, \ldots, n\}$ we have $\pi$ in $G$ such that $\pi i_1 = j_1$ and $\pi i_2 = j_2$. (For example, we could take $G = \text{Sym}(n)$ itself.) Then $G$ has an irreducible complex representation of degree $n - 1$. (Hint: The regular representation affords a character $\mathcal{X}$ which is the sum of the unit character and an irreducible character. To see this write $\mathcal{X} = \sum_{i=1}^t m_i \mathcal{X}_i$ for $\mathcal{X}_i$ irreducible. Letting $G$ act on a base $v_1, \ldots, v_n$ of $V = F^{(n)}$ by permuting the subscripts, one notes $\text{End}_{C[G]} V$ is a direct product of $\sum_{i=1}^t m_i$ copies of $\mathbb{C}$. To show $t = 2$ suppose $\tau \in \text{End}_{C[G]} V$; writing $\tau v_i = \sum \alpha_{ij} v_j$ show $\sum \alpha_{ij} v_{\pi j} = \pi \tau v_i = \tau v_{\pi i} = \sum \alpha_{\pi i, \pi j} v_{\pi j}$ for all $\pi$ in $G$, implying $\alpha_{ij} = \alpha_{12}$ and $\alpha_{ii} = \alpha_{11}$ for all $i \neq j$, and thus $\tau = \alpha_{11} 1_V + \alpha_{12} \rho$ where $\rho v_i = \sum_{j \neq i} v_j$.)

## The Kaplansky Trace Map and Idempotents

8. (Kaplansky) Suppose $\text{char}(F) = 0$ and $0 \neq e \in F[G]$ is idempotent. Then $\text{tr}(e)$ is real and algebraic over $\mathbb{Q}$, and $0 < \text{tr } e < 1$. (Hint: It suffices to show $\text{tr } e$ is real

and positive; if so then $1 - \mathrm{tr}\,e > 0$ since $1 - e$ is also idempotent, and $\mathrm{tr}\,e$ is algebraic since if transcendental its image under a suitable automorphism would be nonreal. Replacing $F$ by the field extension generated by $\mathrm{supp}(e)$, one may assume $F$ is a subfield of $\mathbb{C}$. Define the inner product $\langle,\rangle$ on $\mathbb{C}[G]$ by $\langle r_1, r_2 \rangle = \mathrm{tr}(r_1^* r_2)$, cf., remark 8.1.5. If $L = \mathbb{C}[G]e$ then $L^{\perp}$ is also a left ideal of $\mathbb{C}[G]$, so there is a decomposition $1 = f + f'$ where $f \in L$ and $f' \in L^{\perp}$ are idempotents. $f^*(1 - f) = 0$ so $f^* = f^*f$ implying $f^* = f^{**} = f$ and $\mathrm{tr}\,f = \mathrm{tr}(f^*f) > 0$. But $\mathrm{tr}(e - f) = \mathrm{tr}[f, e] = 0$, so $\mathrm{tr}\,e = \mathrm{tr}\,f$.)

9. (Kaplansky) If $\mathrm{char}(F) = 0$ then $F[G]$ is weakly 1-finite. (Hint: If $ab = 1$ then $\mathrm{tr}(ba) = 1$.)

10. (Zaleskii [72]) If $e \in F[G]$ is idempotent then $\mathrm{tr}\,e$ is in the characteristic subfield of $F$. (Hint: Reduce to the case $\mathrm{char}(F) > 0$ using remark 8.1.25'. Write $e = \sum \alpha_g g$ and let $S = \{1\} \cup p\text{-}\mathrm{supp}(e)$. Taking $q$ large enough one has $\mathrm{tr}\,e = \mathrm{tr}\,e^q = \sum \{\alpha_g^q : g \in S\}$. Then $(\mathrm{tr}\,e)^p = \mathrm{tr}\,e$, as desired.)

## The Center of the Group Algebra

11. $Z(K[G]) \approx K \otimes_C Z(C[G])$ for any commutative $C$-algebra $K$. If $A \lhd C$ then $Z((C/A)[G])$ is the image of $Z(C[G])$ under the canonical map $C[G] \to (C/A)[G]$. (Hint: Compare bases, using IBN.)

12. Suppose $C$ is Noetherian and $H$ is a finitely generated subgroup of $\Delta(G)$ which is normal in $G$. Then $Z(C[G]) \cap C[H]$ is an affine $C$-algebra. (Hint: Let $Z = Z(H)$. Then $[H:Z] < \infty$ so $Z$ is finitely generated. Hence $C[Z]$ is commutative Noetherian, over which $C[H]$ is f.g. and thus Noetherian. Thus $T = Z(C[H])$ is affine by Artin-Tate, and the fixed ring $T^G$ is affine by exercise 6.2.2, where $G$ acts by conjugation on $C[H]$. But $C[H]^G = Z(C[G]) \cap C[H]$ is a $T^G$-submodule of $T$.)

13. $\mathrm{Jac}(Z(C[G]))$ is nil whenever $C$ is Noetherian. (Hint: Write $Z = Z(C[G])$. If $r \in \mathrm{Jac}(Z)$ then taking $H = \langle \mathrm{supp}(r) \rangle \subseteq \Delta(G)$ note $Z \cap C[H]$ is affine over $C$ by exercise 12. Hence $\mathrm{Jac}(Z \cap C[H])$ is nil, implying $r$ is nilpotent.)

14. Suppose $\mathrm{char}(F) = p > 0$. The following are equivalent: (i) $F[G]$ is semiprime. (ii) $Z(F[G])$ is semiprime. (iii) $Z(F[G])$ is semiprimitive. (iv) The order of every normal finite subgroup of $G$ is prime to $p$. (v) $\Delta(G)$ is a $p'$-group. (vi) $Z(F[\Delta(G)])$ is semiprime. (Hint: (ii) $\Rightarrow$ (iii) by exercise 13. (iii) $\Rightarrow$ (iv) $\Rightarrow$ (v) and (vi) $\Rightarrow$ (i) as in theorem 8.1.39.)

## The Augmentation Ideal

15. If $H = \langle h_i : i \in I \rangle \subset G$ then $\omega(C[H])C[G] = \sum_{i \in I}(h_i - 1)C[G]$. (Hint: Let $H_0 = \{h \in H : h - 1 \in \sum(h_i - 1)C[G]\}$. Show $h_i h \in H_0$ and $h_i^{-1}h \in H_0$ for any $h$ in $H_0$; thus $HH_0 = H_0$. Since $1 \in H_0$ conclude $H = H_0$.)

16. A subgroup $H$ of $G$ is finite iff $\mathrm{Ann}\,\omega(C[H]) \neq 0$, in which case $\mathrm{Ann}\,\omega(C[H]) = C[G]\sum_{h \in H} h$. (Hint: Suppose $0 \neq r = \sum_{i=1}^{n} c_i g_i \in \mathrm{Ann}\,\omega(C[H])$. Then $r = rh$ for all $h$ in $H$, implying $g_1 h \in \{g_1, \ldots, g_n\}$ and thus $H$ is finite. Continue along these lines.)

17. If $\omega(F[G])$ is nilpotent then $\mathrm{char}(F) = p > 0$ and $G$ is a finite group.

18. (Auslander-Villamayor) $F[G]$ is (von Neumann) regular iff $G$ is locally finite and is a $\mathrm{char}(F)'$-group. (Hint: ($\Rightarrow$) If $H = \langle h_1, \ldots, h_n \rangle$ then $\omega(F[H])F[G]$ is an f.g.

right ideal and thus has nonzero annihilator, implying $H$ is finite. To check $|H| \neq 0$ in $F$ let $a = \sum_{h \in H} h$ and take $b$ such that $1 - ab \in \text{Ann}' a = C[G]\omega(F[H])$; apply the augmentation.)

## Injective Modules

19. $F[G]$ is a symmetric algebra and thus Frobenius for any finite group $G$. (Hint: Apply the trace map to exercise 3.3.30.) Renault showed $F[G]$ is self-injective iff $G$ is finite (cf., Passman [77B, p. 79] for Farkas' short proof).
20. If $E$ is an injective $F[G]$-module and $\omega(F[G])E \neq E$ then $G$ is locally finite. (Hint: Suppose $H = \langle h_1, \ldots, h_n \rangle$. Then $E \neq \sum (h_i - 1)E$, implying $\text{Ann}\{h_1 - 1, \ldots, h_n - 1\} \neq 0$. Thus $\text{Ann}\,\omega(F[H]) \neq 0$ so $H$ is finite by exercise 16.)
21. (Farkas-Snider [74]) View $F$ as an $F[G]$-module by means of the augmentation map. $F$ is injective iff $F[G]$ is a regular ring. (Hint: ($\Leftarrow$) By exercise 2.11.21. ($\Rightarrow$) To verify the criteria of exercise 18, first note $G$ is locally finite by exercise 20. Furthermore, if $H$ is a finite subgroup and $a = \sum_{h \in H} h$ then there is a map $f: F[G]a \to F$ sending $a$ to 1 (in view of exercise 16); thus $f$ extends to $\hat{f}: F[G] \to E$ and $|H|\hat{f}1 = a\hat{f}1 = \hat{f}a \neq 0$.) Conclude that if $F[G]$ is a $V$-ring then it is a regular ring.

## Primitive and Semiprimitive Group Rings

22. Suppose $G$ is the group free product $G_1 \coprod G_2$. Then $F[G]$ is the free product (as $F$-algebras) of $F[G_1]$ and $F[G_2]$.
23. (Formanek) If $|G_1| > 2$ and $G = G_1 \coprod G_2$ then $F[G]$ is primitive (cf., exercise 2.1.15).
24. (Formanek) Suppose $G = \mathbb{Z} \times (A \coprod B)$ where $|A| > 2$ and $|B| > 1$. Then $F[G] \approx F[\mathbb{Z}] \otimes F[A \coprod B]$ is primitive.
25. (Farkas-Snider [81]) If char$(F) = 0$ then $F[G]$ cannot be primitive with non-zero socle. (Hint: It suffices to show there are only a finite number of independent minimal left ideals, for then $F[G]$ would be simple Artinian, impossible. So assume there are an infinite number. This yields an infinite number of orthogonal primitive idempotents $e_1, e_2$, all with the same Kaplansky trace $\alpha$, since minimal left ideals are isomorphic. But $\alpha$ is real $> 0$; taking $n > 1/\alpha$ yields $1 < n\alpha = \sum \text{tr}\,e_i \leq 1$, contradiction.) Their paper also deals with characteristic $p$.
26. (Passman) Suppose $p = \text{char}(F) > 0$ and $G$ is a $p'$-group. If $F$ is not algebraic over its characteristic subfield then $F[G]$ is semiprimitive. (Hint: As in theorem 8.1.47 first show there is a subfield $F_1$ such that $\text{Jac}(F_1[G]) = 0$ and $F$ is algebraic over $F_1$. To be able to conclude $\text{Jac}(F[G]) = 0$ one may assume $[F:F_1] < \infty$.)
27. If $G$ is a char$(F)'$-group and $F[G]$ satisfies a polynomial identity then $F[G]$ is semiprimitive. (Hint: Reduce to the case $G$ is finitely generated; but then $F[G]$ is affine.)

## Polynomial Identities of Group Rings

28. (Passman) Suppose $H$ is a nonabelian group of order $p^3$. Let $z_i$ be the element of $H^{(k)}$ whose $i$-th component is the generator of $z(H_i)$ and whose other com-

ponents are 1, and let $N = \langle z_i z_1^{-1} : 1 \leq i \leq k \rangle \subseteq Z(H^{(k)})$. Let $G = H^{(k)}/N$. Then $G' = Z(H^{(k)})/N = Z(G)$. Also if $A < G$ is abelian then $[G:A] \geq p^k$. (Hint: The first equality is clear by checking components, as is $G' \subseteq Z(G)$; to complete the equality note that if $h = (h_i) \in H^{(k)} - Z(H^{(k)})$ then $[h_i, x_i] \neq 1$ for some $x = (x_i)$ in $H^{(k)}$ so $[h_i, x_i] \notin N$ since $N \cap H_i = (1)$. Now suppose $A < G$ is abelian of index $p^t$. To show $t \geq k$ let $B = AZ(G)$. Then $B \supseteq G'$ so $B \triangleleft G$ and $G/B$ is a product of $\leq t$ cyclic groups $\langle \bar{g}_1 \rangle, \ldots, \langle \bar{g}_t \rangle$ for suitable $g_i$ in $G$. Each $g_i$ has at most $p$ conjugates since $g g_i g^{-1} = [g, g_i] g_i \in G' g_i$ and $|H_i'| = p$. Hence $Z(G) = B \cap \left(\bigcap C_G(g_i)\right)$ has index $\leq p^t p \cdots p = p^{2t}$ in $G$. But $[G:Z(G)] = p^{2k}$.)

29. A group $G$ is $p$-*abelian* if $G'$ is a finite $p$-group. Suppose $G$ has a $p$-abelian subgroup $H$ of index $n$, such that $|H'| = m = p^t$. If $\text{char}(F) = p$ and $f_1, \ldots, f_m$ are identities of $M_n(\mathbb{Z})$ then $f_1 \cdots f_m$ is an identity of $F[G]$. (Hint: Each $f_i$ is an identity of $M_n(F[H/H'])$ so $f_i(M_n(F[H])) \subseteq \ker \varphi$ where $\varphi = M_n(\rho_{H'}) : M_n(F[H]) \to M_n(F[H/H'])$. But $(\ker \rho_{H'})^m = \omega(F[H'])^m F[H] = 0$ so $(\ker \varphi)^m = 0$.)

30. Applying exercise 29 to exercise 28, show for any $k$ there is a finite $p$-group $G$ such that $F[G]$ satisfies the identity $[X_1, X_2][X_3, X_4] \cdots [X_{2p-1}, X_{2p}]$ but $G$ has no abelian subgroup of index $k$. Applying ultraproducts find a group $G$ having no abelian group of finite index, but satisfying this PI. (Hint: Note $|G'| = p$.)

In view of exercise 30 the Isaacs-Passman theorem is different for $\text{char}(F) = p > 0$; the correct conclusion is that $G$ has a $p$-abelian subgroup of finite index bounded by the PI-degree of $F[G]$ (cf., Passman [77B, corollary 5.3.10].

31. (Bercov-Moser) The maximal order of an abelian subgroup $H$ of $\text{Sym}(n)$ is $t(n)$ = the largest number which is a product of positive integers whose sum is $n$. (Hint: Let $T_1 = \{\sigma 1 : \sigma \in H\}$ and inductively taking $i_m$ to be the smallest number not in $T_1 \cup \cdots \cup T_{m-1}$ let $T_m = \{\sigma i_m : \sigma \in H\}$. Then $H$ acts transitively on each $T_i$. For $\sigma$ in $H$ and $1 \leq u \leq m$ define $\sigma_u$ by $\sigma_u i = \sigma i$ for $i \in T_u$ and $\sigma_u i = i$ otherwise. Now $H_u = \{\sigma_u : \sigma \in H\}$ has order $|T_u|$, and there is an injection $H \to \prod H_u$ given by $\sigma \to (\sigma_u)$, so $|H| \leq \prod |H_u| \leq t(n)$.)

32. Compute $t(n)$ of exercise 31 to be $3^m$ if $n = 3m$; $t(n) = 4 \cdot 3^{m-1}$ if $n = 3m + 1$; $t(n) = 2 \cdot 3^m$ if $n = 3m + 2$.

33. Using the Young-Frobenius formula compare the PI-degree of $\mathbb{Q}[\text{Sym}(n)]$ with the index of a maximal abelian subgroup.

34. If $G$ is finite and $\mathbb{C}[G]$ has PI-degree $n$ then there is $g$ in $G - \{1\}$ such that $[G : C_G(g)] \leq n^2$. (Hint: Write $|G| - 1 = (\sum n_i^2) - 1 \leq (t-1)n^2$ where $t$ is the number of conjugacy classes and, letting $k = \min_{1 \neq g \in G} [G : C_G(g)]$ note $|G| - 1 \geq (t-1)k$. Thus $k \leq n^2$.)

35. (Crossed products) Suppose $G$ is a group each of whose elements acts as an automorphism of $R$. Given a function $t : G \times G \to \text{Unit}(R)$ we define the *crossed product* $R * G$ to be the free $R$-module with base indexed by $G$, endowed with multiplication $\left(\sum r_g g\right)\left(\sum s_h h\right) = \sum_{a \in G}\left(\sum_{gh=a} r_g(gs_h) t(g, h)\right) a$. Determine necessary and sufficient conditions on $t$ such that $R * G$ is an associative ring. (Hint: These resemble the Noether factor set conditions.) When $t(g, h) = 1$ for all $g, h$, $R * G$ is just the skew group ring; at the other extreme, when $G$ acts trivially on $R$ then $R * G$ is called the *twisted group ring*. Letting $U = \text{Unit}(R)$ and $\mathcal{G} = \{ug : u \in U, g \in G\}$ show $\mathcal{G} \subseteq \text{Unit}(R * G)$ and $G \approx \mathcal{G}/U$.

# §8.2

## Noetherian Group Rings

1. If $G$ is polycyclic-by-finite with Hirsch number $t$ then K-dim $F[G] \leq t$. (Hint: As in theorem 8.2.2.)
2. (Formanek [73]) If $G$ is torsion-free and $F[G]$ is left Noetherian then $F[G]$ has no nontrivial idempotents. (Hint: By proposition 8.2.24 the Kaplansky trace of an idempotent of $F[G]$ is an integer; apply exercise 8.1.8.) Compare to exercise 16.

## Ordered Groups

3. If $P$ is the positive cone of a left ordered group $G$ then make $G$ right ordered by using $P^{-1}$.
4. The direct limit of left ordered groups is left ordered. In particular, every torsion-free abelian group is left ordered, and corollary 8.2.32 can be improved accordingly.
5. We say $F[G]$ has *only trivial units* if every unit is of the form $\alpha g$ for $0 \neq \alpha \in F$ and $g \in G$. Show if $G$ is left ordered then $F[G]$ has only trivial units. (Hint: Suppose $u = \sum \alpha_g g$ is a unit, where $|\text{supp}(u)| \geq 2$. As in the proof of proposition 8.2.33 $u^{-1}u = 1$ has an element in its support which is a unique product of elements from $\text{supp}(u^{-1})$ and $\text{supp}(u)$. But there is another one, obtained by taking *minimal* elements in the support. Hence $|\text{supp}(u^{-1}u)| \geq 2$, contrary to $u^{-1}u = 1$.)
6. If $G$ is a torsion-free group and $F[G]$ has only trivial units then $F[G]$ is a domain. (Hint: Since $F[G]$ is prime it suffices to show $F[G]$ is reduced. But if $r^2 = 0$ then $(1 - r)(1 + r) = 1$.) For further discussion of the connection between trivial units and the zero-divisor problem see Passman [77B, p. 591].

**The Generic Construction of a $G$-Crossed Product** Based on Farkas [75], both Rosset [78] and Snider [79] have found very interesting construction of division rings which are crossed products with respect to a Galois group $G$. It is possible to bypass the Farkas-Snider theorem and utilize the easier results on left-ordered groups. The basic set-up is as follows: Given a group $G$ of order $n$, take an epic $\varphi: \Lambda \to G$ where $\Lambda$ is a finitely generated free group. Let $N = \ker \varphi$, and $N'$ be the commutator subgroup. There is an exact sequence $1 \to \bar{N} \to \bar{\Lambda} \to G \to 1$ where $\bar{N} = N/N'$ is abelian, so $\bar{\Lambda} = \Lambda/N'$ is abelian-by-finite.

7. If $G$ is solvable then $\bar{\Lambda}$ is torsion-free polycyclic. (Hint: There is a subnormal series $\Lambda = \Lambda_0 \rhd \Lambda_1 \rhd \cdots \rhd \Lambda_t = N$ with $\Lambda_i$ f.g. and each factor abelian; by the Nielssen-Schreier theorem (cf., Hall [59B]) each $\Lambda_i$ is free, so $\Lambda'_{i-1}/\Lambda'_i$ is a subgroup of the torsion-free abelian group $\Lambda_i/\Lambda'_i$. Hence the series $\Lambda \rhd \Lambda'_0 \rhd \cdots \rhd \Lambda'_t = N'$ has factors which are f.g. torsion-free abelian.)
8. For arbitrary $G$ and for any field $F$, show $\bar{\Lambda}$ is torsion-free and $F[\bar{\Lambda}]$ is a PI-domain whose division ring of fractions $D$ is a crossed product with respect to the Galois group $G$. (Hint: Suppose $h \in \bar{\Lambda}$ were torsion. Letting $\Lambda_1$ be the pre-image of $\langle h \rangle$ in $\Lambda$ and letting $G_1$ be the image of $\Lambda_1$ in $G$, one gets $1 \to N \to$

$\Lambda_1 \to G_1 \to 1$ and $\bar{\Lambda}_1 = \langle h \rangle$ is torsion, contrary to exercise 7. At this point one could apply Farkas-Snider, but the following easier argument is available. Certainly $F[\bar{\Lambda}]$ is prime of PI-degree $\leq n$, so by central localization one gets a central simple algebra $R$ of degree $\leq n$. On the other hand, the field of fractions $F[\bar{N}]$ is a subfield on which $G$ acts as a group of automorphisms via the $G$-module action on $\bar{N}$ of example 5.2.13. Thus $[F[\bar{N}]:Z(R)] = n$ so $F[\bar{N}]$ is a maximal subfield of $R$, and $R$ is a crossed product with respect to $G$. Furthermore, if $\operatorname{char}(F) \nmid |G|$ then $\bar{\Lambda}$ is left ordered by exercise 4, so $F[\bar{\Lambda}]$ is a domain and $R$ is a division ring. Appealing to exercise 7.2.5 conclude in general that $R$ is a division ring.)

9. $R$ of exercise 8 is "generic" in the following sense: If $R_1$ is a crossed product with respect to $G$ then there is a maximal subfield $K$ and factor set $\beta: G \times G \to K^*$ where $K^*$ denotes $K - \{0\}$. Taking $r_\sigma$ in $R_1$ such that $r_\sigma^{-1} a r_\sigma = \sigma a$ for all $a$ in $K$, let $E$ be the multiplicative subgroup of $R$ generated by $K^*$ and the $r_\sigma$. The factor set conditions give an exact sequence $1 \to K^* \to E \to G \to 1$ sending $r_\sigma$ to $\sigma$, and any group homomorphism $\bar{\Lambda} \to E$ gives rise to a ring homomorphism $R \to R_1$.

10. Using exercise 9, show that if $R$ is similar to a tensor product of cyclic algebras then every crossed product with respect to $G$ is similar to a tensor product of cyclics in the same way.

11. Rosset [78, lemma 9] proved $R$ has exponent $n$ by showing $H^2(G, \bar{N}) \approx (\mathbb{Z}/n\mathbb{Z}, +)$.

## Polycyclic-by-Finite Group Algebras

12. (Snider) Suppose $R$ is a prime ring of finite gl. dim, which is free as a module over a Noetherian domain $W \subseteq R$, with normalizing base $a_1, \ldots, a_n$. Then $R$ is a domain if $n \mid u$-dim $P$ for every f.g. projective $R$-module $P$. Here $u$-dim denotes uniform dimension. (Hint: As in the beginning of the proof of theorem 8.2.35.)

13. Exercises are from Strojnowski [86]: A group $H$ is called an $E(F)$-*group* if every free module $Q = F[H]^{(n)}$ satisfies the property that dependent elements remain dependent modulo the augmentation ideal, i.e., $\{x_i + \omega(F[H])Q: 1 \leq i \leq n\}$ are $F$-dependent in $Q/\omega(F[H])Q$ if $\{x_1, \ldots, x_n\}$ are $F$-dependent in $Q$. Suppose $S \subset F[H] - \omega(F[H])$ is a denominator set of regular elements. Then for any f.g. projective $S^{-1}F[H]$-module $P$ one has $u$-dim $P = [P/\omega(F[H])P:F]$. (Hint: Take $P \oplus P'$ free.)

14. Suppose the $E(F)$-group $H$ is a subgroup of index $n$ in $G$, and suppose $R = F[G]$ and $W = F[H]$ satisfy the hypothesis of exercise 12. Then $R$ is a domain if $n \mid [P/\omega(F[H])P:F]$ for every f.g. projective $F[G]$-module $P$.

15. A group $G$ is an $E(F)$-group under any of the following conditions: (i) $G$ is infinite cyclic. (ii) $G$ is a subgroup of an $E(F)$-group. (iii) Every nontrivial f.g. subgroup has a homomorphic image which is an $E(F)$-group $\neq (1)$. (iv) $G$ has a normal subgroup $N$ such that $N$ and $G/N$ are $E(F)$-groups. (v) $G$ is solvable with torsion free abelian factors. (Hint: (i) If $G = \langle g \rangle$ then cancel a suitable power of $(1 - g)$ from any dependence relation. (ii) Write $F[G]$ as a summand of $F[H]$ where $H$ is an $E(F)$-group, and match components. (iii) Induct on the support of a dependence relation. (iv) by (ii) and (iii).)

**Idempotents**

16. If $G$ is an $E(F)$ group then $F[G]$ has no nontrivial idempotents. (Hint: Clearly $\omega(F[G])$ contains no projective left ideal $\neq 0$ and thus no nontrivial idempotent. Now let $\pi\colon F[G] \to F$ be the augmentation map. If $e \neq 0$ is idempotent then $\pi e$ is idempotent implying $\pi e = 1$; hence $\pi(1 - e) = 0$ implying $1 - e = 0$.)

17. (Farkas) Say idempotents $e, e'$ of $R$ are *equivalent* if $e = r_1 r_2$ and $e' = r_2 r_1$ for $r_1, r_2$ in $R$. Suppose $R$ is an $F$-algebra with a nonfree projective module $P$ generated by $n$ elements, and $W_1$ is another $F$-algebra. Let $W = M_n(F) \times W_1$. Then $W \otimes_F R$ has an idempotent not equivalent to any idempotent in $W$. (Hint: $(e, 0) \in M_n(R) \times (W_1 \otimes R) \approx W \otimes R$ is not equivalent to any idempotent in $M_n(F) \times (W_1 \otimes F) \approx W \otimes F \approx W$.

18. (Farkas' surpise) If $G$ is nonabelian torsion-free polycyclic-by-finite then for some finite group $G_1$ the group algebra $\mathbb{C}[G \times G_1]$ has an idempotent *not* equivalent to any idempotent whose support lies in a finite subgroup of $G \times G_1$. (Hint: This relies on the result that there is a nonfree projective module spanned by $n$ elements. When $G$ is not nilpotent this is a result of Lewin [82]; when $G$ is nilpotent this is a result of Artamov and has a relatively easy proof given below in §8.4. Let $G_1 = \text{Sym}(n + 1)$, which has an irreducible character of degree $n$, cf., exercise 8.1.7. By exercise 17 there is an idempotent in $\mathbb{C}[G \times G_1] \approx \mathbb{C}[G] \otimes \mathbb{C}[G_1]$ not equivalent to any idempotent in $\mathbb{C}[G_1]$. But all finite subgroups of $G \times G_1$ lie in $G_1$.) Farkas' surpise shoots down one natural approach to the Goldie rank theorem.

# §8.3

1. $U(\varinjlim L_i) \approx \varinjlim U(L_i)$ canonically.

2. If $L_i$ are Lie subalgebras of $L$ such that $L_1 + L_2 = L$ then taking $H = L_1 \cap L_2$ one has a unique bijective map $f\colon U(L_1) \otimes_{U(H)} U(L_2) \to U(L)$ such that $f(r_1 \otimes r_2) = r_1 r_2$ for $r_i$ in $U(L_i)$. (Hint: To prove $f$ exists, view $U_i = U(L_i)$ as $U(H)$-modules and define the balanced map $U_1 \times U_2 \to U(L)$ by $(r_1, r_2) \mapsto r_1 r_2$. Show $f$ is an isomorphism by matching bases.)

3. Let $L$ be the Lie algebra with base $w, y, z$ satisfying $[wx] = x$, $[wy] = -y$, $[xy] = z$, and $[zL] = 0$. Letting $\psi\colon S(L) \to U(L)$ be the canonical map note $a = wz + xy \in \text{Ann}'_{S(L)} L$ but $(\psi a)^2 \neq \psi(a^2)$.

4. Define the Heisenberg Lie algebra $L$ with base $\{x_1, \ldots, x_n, y_1, \ldots, y_n, z\}$ satisfying $[x_i y_i] = z$ and $[x_i z] = [y_i z] = 0$ for all $i$. Show $L$ is solvable, and the Weyl algebra $A_n(F)$ is a homomorphic image of $U(L)$.

5. Every derivation of the nonabelian 2-dimensional Lie algebra is inner. (Hint: Suppose $\delta$ is a derivation of $L = Fx + Fy$, where $[xy] = x$. Then $\delta x = \delta[xy] \subseteq L' = Fx$, so $\delta x = [ax]$ for some $a$ in $Fy$. Replacing $\delta$ by $\delta - \text{ad } a$ one may assume $\delta x = 0$. Then $[x\delta y] = \delta[xy] = 0$ so $\delta y \in Fx$, and repeat the process.)

6. Describe the Lie algebra of upper triangular $3 \times 3$ matrices in terms of example 8.3.42.

7. Describe the Lie algebra with base $x_1, x_2, x_3$ such that $[x_i x_{i+1}] = x_{i+2}$ subscripts mod 3, in terms of example 8.3.42.

8. (BGR, example 5.9) Let $L$ be as in exercise 3. Localizing at $z$ show $\mathrm{Spec}(U(L))$ is isomorphic to $\mathrm{Spec}\, F[z, z^{-1}, \lambda]$ where $\lambda = w + z^{-1} yx$.

9. Suppose $L$ acts as a Lie algebra of derivations on a ring $R$. Then we can define the *Lie algebra of differential polynomials over* $R$ as the $F$-vector space $R \otimes_F V(L)$ with multiplication agreeing with that in $R$ and in $U(L)$, and satisfying the rule $ar = ra + \delta_a r$ for every $a$ in $L$ and $r$ in $R$. Show this defines an algebra structure and generalize proposition 8.3.32 to this setting.

10. If $[L:F] = n$ then $K$-dim $U(L) \geq n$. (Hint: Define a lattice injection from $\mathcal{L}(U(L))$ to $\mathcal{L}(G(L))$ by $I \mapsto \bigoplus_{n \geq 1}(I \cap U_n(L))/(I \cap U_{n-1}(L))$. Since $G(L)) \approx F[\lambda_1, \ldots, \lambda_n]$, conclude $K$-dim $U(L) \leq$ $K$-dim $F[\lambda_1, \ldots, \lambda_n] = n$.)

11. Show every chain in $\mathrm{Spec}(U(L))$ has length $\leq [L:F]$ by using exercise 10.

## Growth of Infinite Dimensional Lie Algebras

12. Define "growth" of a Lie algebra as in §6.2. If $L$ has subexponential growth then $U(L)$ also has subexponential growth (by PBW).

13. (M. Smith [76]) If $L$ is infinite dimensional but finitely generated as Lie algebra then $U(L)$ does not have polynomial bounded growth. (Hint: The growth function in $L$ is strictly increasing, so the growth function in $U(L)$ increases at least as fast as $p(n)$, the number of ways of writing $n$ as a sum of nondecreasing positive integers. But $p(n)$ is not polynomially bounded.)

14. (M. Smith [76]) Let $L$ be a Lie algebra with base $a_0, a_1, \ldots$ such that $[a_i a_j] = 0$ and $[a_0 a_j] = a_{j+1}$ for $i, j \geq 1$. Then $U(L)$ is subexponential by exercise 12 but not polynomial bounded, by exercise 13. As observed by Bergen-Montgomery-Passman [86], this example can be altered to give an example in which the differential polynomial ring $E[\lambda; \delta]$ does not have polynomial bounded growth, where $E$ is the infinite dimensional exterior algebra and thus has GK dim 0. ($\lambda$ replaces $a_0$, and $x_i$ replace $a_i$.)

## The Ore Condition for Infinite Dimensional Lie Algebras

15. If $L$ has subexponential growth then $U(L)$ is Ore, by exercise 12 and proposition 6.2.15.

16. Suppose $L_1$ is a Lie ideal of a Lie algebra $L$ and $L/L_1$ is solvable. If $U(L_1)$ is Ore then $U(L)$ is Ore. (Hint: One may assume $L/L_1$ is abelian; furthermore, since the Ore condition involves a finite number of elements one may assume $L/L_1$ is finitely generated. But $U(L)$ is a free $U(L_1)$-module and matching coefficients in the base one sees $U(L_1)$ is a left denominator set for $U(L)$; localizing puts one in a position to use the Hilbert basis theorem.)

The next exercises, due to Borho-Rentschler [75], show how to obtain the Ore condition for a subalgebra of finite index.

17. Suppose $L_1$ is a Lie subalgebra of $L$. For any $a$ in $L$ we have $aU(L_1) \subseteq U(L_1)L$. (Hint: Show $aU_n(L_1) \subseteq U_n(L_1)$ by induction.)

18. Suppose $L_1$ is a Lie subalgebra of $L$ of finite index $n$, and $U(L)$ and $U(L_1)$ are both Ore, e.g., $L$ is locally finite. Letting $D_1$ be the division ring of fractions of

$U(L_1)$ show $S = U(L_1) - \{0\}$ is a left and right denominator set of $U(L)$, and $S^{-1}U(L) = D_1 U(L) = U(L)D_1$, viewed in the division ring $D$ of fractions of $U(L)$. (Hint: Let $b_1, \ldots, b_k$ be a complementary base for $L_1$ in $L$. By exercise 17 for any $s$ in $S$ one has $b_i s = \sum_j u_{ij} b_j + s_i$ for suitable $u_{ij}, s_i$ in $U(L_1)$. The $n \times n$ matrix $(u_{ij})$ is invertible in $M_n(D_1)$; indeed, if $a_1, \ldots, a_n \in U(L_1)$ with $\sum_i a_i u_{ij} = 0$ for $1 \le j \le n$ then $(\sum a_i b_i)s = \sum a_i s_i \in U(L_1)$ implying by PBW that $\sum a_i b_i \in U(L_1)$ and so each $a_i = 0$. But now the matrix equation $(b_i)s = (u_{ij})(b_i) + (s_i)$ implies $(u_{ij})^{-1}(b_i) = (b_i)s^{-1} + (u_{ij})^{-1}(s_i)s^{-1}$; thus $(b_i)s^{-1} = (u_{ij})^{-1}(b_i) - (u_{ij})^{-1}(s_i)^{-1}s^{-1} \in (D_1 L)^{(n)}$, proving $LD_1 \subseteq D_1 L$, and thus $S$ is a left denominator set.)

19. A "relative" Poincare-Birkhoff-Witt theorem. Assumptions and notations as in exercise 18, let $\tilde{L}$ be the semidirect product of $L_1$ and $(L/L_1)$, where $L_1$ acts naturally on $L/L_1$, i.e., $[a, b_i + L_1] = [ab_i] + L_1$. Then the algebra $R = D_1 U[L]$ has a filtration whose associated graded ring $G(R)$ is isomorphic to $D_1 U(\tilde{L})$. (Hint: There is a Lie map $\varphi: \tilde{L} \to G(R)$ given by $\varphi(a, b_i) = (a + R(1)) + (b_i + R(2))$, so this extends to an associative homomorphism $U(\tilde{L}) \to G(R)$ and thus to a homomorphism $D_1 U(\tilde{L}) \to G(R)$. Since base is sent to base this is an isomorphism.)

20. (Lichtman) If $L$ is a locally finite dimensional Lie algebra over a field of characteristic 0 then every PI-subalgebra of $U(L)$ is commutative. (Hint: Consider the $\mathbb{Z}$-subalgebra $R$ generated by two noncommuting elements $a, b$, and pass to characteristic $p$ for $p$ sufficiently large. Then PI-deg$(R)$ divides a power of $p$; but this is true for differing $p$, contradiction.)

## §8.4

1. (Amitsur-Small [78]) Suppose $D$ is a division ring. The polynomial ring $R = D[\lambda_1, \ldots, \lambda_t]$ is primitive iff $M_n(D)$ contains a subfield of transcendence degree $t$ over $F = Z(D)$ for some $n$. (Hint: ($\Leftarrow$) is exercise 2.3.5. ($\Rightarrow$) Suppose $R$ has maximal left ideal $L$ with core 0. Then $F[\lambda_1, \ldots, \lambda_t]$ can be embedded into $\text{End}_R(R/L)$, so conclude by means of the Nullstellensatz.) This result is used to prove that there are division rings $D$ for which $D[\lambda_1, \ldots, \lambda_t]$ is primitive but $D[\lambda_1, \ldots, \lambda_{t+1}]$ is not primitive.

### Skew Polynomial Rings (cf., Exercise 1.6.13ff)

2. Suppose $\sigma$ is an automorphism of $R$ and $T = R[\lambda; \sigma]$. Then $\sigma$ extends to an automorphism of $T$ when we put $\sigma\lambda = \lambda$. Suppose $P \lhd R$. If $P \in \text{Spec}(R)$ then $P + T\lambda \in \text{Spec}(T)$ with $T/(P + T\lambda) \approx R/P$; if $P$ is $\sigma$-prime with $\sigma P = P$ then $TP \in \text{Spec}(T)$, with $T/TP \approx (R/P)[\lambda; \sigma]$. Conversely, if $P' \in \text{Spec}(T)$ then either (i) $\lambda \in P'$ and $P' = P + T\lambda$ where $P = P' \cap R \in \text{Spec}(R)$, or (ii) $\lambda \notin P'$, $\sigma P' = P'$, and $P = P' \cap R$ is a $\sigma$-prime ideal with $\sigma P = P$ and $TP \in \text{Spec}(T)$. (Hint: The first assertion is straightforward as is (i) of the second assertion since $T/(P + T\lambda) \approx R/P$. Thus suppose $\lambda \notin P'$. If $a \in T$ and $\lambda a \in P'$ then $\lambda Ta \in P'$ so $a \in P'$, implying $\lambda$ is regular modulo $P'$; thus $(\sigma a)\lambda = \lambda a \in P'$ for any $a$ in $P'$, yielding $\sigma P' = P'$. $P$ is $\sigma$-prime as in exercise 1.6.14 and $\sigma P = P$ so $TP \lhd T$. There is a natural map $T \to (R/P)[\lambda; \sigma]$ taking coefficients mod $P$, with kernel $TP$.)

Assume for exercises 3 through 10 that $R$ is a left Noetherian ring, and $T = R[\lambda; \sigma]$ where $\sigma$ is an automorphism. Note $T$ also is left Noetherian. We aim for the theorem of Goldie-Michler [75] that $T$ is Jacobson if $R$ is Jacobson. Write $S$ for {regular elements of $R$}.

3. If $B$ is a $\sigma$-prime ideal of $R$ (or of $T$) then $\sigma$ permutes the minimal prime ideals over $B$; explicitly there is a prime ideal $P$ of $R$ with $B = \bigcap_{i=1}^{n} \sigma^i P$.

4. $S$ is a left denominator set for $T$. (Hint: Iterate the Ore condition on the coefficients.)

5. Any annihilator ideal in $T$ is a $\sigma$-ideal.

6. If $R$ is $\sigma$-prime then any nonzero $\sigma$-ideal $A$ of $T$ intersects $S$. (Hint: $\text{Ann}_R A$ is a $\sigma$-ideal and is thus 0, so $A$ is large.)

7. If $R$ is semiprime and $P' \in \text{Spec}(T)$ with $P' \cap R = 0$ then every element of $S$ is regular modulo $P'$ in $T$. (Hint: Let $A = \{a \in T : sa \in P'$ for some $s$ in $S\} \lhd T$ by exercise 4. Then $A \subseteq P'$ so $S$ is right regular modulo $P'$. To get left regular consider the ascending chain $\{P's^{-1} : i \in \mathbb{N}\}$; this terminates.)

8. Suppose $R$ is $\sigma$-prime and $a, b \in T$. Then there is $s$ in $S$ and $q, r$ in $T$ such that $sb = qa + r$ where $r = 0$ or $\deg(r) < \deg(a)$. (This is proved just as the usual Euclid algorithm.)

9. If $T$ is prime and $0 \neq P' \in \text{Spec}(T)$ with $P' \cap R = 0$ then every ideal properly containing $P'$ intersects $S$. (Hint: $\sigma P' = P'$ by exercise 2. Take an ideal $A \supseteq P'$ maximal with respect to $A \cap S = \varnothing$. Then $\sigma A = A$. Take $a \in A$ of minimal degree and $b \in P'$ of minimal degree. If $A \supset P'$ then $A$ would contain an element regular modulo $P'$, and exercises 6 and 7 then show $a$ is regular modulo $P$. Now using exercise 8 show $sb = qa \in P'$ implies $q \in P'$ and thus $\deg a = 0$, implying $0 \neq A \cap R$ so $A$ intersects $S$.)

10. If $R$ is Jacobson then $T$ is Jacobson and $\langle \lambda \rangle^{-1} T$ is Jacobson. (Hint: The second assertion follows from the first since $\langle \lambda \rangle^{-1} T \approx T[\mu; \sigma]/A$ where $A$ is the ideal generated by $\lambda\mu - 1$. To prove the first assertion it suffices to prove $\text{Jac}(T/P') = 0$ for all $P'$ in $\text{Spec}(T)$. Let $P = R \cap P'$. If $\lambda \in P'$ then $T/P' \approx R/P$ is Jacobson, so assume $\lambda \notin P$. Replacing $R$ by $R/P$ and $T$ by $T/P'$ one may assume $T$ is prime and $P = 0$, cf., exercise 2. $\text{Jac}(T) = 0$ by §2.5, so we may assume $P' \neq 0$. Let $J' = \bigcap\{$primitive ideals containing $P'\}$, and $J = R \cap J'$. If $P' \subset J'$ then $J \neq 0$ by exercise 9. Let $t$ be the minimal degree of a nonzero element of $P'$, and let $Q = \{$leading coefficients of elements of $P'$ having degree $t\}$.

The strategy is to show $Q \subseteq L$ for every maximal left ideal $L$ of $R$ not containing $J$, for then $Q \subseteq \bigcap\{$cores of maximal left ideals not containing $J\} = 0$ contrary to $P' \neq 0$. Pick $q \in Q$. By an iterative procedure show that if $a \in T$ has degree $n$ there is some $b \in T$ of degree $< t$ such that $q^n a - b \in P'$.

Now take a maximal left ideal $L$ of $R$ with $J \nsubseteq L$. Then $1 = x + y$ for $x$ in $J$ and $y$ in $L$. But there is $a \in T$ with $a(1 - x) - 1 \in P'$, and thus $by - q^n \in P'$ so by minimality of $t$ we see $by - q^n = 0$ implying $q^n \in Ry \subseteq L$. If $q \notin L$ then $Rq + L = R$ so for some $r$ in $R$ and $z$ in $L$ we would have $1 = rq + z$. By the above argument $(rq)^m \in L$ for some $m$; taking $m$ minimal yields $(rq)^{m-1} = (rq)^{m-1}(z + rq) = (rq)^{m-1}z + (rq)^m \in L$, contradiction. Thus $q \in L$ as desired.)

11. (Irving [79]) An unexpected primitive ring. Let $F$ be an absolute field and $C_0 = F[\mu_1, \mu_1^{-1}, \mu_2, \mu_2^{-1}]$. Let $\sigma$ be the automorphism of $C_0$ fixing $F[\mu_1, \mu_1^{-1}]$ satisfying $\sigma\mu_2 = \mu_1\mu_2 + 1$. Let $C = S^{-1}C_0$ where $S$ is the submonoid of $C_0$ generated by $\sigma^n\mu_2$ for all $n$ in $\mathbb{Z}$, and $R$ be the localization of $C[\lambda; \sigma]$ at $<\lambda>$, where $\sigma$ is extended naturally to $C$. Then $R$ is Noetherian Jacobson (by exercise 10) and is affine over $F$, generated by $\lambda, \mu_1, \mu_2$ and their inverses. $R$ also is primitive and thus fails the Nullstellensatz for its faithful simple module $M$. (Hint: Let $M$ be the vector space over $F(\mu_1)$ with base $\{v_n : n \in \mathbb{N}\}$ made into an $R$-module by putting $\mu_2 v_0 = v_0$ and $\lambda v_n = v_{n+1}$ for all $n$. This determines the scalar multiplication since then $\mu_2 v_n = \mu_2 \lambda^n v_0 = \lambda^n \sigma^{-n} \mu_2 v_0$. $M$ is faithful for if $0 \neq r \in$ Ann $M$ then writing $r$ in the form $\sum \lambda^i c_i$ and matching components of $v_m$ in $0 = \sum \lambda^i c_i v_n$ shows each $c_i \in$ Ann $M$, so one may assume $r \in C$, and comparing powers of $\mu_1$ shows $r = 0$. To show $M$ is simple it suffices to show $v_0 \in Rv$ for all $v \neq 0$. If $v = \sum r_i v_i$ with $v_j \neq 0$ then $(\mu_2 - 1)\lambda^{-j}v$ has fewer nonzero terms, and thus $Rv \cap F(\mu_1)v_0 \neq 0$; so assume $v \in F[\mu_1]v_0$. But $\lambda^n \mu_2 \lambda^{-n} v_0 = (\sigma^n \mu_2)v_0 = \sum_{i=0}^{n-1} \mu_1^i v_0$. $F$ being absolute, every irreducible element of $F[\mu_1]$ divides some polynomial $\mu_1^{n-1} + \cdots + 1$ by the theory of finite fields. Hence $(\lambda^n \mu_2 \lambda^{-n})^k v_0 \in Rv$ for some $n$ and $k$, implying $v_0 \in Rv$.)

## F.g. Overrings of Left Noetherian Rings

12. (Small) Suppose $R$ is left Noetherian and $T \supseteq R$ is a prime ring which is f.g. (as $R$-module). Then $A \cap R \neq 0$ for every nonzero $A \triangleleft T$. (Hint: $T$ is a Noetherian $R$-module so any $a$ in $T$ satisfies $\sum_{i \leq n} Ra^i = \sum_{i \leq n+1} Ra^i$ for suitable $n$; thus $a^{n+1} \in \sum_{i \leq n} Ra^i$, so take $a$ regular in $A$.) Using a similar argument show every element of $R$ invertible in $T$ is invertible in $R$, and thus $R \cap \mathrm{Jac}(T) \subseteq \mathrm{Jac}(R)$ by proposition 2.5.17.

12'. (Resco) If $R$ is left Noetherian and $T \supseteq R$ is f.g. then INC holds from $R$ to $T$. (Hint: By exercise 12.)

13. (Small) If $R$ is left Noetherian, Jacobson and $T \supseteq R$ is f.g. then $T$ is left Noetherian, Jacobson. (Hint: One may assume $T$ is prime. Let $A = R \cap \mathrm{Jac}(T)$. Exercise 12 implies $0 \neq A \subseteq \mathrm{Jac}(R) = \mathrm{Nil}(R)$. Then

$$\text{K-dim } T = \text{K-dim } R = \text{K-dim } R/A = \text{K-dim } T/\mathrm{Jac}(T)$$

so $\mathrm{Jac}(T)$ is nilpotent.)

14. (Small) Suppose $R$ is affine over a field $F$ such that, for every algebraic field extension $K$ of $F$, $R \otimes_F K$ is Noetherian, and $\mathrm{End}_{R \otimes K} N$ is algebraic over $K$ for every $R \otimes K$-module $N$ of finite composition length. Then $\mathrm{End}_R M$ is finite dimensional over $F$ for every simple $R$-module $M$. (Sketchy hint: Let $D = \mathrm{End}_R M$. Then $K \otimes_F D \approx \mathrm{End}_{K \otimes R} K \otimes M$ by exercise 2.11.15. But for $[K:F] < \infty$ one has $K \otimes M$ of finite composition length over $K \otimes R$ so taking direct limits note $\mathrm{End}_{K \otimes R} K \otimes M$ is algebraic over $K$ for $K$ the algebraic closure of $F$. Taking a maximal subfield $L$ of $D$ note $L \otimes L \subset L \otimes D \approx \mathrm{End} L \otimes M$ is Noetherian since $L$ is algebraic; thus $L \otimes L$ is Noetherian so $L$ is affine over $F$ by exercise 7.1.32. Hence $[L:F] < \infty$; conclude $[D:F] < \infty$.)

**Almost Commutative Algebras**  A filtered algebra $R$ over a field $F$ is *almost commutative* if $R_1$ generates $R$ as an algebra, and the associated graded algebra of $R$ is commutative. This yields a ring-theoretic approach to studying enveloping algebras in view of the following result.

15. (Duflo) $R$ is almost commutative iff it is a homomorphic image of $U(L)$ for a suitable Lie algebra $L$. (Hint: ($\Rightarrow$) If $r_1, r_2 \in R_1$ then $r_1 r_2 - r_2 r_1 \in R_1$ so $R_1$ has a Lie algebra structure $L$, yielding a homomorphism $U(L) \to R_1$ by PBW. ($\Leftarrow$) Any homomorphism $U(L) \to R$ induces a map of the associated graded algebras; check this is an algebra homomorphism.)

16. If $R$ is almost commutative then the GK-module dimension over $R$ is exact. The connection between K-dim and GK-dim and the Hilbert series are carried out in Krause-Lenagan [85B, Chapter 7].

**Stably Free Nonfree Left Ideals (Following Stafford [85])**

17. If $L$ is an f.d. solvable nonabelian Lie algebra with an abelian Lie subalgebra $L_1$ of codimension 1 then the enveloping algebra $U(L)$ has a nonfree stably free left ideal. (Hint: Let $R = U(L_1)$, and $T = U(L)$. Then $T \approx R[\lambda; \delta]$, where $\lambda$ can be identified with an element in $L$. Take $a_0 \in L$ with $[\lambda a_0] \neq 0$ and inductively let $a_i = [\lambda a_{i-1}] \in L_1$. Take $n$ maximal such that $a_0, \ldots, a_n$ are linearly independent. If $a_{n+1} = 0$ then take $r = a_{n-1} a_n + 1$ and apply corollary 8.4.44, noting $a_{n-1}^2 \delta r - (a_{n-1} a_n - 1) r = 1 + a_{n-1}^2 a_n^2 - a_{n-1}^2 a_n^2 = 1$. If $a_{n+1} \neq 0$ then write $a_{n+1} = \sum_{i=0}^{n} \alpha_i a_i$; assume $\alpha_0 \neq 0$ and again show $\sum R \delta^i r = R$.)

18. The following assertions are equivalent for a finite dimensional Lie algebra $L$: (i) $L$ is abelian. (ii) Every stably free left ideal of $U(L)$ is free. (iii) Every projective left ideal of $U(L)$ is free. (Hint: (i) $\Rightarrow$ (iii) by theorem 5.1.53, and (iii) $\Rightarrow$ (ii). So it suffices to assume $L$ is nonabelian, and prove there is a nonfree stably free left ideal.

*Case I.* $L$ has a nonabelian solvable subalgebra. Then $L$ has a subalgebra $L_1$ as in exercise 17, so conclude via proposition 8.4.46 which is applicable by exercise 8.3.19.

*Case II.* Every solvable subalgebra is abelian. As in case I one can assume $L$ is minimal with respect to being nonabelian, so in particular $L = [LL]$. Pick $a_i$ in $L$ with $a_3 = [a_1 a_2] \neq 0$. Then $a_1, a_2$ generate $L$ as a Lie algebra by the minimality assumption, so $a_2 \in Ta_1 + T(1 + a_2)$ where $T = U(L)$. $I = T(a_2 + 1)a_1^{-1}$ is stably free so one wants $I$ nonprincipal. By exercise 8.3.18 one has $I \cap F[a_2] \neq 0$; likewise $a_3' a_1 = a_1' a_3$ for suitable $a_1' \in F[a_1]$ and $a_3'$ in $T$, implying $a_1'(a_2 + 1) - a_3' \in I$. If $I = Ta$ one could assume $a \in F[a_2]$ using proposition 8.4.46, and $a$ has degree 1 so $a = a_2 + \alpha$, yielding $(a_2 + \alpha)a_1 \in T(a_2 + 1)$ and $a_3 \in Fa_1 + Fa_2$, so $L$ is solvable, contradiction.)

**Noncommutative Unique Factorization Domains**  An element $p$ of a prime ring $R$ is called a *prime* if $pR = Rp$ is a height 1 prime ideal and $R/pR$ is a domain. A (not necessarily commutative) domain $R$ is a UFD (*unique factorization domain*) if every nonzero element has the form $sp_1 \cdots p_n$ for suitable prime elements $p_1, \ldots, p_n$ of $R$, where $s$ is not in any height 1 prime ideal of $R$.

**19.** Suppose $R$ is prime Noetherian. If $P = Rp = pR$ and $R/P$ is a domain then $\bigcap_{i \in \mathbb{N}} P^i = 0$ and $R$ is a domain. (Hint: Note $p$ is regular. If $\bigcap P^i \neq 0$ then it contains a regular element $a$; writing $a = p^i a_i$ yields $Ra_1 \subset Ra_2 \subset \cdots$, contradiction. Now any element can be written as $p^i a$ where $a \notin P$, which implies $R$ is a domain since $Rp = pR$.)

**20.** (Chatters [84]). Suppose $R$ is prime Noetherian and has a height 1 prime ideal. $R$ is a UFD iff every height 1 prime ideal $P$ of $R$ has the form $Rp$ for a suitable prime element $p$ of $R$. (Hint: ($\Rightarrow$) If $sp_1 \cdots p_n \in P$ then some $p_i \in P$ so $P = Rp_i$. ($\Leftarrow$) Use exercise 19. If $r \in R$ then $R/RrR$ has a finite number of minimal primes, so $R$ has a finite number of height 1 primes containing $r$; call these $P_1, \ldots, P_t$. Write $P_i = Rp_i = p_i R$ and

$$r = p_1^{n(1)} r_2 = p_1^{n(1)} p_2^{n(2)} r_2 \cdots = \cdots = p_1^{n(1)} \cdots p_t^{n(t)} r_t.$$

Clearly $r_t$ is in no height 1 prime ideal of $R$.)

Examples of Noetherian UFD's include enveloping algebras of solvable or semi-simple Lie algebras over $\mathbb{C}$, and group algebras of certain polycyclic-by-finite group algebras, cf., Chatters [84] and Gilchrist-Smith [85]. The verifications are via the condition of exercise 20, noting that if $R$ has finite GK dimension then $R$ has height 1 primes. On the other hand, the next few exercises show the class of Noetherian UFD's is rather restricted. Surprisingly, the noncommutative case is more malleable than the commutative case.

**21.** (Gilchrist-Smith) if $R$ is a left Noetherian UFD which is *not* commutative then every prime $P$ of height $> 1$ contains an element which does not belong to any height 1 prime. (Hint: Assume, on the contrary, that every element of $P$ is contained in a height 1 prime. Then every element $a$ of $P$ is contained in a height 1 prime $\subset P$; for if $a = sp_1 \cdots p_n \in P$ then $s \notin P$ so some $p_i \in P$.

In particular, $P$ contains distinct height 1 primes $P_0 = Rp = pR$ and $P_0' = Rp' = p'R$. Given $r$ in $R$ define $c_n = p' + p(r + p^n) \in P$. By assumption each $c_n$ is in a suitable height 1 prime $P_n \subset P$. But $p \notin P_n$ since, otherwise, $p' \in P_n$, contrary to $P_0 \neq P_0'$. Hence $c_n - c_m \notin P_n$ implying $P_n \neq P_m$ for each $n \neq m$. Thus $\bigcap P_n = 0$, by exercise 3.2.1. This is contradicted by the following choice of $r$: (i) If $p, p' \in Z(R)$ take any $r, r'$ with $rr' \neq r'r$; then $[r', c_n] = p[r', r] \in P_n$ so $[r', r] \in P_n$ for each $n$. (ii) If $pp' = p'p$ but $p \notin Z(R)$ then take $r$ with $pr \neq rp$ and write $rp = pr'$ for suitable $r'$; then $p^2(r - r') = [c_n, p] \in P_n$ so $r - r' \in P_n$. (iii) If $pp' \neq p'p$ then $[p, p'] = [p, c_n] \in P_n$.)

**22.** (Making the UFD nicer) Suppose $R$ is a left Noetherian UFD which is not commutative. Let $S = \{$elements not in any height 1 prime ideals of $R\}$. Then $S$ is a left and right denominator set of $R$. $S^{-1}R$ is a left Noetherian UFD, and every nonunit of $S^{-1}R$ belongs to a height 1 prime. Finally every one-sided ideal of $S^{-1}R$ is principal and spanned by a normalizing element, so is a two-sided ideal. (Hint: Suppose $r \in R$ and $s \in S$. Then $ar = bs$ for suitable $0 \neq a, b \in R$. Write $a = p_1 \cdots p_n c$ for suitable $c$ in $S$, and $p_i$ prime, then $bs = p_1 \cdots p_n cr \in p_1 R$ so $b \in p_1 R$; iterating shows $b = p_1 \cdots p_n b'$ so $cr = b's$, proving $S$ is a left denominator set; thus $S^{-1}R$ is left Noetherian and is a UFD by means of exercise 20. Suppose $s^{-1}r$ is a nonunit of $S^{-1}R$. Then $r \notin S$ so $r$ is in a height 1 prime ideal of $R$, implying $s^{-1}r$ is in a height 1 prime ideal of $S^{-1}R$. In fact if $r = cp_1 \cdots p_n$ then $s^{-1}r \in (S^{-1}R)p_1 \cdots p_n \lhd R$ and $p_1 \cdots p_n$ is normalizing, so it remains only to show each left ideal $L$ of $S^{-1}R$ is principal; one may assume $L = S^{-1}Rr_1S^{-1}Rr_2$. It was just shown $L \lhd R$ so is contained in a maximal ideal $P$. $P \cap R \in \operatorname{Spec}(R)$ so has height 1 by exercise 21; hence $P$ has height 1 and may be assumed to contain $p_1$. Conclude by pulling out $p_1$ and iterating the procedure.)

23. If $R$ is a bounded left Noetherian UFD which is not commutative, and such that every prime ideal of $R$ contains a height 1 prime, then $R$ is a PLID. (Hint: Suppose $a \in R$ is a nonunit. Then $Ra$ contains a two-sided ideal $A \neq 0$. There are prime ideals $P_1, \ldots, P_t$ of $R$ with $P_1 \cdots P_t = A$; taking $p_i$ prime elements in $P_i$ yields $p_1 \cdots p_t \in A \subseteq Ra$. Thus $p_1 \cdots p_t = ra$. Hence $r = p_1 \cdots p_i c$ for some $i < t$ and $c$ not in any height 1 prime ideal, implying $a \in Rp_{i+1}$. In the notation of exercise 22, this proves $R = S^{-1}R$ is a PLID.

24. If $R$ is a PI-ring which is a noncommutative UFD then $R$ is a PLID. (Hint: Verify the conditions of exercise 23.)

**Lie Superalgebras** In exercises 25, 26 let $L = L_0 \oplus L_1$ be a finite dimensional Lie superalgebra over $\mathbb{C}$.

25. $U(L_0) - \{0\}$ is a left and right denominator set for $U(L)$; localizing yields an Artinian ring.

26. Viewing $U(L) \subseteq M_n(U(L_0))$ show that the intersection of the ideals of finite codimension is 0, and $U(L)$ has an ideal of finite codimension which misses $L$. (Hint: The first assertion is a consequence of Harish-Chandra's theorem, cf., Jacobson [62B]. To obtain the second assertion, now take an ideal $A$ of finite codimension for which $A \cap L$ has minimal dimension over $\mathbb{C}$. Then $A \cap L = 0$.)

### Hopf Algebras (See SW)

27. Any Hopf algebra has IBN.
28. If $A, B$ are coalgebras then so is $A \otimes B$. (Hint: Take $\Delta$ to be the composite $A \otimes B \to (A \otimes A) \otimes (B \otimes B) \to (A \otimes B) \otimes (A \otimes B)$.)
29. If $C = (C, \Delta, \varepsilon)$ is a coalgebra and $A = (A, \mu, u)$ is an algebra then $\operatorname{Hom}(C, A)$ is an algebra under the *convolution product*, where the convolution product $h$ of $f$ and $g$ is given by $hc = \mu \sum fc_{(1)} \otimes gc_{(2)}$, i.e., $h = \mu(f \otimes g)\Delta$. The unit element of this algebra is $u\varepsilon$. Taking $C = A$ in case $A$ is a bialgebra, show an antipode $S$ is an inverse of $1_A$ under the convolution, and thus unique (if it exists).
30. The antipode $S$ of a Hopf algebra $A$ is an anti-homomorphism and is an involution if $A$ is cocommutative. (Hint: define maps $\eta, v: A \otimes A \to A$ by $\eta(a \otimes b) = S(ab)$ and $v(a_1 \otimes a_2) = SbSa$. Taking $A \otimes A$ as a coalgebra as in exercise 28 and letting $\cdot$ denote the convolution product in $\operatorname{Hom}(A \otimes A, A)$ show $\eta \cdot \mu = \mu \cdot v$; in fact, each evaluation on $a \otimes b$ is $\varepsilon a \varepsilon b$ for all $a, b$ in $A$. This implies $S(ab) = SbSa$. Next note $S1 = 1$, implying $Su = u$. Finally, to prove $S^2 = 1_A$ it is enough to show the convolution product in $\operatorname{Hom}(A, A)$ of $S^2$ and $S$ is $u\varepsilon$. But this is $\sum S^2 a_{(1)} Sa_{(2)} = S(\sum a_{(2)} Sa_{(1)}) = Su\varepsilon a = u\varepsilon a$; note cocommutativity was used.)

**31.** An element $a$ of a Hopf algebra $A$ is *group-like* if $\Delta a = a \otimes a$; $a$ is *primitive* if $\Delta a = a \otimes 1 + 1 \otimes a$. Show $G = \{\text{group-like elements of } A\}$ is a subgroup of Unit($A$), where $a^{-1} = Sa$, and $F[G]$ is canonically isomorphic to the $F$-subspace of $A$ spanned by $G$. Also $L = \{\text{primitive elements of } A\}$ is a Lie algebra. (Hint: If $g \in G$ then $\varepsilon g = 1$. It remains to show that $\{g \in G\}$ are linearly indpendent over $F$. Otherwise, take a dependence relation $\sum_{i=1}^{n} \alpha_i g_i = 0$ with $n$ minimal such that $\alpha_n \neq 0$. Assuming $n \neq 1$ get $\sum_{i=1}^{n-1} \alpha_i g_i \otimes g_i = \sum \alpha_i \Delta g_i = -\Delta g = -g \otimes g = -\sum_{i,j=1}^{n-1} \alpha_i \alpha_j g_i \otimes g_j$, implying $n - 1 = 1$, i.e., $g_2 = -\alpha_1 g_1$. Then $-\alpha_1 = \varepsilon(g_2) = 1$, contradiction.)

**32.** (Hopf version of Maschke's theorem) A finite-dimensional Hopf algebra $A$ is semisimple Artinian iff $\varepsilon \int \neq 0$. (Hint: ($\Rightarrow$) Write $A = L \oplus \ker \varepsilon$. Then $(\ker \varepsilon)L = 0$ so $L \subseteq \int$, implying $L = \int$. ($\Leftarrow$) Pick an integral $a$ of $A$ with $\varepsilon a = 1$. Given any left ideal $L$ of $A$ take a projection $\pi : A \to L$ as $F$-vector spaces: then there is an $A$-module projection $\pi' : A \to L$ given by $\pi' ab = \sum a_{(1)} \pi((Sa_{(2)})b)$. This proof uses an important averaging process arising from the integral; however, for $R$ cocommutative and $F = \mathbb{C}$ there is an easier proof.)

**33.** Any finite dimensional Hopf algebra is a symmetric algebra and thus Frobenius. (Hint: as in exercise 8.1.19.)

**34.** (The fundamental theorem of coalgebras, generalized) Any f.d. subspace $V$ of a coalgebra is contained in an f.d. subcoalgebra of $A$. (Hint, following notes of Kaplansky: Given any subspace $W$ of $A$ let $LW$ (resp. $RW$) denote the subspace of $A$ generated by all $a_{(1)}$ (resp. by all $a_{(2)}$) for each $a$ in $W$, writing $\Delta a = \sum a_{(1)} \otimes a_{(2)}$. Note $a = \sum (\varepsilon a_{(2)}) a_{(1)} \in LW$, so $W \subseteq LW$ and likewise $W \subseteq RW$. Applying coassociativity and matching components shows $LLV = LV$, and $LRV = RLV$. Hence $LRV$ is the desired subcoalgebra.)

# Dimensions for Modules and Rings

(Parenthetical numbers refer to results in text)

## Module Dimensions

*Note:* There are two important possible properties of a module dimension "dim":

*Additive:* $\dim M = \dim N + \dim M/N$ for $N \leq M$;
*Exact:* $\dim M = \max(\dim N, \dim M/N)$ for $N \leq M$

| Name | Abbr. | Kind of module with minimal dimension | Properties |
|------|-------|---------------------------------------|------------|
| composition length | $\ell(M)$ | simple | additive (2.3.4) |
| uniform = Goldie dim. | $u$-dim $M$ | uniform | additive (3.2.19) |
| reduced rank | $\rho(M)$ | torsion | additive (3.5.6) |
| Krull dimension | $K$-dim $M$ | Artinian | exact (3.5.41) |
| basic dimension | $B$-dim $M$ | unfaithful (for $R$ prime) | exact (3.5.59) |
| Gabriel dimension | $G$-dim $M$ | completely reducible | exact (3.5.96f) |
| projective dimension | pd $M$ | projective | see summary 5.1.13 |
| injective dimension | id $M$ | injective | |
| Gelfand-Kirillov dim. | GK dim $M$ | | exact over Noetherian PI-ring (6.3.46) |

# Ring Dimensions

| Name | Abbr. | Ring of minimal dim. | Remarks |
|---|---|---|---|
| Gelfand-Kirillov dim. | GK dim | finite dim. algebra | integer valued for certain classes of rings |
| Krull dimension | $K$-dim | Artinian | $\leq$ gl. dim for Noetherian PI-rings (6.3.46) |
| classical Krull | cl. $K$-dim | every prime is maximal | $\leq k$ dim |
| little Krull dimension | $k$ dim | every prime is maximal | $\leq$ K-dim |
| Gabriel dimension | G-dim | direct prod. of Artinian | $\leq K$-dim $+ 1$ |
| global dimension | gl. dim. | semisimple Artinian | $=\sup\{\text{pd } M\}$ $=\sup\{\text{id } M\}$ |
| weak dimension | w. dim. | von Neumann regular | $\leq$ gl. dim. |
| PI degree | PI deg | commutative | $=$ degree, for f.d. division algebra |
| Goldie rank | | Ore domain | $=$ length, for simple Artinian ring |

*Sketch of theory for an affine PI-algebra over a field F* (cf., 6.3.40 ff). GK dim $R$ = cl. K-dim $R$ = tr. deg $Z(R)/F$, and equals K-dim $R$ if the latter exists!; although K-dim $R$ may fail to exist, G-dim $R$ always exists. Also cl. K-dim $R$ = cl. K-dim $R/P$ + height $P$ for any prime ideal $P$ of $R$.

# Major Theorems and Counterexamples for Volume II

E before number denotes "exercise"; otherwise the result is from the main text.

## Theorems

The vast majority of these results are proved in the text. In this list, precision may be sacrificed for conciseness; please check the text for the full hypotheses.

### §5.1

5.1.13.  The relationships of pd among modules of an exact sequence.

5.1.25, E5.1.4.  (See also E5.1.8, E5.1.16) $\operatorname{gl.dim} R[\lambda; \sigma] = \operatorname{gl.dim} R + 1$.

5.1.34, 5.1.41.  (Serre-Quillen, see also E5.1.13 and the appendix to Ch. 5). For $R$ homologically regular and filtered, if $K_0(R_0)$ is trivial then so is $K_0(R)$. (In fact $K_0(R) \approx K_0(R_0)$ in general.)

5.1.45  (Walker). A prime, homologically regular ring with trivial $K_0$ is a domain.

5.1.53  (Quillen-Suslin). f.g. projectives over $F[\lambda_1, \ldots, \lambda_n]$ are free.

5.1.59, 5.1.60, 5.1.61  (Stafford-Coutinho, generalizing results of Bass and Swan). The main tool for proving a f.g. R-module has a summand isomorphic to $R$, and for proving cancellation.

5.1.67  (Bass, see also E5.1.28). "Uniformly big" projectives are free.

### §5.2

5.2.10, 5.2.11.  Existence of the "long exact" homology and cohomology sequence.

5.2.22, 5.2.25.   Existence of the "long exact" sequence for the derived functors of a left or right exact functor. (This is applied later to Tor
and Ext.)

5.2.34, 5.2.37.   The definitions of $\text{Ext}(M, N)$ and $\text{Tor}(M, N)$ are well-defined.

5.2.40.     The following are equivalent for a module $M$:

(i) pd $M \leq n$
(ii) $\text{Ext}^k(M, \underline{\phantom{x}}) = 0$ for all $k > n$
(iii) $\text{Ext}^{n+1}(M, \underline{\phantom{x}}) = 0$
(iv) The $(n - 1)$ syzygy of any projective resolution of M is projective.

5.2.40'.    The following are equivalent for a module $N$:

(i) id $N \leq n$
(ii) $\text{Ext}^k(\underline{\phantom{x}}, N) = 0$ for all $k > n$
(iii) $\text{Ext}^{n+1}(\underline{\phantom{x}}, N) = 0$
(iv) The $(n - 1)$ cosyzygy of every injective resolution is injective.

5.2.42      (Auslander). gl. dim $R = \sup\{\text{pd } R/L: L < R\}$.

5.2.48.     Any pair of right exact functors that commutes with Hom also
commutes with $\text{Ext}^n$ for all $n \in \mathbb{N}$.

5.2.49.     Any pair of left exact functors that commutes with $\otimes$ also commutes with $\text{Tor}_n$ for all $n \in \mathbb{N}$.

E5.2.21.    Structure of homomorphic images of hereditary semiprimary
rings.

# §5.3

5.3.7.      Equivalent conditions for an algebra to be separable.

5.3.18.     The "modern" definition of "separable" coincides with the classical
definition for algebras over a field.

5.3.20.     Cohomological setting of Wedderburn's principal theorem. ($R =
S \oplus N$ where $S$ is semisimple Artinian and $N = \text{Jac}(R)$.)

5.3.21.     Uniqueness of $S$ in 5.3.20.

5.3.24      (See also E5.3.9, E5.3.11). Equivalent conditions for an algebra to be
Azumaya.

# §6.1

6.1.20      (See also 6.1.23, E6.1.22). A multilinear central polynomial for
matrices.

6.1.25    (Kaplansky's Theorem, see also E6.1.17′). Primitive PI-rings are central simple.

6.1.26.   $R/N(R)$ is embeddible in $n \times n$ matrices for $R$ PI. (cf., definition 2.6.25.)

6.1.27, 6.1.28.   If $R$ is semiprime PI then $\mathrm{Nil}(R) = 0$ and every ideal of $R$ intersects the center nontrivially.

6.1.30.   The ring of central fractions of a prime PI-ring is central simple.

6.1.33.   Suppose $R$ has PI-degree $n$. If an $n^2$-normal central polynomial takes on the value 1 then $R$ is a free $Z(R)$-module.

6.1.35, 6.1.35′   ("Artin-Procesi"). Various PI-conditions equivalent to being Azumaya of unique rank.

6.1.38.   $\mathrm{Spec}_n(R)$ is intimately related to the center.

6.1.44.   $R$ is PI-equivalent to $R \otimes_C H$ for $H$ commutative.

6.1.49    (see also E6.1.24′, E6.1.25). Jacobson's conjecture is true for Noetherian PI-rings.

6.1.50.   In a left Noetherian PI-ring every prime ideal has finite height.

6.1.53.   Technical criterion used to prove a ring is PI.

6.1.59, 6.1.61.   (*)-version of 6.1.27 and 6.1.28.

E6.1.10   (Amitsur-Levitzki). The standard polynomial $S_{2n}$ is an identity of $n \times n$ matrices.

E6.1.31   (Amitsur). A primitive ring is strongly primitive iff it satisfies a proper generalized identity.

# §6.2

6.2.5    (Artin-Tate, see also 6.3.1). If $R$ is affine and f.g. over $Z$ then $Z$ is affine.

6.2.17   (Irving-Small, see also E6.2.8). Any affine, semiprime $F$-algebra of subexponential growth and satisfying ACC on left annihilators, is Goldie.

6.2.25.   GK dim$(R) \geq$ GK dim$(R/P)$ + height$(P)$ for any $P$ in Spec$(R)$, if prime images of $R$ are Goldie.

6.2.27, 6.2.33   (Bergman). If GK dim$(R) \leq 2$ then GK dim$(R)$ is an integer.

6.2.38.   GK module dimension is exact over any filtered affine F-algebra whose associated graded algebra is affine and left Noetherian.

# §6.3

6.3.3     (Amitsur-Procesi). Any affine PI-algebra is Jacobson and satisfies the Nullstellensatzes.

6.3.11.   Embedding a ring into a generalized triangular matrix ring, by means of "universal derivations."

6.3.14   (Lewin). If $F\{X\}/A$ and $F\{X\}/B$ are embeddible, respectively, in $m \times m$ and $n \times n$ matrices then $F\{X\}/AB$ is embeddible in $(m + n) \times (m + n)$ matrices.

6.3.16   (Lewin). If $R$ is PI and $\mathrm{Nil}(R)$ is nilpotent then $R$ satisfies the identities of $k \times k$ matrices for some $k$.

6.3.23.   (Levitzki-Kaplansky-Shirshov, see also E6.3.12, E6.3.13). Suppose $R = C\{r_1, \ldots, r_t\}$ satisfies a multilinear polynomial identity, and each monomial in the $r_i$ of length $< m$ is integral over $C$. Then $R$ is f.g. as C-module, spanned by the monomials in the $x_i$ of a certain bounded length.

6.3.24   (Shirshov). Technical but useful related result, which holds for arbitrary affine PI-algebras.

6.3.25.   Any affine PI-algebra has finite Gelfand-Kirillov dimension.

6.3.30.   If a PI-ring $R$ is integral over $C$ then GU, LO and INC hold from $C$ to $R$.

6.3.31   (See also E6.3.14). Versions of LO, GU, and INC holding from a prime PI-ring to its trace ring.

6.3.34   (Schelter, see also 6.3.36'). Any affine PI-algebra over a Noetherian ring satisfies ACC (prime ideals).

6.3.35, 6.3.36, 6.3.39   (Razmyslov-Braun). The Jacobson radical of an affine PI-algebra is nilpotent.

6.3.41   (See also 6.3.42, 6.3.44). If $R$ is prime affine PI over a field $F$ then GK dim $R$ = cl. K-dim $R$ = tr deg $Z(R)/F$.

6.3.43   (Schelter). Catenarity of affine PI-algebras.

E6.3.17.   Reduction of Bergman-Small theorem (describing the PI-degree in terms of its prime images) to trace ring.

E6.3.18.   "Going down" from a prime PI-ring to its center.

E6.3.19.   Affine PI-algebras have Gabriel dimension.

## §6.4

6.4.5.   $A$ is a T-ideal of $C\{X\}$ iff $C\{X\}/A$ is relatively free.

6.4.7.   (*)-version of 6.4.5.

6.4.8.   (*)-version of 6.1.44.

6.4.19   (Regev, see also 6.4.18, E6.4.4-E6.4.7). The tensor product of PI-algebras is PI.

E6.4.2   (Nagata-Higman). In characteristic 0 nil of bounded degree implies nilpotent.

*Note:*   We write "csa" below for "central simple algebra."

## §7.1

7.1.9     (Double centralizer theorem). $C_R(C_R(A)) = A$ for any simple subalgebra $A$ of the csa $R$.

7.1.10    (Skolem-Noether, see also 7.1.10′, E7.1.5). Any isomorphism of simple F-subalgebras of $R$ can be extended to an inner automorphism of $R$.

7.1.11    (Wedderburn's theorem). Every finite division ring is commutative.

7.1.12.   {separable elements of maximal degree} is dense in the Zariski topology.

7.1.12′   (Koethe). Any separable subfield of a central division algebra is contained in a separable maximal subfield.

7.1.18    (Wedderburn's criterion, see also E7.1.12). The symbol $(\alpha, \beta)_n$ is split iff $\alpha$ is a norm from $F(\beta^{1/n})$ to $F$.

7.1.24.   $UD(F, n)$ is a division algebra of degree $n$.

7.1.37    (See also E7.1.10). The correspondence between central simple algebras and Brauer factor sets.

7.1.40    (See also E7.1.11). Central simple algebras are isomorphic iff their Brauer factor sets are associates.

7.1.44.   (Wedderburn). Every division algebra of degree 2 or 3 is cyclic.

7.1.45.   (Albert). Every csa of index 4 is split by the Galois group $(\mathbb{Z}/2\mathbb{Z})^{(2)}$.

E7.1.18   (Wedderburn's factorization theorem). The minimal polynomial of $d$ over $Z(D)$ splits into linear factors over $D$.

E7.1.23   (Cartan-Brauer-Hua). Any proper division subring, invariant with respect to all inner automorphisms, is central.

## §7.2

7.2.3.    The effect on the index of tensoring by a field extension.

7.2.6.    Identification of the Brauer group with the second cohomology group.

7.2.10.   $\exp(R)$ divides $\mathrm{index}(R)$.

7.2.12.   If $n, m$ are the respective index and exponent then

(i) $m \mid n$ and
(ii) every prime divisor of $n$ also divides $m$.

7.2.13.   Any csa decomposes as a tensor product of algebras of prime power index.

7.2.20.    The explicit algebraic description of $\text{cor}_{L/F} R$ is well defined.

7.2.26.    $\text{cor} \circ \text{res} = [L:F]$.

7.2.32    ("Projection formula"). $\text{cor}_{L/F}(\alpha, \beta; L) \sim (\alpha, N_{L/F}\beta; F)$.

7.2.37    (Rosset-Tate). $\text{cor}_{L/F}(a, b)$ is a tensor product of $\leq [L:F]$ symbols.

7.2.43    (mostly Rosset). Any csa of prime degree $p$ is similar to a tensor product of $(p-1)!/2$ cyclic algebras.

7.2.44.    A csa $R$ has exponent 2 iff $R$ has an involution of the first kind.

7.2.48    (Albert-Riehm-Scharlau). Description of csa with anti-automorphism, in terms of the corestriction.

7.2.49'.    Suppose $[L:F] = 2$. A csa $R$ (over $L$) has an involution (*) of second kind over $F$ iff $\text{cor}_{L/F} R \sim 1$.

7.2.50.    A csa $R$ containing a quadratic field extension $K$ of $F$ has exponent 2, iff $C_R(K)$ has an involution of second kind over $F$.

7.2.57    (Albert). Every division algebra of degree 4 and exponent 2 is a tensor product of quaternion subalgebras.

7.2.60    (Tignol). If $R$ has degree 8 and exponent 2 then $M_2(R)$ is a tensor product of quaternions.

7.2.61.    If $\sqrt{-1} \in F$ then any csa of index 4 is similar to a tensor product of a cyclic of degree 4 and four quaternion algebras.

7.2.62    (Merkurjev-Suslin, see also Appendix B, E7.2.18, and E7.2.33.). Every csa is similar to a tensor product of cyclic algebras. (Proved modulo "Hilbert's theorem in $K_2$".)

E7.2.32    (Rowen, see also E7.2.43). Every csa of index 8 and exponent 2 is split by $(\mathbb{Z}/2\mathbb{Z})^{(3)}$.

## Appendices to Chapter 7

B2 (The real Merkurjev-Suslin Theorem). The $m$-torsion part of the Brauer group is isomorphic to a $K_2$-group.

C3. Properties of generics.

## §8.1

8.1.8'    (Maschke's theorem). $F[G]$ is semiprime if $\text{char}(F) = 0$.

8.1.14    (Maschke's theorem for arbitrary fields, $G$ finite, see also E8.4.32). $F[G]$ is semisimple Artinian iff $|G| \neq 0$ in $F$.

8.1.26.    $F[G]$ has no nonzero left ideals if $G$ is a $\text{char}(F)'$-group.

8.1.32.    If $Z(G)$ is of finite index in $G$ then $G'$ is finite.

8.1.39    (Connell, see also E8.1.14). Characterization of prime group rings.

8.1.41   (Passman, see also E8.1.25). Criteria for $K[G]$ to be primitive if $K \supseteq F$ and $F[G]$ is primitive.

8.1.45   (Amitsur, Herstein). $\mathrm{Jac}(F[G])$ is nil for $F$ uncountable.

8.1.47.   $F[G]$ is semiprimitive when $F$ is a $\mathbb{Q}$-algebra not algebraic over $\mathbb{Q}$.

8.1.52   (See also 8.1.56). If $C[G]$ is semiprime PI then $\Delta(G)$ is of finite index in $G$. (This is part of the Isaacs-Passman Theorem in char 0, that $G$ has an Abelian subgroup of finite index).

## §8.2

8.2.2.   If $G$ is polycyclic-by-finite then $F[G]$ is Noetherian.

8.2.16   (Milnor-Wolf-Gromov). A finitely generated group has polynomial growth iff it is nilpotent-by-finite (not proved in full here).

8.2.20.   If $G$ is polycyclic-by-finite and is a char$(F)'$-group then gl. dim. $F[G] < \infty$.

8.2.21   (Serre). If $G$ is a char$(F)'$-group with a subgroup $H$ of finite index such that gl. dim. $F[H] < \infty$, then gl. dim. $F[G] < \infty$.

8.2.27.   If $G$ is torsion-free and $F[G]$ is Noetherian then $\mathrm{Tr}_G P$ is a well defined integer for each f.g. projective $P$.

8.2.35.   (Farkas-Snider, see also appendix). If $G$ is torsion-free polycyclic-by-finite and char$(F) = 0$ then $F[G]$ is a domain.

## §8.3

8.3.3   (Poincare-Birkhoff-Witt $=$ PBW theorem, see also 8.3.51). $u: L \to U(L)$ is monic.

8.3.14, 8.3.15.   $U(L)$ is a Noetherian domain, and its subrings are semi-primitive.

8.3.23   (Quillen). $U(L)$ is Jacobson.

8.3.27   (McConnell). If $L$ is a nilpotent Lie algebra then every ideal of $U(L)$ is polycentral; hence any $A$-adic completion of $U(L)$ is Noetherian.

8.3.36.   If $L$ is solvable then every prime ideal of $U(L)$ is completely prime.

8.3.39.   If $L$ is solvable then $[L:F] = \mathrm{height}(P) + \mathrm{GK\ dim}\ U(L)/P$ for any prime ideal $P$ of $U(L)$. (In particular, GK dim $U(L) = [L:F]$.)

## §8.4

8.4.4.   Any almost normalizing extension of a left Noetherian ring is left Noetherian.

8.4.11 (McConnell). Conditions for vector generic flatness to pass up from a subring.

8.4.12, 8.4.15 (See also E8.4.3–E8.4.10, E8.4.13, E8.4.14). Theorems concluding that a ring is Jacobson and/or satisfies the Nullstellensatz.

8.4.18, 8.4.20, 8.4.22. Various ring-theoretic implications of Dixmier's conditions.

8.4.31 (Irving-Small). Reduction of the proof of the equivalence of Dixmier's conditions to the algebraically closed case.

8.4.32ff (Warfield). The Goldie rank of $R$ in terms of its prime images.

8.4.38. If $R$ is the enveloping algebra of a finite dimensional Lie algebra or the group algebra of a poly-{infinite cyclic} group then every f.g. projective R-module is stably free.

E8.4.34 (Fundamental theorem of coalgebras). Any element of a coalgebra is contained in a finite dimensional subcoalgebra. Consequently, any simple coalgebra $A$ is finite dimensional.

## Appendix to Chapter 8

**Moody's theorem:** (Verification of the Goldie rank conjecture): *For G polycyclic-by-finite, the Goldie rank of the ring of fractions of $F[G]$ is the least common multiple of the orders of the finite subgroups of G. (Highlights of the proof are given.)*

## Counterexamples

### Chapter 5

E5.1.1. Schanuel's lemma fails for flat modules.

E5.1.9. See also 8.4.43. A f.g. projective $D[\lambda_1, \lambda_2]$ that is not free.

### Chapter 6

E6.1.12. A PI-ring failing $\text{Nil}(R)^{n-1} \subseteq \text{Nil}(R)$.

E6.1.26. A PI-ring with an idempotent maximal ideal.

E6.1.27. A Noetherian PI-ring whose center fails the "principal ideal theorem."

6.2.4. An affine algebra failing the Nullstellensatz.

6.2.9 (Golod). A nil algebra $R_0$ which is not locally nilpotent, but satisfying $\bigcap_{n \in \mathbb{N}} R_0^n = 0$. (Also $M_2(R_0)$ is nil, see 6.2.9'.)

E6.2.4. An affine domain that is left and right Ore, and left but not right Noetherian.

## Chapter 7

## Chapter 8

E8.1.30.   A group $G$ lacking any Abelian group of finite index, but such that $F[G]$ is PI (in characteristic $p$, of course).

E8.2.18   (Farkas). A group ring whose idempotents display a surprising behavior.

8.3.35.   If $L$ is not solvable then $U(L)$ has a primitive ideal that is not completely prime.

8.3.41.   A solvable non-nilpotent Lie algebra.

8.3.49.   A Noetherian ring not embeddible in an Artinian ring; also a Noetherian ring that is not weakly ideal invariant, and with a peculiar simple module.

E8.3.3.   The canonical bijection $S(L) \to U(L)$ is not an algebra homomorphism.

E8.3.13, E8.3.14   (M. Smith). An enveloping algebra with subexponential but not polynomially bounded growth.

8.4.40.   Conditions for $T = R[\lambda; \sigma, \delta]$ to contain a non-free left ideal $L$ satisfying $L \oplus T \approx T^{(2)}$.

8.4.41.   As in 8.4.40, for a skew Laurent extension.

8.4.48.   If $G$ is nonabelian and poly-{infinite cyclic} then the group algebra $F[G]$ has a stably free, non-free left ideal.

E8.4.11   (Irving). A primitive, Noetherian Jacobson ring that fails the Nullstellensatz.

E8.4.18   (Stafford). If $L$ is nonabelian then $U(L)$ has a stably free left ideal that is not free.

# References

## Bibliography of Books

This list contains most of the advanced ring theory books, plus books from other subjects which bear heavily on the material.

ABE, E.
 [77] *Hopf Algebras.* Cambridge University Press, Cambridge
ALBERT, A. A.
 [61] *Structure of Algebras.* AMS Colloq. Pub. 24. American Mathematical Society, Providence
ANDERSON, F., and FULLER, K.
 [74] *Rings and Categories of Modules.* Springer-Verlag, Berlin
ARTIN, E., NESBIT, C. J., and THRALL, R.
 [44] *Rings with Minimum Condition.* University of Michigan Press, Ann Arbor
BARBILIAN, D.
 [46] *Teoria Aritmetica a Idealilor* (in inele necomutative). Ed. Acad. Rep. Pop. Romine, Bucaresti
BARWISE, J. (Ed.)
 [78] *Handbook of Logic.* North-Holland, Amsterdam
BASS, H.
 [68] *Algebraic K-Theory.* Benjamin, New York
BJORK, J.
 [79] *Rings of Differential Operators.* North-Holland, Amsterdam
BOREL, A.
 [69] *Linear Algebraic Groups.* Benjamin, New York
BORHO, W., GABRIEL, P., and RENTSCHLER, R.
 [73] *Primideale in einhüllenden auflösbarer Lie-Algebren.* Lecture Notes in Mathematics 357, Springer-Verlag, Berlin
BOURBAKI, N.
 [72] *Commutative Algebra* (transl. from French). Elements de Mathematique. Hermann, Paris
CARTAN, H., and EILENBERG, S.
 [56] *Homological Algebra.* Princeton University Press, Princeton
CHATTERS, A. W., and HAJARNAVIS, C.
 [80] *Rings with Chain Conditions.* Pitman, London
CHEVALLEY, C.
 [68] *Théorie des groupes de Lie.* Hermann, Paris

COHN, P. M.
[74] *Algebra I.* Wiley, London
[77] *Skew Field Construction,* Cambridge University Press, Cambridge
[81] *Universal Algebra* (second edition). Reidel, Dordrecht, Holland
[85] *Free Rings and Their Relations* (second edition), Academic Press, New York
[88] *Algebra II, III* (second edition, in press). Wiley, London
COZZENS, C., and FAITH, C.
[75] *Simple Noetherian Rings.* Cambridge University Press, Cambridge University Press, Cambridge
CURTIS, C., and REINER, I.
[62] *Representation Theory of Finite Groups and Associative Algebras,* Interscience, New York
[81] *Methods of Representation Theory,* Vol. 1. Wiley, New York (Vol. 2 in press)
DEMEYER, F., and INGRAHAM, E.
[71] *Separable Algebras over Commutative Rings.* Lecture Notes in Mathematics 181. Springer-Verlag, Berlin
DEURING, M.
[66] *Algebren,* Springer-Verlag, Berlin
DICKSON, L. E.
[23] *Algebras and Their Arithmetics,* University of Chicago, Chicago
DIVINSKY, N. J.
[65] *Rings and Radicals.* Toronto University, Toronto
DIXMIER, J.
[77] *Enveloping Algebras.* North Holland, Amsterdam (Transl. of *Algebres Enveloppantes,* Gauthier-Villars, Paris)
DRAXL, P. K.
[83] *Skew Fields.* London Mathematical Society Lecture Notes 81. Cambridge University Press, Cambridge
FAITH, C.
[67] *Lectures on Injective Modules and Quotient Rings.* Lecture Notes in Mathematics 49, Springer-Verlag, Berlin
[73] *Algebra: Rings, Modules, and Categories.* Springer-Verlag, Berlin
[76] *Algebra II: Ring Theory.* Springer-Verlag, Berlin
[82] *Injective Modules and Injective Quotient Rings.* Marcel Dekker, New York
FAITH, C., and PAGE, S.
[84] *FPF Ring Theory.* London Mathematical Society Lecture Notes 88. Cambridge University Press, Cambridge
FREYD, P.
[64] *Abelian Categories: An Introduction to the Theory of Functors.* Academic Press, New York
GILLMAN, L., and JERISON, M.
[60] *Rings of Continuous Functions.* Academic Press, New York
GOLAN, J.
[75] *Localization of Noncommutative Rings.* Marcel Dekker, New York
[86] *Torsion Theories.* Longman Scientific & Technical, Harlow, England
GOODEARL, K.
[76] *Ring Theory: Nonsingular Rings and Modules.* Marcel Dekker, New York
[79] *von Neumann Regular Rings.* Pitman, New York
HALL, M.
[59] *The Theory of Groups.* MacMillan, New York
HARTSHORNE, R.
[75] *Algebraic Geometry.* Springer-Verlag, New York

HERSTEIN, I. N.
[64] *Topics in Algebra* (second edition). Xerox, Lexington, Massachussetts
[68] *Noncommutative Rings*. Carus Mathematical Monographs 15. Mathematical Association
of America, Wiley, New York
[69] *Topics in Ring Theory*. University of Chicago Press, Chicago
[76] *Rings with Involutions*. University of Chicago Press, Chicago
HOCHSCHILD, G.
[71] *Introduction to Affine Algebraic Groups*. Holden-Day, San Francisco
ISAACS, M.
[76] *Character Theory of Finite Groups*. Academic Press, New York
JACOBSON, N.
[43] *Theory of Rings*. AMS Surveys 1. American Mathematical Society, Providence
[62] *Lie Algebras*. Wiley, New York
[64] *Structure of Rings* (second edition). AMS Colloq. Pub. 37. American Mathematical
Society, Providence
[75] *PI-Algebras: An Introduction*. Lecture Notes in Mathematics 441. Springer-Verlag, Berlin
[80] *Basic Algebra II*. Freeman, San Francisco
[85] *Basic Algebra I* (second edition). Freeman, San Francisco
JACOBSON, N., and SALTMAN, D.
[88] *Division algebras* (forthcoming)
JANTZEN, J. C.
[83] *Einhüllende Algebren halbeinfacher Lie-Algebren*. Springer Verlag, Berlin
JATEGAONKAR, A. V.
[70] *Left Principal Ideal Rings*. Lecture Notes in Mathematics 123. Springer-Verlag, Berlin
[86] *Localization in Noetherian Rings*. London Mathematical Society Lecture Notes 98.
Cambridge University Press, Cambridge
KAPLANSKY, I.
[68] *Rings of Operators*. Benjamin, New York
[70] *Commutative Rings*. Allyn and Bacon, Boston
[72] *Fields and Rings*. University of Chicago, Chicago
KASCH, F.
[82] *Modules and Rings* (transl. from German). London Mathematical Society Monographs
17. Academic Press, New York
KELLEY, J.
[55] *General Topology*. van Nostrand, New York
KNUS, M.-A., and OJANGUREN, M.
[74] *Théorie de la descente et algèbres d'Azumaya*. Lecture Notes in Mathematics 389,
Springer-Verlag, Berlin
KRAUSE, G. R., and LENAGAN, T. H.
[85] *Growth of Algebras and Gelfand-Kirillov Dimension*. Research Notes in Mathematics 116.
Putnam, London
KRUSE, R. L., and PRICE, D.
[69] *Nilpotent Rings*. Gordon & Breach, New York
LAM, T. Y.
[78] *Serre's Conjecture*. Lecture Notes in Mathematics 635. Springer-Verlag, Berlin
LAMBEK, J.
[66] *Lectures on Rings and Modules*. Blaisdell, Waltham, Massachussetts
LANG, S.
[84] *Algebra* (second edition). Addison-Wesley, Reading, Massachussetts
MACLANE, S.
[75] *Homology*, Springer-Verlag, Berlin

MATIS, E.
   [72] *Torsion-Free Modules*. University of Chicago Press, Chicago
MCCONNELL, J. and ROBSON, J. C.
   [88] *Noetherian Rings*. Wiley, London
MCDONALD, B.
   [74] *Finite Rings with Identity*. Marcel Dekker, New York
MILNOR, J.
   [71] *Introduction to Algebraic K-Theory*. Annals of Mathematical Studies. Princeton University Press, Princeton
MITCHELL, B.
   [65] *Theory of Categories*. Pure and Applied Mathematics 17. Academic Press, New York.
MONTGOMERY, S.
   [80] *Fixed Rings of Finite Automorphism Groups of Associative Rings*. Lecture Notes in Mathematics 818, Springer-Verlag, Berlin
NATASESCU, C., and VAN OYSTAEYEN, F
   [82] *Graded Ring Theory*. Mathematical Library 28. North-Holland, Amsterdam
OSOFSKY, B.
   [72] *Homological Dimension of Modules*. AMS Regional Conference Series in Mathematics 12, American Mathematical Society, Providence
PASSMAN, D.
   [77] *The Algebraic Structure of Group Rings*. Wiley, New York
PIERCE, R. S.
   [82] *Associative Algebras*. Springer-Verlag, Berlin
PROCESI, C.
   [73] *Rings with Polynomial Identities*. Marcel Dekker, New York
REINER, I.
   [75] *Maximal Orders*. Academic Press, London
RENAULT, J.
   [75] *Algèbre Noncommutative*. Gauthier-Villars, Paris
ROTMAN, J.
   [79] *An Introduction to Homological Algebra*. Academic Press, New York
ROWEN, L. H.
   [80] *Polynomial Identities in Ring Theory*. Academic Press, New York
RUDIN, W.
   [66] *Real and Complex Analysis*. McGraw Hill, New York
SCHOFIELD, A. H.
   [85] *Representations of Rings over Skew Fields*. London Mathematical Society Lecture Notes 92. Cambridge University Press, Cambridge
SERRE, J. P.
   [79] *Local Fields*. Springer-Verlag, Berlin (transl. of *Corps Locaux*)
SHARPE, D. W., and VAMOS, P.
   [72] *Injective Modules*. Cambridge University Press, Cambridge
SMALL, L. (Ed.)
   [81] *Reviews in Ring Theory*. American Mathematical Society, Providence
   [86] *Reviews in Ring Theory II*. American Mathematical Society, Providence
STENSTRÖM, B.
   [75] *Rings of Quotients: An Introduction to Methods of Rings Theory*. Springer-Verlag, Berlin
SWEEDLER, M.
   [69] *Hopf Algebras*. Benjamin, New York
VAN DER WAERDEN, B. L.
   [49] *Modern Algebra*. Ungar, New York

VAN OYSTAEYEN, F. M. J., and VERSCHOREN, A. H.
[81] *Noncommutative Algebraic Geometry: An Introduction.* Springer-Verlag, Berlin
VASCONCELOS, W.
[76] *The Rings of Dimension Two.* Marcel Dekker, New York
WEHRFRITZ, B. A. F.
[73] *Infinite Linear Groups.* Ergebnisse der Mathematik 76. Springer-Verlag, Berlin
ZARISKI, O., and SAMUEL, P.
[58] *Commutative Algebra.* Van Nostrand-Reinhold, Princeton (reprinted Springer-Verlag, Berlin)
ZHEVLAKOV, K. A., SLINKO, A. M., SHESTAKOV, I. P., and SHIRSHOV A. I.
[82] *Rings That Are Nearly Associative.* Academic Press, New York

# Collections of Papers

Collections which are in print, by year of publication.

1972
*Proceedings of Ring Theory Conference, Tulane University: Lectures on the Applications of Sheaves to Ring Theory.* Lecture Notes in Mathematics 248. Springer-Verlag, Berlin
*Proceedings of Ring Theory Conference, Tulane University: Lectures on Rings and Modules.* Lecture Notes in Mathematics 246. Springer-Verlag, Berlin
*Conference on Orders, Group Rings, and Related Topics.* Lecture Notes in Mathematics 353. Springer-Verlag, Berlin
R. Gordon (Ed.). *Ring Theory.* Academic Press, New York

1973
A. Kertesz (Ed.). *Rings Modules and Radicals.* Colloquium Mathematica Societatis Janos Bolyai 6. North-Holland, Amsterdam

1974
B. R. McDonald, A. Magid, and K. Smith (Eds.). *Ring Theory: Oklahoma University Conference.* Lecture Notes in Mathematics 7. Marcel Dekker, New York
V. Dlab and P. Gabriel (Eds.). *Representations of Algebras: University of Ottawa.* Lecture Notes in Mathematics 488. Springer-Verlag, Berlin

1975
B. R. McDonald and R. Morris (Eds.). *Ring Theory II: Second Oklahoma University Conference.* Lecture Notes in Mathematics 26. Marcel Dekker, New York
D. Zelinsky (Ed.). *Brauer Group Conference.* Lecture Notes in Mathematics 549. Springer-Verlag, Berlin
J. H. Cozzens and F. L. Sandomierski (Eds.). *Noncommutative Ring Theory: Kent State.* Lecture Notes in Mathematics 545. Springer-Verlag, Berlin

1977
S. K. Jain (Ed.). *Ring Theory: Ohio State University Conference, 1976.* Lecture Notes in Mathematics 25. Marcel Dekker, New York
M. P. Malliavin (Ed.). *Séminaire d'Algèbre Paul Dubreil, Paris (1975–76),* Lecture Notes in Mathematics 586. Springer-Verlag, Berlin

**1978**

F. Van Oystaeyen (Ed.). *Ring Theory: Antwerp Conference, 1977.* Lecture Notes in Mathematics 40. Marcel Dekker, New York

M. P. Malliavin (Ed.). *Séminaire d'Algèbre Paul Dubreil, Paris (1976-77).* Lecture Notes in Mathematics 641. Springer-Verlag, Berlin

R. Gordon (Ed.). *Representation Theory of Algebras: Temple University Conference, 1976.* Lecture Notes in Mathematics 37. Marcel Dekker, New York

**1979**

D. Handleman and J. Lawrence (Eds.). *Ring Theory: Waterloo Conference, 1978.* Lecture Notes in Mathematics 734. Springer-Verlag, Berlin

F. Van Oystaeyen (Ed.). *Ring Theory: Antwerp Conference, 1978.* Lecture Notes in Mathematics 51. Marcel Dekker, New York

M. P. Malliavin (Ed.). *Séminaire d'Algèbre Paul Dubreil, Paris (1977-78).* Notes in Mathematics 740. Springer-Verlag, Berlin

**1980**

V. Dlab (Ed.). *Representation Theory I: Carleton University Conference, 1979.* Lecture Notes in Mathematics 831. Springer-Verlag, Berlin.

V. Dlab (Ed.). *Representation Theory II: Carleton University Conference, 1979.* Lecture Notes in Mathematics 832. Springer-Verlag, Berlin

B. R. McDonald (Ed.). *Ring Theory and Algebra III: Third Oklahoma University Conference.* Lecture Notes in Mathematics 55. Marcel Dekker, New York

M. P. Malliavin (Ed.). *Séminaire d'Algèbre Paul Dubreil et Marie-Paule Malliavin, Paris (1978-79).* Lecture Notes in Mathematics 795. Springer-Verlag, Berlin.

F. Van Oystaeyen (Ed.). *Ring Theoy: Antwerp Conference, 1980.* Lecture Notes in Mathematics 825. Springer-Verlag, Berlin

**1981**

M. Kervaire and M. Ojanguren (Eds.). *Groupe de Brauer: Séminaire, Les Plans-Sur-Bex.* Lecture Notes in Mathematics 844. Springer-Verlag, Berlin

M. P. Malliavin (Ed.). *Séminaire d'Algèbre Paul Dubreil et Marie-Paule Malliavin, Paris (1979-80).* Lecture Notes in Mathematics 867. Springer-Verlag, Berlin

**1982**

S. A. Amitsur, G. Seligman, and D. Saltman (Eds.). *N. Jacobson: An Algebraists's Homage: Yale University 1980.* Contemporary Mathematics 13. American Mathematical Society, Providence

P. J. Fleury (Ed.). *Advances in Noncommutative Ring Theory: Plattsburgh Conference, 1981.* Lecture Notes in Mathematics 951. Springer-Verlag, Berlin

M. P. Malliavin (Ed.). *Séminaire d'Algèbre Paul Dubreil et Marie-Paule Malliavin, Paris (1981-82).* Lecture Notes in Mathematics 924. Springer-Verlag, Berlin

*Representations of Algebras (Pueblo).* Lecture Notes in Mathematics 944. Springer-Verlag, Berlin

F. Van Oystaeyen and A. Verschoren (Eds.). *Brauer Groups in Ring Theory and Algebraic Geometry.* Lecture Notes in Mathematics 917. Springer-Verlag, Berlin

**1983**

M. P. Malliavin (Ed.). *Séminaire d'Algèbre Paul Dubreil et Marie-Paule Malliavin, Paris (1982-83).* Lecture Notes in Mathematics 1029, Springer-Verlag, Berlin

**1984**

F. Van Oystaeyen (Ed.). *Methods in Ring Theory.* NATO Advanced Science Institutes, Reidel, Dordrecht

B. Srinivasan and J. Sally (Eds.). *Emmy Noether in Bryn Mawr.* Springer-Verlag, Berlin

1985
M. P. Malliavin (Ed.). *Séminaire d'Algèbre Paul Dubreil et Marie-Paule Malliavin, Paris (1983–84)*. Lecture Notes in Mathematics 1146. Springer-Verlag, Berlin
S. Montgomery (Ed.). *Group Actions on Rings*. Contemporary Mathematics 43. American Mathematical Society, Providence
L. Marki and R. Wiegandt (Eds.). *Radical Theory, Colloquium Mathematica Societatis Janos Bolyai 38*. North-Holland, Amsterdam

1986
V. Dlab, P. Gabriel, and G. Michler (Eds.) *Representation Theory I: Carleton University Conference, 1984*. Lecture Notes in Mathematics 1177. Springer-Verlag, Berlin.
F. Van Oystaeyen (Ed.). *Ring Theory: Antwerp Conference, 1985*. Lecture Notes in Mathematics 1197. Springer-Verlag, Berlin

1987
L. W. Small (Ed.). *Noetherian Rings and Their Applications*. Surveys and Monographs 24. American Mathematical Society, Providence

# Bibliography of Articles

*General note:* This is *not* intended as a comprehensive bibliography of ring theory, for which the reader could turn to Small's compilations. Rather, I have selected those articles most relevant to the areas covered in this book. Articles wholly included in books by the same author have sometimes been deleted.

MR denotes the location of the review in *Mathematical Reviews*. The parenthetical number(s) at the end indicate the page number(s) in this book where the article is cited.

AMITSUR, S. A.
[54] A general theory of radicals. II. Radicals in rings and bicategories. III. Applications. *Amer. J. Math.* **76**, 100–136. MR 15 #499b,c.
[55] Generic splitting fields of central simple algebras. *Ann of Math.* **62**, (2) 8–43. MR 17 #9d.
[56] Algebras over infinite fields. *Proc. Amer. Math. Soc.* **7**, 35–48. MR 17 #822b.
[57] Derivations in simple rings. *Proc. London Math. Soc.* (3) **57**, 87–112. MR 19 #525.
[59] Simple algebras and cohomology groups of arbitrary fields. *Trans Amer. Math. Soc.* **90**, 73–112. MR 21 #78.
[65] Generalized polynomial identities and pivotal monomials. *Trans Amer. Math. Soc.* **114**, 210–226. MR 30 #3117.
[66] Rational identities and applications to algebra and geometry. *J. Algebra* **3**, 304–359. MR 33 #139.
[67] Prime rings having polynomial identities with arbitrary coefficients. *Proc. London Math. Soc.* **17**, 470–486. MR 36 #209.
[69] Identities in rings with involution. *Israel J. Math.* **7**, 63–68. MR 39 #4216.
[71] Nil radicals: Historical notes and some new results. *Rings, Modules, and Radicals. Proc. Colloq. Keszthely*, 47–65. North-Holland, Amsterdam MR 50 #374.
[71a] A note on PI-rings. *Israel J. Math.* **10**, 210–211. MR 48 #332.
[71b] Ring of quotients and Morita contexts. *J. Algebra* **17**, 273–298. MR 54 #2704.
[71c] Embeddings in matrix rings. *Pacific J. Math* **36**, 21–29. MR 43 #2017.
[72] On central division algebras. *Israel J. Math* **12**, 408–420. MR 47 #6763.
[72a] On rings of quotients. *Symp Math. Inst. Naz. Alt. Matematica*, 149–164. Academic Press, London MR 48 #11180.

[73] On universal embeddings in matrix rings. *J. Math. Soc. Japan* **25**, 322–328. MR 47 # 3444.

[75] Identities and linear dependence.*Israel J. Math.* **22**, 127–137. MR 52 # 8184.

[80] On the characteristic polynomial of a sum of matrices. *Linear and Mutilinear Algebra* **8**, 177–182. MR 82 # 150.

[82] Division algebras. A survey. *Contemp. Math.* **13**, 3–26. MR 84b # 16021.

[84] The sequence of codimensions of PI-algebras. *Israel J. Math.* **47**, 1–22.

AMITSUR, S. A., and LEVITZKI, J.

[50] Minimal identities for algebras. *Proc. Amer. Math. Soc.* **1**, 449–463. MR 12 # 155d.

AMITSUR, S. A., and REGEV, A.

[82] PI- algebras and their cocharacters. *J. Algebra* **78**, 248–254. MR 84b # 16020.

AMITSUR, S. A., ROWEN, L. H., and TIGNOL, J.-P.

[79] Division algebras of degree 4 and 8 with involution. *Israel J. Math.* **33**, 133–148. MR 81h # 16029.

AMITSUR, S. A., and SALTMAN, D.

[78] Generic Abelian crossed products and p-algebras. *J. Algebra* **51**, 76–87. MR 58 # 10988.

AMITSUR, S. A., and SMALL, L. W.

[78] Polynomials over division rings. *Israel J. Math.* **31**, 353–358. MR 80f # 116022.

[80] Prime ideals in PI-rings. *J. Algebra* **62**, 358–383. MR 81e # 16027.

ANAN'IN, A. Z.

[79] Embedding of algebras in algebras of triangular matrices. *Mat. Sbornik* **108**, 168–186. MR 81c # 16031.

ANAN'IN, A. Z., and ZJABKO, E. M.

[74] On a question due to faith. *Algebra i Logika* **13**, 125–131. MR 50 # 13168.

ANDRUNAKIEVIC, V. A., and RJABUHIN, JU. M.

[68] Rings without nilpotent elements and completely prime rings. *Soviet Math. Dokl.* **9**, 565–568. MR 37 # 6320.

ARMENDARIZ, E. P., FISHER, J. W., and SNIDER, R. L.

[78] On injective and surjective endomorphisms of finitely generated modules. *Comm. Alg.* **6**, 659–672. MR 57 # 9754.

ARTIN, M.

[69] On Azumaya algebras and finite dimensional representation of rings. *J. Algebra* **11**, 532–563. MR 39 # 4217.

ARTIN, M., and SCHELTER, W.

[79] A version of Zariski's main theorem for rings with polynomial identity. *Amer. J. Math.* **101**, 301–330. MR 80e # 14003.

[81] Integral ring homomorphisms. *Advances in Math.* **39**, 289–329. MR 83e # 16015.

[81a] On two-sided modules which are left projective. *J. Algebra* **71**, 401–421. MR 84b # 16026.

AUSLANDER, M.

[55] On the dimension of modules and algebras. III. Global dimension. *Nagoya Math. J.* **9**, 67–77. MR 17 # 579a.

[74] Representation theory of Artin Algebras I, II. *Comm. Algebra* **1**. 177–310. MR 50 # 2240.

[75] Existence theorems for almost split sequences. *Ring Theory II (Oklahoma)*, 1–44. Marcel Dekker, New York. MR 55 # 12764.

[80] A functorial approach to representation theory. *Lecture Notes in Mathematics* **944**, 105–179. Springer-Verlag, Berlin. MR 83m # 16027.

[82] Representation theory of finite dimensional algebras. *Contemp. Math.* **13**, 27–39. MR 84b # 16031.

AUSLANDER, M., AND BUCHSBAUM, D.

[57] Homological dimension in local rings. *Trans. Amer. Math. Soc.* **85**, 390–405. MR 19 # 249.

AUSLANDER, M., and GOLDMAN, O.
[60] The Brauer group of a commutative ring. *Trans. Amer. Math. Soc.* **97**, 367–409. MR 22 #12130.
AUSLANDER, M., and REITEN, I.
[75] Representation theory of Artin algebras. III. Almost split sequences. *Comm. Alg.* **3**, 239–294. MR 52 #504.
AZUMAYA, G.
[48] On generalized semi-primary rings and Krull-Remak-Schmidt's theorem. *Jap. J. Math.* **19**, 524–547. MR 11 #316b.
[50] Corrections and supplements to my paper concerning Krull-Remak-Schmidt's Theorem. *Nagoya Math. J.* **1**, 117–124. MR 12 #314e.
[51] On maximally central algebras. *Nagoya Math. J.* **2**, 119–150. MR 12 #669g.
BASS, H.
[60] Finitistic dimension and a homological generalization of semiprimary rings. *Trans. Amer. Math. Soc.* **95**, 466–488. MR 28 #1212.
[61] Projective modules over algebras. *Ann. Math.* **73**, 532–542.
[62] Injective dimension in Noetherian rings. *Trans. Amer. Math. Soc.* **102**, 18–29. MR 25 #2087.
[63] Big projective modules are free. *Ill. J. Math.* **7**, 24–31. MR 26 #1341
[64] Projective modules over free groups are free. *J. Algebra* **1**, 367–373. MR 31 #2290.
[72] The degree of polynomial growth of finitely generated nilpotent groups. *Proc. London Math. Soc.* (3) **25**, 603–614.
[76] Euler characteristics and characters of discrete groups. *Inventiones Math.* **35**, 155–196. MR 55 #5764.
BAUTISTA, R., GABRIEL, G., ROITER, A. V., and SALMERÓN, L.
[85] Representation-finite algebras and multiplicative bases. *Invent. Math.* **81**, 217–285. MR 87g #16031.
BEACHY, J., and BLAIR, W.
[86] Examples in noncommutative localization. *J. Algebra* **99**, 108–113. MR 87e #16007.
BECK, I.
[72] Projective and free modules. *Math. Z.* **129**, 231–234. MR 47 #1868.
[78] On modules whose endomorphism ring is local. *Israel J. Math.* **29**, 393–407. MR 58 #11020.
BECK, I., and TROSBORG, P. J.
[78] A note on free direct summands. *Math. Scand.* **42**, 34–38. MR 53 #540.
BEIDAR, K. I.
[81] Radicals of finitely generated algebras. (Russian) *Uspekhi Mat. Nauk* **36**, 203–204. MR 83e #16009.
BEIDAR, K. I., MIKHALEV, A. V., and SALAVOVA, K.
[81] Generalized identities and semiprime rings with involutions. *Math. Z.* **178**, 37–62. MR 83b 16012.
BELJAEV, V. YA., and TAITSLIN, M. A.
[79] On elementary properties of existentially closed systems. *Russian Math. Surveys* **34**, 43–107. MR 82a #03028.
BENARD, M., and SCHACHER, M. M.
[72] The Schur subgroup II. *J. Algebra* **22**, 378–385. MR 46 #1890.
BENKART, G.
[76] The Lie inner structure of associative rings. *J. Algebra* **43**, 561–584. MR 55 #8110.
BERELE, A.
[82] Homogeneous polynomial identities. *Israel J. Math.* **42**, 258–272. MR 84b #16018.
BERELE, A., and REGEV, A.
[83] Applications of hook Young diagrams to P. I.-algebras. *J. Algebra* **82** 559–567. MR 84g #16012.

BERGEN, J., MONTGOMERY, S., and PASSMAN, D.
[88] Radicals of crossed products of enveloping algebras. *Israel J. Math.* (in press).

BERGMAN, G. M.
[64] A ring primitve on the right but not on the left. *Proc. Amer. Math. Soc.* **15**, 473–475. MR 29 #4770.
[71] Groups acting on hereditary rings. *Proc. London Math. Soc.* (3) **23**, 70–82; corrigendum ibid. **24** (3), 192 (1972). MR 45 #293.
[71a] Hereditary commutative rings and centers of hereditary rings. *Proc. London Math. Soc.* (3) **23**, 214–236. MR 46 #9022.
[73] On Jacobson radicals of graded rings. Preprint.
[74] Lifting prime ideals to extensions by centralizing elements. More on centralizing elements. Preprints.
[74a] Coproducts and some universal ring constructions. *Trans. Amer. Math. Soc.* **200**, 33–88. MR 50 #9771.
[74b] Some examples in PI ring theory. *Israel J. Math.* **18**, 257–277. MR 50 #9956.
[75] Zero-divisors in tensor products. *Lecture Notes in Mathematics* 545, pp. 32–82. Springer-Verlag, Berlin. MR 55 #12752.
[76] Rational relations and rational identities in division rings I, II. *J. Algebra* **43**, 253–297. MR 55 #5683a.
[78] The diamond lemma for ring theory. *Adv. Math.* **29**, 178–218. MR 81b #16001.
[81] Gel'fand-Kirillov dimension can go up in extension modules. *Comm Alg.* **9**, 1567–1570. MR 83e #16030.
[83] S fields finitely right generated over subrings. *Comm. Alg.* **11**, 1893–1902. MR 85e #16032.
[85] Everybody knows what a Hopf algebra is. *Contemp. Math.* **43**, 25–49. MR 87e #16024.
[86] Radicals, tensor products, and algebraicity. Preprint.

BERGMAN, G. M., and DICKS, W.
[75] On universal derivations. *J. Algebra* **36**, 193–211. MR 52 #8196.

BERGMAN, G. M., and ISAACS, I. M.
[73] Rings with fixed-point-free group actions. *Proc. London Math. Soc.* **27**, 69–87. MR 49 #10743.

BERGMAN, G. M., and OGOUS, A.
[72] Nakayama's lemma for half-exact functors. *Proc. Amer. Math. Soc.* **31**, 67–74. MR 46 #1777.

BERGMAN, G. M., and SMALL, L.
[75] PI degrees and prime ideals. *J. Algebra* **33**, 435–462. MR 50 #13129.

BERGMAN, G. M., VOVSKI, S. M., BRITTEN, D. J., and LEMIRE, F. W.
[83] Embedding rings in completed graded rings. I. Triangular embeddings. II. Algebras over a field. III. Algebras over general *k*. IV. Commutative algebras. *J. Algebra* **84**, 14–106. MR 851 #16001.

BERNSTEIN, I. N., GELFAND, I. M., and PONOMAREV, V. A.
[73] Coxeter functors and Gabriel's theorem. *Russian Math. Surveys* **28**, 17–32. MR 52 #13876.

BJORK, J.-E.
[71] Conditions which imply that subrings of Artinian rings are Artinian. *J. Reine Angew. Math.* **247**, 123–138. MR 43 #6249.
[71a] Conditions which imply that subrings of semiprimary rings are semiprimary. *J. Algebra* **19**, 384–395. MR 44 #1686.
[85] Noncommutative Noetherian rings and the use of homological algebra. *J. Pure Appl. Algebra* **38**, 111–119. MR 87a #16050.

BLATTNER, R., COHEN, M., and MONTGOMERY, S.
[86] Crossed products and inner actions of Hopf algebras. *Trans. Amer. Math. Soc.* **298**, 671–711.

BLOCH, S.
[80] Torsion algebraic cycles, $K_2$, Brauer groups of function fields. *Lecture Notes in Mathematics* 844, pp. 75–102. Springer-Verlag, Berlin. MR 82k #14018.

BLOCK, R. E.
[81] The irreducible representations of the Lie algebra sl(2) and of the Weyl algebra. *Adv. Math.* **39**, 69–110. MR 83c #17010.

BOKUT, L. A.
[69] The problem of Mal'cev. *Siberian Math. J.* **10**, 706–739. MR 41 #267b.
[69a] Groups of fractions of multiplicative semigroups of certain rings, I,II,III. *Siberian Math. J.* **10**, 172–203, 541–600. MR 41 #267a.

BONGARTZ, K., and GABRIEL, P.
[82] Covering spaces in representation theory. *Invent. Math.* **65**, 331–378.

BORHO, W.
[77] Definition einer Dixmer-Abildung für sl(n,ℂ). *Invent. Math.* **40**, 163–169. MR 56 #434.
[82] Invariant dimension and restricted extension of Noetherian rings. *Lecture Notes in Mathematics* 924. Springer-Verlag, Berlin. MR 83h #16018.
[82a] On the Joseph-Small additivity principal for Noetherian ring. *Comp. Math.* **47**, 3–29. MR 84a #17007.
[86] A survey on enveloping algebras of semisimple Lie algebras I. *Canad. Math. Soc. Conference Proceedings* 5, pp. 19–50. MR 87g #17013.

BORHO, W., and JANTZEN, J.-C.
[77] Über primitive Ideale in der Einhüllenden einer halbeinfachen Lie-Algebra. *Invent. Math.* **39**, 1–53.

BORHO, W., and KRAFT, H.
[76] Über die Gelfand-Kirillov-Dimension. *Math. Ann.* **220**, 1–24. MR 54 #367.

BORHO, W., and RENTSCHLER, R.
[75] Oresche Teilmengen in Einhüllenden Algebren. *Math Ann.* **217**, 201–210. MR 53 #5680.

BOWTELL, A.J.
[67] On a question of Mal'cev. *J. Algebra* **7**, 126–139. MR 37 #6310.

BRAUN, A.
[79] Affine polynomial identity rings and their generalizations. *J. Algebra* **58**, 481–494. MR 80f #16021.
[81] A note on Noetherian PI-rings. *Proc. Amer. Math. Soc.* **83**, 670–672. MR 82m #16013.
[82] On Artin's theorem and Azumaya algebras. *J. Algebra* **77**, 323–332. MR 83m #16005.
[84] The nilpotency of the radical in a finitely generated PI-ring. *J. Algebra* **89**, 375–396. MR 85m #16007.
[85] An additivity principle for PI-rings. *J. Algebra* **96**, 433–441.
[86] On the defining axioms of Azumaya algebras. Preprint.

BRAUN, A., and WARFIELD, R. B., JR.
[86] The localization problem in PI-rings. Preprint.

BROCKHAUS, P.
[85] On the radical of a group algebra. *J. Algebra* **95**, 454–472.

BROOKES, C. J. B., and BROWN, K. A.
[85] Primitive group rings and Noetherian rings of quotients. *Trans. Amer. Math. Soc.* **288**, 695–623. MR 86d #16014.

BROWN, B., and McCOY, N. H.
[50] The maximal regular ideal of a ring. *Proc. Amer. Math. Soc.* **1**, 165–171. MR 11 #638.

BROWN, K. A.
[76] On zero divisors in group rings. *Bull London Math. Soc.* **8**, 251–256. MR 54 #2716.
[81] Module extensions over Noetherian rings. *J. Algebra* **69**, 247–260. MR 83g #16027.
[82] The Nullstellensatz for certain group rings. *J. London Math. Soc.* **26** (2), 425–434, MR 84c #16013.

[84] Ore sets in enveloping algebras. *Comp. Math.* **53**, 347–367. MR 86d #17008.

[85] Ore sets in Noetherian rings. *Lecture Notes in Mathematics* 1146, pp. 355–366. Springer-Verlag, Berlin.

BROWN, K. A., and LENAGAN, T. H.

[82] A note on Jacobson's conjecture for right Noetherian rings. *Glasgow Math. J.* **23**, 7–8. MR 83b #16010.

BROWN, K. A., LENAGAN, T. H., and STAFFORD, J. T.

[80] Weak ideal invariance and localization. *J. London Math. Soc.* **21**, 53–61. MR 81i #16016.

[81] K-theory and stable structure of some Noetherian group rings. *Proc London Math. Soc.* **42** (3), 193–230. MR 82g #16017.

BROWN, K. A., and WARFIELD, R. B., JR.

[84] Krull and Gabriel dimensions of fully bounded Noetherian rings. *Proc. Amer. Math. Soc.* **92**, 169–174. MR 86d #16019.

BROWN, K. S., and DROR, E.

[75] The Artin-Rees property and homology. *Israel J. Math.* **22**, 93–109. MR 52 #10809.

BURGESS, W. D., and STEPHENSON, W.

[76] Pierce sheaves of non-commutative rings. *Comm. Alg.* **4**, 51–75. MR 53 #8122.

CAMILLO, V. P.

[84] Morita equivalence and infinite matrix rings. *Proc. Amer. Math. Soc.* **90**, 186–188. MR 85a #16045.

CAMILLO, V. P., and FULLER, K. R.

[74] On the Loewy length of rings. *Pacific J. Math.* **53**, 347–354. MR 52 #10798.

CAUCHON, G.

[76] Anneaux semipremiers Noethériens à identités polynômiales. *Bull. Soc. Math. France* **104**, 99–111. MR 53 #10859.

[76a] Les T-anneaux, la condition (H) de Gabriel, et ses consequences. *Comm. Algebra* **4**, 11–50. MR 49 #2836.

[79] Idéaux bilatères et centre des anneaux de polynômes de Ore sur les anneaux quasi-simples. *Lecture Notes in Mathematics* 740, pp. 397–407. Springer-Verlag, Berlin. MR 82e #16001.

[86] Centralisateurs dans les corps de Weyl. *Comm. Alg.* **14**, 1403–1428.

CAUCHON, G., and LESIEUR, L.

[78] Localisation classique en un idéal premier d'un anneau Noetherian à gauche. *Comm. Alg.* **6**, 1091–1108. MR 58 #10985.

CHASE, S. U.

[60] Direct products of modules. *Trans. Amer. Math. Soc.* **97**, 457–473. MR 22 #11017.

[61] A generalization of the ring of triangular matrices. *Nagoya Math J.* **18**, 13–25. MR 23A #919.

[84] Two results on central simple algebras. *Comm. Algebra* **12**, 2279–2289.

CHASE, S. U., and FAITH, C.

[65] Quotient rings and direct products of full linear rings. *Math. Z.* **88**, 250–264. MR 31 #2281.

CHATTERS, A. W.

[76] Two results on P. P. rings. *Comm. Algebra* **4**, 881–891. MR 54 #7522.

[84] Noncommutative unique factorization domains. *Math. Proc. Cambridge Philos. Soc.* **95**, 49–54. MR 85b #16001.

CHATTERS, A. W., GOLDIE, A. W., HAJARNAVIS, C. J., and LENAGAN, T. H.

[79] Reduced rank in Goldie rings. *J. Algebra* **61**, 582–589. MR 8li #16041.

CHATTERS, A. W., and JONDRUP, J.

[83] Hereditary finitely generated PI-algebras. *J. Algebra* **82**, 40–52. MR 84m #16012.

[84] Quotient rings of finitely generated P. I. Algebras. *Comm. Alg.* **12**, 271–285. MR 85j #16006.

CLIFF, G. H.
[80] Zero divisors and idempotents in group rings. *Canad. J. Math.* **32** (3), 596–602. MR 81k #16011.
[85] Ranks of projective modules of group rings. *Comm. Alg.* **13**, 1115–1130.

COHEN, M.
[78] Centralizers of algebraic elements. *Comm. Algebra* **6**, 1505–1519. MR 80a #16028.
[85] Hopf algebras acting on semiprime algebras. *Contemp. Math.* **43**, 49–61. MR 87a #16016.

COHEN, M., and MONTGOMERY, S.
[84] Group-graded rings, smash products, and group actions. *Trans. Amer. Math. Soc.* **282**, 237–258. MR 85i #16002.

COHN, P. M.
[56] The complement of a f.g. direct summand of an Abelian group. *Proc. Amer. Math. Soc.* **7**, 520–521. MR 17 #1182.
[59] On the free product of associative rings. *Math. Z.* **71**, 380–398. MR 21 #5648.
[60] On the free product of associative rings II. The case of (skew) fields. *Math. Z.* **73**, 433–456. MR 22 #4747.
[61] On the embedding of rings in skew fields. *Proc. London Math. Soc.* (3) **11**, 511–530. MR 25 #100.
[66] Some remarks on the invariant basis property. *Topology* **5**, 215–228. MR 33 #5676.
[68] On the free product of associative rings III. *J. Algebra* **8**, 376–383, correction ibid. **10**, 23 (1968). MR 36 #5170.
[71] The embedding of firs into skew fields. *Proc. London Math. Soc.* (3) **23**, 193–213. MR 45 #6866.
[74] Progress in free associative algebras. *Israel J. Math.* **19**, 109–151. MR 52 #460.
[82] The universal field of fractions of a semifir 1. *Proc. London Math. Soc.* (3) **44**, 1–32. MR 84h #16002.
[84] Fractions. *Bull. London Math. Soc.* **16**, 561–574. MR 86f #16016.

CONNELL, I. G.
[63] On the group ring. *Canad. J. Math.* **15**, 650–685. MR 27 #3666.

CONZE-BERLINE, N.
[75] Espace des idéaux primitifs de l'algèbre enveloppante d'une algèbre de Lie nilpotente. *J. Algebra* **34**, 444–450. MR 51 #8191.

CORNER, A. L. S., and GÖBEL, R.
[85] Prescribing endomorphism algebras, a uniform treatment. *Proc. London Math. Soc.* (3) **50**, 447–479. MR 86h #16031.

COUTINHO, S. C.
[85] K-theoretic properties of generic matrix rings. *J. London Math. Soc.* **32**, 51–56.
[86] Dissertation, University of Leeds.

CRAWLEY-BOEREY, W. W., KROPHOLLER, P. H., and LINNELL, P. A.
[87] Torsion-free soluble groups and the zero-divisor conjecture. *J. Pure Applied Algebra*.

CURTIS, C. W., and JANS, J. P.
[65] On algebras with a finite number of indecomposable modules. *Trans. Amer. Math. Soc.* **114**, 122–132. MR 31 #2270.

DEAN, C., and STAFFORD, J. T.
[86] A nonembeddable Noetherian ring. Preprint.

DEHN, M.
[22] Über die Grundlagen der projektiven Geometrie und allgemeine Zahlsysteme. *Math. Ann.* **85**, 184–193.

DICKS, W.
[83] The HNN construction for rings. *J. Algebra* **81**, 434–487. MR 85c #16005.
[85] Automorphisms of the free algebra of rank two. *Contemp. Math.* **43**, 63–68. MR 86j #16007.

DIXMIER, J.
[68] Sur les algèbres de Weyl. *Bull. Soc. Math. France* **96**, 209–242. MR 39 #4224.
[70] Sur les algèbres de Weyl II. *Bull. Sci. Math.* (2) **94**, 289–301. MR 45 #8680.
[77] Idèaux primitifs des algèbres enveloppantes. *J. Algebra* **48**, 96–112. MR 56 #5673.
DLAB, V., and RINGEL, C. M.
[75] On algebras of finite representation type. *J. Algebra* **33**, 306–394. MR 50 #9774.
[76] Indecomposable representations of graphs and algebras. *Mem Amer. Math. Soc.* **173**. MR 56 #5657.
DOMANOV, O. I.
[77] A prime but not primitive regular ring. (Russian) *Uspehi Mat. Nauk* **32**, 219–220. MR 58 #28058.
[78] Primitive group algebras of polycyclic groups. *Siber. Math. J.* **19**, 25–30. MR 57 #6075.
DRENSKY, V. S.
[82] A minimal basis for identities of a second-order matrix algebra over a field of characteristic 0. *Algebra and Logic* **22**, 188–194. MR 83g #16031.
[84] Codimensions of T-ideals and Hilbert series of relatively free algebras. *J. Algebra* **91**, 1–17. MR 86b #16010.
DRENSKY, V. S., and KASPARIAN, A. K.
[83] Some polynomial identities of matrix algebras. *C. R. Acad. Bulgare Sci.* **36**, 565–568. MR 85e #16036.
DUFLO, M.
[73] Certaines algèbres de type fini sont algèbres de Jacobson. *J. Algebra* **27**, 358–365. MR 49 #9018.
[75] Construction of primitive ideals in an enveloping algebras. *Lie Groups and Their Representations.* Adam Hilger, London. MR 53 #3045.
[77] Sur la classification des idéaux primitifs dans l'algèbre enveloppante d'une algèbre de Lie semisimple. *Ann. Math.* (2) **105**, 107–120. MR 55 #3013.
[82] Sur les idéaux induits dans les algèbres enveloppantes. *Invent Math.* **67**, 385–392. MR 83m #17005.
[82a] Théorie de Mackey pour les groupes de Lie algébriques. *Acta Mathematica* **110**, 154–213. MR 85h #22022.
ECKMANN, B., and SHOPF, A.
[53] Über injektive Moduln. *Archiv Math.* **4**, 75–78. MR 15 #5.
ECKSTEIN, F.
[69] Semigroup methods in ring theory. *J. Algebra* **12**, 177–190. MR 39 #2808.
EILENBERG, S., and MACLANE, S.
[45] General theory of natural equivalences. *Trans. Amer. Math. Soc.* **58**, 231–294. MR 7 #109.
EILENBERG, S., NAGAO, H., and NAKAYAMA, T.
[56] On the dimension of modules and algebras IV: Dimension of residue rings of hereditary rings. *Nagoya Math. J.* **10**, 87–95. MR 18 #9d.
EILENBERG, S., and NAKAYAMA, T.
[55] On the dimension of modules and algebras II: Frobenius algebras and quasi-Frobenius rings. *Nagoya Math. J.* **9**, 1–16. MR 16 #993a.
[57] On the dimension of modules and algebras V: Dimension of residue rings. *Nagoya Math. J.* **11**, 9–12. MR 19 #118c.
EILENBERG, S., ROSENBERG, A., and NAKAYAMA, T.
[57] On the dimension of modules and algebras VIII: Dimension of tensor products. *Nagoya Math. J.* **12**, 71–93. MR 20 #5229.
EISENBUD, D.
[70] Subrings of Artinian and Noetherian rings. *Math. Ann.* **185**, 247–249. MR 41 #6885.

EISENBUD, D., and ROBSON, J. C.
[70] Hereditary Noetherian prime rings. *J. Algebra* **16**, 86–104. MR 45 #316.
ELIZAROV, V. P.
[60] Rings of quotients for associative rings. (Russian) *Izv. Akad. Nau SSSR ser. Mat.* **24**, 153–170. MR 22 #9513.
[66] On the modules of quotients. (Russian) *Sibirsk. Mat. Z.* **7**, 221–226. MR 33 #2675.
[69] Ring of quotients. *Algebra and Logic* **8**, 219–293 (some corrections in *Siberian Math. J.* **13**, 998–1000 (1972). MR 43 #3302.
EVANS, E. G., JR.
[73] Krull-Schmidt and cancellation over local rings. *Pacific J. Math.* **46**, 115–121. MR 48 #2170.
FAITH, C.
[58] On conjugates in division rings. *Canad. J. Math.* **10**, 374–380. MR 22 #5649.
[67] A general Wedderburn theorem. *Bull. Amer. Math. Soc.* **73**, 65–67. MR 34 #2623.
[74] On the structure of indecomposable injective modules. *Comm. Algebra* **2**, 559–571. MR 50 #13150.
FAITH, C., and UTUMI, Y.
[65] On Noetherian prime rings. *Trans. Amer. Math. Soc.* **114**, 53–60. MR 30 #3111.
FAITH, C., and WALKER, E.
[67] Direct-sum representations of injective modules. *J. Algebra* **5**, 203–221. MR 34 #7575.
FARKAS, D. R.
[75] Miscellany on Bieberbach group algebras. *Pacific J. Math.* **59**, 427–435. MR 52 #8257.
[79] Baire category and Laurent extensions. *Canad. J. Math.* **31**, 824–830. MR 81a #16009.
[80] Group rings: An annotated questionnaire. *Comm. Alg.* **8**, 585–602. MR 82f #16008.
[87] Noetherian group rings: An exercise in creating folklore intuition in: *Noetherian rings and their applications*, AMS surveys and monographs 24.
FARKAS, D. R., and PASSMAN, D. S.
[78] Primitive Noetherian group rings. *Comm. Alg.* **6**, 301–315. MR 57 #9741.
FARKAS, D. R., SCHOFIELD, A. H., SNIDER, R. L., and STAFFORD, J. T.
[82] The isomorphism question for division rings of group rings. *Proc. Amer. Math. Soc.* **85**, 327–330. MR 83g #16022.
FARKAS, D. R., and SNIDER, R. L.
[74] Group algebras whose simple modules are injective. *Trans.* Amer. Math. Soc. **194**, 241–248. MR 50 #9943.
[76] Ko and Noetherian group rings. *J. Algebra* **42**, 192–198. MR 54 #10318.
[79] Induced representations of polycyclic groups. *Proc. London Math. Soc.* **39** (3), 193–207. MR 81k #12025.
[81] Group rings which have a nonzero socle. *Bull. London Math. Soc.* **13**, 392–396. MR 83c #16009.
FEIN, B., and SCHACHER, M.
[80] Brauer groups of fields. *Lecture Notes in Mathematics* 55, pp. 345–356. Marcel Dekker, New York. MR 81k #12025.
FEIN, B., SCHACHER, M., and SONN, J.
[86] Brauer groups of algebraic function fields of genus 0 over number fields. *J. Algebra*. (forthcoming).
FIELDS, K. L.
[69] On the global dimension of skew polynomial rings. *J. Algebra* **13**, 1–4; addendum ibid. **14**, 528–530. MR 39 #6929.
FISCHER, I., and STRUIK, R.
[68] Nil algebras and periodic subgroups. *Amer. Math. Monthly* **75**, 611–623. MR 38 #179.
FISHER, J. W.
[72] Nil subrings of endomorphism rings of modules. *Proc. Amer. Math. Soc.* **34**, 75–78. MR 45 #1960.

[73] Von Neumann regular rings versus V-rings. *Lecture Notes in Mathematics* 7, pp. 101–119. Marcel Dekker, New York. MR 48 #11191.

[80] Finite group actions on noncommutative rings: A survey since 1970. *Lecture Notes in Mathematics* 55, pp. 345–356. Marcel Dekker, New York. MR 88li #16001.

FISHER, J. W., and OSTERBURG, J.

[74] On the von Neumann regularity of rings with regular prime factor rings. *Pacific J. Math.* **54**, 135–144. MR 50 #13126.

FISHER, J. W., and SNIDER, R. L.

[76] Rings generated by their units. *J. Algebra* **42**, 363–368. MR 54 #7531.

FORMANEK, E.

[72] Central polynomials for matrix rings. *J. Algebra* **32**, 129–132. MR 46 #1833.

[73] Idempotents in Noetherian group rings. *Canad. J. Math.* **25** (2), 366–369. MR 47 #5041.

[73a] Group rings of free products are primitive. *J. Algebra* **26**, 508–511. MR 48 #329.

[73b] The zero divisor question for supersolvable groups. *Bull. Austr. Math. Soc.* **9**, 67–71. MR 48 #4017.

[73c] A problem of Herstein on group rings. *Canad. Math. Bull.* **17**, 201–202. MR 50 #13118.

[74] Maximal quotient rings of group rings. *Pacific J. Math.* **53**, 109–116. MR 53 #3210.

[82] The polynomial identities of matrices. *Contemp. Math.* **13**, 41–79. MR 84b #16019.

[84] Invariants and the ring of generic matrices. *J. Algebra* **89**, 178–223.

[85] Noncommutative invariant theory. Contemp. Math. **43**, 87–120.

[86] A conjecture of Regev about the Capelli polynomial.

[86a] Functional equations for character series associated with $n \times n$ matrices. *Trans. Amer. Math. Soc.* **294**, 647–663.

FORMANEK, E., and JATEGAONKAR, A. V.

[74] Subrings of Noetherian rings. *Proc. Amer. Math. Soc.* **46**, 181–186. MR 54 #2725.

FORMANEK, E., and LICHTMAN, A. I.

[78] Ideals in group rings of free products. *Israel J. Math.* **31**, 101–104. MR 58 #16752.

FORMANEK, E., and SCHOFIELD, A. H.

[85] Groups acting on the ring of two $2 \times 2$ generic matrices and a coproduct decomposition of its trace ring. *Proc. Amer. Math. Soc.* **95**, 179–183.

FUCHS, L.

[71] On a substitution property of modules. *Monatsh. Math.* **75**, 198–204. MR 45 #5157.

[72] The cancellation property for modules. Lectures on rings and modules. *Lecture Notes in Mathematics* 246, pp. 191–212. Springer-Verlag, Berlin. MR 49 #2846.

FUELBERTH, J., and KUZMANOVICH, J.

[75] The structure of semiprimary and Noetherian hereditary rings. *Trans. Amer. Math. Soc.* **212**, 83–111. MR 51 #12929.

GABRIEL, P.

[62] Des categories Abéliennes. *Bull. Soc. Math. France* **90**, 323–448. MR 38 #1144.

[72] Unzerlegbare Darstellungen 1. *Manuscripta Math.* **6**, 71–103; corr. ibid. **6**, 309. MR 48 #11212.

[73] Indecomposable representations II. *Symposia Mathematica*, Vol. XI, pp. 81–104. Academic Press, New York. MR 49 #5132.

[80] Auslander-Reiten sequences and representation-finite algebras. *Lecture Notes in Mathematics* 831, pp. 1–71. Springer-Verlag, Berlin. MR 82i #16030.

[81] The universal cover of a representation-finite algebra. *Lecture Notes in Mathematics 903*, pp. 68–105. Springer-Verlag, Berlin. MR 83f #16036.

GEL'FAND, I. M., and KIRILLOV, A. A.

[66] Sur les corps Lies aux algèbres enveloppantes des algèbres de Lie. *Inst. Hautes Études Sci. Publ. Math.* **31**, 5–19. MR 34 #7731.

GERASIMOV, V. N., and SAKHAEV, I. I.

[84] A counterexample to two conjectures on projective and flat modules. *Siberian Math. J.* **25**, 855–859. MR 86f #16021.

GERSTE: IHABER, M.

[59] On nilalgebras and linear varieties of nilpotent matrices III. *Ann Math.* **70** (2), 167–205. MR 22 #4743.

GILCHRIST, M. P., and SMITH, M. K.

[84] Noncommutative UFD's are often PID's. *Math. Proc. Camb. Philos. Soc.* **95**, 417–419. MR 85h #16002.

GOLAN, J. S.

[80] Structure sheaves of noncommutative rings. *Lecture Notes in Mathematics* 55, pp. 345–356. Marcel Dekker, New York. MR 81j #16032.

GOLAN, J. S., and RAYNAUD, J.

[74] Dimension de Gabriel et TTK-dimension de modules. *C. R. Acad. Sci. Paris* Ser. A **278**, 1603–1606. MR 49 #7316.

GOLDIE, A. W.

[58] The structure of prime rings under ascending chain conditions. *Proc. London Math. Soc.* **8** (3), 589–608. MR 21 #1988.

[60] Semiprime rings with maximum condition. *Proc. London Math Soc.* (3) **10**, 201–220. MR 22 #2627.

[62] Noncommutative principal ideal rings. *Arch. Math.* **13**, 213–221. MR 29 #2282.

[64] Torsion-free modules and rings. *J. Algebra* **1**, 268–287. MR 25 #3591.

[67] Localization in noncommutative Noetherian rings. *J. Algebra* **5**, 89–105. MR 34 #7562.

[69] Some aspects of ring theory. *Bull. London Math. Soc.* **1**, 129–154. MR 42 #299.

[72] The structure of Noetherian rings. *Lecture Notes in Mathematics* 246, pp. 213–321. Springer-Verlag, Berlin. MR 52 #3226.

[74] Some recent developments in ring theory. *Israel J. Math.* **19**, 153–168. MR 50 #13098.

[80] The reduced rank in Noetherian ring. *Lecture Notes in Mathematics* 867, pp. 396–405. Springer-Verlag, Berlin. MR 83d #16015.

GOLDIE, A. W., and MICHLER, G.

[75] Ore extensions and polycyclic group rings. *J. London Math. Soc.* (2) **9**, 337–345. MR 50 #9968.

GOLDMAN, O.

[69] Rings and modules of quotients. *J. Algebra* **13**, 10–47. MR 39 #6914.

GOLOD, E. S.

[64] On nil-algebras and finitely approximable p-groups (Russian). *Izv. Akad. Nauk SSR Ser Mat.* **28**, 273–276. MR 28 #5082.

[66] Some problems of Burnside type. (Russian) *Proc. Internat. Congr. Math.*, pp. 284–289. Izdat Mis, Moscow. (Transl. in *Amer. Math. Soc. Trans.* (2) **82**, 83–88 (1978). MR 39 #240.

GOODEARL, K. R.

[72] Singular torsion and the splitting properties. *Memoirs of the AMS* 124. American Mathematical Society, Providence. MR 49 #5090.

[73] Prime ideals in regular self-injective rings. *Canad. J. Math.* **25**, 829–839. MR 48 #4037.

[73a] Prime ideals in regular self-injective rings II. *J. Pure Appl. Algebra* **3**, 357–373. MR 48 #8561.

[74] Global dimension of differential operator rings. *Proc. Amer. Math. Soc.* **45**, 315–322. MR 52 #3243.

[75] Global dimension of differential operator rings II. *Trans. Amer. Math. Soc.* **209**, 65–85. MR 52 #3244.

[76] Power-cancellation of groups and modules. *Pacific J. Math.* **64**, 387–411. MR 56 #8630.

[78] Global dimension of differential operator rings III. *J. London Math. Soc.* (2) **17**, 397–409. MR 58 #28072.

[79] Simple Noetherian rings—the Zaleskii-Neroslavskii examples. *Lecture Notes in Mathematics* 734, pp. 118–130. Springer-Verlag, Berlin. MR 81b #16009.

[84] Simple Noetherian rings not isomorphic to matrix rings over domains. *Comm. Algebra* **12**, 1421–1434. MR 85j #16027.

GOODEARL, K. R., and HANDLEMAN, D.

[75] Simple self-injective rings need not be Artinian. *Comm. Alg.* **3**, 797–834. MR 52 #498.

GOODEARL, K. R., HODGES, T. J., and LENAGAN, T. H.

[84] Affine PI-rings have Gabriel dimension. *Comm. Alg.* **12**, 1291–1300. MR 85k #16033. MR 86d #16034.

GOODEARL, K. R., and LENAGAN, T. H.

[83] Krull dimension of differential operator rings. III. Noncommutative coefficients. *Trans. Amer. Math. Soc.* **275**, 833–859. MR 85f #16025.

[83a] Krull dimension of differential operator rings. IV. Multiple derivations. *Proc. London Math. Soc.* (3) **47**, 306–336. MR 85f #16026.

[85] Krull dimension of skew Laurent series. *Pacific J. Math.* **114**, 109–117. MR 86a #16001.

GOODEARL, K. R., and SMALL, L. W.

[84] Krull vs. global dimension in Noetherian PI-rings. *Proc. Amer. Math. Soc.* **92**, 175–177.

GOODEARL, K. R., and WARFIELD, R. B., JR.

[81] State spaces of $K_0$ of Noetherian rings. *J. Algebra* **71**, 322–378. MR 83h #16019.

[82] Krull dimension of differential operator rings. *Proc. London Math. Soc.* (3) **45**, 49–70. MR 83j #16032.

GORDON, R.

[74] Gabriel and Krull dimension. *Lecture Notes in Mathematics* 7, pp. 241–295. Marcel Dekker, New York. MR 51 #10388.

[75] Some aspects of noncommutative Noetherian rings. *Lecture Notes in Mathematics* 545, pp. 105–127. Springer-Verlag, Berlin. MR 55 #2990.

GORDON, R., and GREEN, E. L.

[77] Modules with cores and amalgamations of indecomposable modules. MEMOIRS OF THE AMS 187. American Mathematical Society, Providence. MR 56 #12068.

GORDON, R., and ROBSON, J. C.

[73] Krull dimension. *Memoirs of the AMS* 133. American Mathematical Society, Providence. MR 50 #4664.

GORDON, R., and SMALL, L.

[72] Piecewise domains. *J. Algebra* **23**, 553–564. MR 46 #9087.

[84] Affine PI-rings have Gabriel dimension. *Comm. Alg.* **12**, 1291–1300. MR 85k #16033.

GOURSAUD, J.-M., PASCAUD, J.-L., and VALETTE, K.

[82] Sur les traveaux de V. K. Harchenko. *Lecture Notes in Mathematics* 924, pp. 322–355. Springer-Verlag, Berlin. MR 83j #16006.

GREEN, E.L.

[75] Diagrammatic techniques in the study of indecomposable modules. *Lecture Notes in Mathematics* 26, pp. 149–169. Marcel Dekker, New York. MR 56 #12069.

GROTHENDIECK A.

[57] Sur quelque points d'algèbre homologique. *Tohoku Math. J.* **9**, 119–221. MR 21 #1328.

GRUSON, L., and RAYNAUD, M.

[71] Critères de platitude et de projectivité. *Invent. Math.* **13**, 1–81. MR 46 #7219.

GRZESZCZUK, P.

[85] On F-systems and G-graded rings. *Proc. Amer. Math. Soc.* **95**, 348–352. MR 87a #16003.

GUSTAFSON, W. H.

[76] On maximal commutative algebras of linear transformations. *J. Algebra* **42**, 557–563. MR 54 #10331.

HALPIN, P.

[83] Central and weak identities for matrices. *Comm. Algebra* **11**, 2237–2248. MR 85b #16014.

HANDELMAN, D., and LAWRENCE, J.

[75] Strongly prime rings. *Trans. Amer. Math. Soc.* **211**, 209–223. MR 52 #8175.

HANDELMAN, D., LAWRENCE, J., and SCHELTER, W.
[78] Skew group rings. *Houston J. Math.* **4**, 175–198. MR 58 #5771.

HARADA, M.
[63] Hereditary orders. *Trans. Amer. Math. Soc.* **107**, 273–290. MR 27 #1474.
[66] Hereditary semiprimary rings and triangular matrix rings. *Nagoya Math. J.* **27**, 463–484. MR 34 #4300.
[75] On the exchange property in a direct sum of indecomposable modules. *Osaka J. Math.* **12**, 719–736. MR 52 #13944.

HARCENKO, V. K.
[74] Galois extensions and rings of quotients. *Algebra and Logic* **13**, 265–281. MR 53 #498.
[75] Galois subrings of simple rings. (Russian) *Math. Notes* **17**, 533–536. MR 52 #10795.
[78] Algebras of invariants of free algebras. *Algebra and Logic* **17**, 316–321. MR 803 #16003.
[84] Noncommutative invariants of finite groups and Noetherian varieties. *J. Pure Appl. Algebra* **31**, 83–90.

HART, R.
[67] Simple rings with uniform right ideals. *J. London Math. Soc.* **42**, 614–617. MR 36 #1477.

HART, R., and ROBSON, J. C.
[70] Simple rings and rings Morita equivalent to Ore domains. *Proc. London Math. Soc.* (3) **21**, 232–242. MR 43 #4857.

HATTORI, A.
[60] A foundation of torsion theory for modules over general rings. *Nagoya Math. J.* **17**, 147–158. MR 25 #1194.
[65] Rank element of a projective module. *Nagoya Math. J.* **25**, 113–120. MR 31 #226.

HEINICKE A. G., and ROBSON, J. C.
[81] Normalizing extensions: Prime ideals and incomparability. *J. Algebra* **72**, 237–268. MR 83f #16004.
[82] Normalizing extensions: Nilpotency. *J. Algebra* **76**, 459–470. MR 83i #16030.
[84] Normalizing extensions: Prime ideals and incomparability II. *J. Algebra* **91**, 142–165. MR 86c #16001.
[83] Intermediate normalizing extensions. *Trans. Amer. Math. Soc.* **282**, 645–667. MR 85h #16030.

HELLING, H.
[74] "Eine Kennzeichnung von Charakteren auf Gruppen and Assoziativen Algebren. *Comm. Alg.* **1**, 491–501. MR 50 #381.

HERSTEIN, I. N.
[53] A generalization of a theorem of Jacobson III. *Amer. J. Math.* **75**, 105–111. MR 14 #613e.
[55] On the Lie and Jordan rings of a simple associative ring. *Amer. J. Math.* **77**, 279–285. MR 16 #789e.
[56] Jordan homomorphisms. *Trans. Amer. Math. Soc.* **81**, 311–341. MR 17 #938f.
[65] A counterexample in Noetherian rings. *Proc. Natl. Acad. Sci. U.S.A.* **54**, 1036–1037. MR 32 #5692.
[70] Notes from a ring theory conference. *Conference Board of the Mathematical Sciences* 9. American Mathematical Society, Providence. MR 47 #1840.
[70a] On the Lie structure of an associative ring. *J. Algebra* **14**, 561–571. MR 41 #270.
[75] On the hypercenter of a ring. *J. Algebra* **36**, 151–157. MR 51 #8179.
[76] A commutativity theorem. *J. Algebra* **38**, 112–118. MR 53 #549.

HERSTEIN, I. N., and SMALL, L.
[64] Nil rings satisfying certain chain conditions. *Canad. J. Math.* **16**, 771–776; addendum ibid. **18**, 300–302 (1966). MR 29 #3497.

HIGMAN, G.
[54] Indecomposable representations at characteristic p. *Duke Math. J.* **7**, 377–381.

[56] On a conjecture of Nagata. *Proc. Cambridge philos. Soc.* **52**, 1–4. MR 17 #453e.

HINOHARA, Y.

[60] Note on noncommutative semilocal rings. *Nagoya Math. J.* **17**, 161–166. MR 24A #2595.

HIRANO, Y., and YAMAKAWA, H.

[84] On nil and nilpotent derivations. *Math. J. Okayama Univ.* **26**, 137–141.

HOCHSCHILD, G.

[45] On the cohomology groups of an associative algebra. *Ann. Math.* (2) **46**, 56–87. MR 6 #114f.

[46] On the cohomology theory for associative algebras. *Ann. Math.* (2) **47**, 568–579. MR 8 #64c.

[47] Cohomology and representations of associative algebras. *Duke Math. J.* **14**, 921–948. MR 9 #267b.

[56] Relative homological algebra. *Trans. Amer. Math. Soc.* **82**, 246–269. MR 18 #278a.

HODGES, T.

[84] An example of a primitive polynomial ring. *J. Algebra* **90**, 217–219. MR 86i #16003.

HOPKINS, C.

[39] Rings with minimal condition for left ideas. *Ann. Math.* **40**, 712–730. MR 1 #2d.

HUDRY, A.

[84] Dimension de Gel'fand-Kirillov et décomposition primaire bilatère dans certaines algèbres filtrées. *Comm. Alg.* **12**, 1547–1565. MR 85k #16025.

[86] Sur le principe d'additivité du rang pour des bimodules, II. *Comm. Alg.* **14**, 1429–1438.

HUPERT, L. E. P.

[81] Homological characteristics of rings having Jacobson radical squared zero. *J. Algebra* **69**, 32–42. MR 83b #16018b.

INGRAHAM, E. C.

[74] On the existence and conjugacy of inertial subalgebras. *J. Algebra* **31**, 547–556. MR 50 #2245.

IRVING, R. S.

[78] Some primitive differential operator rings. *Math Z.* **160**, 241–247. MR 58 #16804.

[79] Generic flatness and the Nullstellensatz for Ore extensions. *Comm. Alg.* **7**, 259–277. MR 80b #13005.

[79a] Noetherian algebras and the Nullstellensatz. *Lecture Notes in Mathematics* 740, pp. 80–87. Springer-Verlag, Berlin. MR 81c #16018.

[79b] On the primitivity of certain Ore extensions. *Math. Ann.* **242**, 177–192. MR 80i #16003.

[79c] Prime ideals of Ore extensions over commutative rings II. *J. Algebra* **58**, 399–423. MR 80m #16003.

[79d] Primitive ideals of certain Noetherian algebras. *Math. Z.* **169**, 77–92. MR 81d #16006.

[79e] Some primitive group rings. *J. Algebra* **56**, 274–281. MR 81c #16018.

[84] Primitive ideals of enveloping algebras over commutative rings. *Amer. J. Math.* **106**, 113–135. MR 85h #16033.

[84a] Algebras with bounded growth and endomorphism rings of simple modules. Preprint.

IRVING, R. S., and SMALL, L. W.

[80] On the characterization of primitive ideals in enveloping algebras. *Math. Z.* **173**, 217–221. MR 82j #17015.

[83] The Goldie conditions for algebras with bounded growth. *Bull London Math. Soc.* **15**, 596–600. MR 85a #16017.

[87] The embeddability of affine PI-algebras in rings of matrices. *J. Algebra* **103**, 708–716.

JACOB, B., and WADSWORTH, A. R.

[86] A new construction of noncrossed product algebras. *Trans. Amer. Math. Soc.* **293**, 693–721.

JACOBSON, N.

[45] A topology for the set of primitive ideals in an arbitrary ring. *Proc. Natl. Acad. Sci. U.S.A.* **31**, 333–338. MR 7 #110f.

[45a] The radical and semisimplicity for arbitrary rings. *Amer. J. Math.* **67**, 300–320. MR 7 #2f.

[45b] Structure theory for algebraic algebras of bounded degree. *Ann. Math.* **46** (2), 695–707. MR 7 #238c.

[45c] Structure theory of simple rings without finiteness assumptions. *Trans. Amer. Math. Soc.* **57**, 288–245. MR 6 #200a.

[47] A note on division rings. *Amer. J. Math.* **69**, 27–36. MR 9 #4c.

[47a] On the theory of primitive rings. *Ann. Math.* **48** (2), 8–21. MR 8 #433a.

[50] Some remarks on one-sided inverses. *Proc. Amer. Math. Soc.* **1**, 352–355. MR 12 #75e.

[64] Clifford algebras for algebras with involution of type D. *J. Algebra* **1**, 288–300. MR 29 #5849.

[83] Brauer factor sets, Noether factor sets, and crossed products. *Emmy Noether in Bryn Mawr*, pp. 1–19. Springer-Verlag, New York. MR 84k #16020.

JAIN, S. K.

[71] Prime rings having a one-sided ideal with a polynomial identity coincide with special Johnson rings. *J. Algebra* **19**, 125–130. MR 36 #6451.

JANS, J. P.

[59] On Frobenius algebras. *Ann. Math.* **69** (2), 392–407. MR 21 #3464.

[60] Verification of a conjecture of Gerstenhaber. *Proc. Amer. Math. Soc.* **11**, 335–336. MR 22 #5654.

[66] Some aspects of torsion. *Pacific J. Math.* **15**, 1249–1259. MR 33 #162.

JANUSZ, G. J.

[69] Indecomposable modules for finite groups. *Ann. Math.* **89**, 209–241. MR 39 #5622.

[72] The Schur index and root of unit. *Proc. Amer. Math. Soc.* **35**, 387–388. MR 46 #1891.

JATEGAONKAR, A. V.

[68] Left principal ideal domains. *J. Algebra* **8**, 148–155. MR 36 #1474.

[69] A counterexample in ring theory and homological algebra. *J. Algebra* **12**, 418–440. MR 39 #1485.

[72] Skew polynomial rings over orders in Artinian rings. *J. Algebra* **21**, 51–59. MR 45 #8672.

[74] Jacobson's conjecture and modules over fully bounded Noetherian rings. *J. Algebra* **30**, 103–121. MR 50 #4657.

[74a] Relative Krull dimension and prime ideals in right Noetherian rings. *Comm. Alg.* **2**, 429–468; addendum ibid. **10**, 361–366 (1982). MR 50 #9962; 84i #16035.

[74b] Injective modules and localization in noncommutative Noetherian rings. *Trans. Amer. Math. Soc.* **190**, 109–123. MR 50 #2220.

[75] Certain injectives are Artinian. *Lecture Notes in Mathematics* 545, pp. 128–139. Springer-Verlag, Berlin. MR 55 #5688.

[81] Noetherian bimodules, primary decomposition, and Jacobson's conjecture. *J. Algebra* **71**, 379–400. MR 83d #16016.

[82] Solvable Lie algebras, polycyclic-by-finite groups, and bimodule Krull dimension. *Comm. Algebra* **10**, 19–69. MR 84i #16014.

JENSEN, C. U., and JONDROP, S.

[73] Centres and fixed-point rings of Artinian rings. *Math. Z.* **130**, 189–197. MR 47 #6769.

JESPERS, E., KREMPA, J., and PUCZYLOWSKI, E. R.

[82] On radicals of graded rings. *Comm. Alg.* **10**, 1849–1854. MR 84c #16003b.

JOHNSON, R. E.

[51] The extended centralizer of a ring over a module. *Proc. Amer. Math. Soc.* **2**, 891–895. MR 13 #618c.

[61] Quotient rings of rings with zero singular ideal. *Pacific J. Math.* **11**, 1385–1392. MR 26 #1331.

[63] Principal right ideal rings. *Canad. J. Math.* **15**, 297–301. MR 26 #3741.

[65]  Unique factorization in principal right ideal domains. *Proc. Amer. Math. Soc.* **16**, 526–528. MR 31 #203.

JONAH, D.

[70]  Rings with the minimum condition for principal right ideals have the maximum condition for principal left ideals. *Math. Z.* **113**, 106–112. MR 21 #5402.

JONDRUP, S.

[81]  On the centre of a hereditary P. I. algebra. *Comm. Algebra* **9**, 1673–1679. MR 83e #16017.

JONDRUP, S., and SIMSON, D.

[81]  Indecomposable modules over semiperfect rings. *J. Algebra* **73**, 23–29. MR 83g #16045.

JOSEPH, A.

[74]  Proof of the Gelfand-Kirillov conjecture for solvable Lie algebras. *Proc. Amer. Math. Soc.* **45**, 1–10. MR 52 #522.

[80]  Goldie rank in the enveloping algebra of a semisimple Lie algebra I,II. *J. Algebra* **65**, 269–306. MR 82f #17009.

[81]  Goldie rank in the enveloping algebra of a semisimple Lie algebra III. *J. Algebra* **73**, 295–326. MR 83k #17010.

[83]  On the classification of primitive ideals in the enveloping algebra of a semisimple Lie algebra. *Lecture Notes in Mathematics* 1024, pp. 30–76. Springer-Verlag, Berlin. MR 85b #17008.

[84]  Primitive ideals in enveloping algebras. *Proceedings of the International Congress of Warsaw* (1983), pp. 403–414. MR 86m #17013.

[85]  On the associated variety of a primitive ideal. *J. Algebra* **93**, 509–523. MR 86m #17014.

[86]  Rings which are modules in the Bernstein-Gelfand O category.

JOSEPH, A., and SMALL, L. W.

[78]  An additivity principle for Goldie rank. *Israel J. Math.* **31**, 105–114. MR 80j #17005.

KAC, V.

[84]  Infinite root systems, representations of quivers, and invariant theory. *Lecture Notes in Mathematics* 996, pp. 74–108. Springer-Verlag, Berlin.

KAPLANSKY, I.

[46]  On a problem of Kurosch and Jacobson. *Bull. Amer. Math. Soc.* **52**, 496–500. MR 8 #63b.

[48]  Rings with a polynomial identity. *Bull. Amer. Math. Soc.* **54**, 575–580. MR 10 #7a.

[50]  Topological representations of algebras II. *Trans. Amer. Math. Soc.* **68**, 62–75. MR 11 #317.

[51]  A theorem on division rings. *Canad. J. Math.* **3**, 290–292. MR 13 #101f.

[5a]  The structure of certain operator algebras. *Trans. Amer. Math. Soc.* **70**, 219–255. MR 13 #48.

[58]  Projective modules. *Ann. Math* (2) **68**, 372–377. MR 20 #6453.

[70]  Problems in the theory of rings revisited. *Amer. Math. Monthly* **77**, 445–454. MR 41 #3510.

[75]  Bialgebras. *University of Chicago Lecture Notes.* MR 55 #8087.

KEMER, A. R.

[78]  A remark on the standard identity. *Math. Notes* **23**, 414–416. MR 58 #11005.

[84]  Varieties and Z2-graded algebras. *Izv. Akad. Nauk. SSSR Ser. Mat.* **48**, 1042–1059. Isvestija MR 86f #16015.

KERR, J. W.

[83]  Very long chains of annihilator ideals. *Israel J. Math.* **46**, 197–204. MR 85f #16016.

KIRKMAN, E. and KUZMANOVICH, J.

[87]  On the global dimension of a ring modulo its nilpotent radical. *Proc. Amer. Math. Soc.*

KLEIN, A. A.

[67]  Rings nonembeddable in fields with multiplicative semigroups embeddable in groups. *J. Algebra* **7**, 100–125. MR 37 #6309.

[69] *Necessary conditions for embedding rings into fields. Trans. Amer. Math. Soc.* **137**, 141–151. MR 38 #4510.

[80] A simple proof of a theorem of reduced rings. *Canad. Math. Bull.* **23**, 495–496. MR 82b #16019.

KOH, K.

[65] A note on a prime ring with a maximal annihilator right ideal. *Canad. Math. Bull.* **8**, 109–110. MR 30 #4781.

[65a] A note on a self-injective ring. *Canad. Math. Bull.* **8**, 29–32. MR 30 #4780.

[67] On simple rings with uniform right ideals. *Amer. Math. Monthly* **74**, 685–687. MR 36 #3818.

KOH, K., and MEWBORN, A. C.

[65] Prime rings with maximal annihilator and maximal complement right ideals. *Proc. Amer. Math. Soc.* **16**, 1073–1076. MR 36 #2653.

KOSTANT, B.

[58] A theorem of Frobenius, a theorem of Amitsur-Levitzki, and cohomology theory. *Indiana J. Math.* **7**, 237–264. MR 19 #1153.

KRAKOWSKI, D., and REGEV, A.

[73] The polynomial identities of the Grassman algebra. *Trans. Amer. Math. Soc.* **181**, 429–438. MR 48 #4005.

KRAUSE, G.

[78] Some recent developments in the theory of Noetherian rings. *Lecture Notes in Mathematics* 641, pp. 209–219. Springer-Verlag, Berlin. MR 58 #5778.

KREMPA, J.

[72] Logical connections between some open problems concerning nil rings. *Fundamenta Mathematicae* **76**, 121–130. MR 46 #5377.

KREMPA, J., and OKNINSKI, J.

[86] Gelfand-Kirillov dimensions of a tensor product. Preprint.

KUROSCH, A. G.

[41] Ringtheoretische Probleme, die mit dem Burnsideschen Problem über periodische Gruppen in Zusammenhang stehen. *Bull. Akad. Sci. URSS Ser. Math.* **5**, 233–240. MR 3 #194.

[53] Radicals of rings and algebra. (Russian) *Mat. Sbornik N. S.* **33** (75), 13–26. (Translated in *Rings, Modules, and Radicals. Proc. Colloq. Keszthely*, 297–314, 1971) MR 15 #194.

L'VOV, I. V.

[69] Certain imbeddings of rings I and II. (Russian) *Algebra i Logika* **8**, 335–356. MR 41 #6883.

[80] Representation of nilalgebras in over matrices. (Russian) *Sibirsk. Math. J.* **21**, 158–161. MR 82j #16053.

LAMBEK, J.

[63] On Utumi's ring of quotients. *Canad. J. Math.* **15**, 363–370. MR 26 #5024.

[65] On the ring of quotients of a Noetherian ring. *Canad. Math. Bull.* **8**, 279–290. MR 33 #5677.

[73] Noncommutative localization. *Bull. Amer. Math. Soc.* **79**, 857–872. MR 48 #4009.

LAMBEK, J., and MICHLER, G. O.

[73] The torsion theory at a prime ideal of a right Noetherian ring. *J. Algebra* **25**, 364–389. MR 47 #5034.

[74] Localization of right Noetherian rings at semiprime ideals. *Canad. J. Math.* **26**, 1069–1085. MR 50 #13105.

LATYSHEV, V. N.

[72] On Regev's theorem on identities in a tensor product of PI-algebras. (Russian) *Ups. Mast. Nauk.* **27**, 213–214. MR 52 #13924.

LAWRENCE, J.
[74] A singular primitive ring. *Proc. Amer. Math. Soc.* **45**, 59–62. MR 50 #9934.

LAZARD, D.
[64] Sur les modules plats. *C. R. Acad. Sci. Paris* **258**, 6313–6316. MR 29 #5883.

LE BRUYN, L.
[86] The Artin-Schofield theorem and some applications. *Comm. Alg.* **14**, 1439–1456.

LE BRUYN, L. and OOMS ALFONS, I.
[85] The semicenter of an enveloping algebra is factorial. *Proc. Amer. Math. Soc.* **93**, 397–400.
MR 86h #17008.

LENAGAN, T. H.
[74] Nil ideals in rings with finite Krull dimension. *J. Algebra* **29**, 77–87. MR 49 #5074.
[77] Noetherian rings with Krull dimension one. *J. London Math. Soc.* (2) **15**, 41–47. MR 56 #397.
[82] Gel'fand-Kirillov dimension and affine PI-rings. *Comm. Algebra* **10**, 87–92. MR 83m #16022.
[84] Gel'fand-Kirillov dimension is exact for Noetherian PI-algebras. *Canad. Math. Bull.* **27**, 247–250. MR 85c #16023.

LESIEUR, L., and CROISOT, R.
[59] Sur les anneàux premiers Noetheriens gauche. *Ann. Sci. Ecole Norm. Sup.* (3) **76**, 161–183.
MR 22 #54.
[63] Algèbre Noethérienne noncommutative. *Memoire Sci. Math. Fasc.* CLIV. Gauther-Villars & Co., Paris. MR 27 #5795.

LESTMAN, P.
[78] Simple going down in PI-rings. *Proc. Amer. Math. Soc.* **63**, 41–45. MR 55 #5606.

LEVASSEUR, T.
[79] Idéaux premiers et complétion dans les algébres enveloppantes d'algèbres de Lie nilpotentes." *Lecture Notes in Mathematics* 795, pp. 116–160. Springer-Verlag, Berlin.
MR 84j #17010.
[82] Equidimensionalité de la variété caractéristique. Preprint.

LEVI, F. W.
[49] Über der Kommutativitätsrang in einem Ring. *Math. Am.* **121**, 184–190. MR 11 #311.

LEVITZKI, J.
[39] On rings which satisfy the minimum condition for right-handed ideals. *Compositio Math.* **7**, 214–220. MR1 #100c.
[53] On the structure of algebraic algebras and related rings. *Trans. Amer. Math. Soc.* **74**, 384–409. MR 14 #720a.
[63] On nil subrings. *Israel J. Math.* **1**, 215–216. MR 19 #1230.

LEVY, L.
[63] Torsion-free and divisible modules over non-integral domains. *Canad. J. Math.* **15**, 132–151. MR 26 #155.
[83] Krull-Schmidt uniqueness fails dramatically over subrings of $Z \times Z \times \cdots \times Z$. *Rocky Mountain J. Math.* **13**, 659–678. MR 85c #13008.
[85] Modules over $ZG_n$, $G_n$ cyclic of square-free order $n$, and over related Dedekind-like rings. *J. Algebra* **93**, 354–375.

LEWIN, J.
[74] A matrix representation for associative algebras, I and II. *Trans. Amer. Math. Soc.* **188**, 293–317. MR 49 #2848.
[82] Projective modules over group algebras of torsion-free groups. *Mich. Math. J.* **29**, 59–64.
MR 83m #20014.

LEWIN, J., and LEWIN, T.
[75] An embedding of the group algebra of a torsion-free one-relator group in a field. *J. Algebra* **52**, 39–74. MR 58 #5764.

LICHTMAN, A. I.
[76] On embedding of groups rings in division rings. *Israel J. Math.* **23**, 288–297. MR 55 #5678.
[78] The primitivity of free products of associative algebras. *J. Algebra* **54**, 153–158. MR 80g #16003.
[78a] On matrix rings and linear groups over a field of fractions of enveloping algebras and group rings I. *J. Algebra* **88**, 1–37.
[82] On normal subgroups of multiplicative groups of skew fields generated by a polycyclic-by-finite group, *J. Algebra* **78**, 548–577. MR 84f #16016.
[83] Localization in enveloping algebras. *Math. Proc. Cambridge Philos. Soc.* **93**, 467–475. MR 85b #17009,
[84] Growth in enveloping algebras. *Israel J. Math.* **47**, 296–304. MR 86b #17006.
[84a] On matrix rings and linear groups over fields of fractions of group rings and enveloping algebras II. *J. Algebra* **90**, 516–527. MR 86a #16019.

LINNELL, P. A.
[77] Zero divisors and idempotents in group rings. *Math. Proc. Camb. Phil. Soc.* **81**, 365–368. MR 55 #2049.

LORENZ, M.
[77] Primitive ideals of group algebra of supersolvable groups. *Math. Ann.* **225**, 115–122. MR 54 #12820.
[81] Chains of prime ideals in enveloping algebras of solvable Lie algebras. *J. London Math. Soc.* **24** (2), 205–210. MR 83e #17011.
[81a] Finite normalizing extensions of rings. *Math. Z.* **176**, 447–484. MR 82k #16043.
[81b] Prime ideals in group algebras of polycyclic-by-finite groups. *Lecture Notes in Mathematics* 867, pp. 406–420. Springer-Verlag, Berlin. MR 83j #20017.
[81c] The Goldie rank of prime supersolvable group algebras. *Mitt. Math. Sem. Giessen* **149**, 115–129. MR 82j #16023.
[82] On the Gelfand-Kirillov dimension of skew polynomial rings. *J. Algebra* **77**, 186–188. MR 83g #16004.
[84] Group rings and division rings. *Methods in Ring Theory*, pp. 265–280. D. Reidel, Dordrecht. MR 86d #16016.
[85] On affine algebras. *Lecture Notes in Mathematics* 1197, pp. 121–126. Springer-Verlag, Berlin. MR 87i #16032.

LORENZ, M. and PASSMAN, D. S.
[79] Integrality and normalizing extensions of rings. *J. Algebra* **61**, 289–297. MR 81e #16022.
[81] Prime ideals in group algebras of polycyclic-by-finite groups. *Proc. London Math. Soc.* **43** (3), 520–543. MR 83j #20017.

LORENZ, M., and SMALL, L. W.
[82] On the Gel'fand-Kirillov dimension of Noetherian P. I.-algebras. *Contemp. Math.* **13**, 199–205. MR 84d #16022.

LUMINET, D.
[86] A functional calculus for Banach algebras. *Pacific J. Math.* **125**, 127–160.

MCCONNELL, J. C.
[68] Localisation in enveloping rings. *J. London Math. Soc.* **43**, 421–428; errat. and addend. ibid. **3** (2), 409–410. MR 37 #4112.
[69] The Noetherian property in complete rings and modules. *J. Algebra* **12**, 143–153. MR 39 #1487.
[74] Representations of solvable Lie algebras and the Gelfand-Kirillov conjecture. *Proc. London Math. Soc.* **29** (3), 454–484. MR 50 #9997.
[75] Representations of solvable Lie algebras. II. Twisted group rings. *Ann. Sci. École Norm. Sup.* **8** (4), 157–178. MR 51 #12966.
[77] On the global dimension of some rings. *Math. Z.* **153**, 253–254. MR 56 #15703.

[77a] Representations of solvable Lie algebras. IV. An elementary proof of the (U/P)E-structure theorem. *Proc. Amer. Math. Soc.* **64**, 8–12. MR 56 #12084.

[79] I-adic completions of non-commutative rings. *Israel J. Math.* **32**, 305–310. MR 81i #16015.

[82] Representations of solvable Lie algebras. V. On the Gelfand-Kirillov dimension of simple modules. *J. Algebra* **76**, 489–493. MR 83m #16023.

[82a] The Nullstellensatz and Jacobson properties for rings of differential operators. *J. London Math. Soc.* (2) **26**, 37–42. MR 84a #16006.

[84] On the global and Krull dimensions of Weyl algebras over affine coefficient rings. *J. London Math. Soc.* (2) **29**, 249–253.

[84] The Nullstellensatz and extensions of affine PI-rings. *J. London Math. Soc.* (2) **29**, 254–256. MR 85i #16019.

[85] The K-theory of filtered rings and skew Laurent extensions. *Lecture Notes in Mathematics* 1146, pp. 288–298. Springer-Verlag, Berlin.

MAKAR-LIMANOV, L.

[83] The skew field of fractions of the Weyl algebra contains a free noncommutative subalgebra. *Comm. Algebra* **11**, 2003–2006. MR 84j #16012.

[74] On free subsemigroups of skew fields. *Proc. Amer. Math. Soc.* **91**, 189–191. MR 85j #16022.

MAL'CEV, A. I.

[37] On the immersion of an algebraic ring into a field. *Math. Ann.* **117**, 686–691.

[42] On the representation of an algebra as a direct sum of the radical and a semi-simple subalgebra. *C.R. Acad. Sci. USSR (N.S.)* **36**, 42–45. MR 4 #130c.

[48] On the embedding of group algebras in division algebras. (Russian) *Dokl. Akad. Nauk. SSR* **60**, 1499–1501. MR 10 #8.

MALLIAVIN, M. P.

[79] Caténarité et théorème d'intersection en algèbre noncommutative. *Lecture Notes in Mathematics* 740, pp. 408–431. Springer-Verlag, Berlin. MR 82f #17010.

[76] Dimension de Gelfand-Kirillov des algèbres à identités polynomiales. *C.R. Acad. Sci.*, Paris sér. A-B, **282**, A679–A681. MR 53 #5654.

MAMMONE, P.

[86] Sur la corestriction des p-symboles. *Comm. Alg.* **14**, 517–530.

MARTINDALE, W. S. III

[69] Lie isomorphisms of prime rings. *Trans. Amer. Math. Soc.* **142**, 437–455. MR 40 #4308.

[69a] Prime rings satisfying a generalized polynomial identity. *J. Algebra* **12**, 576–584. MR 39 #257.

[72] Prime rings with involution and generalized polynomial identities. *J. Algebra* **22**, 502–516. MR 46 #5371.

[73] On semiprime PI-rings. *Proc. Amer. Math. Soc.* **40**, 364–369. MR 47 #6762.

[78] Fixed rings of automorphisms and the Jacobson radical. *J. London Math. Soc.* (2) **17**, 42–46. MR 57 #6107.

[82] Lie and Jordan mappings. *Contemp. Math.* **13**, 173–177. MR 84a 00007.

MARTINDALE, W. S. III, and MONTGOMERY, S.

[77] Fixed rings of Jordan automorphisms of associative rings. *Pacific J. Math.* **72**, 181–196. MR 58 #22183.

[83] The normal closure of coproducts of domains. *J. Algebra* **82**, 1–17. MR 84k #16003.

MATLIS, E.

[58] Injective modules over Noetherian rings. *Pacific J. Math.* **8**, 511–528. MR 20 #5800.

MERKURJEV, A. S.

[81] On the norm residue symbol of degree 2. *Soviet Math. Doklady* **24**, 546–551. MR 84i #12007.

[83] Brauer groups of fields. *Comm. Alg.* **11**, 2611–2624. MR 85c #11121.

[86] On the Brauer groups of fields. Preprint.

MERKURJEV, A. S., and SUSLIN, A. A.
[82] K-cohomology of Severi-Brauer varieties and norm residue homomorphism. *Izv. Akad. Nauk.* USSR **46**, 1011–1046.
[86] On the structure of the Brauer groups of fields. *Math. USSR Izvestiya* **27**, 141–155.

MEWBORN, A. C., and WINTON, C. N.
[69] Orders in self-injective semiperfect rings. *J. Algebra* **13**, 5–9. MR 39 #6925.

MICHLER, G. O.
[66] Radicals and structure spaces. *J. Algebra* **4**, 199–219. MR 33 #5672.

MICHLER, G. O., and VILLAMAYOR, O. E.
[73] On rings whose simple modules are injective. *J. Algebra* **25**, 185–201. MR 47 #5052.

MILLER, R. W., and TURNIDGE, D. R.
[73] Some examples from infinite matrix rings. *Proc. Amer. Math. Soc.* **38**, 65–67. MR 46 #9104.

MILNOR, J. W.
[68] Growth of finitely generated solvable groups. *J. Differential Geometry* **2**, 447–449. MR 38 #636.

MILNOR, J. W., and MOORE, J. C.
[64] On the structure of Hopf algebras. *Ann. Math.* **81**, 211–264. MR 30 #4259.

MITCHELL, B.
[68] On the dimension of objects and categories. I. Monoids. II Finite ordered sets. *J. Algebra* **9**, 314–368. MR 38 #5885.

MOEGLIN, C.
[80] Idéaux primitifs des algèbres enveloppantes. *J. Math. Pures Appl.* **59**, 265–336. MR 83i #17008.
[80a] Ideaux bilatères des algèbres enveloppantes. *Bull. Soc. Math. Fr.* **108**, 143–186. MR 84e #17012.

MOEGLIN, C., and RENTSCHLER, R.
[81] Orbites d'un groupe algébrique dans l'espace des idéaux rationnels d'une algèbre enveloppante. *Bull. Soc. Math. Fr.* **109**, 403–426. MR 83i #17009.
[84] Sur la classification des idéaux primitifs des algèbres enveloppantes. *Bul. Soc. Math. France* **112**, 3–40. MR 86e #17006.

MONTGOMERY, S.
[76] The Jacobson radical and fixed rings of automorphisms. *Comm. Algebra* **4**, 459–465. MR 53 #2995.

MONTGOMERY, S., and PASSMAN, D. S.
[84] Galois theory of prime rings. *J. Pure Appl. Algebra* **31**, 139–184. MR 85i 16047.

MONTGOMERY, S., and SMALL, L.
[84] Integrality and prime ideals in fixed rings of prime PI-rings. *J. Pure Appl. Algebra* **31**, 185–190. MR 85i 16020.

MODDY, J.
[86] Dissertation, Columbia University.

MORITA, K.
[58] Duality for modules and its application to the theory of rings with minimum condition. *Sci. Rep. Tokyo Kyoiku Daigaku* (Sect. A) **6**, 83–142. MR 20 #3183.
[62] Category-isomorphisms and endomorphism rings of modules. *Trans. Amer. Math. Soc.* **103**, 451–469. MR 25 #3978.
[72] Quotient rings. *Ring Theory*, 257–285. Academic Press, New York. MR 49 #7290.

MÜLLER, B. J.
[74] The quotient category of a Morita context. *J. Algebra* **28**, 389–407. MR 56 #5649.
[74a] The structure of quasi-Frobenius rings. *Canad. J. Math.* **26**, 1141–1151. MR 52 #482.
[76] Localization in non-commutative Noetherian rings. *Canad. J. Math.* **28**, 600–610. MR 53 #13290.
[79] Ideal invariance and localization. *Comm. Algebra* **7**, 415–441. MR 80i #16007.

[84] Morita duality-a survey, Abelian groups and modules. *CISM Courses and Lectures* 287. Springer-Verlag, Berlin. MR 86i #16019.

[85] Affine Noetherian PI-algebras have enough clans. *J. Algebra* 97, 116–129.

MULLER, W.

[74] On Artin rings of finite representation type. *Mathematical Lecture Notes* 9 (paper 19). Carleton University, Ottawa. MR 54 #12834.

[74a] Unzerlegbare Moduln uber Artinschen Ringen. *Math. Z.* 137, 197–226. MR 50 #4665.

MUSSON, I. M.

[82] Some examples of modules over Noetherian rings. *Glasgow Math. J.* 23, 9–13, MR 83g #16029.

NAGARAJAN, K.

[68] Groups acting on Noetherian rings. *Nieuw. Arch. Wisk.* 16, 25–29. MR 37 #5202.

NAKAYAMA T.

[51] A remark on finitely generated modules. *Nagoya Math. J.* 3, 139–140. MR 13 #313f.

[55] Über die Kommutativität gewisser Ringe. *Abh. Math. Sem. Hamburg* 20, 20–27. MR 17 #341.

[59] A remark on the commutativity of algebraic rings. *Nagoya Math. J.* 14, 39–44. MR 21 #74.

NAZAROVA, L. A., and ROITER, A. V.

[71] Matrix questions and the Brauer-Thrall conjectures on algebras with an infinite number of indecomposable representations. *Amer. Math. Soc. Proc. Sym. Pure Math.* XXI, 111–115. MR 47 #3457.

[73] Categorical matrix problems and the Brauer-Thrall problem. *Izdat. Naukova Dumka* (preprint in Russian), Kiev. (Transl. in *Mitt. Math. Sem. Giessen Heft* 115, (1975)). MR 54 #360.

NEUMANN, B. H.

[49] On ordered division rings. *Trans Amer. Math. Soc.* 66, 202–252. MR 11 #311.

NOETHER, E.

[211] Idealtheorie in Ringbereichen. *Math. Ann.* 83, 24–66.

NOUAZE, Y., and GABRIEL, P.

[67] Ideaux premiers de l'algèbre enveloppante d'une algèbre de Lie nilpotente. *J. Algebra* 6, 77–99. MR 34 #5889.

[67a] Remarques sur "Ideaux premiers de l'algèbre enveloppante d'une algèbre de Lie nilpotente." *Bull. Sci. Math.* 91 (2), 117–124. MR 37 #5262.

ORE, O.

[33] Theory of noncommutative polynomials. *Ann. Math.* 34, 480–508.

[35] On the foundation of abstract algebra I. *Ann. Math.* 36, 406–437.

[36] On the foundation of abstract algebra II. *Ann. Math.* 37, 265–292.

OSOFSKY, B. L.

[64] On ring properties of injective hulls. *Canad. Math. Bull.* 7, 405–413. MR 29 #3504.

[66] A generalization of quasi-Frobenius rings. *J. Algebra* 4, 373–387; correction ibid. 9, 20 (1968). MR 34 #4305.

[68] Noninjective cyclic modules. *Proc. Amer. Math. Soc.* 19, 1383–1384. MR 38 #185.

[68a] Homological dimension and the continuum hypothesis. *Trans. Amer. Math. Soc.* 132, 217–230. MR 37 #205.

[68b] Upper bounds on homological dimensions. *Nagoya Math. J.* 32, 315–322. MR 38 #1128.

[70] A remark on the Krull-Schmidt-Azumaya theorem. *Canad. Math. Bull.* 13, 501–505. MR 43 #281.

[84] A semiperfect one-sided injective ring. *Comm. Algebra* 12, 2037–2041. MR 85c #16034.

PAGE, A.

[84] On the center of hereditary PI-rings. *J. London Math. Soc.* 30, 193–196.

PASSMAN, D. S.

[71] Group rings satisfying a polynomial identity II. *Pacific J. Math.* **39**, 425–438. MR 44 #6849.

[71a] Group rings satisfying a polynomial identity. *J. Algebra* **20**, 103–117. MR 44 #2849.

[71b] Linear identities in group rings, I and II. *Pacific J. Math.* **36**, 457–505. MR 44 #333.

[79] The Jacobson radical of a group ring of a locally solvable group. *Proc. London Math. Soc.* **38** (3), 169–192. MR 81j #16017a.

[81] Prime ideals in normalizing extensions. *J. Algebra* **73**, 556–572. MR 83e #16005.

[82] Universal fields of fractions for polycyclic group algebras. *Glasgow Math. J.* **23**, 103–113. MR 84a #16021.

[83] It's essentially Maschke's theorem. *Rocky Mountain J. Math.* **1**, 37–54. MR 84e #16023.

[84] Group rings of polycyclic groups. *Group Theory–Essays for Philip Hall.* Academic Press, New York. MR 85e #16019.

[86] Group rings, crossed products, and Galois theory. CBMS pub. #64. American Mathematical Society, Providence. MR 87e #16033.

PEARSON, K. R., and STEPHENSON, W.

[77] A skew polynomial ring over a Jacobson ring need not be a Jacobson ring. *Comm. Algebra* **5**, 783–794. MR 55 #12754.

PERLIS, S.

[69] Cyclicity of division algebras of prime degree. *Proc. Amer. Math. Soc.* **21**, 409–411. MR 39 #261.

PIERCE, R. S.

[67] Modules over commutative regular rings. *Memoirs of the AMS* 70. American Mathematical Society, Providence. MR 36 #151.

PLATONOV, V. P.

[75] The Tannaka-Artin problem and groups of projective conorms. *Soviet Math. Dokl.* **16**, 468–473, 782–786. MR 52 #5728b.

[76] The Tannaka-Artin problem and reduced K-theory. (Russian) *Izv. Akad. Nauk. SSR Ser. Mat.* **40**, 227–261, 469. MR 53 #10865.

PROCESI, C.

[67] Noncommutative affine rings. *Atti Accad. Naz. Lincei* **8**, 239–255. MR 37 #256.

[72] On a theorem of M. Artin. *J. Algebra* **22**, 309–315. MR 46 #1825.

[72a] On the identities of Azumaya algebras. *Ring Theory.* Academic Press, New York. MR 49 #2828.

[74] Finite-dimensional representations of algebras. *Israel J. Math.* **19**, 169–182. MR 50 #9975.

[76] The invariant theory of $n \times n$ matrices. *Advances in Math.* **198**, 306–381. MR 54 #7512.

[84] Computing with $2 \times 2$ matrices. *J. Algebra* **87**, 342–359. MR 86g #16022.

PROCESI, C., and SCHACHER, M.

[76] A noncommutative real Nullstellensatz and Hilbert's 17th problem. *Ann. Math.* **104**, 395–406. MR 55 #5599.

PUCZYLOWSKI, E.

[84] On fixed rings of automorphisms. *Proc. Amer. Math. Soc.* **90**, 517–518. MR 85f #16046.

QUILLEN, D.

[69] On the endomorphism ring of a simple module over an enveloping algebra. *Proc. Amer. Math. Soc.* **21**, 171–172. MR 39 #252.

[72] Higher algebraic K-theory: I. *Lecture Notes in Mathematics* 341, pp. 85–139. Springer-Verlag, Berlin. MR 49 #2895.

[76] Projective modules over polynomial rings. *Inventiones Math.* **36**, 167–171. MR 55 #337.

RADFORD, D. E.

[76] The order of the antipode of a finite dimensional hopf algebra is finite. *Amer. J. Math.* **98**, 333–355. MR 53 #10852.

[77]   Finiteness conditions for a Hopf algebra with a nonzero integral. *J. Algebra* **46**, 189–195. MR 56 #5628.

RAZMYSLOV, YU. P.
[73]   A certain problem of Kaplansky. *Math. USSR-Isv.* **7**, 479–496. MR 49 #2830.
[73a]  Finite basing of the identities of a matrix algebra of second order over a field of characteristic 0. *Algebra and Logic* **12**, 47–61. MR 49 #5103.
[74]   The Jacobson radical in PI-algebras. *Algebra and Logic* **13**, 193–204. MR 54 #7536.
[74a]  The finite basis property for certain varieties of algebras. *Algebra and Logic* **13**, 394–399. MR 53 #548.
[74b]  Trace identities of full matrix algebras over a field of characteristic 0. *Math. USSR-Isv.* **8**, 727–760. MR 58 #22158.
[85]   Trace identities and central polynomials in matrix superalgebras $M_{n,k}$. (Russian) *Mat. Sb.* (*N.S.*) **128**(170), 194–215, 287. MR 87f #16014.

REGEV, A.
[72]   Existence of identities in A × B. *Israel J. Math.* **111**, 131–152. MR 44 #1694.
[78]   The representations of $S_n$ and explicit identities for PI-algebras. *J. Algebra* **51**, 25–40. MR 57 #9745.
[80]   The polynomial identities of matrices in characteristic zero. *Comm. Alg.* **8**, 1417–1467.
[84]   Codimensions and trace codimensions of matrices are asymptotically equal. *Israel J. Math.* **47**, 246–250. MR 85j #16024.

REITEN, I.
[77]   Almost split sequences for group algebras of finite representation type. *Trans. Amer. Math. Soc.* **233**, 125–136. MR 58 #28056.
[80]   An overview of integral representation theory. *Lecture Notes in Mathematics* 55, pp. 345–356. Marcel Dekker, New York. MR 82c #16001.
[82]   The use of almost split sequences in the representation theory of Artin algebras. *Lecture Notes in Mathematics* 944, pp. 29–104. Springer-Verlag, Berlin. MR 83j #16040.

RENAULT, G.
[73]   Anneaux reguliers auto-injectifs à droite. *Bull. Soc. Math. France* **101**, 237–254. MR 51 #8171.

RENTSCHLER, R.
[87]   Primitive ideals in enveloping algebras (general case). *Mathematical Surveys and Monographs* 24, pp. 37–57. American Mathematical Society, Providence.

RENTSCHLER, R., and GABRIEL, P.
[67]   Sur la dimension des anneaux et ensembles ordonnés. *C.R. Acad. Sci. Paris* **265**, A712–A715. MR 37 #243.

RESCO, R.
[79]   Transcendental division algebras and simple Noetherian rings. *Israel J. Math.* **32**, 236–256. MR 80i #16027.
[81]   Radicals of finite normalizing extensions. *Comm. Alg.* **9**, 713–725. MR 82f #16005.
[84]   Affine domains of finite Gel'fand-Kirillov dimension which are right, but not left, Noetherian. *Bull. London Math. Soc.* **16**, 590–594. MR 85k #16047.

RESCO, R., SMALL, L., and STAFFORD, T.
[82]   Krull and global dimensions of semiprime Noetherian PI-rings. *Trans. Amer. Math. Soc.* **274**, 285–295.

RESCO, R., SMALL, L. W., and WADSWORTH, A. R.
[79]   Tensor products of division rings and finite generation of subfields. *Proc. Amer. Math. Soc.* **77**, 7–10. MR 80g #16020.

RIEDTMANN, C.
[85]   Algèbres de type de représentation fini d'après Bongartz, Gabriel, Roiter, et d'autres." *Séminaire Bourbaki Astérisque* **133–134**, 335–350. MR 87g #16032.

RIEHM, C.
[70] The corestriction of algebraic structures. *Inventiones Math.* **11**, 73–98. MR 45 #8736.
RINEHART, G. S., and ROSENBERG, A.
[76] The global dimension of Ore extensions and Weyl algebras. *Algebra, Topology, and Category Theory*, pp. 169–180. Academic Press, New York. MR 53 #13317.
RINGEL, C. M.
[80] Report on the Brauer-Thrall conjectures. *Lecture Notes in Mathematics* 831, pp. 104–136. Springer-Verlag, Berlin. MR 82j #16055.
[80a] Tame algebras. *Lecture Notes in Mathematics* **831**, 137–287. Springer-Verlag, Berlin. MR 82j #16056.
RISMAN, L.
[77] Non-cyclic division algebras. *J. Pure Appl. Algebra* **11**, 199–216. MR 58 #11007.
RITTERE, J., and SEHGAL, S.
[83] Isomorphism of group rings. *Arch. Math.* **40**, 32–39. MR 84k #16015.
ROBSON, J. C.
[67] Artinian quotient rings. *Proc. London Math Soc.* (3) **17**, 600–616. MR 36 #199.
[74] Decomposition of Noetherian rings. *Comm. Algebra* **1**, 345–349. MR 49 #7310.
[85] Some constructions of rings of finite global dimension. *Glasgow Math. J.* **26**, 1–11. MR 86i #16026.
ROBSON, J. C., and SMALL, L. W.
[74] Hereditary prime PI-rings are classical hereditary orders. *J. London Math. Soc.* **8**, 499–503. MR 50 #2236.
[81] Liberal extensions. *Proc. London Maths. Soc.* (3) **42**, 87–103. MR 82c #16025.
ROITER, A. V.
[68] Unboundedness of the dimensions of the indecomposable representations of an algebra which has infinitely many indecomposable representations. (Russian) *Izv. Akad. Nauk. SSR Ser. Mat.* **32**, 1275–1282. MR 39 #253.
[80] Matrix problems and representations of BOC's. *Lecture Notes in Mathematics* 831, pp. 288–324. Springer-Verlag, Berlin. MR 83e #16034.
ROOS, J.-E.
[72] Détermination de la dimension homologique des algèbres de Weyl. *C.R. Acad. Sci. Paris* Sér. A. **274**, 23–26. MR 45 #1995.
ROQUETTE, P.
[63] On the Galois cohomology of the projective linear group and its applications to the construction of generic splitting fields of algebras. *Math. Ann.* **150**, 411–439. MR 27 #4832.
ROSEBLADE, J. E.
[73] Group rings of polycyclic groups. *J. Pure Appl. Algebra* **3**, 307–328. MR 48 #11269.
[76] Applications of the Artin-Rees lemma to group rings. *Symposia Matematica* XVII, pp. 471–478. Academic Press, New York. MR 53 #10902.
[78] Prime ideals in group rings of polycyclic groups. *Proc. London Math. Soc.* (3) **36**, 385–447; correction ibid. **38**, 216–218 (1979). MR 58 #10996.
ROSENBERG, A.
[82] Homological dimension and transcendence degree. *Comm. Alg.* **10**, 329–338. MR 84c #16026.
ROSENBERG, A., and STAFFORD, J. T.
[76] Global dimension of Ore extensions. *Algebra, Topology, and Category Theory*, pp. 181–188. Academic Press, New York. MR 53 #13318.
ROSENBERG, A., and ZELINSKY, D.
[56] Cohomology of infinite algebras. *Trans. Amer. Math. Soc.* **82**, 85–98. MR 17 #1181b.
[58] Finiteness of the injective hull. *Math. Z.* **70**, 372–380. MR 21 #4176.

ROSSET, S.

[76] A property of groups of nonexponential growth. *Proc. Amer. Math. Soc.* **54**, 24–26.

[77] Abelian splitting of division algebras of prime degrees. *Comment. Math. Helvetici* **52**, 519–523. MR 58 #27911.

[78] Group extensions and division algebras. *J. Algebra* **53**, 297–303. MR 58 #16627.

[84] A vanishing theorem for Euler characteristics. *Math. Z.* **185**, 211–215. MR 85i #55002.

ROSSET, S., and TATE, J.

[83] A reciprocity law for $K_2$-traces. *Comment. Math. Helv.* **58**, 38–47. MR 85b #11105.

ROWEN, L. H.

[73] Some results on the center of a ring with polynomial identity. *Bull. Amer. Math. Soc.* **79**, 219–223. MR 46 #9099.

[74] On rings with central polynomials. *J. Algebra* **31**, 393–426. MR 50 #2237.

[74a] Maximal quotients of semiprime PI-algebras. *Trans. Amer. Math. Soc.* **196**, 127–135. MR 50 #388.

[74b] Universal PI-algebras and algebras of generic matrices. *Israel J. Math.* **18**, 65–74. MR 51 #596.

[75] Generalized polynomial identities. *J. Algebra* **34**, 458–490. MR 51 #8162.

[76] Generalized polynomial identities II. *J. Algebra* **38**, 380–392. MR 57 #3189a.

[78] Central simple algebras. *Israel J. Math.* **29**, 285–301. MR 58 #11008.

[82] Cyclic division algebras. *Israel J. Math.* **41**, 213–214; corr. ibid. **43**, 277–280 (1982). MR 84d #12021.

[82a] A simple proof of Kostant's theorem and an analogue for the symplectic involution. *Contemp. Math.* **13**, 207–215. MR 84d #16024.

[84] Brauer factor sets and simple algebras. *Trans. Amer. Math. Soc.* **282**, 767–772.

[86] Finitely presented modules over semiperfect rings. *Proc. Amer. Math. Soc.* **97**, 1–7.

ROWEN, L. H., and SALTMAN, D.

[82] Dihedral algebras are cyclic. *Proc. Amer. Math. Soc.* **84**, 162–164. MR 83c #16013.

ROY, A.

[65] A note on filtered rings. *Arch. Math.* **16**, 421–427. MR 32 #7598.

SALTMAN, D. J.

[77] Splittings of cyclic p-algebras. *Proc. Amer. Math. Soc.* **62**, 223–228. MR 55 #8006.

[78] Noncrossed product p-algebras and Galois extensions. *J. Algebra* **52**, 302–314. MR 58 #796.

[78a] Azumaya algebras with involution. *J. Algebra* **52**, 526–539. MR 80g #16021.

[79] Indecomposable division algebras. *Comm. Alg.* **7**, 791–817. MR 80g #16021.

[79a] Retract rational fields and cyclic Galois extensions. *Israel J. Math.* **47**, 165–215.

[81] The Brauer group is torsion. *Proc. Amer. Math. Soc.* **81**, 385–387.

[82] Generic structures. *Lecture Notes in Mathematics* 917, pp. 96–117. Springer-Verlag, Berlin. MR 83f #16028.

[82a] Generic structures and field theory. *Contemp. Math.* **13**, 127–134. MR 84a #00007.

[85] The Brauer group and the center of generic matrices. *J. Algebra* **97**, 53–67.

SANDOMIERSKI, F. L.

[73] Homological dimension under change of rings. *Math. Z.* **130**, 55–65. MR 47 #6772.

SARRAILLÉ, J. J.

[82] Noetherian PI-rings not module-finite over any commutative subring. *Proc. Amer. Math. Soc.* **84**, 457–463. MR 83h #16025.

SASIADA, E., AND COHN, P. M.

[67] An example of a simple radical ring. *J. Algebra* **5**, 373–377. MR 34 #2629.

SCHACHER, M. M.

[68] Subfields of division rings. *J. Algebra* **9**, 451–477. MR 37 #2809.

[77] The crossed-product problem. *Lecture Notes in Mathematics* 26, pp. 237–245. Marcel Dekker, New York. MR 55 #8100.

SCHELTER, W.
[75] Essential extensions and intersection theorems. *Proc. Amer. Math. Soc.* **53**, 328–330. MR 52 #10800.
[76] Integral extensions of rings satisfying a polynomial identity. *J. Algebra* **40**, 245–257; correction ibid. **44**, 576 (1976). MR 54 #5295; 55 #405.
[77] Azumaya algebras and Artin's theorem. *J. Algebra* **46**, 303–304. MR 56 #3044.
[78] Noncommutative affine PI-rings are catenary. *J. Algebra* **51**, 12–16. (delayed publ.) MR 58 #5772.
[82] On the geometry of affine PI-rings. *Contemp. Math.* **13**, 217–221. MR 84c #16020.
[84] Smooth affine PI-algebras. *Methods in Ring Theory*, pp. 483–488. D. Reidel, MR 86g #16024.
SCHELTER, W., and SMALL, L.
[76] Some pathological rings of quotients. *J. London Math. Soc.* **14**, 200–202. MR 55 #2981.
SCHOFIELD, A. H.
[84] Questions on skew fields. *Methods in Ring Theory*, pp. 489–496. Reidel, Dordrecht.
[85] Artin's problem for skew field extensions. *Proc. Camb. Phil. Soc.* **97**, 1–6. MR 86e #16029.
[86] Hereditary Artinian rings of finite representation type. Preprint.
[86a] Universal localisation for hereditary rings and quivers. *Lecture Notes in Mathematics* 1197, pp. 149–165. Springer-Verlag, Berlin.
[86b] Generic representations of quivers. Preprint.
[87] Untitled manuscript.
[87a] Skew rational extensions of commutative fields and division rings of group algebras.
SERRE, J. P.
[71] Cohomologie des groups discrètes. *Prosp. in Mathematics Annals of Mathematical Studies* 70. MR 52 #5876.
SEXAUER, N. E., and WARNOCK, J. E.
[79] The radical of the row-finite matrices over an arbitrary ring. *Trans. Amer. Math. Soc.* **139**, 287–295. MR 39 #249.
SHIRSOV, A. I.
[57] On rings with identity relations. *Mat. Sb.* **43**, 277–283. MR 20 #1698.
[57a] On some nonassociative nil-rings and algebraic algebras. *Mat. Sb.* **41**, 381–394. MR 19 #727.
[66] On certain identity relationships in algebra. *Sibirsk. Mat. Z.* 7, 963–968. MR 33 #4097.
SHOCK, R. C.
[71] Nil subrings in finiteness conditions. *Amer. Math. Monthly* **78**, 741–748. MR 44 #6743.
SKORNJAKOV, L. A.
[67] Homological classification of rings. (Russian) *Mat. Vesnik* **4**, 415–434. MR 37 #2815.
SLOVER, R.
[69] The Jacobson radical of row-finite matrices. *J. Algebra* **12**, 345–359. MR 39 #250.
[72] A note on the radical of row-finite matrices. *Glasgow Math. J.* **13**, 80–81. MR 46 #1834.
SMALL, L. W.
[64] Orders in Artinian rings. *J. Algebra* **4**, 13–41. MR 34 #199.
[65] An example in Noetherian rings. *Proc. Natl. Acad. Sci. U.S.A.* **54**, 1035–1036. MR 32 #5691.
[66] Hereditary rings. *Proc. Natl. Acad. Sci. U.S.A.* **55**, 25–27. MR 32 #4178.
[67] Semihereditary rings. *Bull. Amer. Math. Soc.* **73**, 656–658. MR 32 #4178.
[68] A change of rings theorem. *Proc. Amer. Math. Soc.* **19**, 662–666. MR 36 #6460.
[68a] Orders in Artinian rings II. *J. Algebra* **9**, 268–273. MR 37 #6315.
[79] The embedding problem for Noetherian rings. *Bull Amer. Math. Soc.* **75**, 147–148. MR 38 #1120.

SMALL, L. W., and STAFFORD, J. T.
[81] Localization and completion of Noetherian PI-algebras. *J. Algebra* **70**, 245–257. MR 84d #16022.
[82] Regularity of zero divisors. *Proc. London Math. Soc.* (3) **44**, 405–419. MR 84b #16014.
[85] Homological properties of generic matrix rings. *Israel J. Math.* **51**, 27–32. MR 87a #16031.

SMALL, L. W., STAFFORD, J. T., and WARFIELD, R. B.
[84] Affine algebras of Gel'fand-Kirillov dimension one are PI. *Math. Proc. Cambridge Philos. Soc.* **97**, 407–414. MR 86g #16025.

SMALL, L. W., and WADSWORTH, A. R.
[81] Some examples of rings. *Commun. Algebra* **9**, 1105–1118. MR 82h #16012.

SMALL, L. W., and WARFIELD, R. B.
[84] Prime affine algebras of Gel'fand-Kirillov dimension one. *J. Algebra* **91**, 386–389. MR 86h #16006.

SMITH, M. K.
[71] Group algebras. *J. Algebra* **18**, 477–499. MR 43 #3358.
[76] Universal enveloping algebras with subexponential but not polynomially bounded growth. *Proc. Amer. Math. Soc.* **60**, 22–24. MR 54 #7555.
[77] Growth of algebras. *Lecture Notes in Mathematics* 26, pp. 247–259. Marcel Dekker, New York. MR 55 #10517.

SMITH, P. F.
[71] Localization and the AR property. *Proc. London Math. Soc.* (3) **22**, 39–58. MR 45 #3453.
[71a] Localization in group rings. *Proc. London Math Soc.* (3) **22**, 69–90. MR 4 #3443.
[76] On non-commutative regular local ring. *Glasgow Math. J.* **17**, 98–102. MR 54 #10309.

SMITH, S. P.
[81] An example of a ring Morita equivalent to the Weyl algebra $A_1$. *J. Algebra* **73**, 552–555. MR 83a #16004.
[81a] The primitive factor rings of the enveloping algebra of $sl(2, C)$. *Proc. London Math. Soc.* **24**, 97–108. MR 82i #17016.
[84] Gel'fand-Kirillov dimension of rings of formal differential operators on affine varieties. *Proc. Amer. Math. Soc.* **90**, 1–8. MR 85d #16020.
[85] Central localization and Gel'fand-Kirillov dimension. *Israel J. Math.* **46**, 33–39. MR 85k #16048.

SNIDER, R. L.
[76] On the singular ideal of a group algebra. *Comm. Alg.* **4**, 1087–1089. MR 54 #2721.
[76a] Primitive ideals in group rings of polycyclic groups. *Proc. Amer. Math. Soc.* **57**, 8–10. MR 54 #2722.
[77] The zero divisor conjecture. *Lecture Notes in Mathematics* 26, pp. 261–295. Marcel Dekker, New York. MR 55 #10508.
[79] Is the Brauer group generated by cyclic algebras? *Lecture Notes in Mathematics* 734, pp. 279–301. Springer-Verlag, Berlin. MR 80k #16010.
[80] Representation theory of polycyclic group. *Lecture Notes in Mathematics* 55, pp. 345–356, Marcel Dekker, New York. MR 81k #20011.
[83] The division ring of fractions of a group ring. *Lecture Notes in Mathematics* 1029, pp. 325–339. Springer-Verlag, Berlin. MR 85g #16009.

SPECHT, W.
[50] Gesetze in Ringen I. *Math. Z.* **52**, 557–589. MR 11 #711.

STAFFORD, J. T.
[77] Stable structure of noncommutative Noetherian rings. *J. Algebra* **47**, 244–267. MR 56 #5648.
[77a] Weyl algebras are stably free. *J. Algebra* **48**, 297–304. MR 56 #8622.

[78] A simple Noetherian ring not Morita equivalent to a domain. *Proc. Amer. Math. Soc.* **68**, 159–160. MR 57 #6090.

[79] K-theory of Noetherian group rings. *Lecture Notes in Mathematics* 734, pp. 302–322. Springer-Verlag, Berlin. MR 80k #16018.

[81] Generating modules efficiently: Algebraic K-theory for noncommutative Noetherian rings. *J. Algebra* **69**, 312–346; corr. ibid **82**, 294–296 (1983). MR 82k #16018; 84d #16019.

[81a] On the stable range of right Noetherian rings. *Bull. London Math. Soc.* **13**, 39–41. MR 82f #16017.

[82] Generating modules efficiently over noncommutative rings. *Lecture Notes in Mathematics* 924, pp. 72–88. Springer-Verlag, Berlin. MR 83g #16044.

[82a] Noetherian full quotient rings. *Proc. London Math. Soc.* **44**, 385–405. MR 84b #16015.

[83] Dimensions of division rings. *Israel J. Math.* **45**, 33–40. MR 85d #16001.

[85] Stably free projective right ideals. *Compositio Math.* **54**, 63–78. MR 86i #16026.

[85a] Non-holonomic modules over Weyl algebras and enveloping algebras. *Invent. Math.* **79**, 619–638. MR 86h #17009.

[85b] On the ideals of a Noetherian ring. *Trans. Amer. Math. Soc.* **289**, 381–392. MR 86e #16023.

STAFFORD, J. T., and WARFIELD, R. B., JR.

[84] Hereditary orders with infinitely many idempotent ideals. *J. Pure Appl. Algebra* **31**, 217–225. MR 86c #16002.

[85] Constructions of hereditary Noetherian rings and simple rings. *Proc. London Math. Soc.* **51**, 1–20. MR 86j #16016.

STALLINGS, J. R.

[65] Centerless groups—an algebraic formulation of Gottlieb's theorem. *Topology* **4**, 129–134. MR 34 #2666.

STEWART, P. N.

[83] Nilpotence and normalizing extensions. *Comm. Alg.* **11**, 109–110. MR 84c #16012.

STROJNOWSKI, A.

[86] Idempotents and zero divisors in group rings. *Comm. Algebra* **14**, 1171–1186.

STROOKER, J. R.

[66] Lifting projectives. *Nagoya Math. J.* **27**, 747–751. MR 33 #5679.

SUSLIN, A. A.

[76] Projective modules over polynomial rings are free. *Soviet Math. Dokl.* **17**, 1160–1164. MR 57 #9685.

[77] A cancellation theorem for projective modules over algebras. *Soviet Math. Dokl.* **18**, 1281–1284. MR 57 #5986.

[79] The cancellation problem for projective modules and related topics. *Lecture Notes in Mathematics* 734, pp. 323–338. Springer-Verlag, Berlin. MR 81c #13011.

[82] Torsion in $K_2$ of fields. Lomi preprint E-2-82, Leningrad.

SWAN, R. G.

[60] Induced representations and projective modules. *Ann. Math.* (2) **71**, 552–578. MR 25 #2131.

[62] Projective modules over group rings and maximal orders. *Ann. Math.* (2) **76**, 55–61. MR 25 #3066.

[83] Projective modules over binary polyhedral groups. *J. Reine Angew. Math.* **342**, 66–172. MR 84j #16003.

SWEEDLER, M. E.

[75] Something like the Brauer group. *Proceedings of the International Congress of Mathematicians, Vancouver B. C.*, Vol. 1, pp. 337–341. Canad. Math. Congress, Montreal. MR 54 #10312.

[78] Two classical splitting theorems: Easy proofs. *J. Pure Appl. Algebra* **13**, 33–35. MR 58 #22160.

[82] Preservation of flatness for the product of modules over the product of rings. *J. Algebra* **74**, 159–205. MR 83f #16030.

TACHIKAWA, H.

[60] A note on algebras of unbounded representation type. *Proc. Japan Acad.* **36**, 59–61. MR 22 #4748.

[71] Localization and Artinian quotient rings. *Math. Z.* **119**, 239–253. MR 46 #7284.

TAKEUCHI, M.

[71] A simple proof of Gabriel and Popesco's theorem. *J. Algebra* **18**, 112–113. MR 43 #2048.

TATE, J.

[76] Relations between *K*2 and Galois cohomology. *Inventiones Math.* **36**, 257–274. MR 55 #2847.

TAUVEL, P.

[82] Sur la dimension de Gelfand-Kirillov. *Comm. Alg.* **10**, 939–963. MR 83j #16033.

TIGNOL, J.-P.

[78] On similarity classes of involutorial division algebras of degree 8. *C. R. Acad. Sci. Paris Ser.* A–B **286**, A876–A877.

[81] Produits croisés abéliens. *J. Algebra* **70**, 420–436. MR 84f #16026.

[83] Cyclic algebras of small exponent. *Proc. Amer. Math. Soc.* **87**, 330. MR 85m #11082.

[84] On the length of decompositions of central simple algebras in tensor products of symbols. *Methods in Ring Theory*, pp. 505–516. D. Reidel, Dordrecht. MR 86d #16029.

[85] On the corestriction of central simple algebras. Preprint.

TIGNOL, J.-P., and AMITSUR, S. A.

[85] Kummer subfields of Mal'cev-Neumann division algebras. *Israel J. Math.* **50**, 114–144.

[86] Totally ramified splitting fields of central simple algebras over Henselian field. *J. Algebra* **98**, 95–101. MR 86m #16007.

[86a] Symplectic modules. *Israel J. Math.* **54**, 266–290.

TIGNOL, J.-P., and WADSWORTH, A. R.

[86] Totally ramified valuations on finite dimensional division algebras. Preprint.

TITS, J.

[81] Groups of polynomial growth. *Lecture Notes in Mathematics* 901, pp. 176–188. Springer-Verlag, Berlin. MR 83i #53065.

TREUR, J.

[77] The Cartan-Brauer-Hua theorem. *Neder. Akad. Wetensch. Proc. Ser.* A 80 = Indag. Math. **39**, 453–454. MR 56 #12057.

TUGANBAEV, A. A.

[84] Distributive rings. *Math. Notes* **35**, 173–178. MR 86j #16031.

UTUM, Y.

[56] On quotient rings. *Osaka Math. J.* **8**, 1–18. MR 18 #7c.

VAMOS, P.

[77] Semi-local Noetherian PI-rings. *Bull London Math. Soc.* **9**, 251–256. MR 57 #376.

VAN OYSTAEYEN, F.

[84] Some constructions of rings. *J. Pure Appl. Algebra* **31**, 241–251. MR 85f #16051.

VILLAMAYOR, O. E.

[58] On the semisimplicity of group algebras. *Proc. Amer. Math. Soc.* **9**, 621–627. MR 20 #5224.

[59] On the weak dimension of algebras. *Pacific J. Math.* **9**, 941–951. MR 21 #7243.

VINBERG, E. B.

[65] On the theorem concerning the infinite-dimensionality of an associative algebra. (Russian) *Izv. Akad. Nauk. SSR Ser. Mat.* **29**, 209–214. MR 30 #3108.

WADSWORTH, A. R.

[82] Merkurjev's elementary proof of Merkurjev's theorem. *Contemp. Math.* **55**, 741–776.

[83] *p*-Henselian fields: *K*-theory, Galois cohomology, and graded Witt rings. *Pacific J. Math.* **105**, 473–495. MR 84m #12026.

[86] Extending valuations to finite dimensional division algeberas. *Proc. Amer. Math. Soc.* **98**, 20–22. MR 87i #16025.

[37] Über die Grundlagen der projektiven Geometrie und allgemeine Zahlsysteme. *Math. Z.* **113**, 528–567.

WALKER, C. L., AND WALKER, E. A.

[72] Quotient categories and rigns of quotient. *Rocky Mountain J.* Math. **2**, 513–555. MR 49 #2812.

WALKER, R. G.

[72] Local rings and normalizing sets of elements. *Proc. London Math. Soc.* (3) **24**, 27–45. MR 45 #3469.

WARE, R., and ZELMANOWITZ, J.

[70] The Jacobson radical of the endomorphism ring of a projective module. *Proc. Amer. Math. Soc.* **26**, 15–20. MR 41 #6891.

WARFIELD, R. B., JR.

[69] A Krull-Schmidt theorem for infinite sums of modules. *Proc. Amer. Math. Soc.* **22**, 460–465. MR 39 #4213.

[69a] Decompositions of injective modules. *Pacific J. Math.* **31**, 263–276. MR 40 #2712.

[72] Exchange rings and decompositions of modules. *Math. Ann.* **199**, 31–36. MR 48 #11218.

[76] Large modules over Artinian rings. *Lecture Notes in Mathematics* 37, pp. 451–463. Marcel Dekker, New York. MR 58 #5793.

[79] Modules over fully bounded Noetherian rings. *Lecture Notes in Mathematics* 734, pp. 339–352. Springer-Verlag, Berlin. MR 80m #16017.

[83] Prime ideals in ring extensions. *J. London Math. Soc.* (2) **28**, 453–460. MR 85e #16006.

[84] The Gel'fand-Kirillov dimension of a tensor product. *Math. Z.* **185**, 441–447. MR 85f #17006.

WATERHOUSE, W. C.

[75] Antipodes and group-likes in finite Hopf algebras. *J. Algebra* **37**, 290–295. MR 52 #473.

WEBBER, D. B.

[70] Ideals and modules of simple Noetherian hereditary rings. *J. Algebra* **16**, 239–242. MR 42 #305.

WEDDERBURN, J. H. M.

[21] On division algebras. *Trans. Amer. Math. Soc.* **22**, 129–135.

WONG, E. T., and JOHNSON, R. E.

[59] Self-injective rings. *Canad. Math. Bull.* **2**, 167–173. MR 21 #5652.

WOODS, S. M.

[74] Existence of Krull dimension in group rings. *J. London Math. Soc.* (2) **9**, 406–410. MR 50 #13120.

XU, Y. H.

[83] On the Koethe problem and the nilpotent problem. *Sci Sinica Ser. A* **26**, 901–908. MR 85k #16007.

YAMAGATA, K.

[78] On Artinian rings of finite representation type. *J. Algebra* **50**, 276–283. MR 57 #9758.

YOSHII, T.

[56] On algebras of bounded representation type. *Osaka Math. J.* **8**, 51–105 (contains error). MR 18 #462.

ZAKS, A.

[67] Residue rings of semi-primary hereditary rings. *Nagoya Math. J.* **30**, 279–283. MR 37 #1409.

[71] Some rings are hereditary rings. *Israel J. Math.* **10**, 442–450. MR 45 #5168.

[71a] Semi-primary hereditary algebras. *Trans. Amer. Math. Soc.* **154**, 129–135. MR 43 #2024.

[74] Hereditary Noetherian rings. *J. Algebra* **29**, 513–527. MR 50 #2243.

ZALESSKII, A. E.

[71] Irreducible representations of finitely generated nilpotent torsion-free groups. *Math. Notes* **9**, 117–123. MR 46 #1912.

[72] On a problem of Kaplansky. *Soviet Math.* **13**, 449–452. MR 45 #6947.

ZALESSKII, A. E., and MIHALEV, A. V.

[73] Group rings. (Russian) *Current Problems in Mathematics*, Vol. 2, pp. 5–118. Akad. Nauk. SSSR Vesojuz Inst. Naucn. I Techn. Informacii, Moscow. MR 54 #2723.

ZALESSKII, A. E., and NEROSLAVSKII, O. M.

[77] There exist simple Noetherian rings with zero divisors but without idempotents. (Russian with English summary) *Comm. Algebra* **5**, 231–244. MR 55 #12761.

ZELINSKY, D

[54] Raising idempotents. *Duke Math. J.* **21**, 315–322. MR 15 #928b.

[64] Berkson's theorem. *Israel J. Math.* **2**, 205–209. MR 32 #7608.

[77] Brauer groups. *Lecture Notes in Mathematics* 26, pp. 69–102. Marcel Dekker, New York.

ZELMANOWITZ, J. M.

[67] Endomorphism rings of torsionless modules. *J. Algebra* **5**, 325–341. MR 34 #2626.

[69] A shorter proof of Goldie's theorem. *Canad. Math. Bull.* **12**, 597–602. MR 40 #7297.

[71] Injective hulls of torsion free modules. *Canad. J. Math.* **23**, 1094–1101. MR 44 #6755.

[82] On the Jacobson density theorem. *Contemp. Math.* **13**, 155–162. MR 84c #16015.

# Index of References
# According to Section

## §5.1

Bass [61, 63, 76]; Beck [72]; Beck and Trosborg [78]; Brown-Lenagan-Stafford [81]; Coutinho [86]; Dicks [83]; Eilenberg-MacLane [45]; Eilenberg-Nagao-Nakayama [56]; Fields [69]; Jategaonkar [69]; Kirkland-Kuzmanovich [86]; McConnell [85]; Quillen [72, 76]; Rosset [84]; Roy [65]; Small [66, 68]; Stafford [77, 81, 82]; Stallings [65]; Suslin [76, 77, 79]

## §5.2

Auslander [55]; Auslander-Buchsbaum [57]; Bass [62]; Bergman-Ogous [72]; Bjork [85]; Brown-Dror [75]; Chase [61]; Eilenberg-Nakayama [55, 57]; Eilenberg-Rosenberg-Nakayama [57]; Fuelberth-Kuzmanovich [75]; Goodearl [74, 75, 78]; Grothendieck [57], Harada [63, 66]; Hupert [81]; Mal'cev [37]; McConnell [77, 84]; Osofsky [66, 68]; Resco [79]; Resco-Small-Stafford [82]; Rinehart-Rosenberg [76]; Robson [85]; Ross [72]; Rosenberg [82]; Rosenberg-Stafford [76]; Sandomierski [73]; Skornjakov [67]; Strooker [66]; Takeuchi [71]; Walker [72]; Zaks [67, 74]

## §5.3

Auslander-Goldman [60]; Azumaya [51]; Braun [86]; Hochschild [45–47, 56]; Ingraham [74]; Rosenberg-Zelinsky [56]; Saltman [78]; Villamayor [59]; Zaks [71]; Zelinsky [64]

## §6.1

Amitsur [65–67, 69, 71, 75]; Amitsur-Levitzki [50]; Artin [69]; Beidar-Mikhalev-Salavova [81]; Bergman [74b, 76]; Braun [82]; Cauchon [76]; Dehn [22]; Eisenbud [70]; Formanek [72]; Halpin [83]; Herstein [65, 70]; Jain [71]; Jategaonkar [75]; Jensen-Jondrop [73]; Jondrup [81]; Kaplansky [48]; Kostant [58]; Levi [49]; Martindale [69, 72, 73]; Nagarajan [68]; Page [84]; Procesi [72]; Razmyslov [73]; Robson-Small [74]; Rowen [73–76]; Schelter [77]; Vamos [77]; Wagner [37]

*447*

## §6.2

Beidar [81]; Bergman [81, 86]; Borho-Kraft [76]; Fischer-Struik [68]; Gel'fand-Kirillov [66]; Golod [64, 66]; Hudry [84]; Irving [84a]; Irving-Small [83]; Jacobson [45b]; Krempa-Okninski [86]; Kurosch [41]; Lorenz [82, 85]; Resco [84]; M. Smith [77]; S. P. Smith [84]; Tauvel [82]; Vinberg [65]; Warfield [84].

## §6.3

Amitsur [80]; Amitsur-Small [80]; Artin-Schelter [79, 81, 81a]; Beachy-Blair [86]; Bergman-Dicks [75]; Bergman-Small [75]; Braun [79, 81, 84, 85]; Braun-Warfield [86]; Brown-Warfield [84]; Chatters-Jondrup [83, 84]; Duflo [73]; Goodearl-Small [84]; Gordon-Small [84]; Irving-Small [85]; Kaplansky [46, 50]; Lenagan [82, 84]; Lestman [78]; Levitzki [53]; Lewin [74]; Lorenz-Small [82]; Malliavin [76]; Müller [85]; Procesi [67]; Procesi-Schacher [76]; Razmyslov [74]; Sarraillé [82]; Schelter [76, 78, 82, 84]; Schelter-Small [76]; Shirsov [57, 57a, 66]; Small-Stafford [81]; Small-Stafford-Warfield [84]; Small-Warfield [84]

## §6.4

Amitsur [84]; Amitsur-Regev [82]; Berele [82]; Berele-Regev [83]; Coutinho [85]; Drensky [82, 84]; Drensky-Kasparian [83]; Formanek [82, 84–86a]; Formanek-Schofield [85]; Harcenko [78, 84]; Helling [74]; Higman [56]; Kemer [78, 84]; Krakowski-Regev [73]; Latyshev [72]; Le Bruyn [86]; Luminet [86]; Procesi [72, 74, 76, 84]; Razmyslov [73, 74a, 74b, 85]; Regev [72, 78, 80, 84]; Rowen [74b, 82a]; Specht [50]

## §7.1

Amitsur [72, 82]; Amitsur-Saltman [78]; Amitsur-Small [78]; Amitsur-Tignol [84, 85]; Bergman [75, 83]; Chase [84]; Faith [58]; Gustafson [76]; Jacob-Wadsworth [86]; Jacobson [47a, 83]; Makar-Limanov [83, 84]; Resco-Small-Wadsworth [79]; Risman [77]; Schacher [68, 77]; Schofield [84, 85, 87a]; Stafford [83]; Sweedler [78]; Tignol [81]; Treur [77]; Wadsworth [86]; Wedderburn [21]

## §7.2

Amitsur [59]; Amitsur-Rowen-Tignol [79]; Benard-Schacher [72]; Bloch [80]; Fein-Schacher [80]; Fein-Schacher-Sonn [86]; Jacobson [64]; Janusz [72]; Mammone [86]; Merkurjev [81, 83, 86]; Merkurjev-Suslin [82, 86]; Perlis [69]; Platonov [75, 76]; Riehm [70]; Rosset [77, 78]; Rosset-Tate [83]; Rowen [78, 84]; Rowen-Saltman [82]; Saltman [77–81, 85]; Snider [79]; Suslin [82]; Sweedler [67]; Tate [76]; Tignol [78, 83–85]; Tignol-Wadsworth [86]; Wadsworth [82, 83]; Zelinski [77]

## §7.3

Amitsur [55]; Rowen [82]; Saltman [82, 82a]

# §8.1

Brockhaus [85]; Brookes-Brown [85]; Connell [63]; Domanov [77]; Farkas-Snider [74, 81]; Formanek [73a, 73c, 74]; Formanek-Lichtman [78]; Handelman-Lawrence-Schelter [78]; Irving [79]; Kaplansky [70]; Lichtman [76]; Passman [82, 83]; Reiner [80]; Rittere-Sehgal [83]; M. Smith [71]; Snider [76]; Villamayor [58]; Woods [74]; Zalesskii [72]; Zalesskii-Mihalev [73]

# §8.2

Bass [64, 72]; Brown [76, 82a]; Cliff [80, 85]; Crawley-Boerey-Kropholler-Linnell [87]; Domanov [78]; Farkas [75, 80, 85]; Farkas-Passman [78]; Farkas-Snider [76, 79]; Farkas-Schofield-Snider-Stafford [82]; Formanek [73, 73b]; Hattori [65]; L. Levy [85]; Lewin [82]; Lewin-Lewin [75]; Lichtman [82]; Linnell [77]; Lorenz [77, 81]; Lorenz-Passman [81]; Milnor [68]; Moody [86]; Passman [71, 79, 82, 84, 86]; Roseblade [73, 76, 78]; Rosset [76]; Serre [71]; P. F. Smith [71a]; Snider [76, 77, 80, 83]; Stafford [79]; Strojnowski [86]; Swan [60, 62, 83]; Tits [81]; Zalesskii [71]

# §8.3

Block [81]; Borho [77, 86]; Borho-Jantzen [77]; Borho-Rentschler [75]; Brown [84]; Conze-Berline [75]; Dixmier [68, 70, 77]; Duflo [77, 82]; Gel'fand-Kirillov [66]; Irving [84]; Joseph [74, 80, 81, 83, 83a, 85, 86]; Le Bruyn-Ooms [85]; Levasseur [79, 82]; Lichtman [78, 83, 84]; Lorenz [81]; Malliavin [79]; McConnell [74, 75, 77]; Moeglin [80, 80a]; Moeglin-Rentschler [81, 84]; Nouaze-Gabriel [67]; Quillen [69]; Rentschler [87]; M. K. Smith [76]; P. F. Smith [76]; S. P. Smith [81]

# §8.4

Bergen-Montgomery-Passman [88]; Bergman [85]; Borho [82]; Brown [81]; Cauchon [86]; Chatters [84]; M. Cohen [85]; Cohen-Montgomery [84]; Dean-Stafford [86]; Farkas [79]; Gilchrist-Smith [84]; Goldie-Michler [75]; Goodearl-Hodges-Lenagan [84]; Goodearl-Lenagan [83, 85]; Hudry [86]; Irving [78, 79]; Irving-Small [80]; Jategaonkar [81, 82]; Joseph-Small [78]; Kaplansky [75]; Lichtman [84]; Lorenz [84]; McConnell [82, 84]; Milnor-Moore [64]; Pearson-Stephenson [77]; Radford [76, 77]; S. P. Smith [85]; Stafford [77a, 85, 85a, 85b]; Warfield [83]; Waterhouse [75]

# Cumulative Subject Index
# for Volumes I and II

Page numbers in normal type are from Vol. I; page numbers in italic type are from Vol. II.